Probability and Its Applications

Michael Demuth
Jan A. van Casteren

Stochastic Spectral Theory for Selfadjoint Feller Operators

A functional integration approach

Springer Basel AG

Authors' addresses

Michael Demuth
Technische Universität Clausthal
Institut für Mathematik
Erzstr. 1
38678 Clausthal-Zellerfeld
Germany

Jan A. van Casteren
University of Antwerp (UIA)
Department of Mathematics
and Computer Science
2610 Antwerp
Belgium

2000 Mathematics Subject Classification: 47D06, 47D02, 47A10, 47A40, 60J25, 60J35, 60J45, 81Q10 (primary); 47B07, 47B38, 47B65, 47G10, 46N50 (secondary)

A CIP catalogue record for this book is available from the Library of Congress, Washington D.C., USA

Deutsche Bibliothek Cataloging-in-Publication Data

Demuth, Michael:
Stochastic spectral theory for selfadjoint feller operators : a functional integration approach / Michael Demuth ; Jan A. van Casteren. - Basel ; Boston ; Berlin : Birkhäuser, 2000
(Probability and its applications)
ISBN 978-3-0348-9577-4 ISBN 978-3-0348-8460-0 (eBook)
DOI 10.1007/978-3-0348-8460-0

ISBN 978-3-0348-9577-4

Originally published by Birkhäuser Verlag in 2000
Softcover reprint of the hardcover 1st edition 2000
Printed on acid-free paper produced from chlorine-free pulp. TCF ∞
ISBN 978-3-0348-9577-4

9 8 7 6 5 4 3 2 1

Contents

Preface

The objective of this book is to investigate the spectral theory of self-adjoint Feller operators using stochastic analysis, in particular the theory of strong Markov processes and of martingales. The free Feller operators are generators of Markov semigroups enjoying the Feller property. More specifically, let E be a second countable locally compact Hausdorff space, which has the interpretation of a *state* or *configuration* space, equipped with a non-negative Radon measure m, defined on its Borel field $\mathcal{E}$. The *free Feller* operator, denoted by K_0, is the L^2-generator of a self-adjoint semigroup $\{\exp(-tK_0) : t \geq 0\}$ satisfying the *Feller* property. By the Feller property we mean that every operator $\exp(-tK_0)$ leaves the space $C_\infty(E)$ invariant where $C_\infty(E)$ denotes the space of continuous complex functions on E that vanish at infinity. (For the Feller property we refer to Chapter 1.B, BASSA, property A2, and to Definition 1.2.) Since we also assume the continuity property A3 of BASSA, every operator $\exp(-tK_0)$ can be represented as a mathematical expectation: $[\exp(-tK_0) f](x) = \mathbb{E}_x[f(X(t))]$, where $\{(X(t), t \geq 0), (\mathbb{P}_x, x \in E)\}$ is a strong Markov process, and where the function f belongs to $C_\infty(E)$. By extension, the notation $[\exp(-tK_0) f](x) = \mathbb{E}_x[f(X(t))]$ can be employed for functions f in $L^2(E, m)$. A major part of the book is devoted to regular perturbations of K_0 given by real-valued functions V, called Kato-Feller potentials, and singular perturbations of K_0 which are given by (zero) Dirichlet boundary conditions on (the boundary of) a closed subset Γ of the state space E, called the singularity region. For regular perturbations the corresponding semigroup is denoted by $\{\exp(-t(K_0 \dot{+} V)) : t \geq 0\}$. In the case of singular perturbations one obtains a semigroup, called Dirichlet semigroup, vanishing on the singularity region. Put $\Sigma = E \setminus \Gamma$. This Dirichlet semigroup, denoted by $\{P_\Sigma(t) : t \geq 0\}$, thus obtained vanishes on Γ; more precisely $P_\Sigma(t)f(x) = 0$ for m-almost all $x \in \Gamma$. The Dirichlet semigroup $\{P_\Sigma(t) : t \geq 0\}$ is strongly continuous in $L^2(\Sigma, m)$. Its generator is written as $(K_0 \dot{+} V)_\Sigma$. Under certain conditions, the semigroups $\{\exp(-t(K_0 \dot{+} V)) : t \geq 0\}$ and $\{\exp(-t(K_0 \dot{+} V)_\Sigma) ; t \geq 0\}$ possess the Feller property as well. Therefore we call the operators $K_0 \dot{+} V$ and $(K_0 \dot{+} V)_\Sigma$ *perturbed* Feller operators.

The aim of the present book is to compare the spectra or spectral data of two Feller operators. The behavior of (the various parts of) the spectra for singular or regular perturbations depends on properties of the corresponding resolvent or semigroup differences. These perturbations of the semigroups can be represented by the Feynman-Kac formulas. Via the Laplace transform the same is true for the resolvents. The Feynman-Kac formula, which makes sense for Kato-Feller potentials, serves for instance as an interface between operator theory and stochastic analysis. By introducing measures pinned at two points at the beginning and at the end of each trajectory, the semigroups and resolvents turn out to be integral operators with kernels that can explicitly be expressed as expectations with respect to these tied measures. Specializing to the case of the Laplace operator, these tied or pinned measures correspond to tied Brownian motion or Brownian bridge. When dealing with singular perturbations, it is helpful to introduce the harmonic extension operator. Intuitively this operator solves the Dirichlet problem for boundary data given on Γ. In this way we establish a link between spectral theory and harmonic functions. Consequently a number of conditions are phrased in terms of capacities. Since we put everything in the context of Markov processes, we get a nice interplay between concepts from stochastic analysis and from operator theory. The theory as a whole is called "stochastic spectral analysis". Its framework is depicted on page 56.

The theory starts with some basic assumptions (abbreviated as BASSA) for the integral kernel of the Feller semigroup $\{\exp(-tK_0) : t \geq 0\}$. The assumptions on the integral kernel, also called the transition density, $p_0(t, x, y)$ are rather weak. So an extensive class of "unperturbed" Feller generators lie within the scope of the present theory. For instance, second order elliptic partial differential operators with variable, not necessarily bounded, diffusion and drift coefficients, are included. The same is true for Laplace-Beltrami operators, for certain pseudo-differential operators, for instance generators of stable processes and versions of relativistic Hamiltonians, and for others. Examples of such operators K_0 are given in section D of the first chapter. This list is however not exhaustive.

The singular and regular perturbations are introduced in such a way that the Hamiltonians $K_0 \dot{+} V$ and $(K_0 \dot{+} V)_\Sigma$ are self-adjoint operators in the L^2-spaces $L^2(E, m)$ and $L^2(\Sigma, m)$ respectively. The Kato-Feller class, which determines the potential V, is taken in analogy with the Kato class for the Wiener process or, what is the same, for the Laplace operator. Except for being closed, there are no other requirements imposed on the singularity region Γ. The results are somewhat more transparent if the collection of its regular points and that of its interior coincide. In such a case the relevant stopping time is the first hitting time of the set Γ. If the process starts outside of the polar set $\Gamma \setminus \Gamma^r$, then this hitting time coincides with the entry time. The set Γ^r is the set of regular points of Γ.

In order to study the discrete, essential or absolutely continuous spectra of $K_0 \dot{+} V$ and of $(K_0 \dot{+} V)_\Sigma$, it is useful to investigate the differences of the associated semigroups and/or resolvents. The operator, Hilbert-Schmidt, and trace norm of these differences are studied in some detail in the chapters 4, 5, and 6, respectively. Several features of Markov process theory are crucial to estimating the norms of these differences from above (and sometimes also from below). Important roles are played by the strong Markov property, the strong time-dependent Markov property, and the martingale property. In fact, the martingales we employ play an essential role in the analysis of the above mentioned differences and for obtaining stochastic representations of the integral kernels of Feynman-Kac semigroups. In addition, knowledge is required of how some quantitative and qualitative properties of the distribution of the first hitting times depend on the set Γ.

We present a compendium of results concerning the semigroup differences.

A characteristic feature is that several conditions depend on integrability conditions in which the unperturbed integral kernel $p_0(t,x,y)$, the potential V, and the behavior of the Markov process on Γ appear together in some explicit form. In concrete examples more specialized applications of these conditions can be established. This is certainly the case if more information on the function $p_0(t,x,y)$ is available. This aspect makes the theory attractive because it is applicable in many situations. It can be used for any operator K_0 that generates a semigroup with a kernel satisfying the basic assumptions BASSA. These general results on resolvent differences are applied in spectral theory for Hamiltonians of the form $K_0 \dot{+} V$ and $(K_0 \dot{+} V)_\Sigma$. It is relatively easy to find sufficient conditions for the invariance of the essential spectra. The existence together with completeness of scattering systems yields the invariance of the absolutely continuous spectra. The finiteness of the capacity of the singularity region Γ implies the completeness of the corresponding scattering systems. Since unbounded regions may have finite capacity, there is a non-local freedom in scattering theory. The corresponding trace class criteria are new even in the well established case of the Wiener process generated by the Laplace operator. The existence of bound states is proven, and the behavior of eigenfunctions is examined.

Another important point is the fact that stochastic spectral analysis contains the possibility of investigating the spectral data quantitatively. For two regular perturbations this is achieved in terms of the Kato-Feller norm of their difference. Amongst other topics we consider the limiting absorption principle, the behavior of spectral measures, wave operators, scattering operators, and scattering matrices. It is obvious that it is possible to treat a whole lot of other examples because, to a large extent, many spectral data are in the end based on resolvent differences. The interested reader can use the compendium of results on resolvent differences in Chapter 4, 5, and 6 for further research activities in spectral theory. The results in Chapter 8 may serve as a possible guideline.

The connection between regular and singular perturbations is given in terms of large coupling limits. In this situation the Hamiltonian $K_0 \dot{+} V$ is additionally perturbed by a potential of the form $\beta 1_\Gamma$, where $\beta \geq 0$ is a large coupling parameter and where 1_Γ denotes the indicator function of the singularity region Γ. The resolvent of the operator $K_0 \dot{+} V + \beta 1_\Gamma$ converges to the resolvent of $(K_0 \dot{+} V)_\Sigma$. In strong resolvent sense this is known for rather general regions Γ. However, stochastic spectral analysis brings in a tool to prove the operator norm convergence in a quantitative fashion. The rate of convergence depends on the Laplace transform of the occupation time or spending time of the process in Γ. At the end of the book, we estimate the scattering matrices in terms of this Laplace transform. This enables us to consider the perturbation of resonances. More precise and detailed information about the book is given in Chapter 1, subsection E.5.

The present book aims to build a bridge between spectral and stochastic analysis. We hope to reach a readership in both of these fields. For the convenience of the reader, and to facilitate access to the material in the book, the appendices are devoted to a short survey on spectral theory, on operator semigroups, on martingale theory, and on Markov processes. The reader should have some mathematical maturity in these areas.

We tried to make this theory more or less self-contained and complete. However we know that the theory can be extended in several directions. For instance non-symmetric semigroups and processes are not covered. First order perturbations (which include e.g. perturbations by quantum mechanical magnetic fields) and the corresponding additive functionals are not included either. However we suspect that a number of results presented for perturbations of order zero can be extended to first order.

A Reader's Guideline

For the convenience of the reader, we give some indications how to use the book. Readers who want to learn about Feynman-Kac semigroups can begin with the basic assumptions (BASSA) as exhibited in section B of Chapter 1. Then they can read section A of Chapter 2. In particular Theorem 2.5 is crucial. At a first reading they may omit proofs. In particular this is true for Chapter 3 which covers the proof of Theorem 2.5. The singular perturbations are introduced via a hitting time, which is explained in section B of Chapter 2. In the same section the integral kernels of the perturbed semigroups are introduced. Since this is done in a probabilistic way, the reader should have some knowledge about martingale theory. The necessary background is explained in Appendix C. On the other hand most of the results in the later chapters (Chapters 4, 5, 6, 7, and 8) can be understood without much knowledge about Markov processes or martingales. In the proofs we often use some probabilistic representation (Feynman-Kac formulas and the like), and we apply some classical inequality (like Hölder or Cauchy-Schwarz). Readers who are mainly interested in semigroup theory are advised to read the basic assumptions (BASSA) in section B of Chapter 1 first, then parts of Chapter 2 (like section A and in particular Theorem 2.5). After that they can read the Chapters 4, 5, or 6 independently. If they want to learn about convergence properties, Chapter 7 gives some information. After that they may read Chapter 8. All the necessary topics from probability theory (like Markov processes, martingale theory, and stopping time techniques) are explained in Appendix C. In section B of Chapter 2, the corresponding concepts are used for describing and defining regular perturbations and Feynman-Kac semigroups (Markov property), singular perturbations and Dirichlet semigroups (Markov property in conjunction with stopping times), and integral kernels of these semigroups (martingale theory). Section E in Chapter 1 gives some relevant properties of Gaussian processes (subsection E.1). This includes a discussion of Brownian motion as a Gaussian as well as a Markov process: see subsection E.2. In subsection E.3. Kato-Feller potentials for the Laplace operator are described and in subsection E.4. part of Theorem 2.5. for the Laplace operator is described. Here the perturbed semigroups are coined Schrödinger semigroups. For more general Markov processes the regularly perturbed generators of Markov processes are called generators of Feynman-Kac semigroups. Since the Laplace operator generates Brownian motion, our theory contains the theory of Schrödinger semigroups as a particular case. To some extent this is explained in subsection E.5. of Chapter 1. In section C of Chapter 1, a number of examples are described to which, in principle, our theory is applicable. In section D of Chapter 1, the reader finds some abstract results on heat kernels. The contents of Chapter 2 is explained above. Chapter 3 is devoted to the proof of the continuity and the symmetry of the integral kernel of the Feynman-Kac semigroup. The Chapters 4, 5, and 6 contain, respectively, estimates for the operator norm, the Hilbert-Schmidt norm, and the trace class norm of differences of Feynman-Kac semigroups and the corresponding resolvent differences. In Chapter 7 we discuss some stability and convergence results for resolvent families of Feynman-Kac semigroups. In Chapter 8 we give some spectral consequences of the previous results. The discussions include phenomena like preservation of essential and/or absolutely continuous spectra, limiting absorption principle, resonances, and scattering.

If we write something like "Theorem 2.21", we mean Theorem 2.21 in Chapter 2. A similar convention is used for references to items in an appendix: C.11 denotes section C.11 in Appendix C.

For the notation of "resolvent operators" and "stopping times" which are in vogue, we refer the reader to the "Index of symbols" (page 441).

Acknowledgement

We started our collaboration in January 1988. Both of us were invited by Heinz Langer to a conference in Gaussig near Dresden. At that time traveling was mainly possible in one direction. Nevertheless after some joint papers we started to think about the project of writing a book. The combination of stochastic and spectral analysis attracted both of us very much. It was around 1989/90 that the plans became more concrete. The product of this stimulating collaboration is realized in the present book. It was and is a very fascinating subject for both of us.

Mainly due to the support of the Max-Planck-Gesellschaft, the Belgian National Fund for Scientific Research (NFWO), the University of Antwerp (UIA), the Deutsche Forschungs-Gemeinschaft (DFG), and the Sonder-Forschungsbereich (SFB) "Differential Geometrie und Quantenphysik" (Berlin), the authors were able to meet on several occasions. In particular we want to thank Bert-Wolfgang Schulze, head of the Max-Planck Research group "Partial Differential Equations and Complex Analysis" at the University of Potsdam, who has consistently encouraged the support of the mathematical sciences and scientists, including our project. The first author wants to take this opportunity to express his great pleasure in cooperating with Bert-Wolfgang Schulze on many projects, thereby fruitfully progressing his research in mathematical analysis. The second author is very much obliged to David Elworthy (University of Warwick, Coventry) for providing stimulation in his scientific work. Some of the stochastic analysis in the present book is, if not directly, certainly indirectly due to him. We want to thank Gunter

Lumer (Université de Hainaut, Mons) for the organization of the meeting in Han-sur-Lesse (October 1991). The second author is also obliged to Albert Badrikian (Université de Clermont-Ferrand). Without his emphasising the use of stochastic methods in mathematical analysis, this book would not have been written. Unfortunately Albert Badrikian died in a tragic accident in the mountains in 1994. The second author would like to dedicate this book to Albert Badrikian.

The regular duties at the universities of Antwerp, Clausthal, and formerly Potsdam, are and always have been very substantial. So we benefitted grately from the use of the facilities at the Mathematical Research Institute at Oberwolfach in the framework of the Research in Pairs (RiP) program. In particular we are obliged to its director Matthias Kreck. We also thank the Volkswagenstiftung for the material support of this program. During four weeks in Oberwolfach, the manuscript was completed.

We are also grateful to Walter Renger for his critical reading of the manuscript, and to Ivor McGillivray who did the same for the preface. They discovered several misprints and gave useful comments for improving the presentation of the material. The programming done by Martin Traupe needed to create the pictures in Chapter 8 is gratefully acknowledged.

The authors sincerely thank the people of Birkhäuser Verlag for their help in the final stage of the publication of the present book.

Chapter 1

Basic Assumptions of Stochastic Spectral Analysis: Free Feller Operators

A. Introduction

In this book we want to introduce and emphasize a new point of view in spectral theory called Stochastic Spectral Analysis.

The basic means are detailed estimates of differences of resolvents and semigroups in order to satisfy certain spectral theoretical criteria. The resolvents and semigroups are represented stochastically via the Feynman-Kac formulae. In this fashion the theory of strong Markov processes can and will be used intrinsically and extensively. The main advantage of this approach is that this point of view in spectral theory and stochastic analysis allows us to consider a large and interesting class of operators. First let us illustrate this class of operators, which henceforth will be called *free Feller operators* (see Definition 1.3), by giving some examples. These examples are (pseudo-)differential operators of degree ≤ 2 that satisfy certain (hypo-)ellipticity conditions. Moreover, since we are mainly interested in self-adjoint operators, we will only consider examples of this type.

EXAMPLE 1. (Kochubeĭ [157]). Take the second order elliptic differential operator

$$K_0 = -\frac{1}{2} \sum_{i,j=1}^{\nu} a_{ij}(x) \frac{\partial^2}{\partial x_i \partial x_j} + \sum_{j=1}^{\nu} b_j(x) \frac{\partial}{\partial x_j} + c(x)$$

in $L^2(\mathbb{R}^\nu)$. It is well-known that sufficiently restrictive conditions on a_{ij}, b_j provide us with an operator K_0 being the generator of a diffusion which is canonical in the sense of Dynkin [94, p 167]. But in [157] Kochubeĭ allows more general coefficients and in general the diffusion process corresponding to the operator K_0 does not possess the properties indicated by Dynkin. Nevertheless, the fact remains that K_0 generates a semigroup $\{\exp(-tK_0) : t \geq 0\}$ possessing the Feller property, i.e. $\exp(-tK_0)\, C_\infty(\mathbb{R}^\nu) \subseteq C_\infty(\mathbb{R}^\nu)$. Here $C_\infty(\mathbb{R}^\nu)$ is the space of continuous functions, vanishing at infinity and the precise definition of Feller semigroup is given in our basic assumptions, called BASSA: see Definition 1.2, Remark 5 following it and Definition 1.3. Moreover, there exists an integral kernel $p_0(t,x,y)$ for this semigroup, i.e.

$$[\exp(-tK_0)\, f](x) = \int_{\mathbb{R}^\nu} p_0(t,x,y) f(y) dy,$$

for all $f \in L^2(\mathbb{R}^\nu)$. On the other hand, $p_0(t,x,y)$ is the corresponding transition density function of a Markov process, possibly constructed via the Kolmogorov's consistency conditions (projective system of cylinder measures): see Proposition 1.14 and C.17 in Apendix C.

EXAMPLE 2. (Davies [66]) Consider operators on a locally finite Riemannian manifold E formally given by

$$K_0 f = -\frac{1}{\sigma^2}\nabla\left(\sigma^2\nabla f\right),$$

where σ is a measurable function, $\sigma(x) > 0$, $\sigma \in L^\infty_{\text{loc}}(E)$ and $\sigma^{-1} \in L^\infty_{\text{loc}}(E)$. In particular, σ is not assumed to be differentiable. The operator K_0 can be defined correctly and unambiguously via the closable Dirichlet form

$$\mathcal{E}^0(f,f) = \int_E |\nabla f(x)|^2\,\sigma(x)^2 dx.$$

Associated to the closure of the quadratic form $\mathcal{E}^0$ is the generator K_0 of a positivity preserving strongly continuous semigroup $\{\exp(-tK_0) : t \geq 0\}$ in $L^p(E)$, $1 \leq p < \infty$,

$$[\exp(-tK_0)\, f](x) = \int_E p_0(t,x,y) f(y)\sigma(y)^2 dy.$$

Further conditions on σ also ensure the Feller property for this example. This will be explained in more detail in Example 2 in section C of this chapter.

EXAMPLE 3. (Jacob [140]) On the Sobolev space $H^\infty(\mathbb{R}^\nu) = \bigcap_{q\geq 0} H^q(\mathbb{R}^\nu)$ one may consider pseudo-differential operators

$$[Lu](x) := [p(x,D)u](x) = \frac{1}{(2\pi)^{\nu/2}}\int_{\mathbb{R}^\nu} e^{i\xi\cdot x} p(x,\xi)\widehat{u}(\xi)d\xi, \quad u \in H^\infty(\mathbb{R}^\nu),$$

with the symbol

$$p(x,\xi) = \sum_{j=1}^{\nu} b_j(x)\left|\xi_j\right|^{2r}, \; 0 < r \leq 1.$$

Sufficient conditions on the coefficients b_j, $1 \leq j \leq \nu$, can be given in order that the operator $K_0 := -L$ has a self-adjoint extension which generates a semigroup with the Feller property. The associated Markov process possesses a symmetric transition function $p_0(t,x,y)$ that can be estimated in the following way:

$$0 \leq p_0(t,x,y) \leq Ct^{-n/(2r)}e^{ct}.$$

In this example $\widehat{u}(\xi)$ denotes the Fourier transform, i.e.

$$\widehat{u}(\xi) = \frac{1}{(2\pi)^{\nu/2}}\int_{\mathbb{R}^\nu} e^{i\xi . x}u(x)dx.$$

EXAMPLE 4. (Carmona, Masters, Simon [48]) Let μ be a non-negative Lévy measure on $\mathbb{R}^\nu$, i.e. one with the property that $\int_{\mathbb{R}^\nu} \min\left(1, |x|^2\right) d\mu(x) < \infty$, let a and b be vectors in $\mathbb{R}^\nu$ and let C be a square $\nu \times \nu$ matrix. In addition, let $h : \mathbb{R}^\nu \to \mathbb{R}^\nu$ be a function of compact support with the property that $h(x) = x$ for all x in a neighborhood of the origin. Define the negative-definite function $F : \mathbb{R}^\nu \to \mathbb{R}^\nu$ by

$$F(p) = a + ib \cdot p + p \cdot Cp - \int_{\mathbb{R}^\nu} \left[e^{ip\cdot x} - 1 - ip \cdot h(x)\right] d\mu(x).$$

A classical result, due to Schoenberg (e.g. see [25, Theorem 7.8, page 41]), says that a function $F : \mathbb{R}^\nu \to \mathbb{C}$ is negative-definite if and only if $F(0) \geq 0$ and if, for every $t > 0$, the function $\exp(-tF)$ is positive-definite. Let $\{m_t : t \geq 0\}$ be a vaguely continuous convolution semigroup of probability measures on $\mathbb{R}^\nu$ such that the Fourier transform is

$$\widehat{m}_t(p) = \exp\left(-tF(p)\right), \quad p \in \mathbb{R}^\nu.$$

By definition, a family of probability measures $\{m_t : t \geq 0\}$ is a *vaguely continuous convolution semigroup* if $m_s * m_t = m_{s+t}$, $s,\ t \geq 0$, $m_0 = \delta_0$ (Dirac measure at zero), and $\lim_{t\downarrow 0} \int f dm_t = f(0)$ for all $f \in C_\infty(\mathbb{R}^\nu)$. Then there exists an operator K_0 that generates a semigroup given by

$$\left[\exp\left(-tK_0\right) f\right](x) = \int_{\mathbb{R}^\nu} f(x+y) dm_t(y), \quad f \in C_\infty(\mathbb{R}^\nu).$$

If $\int_{\mathbb{R}^\nu} \exp\left(-tF(p)\right) dp$ is finite, the operator $\exp\left(-tK_0\right)$ possesses an integral kernel given by

$$p_0(t,x,y) = \frac{1}{(2\pi)^\nu} \int_{\mathbb{R}^\nu} \exp\left(-tF(p) + ip \cdot (x-y)\right) dp.$$

In the special case that $F(p)$ is given by $F(p) = \sqrt{p^2 + m^2} - m$, m fixed, then $K_0 = \sqrt{-\Delta + m^2} - m$ and the kernel can be given explicitly:

$$\begin{aligned} &p_0(t,x,y) \\ &= \frac{1}{(2\pi)^\nu} \frac{t}{\sqrt{|x-y|^2 + t^2}} \int_{\mathbb{R}^\nu} \exp\left(mt - \sqrt{\left(|x-y|^2 + t^2\right)\left(p^2 + m^2\right)}\right) dp. \\ &= \int_0^\infty \frac{\exp(mt - m^2 u)}{(4\pi u)^{\nu/2}} \exp\left(-\frac{|x-y|^2}{4u}\right) \frac{t}{u\sqrt{u\pi}} \exp\left(-\frac{t^2}{4u}\right) du. \end{aligned}$$

Here $p^2 = p \cdot p = |p|^2$. These and further examples will be treated in some more detail after defining our basic assumptions for the whole theory in section C of this chapter.

Although we do not explicitly discuss them, operators defined on certain open sub-domains U of a larger locally compact state space E and satisfying certain Wentzell type boundary conditions, are also included in our theory as long as they generate Feller semigroups on U with a symmetric continuous density. For more details the reader may consult Taira [229], Ishikawa [139] and others like Sato and Ueno [201] and Bony, Courrège and Priouret [36]. For the moment the main point is that one can find a collection of operators generating (self-adjoint) Feller semigroups. These semigroups consist of integral operators. In many cases there exist no explicit formula for their integral kernels, but one knows about their existence and often some properties are known. Of course, it is impossible to classify or even to find all operators that generate Feller semigroups possessing integral kernels which correspond to the one-dimensional distributions of a Markov process. Therefore the fundamental idea is to start rigorously with the semigroup kernels, i.e. the transition density function. This is the common feature of all the examples above and below. The objective is to find assumptions which on the one hand are sufficiently strong for the construction of a spectral theory and which on the other hand are so general that interesting classes of Feller operators, generators of Feller semigroups, are included. For instance the different operators K_0, that correspond to the above examples should be included.

This recently developed theory is called *stochastic spectral analysis*. The aim is to consider and study the spectral behavior of regularly and/or singularly perturbed Feller operators. Usually these perturbations are introduced stochastically and the basic tools for this theory also have the origin in stochastic analysis. The perturbations under consideration mainly consist of multiplication operators by appropriate potential functions. Here "appropriate" means that these potential functions possess a certain *Kato-Feller* property that makes them good candidates for perturbation theory. On the other hand this property can also be employed to introduce the generalized Schrödinger semigroups via the so-called Feynman-Kac formula, a formula from stochastic analysis. For the estimates we often use this kind of representation. For singular perturbations we also have such a Feynman-Kac formula but now with stopping times. All this will be explained in detail in Chapter 2.

B. Assumptions and Free Feller Generators

In what follows the function $p_0(t, x, y)$ defined on $(0, \infty)\times E\times E$ will be a continuous density function. The state space E (or configuration space) will be a second countable locally compact Hausdorff space with Borel field $\mathcal{E}$. A non-negative Radon measure m (*reference measure*) on $\mathcal{E}$ is given. Instead of $dm(x)$ or $m(dx)$ we usually write dx.

Basic Assumptions of Stochastic Spectral Analysis (BASSA)
By hypothesis the *continuous* function $(t, x, y) \mapsto p_0(t, x, y)$ possesses the following properties:

A1. Markov property:
It is non-negative and it verifies the Chapman-Kolmogorov identity, i.e.

$$\int p_0(s, x, z)p_0(t, z, y)dz = p_0(s+t, x, y), \quad s,\ t > 0, \quad x,\ y \in E,$$

and its total mass is less than or equal to 1, i.e.

$$\int p_0(t, x, y)dm(y) \leq 1, \quad t > 0 \quad x \in E.$$

A2. Feller property (see Definition 1.2):
For every $f \in C_\infty(E)$ the function $x \mapsto \int f(y)p_0(t, x, y)dm(y)$ belongs to $C_\infty(E)$.

A3. Continuity:
For every $f \in C_\infty(E)$ and for every $x \in E$ the following identity is true: $\lim_{t\downarrow 0} \int f(y)p_0(t, x, y)dm(y) = f(x)$.

A4. Symmetry:
The function $p_0(t, x, y)$ is symmetric: $p_0(t, x, y) = p_0(t, y, x)$ for all $t > 0$, and for all x and y in E.

1.1. Definition. The semigroup $\{\exp(-tK_0) : t \geq 0\}$ is given by

$$[\exp(-tK_0)\, f]\,(x) = \int p_0(t, x, y)f(y)dm(y),$$

for those functions f for which the integral on the right-hand side makes sense. This semigroup $\{\exp(-tK_0) : t \geq 0\}$ is said to be *L^1-L^∞-smoothing* if for every $t > 0$ the operator $\exp(-tK_0)$ sends the space $L^1(E, m)$ into $L^\infty(E, m)$. In view of our basic assumptions this is the same as saying that for every $t > 0$ the quantity (see Remark 2 below)

$$\|\exp(-tK_0)\|_{\infty,1} = \sup_{x,y\in E} p_0(t, x, y) \quad \text{is finite.}$$

The semigroup $\{\exp(-tK_0) : t \geq 0\}$ is said to have *finite spectral dimension* if there exists finite constants m, a_0 and c_0 such that $0 \leq p_0(t, x, y) \leq c_0 t^{-m/2}\exp(a_0 t)$ for all $t > 0$ and for all x, $y \in E$. By definition the quantity a_0 is an upper bound for the L^1-L^∞ type of the (un-perturbed) semigroup.

1.2. DEFINITION. A continuous function $f : E \to \mathbb{C}$ is said to belong to $C_\infty(E)$ if for every $\epsilon > 0$ there exists a compact subset K of E such that $|f(x)| \leq \epsilon$ for $x \notin K$. The space $C_\infty(E)$ is always equipped with the supremum-norm: $\|f\|_\infty = \sup_{x\in E} |f(x)|$, $f \in C_\infty(E)$. It is complete with respect to this norm.

REMARK 1. We will explicitly mention the L^1-L^∞-smoothing property whenever it is used. The same goes if we employ the finiteness of the spectral dimension. The L^1-L^∞-smoothing property will be carried over to the perturbed semigroups: see Theorem 2.5.(d).

REMARK 2. If T is a continuous linear operator from $L^p(E,m)$ to $L^q(E,m)$, $1 \leq p \leq \infty$, $1 \leq q \leq \infty$, then we write $\|T\|_{q,p}$ for its operator norm:

$$\|T\|_{q,p} = \sup\left\{ \|Tf\|_q : f \in L^p(E,m),\ \|f\|_p = 1 \right\}.$$

Of course, $\|f\|_p$ denotes the L^p-norm of f: $\|f\|_p = \left(\int |f(x)|^p \,dx\right)^{1/p}$.

REMARK 3. Given $p_0(t,x,y)$ satisfying BASSA, it is well known that there exists a strong Markov process

$$\{(\Omega, \mathcal{F}, \mathbb{P}_x), (X(t) : t \geq 0), (\vartheta_t : t \geq 0), (E, \mathcal{E})\}$$

(see e.g. Blumenthal and Getoor [35]) such that (1.1) below is valid. For more details on Markov processes and martingale theory, the reader is referred to Appendix C. The mappings $X(t)$ are called the *state variables*. They are defined on Ω and they attain their values in E or in $\triangle$, the point at infinity. $X(\infty, \omega)$ always has the interpretation $\triangle$. The relation between $X(t)$ and ϑ_s is expressed as follows: $X(t) \circ \vartheta_s = X(s+t)$. If T and S are stopping times, then $X(T) \circ \vartheta_S = X(S + T \circ \vartheta_S)$. The one-dimensional or marginal distributions are given by

$$\mathbb{P}_x(X(t) \in B) = \int_B p_0(t,x,y)dy, \quad t > 0, \quad B \text{ Borel subset of } E. \tag{1.1}$$

The sample paths are $\mathbb{P}_x$-almost surely right continuous and possess $\mathbb{P}_x$-almost sure left limits in E on their life time. In other words the process $\{X(t), \mathbb{P}_x\}$ is *cadlag* on its life time. Moreover, we may assume that the closure of the (random) set $\{X(s) : 0 \leq s < t\}$ is a compact subset of E whenever $X(t-)$ belongs to E. In other words, the process does not re-enter E once it has hit $\triangle$. It is quite useful to include processes of finite life time in our discussion. In general the standard Markov process associated to a Feller semigroup has finite life time; it need not have infinite life time. For example, solutions of stochastic differential equations are often Markov processes for their natural filtration. Whether or not they have infinite life time depends strongly on the coefficients. For more information on stochastic differential equations the reader is referred to e.g. Ikeda and Watanabe [137]. In this context we may introduce a *standard* Markov process as follows. Let Ω be the so-called *Skorohod space*: i.e. Ω consists of all paths

$\omega : [0,\infty] \to E^{\triangle}$, which are cadlag on their life time and which have the property that $\omega(s) = \triangle$ and $t \geq s$ imply $\omega(t) = \triangle$. The state variables $X(t)$ are then defined via the equality $X(t)(\omega) = X(t,\omega) = \omega(t)$ and the translation operators ϑ_s are given by $\vartheta_s(\omega)(t) = \omega(s+t)$. Put $\zeta = \inf\{s > 0 : X(s) = \triangle\}$ (ζ is the *life time* of our Markov process). Then, again by definition, a strong Markov process is *standard* if, for every increasing sequence of $\mathcal{F}_t$-stopping times $(T_n : n \in \mathbb{N})$, with $T = \sup_{n\in\mathbb{N}} T_n$, we have $\lim_{n\to\infty} X(T_n) = X(T)$ on the event $\{T < \zeta\}$, $\mathbb{P}_x$-almost surely. (This property is called *quasi left continuity*.) If we require instead that we have $\lim_{n\to\infty} X(T_n) = X(T)$ on the event $\{T < \infty\}$, $\mathbb{P}_x$-almost surely, then our Markov process is called a *Hunt process*. We also notice $\mathbb{E}_x\left[Y \mid \mathcal{F}_0\right] = \mathbb{E}_{X(0)}\left[Y\right] = \mathbb{E}_x\left[Y\right]$, $\mathbb{P}_x$-almost surely, where we use $\mathbb{P}_x\left[X(0) = x\right] = 1$. In fact Blumenthal and Getoor [35] prove that there exists a standard strong Markov process (in fact a Hunt process) $\{(\Omega,\mathcal{F},\mathbb{P}_x),(X(t):t\geq 0),(\vartheta_t : t \geq 0),(E,\mathcal{E})\}$ with the one-dimensional distributions given as in (1.1). Strong Markov processes associated to Feller semigroups are called *Feller processes*; consequently Feller processes are Hunt processes.

REMARK 4. It is not necessarily true that integral densities $p_0(t,x,y)$ are available. In principle one may formulate the basic assumptions (BASSA) in terms of the transition function $P(t,x,B) := \mathbb{P}_x(X(t) \in B)$, $t \geq 0$, $x \in E$, $B \in \mathcal{E}$, where $\mathcal{E}$ is the collection of Borel subsets of E. This can be done for instance for diffusion processes with a symmetric positive-definite diffusion matrix and Hölder continuous diffusion and drift coefficients. However the assumption of the availability of densities does not seem to be too restrictive: the reader may consult e.g. Azencott et. al. [14].

REMARK 5. In the presence of property A1, the continuity property A3 is equivalent to the following:

$$\lim_{s\to t,s>0} \left\|\exp\left(-sK_0\right)f - \exp\left(-tK_0\right)f\right\|_\infty = 0$$

for all $f \in C_\infty(E)$. This is so because for a bounded measure μ on $\mathcal{E}$, property A3 together with Lebesgue's dominated convergence will imply that

$$\lim_{t\downarrow 0} \int \exp\left(-tK_0\right) f d\mu = \int f d\mu$$

for all $f \in C_\infty(E)$. So property A3 is equivalent to the weak continuity. However, for semigroups weak operator continuity and strong operator continuity coincide: see e.g. Yosida [255, p. 233]. It is noticed that, in general, the space $C_\infty(E)$ cannot be replaced with the space $L^\infty(E,\mathcal{E})$ because, if this were the case, then the semigroup $\{\exp(-sK_0) : s \geq 0\}$ would be strongly continuous in the larger space $L^\infty(E,\mathcal{E})$ consisting of those bounded functions which are Borel measurable. But in such a space, a strongly continuous semigroup is uniformly continuous, which is the same as saying that the operator K_0 is bounded. For more details, the reader is referred to Lotz [169].

It is mentioned that in the Chapters 4, 5, and 6 we will usually employ the standard assumptions A1-A4, whereas in Chapters 7 and 8 we often will require that the unperturbed semigroup $\{\exp(-tK_0) : t \geq 0\}$ is L^1-L^∞-smoothing.

1.3. DEFINITION. Let BASSA be given. Then the function $p_0(t, x, y)$ is the kernel of a Feller semigroup, because of property A2. We denote its L^2-generator with K_0. It is called the *(free) Feller generator*, because it is the generator of a Feller semigroup. Property A2 can be rewritten as

$$\exp(-tK_0)\, C_\infty(E) \subseteq C_\infty(E).$$

By definition the operator K_0 is then the L^2-generator of the Markov process

$$\{(\Omega, \mathcal{F}, \mathbb{P}_x), (X(t) : t \geq 0), (\vartheta_t : t \geq 0), (E, \mathcal{E})\}.$$

For any Borel function g, defined on E, we write

$$[\exp(-sK_0)g](x) = \mathbb{E}_x(g(X(s)) = \int p_0(s, x, y) g(y) dy$$

and for a a strictly positive number

$$\begin{aligned}\left[(aI + K_0)^{-1} g\right](x) &= \int_0^\infty e^{-as} \left[\exp(-sK_0)g\right](x) ds \\ &= \int_0^\infty e^{-as} \int p_0(s, x, y) g(y) dy\, ds\end{aligned}$$

whenever these expressions make sense.

Important note: At a first reading the remaining part of the present chapter may be skipped and the reader can continue with Chapter 2 on page 53.

In order to discuss the properties of BASSA we begin with a definition and a result, due to Varopoulos, on the (spectral) dimension of a semigroup. This explains the usefulness of the boundedness assumption of Definition 1.1.

1.4. DEFINITION. The (kernel of the) semigroup $\{\exp(-tK_0) : t \geq 0\}$ is said to be of *spectral dimension* n, if $\|\exp(-tK_0)\|_{\infty,1} \leq Ct^{-n/2}$ for all (small) $t > 0$.

In [243], Varopoulos shows that this is equivalent to saying that the generator K_0 verifies $\|f\|^2_{2n/(n-2)} \leq C \langle K_0 f, f\rangle$ for all $f \in \mathrm{dom}(K_0)$, provided $n > 2$. In [55] Coulhon gives a simple proof of this equivalence and in [60] the authors, Coulhon and Saloff-Coste, apply this result to semigroups of operators acting on functions defined on a Lie group. Following Nash [184], the authors of [44] use the Dirichlet form associated to a semigroup to characterize the dimension of the semigroup. For more information on this line of thought, the reader should consult Coulhon, Saloff-Coste and Varopoulos [61]. Let $\mathcal{E}^0$ be the Dirichlet form associated to the semigroup $\{\exp(-tK_0) : t \geq 0\}$, i.e.

$$\mathcal{E}^0(f, f) = \lim_{t \downarrow 0} \frac{1}{t} \left[\langle f, f\rangle - \langle f, \exp(-tK_0) f\rangle\right], \quad f \in \mathrm{dom}(\mathcal{E}^0).$$

Then, for some $\delta > 0$,

$$\|\exp(-tK_0)\|_{\infty,1} \leq C_1 e^{\delta t} t^{-n/2},$$

for $t > 0$, if and only if

$$\|f\|_2^{2+4/n} \leq C_2 \left[\mathcal{E}^0(f,f) + \delta \|f\|_2^2\right] \|f\|_1^{4/n}$$

for all $f \in \text{dom}(\mathcal{E}^0)$. The constant C_1 depends on C_2 and on n and the constant C_2 depends on C_1 and n. In [15] and in [16] the authors, Bakry and Bakry and Michel, explain the precise relations that exist between the (weak) logarithmic Sobolev inequalities and the finite dimensionality of the semigroup. In [67] and in [68] this topic is treated as well. In [67] Davies also extends bounds of the type $\|\exp(-tK_0)\|_{\infty,1} \leq Ct^{-n/2}$, which can be considered as bounds on the diagonal of $\exp(-tK_0)(x,y)$, to global bounds for the density $\exp(-tK_0)(x,y)$. For that purpose he employs the so-called square gradient operator or, in French, the *carré du champ* operator. This operator will be explained in Definition 1.7 in section D of this chapter.

In the latter inequalities, the constant δ could be considered as an upper bound for the *L^1-L^∞-type* of the semigroup $\{\exp(-tK_0) : t \geq 0\}$. The type δ determines the large time behavior in the condition on the L^1-L^∞-smoothing.

The next criteria will illustrate Assumption A3. They give sufficient conditions for the continuity of paths. A proof can be found in Blumenthal and Getoor [35, Proposition 9.10]. Markov processes for which the condition in Proposition 1.5. is satisfied are called *diffusions.*

1.5. PROPOSITION. *Suppose* $\lim_{t\downarrow 0} \sup_{x\in K} \frac{1}{t} \int_{E\setminus U} p_0(t,x,y)dy = 0$ *for every open subset U of E together with every compact subset K of U. Then the sample paths in Ω are $\mathbb{P}_x$-almost surely continuous on their life time.*

The following result can de found in Smits [212].

1.6. PROPOSITION. *Let* $\{(\Omega, \mathcal{F}, \mathbb{P}_x), (X(t) : t \geq 0), (\vartheta_t : t \geq 0), (E, \mathcal{E})\}$ *be as above. The following assertions are equivalent:*

(a) $\lim_{t\downarrow 0} \frac{1}{t} \sup_{x\in K} \int_{E\setminus U} p_0(t,x,y)dy = 0$ *for every open subset U of E together with every compact subset K of U.*

(b) *Let T_U be the first hitting time after 0 of the complement of U:* $T_U = \inf\{s > 0 : X(s) \in E \setminus U\}$. *(The infimum over the empty set has the interpretation ∞.) Then* $\lim_{t\downarrow 0} \frac{1}{t} \sup_{x\in K} \mathbb{P}_x(T_U \leq t) = 0$ *for every open subset U of E together with every compact subset K of U.*

Instead of "first hitting time after 0" we usually say just "first hitting time" or even "hitting time".

We want to mention some problems.

PROBLEMS. Suppose that for some $q > 2$ the inequality

$$\|u\|^2_{L^q(E,m)} \leq c_1 \langle K_0 u, u\rangle + c_2 \langle u, u\rangle$$

is valid for all $u \in \mathrm{dom}(K_0) \cap L^q(E,m)$. Then there exists a set N of capacity 0 with the property that for every $x \notin N$ the measure $B \mapsto \mathbb{P}_x(X(t) \in B)$ has a density with respect to m and hence, under this condition, there exists an $m\otimes m$-almost everywhere defined density $p_0(t,x,y)$ such that $\mathbb{P}_x(X(t) \in B) = \int_B p_0(t,x,y)dm(y)$: see Fukushima [107] and also Bouleau [38]. For more details on the notion of capacity, the reader is referred to Definition D.17 in Appendix D. However, we would like to have integral kernels $p_0(t,x,y)$ that are everywhere defined and continuous on $(0,\infty) \times E \times E$. Furthermore, it is not clear to what extent the Dirichlet form can be used to see that the corresponding semigroup acting on $L^2(E,m)$ is in fact a Feller semigroup, so that it leaves the space $C_\infty(E)$ invariant. The authors do not know the precise answers to these two questions (the one on the existence of everywhere defined continuous integral kernels and the other one on the Feller property). See the comments following the Definition 1.21 in section E of this chapter and also Example 3 in section C below. In [176] McGillivray gives a partial answer to these questions in the presence of Muckenhaupt weights. In fact consider the operator

$$R^{r/2} := \Gamma\left(\frac{r}{2}\right)^{-1} \int_0^\infty t^{r/2-1} e^{-t} T(t) dt,$$

which corresponds to the operator $(K_0 + I)^{-r/2}$. The norm

$$\|f\|_{r,2} := \left\|(K_0 + I)^{r/2} f\right\|_2,$$

turns the space $R^{r/2}L^2(E,m)$ into a Banach space. Let $C_b(E)$ denote the space of bounded continuous functions on E. Suppose that the following conditions hold true ($\mathrm{ran}(T) = \{Tx : x \in \mathrm{dom}(T)\}$ as explained in Appendix A):

$$\int_0^\infty t^{r/2-1} e^{-t} \|\exp(-tK_0)\|_{2,\infty} dt < \infty \quad \text{and} \tag{1.2}$$

$$\mathrm{ran}\left((K_0+I)^{-r/2}\right) \cap C_b(E) \quad \text{is dense in} \left(\mathrm{ran}\left((K_0+I)^{-r/2}\right), \|\cdot\|_{r,2}\right). \tag{1.3}$$

Then $T(t)f$ admits a continuous m-version for all $f \in L^2(E,m)$, $t > 0$. As a particular example of the above assertion, let U be an open subset of $\mathbb{R}^\nu$, and suppose that $\rho > 0$, $\rho \in L^1_{\mathrm{loc}}(U, dx)$, belongs to the Muckenhaupt class of order 2, i.e. there exists a $c > 0$ such that

$$\frac{1}{|B|}\int_B \rho(x)dx \times \frac{1}{|B|}\int_B \rho^{-1}(x)dx \leq c,$$

for all balls $B \subset U$. Define

$$\mathcal{E}_\rho(f,f) := \int_U \left| \left[K_0^{1/2} f \right](x) \right|^2 \rho(x) dx,$$

for $f \in C_0^\infty(U)$. Then $(C_0^\infty(U), \mathcal{E}_\rho)$ is closable, with closure $(F, \mathcal{E}_\rho)$, say. Assuming a global Sobolev inequality of the form

$$\|f\|_\eta^2 \leq c\,\mathcal{E}_\rho(f,f), \quad f \in F,$$

for some $\eta > 2$, the conditions in (1.2) and (1.3) of the above result are verified, and so $T(t)f$ has a continuous version for all $f \in L^p(\mathbb{R}^\nu, dx)$, if p is sufficiently large.

C. Examples

Before we introduce regular or singular perturbations of the free Feller operator K_0, we return to the examples given above in section A of this chapter and we collect some others. Moreover, the hypotheses in BASSA are checked for these examples. Not every restriction on $p_0(t,x,y)$ in BASSA is of the same depth. The most important feature is the Feller property. This is a basic assumption to start with the semigroup theory in C_∞ and to extend this to L^p-spaces: see e.g. van Casteren [231]. The next main property is the continuity assumption in property A3. This property selects in fact the underlying Markov process with cadlag paths. We may assume that the paths are right continuous on $[0,\infty]$ and that, on their life time, they possess left limits in the state space E. Thus we discuss the examples in such a way that every assumption in BASSA is considered separately. The intersection of the assumptions is left to the reader.

EXAMPLE 1. (Kochubeĭ [157]) Of course, second order differential operators with variable coefficients are the first obvious generalizations of the Laplace operator. But as long as we deal with canonical diffusions certain aspects of this generalization are almost trivial. More or less following Dynkin [94], we define a *canonical* diffusion in the following way. Let the operator K_0 be given by

$$K_0 = -\frac{1}{2} \sum_{i,j=1}^{\nu} a_{ij}(x) \frac{\partial^2}{\partial x_i \partial x_j} + \sum_{j=1}^{\nu} b_j(x) \frac{\partial}{\partial x_j} + c(x). \tag{1.4}$$

If K_0 is the generator of a diffusion process with transition density function $p_0(t,x,y)$, then the matrix of its *diffusion* coefficients $(a_{ij}(x))_{i,j=1}^{\nu}$ is given by

$$a_{ij}(x) := \lim_{t \downarrow 0} \frac{1}{t} \int_{\{y:|y-x|<\epsilon\}} (y_i - x_i)(y_j - x_j) p_0(t,x,y) dy$$

for any $\epsilon > 0$. Such a process is said to be *canonical* if its transition density can be estimated by the Wiener density, i.e. if

$$p_0(t,x,y) \leq c_1 \frac{1}{(2\pi t)^{\nu/2}} \exp\left(-c_2 \frac{|x-y|^2}{2t}\right), \quad 0 < t < t_0.$$

This assumption implies that the entries $a_{ij}(x)$, $1 \leq i,\, j \leq \nu$, of the diffusion matrix are bounded. The study of processes, which are generated by operators K_0 of the form (1.4) with not necessarily bounded coefficients a_{ij}, $1 \leq i,,\, j \leq \nu$, or b_j, $1 \leq j \leq \nu$, is an old topic studied in the literature for many situations. One of our aims is to allow rather general conditions on the *diffusion* coefficients $a_{ij}(x)$ and the *drift* coefficients $b_j(x)$ and still have generators of *diffusion* processes. In [94, p. 167] Dynkin supposes that the coefficients a_{ij} and b_j, $1 \leq i,\, j \leq \nu$, are bounded and (locally) Hölder continuous of some order. Moreover, he assumes that the diffusion coefficients a_{ij}, $1 \leq i,\, j \leq \nu$, satisfy the following ellipticity condition:

$$\sum_{i,j=1}^{\nu} \lambda_i \lambda_j a_{ij}(x) \geq \gamma \sum_{j=1}^{\nu} |\lambda_j|^2,$$

for all $(\lambda_1, \ldots, \lambda_\nu) \in \mathbb{R}^\nu$. It then follows that the corresponding transition density can be estimated by the Wiener density: see Dynkin [94, Theorem 0.5. page 229, part II]. This ellipticity condition can be relaxed to a condition of Hörmander type: see Example 8 below. Here, we have chosen a rather recent example by Kochubeĭ [157]. We do not intend to give an exhaustive list of all possible examples. Even if this were possible, it is not necessary, because the whole general theory is designed in such a way that new examples and further applications can be accommodated. Up to some extent one aim of the book is to initiate the research for a more or less exhaustive set of applications and examples. The reader is invited to explore the borderlines of the theory. Although we have collected a variety of examples we are sure that the list of interesting examples is by no means complete.

Kochubeĭ [157, Theorem 2] gives rather general conditions on the coefficients of the operator in (1.4) which guarantee that it generates a Feller semigroup (Markov process) with continuous density. In fact the assumptions on the coefficients are the following:

(a) The functions a_{ij} are twice differentiable and the partial derivatives $\dfrac{\partial^2 a_{ij}}{\partial x_k \partial x_\ell}$, belong to $L^1_{\mathrm{loc}}(\mathbb{R}^\nu)$, $1 \leq k, \ell \leq \nu$. The functions $\displaystyle\sum_{j=1}^{\nu} \frac{\partial a_{ij}}{\partial x_j}$, $1 \leq i \leq \nu$, and $\displaystyle\sum_{i,j}^{\nu} \frac{\partial^2 a_{ij}}{\partial x_i \partial x_j}$ are locally Hölder continuous.

(b) The functions b_j, $1 \leq j \leq \nu$, belong to $L^\nu_{\mathrm{loc}}(\mathbb{R}^\nu)$ and the function $\displaystyle\sum_{j=1}^{\nu} \frac{\partial b_j}{\partial x_j}$ belongs to $L^{\max(1,\nu/2)}_{\mathrm{loc}}(\mathbb{R}^\nu)$.

(c) The inequality

$$\sum_{i,j}^{\nu} \frac{\partial^2 a_{ij}(x)}{\partial x_i \partial x_j} \leq 0 \tag{1.5}$$

is valid for all $x \in \mathbb{R}^\nu$. In addition,

$$c(x) \geq \max\left(0, \sum_{j=1}^{\nu} \frac{\partial b_j(x)}{\partial x_j}\right) \geq 0,$$

for all $x \in \mathbb{R}^\nu$.

Under the assumptions (a), (b) and (c) the operator K_0 generates a Feller semigroup and hence a Markov process

$$\{(\Omega, \mathcal{F}, \mathbb{P}_x), (X(t) : t \geq 0), (\vartheta_t : t \geq 0), (E, \mathcal{E})\}$$

with state space $E = \mathbb{R}^\nu$ and with a density. Moreover, the one-dimensional distribution $\mathbb{P}_x(X(t) \in A)$ are given by $\mathbb{P}_x(X(t) \in A) = \int_A p_0(t, x, y)dy$. It is not so difficult to give additional conditions in order that all the assumptions in BASSA are satisfied. For instance for the self-adjointness we refer to Weidmann [252], Berezanskii [24], Schechter [203], Glazman [113], Jörgens and Weidmann [146]. If $c \equiv 0$ and the functions a_{ij} and b_j, $1 \leq i, j \leq \nu$ are uniformly bounded then the corresponding Markov process has infinite lifetime. If we want to include more general coefficients we have to allow processes with only finite life time. Notice that (a) and (b) are local assumptions, whereas the assumption in (c) is something like a condition of dissipativeness.

EXAMPLE 2. (Davies [66]) As mentioned above, Davies considered generators of the form

$$K_0 f = -\frac{1}{\sigma^2} \nabla \left(\sigma^2 \nabla f\right), \tag{1.6}$$

on a locally finite Riemannian manifold E. The conditions on σ are very general. Davies only assumes that σ is a measurable function for which both σ and its reciprocal σ^{-1} belong to $L^\infty_{\text{loc}}(E)$. It need not be differentiable. Under these assumptions Davies shows that $\{\exp(-tK_0) : t \geq 0\}$ is a positivity preserving, strongly continuous contraction semigroup on $L^p(E)$, $1 \leq p < \infty$. This semigroup possesses a kernel $p_0(t, x, y)$. The pointwise upper bounds for $p_0(t, x, y)$ depend on the geometry of the manifold E and also on the metric. Davies needs conditions on the local geometry expressed in terms of Harnack's inequality. Here we shall not repeat the technical condition of the weak bounded geometry. Pointwise estimates for the kernel $p_0(t, x, y)$ allow us to find sufficient conditions on E and σ in order that the semigroup $\{\exp(-tK_0) : t \geq 0\}$ has the Feller property. The semigroup $\{\exp(-tK_0) : t \geq 0\}$ is self-adjoint in $L^2(E, \sigma^2 dm)$.

The results are applicable to elliptic operators on Euclidean space $\mathbb{R}^\nu$. In the self-adjoint case Davies considers

$$K_0 := -B^{-1} \sum_{i,j=1}^{\nu} \frac{\partial}{\partial x_i} \left(A_{ij} \frac{\partial f}{\partial x_j} \right). \tag{1.7}$$

The functions $B(x)$ and $B(x)^{-1}$ are assumed to be measurable and locally bounded on $\mathbb{R}^\nu$. The position dependent matrix (A_{ij}) is real and symmetric. It is of the form

$$A_{ij}(x) = C(x) D_{ij}(x),$$

with $\det(D(x)) = 1$ and with $0 < c^{-1} \leq D(x) \leq c < \infty$ (in form sense), and where the functions $C(x)$ and $C(x)^{-1}$ are locally bounded. This problem is then reformulated for a Riemannian manifold with a metric given in terms of B, C and D. For special ratios $\dfrac{B(x)}{C(x)}$, it will follow that $\exp(-tK_0)$ conserves the probability or is *conservative*, i.e. $\exp(-tK_0)1 = 1$. In that case the life time ζ of the corresponding Markov process is $\mathbb{P}_x$-almost surely infinite for all $x \in E$. Finally this theory is applicable to operators given via the quadratic form

$$\mathcal{E}^0(f,f) := \int_E |\nabla f(x)|^2 \sigma(x)^2 dx, \quad \sigma, \quad \sigma^{-1} \in L^\infty_{\text{loc}}. \tag{1.8}$$

The operator K_0 on $L^2(\mathbb{R}^\nu, \sigma^2(x)dx)$ associated with $\mathcal{E}^0$ is unitarily equivalent to a Schrödinger operator $-\Delta + V$ with a singular potential V. For smooth enough σ we may choose $V = \dfrac{\operatorname{div}\sigma}{\sigma}$. But non-smooth functions σ may be considered as well. For instance if we suppose that there exist strictly positive constants c_1 and r for which

$$0 < c_1^{-1} \leq \frac{\sigma(x)}{\sigma(y)} \leq c_1$$

for all x, y with $|y - x| \leq r$, then the kernel of $\exp(-tK_0)$ can be estimated by

$$0 \leq p_0(t,x,y) \leq c_2 \max\left[\frac{2^\nu}{r^\nu}, \frac{1}{t^{\nu/2}}, \frac{|x-y|^\nu}{t^\nu}\right] \exp\left(-\frac{|x-y|^2}{4t}\right) \tag{1.9}$$

for all $t > 0$ and for all x and $y \in \mathbb{R}^\nu$. The constant c_2 depends on c_1 and on the dimension. If one chooses more special functions σ, then the Feller property will follow. For example, suppose that

$$\sigma(x) \geq c^{-1} \exp\left(-c|x|^2\right), \quad \text{for all } x \in \mathbb{R}^\nu,$$

$$c^{-1} \leq \frac{\sigma(x)}{\sigma(y)} \leq c \text{ for all } x, y \text{ in } \mathbb{R}^\nu \text{ with } |x-y| \leq c^{-1}(1+|x|)^{-c}. \tag{1.10}$$

Then the semigroup $\{\exp(-tK_0) : t \geq 0\}$ possesses the Feller property. Here K_0 is the operator associated with the quadratic form (1.8). For more information the reader may consult Davies' book [67] and his paper [66].

EXAMPLE 3. This example is due to N. Jacob (cfr. [140] and [143]). Jacob considers pseudo-differential operators of the form

$$[p(x,D)u](x) = \frac{1}{(2\pi)^{\nu/2}} \int_{\mathbb{R}^\nu} e^{ix\cdot\xi} p(x,\xi)\widehat{u}(\xi)d\xi \tag{1.11}$$

with certain symbols $p(x,\xi)$. To be more precise, Jacob puts himself in the following context. Let $p : \mathbb{R}^\nu \times \mathbb{R}^\nu \to \mathbb{R}$ be a continuous function such that for fixed $\xi \in \mathbb{R}^\nu$ the function $x \mapsto p(x,\xi)$ is a bounded C^∞-function with bounded derivatives of all orders. Suppose that for fixed $x \in \mathbb{R}^\nu$, the function $\xi \mapsto p(x,\xi)$ is negative definite. In addition, Jacob assumes that there exists a continuous negative-definite function $a : \mathbb{R}^\nu \mapsto \mathbb{R}$ such that, for some $0 < r \leq 2$,

$$c_0c_1\left(1+|\xi|^2\right)^{r/2} \leq c_1\left(1+a(\xi)^2\right)^{1/2} \leq p(x,\xi) \leq c_2\left(1+a(\xi)^2\right)^{1/2}. \tag{1.12}$$

Define the Sobolev spaces $H^{q,a}(\mathbb{R}^\nu)$, $q \geq 0$, as follows:

$$H^{q,a}(\mathbb{R}^\nu) = \left\{u \in L^2(\mathbb{R}^\nu) : \|u\|_{q,a} < \infty\right\}, \tag{1.13}$$

where $\|\cdot\|_{q,a}$ is given by:

$$\|u\|_{q,a}^2 = \int_{\mathbb{R}^\nu} \left(1+a(\xi)^2\right)^q \left|\widehat{u}(\xi)\right|^2 d\xi. \tag{1.14}$$

If $a(\xi) \equiv |\xi|$, then we just write $H^q(\mathbb{R}^\nu)$ instead of $H^{q,a}(\mathbb{R}^\nu)$ and $\|\cdot\|_q$ replaces $\|\cdot\|_{q,a}$. Put $H^\infty(\mathbb{R}^\nu) = \bigcap_{q\geq 0} H^q(\mathbb{R}^\nu)$. Then the pseudo-differential operator $p(x,D)$ is defined on $H^\infty(\mathbb{R}^\nu)$. In order to ensure that this operator generates a Feller semigroup, Jacob makes the following assumptions:

(a) For every $t \geq 0$ there exists $s \geq 0$ such that

$$\|p(x,D)u\|_t \leq c\,\|u\|_s$$

for all $u \in H^\infty(\mathbb{R}^\nu)$.

(b) If u is any function in $H^{1,a}(\mathbb{R}^\nu)$ with the property that $p(x,D)u$ belongs to $C_0^\infty(\mathbb{R}^\nu)$, then u belongs to $H^\infty(\mathbb{R}^\nu)$.

(c) The operator $p(x,D)$ possesses an extension $p(x,D) : H^{1,a}(\mathbb{R}^\nu) \to L^2(\mathbb{R}^\nu)$, that is continuous and that is a closed self-adjoint operator on $L^2(\mathbb{R}^\nu)$ with domain $H^{1,a}(\mathbb{R}^\nu)$. Furthermore it is assumed that, for some $\lambda \geq 0$, the operator $p(x,D) + \lambda I$ has a continuous inverse defined on the whole space $L^2(\mathbb{R}^\nu)$.

Under the assumptions (a), (b) and (c) the operator $-p(x,D)$ generates a Feller semigroup acting on $C_\infty(\mathbb{R}^\nu)$. If, in addition, the inequality

$$\|u\|_{L^q}^2 \leq c\left(\mathcal{E}^0(u,u) + c_0\,\|u\|_2^2\right), \quad u \in \operatorname{dom}(\mathcal{E}^0), \tag{1.15}$$

(for the definition of $\mathcal{E}^0$ see (1.16) below) holds for some $q > 2$ and some constants c and c_0, then the associated Markov process

$$M = \{(\Omega, \mathcal{F}, \mathbb{P}_x), (X(t) : t \geq 0), (\vartheta_t : t \geq 0), (E, \mathcal{E})\},$$

where $E = \mathbb{R}^\nu$, possesses the following property. There exists a Borel set N of capacity zero (see Definition 2.36) such that $\mathbb{R}^\nu \setminus N$ is M-invariant (i.e. for every $x \in \mathbb{R}^\nu$ the probability $\mathbb{P}_x [X(t) \in N$ for some finite $t \geq 0] = 0$) and such that the following hold:

(i) The resolvent kernel $R(\lambda)(x, \cdot)$ is absolutely continuous with respect to the Lebesgue measure for every $\lambda > 0$ and for each $x \in \mathbb{R}^\nu \setminus N$.

(ii) The transition function $\exp(tL)(x, \cdot)$ is absolutely continuous with respect to the Lebesgue measure for every $t > 0$ and for each $x \in \mathbb{R}^\nu \setminus N$.
Here $L = p(x, D)$ and $\mathcal{E}^0$ is the corresponding Dirichlet form:

$$\mathcal{E}^0(u, v) = \lim_{t \downarrow 0} \frac{\langle u, v \rangle - \langle \exp(tL)u, v \rangle}{t}, \quad u, v \in \operatorname{dom}(\mathcal{E}^0). \tag{1.16}$$

The function u belongs to $\operatorname{dom}(\mathcal{E}^0)$ if and only if the limit in the right-hand side of (1.16) exists with $v = u$. Because of (1.16) the semigroup generated by $p(x, D)$ possesses a symmetric density. Of course, the class of pseudo-differential operators considered by Jacob has to satisfy several abstract conditions mentioned above. Nevertheless there are interesting examples of pseudo-differential operators $p(x, D)$ that verify all these assumptions.

In [140] Jacob studies operators $L = p(x, D)$ of the form

$$[Lu](x) = \sum_{j=1}^{\nu} b_j(x) \left[\lambda_j^{2r}(D)u\right](x), \tag{1.17}$$

where $\left[\lambda_j^{2r}(D)\right](x)$ is given by

$$\left[\lambda_j^{2r}(D)\, u\right](x) = \frac{1}{(2\pi)^{\nu/2}} \int_{\mathbb{R}^\nu} e^{ix\cdot\xi} \lambda_j^{2r}(\xi) \hat{u}(\xi) d\xi. \tag{1.18}$$

Here $\lambda_j^{2r}(\xi) = |\xi_j|^{2r}$, $\xi \in \mathbb{R}^\nu$. Moreover, in order to have self-adjoint Feller operators, Jacob assumes that $0 < r \leq 1$ and that the functions b_j do not depend on x_j, i.e. $\dfrac{\partial b_j}{\partial x_j} = 0$. In addition, he supposes that

$$\max_{1 \leq j \leq \nu} \sup_{x \in \mathbb{R}^\nu} |b_j(x) - b_j(x_0)| < \frac{\widetilde{c}_{\nu,2r}}{2\nu} \min_{1 \leq j \leq \nu} \inf_{x \in \mathbb{R}^\nu} b_j(x) \tag{1.19}$$

for some $x_0 \in \mathbb{R}^\nu$. The constant $\widetilde{c}_{\nu,2r}$ may be chosen as

$$\widetilde{c}_{\nu,2r} = \inf_{\xi \in \mathbb{R}^\nu} \frac{1 + \sum_{j=1}^{\nu} \lambda_j^{2r}(\xi)}{\left(1 + |\xi|^2\right)^{2r}}. \tag{1.20}$$

Then $\mathcal{E}^0(u,v)$ is given by

$$\mathcal{E}^0(u,v) = \sum_{j=1}^{\nu} \int b_j(x) \left[\lambda_j^r(D)u\right](x) \overline{\left[\lambda_j^r(D)v\right](x)} dx \tag{1.21}$$

and $\mathcal{E}^0$ satisfies (1.15). To $\mathcal{E}^0$ (or to L) there corresponds a Feller semigroup with symmetric integral kernel. In our notation $K_0 = -L$. Then Jacob shows that $\exp(-tK_0)$ is L^1-L^∞-smoothing, i.e.

$$\|\exp(-tK_0)\|_{\infty,1} \leq c_1 \exp(c_2 t)\, t^{-\nu/(2r)}.$$

In other articles Jacob generalizes the class of examples given in (1.17). In [141] Jacob studies operators

$$p(x,D) = \sum_{j=1}^{\nu} b_j(x) \lambda_j^2(D_j),$$

where $\lambda_j^2 : \mathbb{R} \to \mathbb{R}$ are more general continuous negative-definite functions then $|\xi_j|^{2r}$. Finally this procedure generalizes to pseudo-differential operators, where the symbols are not necessarily products of functions depending on x respectively ξ.

In the following two examples we consider relativistic Hamiltonians, which were introduced by Ichinose (see e.g. [127, 128, 129, 130, 131, 132]). For systems without electromagnetic fields we use the notation of Carmona, Masters and Simon [48].

EXAMPLE 4. (See also the explanation in the beginning of this chapter.) For the sake of completeness, Example 4 is repeated here. This example is taken from Carmona, Masters and Simon [48]. Let μ be a non-negative measure on $\mathbb{R}^\nu$ with the property that $\int_{\mathbb{R}^\nu} \min\left(1, |x|^2\right) d\mu(x) < \infty$, let a and b be vectors in $\mathbb{R}^\nu$ and let C be a square $\nu \times \nu$ matrix. In addition, let $h : \mathbb{R}^\nu \to \mathbb{R}^\nu$ be a function of compact support with the property that $h(x) = x$ for all x in a neighborhood of the origin. Define the negative-definite function $F : \mathbb{R}^\nu \to \mathbb{R}^\nu$ by

$$F(p) = a + ib \cdot p + p \cdot Cp - \int_{\mathbb{R}^\nu} \left[e^{ip\cdot x} - 1 - ip \cdot h(x)\right] d\mu(x). \tag{1.22}$$

Here a function $F : \mathbb{R}^\nu \to \mathbb{C}$ is called *negative-definite*, if for every n-tuple $(\lambda_1, \dots, \lambda_n)$ of complex numbers together with every n-tuple of vectors $(\xi_1, \dots, \xi_n)$ in $\mathbb{R}^\nu$, the inequality

$$\sum_{k,\ell=1}^{n} \lambda_k \overline{\lambda}_\ell \left(F(\xi_k) + \overline{F(\xi_\ell)} - F(\xi_k - \xi_\ell)\right) \geq 0$$

holds. Then there exists an operator K_0 that generates a semigroup

$$\{\exp(-tK_0) : t \geq 0\}$$

with the property that $\exp(-tK_0)$ is given by

$$[\exp(-tK_0)f](x) = \int_{\mathbb{R}^\nu} f(x+y)dm_t(y), \tag{1.23}$$

where f belongs to $C_\infty(\mathbb{R}^\nu)$. Here the family $\{m_t : t \geq 0\}$ is a vaguely continuous convolution semigroup of probability measures on $\mathbb{R}^\nu$ with the property that the Fourier transform $\widehat{m_t}$ of m_t is given by $\widehat{m_t}(p) = \exp(-tF(p))$, $p \in \mathbb{R}^\nu$ (see Example 4 in section A of this chapter). If, in addition, the integrals $\int_{\mathbb{R}^\nu} \exp(-tF(p))dp$ are finite, then the function $p_0(t,x,y)$ defined by

$$p_0(t,x,y) = \frac{1}{(2\pi)^\nu}\int_{\mathbb{R}^\nu} \exp(-tF(p) + ip\cdot(x-y))\,dp \tag{1.24}$$

serves as the density of the corresponding conservative Markov process which is in fact a *Lévy process* (i.e. a process with independent increments). Since $\int p_0(t,x,y)dy = 1$, the process is conservative. Two cases will be considered in more detail. They are special cases of the pseudo-differential operators studied in Example 3. In case $F(p) \equiv |p|^\alpha$, $0 < \alpha \leq 2$ (stable case), the following inequalities can be proved:

$$\frac{c_1}{t^{\nu/\alpha}|x-y|^{\nu+\alpha}} \leq p_0(t,x,y) \leq \frac{c_2}{t^{\nu/\alpha}|x-y|^{\nu+\alpha}}, \quad |x-y| \geq 1 \tag{1.25}$$

if $\alpha \neq 2$. In general there exist no explicit formulas for the densities of the corresponding semigroups. An exception is the case $F(p) = \sqrt{p^2+m^2} - m$, m fixed. This corresponds to the *relativistic Hamiltonian* $K_0 = \sqrt{-\Delta + m^2} - m$. The density of the corresponding semigroup is given by

$$p_0(t,x,y)$$

$$= \frac{1}{(2\pi)^\nu}\frac{t}{\sqrt{|x-y|^2+t^2}}\int_{\mathbb{R}^\nu} \exp\left(mt - \sqrt{\left(|x-y|^2+t^2\right)(p^2+m^2)}\right)dp \tag{1.26}$$

$$= \int_0^\infty \frac{\exp(mt-m^2u)}{(4\pi u)^{\nu/2}}\exp\left(-\frac{|x-y|^2}{4u}\right)\frac{t}{u\sqrt{u\pi}}\exp\left(-\frac{t^2}{4u}\right)du. \tag{1.27}$$

It is obvious that the assumptions in BASSA are satisfied for this integral kernel. Following a suggestion of Ben-Artzi, the spectral dimension of this semigroup can be found via the procedure below. First we notice that the spectral measure $E_F(\lambda)$ corresponding to the operator K_0 is determined by the equality

$$\langle E_F(\lambda)f, g\rangle = \int_{\{\xi\in\mathbb{R}^\nu : F(\xi)\leq\lambda\}} \widehat{f}(\xi)\overline{\widehat{g}(\xi)}d\xi, \quad f,\ g \in L^2(\mathbb{R}^\nu).$$

Here F is the characteristic of K_0: $F(\xi) = \sqrt{|\xi|^2 + m^2} - m$. On the surface $\{F(\xi) = \lambda\}$, the length of the gradient $\nabla F(\xi)$ is equal to $\dfrac{\sqrt{\lambda^2 + 2m\lambda}}{\lambda + m}$. Define the operator $A_F(\lambda)$ via the equality

$$\begin{aligned}\langle A_F(\lambda)f, g\rangle &= \frac{d}{d\lambda}\langle E_F(\lambda)f, g\rangle \\ &= \int_{\{F(\xi)=\lambda\}} \frac{\widehat{f}(\xi)\overline{\widehat{g}(\xi)}}{|\nabla F(\xi)|} d\sigma_\lambda(\xi) \\ &= \frac{\lambda + m}{\sqrt{\lambda^2 + 2m\lambda}} \int_{\{F(\xi)=\lambda\}} \widehat{f}(\xi)\overline{\widehat{g}(\xi)} d\sigma_\lambda(\xi).\end{aligned}$$

So, via the equality

$$\langle \exp(-tK_0)f, g\rangle = \int_0^\infty e^{-t\lambda} \langle A_F(\lambda)f, g\rangle \, d\lambda,$$

we obtain, for $f \in L^1(\mathbb{R}^\nu) \cap L^2(\mathbb{R}^\nu)$ and for $g \in L^2(\mathbb{R}^\nu)$,

$$\begin{aligned}&|\langle \exp(-tK_0) f, g\rangle| \leq \int_0^\infty \exp(-t\lambda)\sqrt{\langle A_F(\lambda)f, f\rangle}\sqrt{\langle A_F(\lambda)g, g\rangle} d\lambda \\ &\leq \left(\int_0^\infty e^{-2t\lambda} \langle A_F(\lambda)f, f\rangle \, d\lambda\right)^{1/2} \left(\int_0^\infty \langle A_F(\lambda)g, g\rangle \, d\lambda\right)^{1/2} \\ &= \left(\int \frac{\lambda + m}{\sqrt{\lambda^2 + 2m\lambda}} e^{-2t\lambda} \int_{\{|\xi|=\sqrt{\lambda^2+2m\lambda}\}} \left|\widehat{f}(\xi)\right|^2 d\sigma_\lambda(\xi)\, d\lambda\right)^{1/2} \|g\|_2 \\ &\leq \left(\int \frac{\lambda + m}{\sqrt{\lambda^2 + 2m\lambda}} e^{-2t\lambda} \int_{\{|\xi|=\sqrt{\lambda^2+2m\lambda}\}} 1 d\sigma_\lambda(\xi)\, d\lambda\right)^{1/2} \|f\|_1 \|g\|_2 \\ &= c_\nu \left(\int_0^\infty (\lambda + m)\left(\sqrt{\lambda^2 + 2m\lambda}\right)^{\nu-2} e^{-2\lambda t} d\lambda\right)^{1/2} \|f\|_1 \|g\|_2 \\ &= c_\nu \left(\int_0^\infty \frac{1}{t}\left(\frac{\lambda}{t} + m\right)\left(\sqrt{\left(\frac{\lambda}{t}\right)^2 + 2m\frac{\lambda}{t}}\right)^{\nu-2} e^{-2\lambda} d\lambda\right)^{1/2} \|f\|_1 \|g\|_2 .\end{aligned}$$

It follows that

$$\|\exp(-tK_0)\|_{\infty,1} \leq \|\exp(-tK_0)\|_{2,1}^2 \leq c_\nu' t^{-\nu}.$$

The previous results can be generalized for negative-definite functions F defined on a locally compact, second countable, abelian group G. In that case the variable p varies over the dual group. It is also noted that these results fit in the theory of Lévy processes.

EXAMPLE 5. The following example is related to the previous one and can be found in Ichinose [127]. Let the Lévy measure μ^m be defined by the equality

$$\sqrt{p^2+m^2} = m - \int_{\{|y|>0\}} [\exp(ip\cdot x) - 1 - ip\cdot y 1_B(y)]\, d\mu^m(y). \tag{1.28}$$

Here $B = \{y \in \mathbb{R}^\nu : |y| < 1\}$. An explicit form of these measures is given by

$$d\mu^m(y) = \begin{cases} 2\dfrac{m^{\frac{1}{2}(m+1)}}{(2\pi\,|y|)^{\frac{1}{2}(\nu+1)}} K_{\frac{1}{2}(\nu+1)}(m\,|y|)\, dy, & m > 0, \quad (1.29) \\[2ex] \Gamma\left(\dfrac{\nu+1}{2}\right) \dfrac{1}{(\pi\,|y|)^{\nu+1}} dy, & m = 0. \quad (1.30) \end{cases}$$

Here $\Gamma(z)$ is the gamma function and $K_r(z)$ is the modified Bessel function of the third kind of order r. Let $A : \mathbb{R}^\nu \to \mathbb{R}^\nu$ be a locally bounded function with the property that the function

$$x \mapsto \int_{\{0<|y|<1\}} \frac{|A(x-y/2) - A(x)|}{|y|^\nu} dy \tag{1.31}$$

is also locally bounded. Define the *Weyl quantized relativistic Hamiltonian* H_A^m via the formula

$$\begin{aligned} &[H_A^m f](x) - mf(x) \qquad (1.32) \\ &= -\int_{\mathbb{R}_0^\nu} \left[\exp(-iyA(x+y/2))\, f(x+y) - f(x) - 1_{B_1}(y) y\, (\partial_x - iA(x))\, f(x)\right] d\mu^m(y) \end{aligned}$$

where $B_1 = \{y \in \mathbb{R}^\nu : |y| < 1\}$ and $\mathbb{R}_0^\nu = \mathbb{R}^\nu \setminus \{0\}$. Under suitable conditions on A, the operator H_A^m is the following pseudo-differential operator:

$$\begin{aligned} &[H_A^m f](x) - mf(x) \\ &= \frac{1}{(2\pi)^\nu} \int_{\mathbb{R}^\nu}\int_{\mathbb{R}^\nu} \exp(i(x-y).p)\, h_A^m\left(p, \frac{x+y}{2}\right) f(y) dy\, dp, \end{aligned} \tag{1.33}$$

where h_A^m is the function

$$h_A^m(p,x) = \sqrt{(p-A(x))^2 + m^2}, \quad p \in \mathbb{R}^\nu,\ x \in \mathbb{R}^\nu. \tag{1.34}$$

The operator $K_0 \equiv H_A^m - mI$ is thean the self-adjoint generator of a strongly continuous semigroup in $L^2(\mathbb{R}^\nu)$. Let the density function $p_0(t,x,y)$ be as in (1.24) of Example 4. Then the integral kernel of the operator $\exp(-t(H_A^m - mI))$ is given by

$$\exp(-t(H_A^m - mI))(x,y) = \lim_{s\uparrow t} \mathbb{E}_x \left(\exp(-S(s))\, p_0(t-s, X(s), y)\right).$$

Here $\{(\Omega, \mathcal{F}, \mathbb{P}_x), (X(t) : t \geq 0), (\vartheta_t : t \geq 0), (E, \mathcal{E})\}$ is the Markov process generated by $H_0^m - mI$ with state space $E = \mathbb{R}^\nu$ and transition density $p_0(t,x,y)$. The process $S(t)$ is given by

$$\begin{aligned} S(t) &= i \int_0^t \int_{\mathbb{R}^\nu \setminus B_1} A\left(X(s-) + y/2\right).y N_X(dsdy) \\ &\quad + i \int_0^t \int_{0<|y|<1} A\left(X(s-) + y/2\right).y \widetilde{N}_X(dsdy) \\ &\quad + i \int_0^t \int_{0<|y|<1} \left[A\left(X(s) + y/2\right) - A(X(s))\right].y \widehat{N}_X(dsdy) \\ &= i \int_0^t \int_{|y|>0} A\left(X(s-) + y/2\right).y \widetilde{N}_X(dsdy) \\ &\quad + i \int_0^t \int_{0<|y|} \left[A\left(X(s) + y/2\right) - A(X(s))\right].y \widehat{N}_X(dsdy) \end{aligned}$$

where the random measures N_X, $\widetilde{N}_X$ and $\widehat{N}_X$ are respectively given by

$$\begin{aligned} &N_X\left((t_1,t_2] \times U\right) = \#\left\{s \in (t_1,t_2] : X(s) \neq X(s-), \quad X(s) - X(s-) \in U\right\}, \\ &\widetilde{N}_X(dsdy) = N_X(dsdy) - \widehat{N}_X(dsdy), \quad \text{and} \\ &\widehat{N}_X(dsdy) = \mathbb{E}_0\left(N_X(dsdy)\right) = dsd\mu^m \end{aligned}$$

where $0 < t_1 < t_2$ and where U is a Borel subset of $\mathbb{R}^\nu \setminus \{0\}$. For more details see Ichinose [127, 131].

Example 6. In this example we consider the generator $K_0 := -\frac{1}{2}\Delta + x \cdot \nabla$ of the so-called Ornstein-Uhlenbeck process in $L^2\left(\mathbb{R}^\nu, \exp\left(-|y|^2\right)\pi^{-\nu/2}dy\right)$: see Theorem 1.19 assertion (d), in section E. Its integral kernel $p_0(t,x,y)$ is given by

$$\begin{aligned} &p_0(t,x,y) \\ &= \frac{1}{(1-\exp(-2t))^{\nu/2}} \exp\left(-\frac{\exp(-2t)|x|^2 + \exp(-2t)|y|^2 - 2\exp(-t)\langle x,y\rangle}{1-\exp(-2t)}\right). \end{aligned}$$

The semigroup in $L^2\left(\mathbb{R}^\nu, \exp\left(-|y|^2\right)\pi^{-\nu/2}dy\right)$ is given by

$$\begin{aligned} \left[\exp(-tK_0)f\right](x) &= \int p_0(t,x,y) f(y) \exp\left(-|y|^2\right) \frac{dy}{\left(\sqrt{\pi}\right)^{\nu/2}} \\ &= \int f\left(\exp(-t)x + \sqrt{1-\exp(-2t)}y\right) \exp\left(-|y|^2\right) \frac{dy}{\left(\sqrt{\pi}\right)^{\nu}}. \end{aligned}$$

For more details the reader is referred to Simon [208].

EXAMPLE 7. In this example we consider the generator $K_0 := -\frac{1}{2}\Delta + \frac{1}{2}|x|^2$ of the oscillator process: see item (c) of Theorem 1.19 in section E of this chapter. The integral kernel of the corresponding semigroup $\exp(-tK_0)(x,y)$ may be written as

$$\begin{aligned}
&\exp(-tK_0)(x,y)\\
&= \exp\left(-\frac{1}{2}|x|^2\right)\frac{1}{(2\pi\sinh t)^{\nu/2}}\exp\left(-\frac{|\exp(-t)x-y|^2}{1-\exp(-2t)}\right)\exp\left(\frac{1}{2}|y|^2\right)\\
&= \exp\left(-\frac{1}{2}|x|^2\tanh t\right)\frac{1}{(2\pi\sinh t)^{\nu/2}}\exp\left(-\frac{1}{2\sinh t}\left|\frac{x}{(\cosh t)^{1/2}} - y(\cosh t)^{1/2}\right|^2\right).
\end{aligned}$$

It follows that the corresponding semigroup $\{\exp(-tK_0) : t \geq 0\}$ is given by

$$\begin{aligned}
&[\exp(-tK_0)f](x)\\
&= \frac{\exp\left(-\frac{1}{2}|x|^2\tanh t\right)}{(2\pi\cosh t)^{\nu/2}}\int \exp\left(-\frac{1}{2}|y|^2\right) f\left(\frac{x}{\cosh t} + \sqrt{\tanh t}\,y\right) dy.
\end{aligned}$$

For more details we refer again to Simon [208].

Before we continue with some more examples we insert another problem.

PROBLEM. Give a characterization of those pseudo-differential operators that generate self-adjoint semigroups for which BASSA is satisfied.

EXAMPLE 8. It is perhaps interesting to recall Theorem 10.3 in Ikeda and Watanabe [137], stating that under appropriate conditions (boundedness of certain vector fields V_j, $0 \leq j \leq \nu$, and Hörmander's hypo-ellipticity condition) the $\mathbb{P}_x$-distribution of the solution $(X(t) : t \geq 0)$ of the stochastic differential equation $dX(t) = \sigma(X(t))dB(t) + b(X(t))dt$, $X(0) = x$, defines a Markov process

$$\{(\Omega, \mathcal{F}, \mathbb{P}_x), (X(t) : t \geq 0), (\vartheta_t : t \geq 0), (E, \mathcal{E})\}$$

with the property that, for every compact subset K, there exists a constant C_K such that, for appropriate n, $\|1_K P(t)\|_{\infty,1} \leq C_K t^{-n/2}$, $t \geq 0$. Here $P(t)f(x) = \mathbb{E}_x\left[f(X(t))\right]$ where f is a bounded Borel measurable function defined on $\mathbb{R}^\nu$. To make this precise we write $V_k(x) = \sum_{i=1}^{\nu}\sigma_{ik}(x)\dfrac{\partial}{\partial x_i}$, $k = 1, \dots, \nu$, and

$$V_0(x) = \sum_{i=1}^{\nu}\left[b_i(x) - \frac{1}{2}\sum_{k=1}^{\nu}\sum_{j=1}^{\nu}\frac{\partial}{\partial x_j}\sigma_{ik}(x)\sigma_{jk}(x)\right]\frac{\partial}{\partial x_i}.$$

The above stochastic differential equation can be rewritten as follows:

$$\begin{cases} dX(t) = \sum_{k=1}^{\nu} V_k(X(t))dB_k(t) + V_0(X(t))dt,\\ X(0) = x. \end{cases}$$

Given a vector field V on $\mathbb{R}^\nu$, define the vector fields (V_k, V), $k = 1, \ldots, \nu$, by $(V_k, V) = [V_k, V]$ and define (V_0, V) by

$$(V_0, V) = [V_0, V] + \frac{1}{2} \sum_{\ell=1}^{\nu} [V_\ell, [V_\ell, V]].$$

The subsets of vector fields Σ_n, $n \in \mathbb{N}$, are given by $\Sigma_0 = \{V_1, V_2, \ldots, V_\nu\}$ and, for $n \geq 1$, Σ_n is given by

$$\Sigma_n = \{(V_k, V) : V \in \Sigma_{n-1}, k = 0, 1, \ldots, \nu\}.$$

The vector fields are V_k, $k = 0, 1, \ldots, \nu$ are said to satisfy the hypo-ellipticity condition of Hörmander at $x \in \mathbb{R}^\nu$ if there exists $m \in \mathbb{N}$ and $A_1, \ldots, A_\nu \in \bigcup_{n=0}^{n=m} \Sigma_k$ such that $A_1(x), \ldots, A_\nu(x)$ are linearly independent. Suppose that all the coefficients σ_{ij} and b_j are bounded and have bounded derivatives of all orders. Also suppose that Hörmander's condition is satisfied. Let L be the operator $L := \frac{1}{2}\sum_{k=1}^{\nu} V_k^2 + V_0$. An extension of $K_0 = -L$ generates a Feller semigroup with C^∞-density $p_0(t, x, y)$. For more details we refer the reader to Ikeda and Watanabe [137, Chapter V]. Again densities are available. Moreover, they are L^1-L^∞-smoothing: see Definition 1.1. A recent treatment can also be found in Fang [100].

EXAMPLE 9. The previous example has a counterpart for Riemannian manifolds. Instead of the Laplace operator on $\mathbb{R}^\nu$, we consider the Laplace-Beltrami operator on a Riemannian manifold. For details we refer the reader to Elworthy [95, 96], Azencott et. al. [14], Bismut [33] and several others. The authors also establish existence results for and bounds on the corresponding heat kernels. A recent and very interesting paper is [68], written by Davies. It provides the reader with much insight into the behavior of heat kernels. Of course, his book [67] should also be consulted.

EXAMPLE 10. In this example we consider so-called hyper-singular integrals. Define the operators $\triangle_h^\ell$ on $C(\mathbb{R}^\nu)$ as follows:

$$\left[\triangle_h^\ell f\right](x) = \sum_{k=0}^{\ell} (-1)^k \binom{\ell}{k} f(x - kh).$$

In [158] Kochubeĭ proves that an operator K_0 of the form

$$\begin{aligned}\left[K_0 f\right](x) = &-\beta \sum_{i,j=1}^{\nu} a_{ij}(x) \frac{\partial^2 f}{\partial x_i \partial x_j}(x) + \frac{1-\beta}{d_{n,\ell}(\gamma)} \int_{\mathbb{R}^\nu} \Omega\left(x, \frac{h}{|h|}\right) \frac{\left[\triangle_h^\ell f\right](x)}{|h|^{\nu+\gamma}} dh \\ &+ \sum_{k=1}^{m} \frac{1}{d_{n,\ell}(\gamma_k)} \int_{\mathbb{R}^\nu} \Omega_k\left(x, \frac{h}{|h|}\right) \frac{\left[\triangle_h^\ell f\right](x)}{|h|^{\nu+\gamma_k}} dh \\ &+ \sum_{j=1}^{\nu} b_j(x) \frac{\partial f}{\partial x_j}\end{aligned}$$

generates a Feller semigroup in $C_\infty(\mathbb{R}^\nu)$ provided that the following conditions are satisfied:

(a) The functions Ω and Ω_k, $1 \leq k \leq m$, are non-negative and continuous on $\mathbb{R}^\nu \times S^{\nu-1}$. They are also even: $\Omega(x,\sigma) = \Omega(x,-\sigma)$, $x \in \mathbb{R}^\nu$, $\sigma \in S^{\nu-1}$.

(b) The orders of homogeneity γ, γ_k, $1 \leq k \leq m$, satisfy $0 < \gamma_k < \gamma \leq 2$. If $\gamma = 2$, then $\beta = 1$ and if $\gamma < 2$, then $\beta = 0$. If $\gamma = 1$, then $b_j \equiv 0$, $1 \leq j \leq \nu$.

(c) Some ellipticity conditions on (a_{ij}) are also required. In particular, the inequality $\operatorname{Re}\sum_{i,j=1}^{\nu} a_{ij}(x)\xi_i\bar{\xi}_j \geq a_0 |\xi|^2$, for $\xi_j \in \mathbb{C}$, $1 \leq j \leq \nu$, is assumed. Here a_0 is some strictly positive real number.

(d) The constants $d_{n,\ell}(\gamma)$ have to be chosen suitably: they are chosen so that the expression

$$\widetilde{\Omega}(x,\xi) = \frac{1}{d_{\nu,\ell}(\gamma)} \int_{\mathbb{R}^\nu} \frac{(1-\exp(-i\xi.h))^\ell}{|h|^{\nu+\gamma}} \Omega\left(x, \frac{h}{|h|}\right) dh,$$

called the *symbol* of the hyper-singular integral $D_\Omega^\gamma f$, does not depend on the particular choice of ℓ, where $\ell > \alpha$.

(e) The characteristics Ω and Ω_k, $1 \leq k \leq m$, are non-negative and are symmetric in the second variable, i.e. $\Omega(x,\sigma) = \Omega(x,-\sigma)$ for all $x \in \mathbb{R}^\nu$ and for all $\sigma \in S^{\nu-1}$.

The conditions (a) through (e) imply that the life time of the corresponding Markov process is ∞. Moreover, in [158] Kochubeĭ proves that the corresponding Markov processes posses transition densities $p_0(t,x,y)$ satisfying inequalities of the form

$$p_0(t,x,y) \leq C\left\{\frac{t}{\left[t^{1/\gamma} + |x-y|\right]^{\nu+\gamma}} + \sum_{k=1}^{m} \frac{t}{\left[t^{1/\gamma} + |x-y|\right]^{\nu+\gamma_k}}\right\}.$$

D. Heat kernels

In the theorems and lemmas below we give bounds for heat kernels, provided the generator K_0 generates a diffusion and the logarithmic Sobolev inequality (1.37) is satisfied. In fact if (1.37) below is true for an appropriate function m, then an estimate of the form

$$\exp(-tK_0)(x,y) \leq \exp(m_c(t/2))\exp\left(-\frac{d(x,y)^2}{2t(1+\eta_c)}\right)$$

holds (see Corollary 1.13). The explanation of this inequality is the aim of this section. Here $d(x,y) = \sup\{|\psi(y)-\psi(x)| : \Gamma_1(\psi,\psi) \leq 1,\ \psi \in \mathcal{A}\}$, and $m_c(t/2)$ and η_c are some positive finite numbers. Definition 1.7 contains the definition of Γ_1, the explanation of $\mathcal{A}$ is to be found in the comments preceding Proposition 1.9, and Corollary 1.13. For the definition of $m_c(t)$ see (1.50), and of η_c see (1.52). The quantities $E_p(f)$ and $\mathcal{E}_p(f)$, $1 \leq p < \infty$, $f > 0$, are defined via the following formulae:

$$E_p(f) = \int_E f(x)^p \log\left(\frac{f(x)^p}{\|f\|_p^p}\right) dx \quad \text{and} \tag{1.35}$$

$$\mathcal{E}_p(f) = \left\langle K_0 f, f^{p-1}\right\rangle, \quad f \in \operatorname{dom}(K_0). \tag{1.36}$$

It is noted that in this section only, section D of Chapter 1, the symbol $\mathcal{E}_2$ replaces the more usual $\mathcal{E}^0$.

REMARK. The form $f \mapsto \mathcal{E}_p(f)$ might be called the *energy* of order p (for $p=2$ we have the usual energy form). For $\|f\|_p = 1$ the quantity $E_p(f)$ could be called the entropy of the measure $B \mapsto \int_B f^p dm$.

These upper bounds will be given in terms of the carré du champ operator, introduced by Roth: see e.g. [200]. For more recent work the reader is referred to Bakry [15].

1.7. DEFINITION. The *carré du champ operator* Γ_1 is is a bilinear operator defined by the following identity:

$$\begin{aligned}\left[\Gamma_1(v,w)\right](x) &= \lim_{s\downarrow 0}\frac{1}{s}\mathbb{E}_x\left[(v(X(s))-v(x))(w(X(s))-w(x))\right] \\ &= -\left[K_0(vw)\right](x) + \left[K_0 v\right](x)w(x) + v(x)\left[K_0 w\right](x),\end{aligned}$$

for whatever functions v and w these limits make sense. If $v \equiv w$ is a real-valued function, then $\Gamma_1(v,v) \geq 0$.

We also observe that $\frac{1}{2}\int \Gamma_1(f,\overline{f})dx = \displaystyle\int \left|\left[K_0^{1/2}f\right](x)\right|^2 dx$ for f belonging to $D\left(K_0^{1/2}\right)$. If $-K_0 = \frac{1}{2}\Delta$, then $\Gamma_1(v,w) = \nabla v \cdot \nabla w$.

Let f, g be functions in the domain of K_0 such that the pointwise product fg also belongs to the domain $\mathrm{dom}(K_0)$. The following statements can be verified (see also Bouleau and Lamberton [38]):

(a) The identity $\Gamma_1(f,g) = -K_0(fg) + f(K_0 g) + (K_0 f)g$ holds.

(b) The process

$$(f(X(t)) - f(X(0)))\,(g(X(t)) - g(X(0))) - \int_0^t \Gamma_1(f,g)(X(u))du$$

is a $\mathbb{P}_x$-martingale for every $x \in E$.

(c) For almost all $x \in E$ the equality

$$[\Gamma_1(f,g)]\,(x) = \lim_{s\downarrow 0} \frac{1}{s}\mathbb{E}_x\Big((f(X(s)) - f(X(0)))\,(g(X(s)) - g(X(0)))\Big)$$

is true.

(d) The inequality $|\Gamma_1(f,g)|^2 \leq \Gamma_1(f,f)\Gamma_1(g,g)$ is valid.

The following definition is taken from Bakry [15].

1.8. DEFINITION. The operator K_0 is said to generate a *diffusion* if for every C^∞-function $\Phi : \mathbb{R}^\nu \to \mathbb{R}$, with $\Phi(0,\ldots,0) = 0$, the following identity is valid:

$$\begin{aligned} &K_0\left(\Phi(f_1,\ldots,f_n)\right) \\ &= \sum_{j=1}^n \frac{\partial\Phi}{\partial x_j}\left(f_1,\ldots,f_n\right) K_0 f_j - \frac{1}{2}\sum_{j,k=1}^n \frac{\partial^2\Phi}{\partial x_j\partial x_k}\left(f_1,\ldots,f_n\right)\Gamma_1(f_j,f_k)\end{aligned}$$

for all functions $f_1,\ldots,f_n$ in a rich enough algebra of functions $\mathcal{A}$, contained in the domain of the generator K_0, described below.

If we take $\Phi(f_1,f_2,f_3) = f_1f_2f_3$ we see that Γ_1 is a derivation in its arguments: $\Gamma_1(f_1f_2,f_3) = f_1\Gamma_1(f_2,f_3) + f_2\Gamma_1(f_1,f_3)$. We also have $\Gamma_1(\Phi(f),g) = \Phi'(f)\Gamma_1(f,g)$.

The hypotheses on $\mathcal{A}$ are the following. It is dense in $L^p(E,m)$ for all $1 \leq p < \infty$ and it is a core for K_0 considered as an operator in $L^2(E,m)$. In addition, it is assumed that $\mathcal{A}$ is stable under composition with C^∞-functions of several variables, that vanish at the origin. Moreover, in order to obtain some nice results a rather technical condition is required: whenever $(f_n : n \in \mathbb{N})$ is a sequence in $\mathcal{A}$ that converges to f with respect to the graph norm of K_0 (in $L^2(E,m)$) and whenever $\Phi : \mathbb{R} \to \mathbb{R}$ is a C^∞-function, vanishing at 0, with bounded derivatives of all orders (including the order 0), then one may extract a subsequence $(\Phi(f_{n_k}) : k \in \mathbb{N})$ that converges to $\Phi(f)$ in $L^1(E,m)$, whereas the sequence $(K_0\Phi(f_{n_k}) : k \in \mathbb{N})$ converges in $L^1(E,m)$ to $K_0\Phi(f)$. Notice that all functions of the form $e^\psi f$ with ψ, $f \in \mathcal{A}$ belong to $\mathcal{A}$. This fact will be used in Theorem 1.10. Also notice that the required properties of $\mathcal{A}$ depend on the generator K_0. In Theorem 1.11 we will assume that the algebra $\mathcal{A}$ is rich enough with respect to another generator K as well. For more details see Bakry [15].

In what follows we shall assume that the measure m is *invariant* in the sense that for all $f \in L^1(E,m)$ the equality $\int \exp(-tK_0) f dm = \int f dm$ is valid. The latter equality is equivalent to the fact that $\int K_0 f dm = 0$ for all functions f in an L^1-dense domain of K_0. From the basic assumptions it then follows that $\langle K_0 f, f\rangle = \frac{1}{2}\int \Gamma_1(f,f)dm$ for all real-valued functions in the domain of K_0.

1.9. PROPOSITION. *Let K_0 be an operator generating a diffusion. Let $m : (0,\infty) \to [0,\infty)$ be a continuous function with the property that for all $f \in \operatorname{dom}(K_0)$, $f > 0$, the following inequality is valid:*

$$E_2(f) \leq t\mathcal{E}_2(f) + m(t) \|f\|_2^2, \tag{1.37}$$

for all $t > 0$. Then for all $f > 0$, $p > 1$ and $t > 0$ the following inequality is true:

$$E_p(f) \leq \frac{tp^2}{4(p-1)}\mathcal{E}_p(f) + m(t) \|f\|_p^p. \tag{1.38}$$

Proof. Consider the equality:

$$\int K_0\Phi(f)dm = \langle K_0 f, \Phi'(f)\rangle - \frac{1}{2}\langle \Gamma_1(f,f), \Phi''(f)\rangle .$$

Here f belongs to $\mathcal{A}$ and $f > 0$. With $\Phi(f) = f^p$, this proves the equality

$$\mathcal{E}_p(f) = \frac{1}{2}(p-1)\left\langle f^{p-2}, \Gamma_1(f,f)\right\rangle .$$

In the same way we get

$$\mathcal{E}_2\left(f^{p/2}\right) = \frac{1}{2}\int \Gamma_1\left(f^{p/2}, f^{p/2}\right) dm = \frac{p^2}{8}\left\langle f^{p-2}, \Gamma_1(f,f)\right\rangle .$$

This can be seen as follows. First employ Definition 1.25 with $\Phi(x) = \frac{1}{2}x^2$ and after that with $\Phi(x) = x^{p/2}$ to obtain

$$\begin{aligned}
&\frac{1}{2}\Gamma_1\left(f^{p/2}, f^{p/2}\right) = -K_0\left(\Phi(f^{p/2})\right) + f^{p/2}K_0\left(f^{p/2}\right) \\
&= -\frac{1}{2}K_0\left(f^{p/2}\right) + f^{p/2}\left[\frac{p}{2}f^{p/2-1}K_0 f - \frac{1}{2}\frac{p}{2}\left(\frac{p}{2}-1\right)f^{p/2-2}\Gamma_1(f.f)\right] \\
&= -\frac{1}{2}K_0\left(f^p\right) + \frac{p}{2}f^{p-1}K_0 f - \frac{p}{4}\left(\frac{p}{2}-1\right)f^{p-2}\Gamma_1(f,f),
\end{aligned}$$

and so

$$\frac{1}{2}\int \Gamma_1\left(f^{p/2}, f^{p/2}\right) dm$$
$$= -\frac{1}{2}\int K_0\left(f^p\right) dm + \frac{p}{2}\int f^{p-1}K_0 f dm - \frac{p}{4}\left(\frac{p}{2}-1\right)\int f^{p-2}\Gamma_1(f,f)dm.$$

From this equality, together with

$$\mathcal{E}_p(f) = \frac{1}{2}(p-1)\left\langle f^{p-2}, \Gamma_1(f,f)\right\rangle,$$

we get the desired result. A combination of the above equalities yields the identity:

$$\mathcal{E}_p(f) = \frac{p^2}{4(p-1)}\mathcal{E}_2\left(f^{p/2}\right).$$

Hence (1.38) follows from (1.37) via the substitution $f = g^{p/2}$.

The proof of inequality (1.40) below was inspired by the proof of Lemma 3.2.1 in Davies [67, p. 83].

1.10. PROPOSITION. *Again let K_0 be an operator that generates a diffusion. Suppose that ψ is a function in* dom(K_0) *with the property that $\Gamma_1(\psi,\psi) \le 1$. Put*

$$\mathcal{E}_{\psi,p}(f) = \left\langle e^{-\alpha\psi}K_0\left(e^{\alpha\psi}f\right), f^{p-1}\right\rangle, \quad e^{\alpha\psi}f \in \mathrm{dom}(K_0). \tag{1.39}$$

The following inequality is valid for $p > 1$, for $\alpha \in \mathbb{C}$, and for $f \ge 0$:

$$\mathcal{E}_p(f) \le 2\mathcal{E}_{\alpha\psi,p}(f) + |\alpha|^2\left(1 + \frac{(p-2)^2}{2(p-1)}\right)\|f\|_p^p. \tag{1.40}$$

Proof. Since K_0 generates a diffusion, the following identity is true:

$$e^{-\alpha\psi}K_0\left(e^{\alpha\psi}f\right) = \alpha f K_0\psi + K_0 f - \frac{\alpha^2}{2}f\Gamma_1(\psi,\psi) - \alpha\Gamma_1(\psi,f).$$

It follows that

$$\begin{aligned}
&\mathcal{E}_{\alpha\psi,p}(f) = \left\langle e^{-\alpha\psi}K_0\left(e^{\alpha\psi}f\right), f^{p-1}\right\rangle\\
&= \alpha\left\langle K_0\psi, f^p\right\rangle + \left\langle K_0 f, f^{p-1}\right\rangle - \frac{\alpha^2}{2}\left\langle \Gamma_1(\psi,\psi), f^p\right\rangle - \alpha\left\langle \Gamma_1(\psi,f), f^{p-1}\right\rangle\\
&= \frac{\alpha}{2}\int \Gamma_1\left(\psi, f^p\right) dm + \left\langle K_0 f, f^{p-1}\right\rangle - \frac{\alpha^2}{2}\left\langle \Gamma_1(\psi,\psi), f^p\right\rangle - \alpha\left\langle \Gamma_1(\psi,f), f^{p-1}\right\rangle\\
&= \frac{\alpha p}{2}\int f^{p-1}\Gamma_1\left(\psi, f\right) dm + \left\langle K_0 f, f^{p-1}\right\rangle\\
&\qquad - \frac{\alpha^2}{2}\left\langle \Gamma_1(\psi,\psi), f^p\right\rangle - \alpha\left\langle \Gamma_1(\psi,f), f^{p-1}\right\rangle\\
&= \frac{\alpha(p-2)}{2}\left\langle \Gamma_1\left(\psi, f\right), f^{p-1}\right\rangle + \left\langle K_0 f, f^{p-1}\right\rangle - \frac{\alpha^2}{2}\left\langle \Gamma_1(\psi,\psi), f^p\right\rangle.
\end{aligned} \tag{1.41}$$

So, if $\Gamma_1(\psi,\psi) \leq 1$, we get

$$\begin{aligned}
\mathcal{E}_{\alpha\psi,p}(f) &= \left\langle e^{-\alpha\psi} K_0\left(e^{\alpha\psi} f\right), f^{p-1}\right\rangle \\
&\geq -\frac{|\alpha(p-2)|}{2}\left|\left\langle \Gamma_1(\psi,f), f^{p-1}\right\rangle\right| + \left\langle K_0 f, f^{p-1}\right\rangle - \frac{\alpha^2}{2}\left\langle \Gamma_1(\psi,\psi), f^p\right\rangle \\
&\geq -\frac{1}{2}|\alpha|\,|p-2|\left\langle \Gamma_1(\psi,\psi)^{1/2}\Gamma_1(f,f)^{1/2}, f^{p-1}\right\rangle + \left\langle K_0 f, f^{p-1}\right\rangle \\
&\quad - \frac{\alpha^2}{2}\left\langle \Gamma_1(\psi,\psi), f^p\right\rangle \\
&\geq -\frac{1}{2}|\alpha|\,|p-2|\left\langle \Gamma_1(f,f)^{1/2}, f^{p-1}\right\rangle + \left\langle K_0 f, f^{p-1}\right\rangle - \frac{\alpha^2}{2}\|f\|_p^p. \qquad (1.42)
\end{aligned}$$

Next we estimate the quantity $\left\langle \Gamma_1(f,f)^{1/2}, f^{p-1}\right\rangle$. This can be done as follows:

$$\begin{aligned}
&\left\langle \Gamma_1(f,f)^{1/2}, f^{p-1}\right\rangle = \frac{2}{p}\left\langle \Gamma_1\left(f^{p/2}, f^{p/2}\right)^{1/2}, f^{p/2}\right\rangle \\
&\leq \frac{2}{p}\left(\int \Gamma_1\left(f^{p/2}, f^{p/2}\right) dm\right)^{1/2}\|f\|_p^{p/2} \\
&= \frac{1}{\sqrt{p-1}}\left(\int \Gamma_1\left(f^{p-1}, f\right) dm\right)^{1/2}\|f\|_p^{p/2} = \sqrt{\frac{2}{p-1}}\left(\mathcal{E}_p(f)\right)^{1/2}\|f\|_p^{p/2}. \qquad (1.43)
\end{aligned}$$

A combination of (1.42) and (1.43) yields the inequality

$$\mathcal{E}_{\alpha\psi,p}(f) \geq \mathcal{E}_p(f) - \frac{1}{2}\alpha^2\|f\|_p^p - \frac{|\alpha|\,|p-2|}{\sqrt{2(p-1)}}\mathcal{E}_p(f)^{1/2}\|f\|_p^{p/2}. \qquad (1.44)$$

The inequality in (1.44) is equivalent to

$$\mathcal{E}_p(f)^{1/2} \leq \frac{|p-2|}{2\sqrt{p-1}}\frac{|\alpha|}{\sqrt{2}}\|f\|_p^{p/2} + \left(\mathcal{E}_{\alpha\psi,p}(f) + \left(1 + \frac{(p-2)^2}{4(p-1)}\right)\frac{\alpha^2}{2}\|f\|_p^p\right)^{1/2}. \qquad (1.45)$$

Since $A \leq \gamma C + \sqrt{B^2 + (1+\gamma^2)C^2}$, A, B and $\gamma \geq 0$, implies the inequality $A^2 \leq 2B^2 + 2C^2\left(1 + 2\gamma^2\right)$, the inequality in (1.40) follows.

The following result can be found in Davies [67], in Coulhon, Saloff-Coste and Varopoulos [61] and in Bakry [15]. Our proof follows that of Bakry.

1.11. THEOREM. *Let K be the generator of a semigroup consisting of operators with non-negative integral kernels. Suppose that the algebra $\mathcal{A}$ also possesses the required properties with respect to the operator K. Fix $1 \leq p < q \leq \infty$. Suppose that there exist continuous functions c, $m : (p,q) \to (0,\infty)$ such that for all $p < u < q$ and for all $f \in \mathcal{A}$ the following Sobolev inequality is true:*

$$E_u(f) \leq c(u)\mathcal{E}_{K,u}(f) + m(u)\|f\|_u^u. \qquad (1.46)$$

Put

$$t = \int_p^q \frac{c(u)}{u^2} du \quad \textit{and} \quad \widehat{m} = \int_p^q m(u) \frac{c(u)}{u^2} du. \tag{1.47}$$

Then

$$\left\|\exp(-tK)\right\|_{q,p} \leq e^{\widehat{m}}.$$

Of course, here $\hat{p}$ (see the beginning of the proof below) and $\widehat{m}$ have nothing to do with Fourier transforms. (Compare with Theorem 3.2 in Bakry [15].) We also wrote $\mathcal{E}_{K,u}$ for the energy form of order u corresponding to the operator K. In Theorem 1.12 we will apply this result for $K = K_0$ and for K of the form $K = e^{-\alpha\psi} K_0 e^{\alpha\psi}$ with $\psi \in \mathcal{A}$. In this case the algebra $\mathcal{A}$ is also rich with respect to K. We remark that $\widehat{m}$ depends in an implicit manner on t. If we change t (i.e. if we change p and/or q), then $\widehat{m}$ changes as well.

Proof. Fix $p_0 \in (p, q)$ and let $\hat{p}(t)$ and $\widehat{m}(t)$, $0 \leq t \leq \displaystyle\int_p^q \frac{c(u)}{u^2} du$, be the solutions to the following differential system:

$$\begin{cases} \dfrac{c(\hat{p}(t))}{\hat{p}(t)^2} \dfrac{d\hat{p}(t)}{dt} = 1, & \hat{p}(0) = p_0; \\ \dfrac{d\widehat{m}(t)}{dt} = \dfrac{m(\hat{p}(t))}{c(\hat{p}(t))}, & \widehat{m}(0) = 0. \end{cases}$$

Let $f > 0$ be a bounded function in $\mathcal{A}$. It suffices to show that the following inequality is valid:

$$\left\|\exp(-tK) f\right\|_{\hat{p}(t)} \leq \exp(\widehat{m}(t)) \left\|f\right\|_{p_0}.$$

Therefore we define $\widehat{f}(t)$ by $\widehat{f}(t) = \exp(-tK) f$ and the quantity $U(t)$ by

$$U(t) = \exp(-\widehat{m}(t)) \left(\int \widehat{f}(t)(x)^{\hat{p}(t)} dx \right)^{1/\hat{p}(t)}.$$

Differentiating $U(t)$ yields:

$$\int \widehat{f}(t)(x)^{\hat{p}(t)} dx \times \frac{dU(t)}{dt}$$
$$= U(t) \frac{\hat{p}'(t)}{\hat{p}(t)^2} \left[E_{\hat{p}(t)} \left(\widehat{f}(t) \right) - \frac{\hat{p}(t)^2}{\hat{p}'(t)} \left\{ \mathcal{E}_{\hat{p}(t)} \left(\widehat{f}(t) \right) + \widehat{m}'(t) \left\| \widehat{f}(t) \right\|_{\hat{p}(t)}^{\hat{p}(t)} \right\} \right]. \tag{1.48}$$

Upon using the differential equations satisfied by $\hat{p}(t)$ and $\widehat{m}(t)$ it follows from inequality (1.46) that the right-hand side of (1.48) is less than or equal to zero, provided that we may apply inequality (1.46) to $\widehat{f}(t)$. This can be arranged via the technical hypothesis on $\mathcal{A}$. We refer the interested reader to Bakry [15] and to Bakry and Michel [16].

In the following theorem we again consider the generator K_0 in conjunction with the algebra $\mathcal{A}$: see explanation in the text preceding Proposition 1.9.

1.12. THEOREM. *Suppose that there exists a continuous function* $m : (0,\infty) \to (0,\infty)$ *with the property that for every* $1 < u < \infty$ *the following logarithmic Sobolev inequality is true:*

$$E_2(f) \le u\mathcal{E}_2(f) + m(u)\,\|f\|_2^2 \tag{1.49}$$

for all $f > 0$, $f \in \mathcal{A}$. *Let* $c(u) > 0$ *be a function in* $L^1([1,\infty))$ *with the property that* $\int_1^\infty c(u)du = 1$. *The quantity* $m_c(t)$ *is defined by*

$$m_c(t) = \int_1^\infty \frac{m\big(4tc(u)(u-1)\big)}{u^2}du. \tag{1.50}$$

(a) *Let* $t > 0$ *be such that the corresponding quantity* $m_c(t)$ *is finite. Then* $\exp(-tK_0)$ *maps* $L^1(E,m)$ *to* $L^\infty(E,m)$ *and*

$$\|\exp(-tK_0)\|_{\infty,1} \le \exp(m_c(t)). \tag{1.51}$$

(b) *Put*

$$\eta = \eta_c = \frac{1}{2}\int_1^\infty c(u)\left(u - 1 + \frac{1}{u-1} - 2\right)du \tag{1.52}$$

and suppose that η *is finite. Let* ψ *be a function in* $\mathcal{A}$ *with the property that* $\Gamma_1(\psi,\psi) \le 1$. *Suppose that* $m_c(t/2)$ *is finite. Then*

$$\exp(-tK_0)(x,y) \le \exp(m_c(t/2))\exp\left(-\frac{|\psi(y)-\psi(x)|^2}{2t(1+\eta)}\right). \tag{1.53}$$

REMARK. Let $c(u)$ be as in the theorem and suppose that $\int_2^\infty c(u)du = 1/2$. Define the function $c_1(v)$, $v > 1$, by

$$\begin{cases} c_1(v) = c\left(\dfrac{v}{v-1}\right)\dfrac{1}{(v-1)^2}, & \text{if } 1 < v \le 2; \\ c_1(v) = c(v), & \text{if } v \ge 2. \end{cases}$$

Then the hypotheses and conclusions of Theorem 1.12 remain valid with $c_1(u)$ instead of $c(u)$.

Proof. (a) From the inequality in (1.38) we get

$$E_p(f) \le tc(p)p^2\mathcal{E}_p(f) + m\Big(4tc(p)(p-1)\Big)\,\|f\|_p^p. \tag{1.54}$$

Since $t = \displaystyle\int_1^\infty \frac{tc(u)u^2}{u^2}du$, the conclusion in (a) follows from (1.47) in Theorem 1.11.

(b) A combination of (1.54) for all $t > 0$, i.e.

$$E_p(f) \leq \frac{\tau p^2}{2}\mathcal{E}_p(f) + m\Big(2\tau(p-1)\Big)\,\|f\|_p^p,$$

and of (1.40), i.e.

$$\mathcal{E}_p(f) \leq 2\mathcal{E}_{\alpha\psi,p}(f) + |\alpha|^2\left(1 + \frac{(p-2)^2}{2(p-1)}\right)\|f\|_p^p$$

for all $f > 0$, $f \in \mathcal{A}$, yields

$$\begin{aligned}
E_p(f) &\leq \frac{\tau p^2}{2}\mathcal{E}_p(f) + m\Big(2\tau(p-1)\Big)\,\|f\|_p^p \\
&\leq \tau p^2\mathcal{E}_{\alpha\psi,p}(f) + \left\{\frac{\tau p^2\alpha^2}{2}\left(1 + \frac{(p-2)^2}{2(p-1)}\right) + m(2\tau(p-1))\right\}\|f\|_p^p
\end{aligned}$$

(put $\tau = tc(p)$)

$$= tc(p)p^2\mathcal{E}_{\alpha\psi,p}(f) + \left\{\frac{tc(p)p^2\alpha^2}{2}\left(1 + \frac{(p-2)^2}{2(p-1)}\right) + m(2tc(p)(p-1))\right\}\|f\|_p^p. \tag{1.55}$$

Notice the equality

$$\int_1^\infty c(u)\left(1 + \frac{(u-2)^2}{2(p-1)}\right)du = 1 + \eta.$$

From (1.55) together with Theorem 1.11 we obtain

$$\left\|e^{-\alpha\psi}\exp\left(-tK_0\right)e^{\alpha\psi}\right\|_{\infty,1} \leq \exp\left(\frac{t}{2}\alpha^2(1+\eta) + m_c(t/2)\right). \tag{1.56}$$

Hence from (1.56) we see

$$\begin{aligned}
&\exp\left(-tK_0\right)(x,y) \qquad\qquad (1.57)\\
&\leq \exp\left(m_c(t/2)\right)\exp\left(\frac{t}{2}\alpha^2(1+\eta)\right)\exp\left(-\alpha\left(\psi(y) - \psi(x)\right)\right).
\end{aligned}$$

So upon optimizing for α, i.e. choosing $\alpha = \dfrac{\psi(y) - \psi(x)}{t(1+\eta)}$, the desired conclusion follows from (1.57).

1.13. COROLLARY. *Let the hypotheses and notation be as in Theorem 1.12. Put* $d(x,y) = \sup\left\{|\psi(y) - \psi(x)| : \Gamma_1(\psi,\psi) \leq 1,\ \psi \in \mathcal{A}\right\}$. *If $m_c(t/2)$ is finite and if η_c is finite as well, then*

$$\exp\left(-tK_0\right)(x,y) \leq \exp\left(m_c(t/2)\right)\exp\left(-\frac{d(x,y)^2}{2t(1+\eta_c)}\right).$$

Proof. The conclusion is an immediate consequence of Theorem 1.11.

REMARK. If $K_0 = -\frac{1}{2}\Delta$, then $\Gamma_1(f,f) = \nabla f \cdot \nabla f$ and so $d(x,y) = |y - x|$.

E. Summary of Schrödinger semigroup theory

In this section of Chapter 1, we recall some basic facts about the Feynman-Kac formula for Brownian motion in $\mathbb{R}^\nu$. We also indicate the improvements and generalizations to be presented in the forthcoming chapters. In this section we mention some fundamental properties of Brownian motion: it is a Gaussian process, it is a Markov process, and it is a martingale.

E.1. Gaussian processes

We begin with an important extension theorem of Kolmogorov, which enables us to construct stochastic processes including Gaussian processes, Lévy processes, Poisson processes and others. It is also useful for the construction of Markov processes. For more details about the topics to be discussed, the reader is referred to Bhattacharya [26], to Karatzas and Shreve [147], and to Simon [208].

1.14. PROPOSITION. *(Kolmogorov) Let I be some index set, and let*

$$\{(\Omega_J, \mathcal{A}_J, \mathbb{P}_J) : J \subseteq I, \quad J \quad \text{finite}\}$$

be a projective system of probability distributions. Here $\Omega_J := \prod_{j\in J}\Omega_j$. Suppose that each Ω_i is a metrizable σ-compact Hausdorff space endowed with its Borel field $\mathcal{A}_i$. Then there exists a unique probability measure $\mathbb{P}_I$ on $(\Omega_I, \mathcal{A}_I)$, such that

$$\mathbb{P}_I\left(p_J^{-1}(A)\right) = \mathbb{P}_J(A)$$

for every $J \subseteq I$, J finite, and for every $A \in \mathcal{A}_J$.

Here p_J is a projection mapping from Ω_I to Ω_J defined by $p_J\left(\omega_i\right)_{i\in I} = \left(\omega_j\right)_{j\in J}$. The family $\{(\Omega_J, \mathcal{A}_J, \mathbb{P}_J) : J \subseteq I, \quad J \quad \text{finite}\}$ of probability measures (or distributions) is called *consistent* or a *projective system* if, for any pair of finite subsets J_1 and J_2, for which $J_1 \subseteq J_2 \subseteq I$, and for every $A \in \mathcal{A}_{J_1}$, the equality

$$\mathbb{P}_{J_1}(A) = \mathbb{P}_{J_2}\left\{\left(\omega_j\right)_{j\in J_2} : \left(\omega_j\right)_{j\in J_1} \in A\right\} = \mathbb{P}_{J_2}\left[\left(p_{J_1}^{J_2}\right)^{-1}(A)\right].$$

Here, as is self-explanatory, $p_{J_1}^{J_2} : \Omega_{J_2} \to \Omega_{J_1}$, is given by $p_{J_1}^{J_2}\left(\omega_j\right)_{j\in J_2} = \left(\omega_j\right)_{j\in J_1}$. Next we recall Bochner's theorem.

1.15. THEOREM. *(Bochner) Let $\varphi : \mathbb{R}^n \to \mathbb{C}$ be a continuous complex function, that is positive-definite in the sense that for all $r \in \mathbb{N}$*

$$\sum_{k,\ell=1}^{r} \lambda_k \overline{\lambda}_\ell \varphi\left(\xi^k - \xi^\ell\right) \geq 0,$$

for all $\lambda_1, \dots, \lambda_r \in \mathbb{C}$ and for all $\xi^1, \dots, \xi^r \in \mathbb{R}^n$. Then there exists a unique non-negative Borel measure μ on $\mathbb{R}^n$ such that its Fourier transform $\int \exp(-i\langle x, \xi\rangle)d\mu(x)$ is equal to $\varphi(\xi)$ for $\xi \in \mathbb{R}^n$. In particular $\mu(\mathbb{R}^n) = \varphi(0)$.

EXAMPLE 1. For every $i \in I$, let $\mathbb{P}_i$ be a probability measures on Ω_i and define $\mathbb{P}_J$ on Ω_J, $J \subseteq I$, J finite, by $\mathbb{P}_J(A) = \mathbb{P}_{j_1} \otimes \cdots \otimes \mathbb{P}_{j_n}(A)$, where A belongs to $\mathcal{A}_J$ and where $J = (j_1, \ldots, j_n)$. Then the collection $\{(\Omega_J, \mathcal{A}_J, \mathbb{P}_J) : J \subseteq I, J \text{ finite}\}$ is a projective system. Here $\mathcal{A}_J$ is the product field on $\prod_{j\in J} \Omega_j$. Put $J = \{j_1, \ldots, j_n\}$. Then $\mathcal{A}_J$ coincides with $\sigma(X_j : j \in J)$ where $X_{j_k}(\omega_{j_1}, \ldots, \omega_n) = \omega_{j_k}$, $1 \le k \le n$. By definition, $\sigma(X_j : j \in J)$ is the σ-field generated by the state variables X_j, $j \in J$.

EXAMPLE 2. Let $\sigma : I \times I \to \mathbb{R}$ be a symmetric function (i.e. $\sigma(i,j) = \sigma(j,i)$ for all i, j in I) such that for every finite subset $J = (j_1, \ldots, j_n)$ of I the matrix $(\sigma(i,j))_{i,j\in J}$ is *positive-definite* in the sense that

$$\sum_{i,j\in J} \sigma(i,j)\xi_i\xi_j \ge 0,$$

for all $\xi_{j_1}, \ldots, \xi_{j_n} \in \mathbb{R}$. In the non-degenerate case we shall assume that the above inequality is strict whenever the vector $(\xi_{j_1}, \ldots, \xi_{j_n})$ is non-zero. Define the process $(i,\omega) \mapsto X_i(\omega)$ by $X_i(\omega) = \omega_i$ where $\omega \in \Omega_I = \mathbb{R}^I$ is given by $\omega = (\omega_i)_{i\in I}$. Let $\mu = (\mu_i)_{i\in I} \in \mathbb{R}^I$ be a map from I to $\mathbb{R}$.

There exists a unique probability measure $\mathbb{P}$ on the σ-field on Ω_I generated by $(X_i)_{i\in I}$ with the following property:

$$\mathbb{E}\left(\exp\left(-i\sum_{j\in J}\xi_j X_j\right)\right) = \exp\left(-i\sum_{j\in J}\xi_j\mu_j\right)\exp\left(-\frac{1}{2}\sum_{k,\ell\in J}\sigma(k,\ell)\xi_k\xi_\ell\right).$$

This measure possesses the following additional properties:

$$\begin{aligned} &\mathbb{E}(X_j) = \mu_j, \quad j \in I; \\ &\operatorname{cov}(X_i, X_j) := \mathbb{E}\left[(X_i - \mathbb{E}[X_i])(X_j - \mathbb{E}[X_j])\right] = \sigma(i,j), \quad i,j \in I. \end{aligned}$$

Notice that $\sum_{u,v=1}^{n} \xi_u\xi_v \operatorname{cov}(X_{j_u}, X_{j_v}) \ge 0$ whenever $\xi_1, \ldots, \xi_n$ belong to $\mathbb{R}$. A proof of this result may be based on Bochner's theorem in conjunction with Kolmogorov's extension theorem. For this purpose, let $J = (j_1, \ldots, j_n)$ be a finite subset of I, let λ_k, $1 \le k \le r$, be complex numbers and let ξ^k, $1 \le k \le r$, be vectors in $\mathbb{R}^n \equiv \mathbb{R}^J$. Put $\lambda'_k = \lambda_k \exp\left(i\sum_{u=1}^n \xi^k_{j_u}\mu_{j_u}\right)$ and let U be an orthogonal matrix with the property that the matrix $\left(U\sigma U^{-1}(u,v)\right)_{u,v=1}^n$ has the diagonal form

$$\left(U\sigma U^{-1}(u,v)\right) = \begin{pmatrix} s_1^2 & 0 & \ldots & 0 \\ 0 & s_2^2 & \ldots & 0 \\ \vdots & \vdots & \ddots & \vdots \\ 0 & 0 & 0 & s_n^2 \end{pmatrix}.$$

We also write $\left(\eta_1^\ell, \dots, \eta_n^\ell\right) = U \begin{pmatrix} \xi_{j_1}^\ell \\ \vdots \\ \xi_{j_n}^\ell \end{pmatrix}$. We may and do suppose that the eigenvalues $s_1, \dots, s_m$, $m \leq n$, are non-zero and the others (if any) are 0. Then we get

$$
\begin{aligned}
&\sum_{k,\ell=1}^{r} \lambda_k \overline{\lambda}_\ell \exp\left(-\frac{1}{2}\sum_{u,v=1}^{n} \sigma(j_u, j_v)\left(\xi_{j_u}^\ell - \xi_{j_u}^k\right)\left(\xi_{j_v}^\ell - \xi_{j_v}^k\right)\right) \\
&\qquad \times \exp\left(-i\sum_{u=1}^{n}\left(\xi_{j_u}^\ell - \xi_{j_u}^k\right)\mu_{j_u}\right) \\
&= \sum_{k,\ell=1}^{r} \lambda_k' \overline{\lambda_\ell'} \exp\left(-\frac{1}{2}\sum_{u,v=1}^{n}\left(U\sigma U^{-1}\right)(u,v)\left(\eta_u^\ell - \eta_u^k\right)\left(\eta_v^\ell - \eta_v^k\right)\right) \\
&= \sum_{k,\ell=1}^{r} \lambda_k' \overline{\lambda_\ell'} \exp\left(-\frac{1}{2}\sum_{u=1}^{n} s_u^2\left(\eta_u^\ell - \eta_u^k\right)^2\right) \\
&= \sum_{k,\ell=1}^{r} \lambda_k' \overline{\lambda_\ell'} \exp\left(-\frac{1}{2}\sum_{u=1}^{m} s_u^2\left(\eta_u^\ell - \eta_u^k\right)^2\right) \\
&= \sum_{k,\ell=1}^{r} \lambda_k' \overline{\lambda_\ell'} \frac{1}{(\sqrt{2\pi})^m} \frac{1}{\prod_{u=1}^{m} s_u} \int \dots \int dx_1 \dots dx_m \\
&\qquad \exp\left(-i\sum_{u=1}^{m}\left(\eta_u^\ell - \eta_u^k\right)x_u\right) \exp\left(-\frac{1}{2}\sum_{u=1}^{m}\frac{x_u^2}{s_u^2}\right) \\
&= \frac{1}{(\sqrt{2\pi})^m}\frac{1}{\prod_{u=1}^{m} s_u} \int \dots \int dx_1 \dots dx_m \left|\sum_{k=1}^{r} \lambda_k' \exp\left(i\sum_{u=1}^{m}\eta_u^\ell x_u\right)\right|^2 \\
&\qquad \times \exp\left(-\frac{1}{2}\sum_{u=1}^{m}\frac{x_u^2}{s_u^2}\right) \\
&\geq 0.
\end{aligned}
$$

From Bochner's theorem it follows that there exists a probability measure Π_J on $\mathbb{R}^J$ such that, for all $\xi \in \mathbb{R}^n$,

$$
\int_{\mathbb{R}^J} \exp\left(-i\sum_{u=1}^{n}\xi_u x_u\right) d\Pi_J = \exp\left(-i\sum_{u=1}^{n}\xi_u\mu_u\right)\exp\left(-\frac{1}{2}\sum_{u,v=1}^{n}\sigma(j_u, j_v)\xi_u\xi_v\right).
$$

Define the probability measure $\mathbb{P}_J$ on $\Omega_J = \prod_{j\in J}\Omega_j$ by $\mathbb{P}_J\left(\left(X_{j_1}, \dots, X_{j_n}\right) \in B\right) = \Pi_J(B)$ where B is a Borel subset of $\mathbb{R}^J$. The collection $(\Omega_J, \mathcal{A}_J, \mathbb{P}_J)$ is a projective system. To see this, let $J \subset I$ contain n indices, let $j_0 \in I\setminus J$, and let $J' := \{j_0\} \cup J$.

Also, let B be a Borel subset of $\mathbb{R}^J$. Then the Fourier transform of the measure $B \mapsto \mathbb{P}_J\left((X_{j_1}, \ldots, X_{j_n}) \in B\right)$ is given by the function

$$(\xi_{j_1}, \ldots, \xi_{j_n}) \mapsto \exp\left(-i\sum_{j\in J} \xi_j \mu_j\right) \exp\left(-\frac{1}{2}\sum_{i,j\in J} \sigma(i,j)\xi_i\xi_j\right),$$

and the Fourier transform of the measure $B \mapsto \mathbb{P}_{J'}\left((X_{j_0}, X_{j_1}, \ldots, X_{j_n}) \in \mathbb{R}\times B\right)$ is given by the function (in what follows $\xi_{j_0} = 0$):

$$\begin{aligned}
(\xi_{j_1}, \ldots, \xi_{j_n}) \mapsto & \int_{X_{j_0}\in\mathbb{R}} \exp\left(-i\sum_{j\in J} \xi_j X_j\right) d\mathbb{P}_{J'} \\
&= \int_{\mathbb{R}^{J'}} \exp\left(-i\sum_{j\in J'} \xi_j X_j\right) d\mathbb{P}_{J'} \\
&= \exp\left(-i\sum_{j\in J'} \xi_j \mu_j\right) \exp\left(-\frac{1}{2}\sum_{i,j\in J'} \sigma(i,j)\xi_i\xi_j\right) \\
&= \exp\left(-i\sum_{j\in J} \xi_j \mu_j\right) \exp\left(-\frac{1}{2}\sum_{i,j\in J} \sigma(i,j)\xi_i\xi_j\right).
\end{aligned}$$

An application of the extension theorem of Kolmogorov yields the desired result.

Suppose that the matrix $\left(\sigma(j_u, j_v)\right)_{u,v=1}^n$ is non-degenerate (i.e. suppose that its determinant is non-zero) and let $\left(\alpha(u,v)\right)_{u,v=1}^n$ be its inverse. Then

$$\begin{aligned}
&\mathbb{P}\left((X_{j_1}, \ldots, X_{j_n}) \in B\right) \\
&= \frac{(\det\alpha)^{1/2}}{(2\pi)^{n/2}} \int\cdots\int dx_1\ldots dx_n 1_B(x_1, \ldots, x_n) \\
&\quad\times \exp\left(-\frac{1}{2}\sum_{u,v=1}^n \alpha(u,v)\left(x_u - \mu_u\right)\left(x_v - \mu_v\right)\right).
\end{aligned}$$

This equality can be proved by showing that the Fourier transforms of both measures coincide.

EXAMPLE 3. We mention some relevant examples of positive-definite functions defined on (subsets of) $\mathbb{R}\times\mathbb{R}$. The functions

$$\begin{aligned}
&\sigma(u,v) := \min(u,v), \quad u,\ v \in [0,\infty); \\
&\sigma_t(u,v) := t\min(u,v) - uv, \quad u,\ v \in [0,t]; \\
&\sigma_{\mathbb{R}}(u,v) = \frac{1}{2}\exp\left(-|u-v|\right) = \frac{1}{2}\exp(-u)\exp(-v)\sigma\left(\exp(2u), \exp(2v)\right), \ u,v\in\mathbb{R};
\end{aligned}$$

and

$$\begin{aligned}\sigma_0(u,v) &:= \frac{1}{2}\exp\bigl(-(u+v)\bigr)\left(\exp\left(2\min(u,v)\right)-1\right)\\ &= \frac{1}{2}\exp(-u)\exp(-v)\sigma\left(\exp(2u)-1,\exp(2v)-1\right)\\ &= \frac{1}{2}\left(\exp\left(-|u-v|\right)-\exp\left(-(u+v)\right)\right), \quad u,\ v\in[0,\infty),\end{aligned}$$

are positive-definite.

These facts can be proved as follows. Let $0 = s_0 < s_1 < s_2 < s_3 < \cdots < s_n < t$ and let $\lambda_1, \ldots, \lambda_n$ be complex numbers. The following identities are valid:

$$\begin{aligned}&\sum_{j=1}^{n}\left|\sum_{k=j}^{n}\lambda_k(t-s_k)\right|^2\frac{t(s_j-s_{j-1})}{(t-s_j)(t-s_{j-1})}\\ &= \sum_{j=1}^{n}\sum_{k_1=j}^{n}\sum_{k_2=j}^{n}\lambda_{k_1}\overline{\lambda}_{k_2}\left(t-s_{k_1}\right)\left(t-s_{k_2}\right)\left\{\frac{s_j}{t-s_j}-\frac{s_{j-1}}{t-s_{j-1}}\right\}\\ &= \sum_{k_1=1}^{n}\sum_{k_2=1}^{n}\sum_{j=1}^{\min(k_1,k_2)}\lambda_{k_1}\overline{\lambda}_{k_2}\left\{\frac{s_j}{t-s_j}-\frac{s_{j-1}}{t-s_{j-1}}\right\}\left(t-s_{k_1}\right)\left(t-s_{k_2}\right)\\ &= \sum_{k_1=1}^{n}\sum_{k_2=1}^{n}\lambda_{k_1}\overline{\lambda}_{k_2}\frac{s_{\min(k_1,k_2)}}{t-s_{\min(k_1,k_2)}}\left(t-s_{k_1}\right)\left(t-s_{k_2}\right)\\ &= \sum_{k_1=1}^{n}\sum_{k_2=1}^{n}\lambda_{k_1}\overline{\lambda}_{k_2}s_{\min(k_1,k_2)}\left(t-s_{\max(k_1,k_2)}\right) = \sum_{k_1,k_2=1}^{n}\lambda_{k_1}\overline{\lambda}_{k_2}\sigma_t\left(s_{k_1},s_{k_2}\right),\end{aligned}$$

and hence the function σ_t is positive-definite. Since

$$\sum_{j,k=1}^{n}\lambda_j\overline{\lambda}_k t\min(s_j,s_k) = \sum_{j,k=1}^{n}\lambda_j\overline{\lambda}_k\sigma_t(s_j,s_k)+\left|\sum_{j=1}^{n}\lambda_j s_j\right|^2,$$

it follows that the function σ is positive-definite as well.

The positive-definiteness of $\sigma_{\mathbb{R}}(u,v)$ can be established as follows. For $\lambda_1, \ldots, \lambda_n$ in $\mathbb{C}$ and for real numbers $s_1, \ldots, s_n$ we write

$$\begin{aligned}&\sum_{k,\ell=1}^{n}\lambda_k\overline{\lambda}_\ell\exp\left(-\left|s_k-s_\ell\right|\right)\\ &= \sum_{k,\ell=1}^{n}\lambda_k\overline{\lambda}_\ell\min\bigl(\exp\bigl(-(s_k-s_\ell)\bigr),\exp\bigl(-(s_\ell-s_k)\bigr)\bigr)\end{aligned}$$

$$= \sum_{k,\ell=1}^{n} \exp(-s_k)\lambda_k \exp(-s_\ell)\overline{\lambda}_\ell \min\left(\exp(2s_k), \exp(2s_\ell)\right)$$

$$= \sum_{k,\ell=1}^{n} \exp\left(-s_k\right)\lambda_k \exp\left(-s_\ell\right)\overline{\lambda}_\ell \exp(-2t)\sigma_{\exp(2t)}\left(\exp\left(2s_k\right), \exp\left(2s_\ell\right)\right)$$

$$+ \exp(-2t)\left|\sum_{k=1}^{n} \exp\left(-s_k\right)\lambda_k \exp\left(2s_k\right)\right|^2 .$$

A similar argument can be used to prove that the function $\sigma_0(u,v)$ is positive-definite.

In the following proposition we mention some elementary facts on Gaussian vectors.

1.16. PROPOSITION. *Let $(\Omega, \mathcal{F}, \mathbb{P})$ be a probability space and let $X^i : \Omega \mapsto \mathbb{R}^{n_i}$, $i = 1, 2$, be random vectors with the property that the random vector $X(\omega) := \left(X^1(\omega), X^2(\omega)\right)$ is Gaussian in the sense that ($n = n_1 + n_2$)*

$$\mathbb{E}\left(\exp\left(-i\sum_{k=1}^{n}\xi_k X_k\right)\right) = \exp\left(-i\sum_{k=1}^{n}\xi_k\mu_k - \frac{1}{2}\sum_{k,\ell=1}^{n}\sigma(k,\ell)\xi_k\xi_\ell\right),$$

where the matrix $\sigma(k,\ell)_{k,\ell=1}^{n}$ is positive-definite and where $(\mu_1, \dots, \mu_n)$ is a vector in $\mathbb{R}^n$. Then the vectors X^1 and X^2 are $\mathbb{P}$-independent if and only if they are uncorrelated in the sense that

$$\mathbb{E}\left(X_i^1 X_j^2\right) = \mathbb{E}\left(X_i^1\right)\mathbb{E}\left(X_j^2\right)$$

for all $1 \le i \le n_1$ and for all $1 \le j \le n_2$.

1.17. PROPOSITION. *Let $(\Omega, \mathcal{F}, \mathbb{P})$ be a probability space.*
(a) Let $Q : \mathbb{R}^m \to \mathbb{R}^n$ be a linear map. If $X : \Omega \to \mathbb{R}^m$ is a Gaussian vector, then so is QX.
(b) A random vector $X : \Omega \to \mathbb{R}^n$ is Gaussian if and only if for every $\xi \in \mathbb{R}^n$ the random variable $\omega \mapsto \langle \xi, X(\omega)\rangle$ is Gaussian.

A proof of the following result is contained in Example 2 of the present section E.

1.18. THEOREM. *Let $\sigma : I \times I \to \mathbb{R}$ be a positive-definite function and let $\mu : I \to \mathbb{R}$ be a map. There exists a probability space $(\Omega, \mathcal{F}, \mathbb{P})$ together with a Gaussian process $(t,\omega) \mapsto X_t(\omega) = X(t,\omega)$, $t \in I$, $\omega \in \Omega$, such that $\mathbb{E}(X_t) = \mu_t$ and such that $\operatorname{cov}(X_s, X_t) = \sigma(s,t)$ for all s, $t \in I$.*

E.2. Brownian motion and related processes

In this subsection we introduce *Brownian motion* and *Brownian bridge* as Gaussian processes. First we note that the function $\sigma : [0,\infty) \times [0,\infty) \to \mathbb{R}$, defined by $\sigma(u,v) = \min(u,v)$, u, $v \in [0,\infty)$, and, for $t \geq 0$ fixed, the function $\sigma_t : [0,t] \times [0,t] \to \mathbb{R}$, defined by $\sigma_t(u,v) = t\min(u,v) - uv$, u, $v \in [0,t]$, are positive-definite. In addition the function $\sigma_0(u,v)$ defined by $\sigma_0(u,v) := \frac{1}{2}\exp(-(u+v))(\exp(2\min(u,v)) - 1)$, for u, $v \geq 0$, is positive-definite: cfr. Example 3 of 1.E.1.

We now give existence theorems for the Wiener process (or Brownian motion), for Brownian bridge, for the oscillator process, and for the Ornstein-Uhlenbeck process. Compare (c) of Theorem 1.19 with Example 7 of section C and (d) with Example 6 of section C of the present chapter.

1.19. THEOREM.

(a) *There exists a probability space $(\Omega, \mathcal{F}, \mathbb{P})$ together with a real-valued Gaussian process $\{B(s) : s \geq 0\}$, called Wiener process or Brownian motion, such that $\mathbb{E}(B(s)) = 0$ and such that $\mathbb{E}(B(s_1)B(s_2)) = \min(s_1, s_2)$ for all s_1, $s_2 \geq 0$.*

(b) *Fix $t > 0$. There exists a probability space $(\Omega, \mathcal{F}, \mathbb{P})$ together with a real-valued Gaussian process $\{X_t(s) : t \geq s \geq 0\}$, called Brownian bridge, such that $\mathbb{E}(X_t(s)) = 0$ and such that $\mathbb{E}(X_t(s_1)X_t(s_2)) = \min(s_1, s_2) - \dfrac{s_1 s_2}{t}$ for all s_1, $s_2 \in [0,t]$.*

(c) *There exists a probability space $(\Omega, \mathcal{F}, \mathbb{P})$ together with a real-valued Gaussian process $\{q(s) : s \in \mathbb{R}\}$, called oscillator process, such that $\mathbb{E}(q(s)) = 0$ and such that $\mathbb{E}(q(s_1)q(s_2)) = \frac{1}{2}\exp(-|s_1 - s_2|)$ for all s_1, $s_2 \in \mathbb{R}$.*

(d) *There exists a probability space $(\Omega, \mathcal{F}, \mathbb{P})$ together with a real-valued Gaussian process $\{X(s) : s \geq 0\}$, called Ornstein-Uhlenbeck process, such that $\mathbb{E}(X(s)) = 0$ and such that*

$$\begin{aligned}\mathbb{E}(X(s_1)X(s_2)) &= \frac{1}{2}\exp(-(s_1+s_2))\big(\exp(2\min(s_1,s_2)) - 1\big)\\ &= \frac{1}{2}\big(\exp(-|s_1 - s_2|) - \exp(-(s_1+s_2))\big),\end{aligned}$$

for all s_1, $s_2 \geq 0$.

In what follows x and y are real numbers and so is μ. Let $\{B(s) : s \geq 0\}$ be Brownian motion (starting in 0) on a probability space $(\Omega, \mathcal{F}, \mathbb{P})$ (i.e. $\mathbb{E}(B(s)) = 0$ and $\mathbb{E}(B(s_1)B(s_2)) = \min(s_1, s_2)$). Then the process

$$\{x + B(s) + \mu s : s \geq 0\}$$

is a *Brownian motion with drift μ starting at x*. Let $\{X_t(s) : 0 \leq s \leq t\}$ be a Brownian bridge on a probability space $(\Omega, \mathcal{F}, \mathbb{P})$. Then the process $s \mapsto \left(1 - \frac{s}{t}\right)x + \frac{s}{t}y + X_t(s)$ $0 \leq s \leq t$ is called *pinned* Brownian motion, namely pinned at x at time 0 and pinned at y at time t. Let $\{B_j(s) : s \geq 0\}$, $1 \leq j \leq \nu$, be ν independent Brownian motions on the probability space $(\Omega, \mathcal{F}, \mathbb{P})$. The process $\{(B_1(s), \dots, B_\nu(s)) : s \geq 0\}$ is called ν-dimensional Brownian motion. A similar definition can be given for ν-dimensional Brownian bridge and for the ν-dimensional oscillator process. Notice that a ν-dimensional process $\{B(s) = (B_1(s), \dots, B_\nu(s)) : s \geq 0\}$ is a ν-dimensional Brownian motion, starting at 0, on the probability space $(\Omega, \mathcal{F}, \mathbb{P})$ if and only if $\mathbb{E}(B_j(s_1)B_k(s_2)) = \delta_{j,k} \min(s_1, s_2)$.

In the following proposition we collect a number of interesting properties of the (finite dimensional) joint distributions of some of the Gaussian processes we introduced hitherto (see Theorem 1.19). More general Markov processes do not have these nice features.

1.20. PROPOSITION. *Let $\{B(s) : s \geq 0\}$ be ν-dimensional Brownian motion and let $\{X_t(s) : 0 \leq s \leq t\}$ be ν-dimensional Brownian bridge. In addition let x and y be vectors in $\mathbb{R}^\nu$ and let $Q : \mathbb{R}^\nu \to \mathbb{R}^\nu$ be an orthogonal linear map. Also fix a strictly positive number a.*

(a) The joint distributions of the processes

$$\{B(s) : s > 0\} \text{ and of } \left\{ sB\left(\frac{1}{s}\right) : s > 0 \right\}$$

coincide.

(b) The joint distributions of the processes

$$\{B(as) : s \geq 0\} \text{ and } \left\{ \sqrt{a}B(s) : s \geq 0 \right\}$$

coincide (scale invariance).

(c) The joint distributions of the processes (see Theorem 1.19.(c))

$$\{q(s) : s \in \mathbb{R}\} \text{ and } \left\{ e^{-s} B\left(e^{2s}/2\right) : s \in \mathbb{R} \right\}$$

coincide.

(d) The joint distributions of the processes (see Theorem 1.19.(d))

$$\{X(t) : t \geq 0\} \text{ and } \left\{ e^{-t} B\left((e^{2t} - 1)/2\right) : t \geq 0 \right\}$$

coincide. The process $\{X(t) : t \geq 0\}$ also possesses the same law (i.e. joint distribution) as the process $\left\{ \int_0^t \exp(-(t-s))\, dB(s) : t \geq 0 \right\}$.

(e) The joint distributions of the following processes also coincide:

$$\left(1 - \frac{s}{t}\right) x + \frac{s}{t} y + X_t(s), \quad 0 < s < t,$$

$$\left(1 - \frac{s}{t}\right) x + \frac{s}{t} y + \left(1 - \frac{s}{t}\right) B\left(\frac{st}{t-s}\right), \quad 0 < s < t,$$

$$\text{and } \left(1 - \frac{s}{t}\right) x + \frac{s}{t} y + B(s) - \frac{s}{t} B(t), \quad 0 < s < t.$$

(f) *The process* $\{QB(s) : s \geq 0\}$ *is* ν*-dimensional Brownian motion and hence its joint distribution coincides with that of* ν*-dimensional Brownian motion* $\{B(s) : s \geq 0\}$ *(orthogonal invariance).*

REMARK 1. Because of assertion (d) the oscillator process and the Ornstein-Uhlenbeck process are not always distinguished.

REMARK 2. Put $X^x(t) = \exp(-t)x + X(t)$. Then the process $\{X^x(t) : t \geq 0\}$ is the *Ornstein-Uhlenbeck process of initial velocity* x. Let $f : \mathbb{R}^\nu \to \mathbb{C}$ be a bounded Borel measurable function. Then $\mathbb{E}\left[f(X^x(t))\right]$ is given by

$$\mathbb{E}\left[f\left(X^x(t)\right)\right] = \int f\left(e^{-t}x + \sqrt{1-e^{-2t}}y\right) \frac{\exp\left(-\left|y\right|^2\right)}{\left(\sqrt{\pi}\right)^\nu} dy.$$

Moreover, the Ornstein-Uhlenbeck process is a strong Markov process. This is also true for Brownian motion and for the oscillator process.

REMARK 3. The stochastic integral $\int_0^t \exp(-(t-s))dB(s)$ in (d) can be defined as the L^2-limit of $\sum_{j=1}^n \exp\left(-(t-s_{j-1})\right)\left(B(s_j) - B(s_{j-1})\right)$, whenever $\max\left(s_j - s_{j-1} : 1 \leq j \leq n\right)$ tends to zero. Here $0 = s_0 < s_1 < \cdots s_n = t$ is a subdivision of the interval $[0, t]$.

REMARK 4. Put $B^x(t) = x+B(t)$. Then, by definition, $\{B^x(t) : t \geq 0\}$ is *Brownian motion that starts in* x (translation invariance). Let $\{B^x(t) : t \geq 0\}$ be Brownian motion that starts at x (and has drift zero). Fix $s > 0$. The processes $\{B^x(s+t) - B^x(s) : t \geq 0\}$ and $\{B^x(t) - x : t \geq 0\}$ possess the same (joint) distribution.

Since in Proposition 1.20 all the indicated processes are ν-dimensional Gaussian (the definition of a ν-dimensional Gaussian process is an immediate generalization of the 1-dimension Gaussian process: in fact in the discussion of Example 2 in subsection E.1 and in Theorem 1.19 the expected value μ should be map from I to $\mathbb{R}^\nu$ and the entries of the diffusion matrix σ should be $\nu \times \nu$-matrices), it suffices to show that the corresponding expectations and covariance matrices are the same for the indicated processes. In most cases this is a simple exercise.

1.21. THEOREM. Let $\{X(t) : t \geq 0\}$ be Brownian motion on some probability space $(\Omega, \mathcal{F}, \mathbb{P})$. Then there exists a $\mathbb{P}$-almost surely continuous process $\{B(t) : t \geq 0\}$ that is also a Brownian motion on the probability space $(\Omega, \mathcal{F}, \mathbb{P})$ and that is indistinguishable from the process $\{X(s) : s \geq 0\}$. This means that $\mathbb{P}(X(t) = B(t)) = 1$ for all $t \geq 0$.

Here we suppose that $\mathcal{F}$ contains the $\mathbb{P}$-zero sets.

In the next theorem we compute the finite dimensional distributions of ν-dimensional Brownian motion starting at 0 with drift μ. For this purpose, we define the Gaussian (or Wiener) transition kernel $p_{0,\nu}(t, x, y)$ by

$$p_{0,\nu}(t, x, y) = \frac{1}{(2\pi t)^{\frac{1}{2}\nu}} \exp\left(-\frac{\left|x-y\right|^2}{2t}\right).$$

This transition kernel appears both in the statement of the next theorem and in the following discussion. The result is an easy consequence of the following identities. Let $0 = s_0 < s_1 < \cdots < s_n < \infty$. Fix the vectors x and $\xi_1, \ldots, \xi_n$ in $\mathbb{R}^\nu$. Put $s_0 = 0$ and $x_0 = x$. The following equalities are valid:

$$\sum_{\ell=1}^{n} (s_\ell - s_{\ell-1}) \left| \sum_{j=\ell}^{n} \xi_j \right|^2 = \sum_{j,k=1}^{n} \langle \xi_j, \xi_k \rangle \min(s_j, s_k), \quad \text{and}$$

$$\int_{\mathbb{R}^\nu} dx_1 \ldots \int_{\mathbb{R}^\nu} dx_n \exp\left(-i \sum_{j=1}^{n} \langle \xi_j, x_j \rangle \right) \prod_{j=1}^{n} p_{0,\nu}(s_j - s_{j-1}, x_{j-1}, x_j)$$
$$= \exp\left(-i \left\langle \sum_{j=1}^{n} \xi_j, x \right\rangle \right) \exp\left(-\frac{1}{2} \sum_{j,k=1}^{n} \langle \xi_j, \xi_k \rangle \min(s_j, s_k) \right).$$

The proof of these equalities proceeds as follows. In order to see the first equality, we write

$$\sum_{\ell=1}^{n} (s_\ell - s_{\ell-1}) \left| \sum_{j=\ell}^{n} \xi_j \right|^2 = \sum_{\ell=1}^{n} (s_\ell - s_{\ell-1}) \sum_{j_1, j_2 = \ell}^{n} \langle \xi_{j_1}, \xi_{j_2} \rangle$$
$$= \sum_{j_1, j_2 = 1}^{n} \sum_{\ell=1}^{\min(j_1, j_2)} (s_\ell - s_{\ell-1}) \langle \xi_{j_1}, \xi_{j_2} \rangle = \sum_{j_1, j_2 = 1}^{n} \left(s_{\min(j_1, j_2)} - s_0 \right) \langle \xi_{j_1}, \xi_{j_2} \rangle$$
$$= \sum_{j_1, j_2 = 1}^{n} s_{\min(j_1, j_2)} \langle \xi_{j_1}, \xi_{j_2} \rangle . = \sum_{j_1, j_2 = 1}^{n} \min(s_{j_1}, s_{j_2}) \langle \xi_{j_1}, \xi_{j_2} \rangle .$$

For the second one,

$$\int_{\mathbb{R}^\nu} dx_1 \ldots \int_{\mathbb{R}^\nu} dx_n \exp\left(-i \sum_{j=1}^{n} \langle \xi_j, x_j \rangle \right)$$
$$\prod_{j=1}^{n} \frac{1}{\left(\sqrt{2\pi (s_j - s_{j-1})} \right)^\nu} \exp\left(-\frac{|x_j - x_{j-1}|^2}{2(s_j - s_{j-1})} \right)$$

(substitute $x_j = x + y_1 + \cdots + y_j$)

$$= \exp\left(-i \sum_{j=1}^{n} \langle \xi_j, x \rangle \right) \int_{\mathbb{R}^\nu} dy_1 \ldots \int_{\mathbb{R}^\nu} dy_n \exp\left(-i \sum_{\ell=1}^{n} \left\langle \sum_{j=\ell}^{n} \xi_j, y_\ell \right\rangle \right)$$
$$\prod_{j=1}^{n} \frac{1}{\left(\sqrt{2\pi (s_j - s_{j-1})} \right)^\nu} \exp\left(-\frac{|y_j|^2}{2(s_j - s_{j-1})} \right)$$

(substitute $y_\ell = (s_\ell - s_{\ell-1})^{1/2} z_\ell$)

$$= \exp\left(-i\sum_{j=1}^{n}\langle\xi_j, x\rangle\right)\int_{\mathbb{R}^\nu} dz_1 \dots \int_{\mathbb{R}^\nu} dz_n \exp\left(-i\sum_{\ell=1}^{n}(s_\ell - s_{\ell-1})^{1/2}\sum_{j=\ell}^{n}\langle\xi_j, z_\ell\rangle\right)$$

$$\prod_{j=1}^{n}\frac{1}{\left(\sqrt{2\pi}\right)^\nu}\exp\left(-\frac{|z_j|^2}{2}\right)$$

$$= \exp\left(-i\sum_{j=1}^{n}\langle\xi_j, x\rangle\right)\exp\left(-\frac{1}{2}\sum_{\ell=1}^{n}(s_\ell - s_{\ell-1})\left|\sum_{j=\ell}^{n}\xi_j\right|^2\right)$$

$$\times\prod_{\ell=1}^{n}\frac{1}{\left(\sqrt{2\pi}\right)^\nu}\int_{\mathbb{R}^\nu}\exp\left(-\frac{1}{2}\left(z_\ell + i\sqrt{s_\ell - s_{\ell-1}}\sum_{j=\ell}^{n}\xi_j\right)^2\right)dz_\ell,$$

where $(a+bi)^2 = |a|^2 + 2i\langle a, b\rangle - |b|^2$, for a and $b \in \mathbb{R}^\nu$. From Cauchy's theorem, it then follows that

$$\int_{\mathbb{R}^\nu} dx_1 \dots \int_{\mathbb{R}^\nu} dx_n \exp\left(-i\sum_{j=1}^{n}\langle\xi_j, x_j\rangle\right)\prod_{j=1}^{n}p_{0,\nu}(s_j - s_{j-1}, x_{j-1}, x_j)$$

$$= \exp\left(-i\sum_{j=1}^{n}\langle\xi_j, x\rangle\right)\exp\left(-\frac{1}{2}\sum_{\ell=1}^{n}(s_\ell - s_{\ell-1})\left|\sum_{j=\ell}^{n}\xi_j\right|^2\right)$$

$$\times\prod_{\ell=1}^{n}\frac{1}{\left(\sqrt{2\pi}\right)^\nu}\int_{\mathbb{R}^d} dz_\ell \exp\left(-\frac{1}{2}|z_\ell|^2\right)$$

$$= \exp\left(-i\sum_{j=1}^{n}\langle\xi_j, x\rangle\right)\exp\left(-\frac{1}{2}\sum_{\ell=1}^{n}(s_\ell - s_{\ell-1})\left|\sum_{j=\ell}^{n}\xi_j\right|^2\right)$$

(first equality)

$$= \exp\left(-i\sum_{j=1}^{n}\langle\xi_j, x\rangle\right)\exp\left(-\frac{1}{2}\sum_{j,k=1}^{n}\langle\xi_j, \xi_k\rangle\min(s_j, s_k)\right).$$

The preceding discussion proves Theorem 1.22 for

$$f_j(y) = \exp\left(-i\langle\xi_j, y\rangle\right), \quad 1 \le j \le n.$$

A density argument will extend this result to bounded Borel functions.

1.22. THEOREM. *Let $\{B(s) : s \geq 0\}$ be ν-dimensional Brownian motion with diffusion matrix identity, with drift μ and which starts in 0. Let $f_1, \ldots, f_n$ be bounded Borel measurable functions on $\mathbb{R}^\nu$ and let $0 = s_0 < s_1 < \cdots < s_n$. Then*

$$\mathbb{E}\left(\prod_{j=1}^{n} f_j(x + B(s_j) + \mu s_j)\right)$$
$$= \int_{\mathbb{R}^\nu} \cdots \int_{\mathbb{R}^\nu} dx_1 \ldots dx_n \prod_{j=1}^{n} f_j(x_j) \prod_{j=1}^{n} p_{0,\nu}\left(s_j - s_{j-1}, x_{j-1} - \mu s_{j-1}, x_j - \mu s_j\right),$$

where $x_0 = x$.

REMARK 1. The equality depicted in Theorem 1.22 determines the joint distribution of the process

$$\{X(s) := x + B(s) + \mu s : s \geq 0\}.$$

This will follow from the monotone class theorem (see Appendix E, Definition E.9). The vector μ is the so-called drift vector and the process $X = \{X(s) : s \geq 0\}$ starts at x in $\mathbb{R}^\nu$.

REMARK 2. Another consequence of the equality in Theorem 1.22 is the fact that the random vector $B(t) - B(s)$, $t > s$ fixed, is independent of the σ-field generated by the process $\{B(\sigma) : 0 \leq \sigma \leq s\}$. In other words a Brownian motion (diffusion matrix identity) is a Gaussian process $\{B(s) : s \geq 0\}$ with independent increments $B(t) - B(s)$, $t > s$, with mean $\mu(t-s)$ and covariance matrix $\operatorname{cov}(B_k(t) - B_k(s), B_\ell(t) - B_\ell(s)) = \delta_{k,\ell}(t-s)$.

Next we want to define standard Brownian motion with drift vector μ that starts at $x \in \mathbb{R}^\nu$. From the monotone class theorem (see Appendix E), it follows that the equalities ($f_j \in L^\infty(\mathbb{R}^\nu, \mathcal{B})$, $1 \leq j \leq n$, $0 < s_1 << \cdots < s_n$, $n \in \mathbb{N}$)

$$\mathbb{E}_x\left(\prod_{j=1}^{n} f_j(B(s_j))\right) := \mathbb{E}\left(\prod_{j=1}^{n} f_j(x + B(s_j))\right) = \mathbb{E}_0\left(\prod_{j=1}^{n} f_j(x + B(s_j))\right)$$

determine the probability measure $\mathbb{P}_x$.

1.23. DEFINITION. The standard Brownian motion starting at $x \in \mathbb{R}^\nu$ with drift μ is defined as the canonical Gaussian process $\{X(s) : s \geq 0\}$ defined on $(\Omega, \mathcal{F}, \mathbb{P}_x)$ with the property that the increments $X(t+h) - X(t)$ are mutually independent and have $\mathbb{P}_x$-expectation μh. Moreover, it starts $\mathbb{P}_x$-almost surely at x, i.e. $\mathbb{P}_x(X(0) = x) = 1$, and

$$\operatorname{cov}\left(X_k(t+h) - X_k(t), X_\ell(t+h) - X_\ell(t)\right) = \delta_{k,\ell} h.$$

The covariance is of course also taken with respect to $\mathbb{P}_x$. The process is canonical because for Ω we take $\Omega = C\left([0,\infty), \mathbb{R}^\nu\right)$, and for $X(t)$ we take $X(t)(\omega) = \omega(t)$,

$\omega \in \Omega$. For $\mathcal{F}$ we take the σ-field on Ω generated by the *state variables* $\{X(s) : s \geq 0\}$. For all this we often write

$$\{(\Omega, \mathcal{F}, \mathbb{P}_x), (X(t) : t \geq 0), (\vartheta_t : t \geq 0), (\mathbb{R}^\nu, \mathcal{B})\}.$$

Here the *shift* or *translation* operators ϑ_t, $t \geq 0$, are defined by $\vartheta_t(\omega)(s) = \omega(s+t)$, $\omega \in \Omega$. We also introduce the filtration $(\mathcal{F}_t : t \geq 0)$ defined as the *full history*: $\mathcal{F}_t$ is the σ-field generated by the variables $X(s)$, $0 \leq s \leq t$. We also shall need the right closure $\mathcal{F}_{t+}$ defined by $\mathcal{F}_{t+} = \bigcap_{s>t} \mathcal{F}_s$.

In the following result we give some interesting *martingale* properties for Brownian motion. More details on martingale theory can be found in Appendix C.

1.24. PROPOSITION. *Let $\{(\Omega, \mathcal{F}, \mathbb{P}_x), (X(t) : t \geq 0), (\vartheta_t : t \geq 0), (\mathbb{R}^\nu, \mathcal{B})\}$ be standard Brownian that starts at $x \in \mathbb{R}^\nu$ and that has drift μ. For $t > s$ the variable $X(t) - X(s)$ does not depend on the σ-field $\mathcal{F}_s$. The following processes are $\mathbb{P}_x$-martingales with respect to the filtration $\mathcal{F}_t$, $t \geq 0$:*

$$t \mapsto X(t) - t\mu, \qquad t \mapsto |X(t) - t\mu|^2 - \nu t, \qquad t \mapsto p_{0,\nu}(t_0 - t, X(t), y), \ 0 \leq t < t_0.$$

So far we have looked at Brownian motion as a Gaussian process. On the other hand it is also a Markov process. We would like to discuss that now. In fact, in mathematical terms, the equality in Theorem 1.22 is an equivalent form of the Markov property.

As above, in the following theorem $\mathcal{F}$ is the σ-field generated by $\{X(s) : s \geq 0\}$ and $\mathcal{F}_t$ is the σ-field generated by the *past* or *full history*, i.e. $\mathcal{F}_t = \sigma\{X(s) : 0 \leq s \leq t\}$. An $(\mathcal{F}_{t+})$-stopping time T is an $\mathcal{F}$-measurable map from Ω to $[0, \infty]$ with the property that $\{T \leq t\}$ belongs to $\mathcal{F}_{t+}$ for all $t \geq 0$. If T is an $(\mathcal{F}_{t+})$-stopping time we write $\mathcal{F}_{T+} = \bigcap_{t \geq 0} \{A \in \mathcal{F} : A \cap \{T \leq t\} \in \mathcal{F}_{t+}\}$. Notice that stopping times may take infinite values. Often this is very interesting. For example let T be an $(\mathcal{F}_{t+})$-stopping time and let the event A belong to $\mathcal{F}_{T+}$. The mapping $S : \Omega \to [0, \infty]$ defined by $S = T$, on A, and $S = \infty$, on $\Omega \setminus A$, is an $(\mathcal{F}_{t+})$-stopping time again.

1.25. THEOREM. *Let $\{(\Omega, \mathcal{F}, \mathbb{P}_x), (X(t) : t \geq 0), (\vartheta_t : t \geq 0), (\mathbb{R}^\nu, \mathcal{B})\}$, $x \in \mathbb{R}^\nu$, be ν-dimensional Brownian motion. Then the following conditions are satisfied:*

(a_1) *For every $\alpha > 0$, for every $t \geq 0$ and for every open subset U of $\mathbb{R}^\nu$, the set $\{x \in \mathbb{R}^\nu : \mathbb{P}_x(X(t) \in U) > \alpha\}$ is open.*

(a_2) *For every $\alpha > 0$, for every $t \geq 0$ and for every compact subset K of $\mathbb{R}^\nu$, the set $\{x \in \mathbb{R}^\nu : \mathbb{P}_x(X(t) \in K) \geq \alpha\}$ is compact.*

(b) *For every open subset U of $\mathbb{R}^\nu$ and for every $x \in U$, the equality*

$$\lim_{t \downarrow 0} \mathbb{P}_x(X(t) \in U) = 1 \quad \textit{is valid.}$$

Moreover, ν-dimensional Brownian motion possesses the following equivalent properties:

(i) *For all $t \geq 0$ and for all bounded random variables $Y : \Omega \to \mathbb{C}$, the equality*

$$\mathbb{E}_x \left(Y \circ \vartheta_t \mid \mathcal{F}_t \right) = \mathbb{E}_{X(t)}(Y)$$

holds $\mathbb{P}_x$-almost surely for all $x \in \mathbb{R}^\nu$.

(ii) *For all finite tuples $0 \leq t_1 < t_2 < \ldots < t_n < \infty$ together with Borel subsets $B_1, \ldots, B_n$ of $\mathbb{R}^\nu$, the equality*

$$\begin{aligned} &\mathbb{P}_x \left(X(t_1) \in B_1, \ldots, X(t_n) \in B_n \right) \\ &= \int_{B_1} \cdots \int_{B_{n-1}} \int_{B_n} P(t_n - t_{n-1}, x_{n-1}, dx_n) P(t_{n-1} - t_{n-2}, x_{n-2}, dx_{n-1}) \\ &\qquad \ldots P(t_2 - t_1, x_1, dx_2) P(t_1, x, dx_1) \end{aligned}$$

is valid for all $x \in \mathbb{R}^\nu$ (here $\mathbb{P}_x(X(t) \in B) = P(t, x, B)$).

(iii) *For every $(\mathcal{F}_{t+})$-stopping time T and for every bounded random variable $Y : \Omega \to \mathbb{C}$, the equality*

$$\mathbb{E}_x \left(Y \circ \vartheta_T \mid \mathcal{F}_{T+} \right) = \mathbb{E}_{X(T)} \left(Y \right),$$

holds $\mathbb{P}_x$-almost surely on $\{T < \infty\}$ for all $x \in \mathbb{R}^\nu$.

(iv) *Let $\mathcal{B}_1$ be the Borel field of $[0, \infty)$. For every bounded function $F : [0, \infty) \times \Omega \to \mathbb{C}$, which is measurable with respect to $\mathcal{B}_1 \otimes \mathcal{F}$, and for every $(\mathcal{F}_{t+})$-stopping time T the equality*

$$\mathbb{E}_x \left(\{ \omega \mapsto F\left(T(\omega), \vartheta_T(\omega)\right) \} \mid \mathcal{F}_{T+} \right) = \left\{ \omega' \mapsto \mathbb{E}_{X(T(\omega'))} \left\{ \omega \mapsto F\left(T(\omega'), \omega\right) \right\} \right\}$$

holds $\mathbb{P}_x$-almost surely $\{T < \infty\}$ for all $x \in \mathbb{R}^\nu$.

Since ν-dimensional Brownian motion satisfies (a_1), (a_2) and (b), the properties in (i), (ii), (iii) and (iv) are all equivalent. From the monotone class theorem (cfr. Appendix E) it follows that Properties (i) and (ii) and Properties (iii) and (iv) are always equivalent. The implication (iii) $\Rightarrow$ (ii) is also clear. For the reverse implication the full strength of (a_1), (a_2) and (b) is employed. The fact that Brownian motion possesses property (ii) is a consequence of Theorem 1.22. In fact the right continuity of paths is very important because the strong Markov property depends on the right continuity of paths. More precisely let T be an $(\mathcal{F}_{t+})$-stopping time. Then we may put $T_n = 2^{-n} \lceil 2^n T \rceil$ (so $2^n T_n$ is smallest integer greater than or equal to $2^n T$). Since $T_n \downarrow T$ and since T_n only takes discrete values, it is relatively easy to pass from (i) to (iii) for T_n instead of T in (iii). Since Brownian motion possesses continuous paths $\mathbb{P}_x$-almost surely, this condition is fulfilled. Property (i) is called the *Markov property* and property (iii) is called the *strong Markov property*. The equality in item (iv) is called the *strong time-dependent Markov property*. We shall not prove this result. It is part of the general theory of Markov processes and their sample path properties. The strong Markov property is also closely connected to the theory of Feller semigroups because to every Feller semigroup $\{P(t) : t \geq 0\}$ acting on $C_\infty(E)$, with E locally compact and second countable, there is associated a strong Markov process

$$\{(\Omega, \mathcal{F}, \mathbb{P}_x), (X(t) : t \geq 0), (\vartheta_t : t \geq 0), (E, \mathcal{E})\},$$

for which $P(t)f(x) = \mathbb{E}_x[f(X(t))]$, $f \in C_\infty(E)$. As in Theorem 1.25, let $\{(\Omega, \mathcal{F}, \mathbb{P}_x), (X(t) : t \geq 0), (\vartheta_t : t \geq 0), (\mathbb{R}^\nu, \mathcal{B})\}$ be Brownian motion starting in x. The family of operators $\{\exp(-tH_0) : t \geq 0\}$ defined by $[\exp(-tH_0) f](x) = \mathbb{E}_x(f(X(t))$, $f \in L^\infty(\mathbb{R}^\nu)$, $t \geq 0$, is a Feller semigroup, because it possesses the properties mentioned in Appendix B. In particular it leaves the space $C_\infty(\mathbb{R}^\nu)$ invariant. The semigroup $\{\exp(-tH_0) : t \geq 0\}$ is also called the Gaussian semigroup. Its generator is a suitable version of $\frac{1}{2}\times$ the Laplace operator. More precisely, its domain depends on the space in which we are working (e.g. $C_\infty(\mathbb{R}^\nu)$, $L^p(\mathbb{R}^\nu)$). For instance, in the space $C_\infty(\mathbb{R}^\nu)$, we have

$$\operatorname{dom}(H_0) = \{f \in C_\infty(\mathbb{R}^\nu) : \Delta f \in C_\infty(\mathbb{R}^\nu)\},$$

$$\text{and} \quad \lim_{t\downarrow 0} \frac{1}{t}(\exp(-tH_0) f - f) = -H_0 f = \frac{1}{2}\Delta f,$$

where the function Δf has to be interpreted as the Laplace operator applied to f in distributional sense. A similar definition can be used for the generator in $L^p(\mathbb{R}^\nu)$, $1 \leq p < \infty$.

E.3. Kato-Feller potentials for the Laplace operator

In this subsection we want to discuss the special features of the class of Kato-Feller potentials in case when $K_0 = -\frac{1}{2}\Delta$ on $\mathbb{R}^\nu$.

1.26. Definition. A function $W : \mathbb{R}^\nu \to [-\infty, \infty]$ is called a *Kato* potential (or is said to belong to the space K_ν), if it possesses the following property:

$$\limsup_{t\downarrow 0} \sup_{x\in\mathbb{R}^\nu} \mathbb{E}_x\left[\int_0^t |W(B(s))|\, ds\right] = 0.$$

It follows from the marginal (or one-dimensional) distribution of Brownian motion that the function W is a Kato potential (or belongs to the space K_ν) if and only if the following identity is true:

$$\lim_{t\downarrow 0} \sup_{x\in\mathbb{R}^\nu} \int_0^t \int_{\mathbb{R}^\nu} p_{0,\nu}(s,x,y)\, |W(y)|\, dy\, ds = 0.$$

Here, as always, we write $p_{0,\nu}(t,x,y)$ for the *Gaussian* transition kernel

$$p_{0,\nu}(t,x,y) = \frac{1}{\left(\sqrt{2\pi t}\right)^\nu} \exp\left(-\frac{|y-x|^2}{2t}\right),$$

where x and y belong to $\mathbb{R}^\nu$ and where t is strictly positive. The symbol $|a| = \sqrt{\sum_{j=1}^\nu a_j^2}$ denotes the Euclidean norm of the vector $a = (a_1, \dots, a_\nu) \in \mathbb{R}^\nu$. The following theorem gives an alternative description of a Kato potential.

1.27. THEOREM. *Let $W : \mathbb{R}^\nu \to [0,\infty]$ be a Borel measurable function. Then*

$$\lim_{t\downarrow 0}\sup_{x\in\mathbb{R}^\nu}\int_0^t\int p_{0,\nu}(s,x,y)W(y)dy\,ds = 0$$

if and only if

$$\lim_{t\downarrow 0}\sup_{x\in\mathbb{R}^\nu}\int_{\{|y-x|<\sqrt{2t}\}}\frac{W(y)}{|y-x|^{\nu-2}}dy = 0, \text{ in case } \nu\geq 3,$$

$$\lim_{t\downarrow 0}\sup_{x\in\mathbb{R}^2}\int_{\{|y-x|<\sqrt{2t}\}}W(y)\log\frac{2t}{|y-x|^2}dy = 0, \text{ in case } \nu = 2, \text{ and}$$

$$\sup_{k\in\mathbb{A}}\int_k^{k+1}W(y)dy<\infty, \text{ in case } \nu=1.$$

Theorem 1.27 is an immediate consequence of Theorem 1.28. Both are proved in Aizenman and Simon [3] and in Simon [206].

1.28. THEOREM. *Let $W : \mathbb{R}^\nu \to [0,\infty]$ be a Borel measurable function. Put*

$$C(\ell)=\int_1^\infty \tau^{\ell-1}e^{-\tau}\,d\tau \quad \text{and} \quad C_1(\ell)=\frac{1}{(2\pi)^\ell}\Gamma(\ell-1)+\left(\frac{256}{2\pi}\right)^\ell\frac{C(\ell)}{16},$$

whenever these expressions make sense.

(a) *If $\nu\geq 3$, then the following inequalities are valid:*

$$\begin{aligned}&\frac{1}{2\left(\sqrt{\pi}\right)^\nu}C(\nu/2-1)\sup_{x\in\mathbb{R}^\nu}\int_{\{|y-x|<\sqrt{2t}\}}\frac{W(y)}{|y-x|^{\nu-2}}dy\\&\leq\sup_{x\in\mathbb{R}^\nu}\int_0^t\int_{\mathbb{R}^\nu}p_{0,\nu}(\sigma,x,y)\,W(y)dy\,d\sigma\\&\leq C_1(\nu/2)\sup_{x\in\mathbb{R}^\nu}\int_{\{|y-x|<\sqrt{2t}\}}\frac{W(y)}{|y-x|^{n-2}}dy.\end{aligned}$$

(b) *If $\nu = 2$, then the following inequalities are valid:*

$$\begin{aligned}&\sup_{x\in\mathbb{R}^2}\left[\frac{1}{2\pi e}\int_{\{|y-x|<\sqrt{2t}\}}W(y)\log\frac{2t}{|y-x|^2}dy\right.\\&\qquad\left.+\frac{1}{2\pi}\int_1^\infty\frac{e^{-\tau}}{\tau}d\tau\int_{\{|y-x|<\sqrt{2t}\}}W(y)\,dy\right]\\&\leq\sup_{x\in\mathbb{R}^2}\int_0^t\int p_{0,2}(\sigma,x,y)\,W(y)\,dy\,d\sigma\\&\leq\frac{1}{2\pi}\sup_{x\in\mathbb{R}^\nu}\int_{\{|y-x|<\sqrt{2t}\}}W(y)\log\frac{2t}{|y-x|^2}dy\\&\quad+\frac{1}{\pi}\left(8+\frac{1}{2e}\right)\sup_{x\in\mathbb{R}^2}\int_{\{|y-x|<\sqrt{2t}\}}W(y)\,dy.\end{aligned}$$

(c) *If $\nu = 1$, then the following inequalities are valid:*

$$\frac{1}{e\sqrt{\pi}} \sup_{x\in\mathbb{R}} \left[\int_{\{|y-x|<\sqrt{2t}\}} \left\{ \sqrt{2t} - |y-x| \right\} W(y)dy \right.$$
$$\left. + \frac{1}{2\sqrt{\pi}} \int_1^\infty \tau^{-3/2} e^{-\tau} d\tau \int_{\{|y-x|<\sqrt{2t}\}} |y-x| \, W(y)\, dy \right]$$
$$\leq \sup_{x\in\mathbb{R}} \int_0^t \int p_{0,1}(\sigma, x, y) W(y) dy \, d\sigma$$
$$\leq \left(1 + \sqrt{\frac{2}{\pi}}\right) \sqrt{t} \sup_{x\in\mathbb{R}} \int_{\{|y-x|<\sqrt{2t}\}} W(y)dy.$$

E.4. Schrödinger semigroups

1.29. THEOREM. *Let $V = V_+ - V_-$ be a Kato potential for Brownian motion. By definition this means that V_- is a Kato potential and that $1_K V_+$ is such a potential for every compact subset K of $\mathbb{R}^\nu$. Then the Feynman-Kac formulas*

$$\exp\left(-t\left(H_0 \dot{+} V\right)\right) f(x) = \mathbb{E}_x \left[\exp\left(- \int_0^t V(X(u))du \right) f(X(t)) \right]$$
$$= \int f(y) \int \exp\left(- \int_0^t V(X(u))du \right) d\mu_{0,x}^{t,y} \, dy,$$

define a strongly continuous, positivity preserving semigroup in any of the spaces $C_\infty(\mathbb{R}^\nu)$ and $L^p(\mathbb{R}^\nu)$, $1 \leq p < \infty$.

Here the measure $\mu_{0,x}^{t,y}$ is the non-normalized Brownian bridge measure. In fact

$$\int \prod_{j=1}^n f_j\left(X(s_j)\right) d\mu_{0,x}^{t,y}$$
$$= p_{0,\nu}(t,x,y)\, \mathbb{E}\left[\prod_{j=1}^n f_j\left(\left(1 - s_j t^{-1}\right) x + s_j t^{-1} y + \left(1 - s_j t^{-1}\right) B\left(s_j t \left(t - s_j\right)^{-1} \right) \right) \right],$$

where $0 < s_1 < s_2 < \cdots < s_n < t$. It may be shown that the integral kernel

$$(t,x,y) \mapsto \int \exp\left(- \int_0^t V(X(u))du \right) d\mu_{0,x}^{t,y}$$

is symmetric on $\mathbb{R}^\nu \times \mathbb{R}^\nu$ and continuous on the product space $(0,\infty) \times \mathbb{R}^\nu \times \mathbb{R}^\nu$. In addition, for $t > 0$, the mapping $\exp\left(-t\left(H_0 \dot{+} V\right)\right)$ is a mapping from $L^1(\mathbb{R}^\nu)$ to $L^\infty(\mathbb{R}^\nu)$. More precisely, if $1 \leq p \leq q \leq \infty$ and if $p < \infty$, then the function $\exp\left(-t\left(H_0 \dot{+} V\right)\right) f$ belongs to $L^q(\mathbb{R}^\nu) \cap C_\infty(\mathbb{R}^\nu)$ whenever f belongs to the space $L^p(\mathbb{R}^\nu)$. So the operators $\exp\left(-t\left(H_0 \dot{+} V\right)\right)$, $t > 0$, are L^1-L^∞-smoothing: see Definition 1.1.

1.30. DEFINITION. The semigroup $\left\{\exp\left(-t\left(H_0\dot{+}V\right)\right) : t \geq 0\right\}$ is called the Schrödinger semigroup.

The generator of the Schrödinger semigroup $H_0\dot{+}V$ extends the operator H_0+V. In $L^2\left(\mathbb{R}^\nu\right)$, this semigroup consists of self-adjoint operators and its generator is a self-adjoint operator. The corresponding quadratic form is the *Schrödinger form* $\mathcal{E}^V$. It has domain

$$\operatorname{dom}\left(\mathcal{E}^V\right) = \left\{f \in L^2(\mathbb{R}^\nu) : \ \nabla f \text{ belongs to } L^2\left(\mathbb{R}^\nu\right) \times \cdots \times L^2\left(\mathbb{R}^\nu\right) \text{ and } V_+\left|f\right|^2 \text{ belongs to } L^1\left(\mathbb{R}^\nu\right)\right\}.$$

The Cartesian product in this definition contains ν copies of $L^2(\mathbb{R}^\nu)$. Its action is given by ($f,\, g \in \operatorname{dom}\left(\mathcal{E}^V\right)$)

$$\mathcal{E}^V(f,g) = \langle \nabla f, \nabla g\rangle + \left\langle V_+^{1/2} f, V_+^{1/2} g\right\rangle - \left\langle V_-^{1/2} f, V_-^{1/2} g\right\rangle.$$

A quick proof of an indication of Feynman-Kac formula can be based on Itô calculus as the following example shows.

EXAMPLE. (Feynman-Kac formula) Let U be an open subset of $\mathbb{R}^\nu$, let f belong to $C_\infty(U)$, let $V : U \to \mathbb{R}$ be an appropriate function, and let $u : [0,\infty) \times U \to \mathbb{R}$ be a solution to the following problem:

$$\begin{cases} \dfrac{\partial}{\partial t} u & = \frac{1}{2}\Delta u - Vu \ \text{ in } [0,\infty) \times U; \\ u & \text{is continuous on } [0,\infty) \times U \text{ and } u(0,x) = f(x). \end{cases}$$

Moreover, we require $\lim_{x\to b, x\in U} u(t,x) = 0$ if b belongs to ∂U. Then $u(t,x) = \mathbb{E}_x\left(\exp\left(-\int_0^t V(B(r))dr\right) f(B(t)) : \tau > t\right)$. Here τ is the exit time from U: $\tau = \inf\left\{s > 0 : B(s) \in \mathbb{R}^\nu \setminus U\right\}$.

For a proof we fix $t > 0$ and consider the process $\{M(s) : 0 \leq s \leq t\}$ defined by $M(s) = u(t-s, B(s)) \exp\left(-\int_0^s V(B(r))dr\right) 1_{\{\tau > s\}}$ and we apply Itô's formula to obtain (see e.g. Bhattacharya and Waymire [26]):

$$\begin{aligned} & M(s) - M(0) \\ & = \int_0^s \left\{-\frac{\partial}{\partial t} u(t-r, B(r)) + \frac{1}{2}\Delta u(t-r, B(r)) - V(B(r)) u(t-r, B(r))\right\} \\ & \qquad \exp\left(-\int_0^r V(B(\rho))d\rho\right) 1_{\{\tau > r\}} dr \\ & \qquad\quad + \int_0^s \nabla u(t-r, B(r)) 1_{\{\tau > r\}} . dB(r) \\ & = \int_0^{\min(s,\tau)} \nabla u(t-r, B(r)) . dB(r). \end{aligned}$$

Here we used the fact that, by assumption, u is a solution of our initial value problem. It follows that the process $\{M(s) : 0 \leq s \leq t\}$ is a martingale. So that $\mathbb{E}_x [M(s)] = \mathbb{E}_x [M(0)]$. Hence we may conclude that

$$\begin{aligned} &u(t,x) \\ &= \mathbb{E}_x(M(0)) = \mathbb{E}_x(M(t)) = \mathbb{E}_x\left(u(0,B(t))\exp\left(-\int_0^t V(B(\rho))d\rho\right) : \tau > t\right) \\ &= \mathbb{E}_x\left(f(B(t))\exp\left(-\int_0^t V(B(\rho))d\rho\right) : \tau > t\right). \end{aligned}$$

This proves the Feynman-Kac representation for the function $u(t,x)$.

E.5. Generalizations and modifications

In the sequel a number of results that exist for the Laplace operator and for Brownian motion will be extended and refined. The extension consists of replacing the Euclidean space $\mathbb{R}^\nu$ with a second countable locally compact Hausdorff E space together with a non-negative Borel measure m which is finite on compact subsets of E. Instead of the Laplace operator we consider a self-adjoint generator of a strong Markov process with state space E. This strong Markov process, which we denote by

$$\{(\Omega, \mathcal{F}, \mathbb{P}_x), (X(t) : t \geq 0), (\vartheta_t, t \geq 0), (E, \mathcal{E})\},$$

will play the role of Brownian motion in $\mathbb{R}^\nu$. In section B of this chapter, we have given the details of the assumptions (BASSA). Here we will mention some of the high-lights in the book and also some of the techniques involved. We will assume that the transition probabilities $B \mapsto \mathbb{P}_x[X(t) \in B]$ are absolutely continuous with respect to the reference measure m. Hence, from Radon-Nikodym's theorem (Theorem E.10 in Appendix E) it follows that there exists a function $p_0(t,x,y)$ such that

$$\mathbb{P}_x[X(t) \in B] = \int_B p_0(t,x,y)dm(y).$$

Here we assume the function $(t,x,y) \mapsto p_0(t,x,y) = p_0(t,y,x)$ is continuous on the space $(0,\infty) \times E \times E$. Instead of potential functions in the Kato-Feller class for Brownian motion we consider such potentials in the Kato-Feller class with respect to our Markov process: see Definition 2.1. The non-normalized Markov bridge measure $\mu_{0,x}^{t,y}$, which ties the process at x at time 0 and at y at time t, will be a substitute for the (non-normalized) Brownian bridge measure or tied Brownian motion. To be more precise we set

$$\mu_{0,x}^{t,y}(A) = \mathbb{E}_x\left[p_0(t-s, X(s), y)1_A\right], \text{ where } A \text{ belongs to the } \sigma\text{-field } \mathcal{F}_s.$$

Since the process $s \mapsto p_0(t-s, X(s), y)$ is a $\mathbb{P}_x$-martingale on the interval $[0,t)$, the measure $\mu_{0,x}^{t,y}$ is well defined: for more details see Proposition 2.7 and Example 5 C.4 of Appendix C. Let V be a Kato-Feller potential and define, for $t > 0$, the operator $\exp\left(-t\left(K_0\dot{+}V\right)\right)$ by the Feynman-Kac formula: see Theorem 2.5. Then the family of operators

$$\left\{\exp\left(-t\left(K_0\dot{+}V\right)\right) : t \geq 0\right\}$$

is a strongly continuous positivity preserving self-adjoint semigroup in $L^2(E,m)$. Each operator $\exp\left(-t\left(K_0\dot{+}V\right)\right)$, $t > 0$, is an integral operator with integral kernel $\exp\left(-t\left(K_0\dot{+}V\right)\right)(x,y)$ given by

$$\int \exp\left(-\int_0^t V(X(s))ds\right) d\mu_{0,x}^{t,y}.$$

This integral is continuous on $(0,\infty) \times E \times E$ and symmetric. Next we consider a singularity region Γ, we put $\Sigma = E \setminus \Gamma$, and we let S be the penetration time of Γ:

$$S = \inf\left\{s > 0 : \int_0^s 1_\Gamma(X(\sigma))d\sigma > 0\right\}.$$

Let $\left\{\exp\left(-t\left(K_0\dot{+}V\right)_\Sigma\right) : t \geq 0\right\}$ be the family of operators with integral kernels:

$$\int_{\{S>t\}} \exp\left(-\int_0^t V(X(s))ds\right) d\mu_{0,x}^{t,y}.$$

Then this family is a strongly continuous semigroup in $L^2\left(E \setminus \Gamma, m\right)$: see e.g. equality (2.40), (2.43), Corollary 2.28 and Proposition 2.29. Let V and W be Kato-Feller potentials. Our aim is to compare, in the sense of spectral and scattering theory, the semigroups $\left\{\exp\left(-t\left(K_0\dot{+}V\right)\right) : t \geq 0\right\}$, $\left\{\exp\left(-t\left(K_0\dot{+}W\right)\right) : t \geq 0\right\}$, and $\left\{\exp\left(-t\left(K_0\dot{+}V\right)_\Sigma\right) : t \geq 0\right\}$. Instead of the comparison of semigroups we often consider differences of (powers of) the corresponding resolvents: see Chapter 4. In Chapter 5, and more precisely in Theorems 5.6 and 5.12, characterizations are given which say that differences of Feynman-Kac semigroups consist of Hilbert-Schmidt operators. In Chapter 6 the same is done for a trace class characterization: see Theorem 6.7 and Theorem 6.11. Some of these results are also new for classical Brownian motion. In the case of singular perturbations an important role is played by Newton (or Dirichlet) capacities: see Definition 2.36, Definition D.17, Propositions D.18, D.19, D.20, and Propositions 2.37 and 2.38. These capacities are also used in Proposition 7.2 to give quantitative bounds for differences of Feynman-Kac semigroups: see the remarks following Proposition 7.2 as well. In principle these methods enable us to give rates of convergence in β for differences of the form

$$\exp\left(-t\left(K_0\dot{+}V + \beta 1_\Gamma\right)\right) - J^* \exp\left(-t\left(K_0\dot{+}V\right)_\Sigma\right) J$$

for the usual operator norm, for the Hilbert-Schmidt norm, and for the trace norm. By definition the operator $\left(K_0\dot{+}V\right)_\Sigma$ is the generator of the singularly perturbed semigroup $\left\{\exp\left(-t\left(K_0\dot{+}V\right)_\Sigma\right) : t \geq 0\right\}$. Notice that $\left(K_0\dot{+}V + \beta 1_\Gamma\right)_\Sigma = \left(K_0\dot{+}V\right)_\Sigma$. Chapter 3, in combination with Appendix D, gives a full proof of Theorem 2.5. In this theorem we present the basic properties of the Feynman-Kac semigroups. Finally Chapter 8 is devoted to spectral and scattering theory.

Chapter 2

Perturbations of Free Feller Operators

There are several ways of introducing semigroups with perturbed generators. The analytic way starts with the unperturbed semigroup and uses the Trotter-product formula to find a Feynman-Kac representation of the perturbed semigroup. The semi-analytic or semi-stochastic manner begins again with the unperturbed semigroup. Then the potentials are introduced stochastically by verifying the sensibility and the semigroup property of the Feynman-Kac formula. Here we employ a purely stochastic approach in the sense that we begin with the transition density function of a strong Markov process. Our aim is to formulate all assumptions on the process or its generator in terms of assumptions on the density. In fact these kernels are the one-dimensional transition densities (marginals) of a strong Markov process. We assume that these transition densities possess a number of basic properties given in BASSA. An advantage, as described in the previous chapter, is that we are able to consider a large class of generators of Feller semigroups. Among others this class contains all the examples of section C in Chapter 1. The regular and singular perturbations are introduced stochastically via the Feynman-Kac formulae. In these formulae we use in an essential way the (strong) Markov processes generated by the unperturbed (free) Feller operator. Other stochastic methods (like martingale theory) will also allow us to describe the integral kernels for the corresponding Feynman-Kac semigroups. This can be done for regular as well as singular perturbations. The final objective is to present and construct a spectral theory for these free and perturbed Feller operators. For the reasons mentioned above this theory is called "Stochastic Spectral Theory".

In the previous chapter we started with the basic assumptions on the transition density function which form the foundations of this newly developed theory. Basically, the word "stochastic" is chosen because we assume that the self-adjoint free Feller operators generate Feller semigroups consisting of integral operators. Their integral kernels are the transition densities of a strong Markov process. On many occasions these integral kernels, or their perturbations, are used as well as the strong Markov property. In addition, many of the basic tools of our theory have their origin in stochastic analysis.

Because the spectrum is the complement of the resolvent set, spectral theory is based on a detailed analysis of resolvent differences. As long as the resolvent values have a large positive real value, the resolvent operators can be represented by semigroups via their Laplace transforms. This is one of the main reasons that an extensive part of the present book is devoted to the investigation of differences of semigroups and resolvents with different regular and/or singular perturbations. In subsection 1.E.5 we described how the present theory extends the classical case of

Schrödinger semigroups. Now we will pursue this matter for more general Markov processes. At first the resolvents are considered for large positive real values, but the final considerations also include the limiting absorption case.

Before we start studying the perturbation theory, we describe the contents of the following chapters in order to illustrate the framework of (our version of) Stochastic Spectral Analysis as illustrated on page 56. A natural class for regular perturbations of the free Feller operator K_0 consists of Kato-Feller potential functions V. For a function V belonging to this class, the perturbed Feller semigroup $\{\exp(-t(K_0\dot{+}V)) : t \geq 0\}$ consists of operators which have an explicit stochastic representation: the Feynman-Kac formula, i.e.

$$\left[\exp\left(-t(K_0\dot{+}V)\right) f\right](x) = \mathbb{E}_x\left[\exp\left(-\int_0^t V(X(s))ds\right) f(X(t))\right].$$

In section A of the present chapter, we give an overview on L^p-L^q-estimates for the operator $\exp(-t(K_0\dot{+}V))$ and we study the relation to the Friedrichs' extension of $K_0 + V$. The operator $\exp(-t(K_0\dot{+}V))$ possesses an integral kernel for which several explicit stochastic representations can be given. The representations are phrased in terms of so-called pinned measures. In our case these pinned measures are non-normalized martingale measures, i.e. they are represented as expectations with martingales as densities. In general these martingales are not closed because they are not uniformly integrable. These kernels are treated in section B of this chapter. Some of them are constructed with the aid of terminal stopping times. This approach allows us to include singular perturbations given by Dirichlet boundary conditions on a singularity region Γ, a closed subset of E. The corresponding semigroup is a strongly continuous semigroup in $L^2(\Sigma, m)$, where $\Sigma = E \setminus \Gamma$ and where m is an appropriate Radon measure on E. For these topics see section C of this chapter. Its generator is denoted by $(K_0\dot{+}V)_\Sigma$. Again this operator is related to the Friedrichs' extension of $(K_0)_\Sigma + (V)_\Sigma$, where $(K_0)_\Sigma$ is the free Feller operator with Dirichlet boundary conditions on (the boundary of) Σ, and where $(V)_\Sigma$ is the restriction of the Kato-Feller potential V to Σ. The semigroup difference

$$\exp\left(-t(K_0\dot{+}V)\right) - J^* \exp\left(-t(K_0\dot{+}V)_\Sigma\right) J,$$

where Jf restricts the function f to Σ, is given by the inverse Laplace transform of the harmonic extension operator. The corresponding resolvent difference is governed by the harmonic extension operator itself.

Basic features of the whole Stochastic Spectral Analysis are the continuity and symmetry of the perturbed semigroup kernels $\exp(-t(K_0 + V))(x, y)$ with respect to x and y. Chapter 3 contains a detailed proof of these properties. For the proof, the unperturbed kernel $p_0(t, x, y)$ will be approximated by kernels of Markov processes killed outside the interior of compact subsets. By doing so we can freely use Dyson expansion (see Appendix B.15) for the corresponding Dirichlet semigroups because of our assumptions on V.

In Chapters 4, 5, and 6 we shall give explicit and quantitative estimates for semigroup and resolvent differences with respect to different operator norms. These estimates are derived from the explicit representations of the semigroup and resolvent differences studied in Chapters 2 and 3. Every chapter is divided into two or three sections A and B and/or C. Whereas the sections 2.A, 4.A, 6.B, and 8.B are devoted to regular perturbations, the sections 2.C, 4.B, 5.B, 6.C, and 8.C take care of the singular perturbations. Section 2.B is completely taken up by integral kernels of the Feynman-Kac semigroups and the corresponding martingales (and martingale measures). Section 6.A treats some general trace properties, and section 8.A deals with spectral theory. Chapter 3 is entirely devoted to the proof of Theorem 2.5. Convergence of resolvent differences are discussed in Chapter 7. In these chapters the following topics have a place: weighted L^2-norms, Hilbert-Schmidt norms and trace norms of differences of semigroups and resolvent families. It turns out that in the context of singular perturbations the 1-equilibrium potential plays a role similar to the one of the potential function in case of regular perturbations.

The estimates for the resolvent differences are based on the explicit stochastic representation of the Feynman-Kac kernels. These explicit formulas have the advantage that they enable us to find quantitative upper and lower bounds. In Chapter 7 this property is used for the large coupling limit, i.e. for the operator $K_0 \dot{+} V + \beta 1_\Gamma$ if β tends to infinity. The estimates from Chapter 4, 5, and 6 for the different operator norms yield bounds in terms of the large coupling parameter β. The Chapters 1 to 7 build the platform for the spectral theory of self-adjoint Feller operators as summarized in Chapter 8. Qualitative spectral results in section A of Chapter 8 ensure the existence of bound states, the stability of essential or absolutely continuous spectra, and the existence and completeness of scattering systems. In section B quantitative estimates for regular potentials are given in terms of Kato-Feller norm. We can determine the shift for spectral measure, for local wave operators, and for scattering matrices. Using the results of Chapter 7 the corresponding subjects are treated quantitatively for large coupling perturbations in section C. The estimates are determined by the weighted Laplace transform of the occupation time, containing the large coupling parameter β.

Framework of stochastic spectral analysis

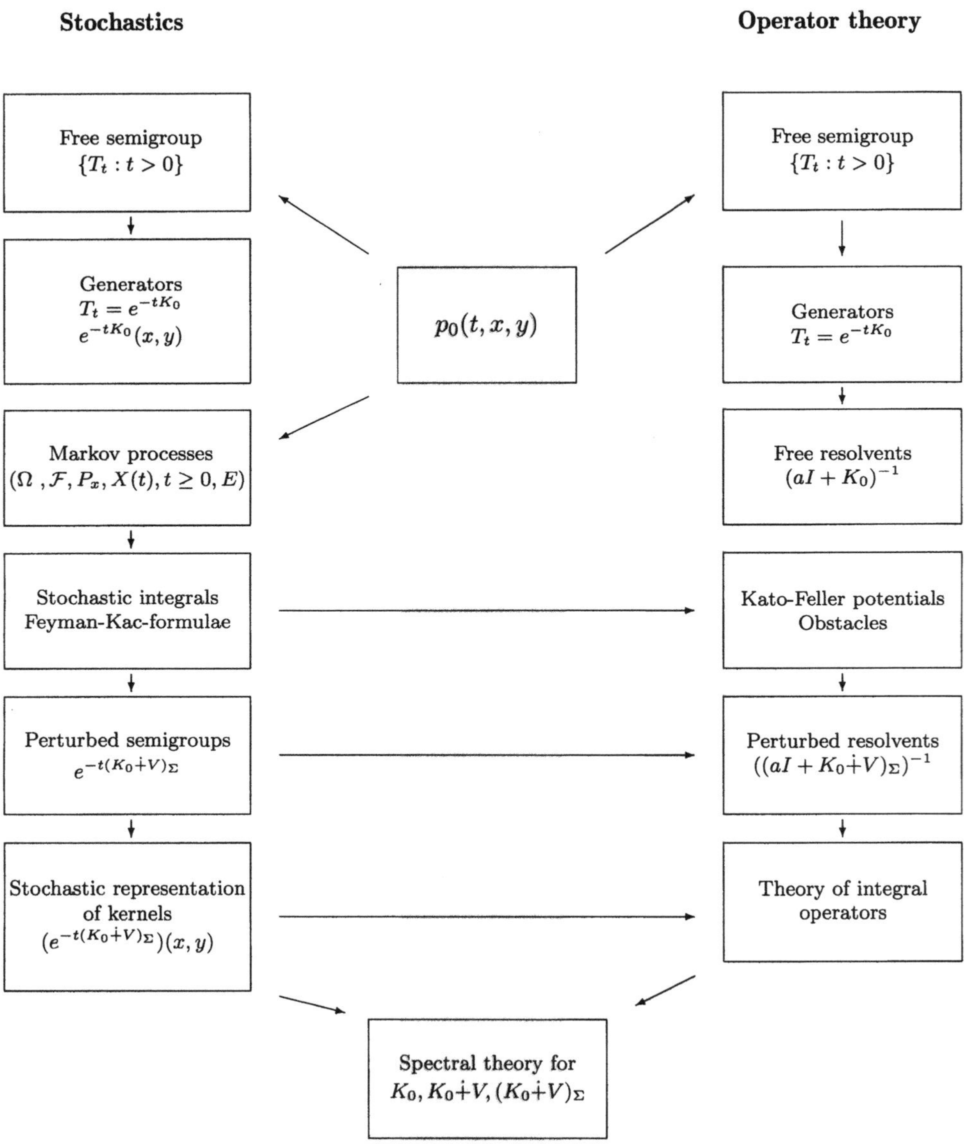

A. Regular Perturbations

In Chapter 1.B we introduced the free Feller operator K_0, defined as the L^2-generator of the Feller semigroup given by the transition density $p_0(t,x,y)$ satisfying BASSA. Here, we consider both regular perturbations, given as a multiplication operator by a real valued function, and singular perturbations, corresponding to infinitely high potential barriers. In [225] Sturm and in [217] Stollmann and Voigt consider perturbations with Kato-Feller measures. Perturbations with measure valued potentials will not be discussed here. However, in this section we shall make a detailed study of regular perturbations. These kinds of perturbations will be called Kato-Feller potentials because they are given in a straightforward manner starting from Kato potentials. The negative (attractive) part of the potential will be required to satisfy a certain uniform condition, whereas the other (repulsive) part only needs to obey a similar condition locally. For a concise formulation of our results we introduce the following definitions (compare with subsection E.4. of Chapter 1).

2.1. Definition. Let $V : E \to [0,\infty]$ be a Borel measurable function on E.
(a) The function V is said to belong to $K(E)$ if

$$\begin{aligned}&\limsup_{t\downarrow 0}\left\|\int_0^t \exp(-sK_0)V\,ds\right\|_{\infty,\infty}\\&=\limsup_{t\downarrow 0}\sup_{x\in E}\int_0^t\left(\int p_0(s,x,y)V(y)dm(y)\right)ds=0.\end{aligned}\tag{2.1}$$

(b) The Borel measurable function $V : E \to [0,\infty]$ belongs to $K_{\mathrm{loc}}(E) = K_{\mathrm{loc}}(E,K_0)$ if $1_B V$ belongs to $K(E)$ for all compact subsets B of E.
(c) The Borel measurable function $V = V_+ - V_-$ is said to be a Kato-Feller potential or is said to belong to the Kato-Feller class if its positive part $V_+ = \max(V,0)$ belongs to $K_{\mathrm{loc}}(E)$ and if its negative part $V_- = \max(-V,0)$ belongs to $K(E)$.

The Kato-Feller class looks somewhat technical, but it is a natural and fairly optimal class in this stochastic context of perturbed semigroups. Saying that a function W belongs to $K(E)$ is equivalent to saying that the equality

$$\lim_{\lambda\to\infty}\left\|(\lambda I+K_0)^{-1}W\right\|_{\infty,\infty}=0,$$

is satisfied. This equality determines the relative form bound of W with respect to K_0: see the proof of Theorem 2.13 and Definition A.18. Moreover, if V_- belongs to $K(E)$, then it follows that

$$\sup_{x\in E}\mathbb{E}_x\left\{\exp\left(\int_0^t V_-(X(u))du\right)\right\}\le M_V e^{b_V t},\tag{2.2}$$

where $V = -V_-$. This is a consequence of Khas'minskii's lemma (Corollary 3.6) which tells us that, if for some (small) $t_0 > 0$

$$\sup_{x\in E} \mathbb{E}_x \left\{ \int_0^{t_0} V_-(X(s))ds \right\} = \alpha < 1,$$

then

$$\sup_{x\in E} \mathbb{E}_x \left\{ \exp\left(\int_0^{t_0} V_-(X(s))ds \right) \right\} \leq \frac{1}{1-\alpha}.$$

The constant b_V in (2.2) could be called an upper bound for the L^∞-L^∞-type of the generalized Schrödinger semigroup $\{\exp(-t(K_0 - V_-)) : t \geq 0\}$ to be defined after the proof of Theorem 2.9: see inequality (2.7). On the other hand one sees immediately from the definition that every function $V \in L^\infty(E, m)$ is a Kato-Feller potential. This property will be interesting for mathematical scattering theory. Generally speaking, the condition that the potential function V be bounded is no real restriction in scattering theory. In the special case of Wiener transition densities the Kato's class is discussed in [3], [63], and [206]. In particular the definition of the Kato-Feller class allows us to study perturbed semigroups.

In order to understand certain aspects of the Kato-Feller norm to be explained later, the following proposition will be useful. In Definition 2.3 the notion of is depicted.

2.2. PROPOSITION. *Let $g : E \to \mathbb{R}$ be a Borel measurable function and let a and η be strictly positive real numbers. Let K_0 be a free Feller generator possessing properties A1-A3 (see BASSA in section B of Chapter 1: Markov property). The following inequalities are valid:*

$$\begin{aligned}&\left(1 - e^{-a\eta}\right) \left\| (aI + K_0)^{-1} |g| \right\|_\infty \\ &\quad\leq \sup_{x\in E} \int_0^\eta \mathbb{E}_x(|g(X(s))|)ds \leq e^{a\eta} \left\| (aI + K_0)^{-1} |g| \right\|_\infty .\end{aligned}$$

Proof. Write $h = |g|$. With the notation and assumptions as in the proposition, the Markov property yields the following string of (in-)equalities:

$$\begin{aligned}&\left[(aI + K_0)^{-1} h\right](x) = \int_0^\infty e^{-as} \mathbb{E}_x\left(h(X(s))\right) ds \\ &= \sum_{k=0}^\infty e^{-ka\eta} \int_0^\eta e^{-as} \mathbb{E}_x\left(h(X(s + k\eta))\right) ds \\ &= \sum_{k=0}^\infty e^{-ka\eta} \mathbb{E}_x \left(\mathbb{E}_{X(k\eta)} \left(\int_0^\eta e^{-as} h(X(s))ds \right) \right) \\ &\leq \sum_{k=0}^\infty e^{-ka\eta} \sup_{y\in E} \mathbb{E}_y \left(\int_0^\eta e^{-as} h(X(s))ds \right)\end{aligned}$$

$$= \frac{1}{1-e^{-a\eta}} \sup_{y\in E} \mathbb{E}_y \left(\int_0^\eta e^{-as} h(X(s))ds \right) \leq \frac{1}{1-e^{-a\eta}} \sup_{y\in E} \mathbb{E}_y \left(\int_0^\eta h(X(s))ds \right)$$
$$\leq \frac{e^{a\eta}}{1-e^{-a\eta}} \sup_{y\in E} \mathbb{E}_y \left(\int_0^\eta e^{-as} h(X(s))ds \right) \leq \frac{e^{a\eta}}{1-e^{-a\eta}} \left\| (aI+K_0)^{-1} h \right\|_\infty .$$

REMARK. We also notice that

$$\|\exp(-tK_0)\|_{\infty,\infty} = \|\exp(-tK_0)\|_{1,1} = \sup_{x\in E} \int p_0(t,x,y)dm(y) \leq 1.$$

Next we define the Kato-Feller norm and we will see that a Kato-Feller function is locally integrable.

2.3. DEFINITION. The Kato-Feller norm $\|g\|_{\mathrm{KF}}$ is defined by the formula

$$\|g\|_{\mathrm{KF}} = \sup_{x\in E} \int_0^1 ds \int_E dy\, p_0(s,x,y) \left| g(y) \right|,$$

whenever this expression is finite. Define the space $KF(E)$ by

$$KF(E) = \{ V \text{ Borel measurable } : \|V\|_{\mathrm{KF}} < \infty \} .$$

REMARK. Let W be a non-negative Borel measurable function. In order to show that W belongs to $K(E)$, the following is sometimes helpful. Put

$$A_k(t,x) = \Big\{ y \in E : (k+1)^2 t \sup_{s>0} \frac{1}{s^2} \int_0^s p_0(\sigma,x,y)d\sigma >$$
$$\sup_{s>0} \frac{1}{s} \int_0^s p_0(\sigma,x,y)d\sigma \geq k^2 t \sup_{s>0} \frac{1}{s^2} \int_0^s p_0(\sigma,x,y)d\sigma \Big\}$$

and notice that, for $0 < t \leq 1$, the following inequality is valid:

$$\int_0^t ds \int dy\, p_0(s,x,y)W(y)$$
$$\leq \int_{A_0(t,x)} dy \int_0^t ds\, p_0(s,x,y)W(y) + \sum_{k=1}^\infty \int_{A_k(t,x)} dy\, t^2 \sup_{s>0} \frac{1}{s^2} \int_0^s p_0(\sigma,x,y)d\sigma W(y).$$

For example, if $p_0(t,x,y)$ is the Gaussian kernel in $\mathbb{R}^\nu$, then the set $A_k(t,x)$ is the annulus $\{ y \in \mathbb{R}^\nu : c_\nu (k+1)\sqrt{t} > |x-y| \geq c_\nu k \sqrt{t} \}$. Here the constant c_ν is such that $c_\nu^2 = \sup_{s>0} s^2 \int_s^\infty \sigma^{\frac{1}{2}\nu - 2} e^{-\sigma} d\sigma / \sup_{s>0} s \int_s^\infty \sigma^{\frac{1}{2}\nu-2} e^{-\sigma} d\sigma$.

We collect some relevant properties of the space $KF(E)$ in the following proposition. The results constitute a generalization of the Erratum in Simon [207].

2.4. Proposition. *A function g that belongs to $KF(E)$ is locally integrable and the space $KF(E)$ is complete.*

Proof. Fix $g \in KF(E)$ and let C be a compact subset of E. Let $(K_n : n \in \mathbb{N})$ be an increasing sequence of compact subsets of E with the following properties: $E = \bigcup_{n=1}^{\infty} K_n$, $K_n \subseteq \operatorname{int}(K_{n+1})$ and $m(K_n) > 0$ for $n \in \mathbb{N}$. From Proposition 2.2 it follows that the supremum

$$\sup_{x \in E} \lambda \int_0^\infty ds\, e^{-\lambda s} \int dy\, p_0(s,x,y)\, |g(y)|$$

is finite for any fixed strictly positive real number λ. An elementary estimate yields the inequality:

$$\begin{aligned} &\sup_n \inf_{y \in C} \frac{\lambda}{m(K_n)} \int_{K_n} dx \int_0^\infty ds\, e^{-\lambda s} p_0(s,x,y) \times \int_C |g(y)|\, dy \\ &\leq \sup_{x \in E} \lambda \int_0^\infty ds\, e^{-\lambda s} \int p_0(s,x,y)\, |g(y)|\, dy. \end{aligned} \tag{2.3}$$

We shall prove that the constant

$$\sup_n \inf_{y \in C} \frac{\lambda}{m(K_n)} \int_{K_n} dx \int_0^\infty ds\, e^{-\lambda s} p_0(s,x,y) \tag{2.4}$$

is strictly positive. Therefore we consider the sequence of functions

$$(h_n : n \in \mathbb{N}), \quad \text{defined by} \quad h_n(y) = \lambda \int_{K_n} dx \int_0^\infty ds\, e^{-\lambda s}\, p_0(s,x,y).$$

Let ζ be the *life time* of the process X, defined by $\zeta = \inf\{s > 0 : X(s) = \triangle\}$. The sequence $(h_n : n \in \mathbb{N})$ increases to

$$\begin{aligned} h(y) &:= \lambda \int_E dx \int_0^\infty ds\, e^{-\lambda s}\, p_0(s,x,y) = \lambda \int_0^\infty ds\, e^{-\lambda s} \mathbb{P}_y\left(X(s) \in E\right) \\ &= \lambda \int_0^\infty ds\, e^{-\lambda s} \mathbb{P}_y\left(\zeta > s\right) = \mathbb{E}_y\left(1 - \exp\left(-\lambda\zeta\right)\right) \\ &= 1 - \mathbb{E}_y\left(\exp\left(-\lambda\zeta\right) : \zeta < \infty\right). \end{aligned}$$

Consequently, the sequence of continuous functions $(h_n : n \in \mathbb{N})$ increases to the continuous function h. From Dini's Lemma (see Lemma E.1 in Appendix E) it follows that this convergence takes place uniformly on the compact subset C. From the right continuity of paths, it follows that $h(y) > 0$ for all $y \in E$ and hence $\inf_{y \in C} h(y) > 0$. From Dini's Lemma (Lemma E.1 in Appendix E) it then follows that $\inf_{y \in C} h_n(y) > 0$ for $n \in \mathbb{N}$ large enough. As a consequence the constant in (2.4) is strictly positive. From (2.3) we then see that the integral

$\int_C |g(y)|\, dy$ is finite. This proves the first statement in Proposition 2.4. Next let $(g_n : n \in \mathbb{N})$ be a Cauchy sequence in $KF(E)$. Then this sequence is a Cauchy sequence in $L^1_{\text{loc}}(E,m)$. So there exists a function $g \in L^1_{\text{loc}}(E,m)$, that serves as a $L^1_{\text{loc}}(E,m)$-limit of the sequence $(g_n : n \in \mathbb{N})$. It then easily follows that this sequence converges to g with respect to the Kato-Feller norm. This completes the proof of Proposition 2.4.

REMARK. The Kato-Feller norm is used later for regular perturbations (see for instance equality (4.29)). It is natural and in several aspects it is also optimal. But this will be explained later. The Kato-Feller norm is equivalent to the norm $g \mapsto \left\|(aI + K_0)^{-1}\, |g|\right\|_\infty$, a norm related to the L^∞-norm of resolvents for Feller operators. This property is based on Proposition 2.2.

A complete proof of the following general result (Theorem 2.5. (a) - (e)) will be given in Chapter 3. Assertion (f) is a consequence of Proposition 2.16. For other approaches in the symmetric case the reader may consult [231], [233], [235], and [236]. For the Gaussian semigroup the reader is also referred to Simon [206], [208] and to Stroock [219, Theorem 4.3.15. p. 227 ff]. From Proposition 2.4 it follows that a function as described in (a) of the following theorem is locally integrable. Also observe that $K(E)$ is the cone of positive (i.e. non-negative) functions in $KF(E)$ and that the space $KF(E) = K(E) - K(E) = \{V - W : V,\ W \in K(E)\}$ is complete for the Kato-Feller norm, defined in Definition 2.3 by

$$\|V\|_{\text{KF}} = \sup_{x\in E} \int dy \int_0^1 ds\, p_0(s,x,y)\, |V(y)|\,, \quad V \in KF(E).$$

Compare this with Definition 2.1.

2.5. THEOREM. *Suppose that $V = V_+ - V_-$ is a Borel measurable function defined on E such that V_- belongs to $K(E)$ and such that V_+ belongs to $K_{\text{loc}}(E)$.*

(a) There exists a closed, densely defined linear operator $K_0 \dot{+} V$ in $C_\infty(E)$, extending $K_0 + V$, which generates a strongly continuous positivity preserving semigroup $\{\exp(-t(K_0 \dot{+} V)) : t \geq 0\}$ in $C_\infty(E)$. As always $C_\infty(E)$ is supplied with the supremum-norm. Every operator $\exp(-t(K_0 \dot{+} V))$, $t > 0$, is of the form

$$\left[\exp(-t(K_0 \dot{+} V))f\right](x) = \int \exp(-t(K_0 + V))(x,y) f(y) dm(y), \quad f \in C_\infty(E),$$

where $\exp(-t(K_0 + V))(x,y)$ is a continuous function in (t,x,y) on $(0,\infty) \times E \times E$ which satisfies the Chapman-Kolmogorov identity:

$$\begin{aligned}&\exp(-(s+t)(K_0+V))(x,y)\\ &= \int \exp(-s(K_0+V))(x,z)\exp(-t(K_0+V))(z,y)\, dz,\end{aligned}$$

for $t > 0$, x, $y \in E$.

(b) The operator $\exp\left(-t\left(K_0\dot{+}V\right)\right)$ *is given via the Feynman-Kac formula:*

$$\left[\exp\left(-t\left(K_0\dot{+}V\right)\right)f\right](x)=\mathbb{E}_x\left[\exp\left(-\int_0^t V(X(u))du\right)f(X(t))\right];$$

Its integral kernel $\exp\left(-t\left(K_0+V\right)\right)(x,y)$ *is given by:*

$$\exp\left(-t\left(K_0+V\right)\right)(x,y)=\lim_{s\uparrow t}\mathbb{E}_x\left[\exp\left(-\int_0^s V(X(u))du\right)p_0(t-s,X(s),y)\right].$$

(c) The semigroup $\{\exp(-t(K_0\dot{+}V)) : t\geq 0\}$ *also acts as a strongly continuous semigroup in* $L^p(E,m)$, $1\leq p<\infty$.

(d) If, for $t>0$, $\exp(-tK_0)$ *maps* $L^1(E,m)$ *into* $L^\infty(E,m)$ *(i.e. if* $\sup\{p_0(t,x,y): x,y\in E\}<\infty$ *for all* $t>0$*), then* $\exp(-t(K_0\dot{+}V))$, $t>0$, *maps* $L^p(E,m)$ *into* $L^q(E,m)$, *for* $1\leq p\leq q\leq\infty$. *If* $t>0$ *and if* $1\leq p\leq q<\infty$, *then* $\exp(-t(K_0\dot{+}V))$ *maps* $L^p(E,m)$ *into* $L^q(E,m)\cap C_\infty(E)$.

(e) In $L^2(E,m)$, *the family* $\{\exp(-t(K_0\dot{+}V)) : t\geq 0\}$ *is a self-adjoint positivity preserving strongly continuous semigroup with a self-adjoint generator.*

(f) The Feynman-Kac semigroup in $L^2(E,m)$ *coincides with the semigroup corresponding to the quadratic form* $\mathcal{E}^V$ *with* $\operatorname{dom}(\mathcal{E}^V)=\operatorname{dom}\left(K_0^{1/2}\right)\cap\operatorname{dom}\left(V_+^{1/2}\right)$ *and defined by*

$$\mathcal{E}^V(f,g)=\left\langle K_0^{1/2}f,K_0^{1/2}g\right\rangle-\left\langle V_-^{1/2}f,V_-^{1/2}g\right\rangle+\left\langle V_+^{1/2}f,V_+^{1/2}g\right\rangle,$$

where f *and* g *belong to* $\operatorname{dom}\left(\mathcal{E}^V\right)$. *(This is proved in Proposition 2.16.)*

REMARK 1. From the assumptions on V_-, it follows that $\operatorname{dom}(K_0^{1/2})$ contains $\operatorname{dom}(V_-^{1/2})$ and that there exist constants $\epsilon(a)$ such that

$$\left\langle V_-^{1/2}f,V_-^{1/2}f\right\rangle\leq\epsilon(a)\left(a\langle f,f\rangle+\left\langle K_0^{1/2}f,K_0^{1/2}f\right\rangle\right),\quad a>0,$$

for all f in $\operatorname{dom}(K_0^{1/2})$. This follows from Theorem 2.12 where we show that

$$\begin{aligned}\left\|(aI+K_0)^{-1/2}V_-^{1/2}\right\|_{2,2}^2&=\left\|V_-^{1/2}(aI+K_0)^{-1}V_-^{1/2}\right\|_{2,2}\\&\leq\left\|(aI+K_0)^{-1}V_-\right\|_\infty<\infty,\end{aligned}$$

and so $\epsilon(a):=\left\|(aI+K_0)^{-1}V_-\right\|_\infty$ will do. The proof of Theorem 2.5 will be given in Chapter 3 and assertion (f) follows from Proposition 2.16.

REMARK 2. Suppose the unperturbed semigroup $\{\exp(-tK_0): t\geq 0\}$ is L^1-L^∞-smoothing. From the general assumptions it then follows that, for $t>0$, the operator $\exp(-tK_0)$ maps $L^1(E,m)$ into $C_\infty(E)$. As indicated in (d), then we may prove that, for $t > 0$, the operator $\exp(-t(K_0\dot{+}V))$ maps $L^p(E,m)$ into $L^q(E,m)\bigcap C_\infty(E)$, provided that $1 \leq p \leq q \leq \infty$, $p \neq \infty$. This is explained in [231], [235], and in [236].

REMARK 3. Other representations for the kernels of the operators

$$\exp\left(-t\left(K_0\dot{+}V\right)\right), \quad t > 0,$$

will be given in Proposition 2.7 and in Proposition 2.17.

2.6. DEFINITION. The formulae in assertion (b) of the Theorem 2.5 are called the *Feynman-Kac formulae.* The formulae that will appear in Proposition 2.7 and in Proposition 2.17 below carry the name Feynman-Kac formula as well.

In the next proposition we give another Feynman-Kac formula for the integral kernel $\exp\left(-t(K_0+V)\right)(x,y)$. It is an alternative description of this kernel. A proof of the equality in (2.5) is contained in the proof of Proposition 3.2.

2.7. PROPOSITION. *The integral kernel of the operator* $\exp\left(-t(K_0\dot{+}V)\right)$ *is given by*

$$\begin{aligned}\exp\left(-t(K_0+V)\right)(x,y) &= \lim_{\tau\uparrow t}\mathbb{E}_x\left(\exp\left(-\int_0^\tau V(X(u))du\right)p_0(t-\tau,X(\tau),y)\right)\\ &= \int \exp\left(-\int_0^t V(X(u))du\right)d\mu_{0,x}^{t,y}\\ &= \sup_{m\in\mathbb{N}}\int \exp\left(-\int_0^t V(X(u))du\right)d\mu_{0,x}^{m,t,y}. \qquad (2.5)\end{aligned}$$

Here the measure $\mu_{0,x}^{t,y}$ pins the motion of the process $\{X(t) : t \geq 0\}$ at x at time 0 and at y at time t. It is determined via the equality (see Definition 2.18.(e))

$$\mu_{0,x}^{t,y}(A) = \mathbb{E}_x\left(p_0(t-s,X(s),y)1_A\right), \qquad (2.6)$$

whenever A is an event in the field $\mathcal{F}_s$ with $s < t$. Since the process $s \mapsto p_0(t-s,X(s),y)$ is a martingale on the interval $[0,t)$ this definition makes sense: see Example 5 in C.4 of Appendix C for this martingale property. One may use Kolmogorov's extension theorem (Proposition 1.14 in Chapter 1.E.1) to prove that the cylinder measure $A \mapsto \mathbb{E}_x\left[1_A p_0(t-s,X(s),y)\right]$, $A \in \bigcup_{0\leq s<t}\mathcal{F}_s$, extends to a genuine measure $\mu_{0,x}^{t,y}$ on $\mathcal{F}_{t-}$. Similarly, the measures $\mu_{0,x}^{m,t,y}$ are associated to the martingale $M_{0,T_m}^{t,y}(s)$, $0 \leq s < t$, given by (see equality (2.26) in Definition 2.18)

$$\begin{aligned}&M_{0,T_m}^{t,y}(s)\\ &= \left[p_0(t-s,X(s),y) - \mathbb{E}_{X(s)}\left(p_0(t-s-T_m,X(T_m),y) : T_m < t-s\right)\right]1_{\{T_m>s\}}.\end{aligned}$$

The sequence $(T_m : m \in \mathbb{N})$ is the increasing sequence of stopping times (in fact they are exit times) given by the identity:

$$T_m = \inf\{s > 0 : X(s) \in E^{\triangle} \setminus \operatorname{int}(K_m)\}.$$

Since T_m is a stopping time, Proposition 2.20 shows that the process $M_{0,T_m}^{t,y}(s)$ is a martingale: see Definition 2.18 as well.

Here the sequence of compact subsets $(K_m : m \in \mathbb{N})$ is supposed to possess the following properties: $E = \bigcup_{m\in\mathbb{N}} K_m$, and $K_m \subseteq \operatorname{int}(K_{m+1})$. The σ-field $\mathcal{F}_t$ contains the information from the past. It is generated by $X(s)$, $s \leq t$, and $\mathcal{F}_{t-}$ contains the information from the genuine past. It is generated by $X(s)$, $s < t$. In Proposition 2.20 it will be shown that the process $s \mapsto M_{0,T_m}^{t,y}(s)$ is a $\mathbb{P}_x$-martingale on the semi-open interval $[0,t)$ with respect to the filtration $\{\mathcal{F}_s : 0 \leq s < t\}$. Essentially speaking, the process $s \mapsto M_{0,T_m}^{t,y}(s)$, $0 \leq s < t$, is the same as the process $s \mapsto p_0(t-s, X(s), y)$, for the process $\{X(s) : s \geq 0\}$, killed outside of the interior of K_m, instead of the global process $\{X(s) : s \geq 0\}$.

2.8. DEFINITION. Let $\{T(t) : t \geq 0\}$ be a strongly continuous positivity preserving semigroup from $L^2(E,m)$ to $L^2(E,m)$ satisfying the Feller property, i.e. $T(t)C_\infty(E) \subseteq C_\infty(E)$. Then we call its generator a *Feller operator*. (This corresponds to Schrödinger operators as generators of Schrödinger semigroups: see Simon [206].)

REMARK. As a consequence of (d) and (e) in Theorem 2.5, the generator of the semigroup $\{\exp\left(-t\left(K_0\dot{+}V\right)\right) : t \geq 0\}$ is a Feller operator. So far we have written it as $K_0\dot{+}V$. In Propositions 2.14 and 2.15, its relation with K_0+V will be explained.

Next we want to include some quantitative estimates for certain norms of the operators $\exp\left(-t\left(K_0\dot{+}V\right)\right)$. One consequence of this theorem is that L^1-L^∞-smoothing is preserved under perturbations by Kato-Feller potentials.

2.9. THEOREM. *Let V be a potential function belonging to the Kato-Feller class. Suppose that the free semigroup $\{\exp(-tK_0) : t \geq 0\}$ is L^1-L^∞-smoothing. Then the following (in-)equalities are true:*

$$(1)\qquad \left\|\exp\left(-t\left(K_0\dot{+}V\right)\right)\right\|_{\infty,1} = \sup_{x,y\in E} \exp\left(-t\left(K_0+V\right)\right)(x,y)$$
$$= \sup_{x\in E} \exp\left(-t\left(K_0+V\right)\right)(x,x);$$

$$(2)\qquad \left\|\exp\left(-t\left(K_0\dot{+}V\right)\right)\right\|_{\infty,2} = \left\|\exp\left(-t\left(K_0\dot{+}V\right)\right)\right\|_{2,1}$$
$$\leq \left\|\exp\left(-t\left(K_0\dot{+}2V\right)\right)\right\|_{\infty,\infty}^{1/2} \left\|\exp\left(-tK_0\right)\right\|_{\infty,1}^{1/2};$$

$$(3)\qquad \left\|\exp\left(-t\left(K_0\dot{+}V\right)\right)\right\|_{\infty,1} \leq \left\|\exp\left(-\frac{t}{2}\left(K_0\dot{+}V\right)\right)\right\|_{\infty,2}^2$$
$$\leq \left\|\exp\left(-\frac{t}{2}\left(K_0\dot{+}2V\right)\right)\right\|_{\infty,\infty} \left\|\exp\left(-\frac{t}{2}K_0\right)\right\|_{\infty,1};$$

$$\text{(4)} \qquad \exp\left(-t\left(K_0+V\right)\right)(x,y) \leq \left\|\exp\left(-\frac{t}{2}\left(K_0\dot{+}4V\right)\right)\right\|_{\infty,\infty}^{1/2}\left\|\exp\left(-\frac{t}{2}K_0\right)\right\|_{\infty,1}^{1/2}\left(p_0(t,x,y)\right)^{1/2};$$

$$\text{(5)} \qquad \left\|\exp\left(-t\left(K_0\dot{+}V\right)\right)\right\|_{\infty,\infty} = \sup_{x\in E}\mathbb{E}_x\left[\exp\left(-\int_0^t V(X(u))du\right)\right] = \left\|\exp\left(-t\left(K_0\dot{+}V\right)\right)1\right\|_{\infty}.$$

Before we prove this theorem we notice that the inequality in (4) may be rewritten in the following fashion:

$$\exp\left(-t\left(K_0+V\right)\right)(x,y) \leq \left(M_{4V}e^{tb_{4V}/2}\right)^{1/2}\left(\sup_{x,y\in E}p_0(t/2,x,y)\right)^{1/2}\left(p_0(t,x,y)\right)^{1/2},$$

where the constants M_{4V} and b_{4V} obey the inequality:

$$\left\|\exp\left(-t\left(K_0\dot{+}4V\right)\right)\right\|_{\infty,\infty} \leq M_{4V}e^{tb_{4V}}$$

for all $t \geq 0$: see inequality (2.7) below.

Proof. The first equality in (1) is an easy consequence of the definition of the norm $\|\cdot\|_{\infty,1}$ and the second one is provided by the following estimate:

$$\begin{aligned}
&\exp\left(-t\left(K_0+V\right)\right)(x,y)\\
&= \int \exp\left(-\frac{t}{2}\left(K_0+V\right)\right)(x,z)\exp\left(-\frac{t}{2}\left(K_0+V\right)\right)(z,y)\,dz\\
&\leq \left(\int \exp\left(-\frac{t}{2}\left(K_0+V\right)\right)(x,z)^2dz\right)^{1/2}\left(\int \exp\left(-\frac{t}{2}\left(K_0+V\right)\right)(z,y)^2dz\right)^{1/2}
\end{aligned}$$

(symmetry and once more Chapman-Kolmogorov)

$$\begin{aligned}
&= \left(\exp\left(-t(K_0+V)\right)(x,x)\right)^{1/2}\left(\exp\left(-t(K_0+V)\right)(y,y)\right)^{1/2}\\
&\leq \sup_{z\in E}\exp\left(-t(K_0+V)\right)(z,z).
\end{aligned}$$

This proves the equalities in (1). The equality $\|\cdot\|_{2,1} = \|\cdot\|_{\infty,2}$ follows from the self-adjointness of the operators $\exp\left(-t\left(K_0\dot{+}V\right)\right)$. Let f be a function in $L^2(E,m)$ and consider the inequalities:

$$\begin{aligned}
\left|\left[\exp\left(-t\left(K_0\dot{+}V\right)\right)f\right](x)\right|^2 &= \left|\mathbb{E}_x\left(\exp\left(-\int_0^t V(X(u))du\right)f(X(t))\right)\right|^2\\
&\leq \mathbb{E}_x\left[\exp\left(-2\int_0^t V(X(u))du\right)\right]\mathbb{E}_x\left(\left|f(X(t))\right|^2\right)\\
&\leq \left\|\exp\left(-t\left(K_0\dot{+}2V\right)\right)\right\|_{\infty,\infty}\sup_{y\in E}p_0(t,x,y)\left\|f\right\|_2^2.
\end{aligned}$$

This inequality easily yields the inequality in (2). The equality

$$\exp\left(-t\left(K_0\dot{+}V\right)\right) = \exp\left(-\frac{t}{2}\left(K_0\dot{+}V\right)\right)\exp\left(-\frac{t}{2}\left(K_0\dot{+}V\right)\right)$$

gives rise to the inequality

$$\left\|\exp\left(-t\left(K_0\dot{+}V\right)\right)\right\|_{\infty,1} \leq \left\|\exp\left(-\frac{t}{2}\left(K_0\dot{+}V\right)\right)\right\|_{\infty,2}\left\|\exp\left(-\frac{t}{2}\left(K_0\dot{+}V\right)\right)\right\|_{2,1}.$$

So (3) follows from (2). In order to prove the pointwise estimate in (4), we use the following representation for $\exp\left(-t\left(K_0+V\right)\right)(x,y)$:

$$\exp\left(-t\left(K_0+V\right)\right)(x,y) = \int \exp\left(-\int_0^t V(X(u))du\right) d\mu_{0,x}^{t,y}.$$

Consider the string of inequalities

$$\begin{aligned}
&\left(\int \exp\left(-\int_0^t V(X(u))du\right) d\mu_{0,x}^{t,y}\right)^2 \\
&\leq \int \exp\left(-2\int_0^t V(X(u))du\right) d\mu_{0,x}^{t,y} \times \mu_{0,x}^{t,y}(\Omega) \\
&\leq \left\|\exp\left(-t\left(K_0\dot{+}2V\right)\right)\right\|_{\infty,1} p_0(t,x,y)
\end{aligned}$$

(use the inequality in (3))

$$\leq \left\|\exp\left(-\frac{t}{2}\left(K_0\dot{+}4V\right)\right)\right\|_{\infty,\infty}\left\|\exp\left(-\frac{t}{2}K_0\right)\right\|_{\infty,1} p_0(t,x,y).$$

This proves (4). The equalities in (5) are easy.

Next we introduce constants M_V, b_V, c_V, a_V and ω_V with the following properties:

$$\left\|\exp\left(-t\left(K_0\dot{+}V\right)\right)\right\|_{\infty,\infty} \leq M_V e^{tb_V}; \tag{2.7}$$

$$\left\|\exp\left(-t\left(K_0\dot{+}V\right)\right)\right\|_{2,2} = e^{t\omega_V}; \tag{2.8}$$

$$\left\|\exp\left(-tK_0\right)\right\|_{\infty,1} \leq c_0 t^{-m/2} e^{ta_0}; \tag{2.9}$$

$$\left\|\exp\left(-t\left(K_0\dot{+}V\right)\right)\right\|_{\infty,1} \leq c_V t^{-m/2} e^{ta_V}. \tag{2.10}$$

The following corollary is implied by the previous theorem. Again we suppose that V is a Kato-Feller function.

2.10. COROLLARY. *(a) Suppose that the unperturbed semigroup $\{\exp(-tK_0) : t \geq 0\}$ is L^1-L^∞-smoothing. Then the function*

$$t \mapsto \|\exp(-tK_0)\|_{\infty,1} = \sup_{x\in E} p_0(t,x,x) = \sup_{x,y\in E} p_0(t,x,y)$$

decreases, and the following inequalities are valid:

$$\limsup_{t\to\infty} \|\exp(-tK_0)\|_{\infty,1}^{1/t} \leq 1;$$

$$\limsup_{t\to\infty} \left\|\exp\left(-t\left(K_0\dot{+}V\right)\right)\right\|_{2,2}^{1/t} \leq \limsup_{t\to\infty} \left\|\exp\left(-t\left(K_0\dot{+}V\right)\right)\right\|_{\infty,\infty}^{1/t};$$

$$\limsup_{t\to\infty} \left\|\exp\left(-t\left(K_0\dot{+}V\right)\right)\right\|_{\infty,1}^{1/t}$$

$$\leq \left(\limsup_{t\to\infty} \left\|\exp\left(-t\left(K_0\dot{+}2V\right)\right)\right\|_{\infty,\infty}^{1/t}\right)^{1/2} \left(\limsup_{t\to\infty} \|\exp(-tK_0)\|_{\infty,1}^{1/t}\right)^{1/2}.$$

(b) Suppose that m is an upper bound for the spectral dimension of the unperturbed semigroup $\{\exp(-tK_0) : t \geq 0\}$. If the function V is a Kato-Feller function, then m is also an upper bound for the spectral dimension of the perturbed semigroup $\left\{\exp\left(-t\left(K_0\dot{+}V\right)\right) : t \geq 0\right\}$.

Proof. (a) From Chapman-Kolmogorov's equality we get, for $t > s > 0$,

$$p_0(t,x,y) = \int p_0(s,x,z)p_0(t-s,z,y)dz \leq \sup_{z\in E} p_0(s,x,z) \int p_0(t-s,z,y)dz$$

(symmetry and Chapman-Kolmogorov once more)

$$\leq \sup_{z\in E} \int p_0(s/2,x,w)p_0(s/2,w,z)dw \int p_0(t-s,y,z)dz$$

(Cauchy-Schwarz together with $\int p_0(t-s,y,z)dz \leq 1$)

$$\leq \sup_{z\in E} \left(\int p_0(s/2,x,w)^2 dw\right)^{1/2} \left(\int p_0(s/2,w,z)^2 dw\right)^{1/2}$$

(symmetry and again Chapman-Kolmogorov)

$$= \left(\int p_0(s,x,x)\right)^{1/2} \sup_{z\in E} \left(\int p_0(s,z,z)^2 dw\right)^{1/2} \leq \sup_{z\in E} p_0(s,z,z).$$

Since $\|\exp(-tK_0)\|_{\infty,1} = \sup_{x,y\in E} p_0(t,x,y)$ the first assertion in (a) follows: compare with Appendix B.12. So the function $t \mapsto \|\exp(-tK_0)\|_{\infty,1}$ is decreasing and hence, for $t > t_0$,

$$\|\exp(-tK_0)\|_{\infty,1}^{1/t} \leq \|\exp(-t_0K_0)\|_{\infty,1}^{1/t}.$$

So that the first assertion in (a) follows. The second and third one follow from the corresponding inequalities in Theorem 2.9.

(b) Choose constants M_{nV}, b_{nV}, c_{nV}, a_{nV} and ω_{nV} as in (2.7) to (2.9) inclusive with nV replacing V. Again from the previous theorem it follows that

$$\left\|\exp\left(-t\left(K_0\dot{+}V\right)\right)\right\|_{\infty,1} \leq c_0 M_{2V} \exp\left(\frac{t}{2}\left(b_{2V}+a_0\right)\right) t^{-m/2}.$$

This proves (b).

REMARK 4. Let K be a self-adjoint generator in a Hilbert space with a lower bound. Let ω_0 be the smallest number ω with the property that $\langle Kf, f\rangle \geq -\omega\langle f, f\rangle$ for all $f \in \mathrm{dom}(K)$. Then ω_0 is called the *L^2-L^2-type* of the semigroup $\{\exp(-tK) : t \geq 0\}$ generated by K and one finds $\|\exp(-tK)\| = \exp(\omega_0 t)$, $t \geq 0$. The corresponding quadratic form $\mathcal{E}$ with domain $\mathrm{dom}(\mathcal{E}) = \mathrm{dom}\,(K+\omega_0 I)^{1/2}$ defined by $\mathcal{E}(f,g) = \lim_{t\downarrow 0} \dfrac{\langle f, g\rangle - \langle \exp(-tK)f, g\rangle}{t}$, f and $g \in \mathrm{dom}(\mathcal{E})$, possesses a lower bound $-\omega_0$. In fact there exists a one-to-one correspondence between the class of self-adjoint operators with largest lower bound ω_0, the class of symmetric closed quadratic forms with largest lower bound ω_0 and the self-adjoint semigroups $\{T(t) : t \geq 0\}$ with the property that $\|T(t)\|_{2,2} = \exp(\omega_0 t)$ for all $t \geq 0$. For these topics the reader may for example consult Chapter VI in [231]. This remark fits into our context of stochastic spectral analysis as follows. Suppose that the boundedness condition

$$p_0(t,x,y) \leq c_0 t^{-m/2} e^{a_0 t}, \quad t > 0, \quad x,\ y \in E$$

is satisfied. Again choose constants M_{nV}, b_{nV}, c_{nV}, a_{nV} and ω_{nV} as in (2.7) through (2.9) with nV replacing V. From Theorem 2.9 and its Corollary 2.10, we conclude that $\omega_V \leq b_V$ and the constant a_V may be chosen in such a way that $a_V \leq \frac{1}{2} b_{2V} + \frac{1}{2} a_0$. Moreover, the constant a_0 may be chosen arbitrarily close to zero (possibly > 0).

In the argument below we use the following inequality (which may be considered as a consequence of the Riesz-Thorin interpolation Theorem (see Appendix A.40):

$$\begin{aligned}\left\|\exp\left(-t(K_0\dot{+}V)\right)\right\|_{2,2} &\leq \left\|\exp\left(-t(K_0\dot{+}V)\right)\right\|_{1,1}^{1/2} \left\|\exp\left(-t(K_0\dot{+}V)\right)\right\|_{\infty,\infty}^{1/2} \\ &= \left\|\exp\left(-t(K_0\dot{+}V)\right)\right\|_{\infty,\infty}.\end{aligned}$$

So the symmetry implies:

$$\begin{aligned}&\left\|\exp\left(-t\left(K_0\dot{+}V\right)\right)\right\|_{2,2} \leq \sup_{x\in E} \int \exp\left(-t\left(K_0+V\right)\right)(x,y)dy \\ &\leq c_0^{1/2} M_{4V}^{1/2} 2^{m/4} t^{-m/4} e^{\frac{t}{4}(b_{4V}+a_0)} \sup_{x\in E} \int \left(p_0(t,x,y)\right)^{1/2} dy.\end{aligned}$$

Hence the L^2-L^2-type ω_V satisfies $\omega_V \le \frac{1}{4}(b_{4V} + a_0)$, provided that the function $t \mapsto t^{-m/4} \sup_{x \in E} \int p_0(t,x,y)^{1/2} dy$ is finite and does not grow exponentially. The constant c_V in (2.9) might be called (an upper bound for) the *quasi-L^1-L^∞-bound* for the semigroup $\{\exp(-t(K_0 \dot{+} V)) : t \ge 0\}$.

In order to make this chapter self-contained we need two additional results related to Hölder's inequality.

2.11. PROPOSITION. *Let $\left(\widetilde{E}, \widetilde{\mathcal{E}}, \mu\right)$ be a measure space and let $\left\{\mu_x : x \in \widetilde{E}\right\}$ be a family of measures on Ω. Then*

$$\left|\int \left(\int F_1(x,\omega) F_2(x,\omega) d\mu_x(\omega)\right) d\mu(x)\right|$$
$$\le \left(\int\int |F_1(x,\omega)|^p \, d\mu_x(\omega) d\mu(x)\right)^{1/p} \left(\int\int |F_2(x,\omega)|^q \, d\mu_x(\omega) d\mu(x)\right)^{1/q},$$

where p and q are conjugate exponents:

$$\frac{1}{p} + \frac{1}{q} = 1, \quad 1 < p,\ q < \infty.$$

Let $\mathcal{F}$ be a sigma-field on Ω. We suppose that the functions F_1 and F_2 are measurable with respect to the σ-field generated by $\mathcal{E} \times \mathcal{F}$. If $p = 1$, then

$$\left|\int \left(\int F_1(x,\omega) F_2(x,\omega) d\mu_x(\omega)\right) d\mu(x)\right|$$
$$\le \int\int |F_1(x,\omega)| \, d\mu_x(\omega) d\mu(x) \operatorname{esssup}_{(x,\omega) \in \widetilde{E} \times \Omega} |F_2(x,\omega)|.$$

Here the essential supremum is taken with respect to the measure, determined by $(B, A) \mapsto \int_B \mu_x(A) d\mu(x)$, $B \in \widetilde{\mathcal{E}}$, $A \in \mathcal{F}$. We assume that, for every $A \in \mathcal{F}$, the mapping $x \mapsto \mu_x(A)$ is μ-measurable.

Proof. The proof for $p = 1$ is left to the reader. We suppose that $1 < p,\ q < \infty$ as well as that the expressions

$$A := \left(\int\int |F_1(x,\omega)|^p \, d\mu_x(\omega) d\mu(x)\right)^{1/p}$$
$$\text{and} \quad B := \left(\int\int |F_2(x,\omega)|^q \, d\mu_x(\omega) d\mu(x)\right)^{1/q}$$

are finite and strictly positive. The inequality $ab \le \frac{a^p}{p} + \frac{b^q}{q}$, $a,\ b \ge 0$, is elementary. Insert $a = \frac{|F_1(x,\omega)|}{A}$ and $b = \frac{|F_2(x,\omega)|}{B}$ to obtain

$$\frac{|F_1(x,\omega)|}{A} \times \frac{|F_2(x,\omega)|}{B} \le \frac{1}{p}\frac{|F_1(x,\omega)|^p}{A^p} + \frac{1}{q}\frac{|F_2(x,\omega)|^q}{B^q}.$$

Integration then yields the desired result.

In the following theorem we interpret $W^{1/p}$ as the function which is identically 1 when $p = \infty$.

2.12. THEOREM. *Let $W \geq 0$ be a function in $K(E)$ and let p and q be conjugate exponents, i.e. $\frac{1}{p} + \frac{1}{q} = 1$, $1 \leq p, q \leq \infty$. For $a > 0$ the operator*

$$f \mapsto W^{1/p} (aI + K_0)^{-1} \left(W^{1/q} f\right)$$

maps $L^p(E, m)$ to $L^p(E, m)$ and

$$\left\| W^{1/p} (aI + K_0)^{-1} W^{1/q} \right\|_{p,p} \leq \left\| (aI + K_0)^{-1} W \right\|_\infty \leq \frac{1}{1 - e^{-a}} \|W\|_{\mathrm{KF}} .$$

REMARK. The operator

$$f \mapsto W^{1/p} (aI + K_0)^{-1} \left(W^{1/q} f\right)$$

is called the *Birman-Schwinger kernel* and can be used to study distributions of eigenvalues: see Chapter 8.

Proof. Let $f \geq 0$ be in $L^p(E, m)$ and let $g \geq 0$ be a function in $L^q(E, m)$. Then we have the following inequalities:

$$\begin{aligned} &\int W(x)^{1/p} \left[(aI + K_0)^{-1} \left(W^{1/q} f\right) \right] (x) \overline{g(x)} dx \\ &= \iint_0^\infty e^{-as} \left(\mathbb{E}_x \left[W(x)^{1/p} f(X(s)) \, W(X(s))^{1/q} g(x) \right] \right) ds \, dx \end{aligned}$$

(Employ Proposition 2.11 with $\widetilde{E} = [0, \infty) \times E$, $\mu_{s,x} = \mathbb{P}_x$ and $\mu(I \times B) = \int_I e^{-as} ds \, m(B)$, $I \subseteq [0, \infty)$, $B \in \mathcal{E}$)

$$\begin{aligned} &\leq \left(\iint_0^\infty e^{-as} \mathbb{E}_x \left(W(x)^{1/p} f(X(s)) \right)^p ds \, dx \right)^{1/p} \\ &\quad \times \left(\iint_0^\infty e^{-as} \mathbb{E}_x \left(W(X(s))^{1/q} g(x) \right)^q ds \, dx \right)^{1/q} \\ &= \left(\iint_0^\infty e^{-as} W(x) \mathbb{E}_x \left(f(X(s))^p \right) ds \, dx \right)^{1/p} \\ &\quad \times \left(\iint_0^\infty e^{-as} \mathbb{E}_x \left(W(X(s)) \right) g(x)^q ds \, dx \right)^{1/p} \end{aligned}$$

$$= \left(\int W(x)\left[(aI+K_0)^{-1}|f|^p\right](x)dx\right)^{1/p}$$

$$\times\left(\int\left[(aI+K_0)^{-1}W\right](x)\,|g(x)|^q\,dx\right)^{1/q}$$

$$\leq \left\|W\,(aI+K_0)^{-1}\right\|_{1,1}^{1/p}\|f\|_p\left\|(aI+K_0)^{-1}W\right\|_\infty^{1/q}\|g\|_q$$

(The operator $(aI+K_0)^{-1}$ is self-adjoint.)

$$= \left\|(aI+K_0)^{-1}W\right\|_{\infty,\infty}^{1/p}\left\|(aI+K_0)^{-1}W\right\|_\infty^{1/q}\|f\|_p\|g\|_q$$

$$= \left\|(aI+K_0)^{-1}W\right\|_\infty^{1/p+1/q}\|f\|_p\|g\|_q = \left\|(aI+K_0)^{-1}W\right\|_\infty\|f\|_p\|g\|_q.$$

This proves Theorem 2.12.

Next we want to discuss the way in which the generator of the Feynman-Kac semigroup $\{\exp(-t(K_0\dot{+}V)) : t\geq 0\}$ is related to the Friedrichs' extension of K_0+V. We also are interested in "core"-type problems. Theorem 2.13 is closely related to the KLMN-theorem: see Reed and Simon [192, Theorem X.17, p. 167]. The fact that $\operatorname{dom}\left(K_0^{1/2}\right)\cap\operatorname{dom}\left(V_+^{1/2}\right)$ is automatically dense implies that the Trotter-Lie product is available: see Kato [150, Theorem 1, p. 694] and also Appendix B.

2.13. THEOREM. *Suppose that for every function* $f\in\operatorname{dom}\left(K_0^{1/2}\right)\cap\operatorname{dom}\left(V_+^{1/2}\right)$ *there exists a sequence* $\{f_n : n\in\mathbb{N}\}$ *in* $\operatorname{dom}(K_0)\cap\operatorname{dom}(V)$ *with the following properties:*

(a) $\lim_{n\to\infty}\|f_n-f\|_2=0$;
(b) $\lim_{m,n\to\infty}\langle K_0(f_n-f_m), f_n-f_m\rangle=0$;
(c) $\lim_{m,n\to\infty}\langle V_+(f_n-f_m), f_n-f_m\rangle=0$.

Then the Feynman-Kac generator $K_0\dot{+}V$ *is the Friedrichs' extension of* K_0+V.

It is noticed that the hypotheses in Theorem 2.13 can be rephrased as "the subspace $\operatorname{dom}(K_0)\cap\operatorname{dom}(V)$ is a core for the operator $K_0^{1/2}+V_+^{1/2}$" or, equivalently, "the subspace $\operatorname{dom}(K_0)\cap\operatorname{dom}(V)$ is a form core for K_0+V_+".

Proof. Fix a number a that is strictly larger then the L^2-L^2-type of the Feynman-Kac semigroup $\{\exp(-t(K_0\dot{+}V)) : t\geq 0\}$. The quadratic form $\mathcal{E}^{a+V}$ associated to the Feynman-Kac semigroup is given by (see Proposition 2.16)

$$\begin{aligned}\mathcal{E}^{a+V}(f,g) &= \left\langle (aI+K_0\dot{+}V)^{1/2}f, (aI+K_0\dot{+}V)^{1/2}g\right\rangle\\ &= \left\langle (aI+K_0)^{1/2}f, (aI+K_0)^{1/2}g\right\rangle\\ &\quad -\left\langle V_-^{1/2}(aI+K_0)^{-1/2}(aI+K_0)^{1/2}f, V_-^{1/2}(aI+K_0)^{-1/2}(aI+K_0)^{1/2}g\right\rangle\\ &\quad +\left\langle V_+^{1/2}f, V_+^{1/2}g\right\rangle,\end{aligned} \tag{2.11}$$

where f and g belong to the domain $\operatorname{dom}(\mathcal{E}^{a+V}) = \operatorname{dom}\left(K_0^{1/2}\right) \cap \operatorname{dom}\left(V_+^{1/2}\right)$. Moreover, for a large enough, by Theorem 2.12 together with the definition of Kato-Feller potential, we have, with $\varepsilon(a) = \left\|(aI + K_0)^{-1} V_-\right\|_\infty$,

$$\begin{aligned}
&\left\|V_-^{1/2}(aI + K_0)^{-1/2} h\right\|_2 \leq \left\|V_-^{1/2}(aI + K_0)^{-1/2}\right\|_{2,2} \|h\|_2 \\
&= \left\|V_-^{1/2}(aI + K_0)^{-1} V_-^{1/2}\right\|_{2,2}^{1/2} \|h\|_2 \leq \left\|(aI + K_0)^{-1} V_-\right\|_\infty^{1/2} \|h\|_2 \\
&\leq \sqrt{\varepsilon(a)}\, \|h\|_2 ,
\end{aligned} \tag{2.12}$$

for h belonging to $L^2(E, m)$. Since the negative part V_- of V is supposed to belong to Kato's class (because V is supposed to be a Kato-Feller potential), Proposition 2.2 implies that $\varepsilon(a) < 1$ for $a > 0$ large enough. From (2.12) it follows that V_- is form bounded with bound ε. In fact we have

$$\langle V_- f, f\rangle \leq \varepsilon(a)\left(a\|f\|_2^2 + \left\|K_0^{1/2} f\right\|_2^2\right), \quad f \in \operatorname{dom}\left(K_0^{1/2}\right).$$

From (a) it follows that the domain of $S := K_0 + V$ is dense in $L^2(E, m)$. This is so because the domain $\operatorname{dom}(\mathcal{E}^{a+V})$ is dense in $L^2(E, m)$. Since, in addition, the Feynman-Kac generator $K_0 \dot{+} V$ extends $K_0 + V$, we see that the operator $K_0 + V$ is closable. Let $\overline{S}$ denote this closure. We also write $\widetilde{S}$ for the Feynman-Kac generator $K_0 \dot{+} V$. Furthermore we define the operators T_1 and T_2 as follows:

$$T_1 = S^* \big|_{\operatorname{dom}\left(\mathcal{E}_{\overline{S}}^{a+V}\right) \cap \operatorname{dom}(S^*)}, \quad T_2 = S^* \big|_{\operatorname{dom}\left(\mathcal{E}_{\widetilde{S}}^{a+V}\right) \cap \operatorname{dom}(S^*)}. \tag{2.13}$$

Here $\mathcal{E}_{\overline{S}}^{a+V}$ is the quadratic form associated to $\overline{S}$ and $\mathcal{E}_{\widetilde{S}}^{a+V}$ is the quadratic form associated to $\widetilde{S}$, the so-called Feynman-Kac or Schrödinger form. Then, from Theorem 5.38 in Weidmann [252, p. 123], it follows that T_1 is the Friedrichs' extension of S. Since K_0 and V are both self-adjoint, the operator S is symmetric and so $S \subset S^*$. We also have $S \subset \widetilde{S}$ and hence $S^* \supset \widetilde{S}^* = \widetilde{S} \supset \overline{S} \supset S$. From the definition T_1 it is clear that $T_1 \subset S^*$ and hence $T_1 = T_1^* \supset \overline{S}$. We also readily see $T_2 \supset \widetilde{S}$ and thus $T_2^* \subset \widetilde{S}$. Since $\operatorname{dom}\left(\mathcal{E}_{\overline{S}}^{a+V}\right) \subseteq \operatorname{dom}\left(\mathcal{E}_{\widetilde{S}}^{a+V}\right)$, we also have $T_1 \subset T_2 \subset S^*$. A combination of these inclusions yields:

$$S \subseteq \overline{S} \subseteq T_2^* \subseteq \widetilde{S} \subseteq T_2 \subseteq S^* \tag{2.14}$$

and

$$S \subseteq \overline{S} \subseteq T_2^* \subseteq T_1 \subseteq T_2 \subseteq S^*. \tag{2.15}$$

From (a), (b) and (c), it follows that $\operatorname{dom}(S)$ forms a core for $\mathcal{E}_{\widetilde{S}}^{a+V}$. For, let $f_0 \in \operatorname{dom}\left(\mathcal{E}_{\widetilde{S}}^{a+V}\right)$ be such that $\left\langle \left(aI + \widetilde{S}\right)^{1/2} f, \left(aI + \widetilde{S}\right)^{1/2} f_0 \right\rangle = 0$ for all $f \in$

dom(S). Then $\langle (aI+S)f, f_0\rangle = 0$ for all $f \in \text{dom}(S)$. By properties (a), (b) and (c), there exists a sequence $(f_n)_{n\in\mathbb{N}}$ in dom(S) such that

$$\lim_{n\to\infty}\left\langle \left(aI+\widetilde{S}\right)^{1/2}(f_0-f_n), \left(aI+\widetilde{S}\right)^{1/2}(f_0-f_n)\right\rangle = 0. \tag{2.16}$$

Consequently

$$\begin{aligned}\left\|\left(aI+\widetilde{S}\right)^{1/2} f_0\right\|_2^2 &= \lim_{n\to\infty}\left\langle \left(aI+\widetilde{S}\right)^{1/2} f_n, \left(aI+\widetilde{S}\right)^{1/2} f_0\right\rangle \\ &= \lim_{n\to\infty}\left\langle \left(aI+\widetilde{S}\right) f_n, f_0\right\rangle = \lim_{n\to\infty}\langle (aI+S)\, f_n, f_0\rangle = 0. \end{aligned}\tag{2.17}$$

Hence $f_0 = 0$. Next we are going to show that

$$\text{dom}\,(S^*)\cap\text{dom}\left(\mathcal{E}_{\widetilde{S}}^{a+V}\right) = \text{dom}\left(\widetilde{S}\right)\cap\text{dom}\left(\mathcal{E}_{\widetilde{S}}^{a+V}\right) = \text{dom}\left(\widetilde{S}\right). \tag{2.18}$$

The following argument shows that this is true. Let $f \in L^2(E,m)$ be such that the functional $g \mapsto \left\langle (aI+S)\,g, \left(aI+\widetilde{S}\right)^{-1/2} f\right\rangle$, $g \in \text{dom}(S)$, is continuous. Since dom(S) is a core for $\mathcal{E}_{\widetilde{S}}^{a+V}$, it follows that the functional $g \mapsto \left\langle \left(aI+\widetilde{S}\right) g, \left(aI+\widetilde{S}\right)^{-1/2} f\right\rangle$, $g\in\text{dom}\left(\widetilde{S}\right)$, is continuous as well. We infer that $\text{ran}\left(\left(aI+\widetilde{S}\right)^{-1/2}\right)\cap\text{dom}\,(S^*) = \text{ran}\left(\left(aI+\widetilde{S}\right)^{-1/2}\right)\cap\text{dom}\left(\widetilde{S}^*\right)$, or stating it differently

$$\text{dom}\left(\mathcal{E}_{\widetilde{S}}^{a+V}\right)\cap\text{dom}\,(S^*) = \text{dom}\left(\mathcal{E}_{\widetilde{S}}^{a+V}\right)\cap\text{dom}\left(\widetilde{S}\right) = \text{dom}\left(\widetilde{S}\right)$$

and hence

$$T_2 = S^*\,\Big|_{\text{dom}\left(\mathcal{E}_{\widetilde{S}}^{a+V}\right)\cap\text{dom}(S^*)} = S^*\,\Big|_{\text{dom}(\widetilde{S})} = \widetilde{S}.$$

It follows that $T_2 = \widetilde{S}$ and hence $T_2 = T_2^* = T_1$. This shows that $\widetilde{S}$ is the Friedrichs' extension of S and proves Theorem 2.13.

2.14. Corollary. *If for all sufficiently large a, $a>0$, the range of the operator $aI+K_0+V$ is dense in $L^2(E,m)$, then the operator K_0+V is essentially self-adjoint on* dom $(K_0)\cap$ dom (V), *and its closure generates the Feynman-Kac semigroup.*

Proof. Let $a>0$ be strictly larger than the type of the semigroup

$$\left\{\exp\left(-t\left(K_0\dot{+}V\right)\right) : t\geq 0\right\}.$$

Since the operator $K_0\dot{+}V$ is bounded from below, with lower bound $-\omega_V$, its L^2-L^2 type, it follows that $\left\|\exp\left(-t(K_0\dot{+}V)\right)\right\|_{2,2}\leq \exp(\omega_V t)$, $t>0$. Let the notation be as in the proof of Theorem 2.13. Then we have $S^* \supseteq \widetilde{S} \supseteq \overline{S}$. Let f_0 belong to the domain of S^*. Since the closure of the range of $aI + K_0 + V$ is dense, it follows that the range of $aI + \overline{S}$ coincides with $L^2(E, m)$ and hence $(aI+S^*)f_0 = (aI+\overline{S})f = (aI+S^*)f$ for some f in the domain of $\overline{S}$. Consequently $(aI + S^*)(f - f_0) = 0$. Since the range of $aI + S$ is dense, it follows that $f = f_0$. Whence $\operatorname{dom}(S^*) \subseteq \operatorname{dom}(\overline{S})$. Altogether this shows that $S^* = \overline{S} = \widetilde{S}$.

2.15. PROPOSITION. *Let* $K_0 \dot{+} V$ *be the Feynman-Kac generator of the semigroup* $\{\exp(-t(K_0 \dot{+} V)) : t \geq 0\}$ *and let* $\mathcal{E}^{a+V}$ *be the corresponding quadratic form. The subspace* $\operatorname{dom}(K_0 \dot{+} V) \cap C_\infty(E)$ *is a core for* $K_0 \dot{+} V$ *and the subspace* $\operatorname{dom}\left(K_0^{1/2}\right) \cap \operatorname{dom}\left(V_+^{1/2}\right) \cap C_\infty(E)$ *is form core for* $\mathcal{E}^{a+V}$.

Proof. Let f be a member of the domain of $K_0 \dot{+} V$. Then there exists a sequence $g_n \in C_\infty(E) \cap L^2(E, m)$, $n \in \mathbb{N}$, such that $(aI + K_0 \dot{+} V)f = \lim_{n\to\infty} g_n$. Next we write g_n in the form $g_n = \left(aI + K_0 \dot{+} V\right) f_n$. Then f_n belongs to $L^2(E, m) \cap C_\infty(E)$. The fact that every f_n belongs to $C_\infty(E)$ follows because the Feynman-Kac semigroup leaves the space $C_\infty(E)$ invariant. Moreover, we have $f = \lim_{n\to\infty} f_n$ and $\left(K_0 \dot{+} V\right) f = \lim_{n\to\infty}(K_0 + V)f_n$. This proves the first statement in Proposition 2.15. Upon replacing $\left(aI + K_0 \dot{+} V\right)$ with $\left(aI + K_0 \dot{+} V\right)^{1/2}$, identity (2.11) will yield the second assertion.

2.16. PROPOSITION. *Let* V *be a Kato-Feller potential and let* $\mathcal{E}^{a+V}(f, g)$, $(f, g) \in \operatorname{dom}\left(aI + K_0 \dot{+} V\right)^{1/2} \times \left(aI + K_0 \dot{+} V\right)^{1/2} \to \mathbb{C}$, *be defined by*

$$\mathcal{E}^{a+V}(f, g) = \left\langle \left(aI + K_0 \dot{+} V\right)^{1/2} f, \left(aI + K_0 \dot{+} V\right)^{1/2} g \right\rangle,$$

be the corresponding Schrödinger form ($a > 0$ *is larger than the* L^2-L^2*-type of the corresponding Feynman-Kac semigroup). Then* $\operatorname{dom}\left(\mathcal{E}^{a+V}\right) = \operatorname{dom}\left(K_0^{1/2}\right) \cap \operatorname{dom}\left(V_+^{1/2}\right)$ *and*

$$\mathcal{E}^{a+V}(f, g) = \left\langle (aI + K_0)^{1/2} f, (aI + K_0)^{1/2} g \right\rangle + \int V(x) f(x) \overline{g(x)} dm(x), \quad (2.19)$$

where f *and* g *belong to* $\operatorname{dom}(\mathcal{E}^{a+V})$.

Proof. Put $V_{m,n} = \max(\min(V, n), -m)$, m, n in $\mathbb{N}$, and fix f in $\operatorname{dom}\left(\mathcal{E}^{a+V}\right)$. Define the functions $f_{m,n} \in \operatorname{dom}\left(K_0^{1/2}\right)$ by the equality

$$\left(aI + K_0 \dot{+} V\right)^{1/2} f = (aI + K_0 + V_{m,n})^{1/2} f_{m,n}.$$

The representation

$$\left(aI + K_0 \dot{+} V\right)^{-1/2} g = \frac{1}{\sqrt{\pi}} \int_0^\infty s^{-1/2} \exp\left(-s\left(K_0 \dot{+} V\right)\right) g\, ds, \quad g \in L^2(E, m),$$

together with the Feynman-Kac formula implies that, in L^2-sense,

$$f = \lim_{m\to\infty} \lim_{n\to\infty} f_{m,n}.$$

In addition, we have

$$\begin{aligned}
&\mathcal{E}^{a+V}(f,f) = \left\langle \left(aI + K_0 \dot{+} V\right)^{1/2} f, \left(aI + K_0 \dot{+} V\right)^{1/2} f\right\rangle \\
&= \left\langle \left(aI + K_0 \dot{+} V_{m,n}\right)^{1/2} f_{m,n}, \left(aI + K_0 \dot{+} V_{m,n}\right)^{1/2} f_{m,n}\right\rangle \\
&= \left\langle \left(aI + K_0\right)^{1/2} f_{m,n}, \left(aI + K_0\right)^{1/2} f_{m,n}\right\rangle \\
&\qquad - \int \left(V_{m,n}\right)_- \left|f_{m,n}\right|^2 dx + \int \left(V_{m,n}\right)_+ \left|f_{m,n}\right|^2 dx \\
&= \left\langle \left(aI + K_0\right)^{1/2} f, \left(aI + K_0\right)^{1/2} f\right\rangle - \int V_- \left|f\right|^2 dx + \int V_+ \left|f\right|^2 dx. \qquad (2.20)
\end{aligned}$$

The ultimate equality in (2.20) follows by letting first n and then m tend to infinity and by observing the following general argument for closed linear operators in Hilbert space. Let B be closed linear operator in a Hilbert space (in our situation we may take $B = (aI + K_0)^{1/2}$ or $B = V_+^{1/2}$ or $B = V_-^{1/2}$ in $L^2(E, m)$). Let (g_n) be a sequence in the domain $\mathrm{dom}(B)$ with the following properties:

(a) $g = \lim_{n\to\infty} g_n$ exists in the weak sense;

(b) $\sup_{n\in\mathbb{N}} \|Bg_n\| < \infty$.

Then the vector g belongs to $\mathrm{dom}(B)$ and some sequence $(\widetilde{g}_n)$ in the convex hull of the sequence (Bg_n) converges in Hilbert space sense to Bg. In our situation we first consider $B = (aI + K_0)^{1/2}$ for $a > 0$ so large that in the inequality

$$\langle V_- f, f\rangle \le \varepsilon(a)\left(a\|f\|_2^2 + \left\|K_0^{1/2} f\right\|_2^2\right), \quad f \in \mathrm{dom}(K_0^{1/2}),$$

the constant $\varepsilon(a)$ is strictly less than 1. As a consequence of (2.12) in the proof of Theorem 2.13, such a choice of $\varepsilon(a)$ is possible. This proves the convergence of the first term in (2.20). For the second term we replace $f_{m,n}$ with $g_{m,n}$ defined by $g_{m,n} = \dfrac{(V_{m,n})_-}{V_-} f_{m,n}$. Then, in L^2-sense, $1_{\{V_-\neq 0\}} f = \lim_{m\to\infty} \lim_{n\to\infty} g_{m,n}$. In order to treat the third term, we replace the function $f_{m,n}$ with $h_{m,n}$ given by $h_{m,n} = \dfrac{(V_{m,n})_+}{V_+}$. Again we have $1_{\{V_+\neq 0\}} f = \lim_{m\to\infty} \lim_{n\to\infty} h_{m,n}$ in L^2-sense. For the second term we use $B = V_-$ and for the third one we make use of $B = V_+$. Of course in both these cases, B is a multiplication operator.

This shows (2.20) and thus Proposition 2.16.

B. Integral kernels, martingales, pinned measures

In this section we give a systematic treatment of integral kernel representations of Feynman-Kac semigroups. A first representation was given in Proposition 2.7. In what follows we shall elaborate on this topic. In Example 5 of Appendix C.4, it is proved that the process $s \mapsto p_0(t-s, X(s), y)$ is a martingale on the interval $[0, t)$. Aside from the martingale $s \mapsto p_0(t-s, X(s), y)$, $0 \leq s < t$, we will introduce some additional martingales $M(s)$, $0 \leq s < t$. It is fairly easy to associate with these martingales the so-called *martingale measures.* These martingale measures will be concentrated on the event $\{X(0) = x, X(t-) = y\}$, contained in the sample space Ω. Let us repeat Proposition 2.7.

2.17. PROPOSITION. *The integral kernel of the operator* $\exp\left(-t(K_0\dot{+}V)\right)$ *is given by*

$$\exp\left(-t(K_0 + V)\right)(x, y) = \lim_{\tau\uparrow t} \mathbb{E}_x\left(\exp\left(-\int_0^\tau V(X(s))ds\right) p_0(t-\tau, X(\tau), y)\right)$$
$$= \int \exp\left(-\int_0^t V(X(u))du\right) d\mu_{0,x}^{t,y} = \mu_{0,x}^{V,t,y}(\Omega)$$
$$= \sup_{m\in\mathbb{N}} \int \exp\left(-\int_0^t V(X(u))du\right) d\mu_{0,x}^{m,t,y}. \tag{2.21}$$

Here the measure $\mu_{0,x}^{t,y}$ pins the motion of the process $\{X(t) : t \geq 0\}$ at x at time 0 and at y at time t. It is determined via the equality

$$\mu_{0,x}^{t,y}(A) = \mathbb{E}_x\left(p_0(t-s, X(s), y)1_A\right), \tag{2.22}$$

whenever A is an event in the field $\mathcal{F}_s$ with $s < t$. Since the process $s \mapsto p_0(t - s, X(s), y)$ is a martingale on the interval $[0, t)$ this definition makes sense. As explained in the text following Proposition 2.7 one may use Kolmogorov's extension theorem (Proposition 1.14 in Chapter 1.E.1), to prove that the cylinder measure $A \mapsto \mathbb{E}_x\left[1_A p_0(t-s, X(s), y)\right]$, $A \in \bigcup_{0\leq s<t} \mathcal{F}_s$, extends to a genuine measure $\mu_{0,x}^{t,y}$ on $\mathcal{F}_{t-}$. See Definition 2.18.(e) as well. Similarly the measure $\mu_{0,x}^{V,t,y}$ is determined via the formula $\mu_{0,x}^{V,t,y}(A) = \mathbb{E}_x\left(M_V^{t,y}(s)1_A\right)$, whenever A is an event in the field $\mathcal{F}_s$ with $s < t$. For the definition of the martingale $M_V^{t,y}$, see (2.23) in Definition 2.18. The measures $\mu_{0,x}^{m,t,y}$ are associated to the martingale $M_{0,T_m}^{t,y}(s)$ given by

$$M_{0,T_m}^{t,y}(s)$$
$$= \left[p_0(t-s, X(s), y) - \mathbb{E}_{X(s)}\left(p_0(t-s-T_m, X(T_m), y) : T_m < t-s\right)\right] 1_{\{T_m > s\}}.$$

This martingale is the one of (2.26) below with $S = T_m$. The sequence $(T_m : m \in \mathbb{N})$ is the increasing sequence of stopping times (in fact they are exit times) given by the identity:

$$T_m = \inf\{s > 0 : X(s) \in E^\triangle \setminus \mathrm{int}(K_m)\}.$$

The sequence of compact subsets $(K_m : m \in \mathbb{N})$ is supposed to possess the following properties: $E = \bigcup_{m\in\mathbb{N}} K_m$, $K_m \subseteq \operatorname{int}(K_{m+1})$. The proof of Proposition 2.17 will be given in Chapter 3: see Proposition 3.2. A proof of the martingale property of the processes in (c) of the next definition will be given in Proposition 2.20, Lemma 3.4: see Appendix C.4 as well.

2.18. DEFINITIONS. (a) Let V be a Kato-Feller potential and let S be a *terminal* stopping time. This means that on the event $\{S > t\}$ the equality $t + S\circ\vartheta_t = S$ is valid $\mathbb{P}_x$-almost surely for all $t > 0$. Since, for terminal stopping times $\mathbb{P}_x(S = t) = 0$ (see Lemma 2.19), it will follow that on the event $\{S \geq t\}$ the equality $t + S\circ\vartheta_t = S$ holds $\mathbb{P}_x$-almost surely.
(b) Fix $t > 0$ and define the functions $\exp\left(-t\left(K_0 + V\right)_S\right)(x, y)$ via the formula

$$\begin{aligned}&\exp\left(-t\left(K_0+V\right)_S\right)(x,y) = \exp\left(-t\left(K_0+V\right)\right)(x,y)\\&\quad - \mathbb{E}_x\left(\exp\left(-\int_0^S V(X(u))du\right)\exp\left(-(t-S)\left(K_0+V\right)\right)(X(S),y) : S < t\right).\end{aligned}$$

(c) Next we introduce the following processes (they all turn out to be martingales on the half-open interval $[0,t)$):

$$M_V^{t,y}(s) = \exp\left(-\int_0^s V(X(u))du\right)\exp\left(-(t-s)\left(K_0+V\right)\right)(X(s),y); \tag{2.23}$$

$$\begin{aligned}M_{V,S}^{t,y}(s) &= \exp\left(-\int_0^s V(X(u))du\right)\exp\left(-(t-s)\left(K_0+V\right)_S\right)(X(s),y)1_{\{S>s\}}\\&= \left\{M_V^{t,y}(s) - \exp\left(-\int_0^s V(X(u))du\right)\mathbb{E}_{X(s)}\left(M_V^{t-s,y}(S) : S < t-s\right)\right\}\\&\qquad\qquad \times 1_{\{S>s\}};\end{aligned} \tag{2.24}$$

$$M_0^{t,y}(s) = p_0(t-s, X(s), y); \tag{2.25}$$

$$\begin{aligned}M_{0,S}^{t,y}(s) &= \exp\left(-(t-s)\left(K_0\right)_S\right)(X(s),y)1_{\{S>s\}}\\&= \left\{p_0(t-s,X(s),y) - \mathbb{E}_{X(s)}\left(p_0(t-s-S,X(S),y) : S<t-s\right)\right\}1_{\{S>s\}}\\&= \left\{M_0^{t,y}(s) - \mathbb{E}_{X(s)}\left(M_0^{t-s,y}(S) : S < t-s\right)\right\}1_{\{S>s\}}.\end{aligned} \tag{2.26}$$

(d) Let M be any of the martingales as described in (c). To M we may associate the *pinned measure* $\mu_{0,x}^{M,t,y}$ on the field $\mathcal{F}_{t-}$, given by the formula

$$\mu_{0,x}^{M,t,y}(A) = \mathbb{E}_x\left(M(s)1_A\right),$$

whenever A is an event in the field $\mathcal{F}_s$ with $s < t$. Since the process $s \mapsto M(s)$ is a martingale on the interval $[0,t)$ this definition makes sense.
(e) In particular we may take $M(s) = p_0(t-s, X(s), y)$, $0 \leq s < t$. Then we get

$$\mu_{0,x}^{M,t,y}(A) = \mathbb{E}_x\left(p_0(t-s,X(s),y)1_A\right),$$

whenever A is an event in the field $\mathcal{F}_s$ with $s < t$. In this case we write $\mu_{0,x}^{t,y}(A)$ instead of $\mu_{0,x}^{M,t,y}(A)$.

REMARK. An application of Kolmogorov's extension theorem (Proposition 1.14 in Chapter 1.E.1) to the family of measures

$$\mathbb{P}_{s_1,s_2,\dots,s_n}(B) := \mathbb{E}_x\left[M(s) : (X(s_1),\dots,X(s_n)) \in B\right],$$

where $\max(s_j : 1 \le j \le n) \le s < t$, yields the existence of the measure $\mu_{0,x}^{M,t,y}$ on the σ-field $\mathcal{F}_{t-}$. That this kind of measure is concentrated on the event $\{X(0) = x,\ X(t-) = y\}$ will be proved in Theorem 2.21. This fact is the reason that these measures are called *pinned measures*.

The following lemma will be used in the proof of Proposition 2.20.

2.19. LEMMA. *Let S be terminal stopping time, let x belong to E and suppose $t > 0$. Then $\mathbb{P}_x(S = t) = 0$.*

Proof. Let $(K_\ell : \ell \in \mathbb{N})$ be a sequence of compact subsets of E with the following properties: $E = \bigcup_{\ell\in\mathbb{N}} K_\ell$ and $K_\ell \subseteq \operatorname{int}(K_{\ell+1})$, $\ell \in \mathbb{N}$. Put

$$\Lambda_{k,\ell} = \left\{s > 0 : \int_{K_\ell} \mathbb{P}_y(S = s)dy \ge \frac{1}{k}\right\}, \quad k,\ \ell \in \mathbb{N}.$$

For any finite subset Λ of $\Lambda_{k,\ell}$, then

$$\frac{1}{k}\#\Lambda \le \sum_{s\in\Lambda}\int_{K_\ell} \mathbb{P}_y(S = s)dy = \int_{K_\ell} \mathbb{P}_y(S \in \Lambda)dy \le \int_{K_\ell} dy = m(K_\ell).$$

Here $\#\Lambda$ denotes the number of elements of Λ. Hence $\#\Lambda \le km(K_\ell)$ and so $\#\Lambda_{k,\ell} \le km(K_\ell)$ for $k,\ \ell \in \mathbb{N}$. Consequently, the set

$$\left\{s > 0 : \int_E \mathbb{P}_y(S = s)dy > 0\right\} = \bigcup_{\ell=1}^{\infty}\bigcup_{k=1}^{\infty}\left\{s > 0 : \int_{K_\ell} \mathbb{P}_y(S = s)dy \ge \frac{1}{k}\right\}$$

is countable. Fix $t > 0$ and let x belong to E. So there exists an s, $0 < s < t$, such that $\mathbb{P}_y(S = s) = 0$ for m-almost all $y \in E$. It follows that

$$\begin{aligned}\mathbb{P}_x(S = t) &= \mathbb{P}_x(S = t,\ S > t - s) = \mathbb{P}_x(S \circ \vartheta_{t-s} = s,\ S > t - s)\\ &\le \mathbb{P}_x(S \circ \vartheta_{t-s} = s) = \int p_0(t - s, x, y)\mathbb{P}_y(S = s)dy = 0.\end{aligned}$$

Here we used the Markov property: $\mathbb{P}_x(S \circ \vartheta_{t-s} = s) = \mathbb{E}_x\left[\mathbb{P}_{X(t-s)}(S = s)\right]$.

In the following proposition we use the notation of the martingales introduced in Definition 2.18.

2.20. Proposition. *Let S be a terminal stopping time. Define the functions* $\exp\left(-t\left(K_0+V\right)_S\right)(x,y)$, $t>0$, *via the formula*

$$\begin{aligned}
&\exp\left(-t\left(K_0+V\right)_S\right)(x,y) \\
&=\exp\left(-t\left(K_0+V\right)\right)(x,y) \\
&-\mathbb{E}_x\left(\exp\left(-\int_0^S V(X(u))du\right)\exp\left(-(t-S)\left(K_0+V\right)\right)(X(S),y) : S<t\right).
\end{aligned} \tag{2.27}$$

Then this family satisfies the Chapman-Kolmogorov identity:

$$\begin{aligned}
&\int \exp\left(-s\left(K_0+V\right)_S\right)(x,z)\exp\left(-t\left(K_0+V\right)_S\right)(z,y)\,dz \\
&=\exp\left(-(s+t)\left(K_0+V\right)_S\right)(x,y),
\end{aligned} \tag{2.28}$$

and the process $s\mapsto M_{V,S}^{t,y}(s)$ is a $\mathbb{P}_x$-martingale on the half-open interval $[0,t)$.

Proof. Put $g(z)=\exp\left(-t(K_0+V)_S\right)(z,y)$. From the Feynman-Kac formula it follows that

$$\begin{aligned}
&\int \exp\left(-s\left(K_0+V\right)_S\right)(x,z)\exp\left(-t\left(K_0+V\right)_S\right)(z,y)dz \\
&=\int \exp\left(-s(K_0+V)\right)(x,z)g(z)dz \\
&-\int \mathbb{E}_x\left[\exp\left(-\int_0^S V(X(u))du\right)\exp\left(-(s-S)\left(K_0+V\right)\right)(X(S),z) : S<s\right] \\
&\qquad\qquad \times g(z)dz \\
&=\mathbb{E}_x\left(\exp\left(-\int_0^s V(X(u))du\right)g(X(s))\right) \\
&\quad -\mathbb{E}_x\left[\exp\left(-\int_0^S V(X(u))du\right)\right. \\
&\qquad \left.\mathbb{E}_{X(S)}\left\{\exp\left(-\int_0^{s-S} V(X(u))du\right)g(X(s-S))\right\} : S<s\right]
\end{aligned}$$

(strong time dependent Markov property)

$$\begin{aligned}
&=\mathbb{E}_x\left(\exp\left(-\int_0^s V(X(u))du\right)g(X(s))\right) \\
&\quad -\mathbb{E}_x\left(\exp\left(-\int_0^s V(X(u))du\right)g(X(s)) : S<s\right) \\
&=\mathbb{E}_x\left(\exp\left(-\int_0^s V(X(u))du\right)g(X(s)) : S\geq s\right)
\end{aligned}$$

$$= \mathbb{E}_x\left\{\exp\left(-\int_0^s V(X(u))du\right)\left[\exp\left(-t(K_0+V)\right)(X(s),y)\right.\right.$$
$$\left.- \mathbb{E}_{X(s)}\left(\exp\left(-\int_0^S V(X(u))du\right)\exp\left(-(t-S)(K_0+V)\right)(X(S),y) : S < t\right)\right]$$
$$\left.\times 1_{\{S\geq s\}}\right\}$$

(Markov property)

$$= \mathbb{E}_x\left\{\exp\left(-\int_0^s V(X(u))du\right)\exp\left(-t(K_0+V)\right)(X(s),y) : S \geq s\right\}$$
$$- \mathbb{E}_x\left\{\exp\left(-\int_0^S V(X(u))du\right)\right.$$
$$\left.\times \exp\left(-(t+s-S)(K_0+V)\right)(X(S),y) : S < t+s, S \geq s\right\}$$
$$= \mathbb{E}_x\left\{\exp\left(-\int_0^s V(X(u))du\right)\exp\left(-t(K_0+V)\right)(X(s),y)\right\}$$
$$- \mathbb{E}_x\left\{\exp\left(-\int_0^S V(X(u))du\right)\exp\left(-(t+s-S)(K_0+V)\right)(X(S),y) :\right.$$
$$\left. S < t+s\right\}$$
$$- \mathbb{E}_x\left\{\exp\left(-\int_0^s V(X(u))du\right)\exp\left(-t(K_0+V)\right)(X(s),y) : S \leq s\right\}$$
$$+ \mathbb{E}_x\left\{\exp\left(-\int_0^S V(X(u))du\right)\exp\left(-(t+s-S)(K_0+V)\right)(X(S),y) : S \leq s\right\}$$

(Chapman-Kolmogorov (Proposition 3.2) together with Doob's optional sampling theorem applied to the martingale $M_V^{t,y}$: see Appendix C.11 and Lemma 3.4)

$$= \int dz \exp\left(-s(K_0+V)\right)(x,z)\exp\left(-t(K_0+V)\right)(z,y)$$
$$- \mathbb{E}_x\left\{\exp\left(-\int_0^S V(X(u))du\right)\exp\left(-(t+s-S)(K_0+V)\right)(X(S),y) :\right.$$
$$\left. S < t+s\right\}$$
$$- \mathbb{E}_x\left\{\exp\left(-\int_0^s V(X(u))du\right)\exp\left(-t(K_0+V)\right)(X(s),y) : S \leq s\right\}$$

$$+\mathbb{E}_x\left\{\exp\left(-\int_0^s V(X(u))du\right)\exp\left(-(t+s-s)(K_0+V)\right)(X(s),y) : S\leq s\right\}$$
$$= \exp\left(-(s+t)\left(K_0+V\right)_S\right)(x,y).$$

This proves the first claim in Proposition 2.20. Next we turn our attention to the martingale property of the process $M_{V,S}^{t,y}(s)$. Let $0<\tau_1<\tau_2<t$ and consider

$$\mathbb{E}_x\left[M_{V,S}^{t,y}(\tau_2)\mid\mathcal{F}_{\tau_1}\right]\quad\text{(formula (2.24))}$$
$$=\mathbb{E}_x\left[\exp\left(-\int_0^{\tau_2}V(X(u))du\right)\exp\left(-\left(t-\tau_2\right)\left(K_0+V\right)_S\right)(X(\tau_2),y)1_{\{S>\tau_2\}}\,\middle|\,\mathcal{F}_{\tau_1}\right]$$
$$=\exp\left(-\int_0^{\tau_1}V(X(u))du\right)\mathbb{E}_x\left[\exp\left(-\int_0^{\tau_2-\tau_1}V(X(u+\tau_1))du\right)\right.$$
$$\left.\times\exp\left(-\left(t-\tau_2\right)\left(K_0+V\right)_S\right)\left(X(\tau_2-\tau_1+\tau_1),y\right)1_{\{S\circ\vartheta_{\tau_1}>\tau_2-\tau_1\}}\,\middle|\,\mathcal{F}_{\tau_1}\right]1_{\{S>\tau_1\}}$$

(Markov property)

$$=\exp\left(-\int_0^{\tau_1}V(X(u))du\right)1_{\{S>\tau_1\}}\mathbb{E}_{X(\tau_1)}\left[\exp\left(-\int_0^{\tau_2-\tau_1}V(X(u))du\right)\right.$$
$$\left.\times\exp\left(-\left(t-\tau_2\right)\left(K_0+V\right)_S\right)\left(X(\tau_2-\tau_1),y\right) : S>\tau_2-\tau_1\right]$$
$$=\exp\left(-\int_0^{\tau_1}V(X(u))du\right)\int dz\exp\left(-\left(\tau_2-\tau_1\right)\left(K_0+V\right)_S\right)\left(X(\tau_1),z\right)$$
$$\times\exp\left(-\left(t-\tau_2\right)\left(K_0+V\right)_S\right)(z,y)\,1_{\{S>\tau_1\}}$$

(Chapman-Kolmogorov)

$$=\exp\left(-\int_0^{\tau_1}V(X(u))du\right)\exp\left(-\left(t-\tau_1\right)\left(K_0+V\right)_S\right)\left(X(\tau_1),y\right)1_{\{S>\tau_1\}}$$
$$=M_{V,S}^{t,y}(\tau_1).$$

This proves the announced martingale property.

2.21. THEOREM. *(a) Let $\mu_{0,x}^{t,y}$ be the pinned measure associated to the martingale $M_0^{t,y}(s)=p_0(t-s,X(s),y)$: see Definition 2.18(e). Then the other pinned measures in Definition 2.18 are absolutely continuous with respect to the measure $\mu_{0,x}^{t,y}$ and all of them are concentrated on the event $\{X(0)=x,\ X(t-)=y\}$. More precisely for a non-negative random variable F, measurable with respect to $\mathcal{F}_{t-}$, the following identities are valid.*

In case $M=M_V^{t,y}$,

$$\int Fd\mu_{0,x}^{M,t,y}=\int F\exp\left(-\int_0^t V(X(u))du\right)d\mu_{0,x}^{t,y},\quad\textit{and}\tag{2.29}$$

in case $M=M_{V,S}^{t,y}$,

$$\int Fd\mu_{0,x}^{M,t,y}=\int F\exp\left(-\int_0^t V(X(u))du\right)1_{\{S\geq t\}}d\mu_{0,x}^{t,y}.\tag{2.30}$$

In addition, the following formulae are valid for the expectation values of the martingales described above:

$$\mathbb{E}_x\left(M_V^{t,y}(s)\right) = \exp\left(-t\left(K_0+V\right)\right)(x,y); \tag{2.31}$$

$$\begin{aligned}\mathbb{E}_x\left(M_{V,S}^{t,y}(s)\right) &= \exp\left(-t\left(K_0+V\right)\right)(x,y) \\ &- \mathbb{E}_x\left(\exp\left(-\int_0^S V(X(u))du\right)\exp\left(-(t-S)\left(K_0+V\right)\right)(X(S),y) : S<t\right).\end{aligned} \tag{2.32}$$

By the martingale property the values in (2.31) and (2.32) do not depend on s as long as $0 \le s < t$. Moreover, (2.31) and (2.32) hold for stopping times between 0 and t. So in general the measure $\mu_{0,x}^{t,y}$ is not absolutely continuous with respect to $\mathbb{P}_x$, because $\mathbb{P}_x\left(X(t-)=y\right)=0$ if the measure m does not charge points.
(b) The following Feynman-Kac formula is valid:

$$\begin{aligned}&\int \exp\left(-t(K_0+V)_S\right)(x,y)f(y)dy \\ &= \mathbb{E}_x\left[\exp\left(-\int_0^t V(X(u))du\right)f(X(t)) : S>t\right].\end{aligned}$$

(c) For every non-negative random variable F that is $\mathcal{F}_{t-}$ measurable and for every non-negative Borel measurable function f the following identities are valid:

$$\iint F d\mu_{0,x}^{M,t,y} f(y)dy = \mathbb{E}_x\left[F\exp\left(-\int_0^t V(X(u))du\right)f(X(t))\right] \tag{2.33}$$

in case $M = M_V^{t,y}$, and

$$\iint F d\mu_{0,x}^{M,t,y} f(y)dy = \mathbb{E}_x\left[F\exp\left(-\int_0^t V(X(u))du\right)f(X(t)) : S>t\right] \tag{2.34}$$

in case $M = M_{V,S}^{t,y}$.

In item (d) of Definition 2.18, the pinned measure $\mu_{0,x}^{M,t,y}$ was introduced.

Proof. (a) We first prove that the measure $\mu_{0,x}^{t,y}$ is concentrated on the event $\{X(0)=x,\ X(t-)=y\}$. Since $\mathbb{P}_x(X(0)=x)=1$, it suffices to show that $\mu_{0,x}^{t,y}(\Omega)=\mu_{0,x}^{t,y}\left(X(t-)=y\right)$. Let U be an open neighborhood of $y\in E$. Then, by virtue of the continuity of the kernel $p_0(t,x,y)$ and by BASSA A3,

$$\begin{aligned}\mu_{0,x}^{t,y}(X(s)\in U) &= \mathbb{E}_x\left[1_U(X(s))p_0(t-s,X(s),y)\right) \\ &= \int_U p_0(s,x,z)p_0(t-s,z,y)dz \to p_0(t,x,y),\end{aligned}$$

as $s\uparrow t$. It follows that $\mu_{0,x}^{t,y}\left(X(t-)\in U\right) = p_0(t,x,y)$. Upon shrinking U, we get $\mu_{0,x}^{t,y}\left(X(t-)=y\right) = p_0(t,x,y)$. A consequence of the absolute continuity of the other pinned measures with respect to $\mu_{0,x}^{t,y}$ will be that the other indicated measures are concentrated on the event $\{X(0)=x,\ X(t-)=y\}$ as well. The fact that the process $M_V^{t,y}$ is a $\mathbb{P}_x$-martingale is proved in Chapter 3, Lemma 3.4. It is a consequence of the Chapman-Kolmogorov identity and the Markov property. Next we prove equality (2.29). The following equality is valid:

$$\mu_{0,x}^{V,t,y}(A) = \int_A \exp\left(-\int_0^t V(X(u))du\right) d\mu_{0,x}^{t,y},\ A\in\mathcal{F}_{t-}.$$

This can be seen as follows. Pick the event $A\in\mathcal{F}_s$, $s<t$. Then

$$\begin{aligned}
&\mathbb{E}_x\left[\exp\left(-\int_0^s V(X(u))du\right)\exp\left(-(t-s)\left(K_0+V\right)\right)(X(s),y)1_A\right]\\
&= \lim_{s'\uparrow t-s}\mathbb{E}_x\left[\exp\left(-\int_0^s V(X(u))du\right)\right.\\
&\qquad \left.\times\,\mathbb{E}_{X(s)}\left\{\exp\left(-\int_0^{s'} V(X(u))du\right)p_0(t-s-s',X(s'),y)1_A\right\}\right]
\end{aligned}$$

(Markov property)

$$\begin{aligned}
&= \lim_{s'\uparrow t-s}\mathbb{E}_x\left[\exp\left(-\int_0^s V(X(u))du\right)\exp\left(-\int_s^{s+s'} V(X(u))du\right)\right.\\
&\qquad \left.\times p_0(t-s-s',X(s+s'),y)1_A\right]\\
&= \lim_{s'\uparrow t-s}\mathbb{E}_x\left[\exp\left(-\int_0^{s+s'} V(X(u))du\right)p_0(t-s-s',X(s+s'),y)1_A\right]\\
&= \int_A \exp\left(-\int_0^t V(X(u))du\right)d\mu_{0,x}^{t,y}.
\end{aligned}$$

This proves the statement in (2.29) above for $F=1_A$, $A\in\mathcal{F}_s$, with $s<t$. The general case follows via an approximation argument.

For a proof of equality (2.30), we proceed as follows. Again we take an event $A\in\mathcal{F}_s$, $s<t$. Then, from (2.24) we get

$$\begin{aligned}
&\mu_{0,x}^{M,t,y}(A) = \mathbb{E}_x\left[M_{V,S}^{t,y}(s),A\right]\\
&= \mathbb{E}_x\left[M_V^{t,y}(s),\ A,\ S>s\right]\\
&\quad - \mathbb{E}_x\left[\exp\left(-\int_0^s V(X(u))du\right)\mathbb{E}_{X(s)}\left\{M_V^{t-s,y}(S): S<t-s\right\},A,\ S>s\right]
\end{aligned}$$

(Markov property)

$$\begin{aligned}
&= \mathbb{E}_x\left[M_V^{t,y}(s),\ A,\ S>s\right]\\
&\quad - \mathbb{E}_x\left[\exp\left(-\int_0^s V(X(u))du\right)\left[M_V^{t-s,y}(S)\right]\circ\vartheta_s : s+S\circ\vartheta_s<t,\ A,\ S>s\right]
\end{aligned}$$

(since on the event $\{S > s\}$ the equality $\exp\left(-\int_0^s V(X(u))du\right)\left[M_V^{t-s,y}(S)\right]\circ\vartheta_s = M_V^{t,y}$ is valid)

$$= \mathbb{E}_x\left[M_V^{t,y}(s),\ A,\ S > s\right] - \mathbb{E}_x\left[M_V^{t,y}(S) : A,\ s < S < t\right]$$

(Doob's optional sampling (Appendix C.11) together with (2.29))

$$= \lim_{\varepsilon\downarrow 0}\left(\mathbb{E}_x\left[M_V^{t,y}(t-\varepsilon) : A,\ S > s\right] - \mathbb{E}_x\left[M_V^{t,y}(t-\varepsilon) : A,\ s < S < t-\varepsilon\right]\right)$$
$$= \lim_{\varepsilon\downarrow 0}\int_{A\cap\{S\geq t-\varepsilon\}}\exp\left(-\int_0^t V(X(u))du\right)d\mu_{0,x}^{t,y}$$
$$= \int_{A\cap\{S\geq t\}}\exp\left(-\int_0^t V(X(u))du\right)d\mu_{0,x}^{t,y}.$$

This proves (2.30).

The equality in (2.31) follows because the left-hand side does not depend on s: take $s = 0$. The equality in (2.32) is a consequence of the same argument together with equality (2.27).

(b) Let x belong to E. We shall show the following Feynman-Kac formula:

$$\int \exp\left(-t(K_0+V)_S\right)(x,y)f(y)dy$$
$$= \mathbb{E}_x\left[\exp\left(-\int_0^t V(X(u))du\right)f(X(t)) : S > t\right],$$

where $\exp\left(-t(K_0+V)_S\right)(x,y)$ is as in formula (2.27). By definition we have

$$\int \exp\left(-t(K_0+V)_S\right)(x,y)f(y)dy$$
$$= \int \exp\left(-t(K_0+V)\right)(x,y)f(y)dy$$
$$- \int \mathbb{E}_x\left[\exp\left(-\int_0^S V(X(u))du\right)\exp\left(-(t-S)\left(K_0+V\right)\right)(X(S),y) : S < t\right]f(y)dy$$

(Feynman-Kac twice: Theorem 2.5)

$$= \mathbb{E}_x\left[\exp\left(-\int_0^t V(X(u))du\right)f(X(t))\right]$$
$$- \mathbb{E}_x\left[\exp\left(-\int_0^S V(X(u))du\right)\right.$$
$$\left.\times\mathbb{E}_{X(S)}\left\{\exp\left(-\int_0^{t-S} V(X(u))du\right)f(X(t-S))\right\} : S < t\right]$$

(strong time dependent Markov property)

$$= \mathbb{E}_x \left[\exp\left(- \int_0^t V(X(u))du \right) f(X(t)) \right]$$

$$- \mathbb{E}_x \left[\exp\left(- \int_0^t V(X(u))du \right) f(X(t)) : S < t \right]$$

(Lemma 2.19)

$$= \mathbb{E}_x \left[\exp\left(- \int_0^t V(X(u))du \right) f(X(t)) : S > t \right].$$

(c) The equalities in (2.33) and (2.34) follow then from the Markov property. In fact we have, for $F \geq 0$ measurable with respect to $\mathcal{F}_s$, $0 \leq s < t$, and $M = M_{V,S}^{t,y}$,

$$\iint F d\mu_{0,x}^{M,t,y} f(y) dy$$
$$= \int \mathbb{E}_x \Big[F \exp\left(- \int_0^s V(X(u))du \right)$$
$$\times \exp\left(-(t-s)\left(K_0 + V\right)_S \right)(X(s), y)\, f(y) : S > s \Big] dy$$

(Feynman-Kac: see (b))

$$= \mathbb{E}_x \left[F \exp\left(- \int_0^s V(X(u))du \right) \right.$$
$$\left. \times \mathbb{E}_{X(s)} \left\{ \exp\left(- \int_0^{t-s} V(X(u))du \right) f(X(t-s)) : S > t-s \right\} : S > s \right]$$

(Markov property)

$$= \mathbb{E}_x \left[F \exp\left(- \int_0^t V(X(u))du \right) f(X(t)) : S \circ \vartheta_s > t-s,\ S > s \right]$$

(S is a terminal stopping time)

$$= \mathbb{E}_x \left[F \exp\left(- \int_0^t V(X(u))du \right) f(X(t)) : S > t \right].$$

This is the identity in (2.34). The identity in (2.33) is similar but easier. Its proof is left to the reader.

The following result will be used with S equal to the penetration time of a singularity region Γ: see Proposition 4.16 and in particular formula (4.70).

2.22. PROPOSITION. *Define the pinned measure $\mu_{0,x}^{t,y}$ on the σ-field generated by $X(s)$, $0 \leq s < t$, via the equality (see Definition 2.18.(e))*

$$\mu_{0,x}^{t,y}(A) = \mathbb{E}_x\left[1_A p_0(t-s, X(s), y)\right], \quad A \in \mathcal{F}_s, \quad 0 \leq s < t.$$

Define the martingale $s \mapsto M_V^{t,y}(s)$, $0 \leq s < t$, by

$$M_V^{t,y}(s) = \exp\left(-\int_0^s V(X(\sigma))d\sigma\right)\exp\left(-(t-s)\left(K_0+V\right)\right)(X(s),y).$$

Then the function

$$D_S(t,x,y) := \int_{\{S<t\}} \exp\left(-\int_0^t V(X(\sigma))d\sigma\right) d\mu_{0,x}^{t,y} \tag{2.35}$$

$$= \mathbb{E}_x\left[M_V^{t,y}(S) : S < t\right] \tag{2.36}$$

serves as an integral kernel for the operator $D_S(t)$ where $D_S(t)$ is the difference of the semigroups

$$\exp\left(-t\left(K_0 \dot{+} V\right)\right) \quad \textit{and} \quad \exp\left(-t\left(K_0 \dot{+} V\right)_S\right),$$

$$\textit{i.e.} \quad D_S(t)f(x) = \mathbb{E}_x\left(\exp\left(-\int_0^t V(X(u))du\right) f(X(t)) : S < t\right),$$

for all $f \in C_\infty(E)$ or $f \in L^p(E,m)$, $1 \leq p \leq \infty$.

Proof. First we establish the equality of the expressions in (2.35) and (2.36). Therefore we consider, for V bounded,

$$\mathbb{E}_x\left[M_V^{t,y}(S) : S < t\right] = \lim_{\epsilon \downarrow 0} \mathbb{E}_x\left[M_V^{t,y}(S) : S \leq t-\epsilon\right]$$

(the process $u \mapsto M_V^{t,y}(u)$ is a $\mathbb{P}_x$-martingale)

$$= \lim_{\epsilon \downarrow 0} \mathbb{E}_x\left[M_V^{t,y}(t-\epsilon) : S \leq t-\epsilon\right]$$

$$= \lim_{\epsilon \downarrow 0} \lim_{\epsilon' \downarrow 0} \mathbb{E}_x\Bigg[\exp\left(-\int_0^{t-\epsilon} V(X(\sigma))d\sigma\right)$$

$$\mathbb{E}_{X(t-\epsilon)}\left(\exp\left(-\int_0^{\epsilon-\epsilon'} V(X(\sigma))d\sigma\right) p_0\left(\epsilon', X(\epsilon-\epsilon'), y\right)\right) : S \leq t-\epsilon\Bigg]$$

(Markov property)

$$= \lim_{\epsilon \downarrow 0} \lim_{\epsilon' \downarrow 0} \mathbb{E}_x\left[\exp\left(-\int_0^{t-\epsilon'} V(X(\sigma))d\sigma\right) p_0\left(\epsilon', X(t-\epsilon'), y\right) : S \leq t-\epsilon\right]$$

$$= \lim_{\epsilon \downarrow 0} \lim_{\epsilon' \downarrow 0} \int 1_{\{S \leq t-\epsilon\}} \exp\left(-\int_0^{t-\epsilon'} V(X(\sigma))d\sigma\right) d\mu_{0,x}^{t,y}$$

(dominated convergence)

$$= \int_{\{S<t\}} \exp\left(-\int_0^t V(X(\sigma))d\sigma\right) d\mu_{0,x}^{t,y}. \tag{2.37}$$

A general Kato-Feller potential is truncated first at level $-m$ and n. Then (2.37) is true for these truncated potentials. Monotone convergence yields the general result. This proves the equalities of the expressions in (2.35) and (2.36). Next we notice the identity

$$\int\int F d\mu_{0,x}^{t,y} f(y) dy = \mathbb{E}_x \left[F f(X(t)) \right]$$

for all non-negative functions $f \in C_\infty(E)$ and for all non-negative random variables F that are measurable with respect to the σ-field $\mathcal{F}_{t-}$. So from (2.35) we get

$$\begin{aligned}
\int D_S(t,x,y) f(y) dy &= \int \int_{\{S<t\}} \exp\left(-\int_0^t V(X(\sigma))d\sigma\right) d\mu_{0,x}^{t,y} f(y) dy \\
&= \mathbb{E}_x \left[\exp\left(-\int_0^t V(X(u))du\right) f(X(t)) : S < t \right] \\
&= \mathbb{E}_x \left[\exp\left(-\int_0^t V(X(u))du\right) f(X(t)) \right] \\
&\quad - \mathbb{E}_x \left[\exp\left(-\int_0^t V(X(u))du\right) f(X(t)) : S \geq t \right] \\
&= \left[D_S(t) f \right](x), \quad \text{by Lemma 2.19.}
\end{aligned} \tag{2.38}$$

This proves Proposition 2.22 for $f \in C_\infty(E)$. The extension to $f \in L^p(E, m)$ is immediate.

C. Singular perturbations

The Feynman-Kac representation in (b) of Theorem 2.5 for regular potentials allows a large variety of (positive) potentials. This variety can be enlarged even more. We can extend the results to potentials that are infinitely high over certain regions $\Gamma \subset E$ and to potentials given as measures as considered by Sturm [222, 225], Albeverio et. al. [4] and Stollmann and Voigt [217]. A potential V is called infinitely high on Γ, if it takes the value ∞ on Γ. The objective of this section is to give the local counterparts of the results in Theorem 2.13, Corollary 2.14 and Proposition 2.15 concerning such singular potentials. First we will discuss the kind of singularity regions we have in mind. The singularity region Γ is supposed to be a closed subset of the second countable locally compact Hausdorff space E. In relation to the set Γ, we shall be employing the following stopping times:

$$S = \inf\left\{ s > 0 : \int_0^s 1_\Gamma(X(\sigma))d\sigma > 0 \right\}, \quad T = \inf\left\{ s > 0 : X(s) \in \Gamma \right\}. \tag{2.39}$$

The quantity $\int_0^s 1_\Gamma(X(\sigma))d\sigma$ is called the *spending time* of the trajectory in Γ up to time s and S is called the *penetration time* of Γ: see e.g. Herbst et. al. [121] or Stroock [218]. The time T is the first *hitting time*, i.e. the first time that the trajectory hits Γ. We note that hitting times and penetration times are so-called *terminal* stopping times. For T this means that it satisfies the equality $t+T\circ\vartheta_t = T$ $\mathbb{P}_x$-almost surely on the event $\{T > t\}$. From Lemma 2.19 it follows that we may replace the event $\{T > t\}$ with the event $\{T \geq t\}$. A similar statement is true for S: see also Definitions 2.18.

It readily follows that $S \geq T$, $\mathbb{P}_x$-almost surely, for all $x \in E$. The following simple examples explain the difference between S and T. Let $E = \mathbb{R}^2$ and let Γ be a strip in $\mathbb{R}^2$, to be more specific let Γ be $[0, d] \times \mathbb{R}$, $d > 0$. Consider Brownian motion (or Wiener process). In this case $S = T$. Now let d tend to zero. Then Γ becomes a straight line in $\mathbb{R}^2$. A straight line has an infinite penetration: two-dimensional Brownian motion never penetrates a straight line. However, $T < \infty$, $\mathbb{P}_x$-almost surely, because a single point in $\mathbb{R}$ is recurrent. If Γ is a finite interval in $\mathbb{R}^2$, then again $S = \infty$ and $T < \infty$.

In quantum mechanical models, Dirichlet decoupling procedures are often advantageous. The decoupling takes place in two different ways. This will explained here in the simplest case of a Laplace operator in $L^2(\mathbb{R})$ with a potential wall given as

$$V(x) = V_0 1_{[a,b]}(x), \quad V_0 > 0.$$

Thus the space $\mathbb{R}$ is decomposed into $\Sigma_1 = (-\infty, a)$, $\Sigma_2 = [a, b]$ and $\Sigma_3 = (b, \infty)$. The first decoupling is to assume Dirichlet conditions in the points $\{a\}$ and $\{b\}$. Consequently the Laplace operator is then decomposed as

$$-\Delta_{\Sigma_1} \oplus -\Delta_{\Sigma_2} + V_0 \oplus -\Delta_{\Sigma_3}.$$

The operators $-\Delta_{\Sigma_i}$ denote the Dirichlet operators in $L^2(\Sigma_i)$ with Dirichlet boundary conditions on $\partial\Sigma_i$, $i = 1, 2, 3$. The corresponding semigroup is given by

$$\exp\left(t\Delta_{\Sigma_1}\right) \oplus \exp\left(t\left(\Delta_{\Sigma_2} - V_0\right)\right) \oplus \exp\left(t\Delta_{\Sigma_3}\right).$$

The Feynman-Kac representation is determined by the set of (Wiener) trajectories which never hit $\{a\}$ nor $\{b\}$. This is the same as saying that, if for instance a trajectory starts in Σ_1, it remains there up to time t. For $i = 1$ or $i = 3$ we have

$$\begin{aligned} \left[\exp\left(t\Delta_{\Sigma_i}\right) f\right](x) &= \mathbb{E}_x \left\{f(X(t)) : X(s) \in \Sigma_i \text{ for all } s \leq t\right\}, \\ \text{and} \quad \left[\exp\left(t\Delta_{\Sigma_2} - V_0\right) f\right](x) &= \mathbb{E}_x \left\{e^{-tV_0} f(X(t)) : X(s) \in \Sigma_2 \text{ for all } s \leq t\right\}. \end{aligned}$$

In this case the singularity region is $\Gamma = \{a\} \cup \{b\}$. For such sets Γ the hitting time T is finite $\mathbb{P}_x$-almost surely, but $S \equiv \infty$, in particular $S \neq T$. The other decoupling procedure (see e.g. Deift and Simon [71] or Baumgärtel and Demuth [20]) consists in comparing this model with the situation where $V_0 \to \infty$. In that case $\Gamma = [a,b]$ and the Laplace operator becomes the Laplace operator on $\Sigma_1 \cup \Sigma_3$ with zero Dirichlet condition on the whole interval $[a,b]$. The semigroup is then decomposed as

$$\exp\left(t\Delta_{\Sigma_1}\right) \oplus 0 \oplus \exp\left(t\Delta_{\Sigma_3}\right).$$

This is the mathematical description of the physical model where the potential height V_0 tends to infinity. In this decoupling the set of regular points for $\Gamma = [a,b]$ coincides with the interval $[a,b]$. Hence we have $S = T$ in this case. These simple examples show that in physically relevant models a distinction has to be made between T and S. Since $T \leq S$, the event $\{T < \infty\}$ contains the event $\{S < \infty\}$, and the inequality (2.42) follows. Moreover, in the following proposition we give a sufficient condition on the singularity region Γ in order that $T = S$. For the sake of simplicity, we shall describe most of our results in terms of the penetration time. In most cases it is a matter of routine to transcribe a result for S to a result for T. There are a few exceptions. The most important one is where we approach a singular potential, identically infinite on some set Γ with $S > T$, via $\beta 1_\Gamma$ with β large. In the limit we automatically get the penetration time. The following proposition gives a sufficient condition on Γ, in order that, for all $x \in E$, $S = T$, $\mathbb{P}_x$-almost surely.

2.23. Definition. A point $x \in E$ is a *regular* point for Γ if $\mathbb{P}_x(T = 0) = 1$. The set of regular points of Γ is denoted by Γ^r. If we want to emphasize that $x \in E$ is regular for the stopping time T, we say x is a T-regular point. If $\mathbb{P}_x[S = 0] = 1$, then x is called an S-regular point. The collection of S-regular points of Γ is denoted by Γ^ρ.

It follows from Doob's theorem that a point x is T-regular if and only if it is regular for the Dirichlet problem: see e.g. Port and Stone [189, Theorem 2.2 page 89] where this result is spelled out for Brownian motion. This means a.o. that $\lim_{x\to x_0, x\in\Sigma} \mathbb{E}_x\left[f(X(T) : T < \infty\right] = f(x_0)$ whenever $f : \Gamma \to \mathbb{C}$ is a bounded Borel measurable function which is continuous at x_0. For example this is true for $f \equiv 1$. See Theorem D.21 in Appendix D as well.

2.24. Proposition. *Suppose* $\Gamma^r = (\mathrm{int}(\Gamma))^r$. *Then* $S = T$, $\mathbb{P}_x$*-almost surely for all* $x \in E$.

Proof. Since $S \geq T$, it suffices to prove that $\mathbb{P}_x(S > T) = 0$, $\mathbb{P}_x$-almost surely for all $x \in E$. Since on the event $\{S > T\}$, $S = T + S \circ \vartheta_T$, $\mathbb{P}_x$-almost surely, we have by the Markov property:

$$\begin{aligned}\mathbb{P}_x(S > T) &= \mathbb{P}_x(S > T, S\circ\vartheta_T > 0) = \mathbb{E}_x\left(\mathbb{P}_{X(T)}(S > 0), S > T\right)\\ &= \mathbb{P}_x\left(S > T, \mathbb{P}_{X(T)}(S > 0) = 1\right)\end{aligned}$$

(On the event $\{T < \infty\}$, $X(T)$ belongs to $\Gamma \bigcup \Gamma^r$ $\mathbb{P}_x$-almost surely)

$$= \mathbb{P}_x\left(S > T, \mathbb{P}_{X(T)}(S > 0) = 1, X(T) \in \Gamma \bigcup \Gamma^r\right)$$

(Because of symmetry, $\Gamma \setminus \Gamma^r$ is a polar set: see Theorem D.14(i) in Appendix D)

$$= \mathbb{P}_x \left(S > T, \mathbb{P}_{X(T)}(S > 0) = 1, X(T) \in \Gamma^r\right)$$

$(\Gamma^r = (\mathrm{int}(\Gamma))^r)$

$$= \mathbb{P}_x \left(S > T, \mathbb{P}_{X(T)}(S > 0) = 1, X(T) \in (\mathrm{int}(\Gamma))^r\right).$$

From the right continuity of paths, it follows that the penetration time of Γ is less than or equal to the (first) hitting time if the interior of Γ. Hence, if y belongs to $(\mathrm{int}(\Gamma))^r$, then $\mathbb{P}_y(S > 0) = 0$, and hence $\mathbb{P}_x(S > T) = 0$.

2.25. DEFINITION. Let S be the penetration time of Γ and let T be its first hitting time. Let V be a Kato-Feller potential as defined in Definition 2.2. Then two families of bounded linear operators will be introduced in the following fashion:

$$[U(t)f](x) := \mathbb{E}_x \left[\exp\left(-\int_0^t V(X(\sigma))d\sigma\right) f(X(t)) : S > t\right]$$

$$\text{and} \quad \left[\widetilde{U}(t)f\right](x) := \mathbb{E}_x \left[\exp\left(-\int_0^t V(X(\sigma))d\sigma\right) f(X(t)) : T > t\right].$$

The operators $U(t)$ and $\widetilde{U}(t)$ are well-defined for all $x \in E$ and for all $f \in L^2(E, m)$. If x is a S-regular point, then $[U(t)f](x) = 0$ and if x is T-regular point, then $\left[\widetilde{U}(t)f\right](x) = 0$. In order to reduce the size of the null spaces, the set $\Sigma := E \setminus \Gamma$, being the complement of Γ, is introduced. The restrictions $U(t) \restriction L^2(\Sigma, m)$ and $\widetilde{U}(t) \restriction L^2(\Sigma, m)$ form strongly continuous self-adjoint and positivity preserving semigroups on $L^2(\Sigma, m)$. Their generators will be denoted by $\left(K_0 \dot{+} V\right)_\Sigma$ and $\left(\widetilde{K_0 \dot{+} V}\right)_\Sigma$ respectively. The operator $(K_0)_\Sigma$ is then the generator of the *free* (i.e. $V \equiv 0$) semigroup, killed in Γ. The semigroup $\left\{\widetilde{U}(t) : t \geq 0\right\}$ is called *Dirichlet* semigroup and some authors (see e.g. Arendt and Batty [11]) call the semigroup $\{U(t) : t \geq 0\}$ a *pseudo-Dirichlet* semigroup. Since we will mainly be using the semigroup $\{U(t) : t \geq 0\}$, we will not make this distinction. Moreover, from the context it will be clear what semigroup we have in mind. The kernels of the semigroups $\left\{\exp\left(-t\left(K_0 \dot{+} V\right)_\Sigma\right) : t \geq 0\right\}$ and $\left\{\exp\left(-t\left(\widetilde{K_0 \dot{+} V}\right)_\Sigma\right) : t \geq 0\right\}$ are given by

$$\begin{aligned} &\exp\left(-t\left(K_0 \dot{+} V\right)_\Sigma\right)(x, y) \qquad (2.40)\\ &= \lim_{t' \uparrow t} \mathbb{E}_x \left(\exp\left(-\int_0^{t'} V(X(\sigma))d\sigma\right) p_0(t - t', X(t'), y) : S > t'\right)\\ &= \int_{\{S > t\}} \exp\left(-\int_0^t V(X(\sigma))d\sigma\right) d\mu_{0,x}^{t,y} \end{aligned}$$

and

$$\begin{aligned}
&\exp\left(-t\left(\widetilde{K_0\dot{+}V}\right)_\Sigma\right)(x,y) \qquad (2.41)\\
&= \lim_{t'\uparrow t}\mathbb{E}_x\left(\exp\left(-\int_0^{t'}V(X(\sigma))d\sigma\right)p_0(t-t',X(t'),y) : T > t'\right)\\
&= \int_{\{T>t\}}\exp\left(-\int_0^t V(X(\sigma))d\sigma\right)d\mu_{0,x}^{t,y},
\end{aligned}$$

respectively.

REMARK. In (2.41) the integration is taken over all trajectories starting in x at time 0, arriving in y at time t and having no values in Γ. The event in (2.40) is somewhat larger, because the trajectories may hit Γ before time S. This means that $T \leq S$.

To be more precise, the "pinned" measure $\mu_{0,x}^{t,y}$ is determined by the equality

$$\mu_{0,x}^{t,y}(A) = \mathbb{E}_x\left(1_A p_0(t-s, X(s), y)\right), \text{ where } A \text{ belongs to } \mathcal{F}_s, \quad s < t.$$

Since the process $s \mapsto p_0(s, X(t-s), y)$ is a $\mathbb{P}_x$-martingale (see Example 4 in Appendix C.4), the total mass of $\mu_{0,x}^{t,y}$ is $p_0(t,x,y)$: see the comments following Proposition 2.17. That such a measure on $\mathcal{F}_{t-} := \sigma\{X(s) : 0 \leq s < t\}$ exists follows from the arguments given in Definition 2.18. So in general the measure $\mu_{0,x}^{t,y}$ is not absolutely continuous with respect to $\mathbb{P}_x$ because $\mathbb{P}_x(X(t-) = y) = 0$, if the measure m does not charge points. Note the inequality $S \geq T$ and hence

$$\exp\left(-t\left(\widetilde{K_0\dot{+}V}\right)_\Sigma\right)(x,y) \leq \exp\left(-t\left(K_0\dot{+}V\right)_\Sigma\right)(x,y). \qquad (2.42)$$

Upon specializing the *Feller property* to these semigroups, we get the following (see Definition 1.2): Let $C_\infty(\Sigma)$ denote the vector space of all continuous functions vanishing at infinity. This means that a function f belongs to $C_\infty(\Sigma)$ if it is continuous on $\Sigma := E \setminus \Gamma$, and if $\lim_{x\to x_0} f(x) = 0$ whenever x_0 is a point in the boundary $\partial\Sigma$ of Σ or if $x_0 = \triangle$, the point at infinity of E. The above semigroups possess the Feller property if they leave the space $C_\infty(\Sigma)$ invariant. This means that

$$\exp\left(-t\left(K_0\dot{+}V\right)_\Sigma\right)C_\infty(\Sigma) \subseteq C_\infty(\Sigma)$$

or that

$$\exp\left(-t\left(\widetilde{K_0\dot{+}V}\right)_\Sigma\right)C_\infty(\Sigma) \subseteq C_\infty(\Sigma),$$

respectively. If Γ is regular for S, then $\exp\left(-t\left(K_0\dot{+}V\right)_\Sigma\right)$ leaves $C_\infty(\Sigma)$ invariant and if every point of Γ is regular for T, then the operator $\exp\left(-t\left(\widetilde{K_0\dot{+}V}\right)_\Sigma\right)$

leaves $C_\infty(\Sigma)$ invariant (also see Doob [92, Chapter 1.VIII] and Theorem D.21 in Appendix D). So for regular sets Γ the semigroups above possess the Feller property.

On account of the inequality (2.42), we mainly confine ourselves to the study of the semigroup $\{\exp\left(-t\left(K_0\dot{+}V\right)_\Sigma\right) : t \geq 0\}$. Moreover, if $\Gamma = (\mathrm{int}(\Gamma))^r$, then the penetration time S may replace the exit time of Σ, which is the same as the hitting time of Γ. This means that in Definition 2.25 we usually consider $U(t) = \exp(-t(K_0\dot{+}V)_\Sigma)$.

On the other hand these semigroups are also related to K_0 via BASSA and to the multiplication operator V as we saw in Definition 2.25. Therefore we want to exhibit the relations that exist between $\left(K_0\dot{+}V\right)_\Sigma$, $K_0\dot{+}V$, $(K_0)_\Sigma$, and K_0. If $\Gamma^r = (\mathrm{int}\Gamma)^r$, then by Proposition 2.24 it suffices to study $\left(K_0\dot{+}V\right)_\Sigma$. The following theorem is due to Baumgärtel and Demuth [20]. The proof will not be repeated here.

2.26. THEOREM. *Let $m(\Gamma) > 0$ and suppose $\Gamma^r = \left(\mathrm{int}(\Gamma)\right)^r$. Let 1_Σ be the projection operator*

$$[1_\Sigma f](x) = 1_\Sigma(x) f(x), \quad f \in L^2(E, m).$$

Assume BASSA and let V be a Kato-Feller potential. Then $K_0\dot{+}V$ is well defined, self-adjoint, bounded from below and it generates the Feynman-Kac semigroup. Let $L^2(\Sigma, m) \cap \mathrm{dom}(K_0\dot{+}V)$ be dense in $L^2(\Sigma, m)$ and assume

$$\left(K_0\dot{+}V\right) 1_\Sigma f = 1_\Sigma \left(K_0\dot{+}V\right) 1_\Sigma f, \quad f \in \mathrm{dom}(K_0\dot{+}V) \cap L^2(\Sigma, m).$$

Then $\left(K_0\dot{+}V\right)_\Sigma$ is the Friedrichs' extension of $K_0\dot{+}V \restriction \mathrm{dom}(K_0\dot{+}V) \cap L^2(\Sigma, m)$, defined in $L^2(\Sigma, m)$.

In the results below we let $\Sigma = E \setminus \Gamma$ be an open subset of E and $\left(K_0\dot{+}V\right)_\Sigma$ denotes the Feynman-Kac generator of the semigroup killed in the complement of Σ, i.e. the semigroup $\{\exp\left(-t\left(K_0\dot{+}V\right)_\Sigma\right) : t \geq 0\}$, defined by

$$\begin{aligned} &\left[\exp(-t(K_0\dot{+}V)_\Sigma)f\right](x) = \mathbb{E}_x\left(\exp\left(-\int_0^t V(X(s))ds\right) f(X(t)) : S > t\right) \\ &= \int \exp\left(-t(K_0 + V)_\Sigma\right)(x, y) f(y) dy. \end{aligned} \tag{2.43}$$

If in Theorem 2.27 the complement of Σ is supposed to be regular, then the Feynman-Kac semigroup, killed on the complement of Σ, leaves $C_\infty(\Sigma)$ invariant. Proofs of Theorem 2.27, its Corollary 2.28, and of Proposition 2.29 are not given; they follow the same lines as the ones given for the regular case: see Theorem 2.13, Corollary 2.14 and Proposition 2.15. Corresponding statements and results are also valid if the stopping time S is replaced with T everywhere. Instead of $\left(K_0\dot{+}V\right)_\Sigma$ we have to write $\left(\widetilde{K_0\dot{+}V}\right)_\Sigma$ in this case. The operator $(W)_\Sigma$ denotes the operator of multiplication by W in the space $L^2(\Sigma, m)$. Here W is some Borel measurable function defined on E.

2.27. THEOREM. *Suppose that for every function f in the intersection*

$$\operatorname{dom}\left((K_0)_\Sigma^{1/2}\right) \cap \operatorname{dom}\left(\left(V_+^{1/2}\right)_\Sigma\right)$$

there exists a sequence $\{f_n : n \in \mathbb{N}\}$ *in the subspace* $\operatorname{dom}((K_0)_\Sigma) \cap \operatorname{dom}((V)_\Sigma)$ *with the following properties:*
(a) $\lim_{n\to\infty} \|f_n - f\|_2 = 0$;
(b) $\lim_{m,n\to\infty} \langle (K_0)_\Sigma (f_n - f_m), f_n - f_m\rangle = 0$;
(c) $\lim_{m,n\to\infty} \langle V_+(f_n - f_m), f_n - f_m\rangle = 0$.
Then the Feynman-Kac generator $\left(K_0 \dot{+} V\right)_\Sigma$ *is the Friedrichs' extension of* $(K_0)_\Sigma + (V)_\Sigma$.

2.28. COROLLARY. *If for all sufficiently large a, $a > 0$, the range of the operator* $aI + (K_0)_\Sigma + (V)_\Sigma$ *is dense in* $L^2(\Sigma, m)$, *then the operator* $(K_0)_\Sigma + (V)_\Sigma$ *is essentially self-adjoint and its closure generates the Feynman-Kac semigroup killed on the complement of* Σ.

2.29. PROPOSITION. *Let* $\left(K_0 \dot{+} V\right)_\Sigma$ *be the Feynman-Kac generator of the semigroup* $\left\{\exp\left(-t\left(K_0 \dot{+} V\right)_\Sigma\right) : t \geq 0\right\}$ *and let* $\mathcal{E}_\Sigma^{a+V}$ *be the corresponding quadratic form. The subspace* $\operatorname{dom}\left(K_0 \dot{+} V\right)_\Sigma \cap C_\infty(\Sigma)$ *is a core for* $\left(K_0 \dot{+} V\right)_\Sigma$ *and the subspace* $\operatorname{dom}\left((K_0)_\Sigma^{1/2}\right) \cap \operatorname{dom}\left(\left(V_+^{1/2}\right)_\Sigma\right) \cap C_\infty(E)$ *is a form core for* $\mathcal{E}_\Sigma^{a+V}$, *provided that* Γ, *defined by* $\Gamma = E \setminus \Sigma$, *is regular in the sense that* $\Gamma = \Gamma^\rho$. *Here* Γ^ρ *is the set of S-regular points of* Γ, *i.e.* $\Gamma^\rho = \{x \in E : \mathbb{P}_x[S = 0] = 1\}$. *The domain of* $\mathcal{E}_\Sigma^{a+V}$ *coincides with the intersection* $\operatorname{dom}\left((K_0)_\Sigma^{1/2}\right) \cap \operatorname{dom}\left((V_+)_\Sigma^{1/2}\right)$ *and*

$$\mathcal{E}_\Sigma^{a+V}(f,g) = a\langle f,g\rangle + \left\langle (K_0)_\Sigma^{1/2} f, (K_0)_\Sigma^{1/2} g\right\rangle - \left\langle V_-^{1/2} f, V_-^{1/2} g\right\rangle + \left\langle V_+^{1/2} f, V_+^{1/2} g\right\rangle, \tag{2.44}$$

where f and g belong to $\operatorname{dom}\left(\mathcal{E}_\Sigma^{a+V}\right)$.

REMARK. Basically we have three kinds of quadratic forms: $\mathcal{E}^0$, corresponding to the unperturbed Hamiltonian K_0; $\mathcal{E}^{a+V}$ (see Proposition 2.16), corresponding to the regularly perturbed Hamiltonian $K_0 \dot{+} V$; and $\mathcal{E}_\Sigma^{a+V}$, corresponding to the singularly perturbed Hamiltonian $\left(K_0 \dot{+} V\right)_\Sigma$. The quadratic form $\mathcal{E}^V$ in assertion (f) of Theorem 2.5 coincides with $\mathcal{E}^{a+V}$ evaluated at $a = 0$. If $\Sigma = E$, then $\mathcal{E}^{a+V}$ coincides with $\mathcal{E}_E^{a+V}$. The quadratic form $\mathcal{E}^0$ was introduced in the comments following Definition 1.30. It usually refers to the quadratic form associated to unperturbed Hamiltonian. Only in section D of Chapter 1 we considered a somewhat different notation. For $f \in \operatorname{dom}(K_0)$, $f \geq 0$, such that f^{p-1} belongs to $L^2(E, m)$, we wrote $\mathcal{E}_p(f,f) = \left\langle K_0 f, f^{p-1}\right\rangle$: see equality (1.36). For the energy form associated to the operator $\exp(-\psi) K_0 \exp(\psi)$, we used the notation $\mathcal{E}_{\psi,p}$: see formula (1.39). So that $\mathcal{E}_2$ of section E in Chapter 1 corresponds to $\mathcal{E}^0$ in the other parts of the book. In Remark 4 following Corollary 2.10 we used the symbol $\mathcal{E}$ to denote the quadratic form associated to a general self-adjoint operator which is bounded from below.

On other occasions the symbol $\mathcal{E}$ denotes the Borel field of the state space E; it will be clear from the context whether or not $\mathcal{E}$ denotes a quadratic form.

Next we want to compare the operators $\left(K_0\dot{+}V\right)_\Sigma$ and $K_0\dot{+}V$. For that we introduce the harmonic extension operator.

2.30. DEFINITION. The *harmonic extension operator* H_Σ^{a+V} is defined as follows. Its domain is $\operatorname{dom}(K_0\dot{+}V)$ and its action is given by

$$\left[H_\Sigma^{a+V} f\right](x) = \mathbb{E}_x\left(\exp\left(-\int_0^S (a+V(X(s)))\,ds\right) f(X(S)) : S < \infty\right), \quad (2.45)$$

where f belongs to $\operatorname{dom}(K_0\dot{+}V)$. For later use we also define the function h_Σ^{a+V} by the formula

$$h_\Sigma^{a+V}(x) = \left[H_\Sigma^{a+V} 1\right](x) = \mathbb{E}_x\left(\exp\left(-\int_0^S (a+V(X(s)))\,ds\right) : S < \infty\right). \quad (2.46)$$

In practice we will use the equality in (2.45) whenever it makes sense. For example if f is a bounded continuous function, then there is no difficulty in the interpretation of this formula, provided V is a Kato-Feller potential and a is large enough: see Proposition 4.20. Intuitively, the function $H_\Sigma^{a+V} f$ is a function that on $\Gamma = E\setminus\Sigma$ coincides with f and that on Σ is "$a+V$-harmonic". The operator H_Σ^{a+V} as defined in (2.45) is a priori an operator defined on bounded continuous functions. It is not clear at all that it is defined on $L^2(E,m)$. In fact the latter does not seem to be true. However, in a natural way it is defined on the domain $\operatorname{dom}(K_0\dot{+}V)$ and a little more thought will show that it has a well-defined meaning on the domain of the corresponding quadratic form given by $\operatorname{dom}\left(\left(aI + K_0\dot{+}V\right)^{1/2}\right)$. For details see Proposition 2.31 and Corollary 2.32 below. It is also noted that $\left(H_\Sigma^{a+V}\right)^2 f = H_\Sigma^{a+V} f$.

The operator J_Σ is the restriction operator: $J_\Sigma f(x) = f(x)$ for $x \in \Sigma$ and f a function defined on E. Sometimes we write J instead of J_Σ; this will be clear from the context. The operator J_Σ^* extends a function f defined on Σ to all of E by setting f equal to zero on Γ, the complement of Σ.

The following result should be compared to the fundamental identity for so-called λ-potentials in Port and Stone [189, p. 41]. It is called *Dynkin's formula*: see Fukushima [108, p. 97].

2.31. PROPOSITION. *Let $a > 0$ be sufficiently large. The following identity holds in $L^2(E,m)$:*

$$\left(aI + K_0\dot{+}V\right)^{-1} - J_\Sigma^*\left(aI + K_0\dot{+}V\right)_\Sigma^{-1} J_\Sigma = H_\Sigma^{a+V}\left(aI + K_0\dot{+}V\right)^{-1}. \quad (2.47)$$

Proof. Let g be a function in the domain of $K_0\dot{+}V$. For $x \in \Sigma$ we have

$$\left[\left\{\left(aI + K_0\dot{+}V\right)^{-1} - J_\Sigma^*\left(aI + K_0\dot{+}V\right)_\Sigma^{-1} J_\Sigma\right\}\left(aI + K_0\dot{+}V\right) g\right](x)$$

$$= \int_0^\infty \mathbb{E}_x\left(\exp\left(-\int_0^s (a + V(X(u)))\,du\right)\left(aI + K_0\dot{+}V\right) g(X(s)) : S \leq s\right) ds$$

$$= \mathbb{E}_x\left[\int_S^\infty \exp\left(-\int_0^s (a + V(X(u)))\,du\right)\left(aI + K_0\dot{+}V\right) g(X(s))ds : S < \infty\right]$$

$$= \mathbb{E}_x\left[\exp\left(-\int_0^S (a + V(X(u)))\,du\right)\right.$$

$$\left.\times \int_0^\infty \exp\left(-\int_0^s (a + V(X(u+S)))\,du\right)\left(aI + K_0\dot{+}V\right) g(X(s+S))ds : S < \infty\right]$$

(Markov property)

$$= \mathbb{E}_x\left[\exp\left(-\int_0^S (a + V(X(u)))\,du\right)\right.$$

$$\left.\times \int_0^\infty \mathbb{E}_{X(S)}\left(\exp\left(-\int_0^s (a + V(X(u)))\,du\right)\left(aI + K_0\dot{+}V\right) g(X(s))\right) : S < \infty\right] ds$$

$$= \mathbb{E}_x\left[\exp\left(-\int_0^S (a + V(X(u)))\,du\right) g(X(S)) : S < \infty\right]. \tag{2.48}$$

Consequently

$$J_\Sigma\left(\left(aI + K_0\dot{+}V\right)^{-1} - J_\Sigma^*\left(aI + K_0\dot{+}V\right)_\Sigma^{-1} J_\Sigma\right) = J_\Sigma H_\Sigma^{a+V}\left(aI + K_0\dot{+}V\right)^{-1} \tag{2.49}$$

and so the equality in (2.47) follows from (2.49) together with the observation that on Γ the equality in (2.48) is also true.

2.32. COROLLARY.

(a) Let $a > 0$ be large enough. The following inclusion holds:

$$\left(aI + K_0\dot{+}V\right)_\Sigma J_\Sigma\left(I - H_\Sigma^{a+V}\right) \supseteq J_\Sigma\left(aI + K_0\dot{+}V\right), \tag{2.50a}$$

in the sense of domains and of equality of operators. If f in $L^2(\Sigma, m)$ is such that $J_\Sigma\left(I - H_\Sigma^{a+V}\right) f$ belongs to dom $\left(\left(aI + K_0\dot{+}V\right)_\Sigma\right)$, *then there exists a function g in $L^2(E, m)$ with the property that $f + H_\Sigma^{a+V} g$ is a member of* dom $\left(aI + K_0\dot{+}V\right)$ *and satisfies the equality:*

$$\left(aI + K_0\dot{+}V\right)_\Sigma J_\Sigma\left(I - H_\Sigma^{a+V}\right) f = J_\Sigma\left(aI + K_0\dot{+}V\right)\left(f + H_\Sigma^{a+V} g\right). \tag{2.50b}$$

(b) The inclusion

$$\left(K_0\dot{+}V\right)_\Sigma \supseteq J_\Sigma\left(K_0\dot{+}V\right) J_\Sigma^* \tag{2.51}$$

is valid in the sense of domains and of equality of operators, i.e.

$$\operatorname{dom}(K_0\dot{+}V)_\Sigma \subseteq \operatorname{dom}(J_\Sigma(K_0\dot{+}V)J_\Sigma^*,$$

and $(K_0\dot{+}V)_\Sigma f = J_\Sigma(K_0\dot{+}V)J_\Sigma^* f$, *for* $J_\Sigma^* f \in \operatorname{dom}(K_0\dot{+}V)$.

(c) The operator $H_\Sigma^{a+V}\left(aI + K_0\dot{+}V\right)^{-1}$ *is self-adjoint and form positive. In fact the following identities are true:*

$$\begin{aligned} &H_\Sigma^{a+V}\left(aI + K_0\dot{+}V\right)^{-1} \\ &= \left(\left(aI + K_0\dot{+}V\right)^{1/2} H_\Sigma^{a+V}\left(aI + K_0\dot{+}V\right)^{-1}\right)^* \left(aI + K_0\dot{+}V\right)^{1/2} H_\Sigma^{a+V}\left(aI + K_0\dot{+}V\right)^{-1} \\ &= H_\Sigma^{a+V}\left(aI + K_0\dot{+}V\right)^{-1/2}\left(H_\Sigma^{a+V}\left(aI + K_0\dot{+}V\right)^{-1/2}\right)^*. \end{aligned} \tag{2.52}$$

Hence the operators

$$H_\Sigma^{a+V}\left(aI + K_0\dot{+}V\right)^{-1}, \qquad H_\Sigma^{a+V}\left(aI + K_0\dot{+}V\right)^{-1/2} \quad \text{and}$$
$$\left(aI + K_0\dot{+}V\right)^{1/2} H_\Sigma^{a+V}\left(aI + K_0\dot{+}V\right)^{-1}$$

are all bounded in $L^2(E,m)$. *Moreover, the operator* H_Σ^{a+V} *is self-adjoint in the space* $\operatorname{dom}(\mathcal{E}^{a+V})$ *equipped with the inner-product*

$$\mathcal{E}^{a+V}(f,g) = \left\langle \left(aI + K_0\dot{+}V\right)^{1/2} f, \left(aI + K_0\dot{+}V\right)^{1/2} g\right\rangle, \tag{2.53}$$

where f *and* g *belong to* $\operatorname{dom}(\mathcal{E}^{a+V}) = \operatorname{dom}\left(\left(aI + K_0\dot{+}V\right)^{1/2}\right)$. *In particular the operator* $\left(aI + K_0\dot{+}V\right)^{1/2} H_\Sigma^{a+V}\left(aI + K_0\dot{+}V\right)^{-1/2}$ *is a self-adjoint projection in* $L^2(E,m)$.

REMARK. We point out that inclusion (2.51) is in agreement with certain examples in Albeverio et. al. in [5, Theorem 3.1.1, page 76]. The relevant example treats a Dirichlet boundary condition in one dimension and in that case the relevant stopping time is the first hitting time.

Proof. (a) The equality (2.50a) in (a) is a reformulation of (2.47). If $\left(I - H_\Sigma^{a+V}\right) f$ belongs to the domain of the operator of $\left(aI + K_0\dot{+}V\right)_\Sigma$, then there exists a function f_0 belonging to the domain of $K_0\dot{+}V$ with the property that

$$J_\Sigma^*\left(aI + K_0\dot{+}V\right)_\Sigma J_\Sigma\left(I - H_\Sigma^{a+V}\right) f = \left(aI + K_0\dot{+}V\right) f_0.$$

Hence, by virtue of (2.50a)

$$\begin{aligned} &\left(aI + K_0\dot{+}V\right)_\Sigma J_\Sigma\left(I - H_\Sigma^{a+V}\right) f = J_\Sigma\left(aI + K_0\dot{+}V\right) f_0 \\ &= \left(aI + K_0\dot{+}V\right)_\Sigma J_\Sigma\left(I - H_\Sigma^{a+V}\right) f_0. \end{aligned}$$

Consequently

$$\left(aI + K_0 \dot{+} V\right)_\Sigma J_\Sigma \left(I - H_\Sigma^{a+V}\right)(f - f_0) = 0.$$

But then we infer that $J_\Sigma \left(I - H_\Sigma^{a+V}\right)(f - f_0) = 0$. The latter implies that $f_0 = f + H_\Sigma^{a+V} g$, with $g = f_0 - f$. This shows (2.50b).

(b) From formula (2.50a) in (a) we infer ($a > 0$ large enough)

$$\begin{aligned}\left(K_0 \dot{+} V\right)_\Sigma &= \left(K_0 \dot{+} V\right)_\Sigma J_\Sigma J_\Sigma^* = \left(aI + K_0 \dot{+} V\right)_\Sigma J_\Sigma J_\Sigma^* - aJ_\Sigma J_\Sigma^* \\ &= \left(aI + K_0 \dot{+} V\right)_\Sigma J_\Sigma \left(I - H_\Sigma^{a+V}\right) J_\Sigma^* - aJ_\Sigma J_\Sigma^* \supseteq J_\Sigma \left(K_0 \dot{+} V\right) J_\Sigma^*.\end{aligned} \tag{2.54}$$

(c) Note the identity $\left[H_\Sigma^{a+V} f\right](x) = f(x)$, for $x \in \Gamma^r$ and for $f \in \operatorname{dom}\left(K_0 \dot{+} V\right)$. Also note the fact that the operator H_Σ^{a+V} is a projection in the sense that its square $H_\Sigma^{a+V} \circ H_\Sigma^{a+V}$ equals H_Σ^{a+V}. The identities in (2.52) follow because the operator $H_\Sigma^{a+V}\left(aI + K_0 \dot{+} V\right)^{-1}$ is self-adjoint. The same argument applies for the proof of the the self-adjointness of the operator H_Σ^{a+V} with respect to the inner-product in (2.53). The latter also implies the final statement in (c) of Corollary 2.32.

In the same spirit as we introduced the *harmonic extension operator* H_Σ^{a+V}, we now want to define more operators of that kind.

2.33. DEFINITION. The family of operators $\left\{T_\Sigma^V(s) : s \geq 0\right\}$ is defined by the formula

$$\left[T_\Sigma^V(s) f\right](x) = \mathbb{E}_x\left(\exp\left(-\int_0^S V(X(u))du\right) f(X(S)) : S \leq s\right). \tag{2.55}$$

The operator $D_\Sigma(t) = D_\Sigma^V(t)$ is defined via the formula

$$D_\Sigma^V(t) = \exp\left(-t\left(K_0 \dot{+} V\right)\right) - J_\Sigma^* \exp\left(-t\left(K_0 \dot{+} V\right)\right) J_\Sigma. \tag{2.56}$$

It goes without saying that the function V is a Kato-Feller function and that the operators $T_\Sigma^V(s)$ are used for whatever functions they make sense. If it is clear from the context which function V is meant in $T_\Sigma^V(t)$ and $D_\Sigma^V(t)$, we suppress it from from the notation. So most of the time we write $T_\Sigma(t)$ and $D_\Sigma(t)$. In the following theorem we exhibit some of the relations that exist between these operators. The formula in (2.61) may be considered as *Duhamel's* formula for singular potentials. A formal proof of the equalities in this theorem will be given in section B of Chapter 4. Assertion (2.57) says that the harmonic extension operator is, up to a sign, the derivative of the *Laplace transform* of the family $\left\{T_\Sigma^V(s) : s \geq 0\right\}$: for a proof see Proposition 4.14.

2.34. THEOREM. *(a) The following equalities are true in the weak operator topology:*

$$H_{\Sigma}^{a+V} f = a \int_0^{\infty} e^{-as} T_{\Sigma}^{V}(s) f ds, \quad f \in \operatorname{dom}(K_0 \dot{+} V); \tag{2.57}$$

$$T_{\Sigma}^{V}(t) T_{\Sigma}^{V}(s) = T_{\Sigma}^{V}(t), \quad t, \ s > 0; \tag{2.58}$$

$$D_{\Sigma}^{V}(t) = T_{\Sigma}^{V}(t) - \int_0^t T_{\Sigma}^{V}(s) \left(K_0 \dot{+} V\right) \exp\left(-(t-s)\left(K_0 \dot{+} V\right)\right) ds; \tag{2.59}$$

$$\int_0^t D_{\Sigma}^{V}(s) ds = \int_0^t T_{\Sigma}^{V}(s) \exp\left(-(t-s)\left(K_0 \dot{+} V\right)\right) ds; \tag{2.60}$$

$$\begin{aligned} D_{\Sigma}^{V}(t) &= 1_{\Gamma} \exp\left(-t\left(K_0 \dot{+} V\right)\right) \\ &\quad + J_{\Sigma}^{*}\left(aI + K_0 \dot{+} V\right)_{\Sigma} \int_0^t du \exp\left(-u\left(K_0 \dot{+} V\right)_{\Sigma}\right) J_{\Sigma} \\ &\quad H_{\Sigma}^{a+V} \exp\left(-(t-u)\left(K_0 \dot{+} V\right)\right). \end{aligned} \tag{2.61}$$

(b) The integral kernel of the operator $D_{\Sigma}^{V}(t)$ is given by the formula

$$\begin{aligned} &D_{\Sigma}^{V}(t, x, y) && (2.62) \\ &= \mathbb{E}_x \left[\exp\left(-\int_0^S V(X(u)) du\right) \exp\left(-(t-S)\left(K_0 + V\right)\right)(X(S), y) : S < t\right] \\ &= \mu_{0,x}^{V,t,y}\{S < t\}. && (2.63) \end{aligned}$$

REMARK. Here the measure $\mu_{0,x}^{V,t,y}$ is one of the pinned measures introduced in Definition 2.18 item (d). It is determined by the equality

$$\mu_{0,x}^{V,t,y}(A) = \mathbb{E}_x\left(M_V^{t,y}(s) 1_A\right), \quad A \in \mathcal{F}_s, \quad s < t. \tag{2.64}$$

The process $s \mapsto M_V^{t,y}(s)$, $s \in [0, t)$, is the martingale given in (c) of Definition 2.18.

Proof. For a proof of assertion (b) of Theorem 2.34, the reader is referred to Theorem 2.21 and also to Proposition 2.22. In fact equality (2.32) corresponds to (2.62) in Theorem 2.21 and (2.36) in Proposition 2.22 coincides with (2.63) The formula in (2.61) above is called the *singular Duhamel's formula*. Proposition 4.15 and its proof is devoted to this formula. To be precise (2.57) corresponds to equality (4.58) in Proposition 4.14, (2.59) to (4.62), (2.60) to the first equality in (4.63), and (2.61) to the second equality in (4.64). The equalities in (4.62), (4.63), and (4.64) form part of Proposition 4.15. The only equality, which is neither covered by Proposition 4.14 nor by 4.15 is equality (2.58). To see this we employ the equality $S \circ \vartheta_S = 0$, $\mathbb{P}_x$-almost surely on the event $\{S < \infty\}$: see equality (D.34) in Theorem D.14 of Appendix D with S instead of $\tau_A = \tau_B = \tau_\Gamma$. That the equality in (D.34) is true for S as well as T is explained in Remark 2 following Theorem D.16 of Appendix D. A proof of (2.58) proceeds as follows ($f \geq 0$, f continuous and bounded):

$$\left[T_\Sigma^V(t)T_\Sigma^V(s)f\right](x) = \mathbb{E}_x\left[\exp\left(-\int_0^S V(X(u))du\right)\left[T_\Sigma^V f\right](X(S)) : S < t\right]$$

$$= \mathbb{E}_x\left\{\exp\left(-\int_0^S V(X(u))du\right)\right.$$

$$\left.\times \mathbb{E}_{X(S)}\left[\exp\left(-\int_0^S V(X(u))du\right) f(X(S)) : S < s\right] : S < t\right\}$$

(strong Markov property)

$$= \mathbb{E}_x\left\{\exp\left(-\int_0^S V(X(u))du\right)\right.$$

$$\left.\times \exp\left(-\int_0^{S\circ\vartheta_S} V(X(u+S))du\right) f\left(X\left(S + S\circ\vartheta_S\right)\right) : S\circ\vartheta_S < s,\ S < t\right\}$$

(on the event $\{S < t\}$, the equality $S\circ\vartheta_S = 0$, $\mathbb{P}_x$-almost surely)

$$= \mathbb{E}_x\left\{\exp\left(-\int_0^S V(X(u))du\right) f\left(X\left(S\right)\right) : S < t\right\} = T_\Sigma^V(s)f(x). \tag{2.65}$$

We want to close this section with some results from stochastic potential theory, which will be particularly useful in the context of trace class and Hilbert-Schmidt conditions.

The following proposition gives some elementary results on λ-equilibrium potentials, i.e. bounded λ-harmonic functions, that are identically 1 on a given set Γ. Notice that the assumption in (c) implies that the sample paths may be taken to be continuous: see Proposition 1.5.

2.35. PROPOSITION. *Suppose $\Gamma = \Gamma^r$, fix $\lambda \geq 0$ and put*

$$v_{\lambda,\Gamma}(x) = \mathbb{E}_x\left(\exp(-\lambda T) : T < \infty\right),$$

where T is the hitting time of Γ. The following assertions hold true (K denotes a compact subset):

(a) The function $v_{\lambda,\Gamma}$ is continuous and $v_{\lambda,\Gamma} \equiv 1$ on Γ.

(b) The function $v_{\lambda,\Gamma}$ is λ-subharmonic, i.e.

$$\limsup_{s\downarrow 0} \frac{\left[\left(\exp(-sK_0) - I\right)v_{\lambda,\Gamma}\right](x)}{s} \leq \lambda v_{\lambda,\Gamma}(x). \tag{2.66}$$

(c) *If* $\lim_{s\downarrow 0}\frac{1}{s}\sup_{x\in K}\int_{E\setminus U} p_0(t,x,y)dy = 0$ *for all open subsets* U *of* E *together with all compact subsets* K *of* U*, then*

$$\limsup_{s\downarrow 0}\sup_{x\in K}\left(\frac{\left[(\exp(-sK_0)-I)\,v_{\lambda,\Gamma}\right](x)}{s} - \lambda v_{\lambda,\Gamma}(x)\right) = 0 \tag{2.67}$$

and for every $b > 0$ *and for every compact subset* K *of* $\{v_{\lambda,\Gamma} < b\}$

$$\lim_{s\downarrow 0}\frac{1}{s}\sup_{x\in K}\mathbb{P}_x(v_{\lambda,\Gamma}(X(s)) > b) = \lim_{s\downarrow 0}\frac{1}{s}\sup_{x\in K}\mathbb{E}_x(\max(v_{\lambda,\Gamma}(X(s)) - b, 0)) = 0, \tag{2.68}$$

Proof. (a) Approximate $v_{\lambda,\Gamma}$, uniformly on compact subsets of $\Sigma = E\setminus\Gamma$, by $v^{\epsilon}_{\lambda,\Gamma} = \exp(-\epsilon K_0)v_{\lambda,\Gamma}$.
(b) Consider the (in-)equalities (λ-excessivity of $v_{\lambda,\Gamma}$)

$$\begin{aligned}\left[\exp(-sK_0)v_{\lambda,\Gamma}\right](x) &= \mathbb{E}_x(v_{\lambda,\Gamma}(X(s))) = \mathbb{E}_x\left(\exp(-\lambda T\circ\vartheta_s) : T < \infty\right)\\ &\leq \mathbb{E}_x\left(\exp(\lambda s - \lambda T) : T < \infty\right) = \exp(\lambda s)v_{\lambda,\Gamma}(x).\end{aligned} \tag{2.69}$$

(c) The following identity is readily verified:

$$\begin{aligned}\left[\exp(-sK_0)v_{\lambda,\Gamma}\right](x) - v_{\lambda,\Gamma}(x) &= \left(e^{\lambda s} - 1\right)\mathbb{E}_x\left(\exp(-\lambda T) : \infty > T > s\right)\\ &\quad + \mathbb{E}_x\left(\exp(-\lambda T) + \exp(-\lambda T\circ\vartheta_s) : T \leq s\right).\end{aligned} \tag{2.70}$$

Hence, by Proposition 1.6.(b), (2.67) follows. Since the function $v_{\lambda,\Gamma}$ is continuous, the first assertion in (2.68) follows from our extra assumption. Since $\mathbb{E}_x\left(\max(v_{\lambda,\Gamma}(X(s)) - b, 0)\right) \leq \mathbb{P}_x\left(v_{\lambda,\Gamma}(X(s)) > b\right)$, the second assertion in (2.68) follows from the first.

REMARK. In particular, we may consider $v_0(x) = \mathbb{P}_x(T < \infty)$. Then v_0 is harmonic on Σ and $v_0\mid_\Gamma \equiv 1$. In addition, we write $\Gamma^* = \{x \in E : v_0(x) = 1\}$. Then, since $\Gamma = \Gamma^r$, we have $\Gamma^* \supseteq \Gamma$. In many cases, $\Gamma^* = \Gamma$, e.g. if Γ is a closed ball in $\mathbb{R}^\nu$, $\nu \geq 3$ and if $\{(\Omega,\mathcal{F},\mathbb{P}_x), (X(t) : t\geq 0), (\vartheta_t : t \geq 0), (E,\mathcal{E})\}$ with $E = \mathbb{R}^\nu$ is ν-dimensional Brownian motion. However, if $\nu = 2$ and Γ is the unit disc, then $\Gamma^* = \mathbb{R}^2$. This stems from the fact that, if $\nu = 2$, then the unit disc is recurrent, and if $\nu \geq 3$, then bounded sets are transient. For these assertions, the reader may consult Port and Stone [189, Proposition 2.10, p. 24, Proposition 1.6, p. 56]. Next, suppose that $\Gamma = \Gamma^r$ and that $E\setminus\Gamma$ is connected. Then either $\Gamma^* = E$ or $\Gamma^* = \Gamma$. This is so because the function v_0 satisfies the maximum principle and hence, if for some $x_0 \in E\setminus\Gamma$, $v_0(x_0) = 1$, then for all $x \in E$, $v_0(x) = 1$, and thus $\Gamma^* = E$. If there are no points $x \in E\setminus\Gamma$ with $v_0(x) = 1$, then $\Gamma^* = \Gamma$.

Next suppose that $\zeta = \infty$, or what is equivalent, $\int p_0(t,x,y)dy = 1$, for all $t > 0$ and for all $x \in E$. So that the equality $\int K_0 f dm = 0$, $f \in \operatorname{dom}(K_0)$, $K_0 f \in L^1(E,m)$, follows. Although it was in Chapter 1 only that we needed the *carré du champ operator* Γ_1, introduced in Definition 1.7, we mention that, $\frac{1}{2}\int\Gamma_1(f,\overline{f})dx = \int\left|\left[K_0^{1/2}f\right](x)\right|^2 dx = \left\langle K_0^{1/2}f, K_0^{1/2}f\right\rangle$, for all functions f belonging to $\operatorname{dom}\left(K_0^{1/2}\right) = \operatorname{dom}\left(\mathcal{E}^\lambda\right)$, $\lambda > 0$.

2.36. DEFINITION. Let Γ be closed subset of E.

(a) The set Γ is said to have *finite λ-Dirichlet capacity* if there exists an open subset U, $U \supseteq \Gamma$, together with a function v, $v \geq 1_U$, in the domain of $\mathcal{E}^\lambda$, for which $\mathcal{E}^\lambda(v,v)$ is finite. Here $\mathcal{E}^\lambda(f,g)$, f, $g \in \operatorname{dom}\left(\mathcal{E}^\lambda\right) = \operatorname{dom}\left(K_0^{1/2}\right)$ is given by

$$\mathcal{E}^\lambda(f,g) = \left\langle K_0^{1/2} f, K_0^{1/2} g\right\rangle + \lambda \langle f, g\rangle .$$

The λ-Dirichlet capacity is given by the expression

$$\inf\left\{\mathcal{E}^\lambda(v,v) : v \geq 1_U,\ v \in \operatorname{dom}\left(\mathcal{E}^\lambda\right),\ U \text{ open } , U \supseteq \Gamma\right\}.$$

(b) The set Γ is said to have *finite λ-Newton capacity* if there exists a non-negative finite Borel measure μ, with the property that $\int g^\lambda(x,y)d\mu(y) = \lambda \int g^\lambda(x,y)dy$ for all $x \in \Gamma^r$. Here $g^\lambda(x,y) = \int_0^\infty e^{-\lambda s} p_0(s,x,y)\, ds$, and for μ we may choose $\mu(B) = \mu_\Gamma^\lambda(B) = \lambda \int \mathbb{E}_x\left[\exp(-\lambda T) : X(T) \in B,\ T < \infty\right] dx$. The *$\lambda$-Newton capacity* of the set Γ is then equal to $\mu_\Gamma^\lambda(E) = \lambda \int \mathbb{E}_x\left[\exp(-\lambda T) : T < \infty\right] dx$.

The connection of both capacities will be explained in Definition 2.36 and Proposition 2.37 below. More details can be found in Definition D.17 and the Propositions D.18, D.19, and D.20 in Appendix D. A function realizing the infimum in Definition 2.36.(a) is given by the function $v = v_{\lambda,\Gamma}$, defined by $v_{\lambda,\Gamma}(x) = \mathbb{E}_x\left[\exp(-\lambda T) : T < \infty\right]$: see Proposition 2.35. Here T is the hitting time of Γ. The function $v_{\lambda,\Gamma}$ is called the *λ-equilibrium* potential of the set Γ. A measure realizing the infimum in Definition 2.36.(b) is μ_Γ^λ. This measure is called the *λ-equilibrium* measure of Γ. From Theorem D.16 it follows that the set $\Gamma \setminus \Gamma^r$ is polar and hence that the measure μ_Γ^λ is concentrated on Γ^r. For Brownian motion the λ-Newton capacity is treated in Port and Stone [189, pp. 41–43]. There it is just called λ-capacity. It turns out that the Dirichlet and Newton capacity coincide. In order to compare these two definitions of capacity we make a note of the following equalities. From Theorem D.4 in Appendix D we obtain the following symmetry: $q_\Gamma(s,x,z) = q_\Gamma(s,z,x) = p_0(s,z,x) - \mathbb{E}_z\left[p_0(s-T, X(T), x) : T < s\right]$. A nice consequence of this equality is the following (see Proposition D.15.(c)):

$$\begin{aligned}\int_0^\infty g^\lambda(x,y)d\mu_\Gamma^\lambda(y) &= \int_0^\infty \int e^{-\lambda s} p_0(s,x,y)d\mu_{\lambda,\Gamma}(y)\, ds \\ &= \mathbb{E}_x\left[\exp(-\lambda T) - \exp(-\lambda\zeta) : T < \infty\right].\end{aligned}$$

So, for $x \in \Gamma^r$, we get

$$\int_0^\infty g^\lambda(x,y)d\mu_\Gamma^\lambda(y) = \mathbb{E}_x\left[1 - \exp(-\lambda\zeta)\right] = \lambda \int g^\lambda(x,y)dy.$$

Since the measure μ_Γ^λ is concentrated on Γ^r, it follows that

$$\lambda \int\int g^\lambda(x,y)d\mu_\Gamma^\lambda(y)\, dx = \int \mathbb{E}_y\left[1 - \exp(-\lambda\zeta)\right] d\mu_\Gamma^\lambda(dy) \leq \mu_\Gamma^\lambda(E).$$

The latter inequality is in fact an equality if $\zeta \equiv \infty$ or, equivalently, if $\int p_0(t, y, z)dz = 1$ for all $y \in E$. For the sake of simplicity assume that $\zeta \equiv \infty$. From the proof of Lemma 3.1.1 in Fukushima [108, p. 62] (and also from assertion (7) in Proposition D.18) it then follows that

$$\left\langle K_0^{1/2} v_{\lambda,\Gamma}, K_0^{1/2} v_{\lambda,\Gamma} \right\rangle + \lambda \left\langle v_{\lambda,\Gamma}, v_{\lambda,\Gamma} \right\rangle = \lambda \int v_{\lambda,\Gamma}(x)dx = \mu_\Gamma^\lambda(E).$$

The above arguments show that in the case when the Markov process is conservative (or equivalently its life time $\zeta \equiv \infty$), the notions of Dirichlet and Newton capacity coincide. So we have proved the following proposition.

2.37. PROPOSITION. *Suppose the Markov process*

$$\left\{ (\Omega, \mathcal{F}, \mathbb{P}_x), (X(t) : t \geq 0), (\vartheta_t : t \geq 0), (E, \mathcal{E}) \right\}$$

has infinite life time. Let Γ be a closed subset of E. Then its λ-Newton-capacity coincides with its λ-Dirichlet capacity. So that if one them is finite (infinite) then the other is so and they are equal.

REMARK. In assertion (7) of Proposition D.18, we shall prove Proposition 2.37 in more generality. So we shall mainly be interested in the λ-Dirichlet capacity. Henceforth we often say λ-capacity (instead of λ-Dirichlet capacity). In Proposition D.19(b), it is shown that the finiteness of the λ-capacity does not depend on λ, as long as λ is strictly positive. Usually we just say Γ has finite capacity to indicate that, for some $\lambda > 0$ (and hence for all $\lambda > 0$), its λ-capacity is finite.

The following proposition states that every compact subset of E has finite capacity. For a proof we refer the reader to Proposition D.20.

2.38. PROPOSITION. *If Γ is a compact subset of E with λ-equilibrium potential $v_{\lambda,\Gamma}$, then $\int v_{\lambda,\Gamma}(x)dx < \infty$. In particular, the following inequality is valid:*

$$\lambda \int v_{\lambda,\Gamma}(x)\, dx \leq \frac{1}{\sup_{\mu \in P(\Gamma^r)} \inf_{x \in \Gamma^r} \int g^\lambda(x, y) d\mu(y)}.$$

Here $v_{\lambda,\Gamma}(x) = \mathbb{E}_x \left[\exp(-\lambda T) : T < \infty\right]$, the symbol $P(\Gamma^r)$ denotes the collection of probability measures on E that do not charge $E \setminus \Gamma^r$ and g^λ is the function (λ-potential) $g^\lambda(x, y) = \int_0^\infty e^{-\lambda s} p_0(s, x, y) ds$.

REMARK. Since $g^\lambda(x, y) = \sup_{\eta > 0} \int_\eta^\infty e^{-\lambda s} p_0(s, x, y) ds$ and since the function $(s, x, y) \mapsto p_0(s, x, y)$ is continuous, we see that the function $g^\lambda(x, y)$ is lower semi-continuous. Notice that the set Γ^r of regular points of a set Γ is a subset of Γ, provided that the set Γ is closed. This follows from the right continuity of paths. Since $\lambda \int g^\lambda(x, y) dx \leq 1$, the function $g^\lambda(x, y)$ is $m \otimes m$-almost everywhere finite. If $m(U) > 0$ for every non-empty open subset of E, then, by lower semi-continuity, we see that $g^\lambda(x, y)$ is finite for all x and y.

In connection with spectral theory (Hilbert-Schmidt and/or trace properties), sets of finite capacity will be of some importance.

Chapter 3

Proof of Continuity and Symmetry of Feynman-Kac Kernels

In this chapter we shall give a self-contained proof of Theorem 2.5. Readers, who are not interested in this proof at the moment, may continue with Chapter 4 where resolvent differences are considered. Because of its length, the proof of Theorem 2.5. got a separate chapter. The proof here is more general than the one in [231] and also the representations of the kernels are described in much more detail. It is also noticed that if the local Kato-Feller property of the positive part of V is replaced with a global property, then the proof of Theorem 2.5. is much simpler. This is so because the property

$$\lim_{t\downarrow 0}\sup_{x\in E}\int_0^t ds\int_E dy\, p_0(s,x,y)V_+(y)=0, \quad \text{which then replaces}$$

$$\lim_{t\downarrow 0}\sup_{x\in E}\int_0^t ds\int_K dy\, p_0(s,x,y)V_+(y)=0,$$

for all compact subsets K of E, enables us to use globally convergent Dyson expansions (see Appendix B, B.15) from the start. If we only require a local Kato-Feller property, then this need not be the case. We feel that a proof of our general result is justified for at least two reasons. The first one being that the present result is more general, than otherwise would be obtained and a second reason being that some of the techniques in our proof are interesting for their own sake.

The proof is divided into three parts. In Proposition 3.1 the existence of integral kernels is established. In Proposition 3.2 we show that they satisfy the Chapman-Kolmogorov identity. In Proposition 3.3 it is shown that these kernels are continuous and that they are symmetric. In the main part of this chapter a proof of Theorem 2.5 is actually given. After the proof of Theorem 2.5 we give a proof of Khas'minskii's lemma with the use of an appropriate sequence of stopping times.

We begin with the existence of the integral kernels.

3.1. PROPOSITION. *Define the operator* $\exp\left(-t\left(K_0\dot{+}V\right)\right)$ *via the Feynman-Kac formula:*

$$\left[\exp\left(-t\left(K_0\dot{+}V\right)\right)f\right](x)=\mathbb{E}_x\left[\exp\left(-\int_0^t V(X(u))du\right)f(X(t))\right].$$

Then its integral kernel $\exp\left(-t\left(K_0+V\right)\right)(x,y)$ *is given by*

$$\exp\left(-t\left(K_0+V\right)\right)(x,y)=\sup_{m\in\mathbb{N}}\int\exp\left(-\int_0^t V(X(u))du\right)d\mu_{0,x}^{m,t,y}.$$

Here the measure $\mu_{0,x}^{m,t,y}$ is defined on $\mathcal{F}_{t-}$ and it is determined by the equality:

$$\mu_{0,x}^{m,t,y}(A)=\mathbb{E}_x\left[1_A p_{0,m}(t-s,X(s),y):T_m>s\right],$$

where the event A belongs to $\mathcal{F}_s$ with $s<t$, where T_m is the exit time from the interior of K_m and where the function $p_{0,m}(t,x,y)$ is given by the formula:

$$p_{0,m}(t,x,y)=p_0(t,x,y)-\mathbb{E}_x\left(p_0(t-T_m,X(T_m),y):T_m<t\right).$$

Notice that the measure $\mu_{0,x}^{m,t,y}$ is the same as the measure $\mu_{0,x}^{M,t,y}$ of Definition 2.18.(d), where M is the martingale given by (see (2.26))

$$\begin{aligned}M(s)&=M_{0,T_m}^{t,y}(s)\\&=\left\{p_0(t,x,y)-\mathbb{E}_{X(s)}\left[p_0(t-s-T_m,X(T_m),y):T_m<t-s\right]\right\}1_{\{T_m>s\}}.\end{aligned}$$

See Proposition 2.17 in Chapter 2.B as well. The sequence $(K_m:m\in\mathbb{N})$ is an increasing sequence of compact subsets of E with the following properties: $K_m\subseteq\operatorname{int}(K_{m+1})$ and $E=\bigcup_{m=1}^{\infty}K_m$. Notice that the process $s\mapsto p_{0,m}(t-s,X(s),y)1_{\{T_m>s\}}$ is a martingale on the interval $[0,t)$: see (3.4) below.

Proof. Put

$$V_{k,\ell}=\max\left(\min\left(V,k\right),-\ell\right)\right)=\min\left(\max\left(V,-\ell\right),k\right). \tag{3.1}$$

For brevity we also write, for k, ℓ and k', ℓ' and for m fixed, $W=V_{k,\ell}1_{K_m}$, $W'=V_{k',\ell'}1_{K_m}$. As above, $\{K_m:m\in\mathbb{N}\}$ is an increasing sequence of compact subsets of the state space E, that exhausts E; i.e. $E=\bigcup_{m=1}^{\infty}K_m$ and $\operatorname{int}(K_{m+1})\supseteq K_m$. We define the functions

$$\exp\left(-t\left(K_0+V_{k,\ell}\right)_{\operatorname{int}(K_m)}\right)(x,y)=\exp\left(-t\left(K_0+V_{k,\ell}1_{K_m}\right)_{\operatorname{int}(K_m)}\right)(x,y)$$

as follows:

$$\begin{aligned}&\exp\left(-t\left(K_0+V_{k,\ell}\right)_{\operatorname{int}(K_m)}\right)(x,y)=\exp\left(-t\left(K_0+V_{k,\ell}1_{K_m}\right)_{\operatorname{int}(K_m)}\right)(x,y)\\&=\lim_{s\uparrow t}\mathbb{E}_x\left(\exp\left(-\int_0^s V_{k,\ell}(X(u))du\right)p_{0,m}\left(t-s,X(s),y\right):T_m>s\right),\end{aligned} \tag{3.2}$$

where T_m is the first exit time from $\operatorname{int}(K_m)$, i.e.

$$T_m=\inf\left\{s>0:X(s)\in E^{\triangle}\setminus\operatorname{int}\left(K_m\right)\right\}$$

and where the function $p_{0,m}(\tau, z, y) = q_{E\setminus \int(K_m)}$ is defined by

$$p_{0,m}(\tau, z, y) = p_0(\tau, z, y) - \mathbb{E}_z\left(p_0\left(\tau - T_m, X(T_m), y\right) : T_m < \tau\right) \tag{3.3}$$
$$= p_0(\tau, z, y) - \lim_{\delta \downarrow 0} \mathbb{E}_z\left(p_0\left(\delta, X(\tau - \delta), y\right) : T_m \leq \tau - \delta\right).$$

(Compare with Theoem D.3 in Appendix D.) The latter equality is a consequence of the fact that the process $\tau \mapsto p_{0,m}\left(t - s, X(s), y\right) 1_{\{T_m > s\}}$ is a $\mathbb{P}_x$-martingale on the interval $[0, t)$. This can be seen as follows. Fix $0 \leq s_1 < s_2 < t$ and consider

$$\mathbb{E}_x\left(p_{0,m}\left(t - s_2, X(s_2), y\right) 1_{\{T_m > s_2\}} \mid \mathcal{F}_{s_1}\right)$$
$$= \mathbb{E}_x\left(p_{0,m}\left(t - s_2, X(s_2 - s_1 + s_1), y\right) 1_{\left\{T_m \circ \vartheta_{s_1} > s_2 - s_1\right\}} \mid \mathcal{F}_{s_1}\right) 1_{\{T_m > s_1\}}$$

(Markov property)

$$= \mathbb{E}_{X(s_1)}\left(p_{0,m}\left(t - s_2, X(s_2 - s_1), y\right) 1_{\{T_m > s_2 - s_1\}}\right) 1_{\{T_m > s_1\}}$$
$$= \int p_{0,m}\left(s_2 - s_1, X(s_1), z\right) p_{0,m}\left(t - s_2, z, y\right) dz \times 1_{\{T_m > s_1\}}$$
$$= p_{0,m}\left(t - s_1, X(s_1), y\right) 1_{\{T_m > s_1\}}. \tag{3.4}$$

Next put

$$\exp\left(-t\left(K_0 + V_{k,\ell}\right)\right)(x, y) = \sup_{m \in \mathbb{N}} \exp\left(-t\left(K_0 + V_{k,\ell}\right)_{\mathrm{int}(K_m)}\right)(x, y)$$
$$= \lim_{m \to \infty} \exp\left(-t\left(K_0 + V_{k,\ell}\right)_{\mathrm{int}(K_m)}\right)(x, y),$$

and set

$$\exp\left(-t\left(K_0 + V\right)\right)(x, y) = \lim_{k \to \infty} \lim_{\ell \to \infty} \lim_{m \to \infty} \exp\left(-t\left(K_0 + V_{k,\ell}\right)_{\mathrm{int}(K_m)}\right)(x, y).$$

Then, for f belonging to $C_\infty(E)$,

$$\int \exp\left(-t\left(K_0 + V\right)\right)(x, y) f(y) dy$$
$$= \lim_{k \to \infty} \lim_{\ell \to \infty} \lim_{m \to \infty} \int \exp\left(-t\left(K_0 + V_{k,\ell}\right)_{\mathrm{int}(K_m)}\right)(x, y)\, f(y) dy$$
$$= \lim_{k \to \infty} \lim_{\ell \to \infty} \lim_{m \to \infty} \int \lim_{s \uparrow t} \mathbb{E}_x\left[\exp\left(-\int_0^s V_{k,\ell}(X(u)) du\right)\right.$$
$$\left. \times\, p_{0,m}\left(t - s, X(s), y\right) : T_m > s\right] f(y) dy$$
$$= \lim_{k \to \infty} \lim_{\ell \to \infty} \lim_{m \to \infty} \lim_{s \uparrow t} \mathbb{E}_x\left[\exp\left(-\int_0^s V_{k,\ell}(X(u)) du\right)\right.$$
$$\left. \times \int p_{0,m}\left(t - s, X(s), y\right) f(y) dy : T_m > s\right]$$

$$= \lim_{k\to\infty} \lim_{\ell\to\infty} \lim_{m\to\infty} \lim_{s\uparrow t} \mathbb{E}_x \left[\exp\left(- \int_0^s V_{k,\ell}(X(u))du \right) \right.$$
$$\left. \times \mathbb{E}_{X(s)} \left(f(X(t-s)) : T_m > t-s \right) : T_m > s \right]$$

(Markov property)

$$= \lim_{k\to\infty} \lim_{\ell\to\infty} \lim_{m\to\infty} \lim_{s\uparrow t} \mathbb{E}_x \left[\exp\left(- \int_0^s V_{k,\ell}(X(u))du \right) f(X(t)) : \right.$$
$$\left. T_m \circ \vartheta_s > t-s, T_m > s \right]$$
$$= \lim_{k\to\infty} \lim_{\ell\to\infty} \lim_{m\to\infty} \lim_{s\uparrow t} \mathbb{E}_x \left[\exp\left(- \int_0^s V_{k,\ell}(X(u))du \right) f(X(t)) : T_m > t \right]$$
$$= \lim_{k\to\infty} \lim_{\ell\to\infty} \lim_{m\to\infty} \mathbb{E}_x \left[\exp\left(- \int_0^t V_{k,\ell}(X(u))du \right) f(X(t)) : T_m > t \right]$$
$$= \lim_{k\to\infty} \lim_{\ell\to\infty} \mathbb{E}_x \left[\exp\left(- \int_0^t V_{k,\ell}(X(u))du \right) f(X(t)) : \zeta > t \right]$$
$$= \mathbb{E}_x \left[\exp\left(- \int_0^t V(X(u))du \right) f(X(t)) : \zeta > t \right]$$
$$= \mathbb{E}_x \left[\exp\left(- \int_0^t V(X(u))du \right) f(X(t)) \right]. \tag{3.5}$$

In these formulae the random variable ζ is the life-time of the process:

$$\zeta = \inf \{ s > 0 : X(s) = \triangle \},$$

so that $f(X(t)) = 0$ on the event $\{\zeta \leq t\}$, whenever f belongs to $C_\infty(E)$. The $\mathbb{P}_x$-almost sure equality $\zeta = \sup_m T_m$ follows from the fact that the process $\{X(t) : \mathbb{P}_x\}$ does not visit E after it has hit $\triangle$: see Blumenthal and Getoor [35, proof of Theorem 9.4]. In fact we have

$$\mathbb{P}_x \left(\zeta < \sup T_m \right) = \sup_m \mathbb{P}_x \left(\zeta < T_m \right) = \sup_m \mathbb{P}_x \left\{ \zeta < T_m, T_m \circ \vartheta_\zeta > 0 \right\}$$
$$= \sup_m \mathbb{E}_x \left(\mathbb{P}_{X(\zeta)} \left(T_m > 0 \right) : \zeta < T_m \right) = \sup_m \mathbb{E}_x \left(\mathbb{P}_\triangle \left(T_m > 0 \right) : \zeta < T_m \right)$$
$$= \sup_m \mathbb{E}_x \left(0 : \zeta < T_m \right) = 0.$$

From the quasi-left continuity, which says that on the event $\left\{ \zeta > \sup_m T_m \right\}$,

$$X(\sup_m T_m) = \lim_{m\to\infty} X(T_m), \quad \mathbb{P}_x\text{-almost surely,}$$

we infer that $X(\sup_m T_m) = \triangle$, $\mathbb{P}_x$-almost surely. Hence, with $T = \sup_m T_m$,

$$\begin{aligned}\mathbb{P}_x(\zeta > T) &= \mathbb{P}_x(\zeta \circ \vartheta_T > 0, \zeta > T) = \mathbb{E}_x\left(\mathbb{P}_{X(T)}(\zeta > 0) : \zeta > T\right) \\ &= \mathbb{E}_x(\mathbb{P}_\triangle(\zeta > 0) : \zeta > T) = \mathbb{P}_\triangle(\zeta > 0)\, \mathbb{P}_x(\zeta > T) = 0.\end{aligned}$$

From the above observations we conclude that the Feynman-Kac semigroup $\left\{\exp\left(-t\left(K_0 \dot{+} V\right)\right) : t \geq 0\right\}$ defined by

$$\left[\exp\left(-t\left(K_0 \dot{+} V\right)\right) f\right](x) = \mathbb{E}_x\left(\exp\left(-\int_0^t V(X(u))du\right) f(X(t))\right)$$

consists of integral operators given by the formulae indicated in Proposition 3.1.

In the following proposition we prove the Chapman-Kolmogorov identity and a somewhat more convenient representation of the kernels

$$\exp\left(-t\left(K_0 + V\right)\right)(x, y).$$

It is emphasized that the equality is proved for all x and y in E.

3.2. PROPOSITION. *The kernels constructed in Proposition 3.1 satisfy the Chapman-Kolmogorov identity, i.e.*

$$\begin{aligned}&\exp\left(-\left(t_1 + t_2\right)\left(K_0 + V\right)\right)(x, y) \\ &= \int dz \exp\left(-t_1\left(K_0 + V\right)\right)(x, z) \exp\left(-t_2\left(K_0 + V\right)\right)(z, y),\end{aligned}$$

for all x and y in E. The kernel $\exp\left(-t\left(K_0 + V\right)\right)(x, y)$ is also given by the formula:

$$\exp\left(-t\left(K_0 + V\right)\right)(x, y) = \int \exp\left(-\int_0^t V(X(u))du\right) d\mu_{0,x}^{t,y}.$$

Here, of course, the measure $\mu_{0,x}^{t,y}$ is determined by the equality $\mu_{0,x}^{t,y}(A) = \mathbb{E}_x\left[1_A p_0(t-s, X(s), y)\right]$, where A belongs to $\mathcal{F}_s$, $0 \leq s < t$.

Proof. The Chapman-Kolmogorov identity can be viewed as follows. First we prove this identity on $\operatorname{int}(K_m)$:

$$\begin{aligned}&\lim_{s_1 \uparrow t_1} \lim_{s_2 \uparrow t_2} \int dz \mathbb{E}_x\left(\exp\left(-\int_0^{s_1} V(X(u))du\right) p_{0,m}\left(t_1 - s_1, X(s_1), z\right) : T_m > s_1\right) \\ &\times \mathbb{E}_z\left[\exp\left(-\int_0^{s_2} V(X(u))du\right) p_{0,m}\left(t_2 - s_2, X(s_2), y\right) : T_m > s_2\right]\end{aligned}$$

$$= \lim_{s_1 \uparrow t_1} \lim_{s_2 \uparrow t_2} \mathbb{E}_x \left[\exp\left(-\int_0^{s_1} V(X(u))du\right) \mathbb{E}_{X(s_1)} \left(\mathbb{E}_{X(t_1-s_1)} \left(\right.\right.\right.$$

$$\exp\left(-\int_0^{s_2} V(X(u))du\right) p_{0,m}\left(t_2 - s_2, X(s_2), y\right) : T_m > s_2 \Big) :$$

$$\left. \left. T_m > t_1 - s_1 \right) : T_m > s_1 \right]$$

(Markov property)

$$= \lim_{s_1 \uparrow t_1} \lim_{s_2 \uparrow t_2} \mathbb{E}_x \left[\exp\left(-\int_0^{s_1} V(X(u))du\right) \right.$$

$$\times \mathbb{E}_{X(t_1)} \left(\exp\left(-\int_0^{s_2} V(X(u))du\right) p_{0,m}\left(t_2 - s_2, X(s_2), y\right) : T_m > s_2 \right) :$$

$$\left. T_m \circ \vartheta_{s_1} > t_1 - s_1, T_m > s_1 \right]$$

(apply the Markov property twice)

$$= \lim_{s_2 \uparrow t_2} \lim_{s_1 \uparrow t_1} \mathbb{E}_x \left[\exp\left(-\int_0^{s_1} V(X(u))du - \int_{t_1}^{s_2+t_1} V(X(u))du\right) \right.$$

$$\left. \times p_{0,m}\left(t_1 + t_2 - (t_1 + s_2), X(t_1 + s_2), y\right) : T_m > t_1 + s_2 \right]$$

$$= \lim_{s \uparrow t_1 + t_2} \mathbb{E}_x \left[\exp\left(-\int_0^{s} V(X(u))du\right) p_{0,m}\left(t_1 + t_2 - s, X(s), y\right) : T_m > s \right]$$

$$= \exp\left(-(t_1 + t_2)\left(K_0 + V\right)_{\mathrm{int}(K_m)}\right)(x, y).$$

This proves the Chapman-Kolmogorov identity on $\mathrm{int}(K_m)$. Upon letting m tend to infinity we obtain the desired result on E.

The representation

$$\exp\left(-t\left(K_0 + V\right)\right)(x, y) = \int \exp\left(-\int_0^t V(X(u))du\right) d\mu_{0,x}^{t,y}$$

remains to be shown. First we take an event $A \in \mathcal{F}_s$ with $0 \le s < t$. Then we have

$$\begin{aligned}
&\mu_{0,x}^{t,y}(A) - \mu_{0,x}^{m,t,y}(A) \\
&= \mathbb{E}_x\left[1_A\left(p_0(t-s, X(s), y) - p_{0,m}(t-s, X(s), y) 1_{\{T_m > s\}}\right)\right] \\
&= \mathbb{E}_x\left[1_A p_0(t-s, X(s), y) 1_{\{T_m \le s\}}\right] \\
&\quad + \mathbb{E}_x\left[\mathbb{E}_{X(s)}\left(p_0(t-s-T_m, X(T_m), y) : T_m < t-s\right) 1_{\{T_m > s\}}\right]
\end{aligned}$$

(Markov property and $s + T_m \circ \vartheta_s = T_m$ on the event $\{T_m > s\}$)

$$= \mathbb{E}_x \left[1_A p_0(t-s, X(s), y) 1_{\{T_m \le s\}}\right] + \mathbb{E}_x \left[1_A p_0(t - T_m, X(T_m), y) : s < T_m < t\right]$$

(the process $s \mapsto p_0(t-s, X(s), y)$ is a martingale on $[0,t)$: see Example 4 in Appendix C.4)

$$\begin{aligned}
&= \mathbb{E}_x \left[1_A p_0(t - T_m, X(T_m), y) 1_{\{T_m \le s\}}\right] \\
&\quad + \mathbb{E}_x \left[1_A p_0(t - T_m, X(T_m), y) : s < T_m < t\right] \\
&= \mathbb{E}_x \left[1_A p_0(t - T_m, X(T_m), y) : T_m < t\right] \\
&= \lim_{\tau \uparrow t} \mathbb{E}_x \left[1_A p_0(t - T_m, X(T_m), y) : T_m \le \tau\right]
\end{aligned}$$

(the process $s \mapsto p_0(t-s, X(s), y)$ is a martingale on $[0,t)$: see Example 4 in Appendix C.4)

$$\begin{aligned}
&= \lim_{\tau \uparrow t} \mathbb{E}_x \left[1_A p_0(t - \tau, X(\tau), y) : T_m \le \tau\right] \\
&= \sup_{\tau < t} \mu_{0,x}^{t,y} \left(A \cap \{T_m \le \tau\}\right) = \mu_{0,x}^{t,y} \left(A \cap \{T_m < t\}\right).
\end{aligned} \tag{3.6}$$

From the symmetry of $p_0(t,x,y)$, it follows that the functions $(x,y) \mapsto \mu_{0,x}^{t,y}(T_m < t)$ are symmetric as well. For Brownian motion this symmetry is explained in Port and Stone [189, Theorem 4.3]. The proof in [189] can be adapted to include our more general situation: see Appendix D, Theorem D.4. Consequently, from this symmetry we get

$$\begin{aligned}
&\mu_{0,y}^{t,x}\left(T_m < \frac{t}{2}\right) - \int dz \mu_{0,x}^{t/2,z}\left(T_m < \frac{t}{2}\right) \mu_{0,z}^{t/2,y}\left(T_m < \frac{t}{2}\right) \\
&= \mathbb{E}_y \left[p_0\left(t - T_m, X(T_m), x\right) : T_m < \frac{t}{2}\right] - \int \mu_{0,x}^{t/2,z}\left(T_m < \frac{t}{2}\right) \mu_{0,z}^{t/2,y}\left(T_m < \frac{t}{2}\right) dz \\
&= \int p_0(t/2, z, x) \mathbb{E}_y \left[p_0\left(t/2 - T_m, X(T_m), z\right) : T_m < \frac{t}{2}\right] dz \\
&\quad - \int dz \mu_{0,x}^{t/2,z}\left(T_m < \frac{t}{2}\right) \mu_{0,z}^{t/2,y}\left(T_m < \frac{t}{2}\right) \\
&= \int p_0(t/2, z, x) \mu_{0,y}^{t/2,z}\left(T_m < t/2\right) dz - \int \mu_{0,x}^{t/2,z}\left(T_m < \frac{t}{2}\right) \mu_{0,z}^{t/2,y}\left(T_m < \frac{t}{2}\right) dz
\end{aligned}$$

(symmetry)

$$\begin{aligned}
&= \int p_{0,m}(t/2, x, z) \mu_{0,z}^{t/2,y}\left(T_m < t/2\right) dz \\
&= \int p_{0,m}(t/2, x, z) \mathbb{E}_z \left(p_0(t/2 - T_m, X(T_m), y) : T_m < t/2\right) dz \\
&= \mathbb{E}_x \left[\mathbb{E}_{X(t/2)} \left(p_0(t/2 - T_m, X(T_m), y) : T_m < t/2\right) : T_m \ge t/2\right]
\end{aligned}$$

(Markov property together with the fact that on the event $\{T_m \geq t/2\}$ the equality $t/2 + T_m \circ \vartheta_{t/2} = T_m$ holds $\mathbb{P}_x$-almost surely)

$$= \mathbb{E}_x \left[p_0(t - T_m, X(T_m), y) : t/2 \leq T_m < t\right] = \mu_{0,x}^{t,y}\left(t/2 \leq T_m < t\right).$$

Hence we obtain

$$\begin{aligned}
&\mu_{0,x}^{t,y}\left(T_m < \frac{t}{2}\right) + \mu_{0,y}^{t,x}\left(T_m < \frac{t}{2}\right) \\
&= \mu_{0,x}^{t,y}\left(T_m < t\right) + \int dz \mu_{0,x}^{t/2,z}\left(T_m < \frac{t}{2}\right)\mu_{0,z}^{t/2,y}\left(T_m < \frac{t}{2}\right) \geq \mu_{0,x}^{t,y}\left(T_m < t\right).
\end{aligned}$$

This shows that the non-negative quantity in (3.6) is dominated by

$$\mu_{0,x}^{t,y}\left(T_m < t\right) \leq \mu_{0,x}^{t,y}\left(T_m < t/2\right) + \mu_{0,y}^{t,x}\left(T_m < t/2\right),$$

and that

$$\begin{aligned}
0 &\leq \mu_{0,x}^{t,y}(A) - \sup_m \mu_{0,x}^{m,t,y}(A) \leq \inf_m \left\{\mu_{0,x}^{t,y}\left(T_m < t/2\right) + \mu_{0,y}^{t,x}\left(T_m < t/2\right)\right\} \\
&= \inf_m \left\{\mathbb{E}_x\left[p_0(t/2, X(t/2), y) : T_m < t/2\right] + \mathbb{E}_y\left[p_0(t/2, X(t/2), x) : T_m < t/2\right]\right\} \\
&\leq \mathbb{E}_x\left[p_0(t/2, X(t/2), y) : \zeta \leq t/2\right] + \mathbb{E}_y\left[p_0(t/2, X(t/2), x) : \zeta \leq t/2\right] = 0 + 0 = 0,
\end{aligned}$$

because we may define $p_0(s, \triangle, y) := 0$ for all $s > 0$. So, for $F \geq 0$ an arbitrary random variable that is measurable with respect to $\mathcal{F}_{t-}$, we approximate F by an increasing sequence $(F_\ell : \ell \in \mathbb{N})$ of elementary non-negative random variables of the form $F_\ell = \sum_{k=1}^{N_\ell} \alpha_{k,\ell} 1_{A_{k,\ell}}$, with $\alpha_{k,\ell} > 0$ and with $A_{k,\ell}$ an event in $\bigcup_{s<t} \mathcal{F}_s$. From what is shown above, we then infer that

$$\begin{aligned}
&\sup_m \int F d\mu_{0,x}^{m,t,y} = \sup_m \sup_\ell \int F_\ell d\mu_{0,x}^{m,t,y} = \sup_\ell \sup_m \int F_\ell d\mu_{0,x}^{m,t,y} \\
&= \sup_\ell \int F_\ell d\mu_{0,x}^{t,y} = \int F d\mu_{0,x}^{t,y}.
\end{aligned}$$

For a random variable $F \geq 0$ that is measurable with respect to $\mathcal{F}_{t-}$ we see that $\int F d\mu_{0,x}^{t,y} = \int_{\{\zeta \geq 0\}} F d\mu_{0,x}^{t,y}$. So, in order to prove the second statement in Proposition 3.2, it suffices to prove that $\exp\left(-\int_0^t V(X(u))du\right) 1_{\{\zeta \geq t\}}$ is $\mathcal{F}_{t-}$ measurable. Therefore we fix $0 < s_1,\ s_2 < t$, and consider

$$\exp\left(-\int_0^{s_1} V_+(X(u))du\right) 1_{\{\zeta > s_1\}} \exp\left(-\int_0^{s_2} V_-(X(u))du\right).$$

We first let s_1 increase to t and then let s_2 increase to t. Monotone convergence then yields the desired result.

Next we want to prove that these integral kernels are continuous on $(0, \infty) \times E \times E$. We also want to show that they are symmetric.

3.3. PROPOSITION. *The function* $(t,x,y) \mapsto \exp(-t(K_0+V))(x,y)$ *is continuous on the set* $(0,\infty)\times E\times E$. *Moreover, it is symmetric in the sense that*

$$\exp(-t(K_0+V))(x,y) = \exp(-t(K_0+V))(y,x)$$

for all $t>0$ *and for all* x *and* y *in* E.

Proof. We prove these statements first for the densities

$$\exp\left(-t\left(K_0+V_{k,\ell}\right)_{\mathrm{int}(K_m)}\right)(x,y), \quad t>0.$$

Expanding the exponential in the definition of such an integral kernel, and employing the Markov property yields the following Dyson expansion (see Appendix B, B.15):

$$\begin{aligned}
&\exp\left(-t\left(K_0+V_{k,\ell}\right)_{\mathrm{int}(K_m)}\right)(x,y)\\
&= \exp\left(-t\left(K_0\right)_{\mathrm{int}(K_m)}\right)(x,y)\\
&\quad+\sum_{n=1}^{\infty}(-1)^n\int_{0<t_1<\dots<t_n<t}\int dt_1\dots dt_n\int\dots\int dx_1\dots dx_n\\
&\quad p_{0,m}(t_1,x,x_1)\,V_{k,\ell}(x_1)p_{0,m}(t_2-t_1,x_1,x_2)\dots V_{k,\ell}(x_n)p_{0,m}(t-t_n,x_n,y)\\
&= p_{0,m}(t,x,y)+\sum_{n=1}^{\infty}(-1)^n\int_{0<t_1<\dots<t_n<t}\int dt_1\dots dt_n\int\dots\int dx_1\dots dx_n\\
&\quad\prod_{j=1}^{n+1}p_{0,m}(t_j-t_{j-1},x_{j-1},x_j)\prod_{j=1}^{n}V_{k,\ell}(x_j),
\end{aligned} \tag{3.7}$$

where in (3.7) the following conventions are used: $t_0=0$, $x_0=x$, $t_{n+1}=t$, $x_{n+1}=y$ and

$$p_{0,m}(t,x,y)=\exp\left(-t\left(K_0\right)_{\mathrm{int}(K_m)}\right)(x,y).$$

Since the density $p_{0,m}(t,x,y)$ is symmetric it will follow that the same is true for the function $\exp\left(-t\left(K_0+V_{k,\ell}\right)_{\mathrm{int}(K_m)}\right)(x,y)$. The continuity of

$$\exp\left(-t\left(K_0+V_{k,\ell}\right)_{\mathrm{int}(K_m)}\right)(x,y) \quad \text{on} \quad (0,\infty)\times\mathrm{int}(K_m)\times\mathrm{int}(K_m)$$

will be a consequence of the continuity of $p_{0,m}(t,x,y)$ on the same domain: see Appendix D, Theorem D.5. A formal proof reads as follows. Let S_n be the set of all strictly increasing sequences of length n in $(0,\infty)$ and define the family

$$\left\{g_{n,t,x,y}^{V_{k,\ell}} : t>0,\ x\in\mathrm{int}(K_m),\ y\in\mathrm{int}(K_m)\right\}$$

on the set $\mathbb{N} \times E^n \times S_n$ via the formula ($t_0 = 0$, $x_0 = x$ and $x_{n+1} = y$)

$$g_{n,t,x,y}^{V_{k,\ell}}(x_1, \ldots, x_n, t_1, \ldots, t_n) = \prod_{j=1}^{n+1} p_{0,m}\left(t_j - t_{j-1}, x_{j-1}, x_j\right) \prod_{j=1}^{n} V_{k,\ell}(x_j) 1_{(0,t)}(t_n).$$

Since the absolute value of $V_{k,\ell}$ is dominated by $\max(k, \ell)$ it follows that the absolute value of $g_{n,t,x,y}^{V_{k,\ell}}\left(x_1, \ldots, x_n, t_1, \ldots t_n\right)$ is dominated by

$$(\max(k, \ell))^n \prod_{j=1}^{n+1} p_{0,m}\left(t_j - t_{j-1}, x_{j-1}, x_j\right).$$

Next let (t_ν, x_ν, y_ν) be a sequence in $(0, \infty) \times \operatorname{int}(K_m) \times \operatorname{int}(K_m)$ converging to (t, x, y) in $(0, \infty) \times \operatorname{int}(K_m) \times \operatorname{int}(K_m)$ and put

$$h_\nu^{V_{k,\ell}}(x_1, \ldots, x_n, t_1, \ldots, t_n) = g_{n,t_\nu,x_\nu,y_\nu}^{V_{k,\ell}}\left(x_1, \ldots, x_n, t_1, \ldots, t_n\right).$$

Then the absolute value of $h_\nu^{V_{k,\ell}}$ is dominated by $h_\nu^{\max(k,\ell)}$. Furthermore notice that

$$\begin{aligned}
&\int_{0<t_1<\ldots t_n<t} \int dt_1 \ldots dt_n \int \cdots \int dx_1 \ldots dx_n \left|h_\nu^{V_{k,\ell}}(t_1, \ldots, t_n, x_1, \ldots, x_n)\right| \\
&\leq \frac{(\max(k, \ell))^n}{n!} p_{0,m}(t_\nu, x_\nu, y_\nu).
\end{aligned} \tag{3.8}$$

Since the function $p_{0,m}(t, x, y)$ is continuous on $(0, \infty) \times \operatorname{int}(K_m) \times \operatorname{int}(K_m)$ it will follow that

$$\begin{aligned}
&\lim_{\nu \to \infty} \int_{S_n} dt_1 \ldots dt_n \int_{E^n} dx_1 \ldots dx_n h_\nu^{\max(k,\ell)}(t_1, \ldots, t_n, x_1, \ldots, x_n) \\
&= \int_{S_n} dt_1 \ldots dt_n \int_{E^n} dx_1 \ldots dx_n \lim_{\nu \to \infty} h_\nu^{\max(k,\ell)}(t_1, \ldots, t_n, x_1, \ldots, x_n).
\end{aligned}$$

From Scheffé's theorem (see Appendix E, Theorem E.3) it will then follow that the sequence $\left\{h_\nu^{\max(k,\ell)} : \nu \in \mathbb{N}\right\}$ is uniformly integrable.

One version of Scheffé's theorem (Theorem E.3) says that a pointwise convergent sequence of non-negative functions $\{f_\nu, \nu \in \mathbb{N}\}$ in some space $L^1(\Omega, \mathcal{A}, \mu)$ is uniformly μ-integrable if and only if

$$\infty > \lim_{\nu \to \infty} \int f_\nu d\mu = \int \lim_{\nu \to \infty} f_\nu d\mu. \tag{3.9}$$

Consequently the sequence $\left\{h_\nu^{V_{k,\ell}} : \nu \in \mathbb{N}\right\}$ is uniformly integrable. It is also pointwise convergent. A moment's thought then shows that each term in the Dyson expansion (3.7) of $\exp\left(-t\left(K_0 + V_{k,\ell}\right)_{\operatorname{int}(K_m)}\right)(x,y)$ is continuous on $(0,\infty) \times \operatorname{int}(K_m) \times (K_m)$. Since the series in the Dyson expansion converges uniformly on compact subsets of $(0,\infty) \times E \times E$ it will follow that

$$\exp\left(-t\left(K_0 + V_{k,\ell}\right)_{\operatorname{int}(K_m)}\right)(x,y)$$

is continuous on $(0,\infty) \times \operatorname{int}(K_m) \times \operatorname{int}(K_m)$. Next we want to show that the same is true for the integral kernel

$$\exp\left(-t\left(K_0 + V\right)_{\operatorname{int}(K_m)}\right)(x,y).$$

The Dyson expansion will also yield the following Duhamel's formula:

$$\begin{aligned}
&\exp\left(-t\left(K_0 + V_{k',\ell'}\right)_{\operatorname{int}(K_m)}\right)(x,y) - \exp\left(-t\left(K_0 + V_{k,\ell}\right)_{\operatorname{int}(K_m)}\right)(x,y)\\
&= \int_0^t ds \int dz \exp\left(-s\left(K_0 + V_{k',\ell'}\right)_{\operatorname{int}(K_m)}\right)(x,z)\left(V_{k,\ell}(z) - V_{k',\ell'}(z)\right)\\
&\qquad \times \exp\left(-(t-s)\left(K_0 + V_{k,\ell}\right)_{\operatorname{int}(K_m)}\right)(z,y). \qquad (3.10)
\end{aligned}$$

For the time being we write $W = V_{k,\ell}1_{K_m}$ and $W' = V_{k',\ell'}1_{K_m}$. Notice that the negative parts W_- and W'_- of respectively W and W' satisfy the inequality: $W_- \leq V_-$ and $W'_- \leq V_-$. Then we see

$$\begin{aligned}
&\exp\left(-t\left(K_0 + W'\right)_{\operatorname{int}(K_m)}\right)(x,y) - \exp\left(-t\left(K_0 + W\right)_{\operatorname{int}(K_m)}\right)(x,y)\\
&= \int_0^t ds \int dz \exp\left(-s\left(K_0 + W'\right)_{\operatorname{int}(K_m)}\right)(x,z)\left(W(z) - W'(z)\right)\\
&\qquad \times \exp\left(-(t-s)\left(K_0 + W\right)_{\operatorname{int}(K_m)}\right)(z,y)\\
&= \int_0^{t/2} ds \int dz \exp\left(-s\left(K_0 + W'\right)_{\operatorname{int}(K_m)}\right)(x,z)\left(W(z) - W'(z)\right)\\
&\qquad \times \exp\left(-(t-s)\left(K_0 + W\right)_{\operatorname{int}(K_m)}\right)(z,y)\\
&\quad + \int_0^{t/2} ds \int dz \exp\left(-s\left(K_0 + W\right)_{\operatorname{int}(K_m)}\right)(z,y)\left(W(z) - W'(z)\right)\\
&\qquad \times \exp\left(-(t-s)\left(K_0 + W'\right)_{\operatorname{int}(K_m)}\right)(x,z). \qquad (3.11)
\end{aligned}$$

Hence, from the Cauchy-Schwarz' inequality, we get

$$\begin{aligned}
&\left|\exp\left(-t\left(K_0+W'\right)_{\mathrm{int}(K_m)}\right)(x,y)-\exp\left(-t\left(K_0+W\right)_{\mathrm{int}(K_m)}\right)(x,y)\right| \\
&\leq \int_0^{t/2} ds \int dz \exp\left(-s\left(K_0+W'\right)_{\mathrm{int}(K_m)}\right)(x,z)\left|W(z)-W'(z)\right| \\
&\qquad \times \sup_{z\in K_m}\sup_{t/2\leq s\leq t}\exp\left(-s\left(K_0+W\right)_{\mathrm{int}(K_m)}\right)(z,y) \\
&\quad + \int_0^{t/2} ds \int dz \exp\left(-s\left(K_0+W\right)_{\mathrm{int}(K_m)}\right)(z,y)\left|W(z)-W'(z)\right| \\
&\qquad \times \sup_{z\in K_m}\sup_{t/2\leq s\leq t}\exp\left(-s\left(K_0+W'\right)_{\mathrm{int}(K_m)}\right)(x,z) \\
&= \int_0^{t/2} ds\mathbb{E}_x\left[\exp\left(-\int_0^s W'(X(u))du\right)\left|W(X(s))-W'(X(s))\right| : T_m>s\right] \\
&\qquad \times \sup_{z\in K_m}\sup_{t/2\leq s\leq t}\exp\left(-s\left(K_0+W\right)_{\mathrm{int}(K_m)}\right)(z,y) \\
&\quad + \int_0^{t/2} ds\mathbb{E}_y\left[\exp\left(-\int_0^s W(X(u))du\right)\left|W(X(s))-W'(X(s))\right| : T_m>s\right] \\
&\qquad \times \sup_{z\in K_m}\sup_{t/2\leq s\leq t}\exp\left(-s\left(K_0+W'\right)_{\mathrm{int}(K_m)}\right)(x,z) \\
&\leq \mathbb{E}_x\left[\exp\left(\int_0^{\min(t/2,T_m)} V_-(X(u))du\right)\int_0^{\min(t/2,T_m)}\left|W(X(s))-W'(X(s))\right| ds\right] \\
&\qquad \times \sup_{z\in K_m}\sup_{t/2\leq s\leq t}\exp\left(-s\left(K_0+W\right)_{\mathrm{int}(K_m)}\right)(z,y) \\
&\quad + \mathbb{E}_y\left[\exp\left(\int_0^{\min(t/2,T_m)} V_-(X(u))du\right)\int_0^{\min(t/2,T_m)}\left|W(X(s))-W'(X(s))\right| ds\right] \\
&\qquad \times \sup_{z\in K_m}\sup_{t/2\leq s\leq t}\exp\left(-s\left(K_0+W'\right)_{\mathrm{int}(K_m)}\right)(x,z)
\end{aligned}$$

(Cauchy-Schwarz)

$$\begin{aligned}
&\leq \left(\mathbb{E}_x\left[\exp\left(2\int_0^{\min(t/2,T_m)}\left(V_-1_{K_m}\right)(X(u))du\right)\right]\right)^{1/2} \\
&\qquad \times \left(\mathbb{E}_x\left[\left(\int_0^{\min(t/2,T_m)}\left|W(X(s))-W'(X(s))\right| ds\right)^2\right]\right)^{1/2} \\
&\qquad \times \sup_{z\in K_m}\sup_{t/2\leq s\leq t}\exp\left(-s\left(K_0+W\right)_{\mathrm{int}(K_m)}\right)(z,y)
\end{aligned}$$

$$+\left(\mathbb{E}_y\left[\exp\left(2\int_0^{\min(t/2,T_m)}(V_-1_{K_m})(X(u))du\right)\right]\right)^{1/2}$$
$$\times\left(\mathbb{E}_y\left[\left(\int_0^{\min(t/2,T_m)}|W(X(s))-W'(X(s))|\,ds\right)^2\right]\right)^{1/2}$$
$$\times\sup_{z\in K_m}\sup_{t/2\le s\le t}\exp\left(-s\left(K_0+W'\right)_{\operatorname{int}(K_m)}\right)(x,z). \tag{3.12}$$

Fix $\epsilon>0$ and define ϵ' by the equality:

$$2\epsilon'\left\|\exp\left(-t/2\left(K_0-2V_-1_{K_m}\right)1_{K_m}\right)\right\|_\infty\sup_{t/2\le s\le t}\sup_{x,y\in K_m}p_0\left(s/2,x,y\right)=\epsilon. \tag{3.13}$$

Next we notice the (in-)equalities (for $y\in K_m$):

$$\left(\mathbb{E}_y\left[\left(\int_0^{\min(t/2,T_m)}|W(X(s))-W'(X(s))|\,ds\right)^2\right]\right)^{1/2}$$
$$=2^{1/2}\left(\mathbb{E}_y\left[\int_0^{\min(t/2,T_m)}dt_2\int_0^{t_2}dt_1\left|W\left(X(t_1)\right)-W'\left(X(t_1)\right)\right|\right.\right.$$
$$\left.\left.\times\left|W\left(X(t_2)\right)-W'\left(X(t_2)\right)\right|\right]\right)^{1/2}$$

(Markov property)

$$=2^{1/2}\left(\mathbb{E}_y\left[\int_0^{\min(t/2,T_m)}dt_2\int_0^{t_2}dt_1\left|W\left(X(t_1)\right)-W'\left(X(t_1)\right)\right|\right.\right.$$
$$\left.\left.\times\,\mathbb{E}_{X(t_1)}\left|W\left(X(t_2-t_1)\right)-W'\left(X(t_2-t_1)\right)\right|\right]\right)^{1/2}$$

(Fubini)

$$\le 2^{1/2}\left(\int_0^{t/2}dt_1\int p_{0,m}\left(t_1,y,z\right)|W(z)-W'(z)|\,dz\right)^{1/2}$$
$$\times\sup_{w\in K_m}\left(\int_0^{t/2}dt_1\int p_{0,m}\left(t_1,w,z\right)|W(z)-W'(z)|\right)^{1/2}$$
$$\le 2^{1/2}\sup_{w\in K_m}\int_0^{t/2}ds\int dzp_0(s,w,z)\,|W(z)-W'(z)|$$
$$\le 2^{1/2}\sup_{w\in K_m}\int_0^{\delta}ds\int dzp_0(s,w,z)\,|W(z)-W'(z)|$$

$$+ 2^{1/2} \sup_{w\in K_m} \int dz_1 p_0 (\delta, w, z_1) \int_0^{t/2-\delta} ds \int dz_2 p_0(s, z_1, z_2) \left|W(z_2) - W'(z_2)\right|$$

$$\leq 2\sqrt{2} \sup_{w\in K_m} \int_0^\delta ds \int dz p_0(s, w, z) \left|V(z) 1_{K_m}(z)\right|$$

$$+ \sqrt{2} \sup_{w\in K_m} \int dz_1 p_0 (\delta, w, z_1) \int_0^{t/2-\delta} ds \int dz_2 p_0(s, z_1, z_2) \left|W(z_2) - W'(z_2)\right|$$

$$= 2\sqrt{2} \sup_{w\in K_m} \int_0^\delta ds \int dz p_0(s, w, z) \left|V(z) 1_{K_m}(z)\right|$$

$$+ \sqrt{2} \sup_{w\in K_m} \left[\exp(-\delta K_0) \int_0^{t/2-\delta} ds \int dz p_0(s, \cdot, z_2) \left|W(z) - W'(z)\right|\right](w)$$

$$\leq \epsilon'/2 + \epsilon'/2 = \epsilon'. \tag{3.14}$$

Here $0 < \delta < t/2$ is chosen in such a way that

$$2\sqrt{2} \sup_{w\in K_m} \int_0^\delta ds \int dz p_0(s, w, z) \left|V 1_{K_m}\right|(z) \leq \frac{1}{2}\epsilon'. \tag{3.15}$$

Then, for such a $\delta > 0$ fixed, choose $k(\epsilon')$ and $\ell(\epsilon')$ in such a way that

$$\sqrt{2} \sup_{w\in K_m} \left[\exp(-\delta K_0) \int_0^{t/2} ds p_0 (s, \cdot, z) \left|V_{k,\ell}(z) - V_{k',\ell'}(z)\right| dz\right](w) \leq \frac{1}{2}\epsilon' \tag{3.16}$$

for k, $k' \geq k(\epsilon')$ and ℓ, $\ell' \geq \ell(\epsilon')$. The latter choices are possible, because the function $w \mapsto \int_0^{t/2-\delta} ds \int p_0(s, w, z) \left|W(z) - W'(z)\right| dz$ is bounded and Borel measurable and the operator $\exp(-\delta K_0)$ maps bounded Borel measurable functions on bounded continuous functions. So Dini's lemma (see Lemma E.1 in Appendix E) is available. Consequently, for x, $y \in K_m$, and for k, $k' \geq k(\epsilon')$, ℓ, $\ell' \geq \ell(\epsilon')$ we obtain

$$\left|\exp\left(-t\left(K_0 + V_{k',\ell'}\right)_{\mathrm{int}(K_m)}\right)(x, y) - \exp\left(-t\left(K_0 + V_{k,\ell}\right)_{\mathrm{int}(K_m)}\right)(x, y)\right|$$

$$\leq \epsilon' \left\|\exp\left(-t/2\left(K_0 - 2V_- 1_{K_m}\right)\right) 1_{K_m}\right\|_\infty^{1/2}$$

$$\times \sup_{z\in K_m} \sup_{t/2\leq s\leq t} \exp\left(-s\left(K_0 + V_{k,\ell} 1_{K_m}\right)_{\mathrm{int}(K_m)}\right)(z, y)$$

$$+ \epsilon' \left\|\exp\left(-t/2\left(K_0 - 2V_- 1_{K_m}\right)\right) 1_{K_m}\right\|_\infty^{1/2}$$

$$\times \sup_{z\in K_m} \sup_{t/2\leq s\leq t} \exp\left(-s\left(K_0 + V_{k',\ell'} 1_{K_m}\right)_{\mathrm{int}(K_m)}\right)(x, z)$$

$$\leq 2\epsilon' \left\|\exp\left(-t/2\left(K_0 - 2V_- 1_{K_m}\right)\right) 1_{K_m}\right\|_\infty^{1/2}$$

$$\times \sup_{z_1, z_2\in K_m} \sup_{t/2\leq s\leq t} \exp\left(-s\left(K_0 - V_-\right)_{\mathrm{int}(K_m)}\right)(z_1, z_2)$$

$$= 2\epsilon' \left\|\exp\left(-t/2\left(K_0 - 2V_- 1_{K_m}\right)\right) 1_{K_m}\right\|_\infty^{1/2} \times \sup_{t/2\le s\le t} \left\|1_{K_m} \exp\left(-s\left(K_0 - V_-\right)_{\mathrm{int}(K_m)}\right) 1_{K_m}\right\|_{\infty,1}. \tag{3.17}$$

We also have

$$\begin{aligned}
&\left\|1_{K_m} \exp\left(-s\left(K_0 - V_-\right)_{\mathrm{int}(K_m)}\right) 1_{K_m}\right\|_{\infty,1} \\
&\le \left\|1_{K_m} \exp\left(-\frac{s}{2}\left(K_0 - V_-\right)_{\mathrm{int}(K_m)}\right\|_{\infty,2} \left\|\exp\left(-\frac{s}{2}\left(K_0 - V_-\right)_{\mathrm{int}(K_m)}\right)\right) 1_{K_m}\right\|_{2,1} \\
&= \left\|1_{K_m} \exp\left(-\frac{s}{2}\left(K_0 - V_-\right)_{\mathrm{int}(K_m)}\right)\right\|_{\infty,2}^2.
\end{aligned} \tag{3.18}$$

For f a function in $L^2(E,m)$, with $\|f\|_2 = 1$, and for $x \in K_m$ we get

$$\begin{aligned}
&\left|\exp\left(-\frac{s}{2}\left(K_0 - V_-\right)_{\mathrm{int}(K_m)}\right) f(x)\right|^2 \\
&= \left|\mathbb{E}_x\left[\exp\left(\int_0^{s/2} V_-(X(u))du\right) f\left(X(s/2)\right) : T_m > \frac{s}{2}\right]\right|^2 \\
&\le \mathbb{E}_x\left[\exp\left(2\int_0^{s/2} V_-(X(u))du\right) : T_m > \frac{s}{2}\right] \mathbb{E}_x\left[\left|f\left(X(s/2)\right)\right|^2 : T_m > \frac{s}{2}\right) \\
&= \mathbb{E}_x\left(\exp\left(2\int_0^{s/2} V_-(X(u))du\right) : T_m > \frac{s}{2}\right] \int dy\, p_{0,m}\left(\frac{s}{2}, x, y\right) |f(y)|^2 \\
&\le \sup_{x,y\in K_m} p_{0,m}\left(\frac{s}{2}, x, y\right) \\
&\quad \times \sup_{x\in K_m} \mathbb{E}_x\left[\exp\left(2\int_0^{s/2} V_-(X(u)) 1_{K_m}(X(u))du\right) : T_m > \frac{s}{2}\right].
\end{aligned} \tag{3.19}$$

Hence, for x, $y \in K_m$, and for k, $k' \ge k(\epsilon')$, ℓ, $\ell' \ge \ell(\epsilon')$ we obtain

$$\begin{aligned}
&\left|\exp\left(-t\left(K_0 + V_{k',\ell'}\right)_{\mathrm{int}(K_m)}\right)(x,y) - \exp\left(-t\left(K_0 + V_{k,\ell}\right)_{\mathrm{int}(K_m)}\right)(x,y)\right| \\
&\le 2\epsilon' \left\|\exp\left(-\frac{t}{2}\left(K_0 - 2V_- 1_{K_m}\right)\right) 1_{K_m}\right\|_\infty \sup_{t/2\le s\le t} \sup_{x,y\in K_m} p_0\left(\frac{s}{2}, x, y\right) = \epsilon.
\end{aligned}$$

It follows that, uniformly on $K_m \times K_m$,

$$\lim_{k\to\infty} \lim_{\ell\to\infty} \exp\left(-t\left(K_0 + V_{k,\ell}\right)_{\mathrm{int}(K_m)}\right)(x,y) = \exp\left(-t\left(K_0 + V\right)_{\mathrm{int}(K_m)}\right)(x,y).$$

The continuity of the functions $\exp\left(-t\left(K_0 + V_{k,\ell}\right)_{\mathrm{int}(K_m)}\right)(x,y)$ on $(0,\infty) \times \mathrm{int}(K_m) \times \mathrm{int}(K_m)$ is inherited by $\exp\left(-t\left(K_0 + V\right)_{\mathrm{int}(K_m)}\right)(x,y)$.

Next we want to let m tend to infinity. Therefore we consider for $m' > m$:

$$\exp\left(-t\left(K_0+V\right)_{\mathrm{int}(K_{m'})}\right)(x,y) - \exp\left(-t\left(K_0+V\right)_{\mathrm{int}(K_m)}\right)(x,y)$$

$$= \lim_{s\uparrow t}\mathbb{E}_x\Big[\exp\left(-\int_0^s V(X(u))du\right)\left(p_{0,m'}\left(t-s,X(s),y\right)1_{\{T_{m'}>s\}}\right.$$
$$\left.-p_{0,m}\left(t-s,X(s),y\right)1_{\{T_m>s\}}\right)\Big]$$

(notice that $p_{0,m}(u,z,y) = p_0(u,z,y) - \mathbb{E}_z\left[p_0\left(u-T_m,X(T_m),y\right) : T_m < u\right]$)

$$= \lim_{s\uparrow t}\mathbb{E}_x\Big[\exp\left(-\int_0^s V(X(u))du\right)\left\{p_0(t-s,X(s),y)1_{\{T_{m'}>s\geq T_m\}}\right.$$
$$+\mathbb{E}_{X(s)}\left(p_0\left(t-s-T_m,X(T_m),y\right) : T_m < t-s\right)1_{\{T_m>s\}}$$
$$\left.-\mathbb{E}_{X(s)}\left(p_0\left(t-s-T_{m'},X(T_{m'}),y\right) : T_{m'} < t-s\right)1_{\{T_{m'}>s\}}\right\}\Big]$$

(Markov property together with the identity $s + T_m \circ \vartheta_s = T_m$ on the event) $\{T_m > s\}$)

$$= \lim_{s\uparrow t}\mathbb{E}_x\Big[\exp\left(-\int_0^s V(X(u))du\right)\left\{p_0(t-s,X(s),y)1_{\{T_{m'}>s\geq T_m\}}\right.$$
$$\left.+p_0\left(t-T_m,X(T_m),y\right)1_{\{s<T_m<t\}} - p_0\left(t-T_{m'},X(T_{m'}),y\right)1_{s<\{T_{m'}<t\}}\right\}\Big]$$

$$= \lim_{s\uparrow t}\mathbb{E}_x\Big[\exp\left(-\int_0^s V(X(u))du\right)\Big\{p_0(t-s,X(s),y)1_{\{T_{m'}>s\geq T_m\}}$$
$$+ p_0\left(t-T_m,X(T_m),y\right)1_{\{T_m<t\}} + p_0\left(t-T_{m'},X(T_{m'}),y\right)1_{\{T_{m'}\leq s\}}$$
$$- p_0\left(t-T_{m'},X(T_{m'}),y\right)1_{\{T_{m'}<t\}} - p_0\left(t-T_m,X(T_m),y\right)1_{\{T_m\leq s\}}\Big\}\Big]$$

(martingale property of the process $s \mapsto p_0(t-s,X(s),y)$: see Example 4 in Appendix C.4)

$$= \lim_{s\uparrow t}\mathbb{E}_x\left[\exp\left(-\int_0^s V(X(u))du\right)p_0\left(t-T_m,X(T_m),y\right) : T_m < t\right]$$
$$- \lim_{s\uparrow t}\mathbb{E}_x\left[\exp\left(-\int_0^s V(X(u))du\right)p_0\left(t-T_{m'},X(T_{m'}),y\right) : T_{m'} < t\right]$$
$$+ \lim_{s\uparrow t}\mathbb{E}_x\left[\exp\left(-\int_0^s V(X(u))du\right)p_0(t-s,X(s),y) : s < T_m \leq T_{m'} \leq s\right]$$
$$= \mathbb{E}_x\left[\exp\left(-\int_0^t V(X(u))du\right)p_0\left(t-T_m,X(T_m),y\right) : T_m < t\right]$$
$$- \mathbb{E}_x\left[\exp\left(-\int_0^t V(X(u))du\right)p_0\left(t-T_{m'},X(T_{m'}),y\right) : T_{m'} < t\right]. \tag{3.20}$$

On the other hand we also have

$$\begin{aligned}
&\mathbb{E}_x\Bigg[\exp\left(-\int_0^{T_m} V(X(u))du\right)\\
&\quad\times\exp\Big(-(t-T_m)\left(K_0+V\right)_{\mathrm{int}(K_{m'})}\Big)\left(X(T_m),y\right):T_m<t\Bigg]\\
&=\lim_{s\uparrow t}\mathbb{E}_x\Bigg[\exp\left(-\int_0^{T_m} V(X(u))du\right)\\
&\qquad\times\mathbb{E}_{X(T_m)}\Bigg\{\exp\left(-\int_0^{s-T_m} V(X(u))du\right)p_{0,m'}\left(t-s,X(s-T_m),y\right):\\
&\qquad\qquad T_{m'}>s-T_m\Bigg\}:T_m<s\Bigg]
\end{aligned}$$

(strong Markov property)

$$\begin{aligned}
&=\lim_{s\uparrow t}\mathbb{E}_x\left[\exp\left(-\int_0^{s} V(X(u))du\right)p_{0,m'}\left(t-s,X(s),y\right):T_{m'}>s>T_m\right]\\
&=\lim_{s\uparrow t}\mathbb{E}_x\Bigg[\exp\left(-\int_0^{s} V(X(u))du\right)\Big(p_0\left(t-s,X(s),y\right)\\
&\quad-\mathbb{E}_{X(s)}\left(p_0\left(t-s-T_{m'},X(T_{m'})<y\right):T_{m'}<t-s\right)\Big):T_{m'}>s>T_m\Bigg]
\end{aligned}$$

(Markov property)

$$\begin{aligned}
&=\lim_{s\uparrow t}\mathbb{E}_x\Bigg[\exp\left(-\int_0^{s} V(X(u))du\right)\Big(p_0\left(t-s,X(s),y\right)\\
&\quad-p_0\left(t-T_{m'},X(T_{m'})<y\right)1_{\{T_{m'}<t\}}\Big):T_{m'}>s>T_m\Bigg]\\
&=\lim_{s\uparrow t}\mathbb{E}_x\left[\exp\left(-\int_0^{s} V(X(u))du\right)p_0\left(t-s,X(s),y\right)1_{\{T_{m'}>s>T_m\}}\right]\\
&-\lim_{s\uparrow t}\mathbb{E}_x\left[\exp\left(-\int_0^{s} V(X(u))du\right)p_0\left(t-T_{m'},X(T_{m'}),y\right)1_{\{T_{m'}<t\}}1_{\{T_{m'}>s>T_m\}}\right]\\
&=\lim_{s\uparrow t}\mathbb{E}_x\left[\exp\left(-\int_0^{s} V(X(u))du\right)p_0\left(t-s,X(s),y\right)1_{\{T_{m'}>s>T_m\}}\right]\\
&=\mathbb{E}_x\left[\exp\left(-\int_0^{t} V(X(u))du\right)p_0\left(t-T_m,X(T_m),y\right):T_m<t\right]\\
&\quad-\mathbb{E}_x\left[\exp\left(-\int_0^{t} V(X(u))du\right)p_0\left(t-T_{m'},X(T_{m'}),y\right):T_{m'}<t\right].
\end{aligned}\tag{3.21}$$

Consequently a combination of (3.20) and (3.21) yields the identity:

$$\begin{aligned}
&\exp\left(-t\left(K_0+V\right)_{\operatorname{int}(K_{m'})}\right)(x,y)-\exp\left(-t\left(K_0+V\right)_{\operatorname{int}(K_m)}\right)(x,y) \qquad (3.22)\\
&=\mathbb{E}_x\Bigg[\exp\left(-\int_0^{T_m}V(X(u))du\right)\exp\left(-(t-T_m)\left(K_0+V\right)_{\operatorname{int}(K_{m'})}\right)(X(T_m),y):\\
&\qquad\qquad T_m<t\Bigg].
\end{aligned}$$

From this equality we infer that

$$\begin{aligned}
&\left(\exp\left(-t\left(K_0+V\right)_{\operatorname{int}(K_{m'})}\right)(x,y)-\exp\left(-t\left(K_0+V\right)_{\operatorname{int}(K_m)}\right)(x,y)\right)^2\\
&\le\left(\mathbb{E}_x\left[\exp\left(-\int_0^{T_m}V(X(u))du\right)\exp\left(-\left(t-T_m\right)\left(K_0+V\right)\right)(X(T_m),y):T_m<t\right]\right)^2\\
&\le\mathbb{E}_x\left[\exp\left(-2\int_0^{T_m}V(X(u))du\right)\exp\left(-\left(t-T_m\right)\left(K_0+2V\right)\right)(X(T_m),y):T_m<t\right]\\
&\quad\times\mathbb{E}_x\left[p_0\left(t-T_m,X(T_m),y\right):T_m<t\right]\\
&\quad\text{(the process } \tau\mapsto\exp\left(-2\textstyle\int_0^\tau V(X(u))du\right)\exp\left(-\left(t-\tau\right)\left(K_0+2V\right)\right)(X(\tau),y)\\
&\quad\text{is a martingale on the interval } [0,t)\text{: see Lemma 3.4 below)}\\
&\le\exp\left(-t\left(K_0+2V\right)\right)(x,y)\mu_{0,x}^{t,y}\left(T_m<t\right), \qquad (3.23)
\end{aligned}$$

where the measure $\mu_{0,x}^{t,y}$ is determined by the equality

$$\mu_{0,x}^{t,y}(A)=\mathbb{E}_x\left(p_0(t-s,X(s),y):A\right),\quad A\in\mathcal{F}_s. \qquad (3.24)$$

Here $s<t$ is arbitrary. Since the process $\tau\mapsto p_0(t-\tau,X(\tau),y)$ is martingale on the interval $[0,t)$, the latter equality defines a measure on $\mathcal{F}_{t-}$ indeed. From the symmetry of $p_0(t,x,y)$ it follows that the functions $(x,y)\mapsto\mu_{0,x}^{t,y}\left(T_m<t\right)$ are symmetric as well. For Brownian motion this symmetry is explained Port and Stone [189, Theorem 4.3]. Their proof can be adapted to include our more general situation: see Appendix D. As in the proof of Proposition 3.2 we see

$$\begin{aligned}
&\mu_{0,x}^{t,y}\left(T_m<\frac{t}{2}\right)+\mu_{0,y}^{t,x}\left(T_m<\frac{t}{2}\right)\\
&=\mu_{0,x}^{t,y}\left(T_m<t\right)+\int dz\,\mu_{0,x}^{t/2,z}\left(T_m<\frac{t}{2}\right)\mu_{0,z}^{t/2,y}\left(T_m<\frac{t}{2}\right)\ge\mu_{0,x}^{t,y}\left(T_m<t\right).
\end{aligned}$$

So that

$$\sup_{x,y\in K}\mu_{0,x}^{t,y}\left(T_m<t\right)\le 2\sup_{x,y\in K}\mu_{0,x}^{t,y}\left(T_m<\frac{t}{2}\right)$$

and also (see the proof of Proposition 3.2)

$$\begin{aligned}
&\mu_{0,x}^{t,y}\left(\bigcap_{m=1}^{\infty}\{T_m < t\}\right) = \lim_{m\to\infty}\mu_{0,x}^{t,y}(T_m < t)\\
&\leq \lim_{m\to\infty}\mu_{0,x}^{t,y}\left(T_m < \frac{t}{2}\right) + \lim_{m\to\infty}\mu_{0,y}^{t,x}\left(T_m < \frac{t}{2}\right)\\
&= \mu_{0,x}^{t,y}\left(\bigcap_{m=1}^{\infty}\left\{T_m < \frac{t}{2}\right\}\right) + \mu_{0,y}^{t,x}\left(\bigcap_{m=1}^{\infty}\left\{T_m < \frac{t}{2}\right\}\right)\\
&\leq \mu_{0,x}^{t,y}\left(\zeta \leq \frac{t}{2}\right) + \mu_{0,y}^{t,x}\left(\zeta \leq \frac{t}{2}\right) = 0 + 0 = 0.
\end{aligned}$$

Hence, by Dini's lemma, the convergence in (3.23) is uniform on compact subsets of $(0,\infty)\times E\times E$ and consequently the integral kernels $\exp\left(-t\left(K_0+V\right)\right)(x,y)$ are continuous on $(0,\infty)\times E\times E$.

The fact that the process

$$\tau \mapsto \exp\left(-2\int_0^{\tau} V(X(u))du\right)\exp\left(-\left(t-\tau\right)\left(K_0+2V\right)\right)\left(X(\tau),y\right)$$

is a martingale on the interval $[0,t)$ is a consequence of the Chapman-Kolmogorov identity, a proof of which follows: compare with Proposition 2.20.

3.4. LEMMA. *The process $s \mapsto M_V^{t,y}(s)$ is a $\mathbb{P}_x$-martingale on the interval $[0,t)$.*

Proof. Let $0 < \tau_1 < \tau_2 < t$ and consider

$$\begin{aligned}
&\mathbb{E}_x\left[\exp\left(-\int_0^{\tau_2} V(X(u))du\right)\exp\left(-\left(t-\tau_2\right)\left(K_0+V\right)\right)\left(X(\tau_2),y\right) \;\middle|\; \mathcal{F}_{\tau_1}\right]\\
&= \exp\left(-\int_0^{\tau_1} V(X(u))du\right)\mathbb{E}_x\left[\exp\left(-\int_0^{\tau_2-\tau_1} V(X(u+\tau_1))du\right)\right.\\
&\qquad \left.\times \exp\left(-\left(t-\tau_2\right)\left(K_0+V\right)\right)\left(X(\tau_2-\tau_1+\tau_1),y\right) \;\middle|\; \mathcal{F}_{\tau_1}\right]
\end{aligned}$$

(Markov property)

$$\begin{aligned}
&= \exp\left(-\int_0^{\tau_1} V(X(u))du\right)\\
&\quad \times \mathbb{E}_{X(\tau_1)}\left[\exp\left(-\int_0^{\tau_2-\tau_1} V(X(u))du\right)\exp\left(-\left(t-\tau_2\right)\left(K_0+V\right)\right)\left(X(\tau_2-\tau_1),y\right)\right]\\
&= \exp\left(-\int_0^{\tau_1} V(X(u))du\right)\\
&\quad \times \int dz \exp\left(-\left(\tau_2-\tau_1\right)\left(K_0+V\right)\right)\left(X(\tau_1),z\right)\exp\left(-\left(t-\tau_2\right)\left(K_0+V\right)\right)(z,y)
\end{aligned}$$

(Chapman-Kolmogorov)

$$= \exp\left(-\int_0^{\tau_1} V(X(u))du\right) \exp\left(-\left(t-\tau_1\right)\left(K_0+V\right)\right)\left(X(\tau_1),y\right).$$

This proves the announced martingale property.

Proof of Theorem 2.5. Next we prove the assertions in Theorem 2.5.
(a) In order to prove (a) we consider the semigroup

$$\left\{\exp\left(-t\left(K_0\dot{+}V\right)\right) : t \geq 0\right\}$$

defined by the Feynman-Kac formula:

$$\left[\exp\left(-t\left(K_0\dot{+}V\right)\right)f\right](x) = \mathbb{E}_x\left(\exp\left(-\int_0^t V(X(u))du\right)f(X(t))\right), \tag{3.25}$$

for f belonging to $C_\infty(E)$ and for $t \geq 0$. From the discussions above it follows that this Feynman-Kac semigroup can be represented in the form as described in assertion (a) of Theorem 2.5. The Chapman-Kolmogorov identity is satisfied as well. We still have to prove that the Feynman-Kac semigroup leaves the space $C_\infty(E)$ invariant. From Khas'minskii's lemma (see Corollary 3.6) together with the continuity of the integral kernel $\exp\left(-t(K_0+V)\right)(x,y)$ we infer that functions of the form $x \mapsto \left[\exp\left(-t\left(K_0\dot{+}V\right)\right)f\right](x)$, where f is any bounded Borel measurable function, are continuous and bounded. Now let f be a function belonging to $C_\infty(E)$. Then the function $x \mapsto \left[\exp\left(-t\left(K_0\dot{+}V\right)\right)f\right](x)$ is continuous. It also vanishes at infinity ($=\triangle$), because we have the following estimate:

$$\begin{aligned} &\left|\left[\exp\left(-t\left(K_0\dot{+}V\right)\right)f\right](x)\right|^2 \leq \left[\exp\left(-t\left(K_0\dot{+}2V\right)\right)1\right](x)\int p_0(t,x,y)\left|f(y)\right|^2 dy \\ &\leq M_{2V}\exp\left(b_{2V}t\right)\int p_0(t,x,y)\left|f(y)\right|^2 dy. \end{aligned} \tag{3.26}$$

Here the constants M_{2V} and b_{2V} are chosen in such a way that

$$\exp\left(-t\left(K_0\dot{+}2V\right)\right)1 \leq M_{2V}\exp\left(b_{2V}t\right). \tag{3.27}$$

The function f belongs to $C_\infty(E)$, and hence the function $|f|^2$ belongs to $C_\infty(E)$ as well. Since the semigroup $\{\exp\left(-tK_0\right) : t \geq 0\}$ leaves the space $C_\infty(E)$ invariant, we conclude that the right-hand side of (3.26) converges to zero, as x tends to infinity.
(b) The Feynman-Kac representation of the semigroup

$$\left\{\exp\left(-t\left(K_0\dot{+}V\right)\right) : t \geq 0\right\}$$

follows from the discussion prior to and in (a). In order to prove the kernel representation we employ the measure $\mu_{0,x}^{W,t,y}$ on the σ-field $\mathcal{F}_{t-}$, which is defined as follows (cfr. Definition 2.18.(d)):

$$\mu_{0,x}^{W,t,y}(A) = \mathbb{E}_x\left(M_W^{t,y}(s)1_A\right). \tag{3.28}$$

For more details see the items (c) and (d) of Definition 2.18. The event A belongs to $\mathcal{F}_s$, $s < t$, and the process $M_W^{t,y}(s)$ is the martingale given by the expression:

$$M_W^{t,y}(s) = \exp\left(-\int_0^s W(X(u))du\right)\exp\left(-(t-s)\left(K_0+W\right)\right)(X(s),y),$$

where W is some Kato-Feller function. As in the case $W \equiv 0$, we have

$$\begin{aligned}&\mu_{0,x}^{W,t,y}\left(T_m < \frac{t}{2}\right) + \mu_{0,y}^{W,t,x}\left(T_m < \frac{t}{2}\right)\\ &= \mu_{0,x}^{W,t,y}\left(T_m < t\right) + \int \mu_{0,x}^{W,t/2,z}\left(T_m < \frac{t}{2}\right)\mu_{0,z}^{W,t/2,y}\left(T_m < \frac{t}{2}\right)dz\\ &\geq \mu_{0,x}^{W,t,y}\left(T_m < t\right). \end{aligned} \tag{3.29}$$

We note the following inequalities:

$$\begin{aligned}0 &\leq \int \exp\left(-\int_0^t V(X(u))du\right)d\mu_{0,x}^{t,y} - \int_{\{T_m\geq t\}} \exp\left(-\int_0^t V(X(u))du\right)d\mu_{0,x}^{t,y}\\ &\leq \mu_{0,x}^{-V_-,t,y}\left\{T_m < t\right\}\\ &\leq \mu_{0,x}^{-V_-,t,y}\left\{T_m < \frac{t}{2}\right\} + \mu_{0,y}^{-V_-,t,x}\left\{T_m < \frac{t}{2}\right\}.\end{aligned}$$

Likewise, as in the case $V \equiv 0$, we infer

$$\int \exp\left(-\int_0^t V(X(u))du\right)d\mu_{0,x}^{t,y} = \lim_{m\to\infty}\int_{\{T_m\geq t\}} \exp\left(-\int_0^t V(X(u))du\right)d\mu_{0,x}^{t,y}. \tag{3.30}$$

This proves the integral representation in (b).

(c) The easiest way to establish (c) is to invoke the Riesz-Thorin interpolation theorem (see Appendix A.40). With $t = 1 - \dfrac{1}{p}$ we have

$$\frac{1}{p_t} = \frac{1-t}{p_0} + \frac{t}{p_1}, \quad \frac{1}{q_t} = \frac{1-t}{q_0} + \frac{t}{q_1},$$

where $q_0 = p_0 = 1$, $q_1 = p_1 = \infty$. Put $T = \exp\left(-t\left(K_0\dot{+}V\right)\right)$. Then T is an operator in $L^\infty(E,m)$ and by symmetry it also acts as a linear operator in $L^1(E,m)$. Moreover, we have $\|T\|_{1,1} = \|T\|_{\infty,\infty}$. Since $L^p(E,m)$ is a vector subspace of $L^1(E,m) + L^\infty(E,m)$, it also acts in $L^p(E,m)$. In addition, we have

$$\|T\|_{q_t,p_t} \leq \|T\|_{q_0,p_0}^{1-t}\|T\|_{q_1,p_1}^{t} \tag{3.31}$$

and hence

$$\begin{aligned}\left\|\exp\left(-t\left(K_0\dot{+}V\right)\right)\right\|_{p,p} &\leq \left\|\exp\left(-t\left(K_0\dot{+}V\right)\right)\right\|_{1,1}^{1/p}\left\|\exp\left(-t\left(K_0\dot{+}V\right)\right)\right\|_{\infty,\infty}^{1-1/p}\\ &= \left\|\exp\left(-t\left(K_0\dot{+}V\right)\right)\right\|_{\infty,\infty}. \end{aligned} \tag{3.32}$$

Next we want to establish the strong continuity of the Feynman-Kac semigroup for $1 \leq p < \infty$, in the sense that

$$\lim_{t\downarrow 0}\left\|\exp\left(-t\left(K_0\dot{+}V\right)\right)f - f\right\|_p = 0, \tag{3.33}$$

for $f \in L^p(E,m)$. From a general argument in operator semigroup theory, it suffices to prove that the Feynman-Kac semigroup is *weakly continuous* in the sense that

$$\lim_{t\downarrow 0}\int dx\left[\exp\left(-t\left(K_0\dot{+}V\right)\right)f\right](x)g(x) = \int dx f(x)g(x) \tag{3.34}$$

for all functions $f \in L^p(E,m)$ and for all functions g belonging to $L^q(E,m)$, where $1 = \dfrac{1}{p} + \dfrac{1}{q}$. For example, the reader may consult Yosida [255, p. 233]. Since on compact time intervals the Feynman-Kac semigroup is equi-continuous, we conclude that it suffices to show (3.34) for functions f and g that are continuous and that have compact support. However for such functions the equality in (3.34) is a consequence of right continuity of paths together with Lebesgue dominated convergence theorem.

(d) We apply the Riesz-Thorin interpolation theorem (Appendix A.40) twice. Put $r = \dfrac{p(q-1)}{q-p}$ and put $t = 1 - \dfrac{1}{q}$. Then

$$\frac{1}{p} = \frac{1-t}{1} + \frac{t}{r} \quad\text{and}\quad \frac{1}{q} = \frac{1-t}{1} + \frac{t}{\infty}. \tag{3.35}$$

With $T = \exp\left(-t\left(K_0\dot{+}V\right)\right)$ we then have:

$$\begin{aligned}\|T\|_{q,p} &\leq \|T\|_{1,1}^{1/q}\|T\|_{\infty,r}^{1-1/q} \leq \|T\|_{1,1}^{1/q}\left(\|T\|_{\infty,1}^{1/r}\|T\|_{\infty,\infty}^{1-1/r}\right)^{1-1/q}\\ &= \|T\|_{1,1}^{1/q}\|T\|_{\infty,1}^{1/p-1/q}\|T\|_{\infty,\infty}^{1-1/p}. \end{aligned} \tag{3.36}$$

Next we estimate $\|T\|_{\infty,1}$. This is achieved in the following fashion:

$$\begin{aligned}\|T\|_{\infty,1} &= \left\|\left(\exp\left(-\frac{t}{2}\left(K_0\dot{+}V\right)\right)\right)^2\right\|_{\infty,1}\\ &\leq \left\|\exp\left(-\frac{t}{2}\left(K_0\dot{+}V\right)\right)\right\|_{\infty,2}\left\|\exp\left(-\frac{t}{2}\left(K_0\dot{+}V\right)\right)\right\|_{2,1}\end{aligned}$$

(symmetry)

$$= \left\| \exp\left(-\frac{t}{2} \left(K_0 \dot{+} V \right) \right) \right\|_{\infty,2}^{2}. \tag{3.37}$$

Next let f be a function in $L^2(E, m)$ and consider

$$\left| \left[\exp\left(-\frac{t}{2} \left(K_0 \dot{+} V \right) \right) f \right] (x) \right| = \left| \mathbb{E}_x \left(\exp\left(-\int_0^{t/2} V(X(u)) du \right) f(X(t/2)) \right) \right|$$

$$\leq \left(\mathbb{E}_x \left(-2 \int_0^{t/2} V(X(u)) du \right) \right)^{1/2} \left(\mathbb{E}_x \left(\left| f(X(t/2)) \right|^2 \right) \right)^{1/2}$$

$$= \left\| \exp\left(-\frac{t}{2} \left(K_0 \dot{+} 2V \right) \right) \right\|_{\infty,\infty}^{1/2} \left\| \exp\left(-\frac{t}{2} K_0 \right) \right\|_{\infty,1}^{1/2} \|f\|_2 . \tag{3.38}$$

Inserting the result of (3.38) into (3.37) yields

$$\|T\|_{\infty,1} \leq \left\| \exp\left(-\frac{t}{2} \left(K_0 \dot{+} 2V \right) \right) \right\|_{\infty,\infty} \left\| \exp\left(-\frac{t}{2} K_0 \right) \right\|_{\infty,1} . \tag{3.39}$$

From (3.36) and (3.39) we obtain:

$$\left\| \exp\left(-t \left(K_0 \dot{+} V \right) \right) \right\|_{q,p} \leq \left\| \exp\left(-t \left(K_0 \dot{+} V \right) \right) \right\|_{\infty,\infty}^{1-(1/p-1/q)}$$

$$\times \left\| \exp\left(-\frac{t}{2} \left(K_0 \dot{+} 2V \right) \right) \right\|_{\infty,\infty}^{1/p-1/q} \left\| \exp\left(-\frac{t}{2} K_0 \right) \right\|_{\infty,1}^{1/p-1/q}$$

$$\leq \left\| \exp\left(-t \left(K_0 - 2V_- \right) \right) 1 \right\|_{\infty} \left(\sup_{x,y \in E} p_0(t/2, x, y) \right)^{1/p-1/q} . \tag{3.40}$$

We still have to prove that, for $f \in L^p(E, m)$, $1 \leq p < \infty$, the function $x \mapsto \left[\exp\left(-t \left(K_0 \dot{+} V \right) \right) f \right](x)$ belongs to $C_\infty(E)$. Fix $\varepsilon > 0$ and choose a function $g \in C_\infty(E)$ with the property that

$$\left\| \exp\left(-t \left(K_0 \dot{+} V \right) \right) \right\|_{\infty,p} \|f - g\|_p < \varepsilon.$$

For $x \in E$ the following inequality is evident:

$$\left| \left[\exp\left(-t \left(K_0 \dot{+} V \right) \right) f \right](x) - \left[\exp\left(-t \left(K_0 \dot{+} V \right) \right) g \right](x) \right|$$

$$\leq \left\| \exp\left(-t \left(K_0 \dot{+} V \right) \right) \right\|_{\infty,p} \|f - g\|_p < \varepsilon.$$

Hence the function $\exp\left(-t \left(K_0 \dot{+} V \right) \right) f$ can be uniformly approximated by continuous functions that vanish at infinity and are of the form $\exp\left(-t \left(K_0 \dot{+} V \right) \right) g$ where g belongs to $C_\infty(E)$. Consequently (d) follows.

(e) This assertion is in fact a part of (d). The symmetry implies the self-adjointness of the operator $\exp\left(-t \left(K_0 \dot{+} V \right) \right)$.

(f) This assertion was proved in Proposition 2.16.

We continue this chapter with a proof of Khas'minskii's lemma.

3.5. PROPOSITION. *Let $W \geq 0$ be a function in $K(E)$. Put $Z(t) = \int_0^t W(X(u))du$ and suppose that, for some strictly positive real constants t_0, a_0, and γ_0, the following inequality is valid:*

$$\sup_{y\in E} \mathbb{P}_y \left(Z(t_0) > a_0\right) \leq \exp\left(-\left(1+\gamma_0\right) a_0\right). \tag{3.41}$$

Then

$$\sup_{x\in E} \mathbb{E}_x \left[\exp\left(\int_0^t W(X(u))du\right)\right] \leq M_{-W} e^{tb_{-W}}, \tag{3.42}$$

where the constants M_{-W} and b_{-W} are given by

$$M_{-W} = e^{a_0}\left(1+\frac{1}{\gamma_0}\right), \tag{3.43}$$

$$b_{-W} = \frac{1}{t_0}\left(a_0 + \log\left(1+\frac{1}{\gamma_0}\right)\right). \tag{3.44}$$

Proof. We shall use a stopping time argument. We write $W(\triangle) = 0$ and we define the sequence of stopping times $(T_k : k \in \mathbb{N})$ as follows: $T_k = \inf\{s>0 : Z(s) > ka_0\}$. (This is called a *stochastic time change*). Then, $\mathbb{P}_x$-almost surely, $T_k = T_{k-1} + T_1 \circ \vartheta_{T_{k-1}}$. For $0 \leq t \leq t_0$ we see for $(k+1)a_0 > \xi \geq ka_0$

$$\begin{aligned}
&\mathbb{P}_x\left(Z(t) > \xi\right) \leq \mathbb{P}_x\left(Z(t_0) > ka_0\right) = \mathbb{P}_x\left(T_k < t_0\right) \\
&= \mathbb{P}_x\left(T_{k-1} + T_1 \circ \vartheta_{T_{k-1}} < t_0\right) = \mathbb{E}_x\left[\mathbb{E}_{X(T_{k-1})}\left(T_1 < t_0 - T_{k-1}\right) : T_{k-1} < t_0\right] \\
&\leq \sup_{y\in E} \mathbb{P}_y\left(T_1 < t_0\right) \mathbb{P}_x\left(T_{k-1} < t_0\right) \leq \sup_{y\in E} \mathbb{P}_y\left(T_1 < t_0\right)^k \\
&= \sup_{y\in E} \mathbb{P}_y\left(Z(t_0) > a_0\right)^k \leq \exp\left(-\left(1+\gamma_0\right) ka_0\right) \\
&\leq \exp\left(-\left(1+\gamma_0\right) \max\left(\xi - a_0, 0\right)\right).
\end{aligned} \tag{3.45}$$

From (3.45) it follows that

$$\mathbb{E}_x\left[\exp\left(Z(t)\right)\right] = 1 + \int_0^\infty \mathbb{P}_x\left(Z(t) > \xi\right) e^\xi d\xi \leq e^{a_0}\left(1+\frac{1}{\gamma_0}\right) \tag{3.46}$$

for $0 \leq t \leq t_0$.

Next let $t > 0$ be arbitrary. Then we write $t = t_1 + \cdots + t_n$, where $t_j = jt_0$, $1 \leq j \leq n-1$, $t_n = t - t_{n-1}$, and where $(n-1)t_0 < t \leq nt_0$. From the additivity of the process $Z(t)$, it follows that $(\tau_{n-1} = t_1 + \cdots + t_{n-1})$

$$\begin{aligned}
\mathbb{E}_x\left[\exp\left(Z(t)\right)\right] &= \mathbb{E}_x\left[\exp\left(Z\left(\tau_{n-1} + Z(t_n) \circ \vartheta_{\tau_{n-1}}\right)\right)\right] \\
\text{(Markov property)} \quad &= \mathbb{E}_x\left[\exp\left(Z\left(\tau_{n-1}\right)\right) \mathbb{E}_{X(\tau_{n-1})}\left\{\exp\left(Z(t_n)\right)\right\}\right] \\
&\leq \mathbb{E}_x\left[\exp\left(Z\left(\tau_{n-1}\right)\right)\right] \sup_{y\in E} \mathbb{E}_y\left[\exp\left(Z(t_n)\right)\right] \\
\text{(induction and (3.46))} \quad &\leq \sup_{y\in E}\left(\mathbb{E}_y\left[\exp\left(Z(t_0)\right)\right]\right)^n \\
&\leq e^{a_0}\left(1+\frac{1}{\gamma_0}\right)^{1+t/t_0} = M_{-W} \exp\left(tb_{-W}\right).
\end{aligned} \tag{3.47}$$

From (3.47) the claim in Proposition 3.5 follows.

The result in the following corollary is usually called Khas'minskii's lemma.

3.6. COROLLARY. *Put $Z(t) = \int_0^t W(X(u))du$, where $W \geq 0$ is a Borel measurable function. If $M_0 := \sup_{y\in E} \mathbb{E}_y [Z(t_0)] < 1$, then there exist finite constants M_{-W} and b_{-W} with the following properties:*

$$\mathbb{E}_x [\exp(Z(t))] \leq M_{-W} \exp(tb_{-W}) \tag{3.48}$$

for all $t \geq 0$ and for all $x \in E$.

Proof. Let W and $Z := \{Z(t) : t \geq 0\}$ be as in Corollary 3.6. Fix $t_0 > 0$ and suppose that $M_0 := \sup_{y\in E} \mathbb{E}_y [Z(t_0)] < 1$. Then choose $\gamma_0 > 0$ such that $(1+\gamma_0) M_0 < 1$ and fix $n \in \mathbb{N}$ such that $n > (1+\gamma_0) M_0 e (n!)^{1/n}$. Put $a_n = eM_0 (n!)^{1/n}$. Then we have the following inequalities:

$$\mathbb{P}_y (Z(t_0) > a_n) \leq \frac{n!}{a_n^n} \mathbb{E}_y \left[\int \!\!\! \int_{0<u_1<\cdots<u_n<t_0} W(X(u_1)) \ldots W(X(u_n)) du_1 \ldots du_n \right]$$

(Markov property)

$$\leq \frac{n!}{a_n^n} \sup_{y\in E} \left(\mathbb{E}_y \left[\int_0^{t_0} W(X(u))du \right] \right)^n = e^{-n} \leq \exp\left(- (1+\gamma_0) eM_0 (n!)^{1/n} \right).$$

For $n_0 = 1$, the following corollary coincides with the previous one. It is often used with $W = V_-$.

3.7. COROLLARY. *Let $Z(t)$ and W be as in Corollary 3.6. If for some $n_0 \in \mathbb{N}$, $n_0 > 0$, and for some $t_0 > 0$,*

$$\sup_{x\in E} \mathbb{E}_x [(Z(t_0))^{n_0}] < n_0!,$$

then there exist finite constants M_{-W} and b_{-W} such that, for all $t > 0$ and for all $x \in E$,

$$\mathbb{E}_x [\exp(Z(t))] \leq M_{-W} \exp(tb_{-W}).$$

Proof. Put $M_0 = \sup_{x\in E} \left(\mathbb{E}_x \left[(Z(t_0))^{n_0} \right] \right)^{1/n_0}$ and choose $\gamma_0 > 0$ such that $(n_0!)^{1/n_0} > (1+\gamma_0) M_0$. Then we choose $k \in \mathbb{N}$ so large that

$$kn_0 \geq (1+\gamma_0) ((kn_0)!)^{1/(kn_0)} \frac{eM_0}{(n_0!)^{1/n_0}} =: (1+\gamma_0) a_{kn_0}.$$

From the Markov property we infer that

$$
\begin{aligned}
&\mathbb{E}_x\left[\left(\int_0^{t_0} W(X(u))du\right)^{kn_0}\right] \\
&= (kn_0)!\,\mathbb{E}_x\left[\int_0^{t_0} W(X(u_1))\int_{u_1}^{t_0} W(X(u_2))\dots\int_{u_{(k-1)n_0-1}}^{t_0} W(X(u_{(k-1)n_0}))\right. \\
&\int_{u_{(k-1)n_0}}^{t_0} W\left(X(u_{(k-1)n_0+1})\right)\dots \\
&\left.\int_{u_{kn_0-1}}^{t_0} W\left(X(u_{kn_0})\right)du_{kn_0}\dots du_{(k-1)n_0+1}\,du_{(k-1)n_0}\dots du_2du_1\right] \\
&= \frac{(kn_0)!}{n_0!}\mathbb{E}_x\left[\int_0^{t_0} W(X(u_1))\int_{u_1}^{t_0} W(X(u_2))\dots W\left(X(u_{(k-1)n_0})\right)\right. \\
&\left.\left(\int_{u_{(k-1)n_0}}^{t_0} W\left(X(u)\right)du\right)^{n_0} du_{(k-1)n_0}\dots du_1\right] \\
&= \frac{(kn_0)!}{n_0!}\mathbb{E}_x\left[\int_0^{t_0} W(X(u_1))\int_{u_1}^{t_0} W(X(u_2))\dots\int_{u_{(k-1)n_0-1}}^{t_0} W\left(X(u_{(k-1)n_0})\right)\right. \\
&\left.\mathbb{E}_{X\left(u_{(k-1)n_0}\right)}\left\{\left(\int_0^{t_0-u_{(k-1)n_0}} W(X(u))du\right)^{n_0}\right\}du_{(k-1)n_0}\dots du_1\right] \\
&\le \frac{(kn_0)!}{n_0!}M_0^{n_0}\mathbb{E}_x\left[\int_0^{t_0} W(X(u_1))\int_{u_1}^{t_0} W(X(u_2))\dots\right. \\
&\left.\int_{u_{(k-1)n_0-1}}^{t_0} W\left(X(u_{(k-1)n_0})\right)du_{(k-1)n_0}\dots du_2du_1\right] \\
&\le \frac{(kn_0)!}{(n_0!)^k}M_0^{kn_0}.
\end{aligned}
$$

Hence

$$
\begin{aligned}
\mathbb{P}_x\left[Z(t_0) > a_{kn_0}\right] &\le \frac{1}{a_{kn_0}^{kn_0}}\mathbb{E}_x\left[\left(Z(t_0)\right)^{kn_0}\right] \le \frac{1}{a_{kn_0}^{kn_0}}\frac{(kn_0)!}{(n_0!)^k}M_0^{kn_0} \\
&= e^{-kn_0} \le \exp\left(-\left(1+\gamma_0\right)a_{kn_0}\right),
\end{aligned}
$$

where γ_0 is strictly positive. Then apply Proposition 3.5.

Chapter 4

Resolvent and Semigroup Differences for Feller Operators: Operator Norms

In section A of this chapter we give some weighted norm estimates for differences of Feynman-Kac semigroups. In section B we shall state and prove some general representations for (differences of) singularly perturbed Feynman-Kac semigroups and resolvent families.

A. Regular Perturbations

4.1. HYPOTHESES. As in Chapter 2 we place ourselves in the setting in which the basic assumptions on stochastic spectral analysis (BASSA) are satisfied. Let K_0 be the generator of the self-adjoint semigroup $\{\exp(-tK_0) : t \geq 0\}$ on $L^2(E, m)$ defined by

$$\exp\left[(-tK_0) f\right](x) = \int p_0(t, x, y) f(y) dm(y). \tag{4.1}$$

Recall that BASSA requires that $p_0(t, x, y)$ is symmetric and continuous on $(0, \infty) \times E \times E$ and that m is some non-negative Borel measure on E, which is bounded on compact subsets. Usually we write dy instead of $dm(y)$. As in Chapter 2 the generator K_0 will be perturbed in two ways. First we will consider a "regular" perturbation, being a multiplication operator V and in section B there will also be a potential barrier on a singularity region Γ. This singularity region Γ will be a closed subset of E.

We begin this chapter with some estimates on the norm of a "sandwiched" semigroup and resolvent difference. These "sandwiched", or more properly, these weighted estimates are important for at least two reasons. The first one being that the trace norm of the product of two operators T and S is dominated by the product of the Hilbert-Schmidt norms of $T\varphi$ and of $\varphi^{-1}S$ or, in a formula, $\|TS\|_{\text{trace}} \leq \|T\varphi\|_{\text{HS}} \left\|\varphi^{-1}S\right\|_{\text{HS}}$. Here φ is a nowhere vanishing Borel function. The second reason is that for certain points λ_0 in the absolutely continuous part of the spectrum of a self-adjoint operator K, the limiting absorption principle is valid. This means that, for certain functions φ and ψ and for a certain interval I containing λ_0 in its interior, the quantity

$$\sup_{\lambda \in \mathbb{C}, \operatorname{Re} \lambda \in I, \operatorname{Im} \lambda \neq 0} \left\|\varphi \left(\lambda I + K\right)^{-1} \psi\right\|$$

is finite. For more details we refer the reader to results based on Mourre estimates as exhibited in Chapter 4 of Cycon et. al. [63], Perry et. al. [188] and Mourre [180].

Closely related results and applications can be found in Agmon [2], Lavine [162], Ben-Artzi and Devinatz [23], Robert and Tamura [196].

It is also observed that among the inequalities proved in this section on regular perturbations, the most interesting are perhaps (4.2), (4.3), and (4.35).

4.2. PROPOSITION. *Let V and W be Kato-Feller potentials and let φ and ψ be non-vanishing Borel measurable functions on E. The following inequalities are valid for any real numbers α and β and $0 \leq \gamma \leq 1$:*

$$\|\varphi\left(\exp\left(-t\left(K_0 \dot{+} V\right)\right) - \exp\left(-t\left(K_0 \dot{+} W\right)\right)\right)\psi\|_{2,2}$$
$$\leq \left\|\int_0^t ds\, |\psi|^{2\beta} \exp\left(-sK_0\right)|V-W|^{2\gamma} \exp\left(-(t-s)\left(K_0 \dot{+} 2V\right)\right)|\varphi|^{2\alpha}\right\|_{\infty}^{1/2}$$
$$\times \left\|\int_0^t ds\, |\varphi|^{2-2\alpha} \exp\left(-sK_0\right)|V-W|^{2-2\gamma} \exp\left(-(t-s)\left(K_0 \dot{+} 2W\right)\right)|\psi|^{2-2\beta}\right\|_{\infty}^{1/2};$$

and (4.2)

$$\left\|\varphi\left(\left(aI + K_0 \dot{+} V\right)^{-1} - \left(aI + K_0 \dot{+} W\right)^{-1}\right)\psi\right\|_{2,2}$$
$$\leq \left\||\psi|^{2\beta}\left(aI + K_0\right)^{-1}|V-W|^{2\gamma}\left(aI + K_0 \dot{+} 2V\right)^{-1}|\varphi|^{2\alpha}\right\|_{\infty}^{1/2}$$
$$\times \left\||\varphi|^{2-2\alpha}\left(aI + K_0\right)^{-1}|V-W|^{2-2\gamma}\left(aI + K_0 \dot{+} 2W\right)^{-1}|\psi|^{2-2\beta}\right\|_{\infty}^{1/2}. \tag{4.3}$$

REMARK. Before proceeding with the proof, let us mention a special case of the above. Choose constants M_V and b_V such that

$$\left\|\exp\left(-t\left(K_0 \dot{+} V\right)\right)1\right\|_{\infty} \leq M_V e^{b_V t}$$

for all $t \geq 0$ and similarly choose M_W and b_W. Suppose that $\psi \equiv \varphi \equiv 1$ in Proposition 4.2. Since

$$\left\|\int_0^t ds \exp\left(-sK_0\right)|V-W| \exp\left(-(t-s)\left(K_0 \dot{+} 2V\right)\right)\right\|_{\infty}$$
$$\leq \left\|\int_0^t ds \exp\left(-sK_0\right)|V-W|\right\|_{\infty} \sup_{0<s<t}\left\|\exp\left(-s\left(K_0 \dot{+} 2V\right)\right)1\right\|_{\infty}$$
$$\leq M_{2V} \sup_{0<s<t} \exp\left(sb_{2V}\right) \int_0^t ds \exp\left(-sK_0\right)\| \, |V-W| \, \|_{\infty},$$

and since a similar inequality is true for $2W$ instead of $2V$, from (4.2) and (4.3) with $\gamma = 1/2$ we get the following:

$$\left\|\left(\exp\left(-t\left(K_0\dot{+}V\right)\right) - \exp\left(-t\left(K_0\dot{+}W\right)\right)\right)\right\|_{2,2} \tag{4.4}$$
$$\leq M_{2V}^{1/2} M_{2W}^{1/2} \sup_{0<s_1,s_2<t/2} \exp\left(s_1 b_{2V} + s_2 b_{2W}\right) \left\|\int_0^t ds \exp\left(-sK_0\right)|V-W|\right\|_\infty$$

and

$$\left\|\left(\left(aI + K_0\dot{+}V\right)^{-1} - \left(aI + K_0\dot{+}W\right)^{-1}\right)\right\|_{2,2} \tag{4.5}$$
$$\leq \left\|(aI+K_0)^{-1}|V-W|\int_0^\infty e^{-as}\exp\left(-s\left(K_0\dot{+}2V\right)\right)1\right\|_\infty^{1/2}$$
$$\times \left\|(aI+K_0)^{-1}|V-W|\int_0^\infty e^{-as}\exp\left(-s\left(K_0\dot{+}2W\right)\right)1\right\|_\infty^{1/2}$$
$$\leq M_{2W}^{1/2} M_{2V}^{1/2} \frac{1}{\sqrt{a-b_{2V}}\sqrt{a-b_{2W}}} \left\|(aI+K_0)^{-1}|V-W|\right\|_\infty .$$

Therefore the $\|\cdot\|_{2,2}$-norm of the operators

$$\exp\left(-t\left(K_0\dot{+}V\right)\right) - \exp\left(-t\left(K_0\dot{+}W\right)\right) \text{ and}$$
$$\left(aI + K_0\dot{+}V\right)^{-1} - \left(aI + K_0\dot{+}W\right)^{-1}$$

can be estimated in terms of the Kato-Feller norm of the function $V - W$: see Definition 2.1 and also Proposition 2.2. In fact we have

$$\left(1-e^{-a}\right)\left\|(aI+K_0)^{-1}|W-V|\right\|_\infty$$
$$\leq \|W-V\|_{\mathrm{KF}} = \sup_{x\in E}\int_0^1 \mathbb{E}_x(|W(X(s)) - V(X(s))|)ds$$
$$\leq e^a\left\|(aI+K_0)^{-1}|g|\right\|_\infty .$$

We continue with a proof of Proposition 4.2.

Proof. Let f and g be functions in $L^2(E,m)$. First we notice the identity:

$$\left\langle\varphi\left(\exp\left(-t\left(K_0\dot{+}V\right)\right) - \exp\left(-t\left(K_0\dot{+}W\right)\right)\right)\psi f, g\right\rangle \tag{4.6}$$
$$= \int_0^t ds\left\langle\varphi\exp\left(-s\left(K_0\dot{+}V\right)\right)(W-V)\exp\left(-(t-s)\left(K_0\dot{+}W\right)\right)\psi f, g\right\rangle .$$

Hence, for any strictly positive Borel measurable function $\chi_s(x)$ on E, to be specified later,

$$\left|\left\langle\varphi\left(\exp\left(-t\left(K_0\dot{+}V\right)\right) - \exp\left(-t\left(K_0\dot{+}W\right)\right)\right)\psi f, g\right\rangle\right|$$
$$\leq \int_0^t ds\left\langle|\varphi|\exp\left(-s\left(K_0\dot{+}V\right)\right)|W-V|\exp\left(-(t-s)\left(K_0\dot{+}W\right)\right)|\psi|\,|f|, |g|\right\rangle$$
$$= \int_0^t ds\int dx\,|W(x)-V(x)|^\gamma\left[\exp\left(-(t-s)\left(K_0\dot{+}W\right)\right)|\psi f|\right](x)$$
$$\times |W(x)-V(x)|^{1-\gamma}\left[\exp\left(-s\left(K_0\dot{+}V\right)\right)|\varphi g|\right](x)$$

(Feynman-Kac formula)

$$= \int_0^t ds \int dx \, \mathbb{E}_x \Big[|W(x) - V(x)|^\gamma |\psi(X(t-s))|^\beta |f(X(t-s))| \chi_{t-s}(x)$$
$$\times |\psi(X(t-s))|^{1-\beta} \exp\left(-\int_0^{t-s} W(X(\sigma))d\sigma\right) \frac{1}{\chi_{t-s}(x)}$$
$$\times |W(x) - V(x)|^{1-\gamma} \left[\exp\left(-s(K_0 \dot{+} V)\right) |\varphi g|\right](x) \Big]$$

(Cauchy-Schwarz inequality)

$$\leq \left(\int_0^t ds \int dx \, \mathbb{E}_x \left[|W(x) - V(x)|^{2\gamma} |\psi(X(t-s))|^{2\beta} |f(X(t-s))|^2 \chi_{t-s}(x)^2 \right] \right)^{1/2}$$
$$\times \Bigg(\int_0^t ds \int dx \, \mathbb{E}_x \Big[|\psi(X(t-s))|^{2-2\beta} \exp\left(-2\int_0^{t-s} W(X(\sigma))d\sigma\right) \frac{1}{\chi_{t-s}(x)^2}$$
$$\times |W(x) - V(x)|^{2-2\gamma} \left[\exp\left(-s(K_0 \dot{+} V)\right) |\varphi g|\right](x)^2 \Big] \Bigg)^{1/2}$$
$$= \left\| \int_0^t ds \chi_s^2 |W - V|^{2\gamma} \exp(-sK_0) \left(|\psi|^{2\beta} |f|^2 \right) \right\|_1^{1/2}$$
$$\times \Big\| \int_0^t ds \frac{1}{\chi_{t-s}^2} \exp\left(-(t-s)(K_0 \dot{+} 2W)\right) |\psi|^{2-2\beta}$$
$$\times |W - V|^{2-2\gamma} \left(\exp\left(-s(K_0 \dot{+} V)\right) |\varphi g| \right)^2 \Big\|_1^{1/2}. \tag{4.7}$$

By the Feynman-Kac formula and another application of the inequality of Cauchy-Schwarz, we have additionally:

$$\left[\exp\left(-s\left(K_0 \dot{+} V\right)\right) |\varphi g|\right](x)^2$$
$$= \left(\mathbb{E}_x \left[\exp\left(-\int_0^s V(X(\sigma))d\sigma\right) |\varphi(X(s)) g(X(s))| \right] \right)^2$$
$$\leq \mathbb{E}_x \left[\exp\left(-2\int_0^s V(X(\sigma))d\sigma\right) |\varphi(X(s))|^{2\alpha} \right] \mathbb{E}_x \left[|\varphi(X(s))|^{2-2\alpha} |g(X(s))|^2 \right]$$
$$= \left[\exp\left(-s(K_0 \dot{+} 2V)\right) |\varphi|^{2\alpha} \right](x) \left[\exp(-sK_0) |\varphi|^{2-2\alpha} |g|^2 \right](x). \tag{4.8}$$

Consequently, from (4.7) and (4.8) it follows that

$$\left| \left\langle \varphi \left(\exp\left(-t\left(K_0 \dot{+} V\right)\right) - \exp\left(-t\left(K_0 \dot{+} W\right)\right) \right) \psi f, g \right\rangle \right|$$
$$\leq \left\| \int_0^t ds \chi_s^2 |W - V|^{2\gamma} \exp(-sK_0) |\psi|^{2\beta} \right\|_{1,1}^{1/2} \|f\|_2$$
$$\times \Big\| \int_0^t ds \frac{1}{\chi_{t-s}^2} |W - V|^{2-2\gamma} \exp\left(-(t-s)(K_0 \dot{+} 2W)\right) |\psi|^{2-2\beta}$$
$$\times \exp\left(-s(K_0 \dot{+} 2V)\right) |\varphi|^{2\alpha} \times \exp(-sK_0) |\varphi|^{2-2\alpha} \Big\|_{1,1}^{1/2} \|g\|_2$$

$$= \left\| \int_0^t ds\, |\psi|^{2\beta} \exp(-sK_0)\, |W-V|^{2\gamma} \chi_s^2 \right\|_\infty^{1/2}$$

$$\times \left\| \int_0^t ds\, |\varphi|^{2-2\alpha} \exp(-sK_0) \left(|W-V|^{2-2\gamma} \frac{1}{\chi_{t-s}^2} \right.\right.$$

$$\left.\left. \exp\left(-(t-s)\left(K_0\dot{+}2W\right)\right) |\psi|^{2-2\beta} \exp\left(-s\left(K_0\dot{+}2V\right)\right) |\varphi|^{2\alpha} \right) \right\|_\infty^{1/2}$$

$$\times \|f\|_2 \|g\|_2 . \tag{4.9}$$

From (4.9) inequality (4.2) in Proposition 4.2 follows upon choosing

$$\chi_s^2 := \exp\left(-(t-s)\left(K_0\dot{+}2V\right)\right) |\varphi|^{2\alpha} .$$

The proof of inequality (4.3) follows the same pattern, therefore it is omitted.

The following corollary is immediate. In the lemmas, propositions, and corollaries, following Corollary 4.3 we will disentangle the implicit conditions on V, W, ψ and φ imposed by the existence of the functions $C_{V,\varphi}(\cdot)$ and $C_{W,\varphi}(\cdot)$ and the constants $C_{V,\varphi}$, and $C_{W,\varphi}$ of Corollary 4.3.

4.3. COROLLARY. *Let the hypotheses be as in Proposition 4.2.*

(a) *Let $C_{V,\varphi}$, $C_{W,\psi} : (0,\infty) \to [0,\infty)$ be Borel measurable functions with the property that the pointwise inequality*

$$\left[\exp\left(-s\left(K_0\dot{+}2V\right)\right) |\varphi|^{2\alpha}\right](x) \le C_{V,\varphi}(s)\, |\varphi(x)|^{2\alpha} , \quad s>0, \tag{4.10}$$

holds for m-almost all x and with the property that the inequality

$$\left[\exp\left(-s\left(K_0\dot{+}2W\right)\right) |\psi|^{2-2\beta}\right](x) \le C_{W,\psi}(s)\, |\psi(x)|^{2-2\beta} \quad s>0, \tag{4.11}$$

is true for m-almost all $x \in E$. Then the following inequality is valid for $0 \le \gamma \le 1$:

$$\left\|\varphi\left(\exp\left(-t\left(K_0\dot{+}V\right)\right) - \exp\left(-t\left(K_0\dot{+}W\right)\right)\right)\psi\right\|_{2,2} \tag{4.12}$$

$$\le \left\| \int_0^t ds C_{V,\varphi}(t-s)\, |\psi|^{2\beta} \exp(-sK_0)\, |V-W|^{2\gamma} |\varphi|^{2\alpha} \right\|_\infty^{1/2}$$

$$\times \left\| \int_0^t ds C_{W,\psi}(t-s)\, |\varphi|^{2-2\alpha} \exp(-sK_0)\, |V-W|^{2-2\gamma} |\psi|^{2-2\beta} \right\|_\infty^{1/2} .$$

(b) *Let $C_{V,\varphi}$ and $C_{W,\psi}$ be constants with the property that*

$$\left[\left(aI+K_0\dot{+}2V\right)^{-1} |\varphi|^{2\alpha}\right](x) \le C_{V,\varphi}\, |\varphi(x)|^{2\alpha} \quad \textit{and} \tag{4.13}$$

$$\left[\left(aI+K_0\dot{+}2W\right)^{-1} |\psi|^{2-2\beta}\right] \le C_{W,\psi}\, |\psi(x)|^{2-2\beta} , \tag{4.14}$$

holds for m-almost all $x \in E$. Then the following inequality is valid for $0 \leq \gamma \leq 1$:

$$\begin{aligned}
&\left\| \varphi \left(\left(aI + K_0 \dot{+} V\right)^{-1} - \left(aI + K_0 \dot{+} W\right)^{-1} \right) \psi \right\|_{2,2} \\
&\leq C_{V,\varphi}^{1/2} C_{W,\psi}^{1/2} \left\| |\psi|^{2\beta} \left(aI + K_0\right)^{-1} |V - W|^{2\gamma} |\varphi|^{2\alpha} \right\|_\infty^{1/2} \\
&\times \left\| |\varphi|^{2-2\alpha} \left(aI + K_0\right)^{-1} |V - W|^{2-2\gamma} |\psi|^{2-2\beta} \right\|_\infty^{1/2}. \qquad (4.15)
\end{aligned}$$

Next we want to take a closer look at the functions $C_{V,\varphi}(\cdot)$ and $C_{W,\psi}(\cdot)$ and the constants $C_{V,\varphi}$ and $C_{W,\psi}$. Let ρ and ρ' be conjugate exponents in the sense that $\dfrac{1}{\rho} + \dfrac{1}{\rho'} = 1$, $1 < \rho, \rho' < \infty$. In the following lemma we shall describe possible functions and constants. The conditions in (4.10), (4.11), (4.13) and (4.14) are rather implicit at the moment. They will be disentangled in the following lemma and proposition.

4.4. LEMMA. *For the functions $C_{V,\varphi}(\cdot)$ and $C_{W,\psi}(\cdot)$ and the constants $C_{V,\varphi}$ and $C_{W,\psi}$ we may choose respectively*

$$\begin{aligned}
C_{V,\varphi}(s) &= \sup_{x \in E} \left(\left[\exp\left(-s(K_0 \dot{+} 2\rho' V)\right) 1\right](x) \right)^{1/\rho'} \\
&\times \sup_{x \in E} \frac{\left(\left[\exp\left(-sK_0\right) |\varphi|^{2\rho\alpha}\right](x) \right)^{1/\rho}}{|\varphi(x)|^{2\alpha}}, \qquad (4.16)\\
C_{W,\psi}(s) &= \sup_{x \in E} \left(\left[\exp\left(-s(K_0 \dot{+} 2\rho' W)\right) 1\right](x) \right)^{1/\rho'} \\
&\times \sup_{x \in E} \frac{\left(\left[\exp\left(-sK_0\right) |\psi|^{2\rho - 2\rho\beta}\right](x) \right)^{1/\rho}}{|\psi(x)|^{2-2\beta}}, \qquad (4.17)\\
C_{V,\varphi} &= \sup_{x \in E} \left(\left[\left(aI + K_0 \dot{+} 2\rho' V\right)^{-1} 1\right](x) \right)^{1/\rho'} \\
&\times \sup_{x \in E} \frac{\left(\left[\left(aI + K_0\right)^{-1} |\varphi|^{2\rho\alpha}\right](x) \right)^{1/\rho}}{|\varphi(x)|^{2\alpha}}, \qquad (4.18)\\
C_{W,\psi} &= \sup_{x \in E} \left(\left[\left(aI + K_0 \dot{+} 2\rho' W\right)^{-1} 1\right](x) \right)^{1/\rho'} \\
&\times \sup_{x \in E} \frac{\left(\left[\left(aI + K_0\right)^{-1} |\psi|^{2\rho - 2\rho\beta}\right](x) \right)^{1/\rho}}{|\psi(x)|^{2-2\beta}}. \qquad (4.19)
\end{aligned}$$

REMARK. It is not clear under what conditions on K_0, φ, and ψ the constants in (4.18) and (4.19) are finite. If $K_0 = -\frac{1}{2}\Delta$, then they are finite if φ and ψ are polynomials, if they are of the form $x \mapsto \exp(-\gamma \cdot x)$, or if they have the form $x \mapsto \exp(-\gamma\,|x|)$. Here γ is a vector in $\mathbb{R}^\nu$ or a positive real number.

Proof. Let V be a Kato-Feller potential and let v be any non-negative Borel function defined on E. Upon employing the Feynman-Kac formula together with Hölder's inequality, the following inequalities are readily established:

$$\begin{aligned}
&\left[\exp\left(-s\left(K_0\dot{+}V\right)\right)v\right](x) &(4.19)\\
&\quad\leq \left(\left[\exp\left(-s\left(K_0\dot{+}\rho' V\right)\right)1\right](x)\right)^{1/\rho'}\left(\left[\exp\left(-sK_0\right)v^\rho\right](x)\right)^{1/\rho};\\
&\left[\left(aI+K_0\dot{+}V\right)^{-1}v\right](x) &(4.20)\\
&\quad\leq \left(\left[\left(aI+K_0\dot{+}\rho' V\right)^{-1}1\right](x)\right)^{1/\rho'}\left(\left[\left(aI+K_0\right)^{-1}v^\rho\right](x)\right)^{1/\rho}.
\end{aligned}$$

From these formulas the statements in Lemma 4.4. follow immediately.

4.5. PROPOSITION. *Suppose that V and W are Kato-Feller potentials and let φ and ψ be Borel measurable functions defined on E. The symbol $C_V(t)$ is defined by*

$$C_V(t) := \sup_{0<s<t}\sup_{x\in E}\left(\mathbb{E}_x\left[\exp\left(-\int_0^s 4V(X(\sigma))d\sigma\right)\right]\right)^{1/4} \tag{4.21}$$

and a similar expression is employed for $C_W(t)$. That these constants are finite follows from the fact that the functions V and W are Kato-Feller potentials together with Khas'minskii's lemma: see Corollary 3.6.

(a) Fix $t>0$ and suppose that the following expressions are finite:

$$\begin{aligned}
c_\varphi(t) &:= \sup_{0<s<t}\sup_{x\in E}\frac{\left(\left[\exp\left(-sK_0\right)|\varphi|^2\right](x)\right)^{1/2}}{|\varphi(x)|},\\
c_\psi(t) &:= \sup_{0<s<t}\sup_{x\in E}\frac{\left(\left[\exp\left(-sK_0\right)|\psi|^2\right](x)\right)^{1/2}}{|\psi(x)|}.
\end{aligned}\tag{4.22}$$

Then the operator norm of $\varphi\left(\exp\left(-t\left(K_0\dot{+}V\right)\right)-\exp\left(-t\left(K_0\dot{+}W\right)\right)\right)\psi$ is finite. In fact the following inequality is valid:

$$\begin{aligned}
&\left\|\varphi\left(\exp\left(-t\left(K_0\dot{+}V\right)\right)-\exp\left(-t\left(K_0\dot{+}W\right)\right)\right)\psi\right\|_{2,2}\\
&\leq C_V(t)C_W(t)c_\varphi(t)^{1/2}c_\psi(t)^{1/2}\left\|\int_0^t ds\,|\psi|\exp\left(-sK_0\right)|V-W|\,|\varphi|\right\|_\infty^{1/2}\\
&\qquad\times\left\|\int_0^t ds\,|\varphi|\exp\left(-sK_0\right)|V-W|\,|\psi|\right\|_\infty^{1/2}.
\end{aligned}\tag{4.23}$$

(b) For $a \in \mathbb{R}$ sufficiently large, we put

$$C'_V := \sup_{x\in E} \left(\left[\left(aI + K_0 \dot{+} 4V \right)^{-1} 1 \right] (x) \right)^{1/4}$$

and we let C'_W stand for a similar expression with W replacing V. Fix $t > 0$ and suppose that the following expressions are finite:

$$c'_\varphi(a) := \sup_{x\in E} \frac{\left(\left[(aI + K_0)^{-1} |\varphi|^2 \right] (x) \right)^{1/2}}{|\varphi(x)|},$$

$$c'_\psi(a) := \sup_{x\in E} \frac{\left(\left[\exp (aI + K_0)^{-1} |\psi|^2 \right] (x) \right)^{1/2}}{|\psi(x)|}. \tag{4.24}$$

Then the operator norm of $\varphi \left(\left(aI + K_0 \dot{+} V \right)^{-1} - \left(aI + K_0 \dot{+} W \right)^{-1} \right) \psi$ is finite. In fact the following inequality is valid:

$$\begin{aligned} &\left\| \varphi \left(\left(aI + K_0 \dot{+} V \right)^{-1} - \left(aI + K_0 \dot{+} W \right)^{-1} \right) \psi \right\|_{2,2} \\ &\leq C'_V(a) C'_W(a) c'_\varphi(a)^{1/2} c'_\psi(a)^{1/2} \left\| |\psi| \left(aI + K_0 \right)^{-1} |V - W| \, |\varphi| \right\|_\infty^{1/2} \\ &\times \left\| |\varphi| \left(aI + K_0 \right)^{-1} |V - W| \, |\psi| \right\|_\infty^{1/2}. \end{aligned} \tag{4.25}$$

Proof. In Lemma 4.4. we set $\rho = \rho' = 2$ and $\alpha = \beta = 1/2$.

In the following proposition we want to compare the norms

$$V \mapsto \left\| |\psi| \left(aI + K_0 \right)^{-1} |V| \, |\varphi| \right\|_\infty \tag{4.26}$$

and

$$V \mapsto \left\| \int_0^t ds \, |\psi| \exp(-sK_0) \, |V| \, |\varphi| \right\|_\infty. \tag{4.27}$$

Here a and t are fixed and strictly positive. For $\varphi = \psi \equiv 1$ this is done in Proposition 2.2.

REMARK. Let the function V be a Kato-Feller potential. The functions $t \mapsto C_V(t)$, $t \geq 0$, and $a \mapsto C'_V(a)$, a sufficiently large, are real valued. Whether or not the functions $t \mapsto c_\varphi(t)$ and $a \mapsto c'_\varphi(a)$ take only finite real values, is not so clear. In the Propositions 4.9 and 4.10, we shall elaborate on that in the case we are dealing with ν-dimensional Brownian motion.

4.6. PROPOSITION. *Let φ and ψ be Borel measurable functions on E. Define the function $D_\psi(a,t)$ by*

$$D_\psi(a,t) = \sup_{x\in E} |\psi(x)|\, e^{-at}\left[\exp(-tK_0)\frac{1}{|\psi|}\right](x). \tag{4.28}$$

Suppose that $a > 0$ and $t > 0$ are chosen in such a way that $D_\psi(a,t) < 1$. Let V be another Borel measurable function. The following inequalities are valid:

$$\begin{aligned}&(1 - D_\psi(a,t))\left\|\left|\psi\right|(aI + K_0)^{-1}|V|\,|\varphi|\right\|_\infty \\ &\leq \left\|\int_0^t |\psi|\exp(-sK_0)\,|V|\,|\varphi|\right\|_\infty \leq e^{at}\left\|\left|\psi\right|(aI + K_0)^{-1}|V|\,|\varphi|\right\|_\infty .\end{aligned} \tag{4.29}$$

REMARK. For the Laplace operator and for the relativistic Hamiltonian the finiteness of the expression in (4.28) is established in Chapter 8, Remark 1 after Proposition 8.21.

Proof. The proof goes along the same lines as the proof of Proposition 2.2 where $\varphi \equiv \psi \equiv 1$. Let $a > 0$, $t > 0$, V, φ and ψ be as in Proposition 4.6. Using the semigroup property together with some trivial inequalities shows the following:

$$\begin{aligned}&|\psi|\,(aI + K_0)^{-1}|V|\,|\varphi| = |\psi|\int_0^\infty dse^{-as}\exp(-sK_0)\,|V|\,|\varphi| \\ &\leq \sum_{k=0}^\infty e^{-akt}\,|\psi|\int_0^t ds\exp(-ktK_0)\exp(-sK_0)\,|V|\,|\varphi| \\ &= \sum_{k=0}^\infty\left(|\psi|\exp(-at - tK_0)\frac{1}{|\psi|}\right)^k\int_0^t ds\,|\psi|\exp(-sK_0)\,|V|\,|\varphi| \\ &\leq \frac{1}{1 - D_\psi(a,t)}\int_0^t ds\,|\psi|\exp(-sK_0)\,|V|\,|\varphi| \\ &\leq \frac{e^{at}}{1 - D_\psi(a,t)}\int_0^\infty dse^{-as}\,|\psi|\exp(-sK_0)\,|V|\,|\varphi| .\end{aligned} \tag{4.30}$$

From (4.30) the inequalities in (4.29) readily follow.

4.7. PROPOSITION. *Let ψ and φ be Borel measurable functions and set*

$$v = (aI + K_0)^{-1}\left(|\psi|^2\left((aI + K_0)^{-1}|\varphi|^4\right)^{1/2}\right).$$

Let V and W be Borel measurable functions defined on E for which

$$\mathbb{E}_x\left[\exp\left(-4\int_0^t \min(V,W)(X(s))ds\right)\right] \leq C\exp(At), \tag{4.31}$$

where $a > A$. Then

$$\left\|\psi\left(\left(aI+K_0\dot{+}W\right)^{-1}-\left(aI+K_0\dot{+}V\right)^{-1}\right)\varphi\right\|$$
$$\leq 2\left(\frac{C}{a-A}\right)^{1/4}\left\|\left((aI+K_0)^{-1}|V-W|\right)^2 v\right\|_{\infty}^{1/2} \tag{4.32}$$
$$\leq 2\left(\frac{C}{a-A}\right)^{1/4}\left\|(aI+K_0)^{-1}\left(|V-W|\max(1,v)\right)\right\|_{\infty}. \tag{4.33}$$

4.8. COROLLARY. *In addition to the assumptions in Proposition 4.7, suppose that there are constants C_1 and C_2 such that, pointwise,*

$$(aI+K_0)^{-1}|\varphi|^4 \leq C_1|\varphi|^4 \quad \textit{and} \quad (aI+K_0)^{-1}|\psi\varphi|^2 \leq C_2|\psi\varphi|^2. \tag{4.34}$$

Then

$$\left\|\psi\left(\left(aI+K_0\dot{+}W\right)^{-1}-\left(aI+K_0\dot{+}V\right)^{-1}\right)\varphi\right\|_{2,2} \tag{4.35}$$
$$\leq 2\left(\frac{C}{a-A}\right)^{1/4} C_1^{1/4}C_2^{1/2}\left\|(aI+K_0)^{-1}\left(|V-W|\max(1,|\psi\varphi|^2)\right)\right\|_{\infty}.$$

Proof of Corollary 4.8. This result immediately follows from Proposition 4.7.

REMARK. In inequality (4.15) of Corollary 4.3, the roles of φ and ψ are entirely separated, whereas in inequality (4.35) their product appears. This could be interesting, if for example the functions φ and ψ are each other's inverse. If $\varphi \equiv \psi \equiv 1$, then the inequality in (4.35) is the one in (4.15) with $\gamma = 1/2$. If $\varphi\psi \equiv 1$, then the right-hand side of (4.35) is dominated by a constant times the Kato-Feller norm of the function $V - W$ and, for $\gamma = 1/2$, the right-hand side of the inequality in (4.15) is dominated by a constant times the square root of the product of Kato-Feller norms of the functions $(V-W)\varphi^2$ and $(V-W)\varphi^{-2}$ (set $\alpha = 1$ and $\beta = 0$).

Proof of Proposition 4.7. We want to estimate the norm of the operator

$$\psi\left(\left(aI+K_0\dot{+}W\right)^{-1}-\left(aI+K_0\dot{+}V\right)^{-1}\right)\varphi. \tag{4.36}$$

Put $R(a,V) = \left(aI+K_0\dot{+}V\right)^{-1}$ and note the identity

$$[\psi\left(R(a,W)-R(a,V)\right)\varphi f](x)$$
$$= \psi(x)\left[R(a,W)(V-W)R(a,V)\varphi f\right](x)$$
$$= \psi(x)\int_0^{\infty} dv e^{-av}\mathbb{E}_x\left[\int_0^v ds \exp\left(-\int_0^s W(X(\sigma))d\sigma - \int_s^v V(X(\sigma))d\sigma\right)\right.$$
$$\left.\times\left(V(X(s))-W(X(s))\right)\varphi(X(v))f(X(v))\right]. \tag{4.37}$$

It follows that

$$\begin{aligned}
&|[\psi (R(a,W) - R(a,V))\,\varphi f]\,(x)| \\
&\quad = |\psi(x)\,[R(a,W)(V-W)R(a,V)\varphi f]\,(x)| \\
&\quad \leq |\psi(x)| \int_0^\infty dv\, e^{-av}\mathbb{E}_x \Bigg[\int_0^v ds\, \exp\left(-\int_0^v \min\left(W(X(\sigma)), V(X(\sigma))\right) d\sigma\right) \\
&\qquad \times |V(X(s)) - W(X(s))|\, |\varphi(X(v)) f(X(v))|\Bigg]. \qquad (4.38)
\end{aligned}$$

From (4.38) we infer that for any nowhere vanishing continuous function χ defined on E and for $0 \leq \alpha,\ \beta \leq 1$,

$$\begin{aligned}
&|\langle\psi (R(a,W) - R(a,V))\,\varphi f, g\rangle|^2 \\
&\leq \Bigg(\int dx \int_0^\infty dv\, e^{-av} \\
&\quad \mathbb{E}_x\Bigg[\chi(x)\,|\psi(x)|^\beta \int_0^v |V(X(s)) - W(X(s))|\, ds\, |\varphi(X(v))|^\alpha\, |f(X(v))| \\
&\qquad \times |\psi(x)|^{1-\beta}\, |\varphi(X(v))|^{1-\alpha} \exp\left(-\int_0^v \min(W,V)(X(s))ds\right) \frac{1}{\chi(x)}\, |g(x)|\Bigg]\Bigg)^2
\end{aligned}$$

(Cauchy-Schwarz inequality)

$$\begin{aligned}
&\leq \int dx_1 \int_0^\infty dv_1\, e^{-av_1} \int dx_2 \int_0^\infty dv_2\, e^{-av_2} \\
&\mathbb{E}_{x_1}\left[\chi(x_1)^2\, |\psi(x_1)|^{2\beta} \left(\int_0^{v_1} |W(X(s)) - V(X(s))|\, ds\right)^2 |\varphi(X(v_1))|^{2\alpha}\, |f(X(v_1))|^2\right] \\
&\times \mathbb{E}_{x_2}\Bigg[|\psi(x_2)|^{2-2\beta}\, |\varphi(X(v_2))|^{2-2\alpha} \\
&\quad \exp\left(-2\int_0^{v_2} \min(W,V)(X(s))ds\right) \frac{1}{(\chi(x_2))^2}\, |g(x_2)|^2\Bigg]
\end{aligned}$$

(Markov property together with the definition of $R(a, 2\min(W,V))$)

$$\begin{aligned}
&= 4\int dx(\chi(x))^2\, |\psi(x)|^{2\beta} \left[\left((aI+K_0)^{-1}\, |W-V|\right)^2 (aI+K_0)^{-1}\, |\varphi|^{2\alpha}\, |f|^2\right](x) \\
&\times \int dx \frac{|\psi(x)|^{2-2\beta}}{(\chi(x))^2} \left[\left(aI + K_0 \dot{+} 2\min(W,V)\right)^{-1} |\varphi|^{2-2\alpha}\right](x)\, |g(x)|^2
\end{aligned}$$

(symmetry)

$$= 4 \int dx \left[\left((aI + K_0)^{-1} |W - V| \right)^2 (aI + K_0)^{-1} \chi^2 |\psi|^{2\beta} \right] (x) |\varphi(x)|^{2\alpha} |f(x)|^2$$
$$\times \int dx \frac{|\psi(x)|^{2-2\beta}}{(\chi(x))^2} \left[\left(aI + K_0 \dot{+} 2 \min(W, V) \right)^{-1} |\varphi|^{2-2\alpha} \right] (x) |g(x)|^2 . \tag{4.39}$$

With

$$\chi^2 = |\psi|^{2-2\beta} \left(aI + K_0 \dot{+} 2 \min(W, V) \right)^{-1} |\varphi|^{2-2\alpha} \tag{4.40}$$

we infer that

$$\begin{aligned} & |\langle \psi \left(R(a, W) - R(a, V) \right) \varphi f, g \rangle|^2 \\ & \leq 4 \int dx \left[\left((aI + K_0)^{-1} |V - W| \right)^2 (aI + K_0)^{-1} |\psi|^2 \right. \\ & \quad \times \left. \left(aI + K_0 \dot{+} 2 \min(V, W) \right)^{-1} |\varphi|^{2-2\alpha} \right] (x) |\varphi(x)|^{2\alpha} |f(x)|^2 \int dx \, |g(x)|^2 \\ & \leq 4 \|u_\alpha\|_\infty \|f\|_2^2 \|g\|_2^2 , \end{aligned} \tag{4.41}$$

where

$$\begin{aligned} u_\alpha = & |\varphi|^{2\alpha} \left((aI + K_0)^{-1} |V - W| \right)^2 \\ & \times (aI + K_0)^{-1} \left(|\psi|^2 \left(aI + K_0 \dot{+} 2 \min(W, V) \right)^{-1} |\varphi|^{2-2\alpha} \right) . \end{aligned} \tag{4.42}$$

Consequently

$$\|\psi \left(R(a, W) - R(a, V) \right) \varphi\|_{2,2} \leq 2 \|u_\alpha\|_\infty^{1/2} . \tag{4.43}$$

We want to have a better control over the function u_α. Let ρ and ρ' be conjugate exponents, i.e. $\frac{1}{\rho} + \frac{1}{\rho'} = 1$, $1 < \rho$, $\rho' < \infty$. By Hölder's inequality the following (in-)equalities are valid for $y \in E$:

$$\begin{aligned} & \left[\left(aI + K_0 \dot{+} 2 \min(V, W) \right)^{-1} |\varphi|^{2-2\alpha} \right] (y) \\ & = \int_0^\infty ds e^{-as} \mathbb{E}_y \left(\exp \left(-2 \int_0^s \min(V, W)(X(\sigma)) d\sigma \right) |\varphi(X(s))|^{2-2\alpha} \right) \\ & \leq \left(\int_0^\infty e^{-as} \mathbb{E}_y \left(\exp \left(-2\rho \int_0^s \min(V, W)(X(\sigma)) d\sigma \right) \right) ds \right)^{1/\rho} \\ & \quad \times \left(\int_0^\infty e^{-as} \mathbb{E}_y \left(|\varphi(X(s))|^{2(1-\alpha)\rho'} \right) ds \right)^{1/\rho'} \\ & \leq \left\| \left(aI + K_0 \dot{+} 2\rho \min(V, W) \right)^{-1} 1 \right\|_\infty^{1/\rho} \\ & \quad \times \left(\left[(aI + K_0)^{-1} |\varphi|^{2(1-\alpha)\rho'} \right] (y) \right)^{1/\rho'} \\ & \leq \left(\frac{C}{a - A} \right)^{1/\rho} \left(\left[(aI + K_0)^{-1} |\varphi|^{2(1-\alpha)\rho'} \right] (y) \right)^{1/\rho'} . \end{aligned} \tag{4.44}$$

Hence, for $\alpha = 0$ and $\rho = \rho' = 2$, we obtain

$$u_0 \le \left(\frac{C}{a-A}\right)^{1/2} \left((aI+K_0)^{-1}|V-W|\right)^2 \times (aI+K_0)^{-1}\left(|\psi|^2\left((aI+K_0)^{-1}|\varphi|^4\right)^{1/2}\right). \tag{4.45}$$

From these discussions, the inequality

$$u_0 \le \left(\frac{C}{a-A}\right)^{1/2} \left((aI+K_0)^{-1}|V-W|\right)^2 v \tag{4.46}$$

is clear. Together with (4.43) this yields inequality (4.32). In order to obtain (4.33), we observe that

$$\begin{aligned} \|u_0\|_\infty &\le \left(\frac{C}{a-A}\right)^{1/2} \left\|(aI+K_0)^{-1}|V-W|\right\|_{\infty,\infty} \left\|(aI+K_0)^{-1}(|V-W|\,v)\right\|_\infty \\ &\le \left(\frac{C}{a-A}\right)^{1/2} \left\|(aI+K_0)^{-1}|V-W|\max(1,v)\right\|_\infty^2. \end{aligned} \tag{4.47}$$

Again we may invoke (4.43) to obtain (4.33).

Next we come back to the remark preceding Proposition 4.6, the case when $-K_0$ is the Laplace operator. Proposition 4.9 below is usually applied with ψ of the form $\psi(x) = \left(1+|x|^2\right)^{-\alpha/2}$ or $\psi(x) = \exp(-|x|)$. The relevant Corollary 4.11 is applicable for these functions.

4.9. Proposition. *Let the operator K_0 denote $-\frac{1}{2}\Delta$ in $\mathbb{R}^\nu$. Let a be a bounded non-negative function defined on $(0,\infty)$ and define, for $\nu \in \mathbb{N}$, the function a_ν by*

$$a_\nu(\tau) = \int_0^\infty a(s)\exp\left(-\frac{\tau}{s}\right)\frac{1}{(\sqrt{2\pi s})^\nu}ds, \quad \tau > 0. \tag{4.48}$$

Furthermore let φ and ψ be Borel measurable functions on $\mathbb{R}^\nu$. Let $C(\nu)$ be the following covering number: $C(\nu)$ is the supremum of all $k \in \mathbb{N}$, $k \ge 1$, of the expression

$$\min_{m\in\mathbb{N}} \frac{\{\textit{the annulus } \{k \le |z| \le k+1\} \textit{ can be covered by } m \textit{ balls of radius } 1\}}{k^{\nu-1}}.$$

It is not so difficult to show that $C(\nu) \le 2\times 16^{\nu-1}$. Put

$$C_\psi(t) = C(\nu)\sup_{x\in\mathbb{R}^\nu}\sum_{k=1}^\infty k^{\nu-1}\frac{a_\nu(k^2t)}{a_\nu(t)}\sup_{k\sqrt{2t}\le|z-x|\le(k+1)\sqrt{2t}}\frac{|\psi(x)|}{|\psi(z)|} \tag{4.49}$$

and suppose that $C_\psi(t)$ is finite. Then

$$\begin{aligned}
&|\psi(x)| \int_0^\infty a(s) \left[\exp(-sK_0) |V\varphi|\right](x) ds \\
&\le (1 + C_\psi(t)) \sup_{x\in\mathbb{R}^\nu} \left(|\psi(x)| \int_{\{|y-x|\le\sqrt{2t}\}} a_\nu \left(|y-x|^2/2\right) |V(y)\varphi(y)| \, dy \right) \\
&\le (1 + C_\psi(t)) \sup_{x\in\mathbb{R}^\nu} \left(|\psi(x)| \int_0^\infty a(s) \left[\exp(-sK_0) |V\varphi|\right](x) ds \right). \qquad (4.50)
\end{aligned}$$

Proof. Let $R_k(t,x)$ be the annulus

$$R_k(t,x) = \left\{ z \in \mathbb{R}^\nu : k\sqrt{2t} \le |z-x| \le (k+1)\sqrt{2t} \right\}.$$

From the explicit expression of the integral kernel of $\exp(-sK_0)$, the (in-) equalities

$$\begin{aligned}
&|\psi(x)| \int_0^\infty a(s) \left[\exp(-sK_0) |V\varphi|\right](x) ds \\
&= |\psi(x)| \int a_\nu \left(|y-x|^2/2\right) |V(y)\varphi(y)| \, dy \\
&= |\psi(x)| \int_{\{|y-x|\le\sqrt{2t}\}} a_\nu \left(|y-x|^2/2\right) |V(y)\varphi(y)| \, dy \\
&\quad + |\psi(x)| \sum_{k=1}^\infty \int_{\{k\sqrt{2t}\le|y-x|\le(k+1)\sqrt{2t}\}} a_\nu \left(|y-x|^2/2\right) |V(y)\varphi(y)| \, dy \\
&\le |\psi(x)| \int_{\{|y-x|\le\sqrt{2t}\}} a_\nu \left(|y-x|^2/2\right) |V(y)\varphi(y)| \, dy \\
&\quad + |\psi(x)| \, C(\nu) \sum_{k=1}^\infty k^{\nu-1} \frac{a_\nu(k^2 t)}{a_\nu(t)} \sup_{z\in R_k(t,x)} \int_{\{|y-z|\le\sqrt{2t}\}} |V(y)\varphi(y)| \, a_\nu(t) dy \\
&\le |\psi(x)| \int_{\{|y-x|\le\sqrt{2t}\}} a_\nu \left(|y-x|^2/2\right) |V(y)\varphi(y)| \, dy \qquad (4.51) \\
&\quad + C(\nu) \sum_{k=1}^\infty k^{\nu-1} \frac{a_\nu(k^2 t)}{a_\nu(t)} \sup_{z\in R_k(t,x)} \frac{|\psi(x)|}{|\psi(z)|} \\
&\quad \times |\psi(z)| \int_{\{|y-z|\le\sqrt{2t}\}} |V(y)\varphi(y)| \, a_\nu \left(|y-z|^2/2\right) dy
\end{aligned}$$

readily follow. Here we used the fact that the function a_ν decreases. Inequality (4.51) easily yields (4.50).

Next we give some estimates for the function a_ν, defined in (4.48).

4.10. PROPOSITION. *Let a be a bounded non-negative function defined on $(0,\infty)$ and let, for $\nu \in \mathbb{N}$, the function a_ν be defined as in equality (4.48) of Proposition 4.9. Put*

$$\|a\|_\delta = \sup_{0<s<\delta} a(s) + \frac{\displaystyle\int_\delta^\infty \frac{a(s)}{s} ds}{\displaystyle\int_1^\infty \exp(-\sigma)\sigma^{\frac{1}{2}\nu-2} d\sigma}.$$

For $0 < \tau \le \delta$ and $\nu \ge 2$, the following inequalities are valid:

$$\begin{aligned} &\inf_{0<s\le\delta} a(s) \frac{\tau \int_{\tau/\delta}^\infty \exp(-\sigma)\sigma^{\frac{1}{2}\nu-2} d\sigma}{(\sqrt{2\pi\tau})^\nu} \\ &\qquad \le a_\nu(\tau) \le \|a\|_\delta \frac{\tau \int_{\tau/\delta}^\infty \exp(-\sigma)\sigma^{\frac{1}{2}\nu-2} d\sigma}{(\sqrt{2\pi\tau})^\nu}. \end{aligned} \tag{4.52}$$

Proof. The proof follows from the observation that

$$\begin{aligned} &(\sqrt{2\pi})^\nu a_\nu(\tau)\tau^{\frac{1}{2}\nu-1} \\ &= \int_{\tau/\delta}^\infty a\left(\frac{\tau}{\sigma}\right) \exp(-\sigma)\sigma^{\frac{1}{2}\nu-2} d\sigma + \int_\delta^\infty \frac{a(s)}{s} \left(\frac{\tau}{s}\right)^{\frac{1}{2}\nu-1} \exp\left(-\frac{\tau}{s}\right) ds \\ &\le \|a\|_\delta \int_{\tau/\delta}^\infty \exp(-\sigma)\sigma^{\frac{1}{2}\nu-2} d\sigma. \end{aligned}$$

REMARK. Note the following inequalities ($\tau \le \delta$):

$$\delta^{1-\frac{1}{2}\nu} \int_1^\infty \exp(-\sigma)\sigma^{\frac{1}{2}\nu-2} d\sigma \le \frac{\displaystyle\int_{\tau/\delta}^\infty \exp(-\sigma)\sigma^{\frac{1}{2}\nu-2} d\sigma}{\tau^{\frac{1}{2}\nu-1}} \le \frac{\delta^{1-\frac{1}{2}\nu}}{1-\dfrac{1}{2}\nu}, \quad \text{for } \nu < 2,$$

$$\frac{1}{e}\log\frac{\delta}{\tau} + \int_1^\infty \frac{e^{-\sigma}}{\sigma} d\sigma \le \frac{\displaystyle\int_{\tau/\delta}^\infty \exp(-\sigma)\sigma^{\frac{1}{2}\nu-2} d\sigma}{\tau^{\frac{1}{2}\nu-1}} \le \log\frac{\delta}{\tau} + \int_1^\infty \frac{e^{-\sigma}}{\sigma} d\sigma, \text{ for } \nu = 2,$$

and

$$\frac{\displaystyle\int_1^\infty \exp(-\sigma)\sigma^{\frac{1}{2}\nu-2} d\sigma}{\tau^{\frac{1}{2}\nu-1}} \le \frac{\displaystyle\int_{\tau/\delta}^\infty \exp(-\sigma)\sigma^{\frac{1}{2}\nu-2} d\sigma}{\tau^{\frac{1}{2}\nu-1}} \le \frac{\Gamma\left(\dfrac{1}{2}\nu - 1\right)}{\tau^{\frac{1}{2}\nu-1}}, \quad \text{for } \nu > 2.$$

The following result is a refinement of Theorem 1.27.

4.11. COROLLARY. *Let $\psi : \mathbb{R}^\nu \to \mathbb{R}$ be a Borel measurable function with the following property:*

$$\sup_{0<t<1}\sup_{x\in\mathbb{R}^\nu}\sum_{k=1}^{\infty} k\int_{k^2}^{\infty} e^{-\sigma}\sigma^{\frac{1}{2}\nu-2}d\sigma \sup_{z:k\sqrt{2t}\leq|z-x|\leq(k+1)\sqrt{2t}}\frac{|\psi(x)|}{|\psi(z)|}<\infty.$$

Then

$$\lim_{t\downarrow 0}\sup_{x\in\mathbb{R}^\nu}|\psi(x)|\int_0^t\left[\exp\left(-sK_0\right)|V\varphi|\right](x)ds=0$$

if and only if

$$\lim_{t\downarrow 0}\sup_{x\in\mathbb{R}^\nu}|\psi(x)|\int_{\{|y-x|\leq\sqrt{2t}\}}\frac{|V(y)\varphi(y)|}{|y-x|^{\nu-2}}dy=0,\quad (\nu\geq 3);$$

$$\lim_{t\downarrow 0}\sup_{x\in\mathbb{R}^\nu}|\psi(x)|\int_{\{|y-x|\leq\sqrt{2t}\}}|V(y)\varphi(y)|\log\frac{\sqrt{2t}}{|y-x|^2}dy=0,\quad (\nu=2);$$

$$\sup_{x\in\mathbb{R}^\nu}|\psi(x)|\int_{x-1}^{x+1}|V(y)\varphi(y)|\,dy<\infty,\quad (\nu=1).$$

Proof. Put $a(s)=1_{[0,t]}(s)$. An elementary calculation shows

$$k^{\nu-1}\frac{a_\nu(k^2t)}{a_\nu(t)}=k\frac{\displaystyle\int_{k^2}^{\infty}e^{-\sigma}\sigma^{\frac{1}{2}\nu-2}d\sigma}{\displaystyle\int_1^{\infty}e^{-\sigma}\sigma^{\frac{1}{2}\nu-2}d\sigma}.$$

Put $C_\psi=\sup_{0<t<1}C_\psi(t)$ where $C_\psi(t)$ is as in (4.49) with $a(s)=1_{[0,t]}(s)$. From our hypothesis it follows that C_ψ is finite. The claim in Corollary 4.11 follows from (4.50) because another calculation shows that

$$a_\nu\left(|y-x|^2/2\right)=\frac{2}{\left(\sqrt{\pi}\right)^\nu}\frac{1}{|y-x|^{\nu-2}}\int_{\frac{|y-x|^2}{2t}}^{\infty}e^{-\sigma}\sigma^{\frac{1}{2}\nu-2}d\sigma,$$

provided $a(s)=1_{[0,t]}(s)$.

B. Singular perturbations

4.12. GENERAL FACTS. As in Chapter 2 the generator K_0 will be perturbed in two ways. The first one will be a "regular" perturbation, being a *multiplication* operator V. That kind of operator was studied in section A of the present Chapter. The other kind of perturbation in Chapter 2 was the "singular" one, i.e. a perturbation by a potential barrier on a closed subset Γ of E. These singular perturbations will be treated in this section. Put $\Sigma := E \setminus \Gamma$ and introduce the restriction operator $J = J_\Sigma$ as follows: $Jf = f \restriction_\Sigma$. Then its adjoint $J^* : L^2(\Sigma, m) \to L^2(E, m)$ is given by the canonical extension: $J^*f(x) = f(x)$ for $x \in \Sigma$ and $J^*f(x) = 0$ for $x \in \Gamma$. Moreover, we have $J^*J = 1_\Sigma$ and JJ^* is the identity in $L^2(\Sigma, m)$. By $\left(K_0 \dot{+} V\right)_\Sigma$ we denote the generator of the semigroup $\left\{\exp\left(-t\left(K_0 \dot{+} V\right)_\Sigma\right) : t \geq 0\right\}$. As explained in Chapter 2 the operator $\exp\left(-t\left(K_0 \dot{+} V\right)_\Sigma\right)$ is given by the formula

$$\left[\exp\left(-t\left(K_0 \dot{+} V\right)_\Sigma\right) f\right](x) = \mathbb{E}_x\left[\exp\left(-\int_0^t V(X(u))du\right) f(X(t)) : S > t\right], \tag{4.53}$$

where S is the *penetration time* of Γ given by

$$S = \inf\left\{s > 0 : \int_0^s 1_\Gamma(X(\sigma))d\sigma > 0\right\}. \tag{4.54}$$

Suppose the set of S-regular points coincides with Γ. Then, for $f \in C_\infty(E)$, the function $g : \Sigma \to \mathbb{C}$, defined by $g(x) = \left[\exp\left(-t\left(K_0 \dot{+} V\right)_\Sigma\right) f\right](x)$ possesses a continuous extension to all of E. In fact the canonical extension J^*f is continuous on all of E. This is a consequence of the fact that the "killed" Feynman-Kac or Dirichlet semigroup $\left\{\exp\left(-t\left(K_0 \dot{+} V\right)_\Sigma\right) : t \geq 0\right\}$ leaves the space $C_\infty(\Sigma)$ invariant: see e.g. Doob [92, Chapter 1.VIII]. For $V = 0$, this is shown in Appendix D, Theorem D.21. Functions in $L^p(E, m)$ functions can be approximated by $C_\infty(E)$ functions in the L^p-norm. Thus in the presence of L^1-L^∞-smoothing, the L^p-spaces $L^p(\Sigma, m)$, $1 \leq p < \infty$, are mapped into $C_\infty(\Sigma)$ by the Feynman-Kac semigroups "killed" on Γ. Their canonical extensions then belong to $C_\infty(E)$: for these results one has to mimic the corresponding proofs of Theorem 2.5 for the singular case. The proof of this theorem was discussed in Chapter 3. From formula (2.51) in Corollary 2.32 item (b), we see that the operator $\left(K_0 \dot{+} V\right)_\Sigma$ extends the operator $J\left(K_0 \dot{+} V\right)J^*$. As in Chapter 2 we will be interested in the *harmonic extension operator* given by the formula (see Definition 2.30)

$$\left[H_\Sigma^{a+V} f\right](x) = \mathbb{E}_x\left[\exp\left(-\int_0^S (a + V(X(u)))\,du\right) f(X(S)) : S < \infty\right], \tag{4.55}$$

for whatever functions f this operator makes sense. From Proposition 2.31 and its Corollary 2.32, it follows that (4.55) has a meaning for functions $f \in \operatorname{dom}(K_0 \dot{+} V)$

or even for functions f in the domain of the generalized Schrödinger form which might also be called Feller form. In Chapter 2 we already discussed the relation between H_Σ^{a+V} and the resolvents $\left(aI + K_0\dot{+}V\right)^{-1}$ and $J_\Sigma^*\left(aI + K_0\dot{+}V\right)_\Sigma^{-1} J_\Sigma$. In fact we have the following Dynkin's formula (see Proposition 2.31):

$$H_\Sigma^{a+V}\left(aI + K_0\dot{+}V\right)^{-1} = \left(aI + K_0\dot{+}V\right)^{-1} - J_\Sigma^*\left(aI + K_0\dot{+}V\right)_\Sigma^{-1} J_\Sigma. \tag{4.56}$$

4.13. Definition. Another family of operators which will play a decisive role consists of the family of projections $\{T_\Sigma(t) : t \geq 0\}$, where $T_\Sigma(t) = T_\Sigma^V(t)$ is defined by (see Definition 2.33)

$$\left[T_\Sigma(t)f\right](x) = \mathbb{E}_x\left[\exp\left(-\int_0^S V(X(u))du\right) f(X(S)) : S \leq t\right].$$

Remark 1. For additional properties of the operators $T_\Sigma(t)$, $t \geq 0$, we refer the reader to Definition 2.33 and to (2.57) in Theorem 2.34 of Chapter 2. We also note the equality

$$\begin{aligned}
&\lim_{t\downarrow 0}\left[T_\Sigma(t)f\right](x) = \mathbb{E}_x\left\{f(X(S)) : S \leq 0\right\} \\
&= \mathbb{E}_x\left\{f(X(0)) : S = 0\right\} = f(x)\mathbb{P}_x\left\{S = 0\right\}.
\end{aligned}$$

Hence, if $\mathbb{P}_x\{S = 0\} = 1$, then $\left[T_\Sigma(0)f\right](x) = f(x)$. In words, if x is an S-regular point for $\Gamma = E \setminus \Sigma$, then $T_\Sigma(0)f(x) = f(x)$. If, for instance x belongs to the open set Σ, then $\left[T_\Sigma(0)f\right](x) = 0$, i.e. $\left[T_\Sigma(0)f\right](x) = 1_\Gamma(x)f(x)$.

Remark 2. In Remark 2 following Theorem D.16 in Appendix D it is explained that $m\left(\Gamma \setminus \Gamma^\rho\right) = 0$. Here Γ^ρ is the set of *S-regular* points of Γ, i.e.

$$\Gamma^\rho := \left\{x \in E : \mathbb{P}_x\left[S = 0\right] = 1\right\}.$$

Blumenthal's zero-one law implies that $\mathbb{P}_x\left[S = 0\right]$ is either 1 or 0: see e.g. Port and Stone [189, Proposition 3.5, p. 8] or Blumenthal and Getoor [35, Proposition 5.17, p. 30]. As a consequence, in the $L^p(E, m)$-context, the operator $T_\Sigma(0)$ is the multiplication operator with 1_Γ. In genuine function spaces, like the space of bounded Borel functions on E, $T_\Sigma(0)$ is the operator of multiplication by 1_{Γ^ρ}.

Remark 3. Because of equality (see (4.61) in Proposition 4.15 below)

$$T_\Sigma(t)f = D_\Sigma(t)f + \int_0^t D_\Sigma(u)(K_0\dot{+}V)f\,du, \quad f \in \operatorname{dom}\left(K_0\dot{+}V\right), \quad \text{where}$$

$$D_\Sigma(t) = \exp\left(-t\left(K_0\dot{+}V\right)\right) - J_\Sigma^*\exp\left(-t\left(K_0\dot{+}V\right)_\Sigma\right) J_\Sigma.$$

It follows that the operators $T_\Sigma(t)$ are well-defined on the domain of $K_0\dot{+}V$. In the following proposition we collect some of the basic properties of the operators $D_\Sigma(t)$, $T_\Sigma(t)$, H_Σ^{a+V} and the resolvents $\left(aI + K_0\dot{+}V\right)^{-1}$ and $J_\Sigma^*\left(aI + K_0\dot{+}V\right)_\Sigma^{-1} J_\Sigma$. Some of the results were stated without proof in Theorem 2.34. Equality (4.58) says that the harmonic extension operator is, up to the factor a, the Laplace transform of the family $\{T_\Sigma(s) : s \geq 0\}$. So we say that the family $\{T_\Sigma(t) : t \geq 0\}$ is the *inverse Laplace transform* of the family $\left\{a^{-1}H_\Sigma^{a+V} : a \geq 0\right\}$. Let $-\omega_{V,\Sigma}$ be the bottom of the spectrum of the operator $\left(K_0\dot{+}V\right)_\Sigma$ and suppose $a > \omega_{V,\Sigma}$. From equality (4.59) below it follows that the domain of the operator $T_\Sigma^{a+V}(t)$, $t > 0$, coincides with that of H_Σ^{a+V}. From equality (4.60) we see that the domain of H_Σ^{a+V} does not depend on a, as long as $a > \omega_{V,\Sigma}$. From Corollary 2.32 we see that, for a large enough, the operator H_Σ^{a+V} is a mapping from dom $\left(aI + K_0\dot{+}V\right)^{1/2}$ into itself. From assertion (f) in Theorem 2.5 it follows that the latter domain coincides with the intersection dom $\left(K_0^{1/2}\right) \cap \text{dom}\left(V_+^{1/2}\right)$. We note that the domain of $K_0\dot{+}V$ is a core for the quadratic form

$$(f,g) \mapsto a\langle f,g\rangle + \left\langle K_0^{1/2}f, K_0^{1/2}g\right\rangle + \left\langle V_+^{1/2}f, V_+^{1/2}g\right\rangle - \left\langle V_-^{1/2}f, V_-^{1/2}g\right\rangle,$$

for f, $g \in$ dom $\left(K_0^{1/2}\right) \cap \text{dom}\left(V_+^{1/2}\right)$. The domain of this generalized Schrödinger form coincides with dom $\left(\left(aI + K_0\dot{+}V\right)^{1/2}\right)$.

REMARK 4. If we want to emphasize that operators like $T_\Sigma(s)$ or $D_\Sigma(s)$ depend on the potential V, we write $T_\Sigma^V(s)$ instead of $T_\Sigma(s)$ and similarly for the operator $D_\Sigma(s)$. Here $D_\Sigma(t)$ is the operator (see also formula (2.56)):

$$D_\Sigma(t) = \exp\left(-t\left(K_0\dot{+}V\right)\right) - J^*\exp\left(-t\left(K_0\dot{+}V\right)_\Sigma\right)J.$$

Here we wrote J instead of J_Σ.

In the following proposition, $-\omega_V$ denotes the bottom of the L^2-spectrum of the operator $K_0\dot{+}V$ and $-\omega_{V,\Sigma}$ that of $\left(K_0\dot{+}V\right)_\Sigma$. We note that $\omega_{V,\Sigma} \geq \omega_V$. This follows from the (in-)equalities

$$\exp\left(t\omega_{V,\Sigma}\right) = \left\|\exp\left(-t\left(K_0\dot{+}V\right)_\Sigma\right)\right\|_{2,2} \leq \left\|\exp\left(-t\left(K_0\dot{+}V\right)\right)\right\|_{2,2} = \exp\left(t\omega_V\right).$$

4.14. PROPOSITION. *Let $a > \omega_{V,\Sigma}$ and $b > \omega_V$. The following identities are true for $f \in$ dom $\left(K_0\dot{+}V\right)$:*

$$T_\Sigma(t_2)f = T_\Sigma(t_1)f + \exp\left(-t_1\left(K_0\dot{+}V\right)_\Sigma\right)T_\Sigma(t_2-t_1)f,\ t_2 > t_1 \geq 0; \tag{4.57}$$

$$H_\Sigma^{a+V}f = a\int_0^\infty e^{-at}T_\Sigma^V(t)f\,dt; \tag{4.58}$$

$$T_\Sigma^{b+V}(t)f = \left(I - J_\Sigma^*\exp\left(-t\left(bI + K_0\dot{+}V\right)_\Sigma\right)J_\Sigma\right)H_\Sigma^{b+V}f; \tag{4.59}$$

$$H_\Sigma^{a+b+V}f = \left(I - aJ_\Sigma^*\left(aI + K_0\dot{+}V\right)_\Sigma^{-1}J_\Sigma\right)H_\Sigma^{b+V}f. \tag{4.60}$$

REMARK 1. From Remark 3 preceding Proposition 4.14, it follows that the equalities (4.57) through (4.60) can also regarded as equalities on $\operatorname{dom}\left(K_0^{1/2}\right) \cap \operatorname{dom}\left(V_+^{1/2}\right)$. Since the expressions for $T_\Sigma^{a+V} f$ and $H_\Sigma^{b+V} f$ have also a meaning for bounded Borel measurable functions $f : E \to \mathbb{C}$, the above equalities are also true for all such functions. In the proof below, we do not specify the function f very precisely. The main point is that all the expressions involved should make sense.

REMARK 2. From (4.57) it follows that the family $\{T_\Sigma(t) : t \geq 0\}$ is continuous. To see this, notice that $\exp\left(-t\left(K_0 \dot{+} V\right)_\Sigma\right) T_\Sigma(0) = \exp\left(-t\left(K_0 \dot{+} V\right)_\Sigma\right) 1_\Gamma = 0$.

Proof. We first establish equality (4.57). Let f be appropriate, e.g. bounded and positive. Then we have

$$\exp\left(-t_1\left(K_0 \dot{+} V\right)_\Sigma\right) T_\Sigma(t_2 - t_1) f(x)$$

$$= \mathbb{E}_x\left[\exp\left(-\int_0^{t_1} V(X(u))du\right) T_\Sigma(t_2 - t_1) f(X(t_1)) : S > t_1\right]$$

$$= \mathbb{E}_x\left[\exp\left(-\int_0^{t_1} V(X(u))du\right)\right.$$

$$\left.\mathbb{E}_{X(t_1)}\left\{\exp\left(-\int_0^S V(X(u))du\right) f(X(S)) : S \leq t_2 - t_1\right\} : S > t_1\right]$$

(Markov property)

$$= \mathbb{E}_x\left[\exp\left(-\int_0^{t_1 + S\circ\vartheta_{t_1}} V(X(u))du\right) f\left(X\left(t_1 + S\circ\vartheta_{t_1}\right)\right) : S > t_1, t_1 + S\circ\vartheta_{t_1} \leq t_2\right]$$

($t_1 + S\circ\vartheta_{t_1} = S$ on the event $\{S > t_1\}$, $\mathbb{P}_x$-almost surely)

$$= \mathbb{E}_x\left[\exp\left(-\int_0^S V(X(u))du\right) f\left(X\left(S\right)\right) : S > t_1,\ S \leq t_2\right]$$

$$= T_\Sigma(t_2) f(x) - T_\Sigma(t_1) f(x).$$

A proof of (4.58) reads as follows. Again we take an appropriate function f and consider

$$a\int_0^\infty e^{-at} T_\Sigma(t) f(x) dt$$

$$= a\int_0^\infty \mathbb{E}_x\left[\exp\left(-at - \int_0^S V(X(u))du\right) f(X(S)) : S \leq t\right] dt$$

(Fubini)

$$= \mathbb{E}_x\left[a\int_S^\infty \exp\left(-at - \int_0^S V(X(u))du\right) f(X(S)) dt : S < \infty\right]$$

$$= \mathbb{E}_x \left[\exp\left(-aS - \int_0^S V(X(u))du \right) f(X(S)) : S < \infty \right]$$
$$= H_\Sigma^{a+V} f(x).$$

In order to show (4.59) we again fix an appropriate function f and notice the following identities:

$$\left(I - J_\Sigma^* \exp\left(-t\left(bI + K_0 \dot{+} V\right)_\Sigma J_\Sigma\right)\right) H_\Sigma^{b+V} f(x)$$
$$= H_\Sigma^{b+V} f(x) - \mathbb{E}_x \left[\exp\left(- \int_0^t (b + V(X(u)))\, du \right) H_\Sigma^{b+V} f(X(t)) : S > t \right]$$
$$= H_\Sigma^{b+V} f(x) - \mathbb{E}_x \left[\exp\left(- \int_0^t (b + V(X(u)))\, du \right) \right.$$
$$\left. \mathbb{E}_{X(t)} \left\{ \exp\left(- \int_0^S (b + V(X(u)))\, du \right) f(X(S)) : S < \infty \right\} : S > t \right]$$

(Markov property)

$$= H_\Sigma^{b+V} f(x) - \mathbb{E}_x \left[\exp\left(- \int_0^S (b + V(X(u)))\, du \right) f(X(S)) : t < S < \infty \right]$$
$$= \mathbb{E}_x \left[\exp\left(- \int_0^S (b + V(X(u)))\, du \right) f(X(S)) : S < \infty \right]$$
$$- \mathbb{E}_x \left[\exp\left(- \int_0^S (b + V(X(u)))\, du \right) f(X(S)) : t < S < \infty \right]$$
$$= T_\Sigma^V(t) f(x).$$

This proves the equality in (4.59). The equality in (4.60) follows upon applying equality (4.59) and then taking Laplace transforms. This proves Proposition 4.14.

Equality (4.64) in the next proposition will among others be used in Theorem 5.12 containing a general result on Hilbert-Schmidt properties and in Theorem 6.11, the corresponding theorem about trace properties. The equalities in (4.64) are called *Duhamel's formula* for singular perturbations: compare with Theorem 2.34 formula (2.61). Taking a rather formal point of view, the second equality in (4.64) below is a consequence of the equalities (4.63) and (4.59).

4.15. PROPOSITION. *(a) Between the families* $\{T_\Sigma(t) : t \geq 0\}$ *and the families* $\{D_\Sigma(t) : t \geq 0\}$ *the following relations exist. Suppose that the function* f *belongs to* $\mathrm{dom}(K_0 \dot{+} V)$. *The following identities are true:*

$$T_\Sigma(t) f = D_\Sigma(t) f + \int_0^t D_\Sigma(u)(K_0 \dot{+} V) f du; \tag{4.61}$$

$$D_\Sigma(t)f = T_\Sigma(t)f - \int_0^t T_\Sigma(u)(K_0\dot{+}V)\exp\left(-(t-u)\left(K_0\dot{+}V\right)\right)f du \tag{4.62}$$

$$= \frac{\partial}{\partial t}\int_0^t T_\Sigma(u)\exp\left(-(t-u)\left(K_0\dot{+}V\right)\right)f\,du \tag{4.63a}$$

$$= \lim_{s\downarrow 0}\int_0^t \exp\left(-u\left(K_0\dot{+}V\right)_\Sigma\right)\frac{T_\Sigma(s)-T_\Sigma(0)}{s}\exp\left(-(t-u)\left(K_0\dot{+}V\right)\right)f du$$
$$+ T_\Sigma(0)\exp\left(-t\left(K_0\dot{+}V\right)\right)f. \tag{4.63b}$$

(b) Suppose that in form sense $K_0\dot{+}V + aI \geq bI > 0$. Then on $\mathrm{dom}(K_0\dot{+}V)$ *the following identities are true:*

$$D_\Sigma(t) = H_\Sigma^{a+V}\exp\left(-t\left(K_0\dot{+}V\right)\right) - J^*\exp\left(-t\left(K_0\dot{+}V\right)_\Sigma\right)JH_\Sigma^{a+V} \tag{4.64}$$
$$+\int_0^t dsJ^*\exp\left(-s\left(\left(K_0\dot{+}V\right)_\Sigma\right)\right)JH_\Sigma^{a+V}\left(aI+K_0\dot{+}V\right)\exp\left(-(t-s)\left(K_0\dot{+}V\right)\right)$$
$$= 1_\Gamma\exp\left(-t\left(K_0\dot{+}V\right)\right)$$
$$+ J^*\left(aI+K_0\dot{+}V\right)_\Sigma\int_0^t ds\exp\left(-s\left(\left(K_0\dot{+}V\right)_\Sigma\right)\right)JH_\Sigma^{a+V}\exp\left(-(t-s)\left(K_0\dot{+}V\right)\right).$$

Proof. (a) Let f belong to $\mathrm{dom}\left(K_0\dot{+}V\right)$. In order to prove (4.61), we take Laplace transforms of both sides. So it suffices to show that the integral

$$a\int_0^\infty e^{-at}\left[D_\Sigma(t)f + \int_0^t D_\Sigma(u)(K_0\dot{+}V)f\,du\right]dt$$

equals H_Σ^{a+V}. Standard manipulations with Laplace transforms will show that the latter integral is equal to the following function:

$$a\left[\left(aI+K_0\dot{+}V\right)^{-1}f - J_\Sigma^*\left(aI+K_0\dot{+}V\right)_\Sigma^{-1}J_\Sigma f\right]$$
$$+\left[\left(aI+K_0\dot{+}V\right)^{-1} - J_\Sigma^*\left(aI+K_0\dot{+}V\right)_\Sigma^{-1}J_\Sigma\right]\left(K_0\dot{+}V\right)f$$
$$= a\,H_\Sigma^{a+V}\left(aI+K_0\dot{+}V\right)^{-1}f + H_\Sigma^{a+V}\left(aI+K_0\dot{+}V\right)^{-1}\left(K_0\dot{+}V\right)f$$
$$= H_\Sigma^{a+V}f.$$

Hence (4.61) follows. Once (4.62) is established the other equalities for $D_\Sigma(t)$ follow. This is so because equality (4.63a) follows via a standard differentiation rule and equality (4.63b) is a consequence of equality (4.57). The equality in (4.62) can be proved via a by now standard manipulation of Laplace transforms. In fact (4.62) is equivalent to the identity

$$a\int_0^\infty e^{-at}D_\Sigma(t)f\,dt = a\int_0^\infty e^{-at}T_\Sigma(t)f dt$$
$$- a\int_0^\infty e^{-at}\int_0^t T_\Sigma(u)(K_0\dot{+}V)\exp\left(-(t-u)\left(K_0\dot{+}V\right)\right)f du\,dt.$$

The right-hand side of the latter equality equals

$$H_{\Sigma}^{a+V} f - H_{\Sigma}^{a+V} \left(K_0 \dot{+} V\right) \left(aI + K_0 \dot{+} V\right)^{-1} f = aH_{\Sigma}^{a+V} \left(aI + K_0 \dot{+} V\right)^{-1} f$$
$$= \text{left-hand side.}$$

Consequently (4.62) is established now.
(b) Use $T_{\Sigma}(s)f(x) = \mathbb{E}_x\left(\exp\left(-\int_0^S V(X(u))du\right) f(X(S)) : S < s\right)$. Notice the equality (see (2.58) in Theorem 2.34) $T_{\Sigma}(s)^2 = T_{\Sigma}(s)$. From the equality in (4.63a) of Proposition 4.15, it follows that

$$\int_0^t D_{\Sigma}(s)ds = \int_0^t T_{\Sigma}(s) \exp\left(-(t-s)\left(K_0 \dot{+} V\right)\right) ds$$
$$= \int_0^t T_{\Sigma}(t-s) \exp\left(-s\left(K_0 \dot{+} V\right)\right) ds. \tag{4.65}$$

Hence, from the equalities (4.62) and (4.61) (in this order), we infer that, on $\operatorname{dom}\left(K_0 \dot{+} V\right)$,

$$D_{\Sigma}(t)$$
$$= T_{\Sigma}(t) - \int_0^t T_{\Sigma}(s) \exp\left(-(t-s)\left(K_0 \dot{+} V\right)\right) (K_0 \dot{+} V) ds \tag{4.66}$$
$$= T_{\Sigma}(t) - \int_0^t D_{\Sigma}(s)(K_0 \dot{+} V) ds$$
$$= T_{\Sigma}(t) - \int_0^t \left(\exp\left(-s\left(K_0 \dot{+} V\right)\right) - J^* \exp\left(-s\left(K_0 \dot{+} V\right)_{\Sigma}\right)\right) (K_0 \dot{+} V) ds \tag{4.67}$$
$$= T_{\Sigma}(t) - I + \exp\left(-t\left(K_0 \dot{+} V\right)\right) + \int_0^t J^* \exp\left(-s\left(K_0 \dot{+} V\right)_{\Sigma}\right) (K_0 \dot{+} V) J\, ds$$

and consequently

$$T_{\Sigma}(t) = D_{\Sigma}(t) + I - \exp\left(-t(K_0 \dot{+} V)\right) - \int_0^t J^* \exp\left(-s\left(K_0 \dot{+} V\right)_{\Sigma}\right) J \left(K_0 \dot{+} V\right)$$
$$= 1_{\Gamma} + 1_{\Sigma} - J^* \exp\left(-t\left(K_0 \dot{+} V\right)_{\Sigma}\right) J - \int_0^t J^* \exp\left(-s\left(K_0 \dot{+} V\right)_{\Sigma}\right) J \left(K_0 \dot{+} V\right)$$
$$= 1_{\Gamma} + J^* \int_0^t ds \exp\left(-s\left(K_0 \dot{+} V\right)_{\Sigma}\right) J \left(J^* \left(K_0 \dot{+} V\right)_{\Sigma} J - \left(K_0 \dot{+} V\right)\right).$$

So from (4.66) we infer that

$$1_{\Gamma} \exp\left(-t\left(K_0 \dot{+} V\right)\right)$$

$$
\begin{aligned}
&+\int_0^t J^* \exp\left(-s\left(K_0\dot{+}V\right)_\Sigma\right) J\left[J^*\left(K_0\dot{+}V\right)_\Sigma J-\left(K_0\dot{+}V\right)\right] \\
&\qquad \exp\left(-(t-s)\left(K_0\dot{+}V\right)\right) ds \\
&= 1_\Gamma \exp\left(-t\left(K_0\dot{+}V\right)\right)+\int_0^t\left(\frac{\partial}{\partial s}T_\Sigma(s)\right)\exp\left(-(t-s)\left(K_0\dot{+}V\right)\right) ds
\end{aligned}
$$

(integration by parts)

$$
\begin{aligned}
&= 1_\Gamma \exp\left(-t\left(K_0\dot{+}V\right)\right)+T_\Sigma(t)-T_\Sigma(0)\exp\left(-t\left(K_0\dot{+}V\right)\right) \\
&\qquad -\int_0^t T_\Sigma(s)\exp\left(-(t-s)\left(K_0\dot{+}V\right)\right)\left(K_0\dot{+}V\right) \\
&= D_\Sigma(t).
\end{aligned}
$$

Hence, for a as indicated, i.e. sufficiently large,

$$
\begin{aligned}
&D_\Sigma(t) \\
&= 1_\Gamma \exp\left(-t\left(K_0\dot{+}V\right)\right) \\
&\quad + J^* e^{at}\left(aI+\left(K_0\dot{+}V\right)_\Sigma\right)\int_0^t ds \exp\left(-s\left(aI+\left(K_0\dot{+}V\right)_\Sigma\right)\right) J \\
&\quad \left(\left(aI+K_0\dot{+}V\right)^{-1}-J^*\left(aI+\left(K_0\dot{+}V\right)_\Sigma\right)^{-1}J\right) \\
&\quad \left(aI+K_0\dot{+}V\right)\exp\left(-(t-s)\left(aI+K_0\dot{+}V\right)\right)
\end{aligned}
$$

(integration by parts)

$$
\begin{aligned}
&= 1_\Gamma \exp\left(-t\left(K_0\dot{+}V\right)\right) \\
&\quad + 1_\Sigma\left(\left(aI+K_0\dot{+}V\right)^{-1}-J^*\left(aI+\left(K_0\dot{+}V\right)_\Sigma\right)^{-1}J\right)\left(aI+K_0\dot{+}V\right) \\
&\qquad\qquad \exp\left(-t\left(K_0\dot{+}V\right)\right) \\
&\quad - J^*\exp\left(-t\left(K_0\dot{+}V\right)_\Sigma\right)J\left(\left(aI+K_0\dot{+}V\right)^{-1}-J^*\left(aI+\left(K_0\dot{+}V\right)_\Sigma\right)^{-1}J\right) \\
&\qquad\qquad \left(aI+K_0\dot{+}V\right) \\
&\quad + J^*\int_0^t ds \exp\left(-s\left(\left(K_0\dot{+}V\right)_\Sigma\right)\right) J \\
&\quad \left(\left(aI+K_0\dot{+}V\right)^{-1}-J^*\left(aI+\left(K_0\dot{+}V\right)_\Sigma\right)^{-1}J\right) \\
&\quad \left(aI+K_0\dot{+}V\right)^2\exp\left(-(t-s)\left(K_0\dot{+}V\right)\right) \\
&= \left(\left(aI+K_0\dot{+}V\right)^{-1}-J^*\left(aI+\left(K_0\dot{+}V\right)_\Sigma\right)^{-1}J\right)\left(aI+K_0\dot{+}V\right) \\
&\qquad\qquad \exp\left(-t\left(K_0\dot{+}V\right)\right) \\
&\quad - J^*\exp\left(-t\left(K_0\dot{+}V\right)_\Sigma\right)J\left(\left(aI+K_0\dot{+}V\right)^{-1}-J^*\left(aI+\left(K_0\dot{+}V\right)_\Sigma\right)^{-1}J\right) \\
&\qquad\qquad \left(aI+K_0\dot{+}V\right)
\end{aligned}
$$

$$+ J^* \int_0^t ds \exp\left(-s\left(\left(K_0 \dot{+} V\right)_\Sigma\right)\right) J$$
$$\left(\left(aI + K_0 \dot{+} V\right)^{-1} - J^*\left(aI + \left(K_0 \dot{+} V\right)_\Sigma\right)^{-1} J\right)$$
$$\left(aI + K_0 \dot{+} V\right)^2 \exp\left(-(t-s)\left(K_0 \dot{+} V\right)\right). \tag{4.68}$$

Since

$$\left(aI + K_0 \dot{+} V\right)^{-1} - J^*\left(aI + \left(K_0 \dot{+} V\right)_\Sigma\right)^{-1} J = H_\Sigma^{a+V}\left(aI + K_0 \dot{+} V\right)^{-1}$$

the equality in (4.64) is a direct consequence of equality (4.68). The second equality in (4.64) follows from the first via partial integration.

The statement and proof of the following proposition is to be found in Chapter 2, Proposition 2.22. In particular the quantities in (2.35) and (2.36) correspond to (4.69) and (4.70) respectively. The stopping time S is the penetration time of Γ. In this case $D_S(t) = D_\Sigma(t)$ for $t \geq 0$.

4.16. PROPOSITION. *Define the pinned measure $\mu_{0,x}^{t,y}$ on the σ-field generated by $X(s)$, $0 \leq s < t$, via the equality:*

$$\mu_{0,x}^{t,y}(A) = \mathbb{E}_x\left[1_A p_0(t-s, X(s), y)\right], \quad A \in \mathcal{F}_s, \quad 0 \leq s < t.$$

Define the martingale $s \mapsto M_V^{t,y}(s)$, $0 \leq s < t$, by

$$M_V^{t,y}(s) = \exp\left(-\int_0^s V(X(\sigma))d\sigma\right)\exp\left(-(t-s)\left(K_0 + V\right)\right)(X(s), y).$$

Then the function

$$\int_{\{S<t\}} \exp\left(-\int_0^t V(X(\sigma))d\sigma\right) d\mu_{0,x}^{t,y} \tag{4.69}$$
$$= \mathbb{E}_x\left[M_V^{t,y}(S) : S < t\right] \tag{4.70}$$

serves as an integral kernel for the operator $D_\Sigma(t)$, $t > 0$.

Instead of (4.69) or (4.70) we often write $D_\Sigma(t, x, y)$.

We want to emphasize the result in the following corollary. It will be used frequently.

4.17. COROLLARY. *Write $D_\Sigma(t)$ for the operator (see Remark 3 following Definition 4.13)*

$$D_\Sigma(t) = \exp\left(-t\left(K_0 \dot{+} V\right)\right) - J_\Sigma^* \exp\left(-t\left(K_0 \dot{+} V\right)_\Sigma\right) J_\Sigma.$$

Its integral kernel is given by

$$D_\Sigma(t,x,y) \tag{4.71}$$
$$= \mathbb{E}_x\left[\exp\left(-\int_0^S V(X(u))du\right)\exp\left(-(t-S)\left(K_0+V\right)\right)(X(S),y) : S < t\right].$$

Proof. The integral kernel $D_\Sigma(t,x,y)$ of $D_\Sigma(t)$ is given by (see the proof of Proposition 2.22)

$$D_\Sigma(t,x,y) \tag{4.72}$$
$$= \lim_{\epsilon\downarrow 0}\mathbb{E}_x\left[\exp\left(-\int_0^{t-\epsilon} V(X(u))du\right)p_0(\epsilon, X(t-\epsilon),y) : S \leq t-\epsilon\right]$$

(strong Markov property together with the fact that $S\circ\vartheta_S = 0$, $\mathbb{P}_x$-almost surely on $\{S<\infty\}$)

$$= \lim_{\epsilon\downarrow 0}\mathbb{E}_x\left[\exp\left(-\int_0^{S} V(X(u))du\right)\right.$$
$$\left.\times\mathbb{E}_{X(S)}\left\{\exp\left(-\int_0^{t-\epsilon-S} V(X(u))du\right)p_0(\epsilon, X(t-\epsilon-S),y)\right\} : S \leq t-\epsilon\right]$$
$$= \mathbb{E}_x\left[\exp\left(-\int_0^S V(X(u))du\right)\exp\left(-(t-S)(K_0+V)\right)(X(S),y) : S < t\right].$$

This proves Corollary 4.17.

4.18. PROPOSITION. *For $t>0$, the integral kernel $D_\Sigma(t,x,y)$ satisfies the following identities ($t>0$):*

$$D_\Sigma(t,x,y) + \int D_\Sigma(t/2,x,z)D_\Sigma(t/2,z,y)dz$$
$$= \mathbb{E}_x\left[\exp\left(-\int_0^S V(X(\sigma))d\sigma\right)\exp\left(-(t-S)\left(K_0+V\right)\right)(X(S),y) : S < \frac{t}{2}\right]$$
$$+ \mathbb{E}_y\left[\exp\left(-\int_0^S V(X(\sigma))d\sigma\right)\exp\left(-(t-S)\left(K_0+V\right)\right)(X(S),x) : S < \frac{t}{2}\right] \tag{4.73}$$

$$= \int D_\Sigma(t/2,x,z)\exp\left(-\frac{t}{2}\left(K_0+V\right)\right)(z,y)dz$$
$$+ \int D_\Sigma(t/2,y,z)\exp\left(-\frac{t}{2}\left(K_0+V\right)\right)(z,x)dz; \tag{4.74}$$

$$D_\Sigma(t,x,y) - \int D_\Sigma(t/2,x,z)D_\Sigma(t/2,z,y)dz$$

$$= \int D_\Sigma(t/2,x,z)\exp\left(-\frac{t}{2}(K_0+V)_\Sigma\right)(z,y)dz$$
$$+ \int D_\Sigma(t/2,y,z)\exp\left(-\frac{t}{2}(K_0+V)_\Sigma\right)(z,x)dz. \tag{4.75}$$

Proof. Again we write

$$M_V^{t,y}(s) = \exp\left(-\int_0^s V(X(\sigma))d\sigma\right)\exp\left(-(t-s)(K_0+V)\right)(X(s),y).$$

From the identity of Chapman-Kolmogorov together with the symmetry we obtain:

$$D_\Sigma(t,x,y) + \int D_\Sigma(t/2,x,z)D_\Sigma(t/2,z,y)dz$$
$$= \exp(-t(K_0+V))(x,y) - \exp(-t(K_0+V)_\Sigma)(x,y)$$
$$+ \exp(-t(K_0+V))(x,y) + \exp(-t(K_0+V)_\Sigma)(x,y)$$
$$- \int dz \exp\left(-\frac{t}{2}(K_0+V)_\Sigma\right)(x,z)\exp\left(-\frac{t}{2}(K_0+V)\right)(z,y)$$
$$- \int dz \exp\left(-\frac{t}{2}(K_0+V)_\Sigma\right)(y,z)\exp\left(-\frac{t}{2}(K_0+V)\right)(z,x)$$

(Feynman-Kac)

$$= 2\exp(-t(K_0+V))(x,y)$$
$$- \mathbb{E}_x\left[\exp\left(-\int_0^{t/2} V(X(\sigma))d\sigma\right)\exp\left(-\frac{t}{2}(K_0+V)\right)(X(t/2),y) : S \geq \frac{t}{2}\right]$$
$$- \mathbb{E}_y\left[\exp\left(-\int_0^{t/2} V(X(\sigma))d\sigma\right)\exp\left(-\frac{t}{2}(K_0+V)\right)(X(t/2),x) : S \geq \frac{t}{2}\right]$$
$$= 2\exp(-t(K_0+V))(x,y)$$
$$- \mathbb{E}_x\left[\exp\left(-\int_0^{t/2} V(X(\sigma))d\sigma\right)\exp\left(-\frac{t}{2}(K_0+V)\right)(X(t/2),y)\right]$$
$$+ \mathbb{E}_x\left[\exp\left(-\int_0^{t/2} V(X(\sigma))d\sigma\right)\exp\left(-\frac{t}{2}(K_0+V)\right)(X(t/2),y) : S < \frac{t}{2}\right]$$
$$- \mathbb{E}_y\left[\exp\left(-\int_0^{t/2} V(X(\sigma))d\sigma\right)\exp\left(-\frac{t}{2}(K_0+V)\right)(X(t/2),x)\right]$$
$$+ \mathbb{E}_y\left[\exp\left(-\int_0^{t/2} V(X(\sigma))d\sigma\right)\exp\left(-\frac{t}{2}(K_0+V)\right)(X(t/2),x) : S < \frac{t}{2}\right]$$

(Feynman-Kac and Chapman-Kolmogorov)

$$
\begin{aligned}
&= 2\exp\left(-t(K_0+V)\right)(x,y) \\
&\quad - \exp\left(-t\left(K_0+V\right)\right)(x,y) \\
&\quad + \mathbb{E}_x\left[\exp\left(-\int_0^{t/2} V(X(\sigma))d\sigma\right)\exp\left(-\frac{t}{2}\left(K_0+V\right)\right)(X(t/2),y) : S < \frac{t}{2}\right] \\
&\quad - \exp\left(-t\left(K_0+V\right)\right)(y,x) \\
&\quad + \mathbb{E}_y\left[\exp\left(-\int_0^{t/2} V(X(\sigma))d\sigma\right)\exp\left(-\frac{t}{2}\left(K_0+V\right)\right)(X(t/2),x) : S < \frac{t}{2}\right]
\end{aligned}
$$

(for $b \in E$, the process $u \mapsto M_V^{t,a}(u)$ is a $\mathbb{P}_b$-martingale)

$$
\begin{aligned}
&= \mathbb{E}_x\left[\exp\left(-\int_0^{S} V(X(\sigma))d\sigma\right)\exp\left(-(t-S)\left(K_0+V\right)\right)(X(S),y) : S < \frac{t}{2}\right] \\
&\quad + \mathbb{E}_y\left[\exp\left(-\int_0^{S} V(X(\sigma))d\sigma\right)\exp\left(-\left(t-S\right)\left(K_0+V\right)\right)(X(S),x) : S < \frac{t}{2}\right].
\end{aligned}
\tag{4.76}
$$

From equality (4.76) the equality in (4.73) follows. The equality in (4.74) is a consequence of (4.73) and so is equality (4.75).

As in (2.46) of Definition 2.30, we define the function h_Σ^{a+V} as follows.

4.19. DEFINITION. The function h_Σ^{a+V} is defined as follows:

$$
h_\Sigma^{a+V}(x) = \mathbb{E}_x\left[\exp\left(-\int_0^S \left(a+V(X(u))\right)du\right) : S < \infty\right] = H_\Sigma^{a+V}1(x). \tag{4.77}
$$

The function h_Σ^{a+V} will play an important role in Hilbert-Schmidt and trace conditions. Intuitively, it is an $a+V$-harmonic function on Σ and it is identically 1 on Γ.

Suppose that a is large enough. The following proposition says that the function

$$
x \mapsto \left[T_\Sigma^{a+V}(t)1\right](x) = \mathbb{E}_x\left[\exp\left(-\int_0^S \left(a+V(X(u))\right)du\right) : S < t\right]
$$

belongs to $L^p(E,m)$, $1 \leq p \leq \infty$, if and only if the function h_Σ^{a+V} belongs to $L^p(E,m)$ as well.

4.20. PROPOSITION. *Let $1 \leq p \leq \infty$. Let V be a Kato-Feller potential and choose $a > 0$ and $t > 0$ in such a way that*

$$
C(t) := e^{-at}\sup_{x\in E}\mathbb{E}_x\left[\exp\left(-\int_0^t V(X(u))du\right)\right] < 1.
$$

If the function $T_\Sigma^{a+V}(t)1$ *belongs to* $L^p(E,m)$*, then the function* $H_\Sigma^{a+V}1$ *is a member of* $L^p(E,m)$ *as well. Moreover, the following inequalities are valid:*

$$\left\|T_\Sigma^{a+V}(t)1\right\|_p \leq \left\|H_\Sigma^{a+V}1\right\|_p \leq \frac{1}{1-C(t)}\left\|T_\Sigma^{a+V}(t)1\right\|_p.$$

In particular, for such a choice of a *the function* $H_\Sigma^{a+V}1$ *is bounded and*

$$H_\Sigma^{a+V}1(x) \leq \frac{1}{1-C(t)}\sup_{x\in E}\mathbb{E}_x\left(\exp\left(\int_0^t V_-(X(u))du\right)\right).$$

That we can choose $a > 0$ and $t > 0$ such that $C(t) < 1$ is a consequence of Khas'minskii's lemma (see Corollary 3.6).

Proof. From equality (4.59) in Proposition 4.14, we get

$$T_\Sigma^{a+V}(t) = \left(I - J_\Sigma^*\exp\left(-t\left(aI+K_0\dot{+}V\right)_\Sigma\right)J_\Sigma\right)H_\Sigma^{a+V}.$$

From the latter it follows that

$$\left\|T_\Sigma^{a+V}(t)1\right\|_p \leq \left\|H_\Sigma^{a+V}1\right\|_p, \tag{4.78}$$

and also

$$\begin{aligned}
&\left\|T_\Sigma^{a+V}(t)1\right\|_p \geq \left(1-\left\|J^*\exp\left(-t\left(aI+K_0\dot{+}V\right)_\Sigma\right)\right\|_{p,p}\right)\left\|H_\Sigma^{a+V}1\right\|_p\\
&\geq \left(1-\left\|J^*\exp\left(-t\left(aI+K_0\dot{+}V\right)_\Sigma\right)\right\|_{\infty,\infty}\right)\left\|H_\Sigma^{a+V}1\right\|_p\\
&\geq (1-C(t))\left\|H_\Sigma^{a+V}1\right\|_p.
\end{aligned}\tag{4.79}$$

From (4.78) and (4.79) the first claim follows. The second assertion follows from the first for $p=\infty$.

In the following lemma, a sufficient condition is given in order that a function of the form $T_\Sigma^W(s)1$ belongs to $L^p(E,m)$ for some $1 \leq p \leq \infty$. It would be of some interest to know a similar result for $0 < p < 1$. In view of trace conditions, it would be particularly nice if Lemma 4.21 is true for $p = \frac{1}{2}$. Lemma 4.21 sometimes replaces Hölder's inequality.

4.21. Lemma. *Let* $1 \leq q \leq \infty$ *and suppose that the function* $W \geq 0$ *possesses the property that* $W^{1/q}(\cdot)\mathbb{P}_{(\cdot)}(S<1)$ *belongs to* $L^q(E,m)$*. Also suppose that the function* $x \mapsto \mathbb{P}_x(S<1)$ *is a member of* $L^q(E,m)$*. In addition, we suppose that* $\inf_{t>0}\sup_{x\in E}\int_0^t\left[\exp\left(-s(K_0)_\Sigma\right)W\right](x)ds < 1$*. Then the functions* $x \mapsto \mathbb{E}_x\left(\exp\left(\int_0^S W(X(u))du\right) : S<t\right)$, $t>0$*, belong to* $L^q(E,m)$*.*

Remark 1. Let $T : L^p(E,m) \to L^q(E,m)$ be a continuous linear operator. Then we write $\|T\|_{q,p}$ for its operator norm. We also freely use the (complex) Stein interpolation theorem: see Appendix A.40. This interpolation result will be applied to the operator valued analytic function $z \mapsto \int_0^s W^{1-z} \exp\left(-\sigma \left(K_0\right)_\Sigma\right) W^z d\sigma$, $0 \leq \operatorname{Re} z \leq 1$. Then the Stein interpolation theorem (see Appendix A.40) says, among other things, that

$$\begin{aligned} \left\|T(p^{-1})\right\|_{q,q} &\leq \sup_{\{\operatorname{Re} z=0\}} \|T(z)\|_{1,1}^{1-1/p} \sup_{\{\operatorname{Re} z=1\}} \|T(z)\|_{\infty,\infty}^{1/p} \\ &= \left\|\int_0^s \exp\left(-\sigma \left(K_0\right)_\Sigma\right) W d\sigma\right\|_\infty . \end{aligned}$$

The reader may find more explanation about interpolation theory in e.g. Reed and Simon [192]. We also freely use the Feynman-Kac formula. An extensive treatment of the Feynman-Kac formula is provided in Chapter 2. It can also be found in Demuth and van Casteren [82, 83] (for the general case) and in Simon [208, 206] for the Brownian motion case.

REMARK 2. Saying that the function $x \mapsto \mathbb{P}_x[S < 1]$ belongs to $L^1(E, m)$ is the same as saying that the set Γ possesses finite Dirichlet (or Newton) capacity, provided its hitting time T coincides with its penetration time S: see Definition 2.36, Definition D.17, and Proposition D.19.(b).

Proof. Let p be the conjugate exponent of q, i.e. $\dfrac{1}{p} + \dfrac{1}{q} = 1$, $1 \leq p, q \leq \infty$. First we fix $1 \geq t > 0$ so small that $\sup_{x\in E} \int_0^t \left[\exp(-sK_0)W\right](x)ds < 1$. We also introduce the operator families $S_{p,q}^W(s)$ and $T_{p,q}^W(s)$, $s \geq 0$, defined by respectively

$$\begin{aligned} S_{p,q}^W(s) &= W^{1/q} \exp\left(-s(K_0)_\Sigma\right) W^{1/p}; \\ T_{p,q}^W(s) &= \int_0^s W^{1/q} \exp\left(-\sigma(K_0)_\Sigma\right) W^{1/p} d\sigma = \int_0^s S_{p,q}^W(\sigma) d\sigma. \end{aligned}$$

For convenience we also write

$$G_t(x) = \mathbb{E}_x\left[\exp\left(\int_0^S W(X(u))du\right) : S < t\right].$$

First note the identity

$$G_t(x) = \mathbb{P}_x(S < t) + \sum_{n=1}^\infty \mathbb{E}_x\left[\frac{\left(\int_0^S W(X(u))du\right)^n}{n!} : S < t\right].$$

Then we fix $n \in \mathbb{N}$ and we consider

$$\mathbb{E}_x\left[\frac{\left(\int_0^S W(X(u))du\right)^{n+1}}{(n+1)!} : S < t\right]$$

(Markov property)

$$= \left[\int_{0<\tau_j,\tau_1+\cdots+\tau_{n+1}<t} \int d\tau_1 \dots d\tau_{n+1}\right.$$

$$\left. \exp\left(-\tau_1 \left(K_0\right)_\Sigma\right) W \dots \exp\left(-\tau_{n+1} \left(K_0\right)_\Sigma\right) W \mathbb{P}_{(\cdot)} \left(S < t - \left(\tau_1 + \cdots + \tau_{n+1}\right)\right)\right](x)$$

$$= \left[\int_{0<\tau_j,\tau_1+\cdots+\tau_{n+1}<t} \int d\tau_1 \dots d\tau_{n+1} \exp\left(-\tau_1 \left(K_0\right)_\Sigma\right) W^{1/p} S_{p,q}^W\left(\tau_2\right) \dots S_{p,q}^W\left(\tau_{n+1}\right)\right.$$

$$\left. \left(W^{1/q} \mathbb{P}_{(\cdot)} \left(S < t - \left(\tau_1 + \cdots + \tau_{n+1}\right)\right)\right)\right](x)$$

$$\leq \left[\int_0^t d\tau \exp\left(-\tau \left(K_0\right)_\Sigma\right) W^{1/p} \left(T_{p,q}^W(t-\tau)\right)^n W^{1/q} \mathbb{P}_{(\cdot)}(S < t-\tau)\right](x)$$

$$\leq \left[\int_0^t d\tau \exp\left(-\tau \left(K_0\right)_\Sigma\right) W^{1/p} \left(T_{p,q}^W(t)\right)^n W^{1/q} \mathbb{P}_{(\cdot)}(S < t)\right](x). \tag{4.80}$$

From the complex Stein interpolation theorem, it follows that

$$\begin{aligned} \left\|T_{p,q}^W(t)\right\|_{q,q} &= \left\|T_{p,q}^W(t)\right\|_{L^q(E,m),L^q(E,m)} \\ &\leq \left\|T_{1,\infty}^W\right\|_{\infty,\infty} = \sup_{x\in E} \int_0^t \left[\exp\left(-\sigma \left(K_0\right)_\Sigma\right) W\right](x) d\sigma < 1. \end{aligned} \tag{4.81}$$

Consequently $\left\|T_{p,q}^W(t)\right\|_{q,q} < 1$ and hence it makes sense to write $\left(I - T_{p,q}^W(t)\right)^{-1}$. From (4.80) together with (4.81) we infer that

$$\begin{aligned} &\sum_{n=1}^\infty \mathbb{E}_x\left(\frac{\left(\int_0^S W(X(u))du\right)^n}{n!} : S < t\right) \\ &\qquad \leq \left[\int_0^t d\tau \exp\left(-\tau \left(K_0\right)_\Sigma\right) W^{1/p} \left(I - T_{p,q}^W(t)\right)^{-1} W^{1/q} \mathbb{P}_{(\cdot)}(S < t)\right](x). \end{aligned}$$

Since

$$G_t(x) \leq \mathbb{P}_x(S < t) + \sum_{n=1}^\infty \mathbb{E}_x\left(\frac{\left(\int_0^S W(X(u))du\right)^n}{n!} : S < t\right),$$

we see that (here we use Stein interpolation again)

$$\begin{aligned} &\left\|G_t\right\|_q - \left(\int \left(\mathbb{P}_x(S < t)\right)^q dx\right)^{1/q} \\ &\leq \frac{\left\|\int_0^t d\tau \exp\left(-\tau \left(K_0\right)_\Sigma\right) W^{1/p}\right\|_{q,q}}{1 - \left\|T_{p,q}^W(t)\right\|_{q,q}} \left(\int W(x) \left(\mathbb{P}_x(S < t)\right)^q dx\right)^{1/q} \end{aligned}$$

$$\leq \frac{\left\| \int_0^t d\tau \max(W,1)^{1/q} \exp\left(-\tau \left(K_0\right)_\Sigma\right) \max(W,1)^{1/p} \right\|_{q,q}}{1 - \left\| T^W_{p,q}(t) \right\|_{q,q}}$$
$$\times \left(\int W(x) \left(\mathbb{P}_x(S < t)\right)^q dx \right)^{1/q}$$
$$\leq C(t) \left(\int W(x) \left(\mathbb{P}_x(S < t)\right)^q dx \right)^{1/q}, \tag{4.82}$$

where the constant $C(t)$ is given by

$$C(t) = \frac{\sup_{x\in\Sigma} \int_0^t \left[\exp\left(-\tau (K_0)_\Sigma\right) \max(W,1)\right](x) d\tau}{1 - \sup_{x\in\Sigma} \int_0^t \left[\exp\left(-\tau \left(K_0\right)_\Sigma\right) W\right](x) d\tau}.$$

From (4.82) the result in the lemma follows for $t > 0$, but sufficiently small. For general $t > 0$ we proceed in the following manner. Fix $1 \geq t_0 > 0$ so small that $\left\| G_{t_0} \right\|_q < \infty$ and define the functions $G^{(k)}(x)$ by

$$\begin{aligned} G^{(k)}(x) &= \mathbb{E}_x \left(\exp\left(\int_0^S W(X(u)) du \right) : kt_0 \leq S < (k+1)t_0 \right) \\ &= \left[\exp\left(-kt_0 \left(K_0\right)_\Sigma\right) G_{t_0}\right](x) \end{aligned} \tag{4.83}$$

and we note that

$$\left\| G^{(k)} \right\|_q \leq \left\| G_{t_0} \right\|_q < \infty \tag{4.84}$$

and also that

$$G_t(x) \leq \sum_{k=0}^{n+1} G^{(k)}(x), \quad \text{where} \quad n = \left\lfloor \frac{t}{t_0} \right\rfloor. \tag{4.85}$$

It is clear that the claim in the lemma follows from (4.84) and (4.85).

Chapter 5

Hilbert-Schmidt Properties of Resolvent and Semigroup Differences

As in Chapter 2 the generator K_0 will be perturbed in two ways. First there will be a "regular" perturbation, being a multiplication operator V and secondly, there will be a potential barrier on Γ, a closed subset of E. The aim of this chapter is to study Hilbert-Schmidt properties for resolvent and/or semigroup differences corresponding to both kinds of perturbations. In part A we consider regular perturbations. The main results and estimates in section A of this chapter can be found in Theorem 5.6 and 5.7, and the most important theorems in section B are Theorem 5.12 and 5.16.

A. Regular perturbations

In the following propositions we collect some of the elementary, but very useful, identities and preliminary inequalities. The first identity in Proposition 5.1 is called *Duhamel's formula.* Strictly speaking this formula cannot be correct as stated because there is a domain problem. If we interpret Duhamel's formula in form sense, it can be given a meaning. More precisely, for $f \in \mathrm{dom}(A)$ and $g \in \mathrm{dom}(B^*)$ formula (5.1) below can be written as

$$\langle D(t)f, g\rangle = \int_0^t \langle W(t-u)f, B^*V(u)^*g\rangle du - \int_0^t \langle AW(t-u)f, V(u)^*g\rangle du.$$

Also if the operator $B-A$ in Proposition 5.1 is not bounded (see B.6. in Appendix B), then the difference has to be interpreted in form sense. The formula is correct on $\mathrm{dom}(B)$ if the domain of B is a subset of the domain of A. The *Lie product* $[A, B]$ of two operators A and B will be denoted by $[A, B] = AB - BA$.

5.1. PROPOSITION. *Let $-A$ and $-B$ be the generators of the semigroups $\{V(t) : t \geq 0\}$ respectively $\{W(t) : t \geq 0\}$ in $L^2(E, m)$. Put $D(t) = V(t) - W(t)$. The following identities are true:*

$$D(t) = \int_0^t V(u)(B-A)W(t-u)du; \tag{5.1}$$

$$D(2t) = D(t)V(t) + W(t)D(t) \tag{5.2}$$

$$= D(t)V(t) + V(t)D(t) - D(t)^2 \tag{5.3}$$

$$= D(t)W(t) + W(t)D(t) + D(t)^2; \tag{5.4}$$

$$D(4t) = V(t)D(2t)V(t) + W(t)D(2t)W(t) + [V(t), W(t)]\, W(t)V(t) - W(t)V(t)\, [V(t), W(t)]. \tag{5.5}$$

Proof. The first formula follows because we have (in form sense)

$$\begin{aligned}
&(aI + A)^{-1} D(t)(aI + B)^{-1} \\
&= \int_0^t du \frac{\partial}{\partial u} (aI + A)^{-1} V(u) W(t-u)(aI + B)^{-1} \\
&= \int_0^t du V(u) \left((aI + A)^{-1} - (aI + B)^{-1} \right) W(t-u).
\end{aligned}$$

In order to avoid domain problems we have used the "resolvent trick" (see B.6. in Appendix B as well). The other identities can be proved via a straight-forward use of the semigroup properties.

Now we come to regular perturbations of K_0 and applications of Duhamel's formula to these Feller semigroups. Instead of $K_0\dot{+}[(1-s)V + sW]$ we write $K_0\dot{+}(1-s)V + sW$. Here V and W are Kato-Feller potentials.

5.2. PROPOSITION. *Let V and W be Kato-Feller potentials and write $K(s) = K_0\dot{+}(1-s)V + sW$. Put*

$$D_s(t) = \int_0^t \exp\left(-uK(s)\right)(W - V)\exp\left(-(t-u)K(s)\right) du = -\frac{\partial}{\partial s} \exp\left(-tK(s)\right), \tag{5.6}$$

and put

$$D(t) = \exp\left(-t\left(K_0\dot{+}V\right)\right) - \exp\left(-t\left(K_0\dot{+}W\right)\right).$$

Then the following identities are valid:

$$D_s(2t) = D_s(t)\exp\left(-tK(s)\right) + \exp\left(-tK(s)\right) D_s(t); \tag{5.7}$$

$$D(t) = \int_0^1 D_s(t) ds. \tag{5.8}$$

Proof. Fix $0 \leq s,\ s_0 \leq 1,\ s \neq s_0$. From Duhamel's formula we get

$$\begin{aligned}
&\exp(-tK(s)) - \exp\left(-tK(s_0)\right) \\
&= -(s - s_0)\int_0^t \exp\left(-uK(s)\right)(W - V)\exp\left(-(t-u)K(s_0)\right) du.
\end{aligned}$$

So that the second equality in (5.6) follows. The equality in (5.7) follows from the first equality in (5.6) and the second equality in (5.6) implies (5.8).

5.3. PROPOSITION. *Let V and W be Kato-Feller potentials and let $D_s(t)$ be as in Proposition 5.2. Let $\mu_{0,x}^{t,y}$ be the pinned Markov measure determined by the equality*

$$\mu_{0,x}^{t,y}(A) = \mathbb{E}_x\left(1_A p_0(t-s, X(s), y)\right)$$

whenever A is an event in $\mathcal{F}_s$, where $0 \leq s < t$: see Definition 2.18, item (e). Let the additive process $A(t)$ be defined by $A(t) = \int_0^t (W(X(u)) - V(X(u)))\,du$. Let $D(t,x,y)$ and $D_s(t,x,y)$ be the integral kernels of the operators $D(t)$ and $D_s(t)$ respectively. Finally put $V_s = (1-s)V + sW$. The following identities are true:

$$D_s(t,x,y) = \int A(t) \exp\left(-\int_0^t V_s(X(\sigma))d\sigma\right) d\mu_{0,x}^{t,y}; \tag{5.9}$$

$$D(t,x,y) = \int_0^1 ds\, D_s(t,x,y); \tag{5.10}$$

$$\begin{aligned}&\int D_s(t_1,x,z) \exp\left(-t_2 K(s)\right)(z,y)dz \\ &\qquad = \int A(t_1) \exp\left(-\int_0^{t_1+t_2} V_s(X(\sigma))d\sigma\right) d\mu_{0,x}^{t_1+t_2,y}.\end{aligned} \tag{5.11}$$

The following inequalities are true:

$$\begin{aligned}&\left|\int D_s(t_1,x,z) \exp\left(-t_2 K(s)\right)(z,y)dz\right|^2 \\ &\leq \int A(t_1)^2 d\mu_{0,x}^{t_1+t_2,y} \exp\left(-\left(t_1+t_2\right)\left(K_0 \dot{+} 2V_s\right)\right)(x,y);\end{aligned} \tag{5.12}$$

$$|D_s(t,x,y)|^2 \leq \int A(t)^2 d\mu_{0,x}^{t,y} \exp\left(-t\left(K_0 \dot{+} 2V_s\right)\right)(x,y); \tag{5.13}$$

$$\begin{aligned}&\int \left|\int D_s(t_1,x,z) \exp\left(-t_2\left(K_0 \dot{+} 2V_s\right)\right)(z,y)dz\right|^2 dx \\ &\leq \left[\exp\left(-t_2 K_0\right) \mathbb{E}_{(\cdot)}\left(A(t_1)^2\right)\right](y) \sup_{x\in E} \exp\left(-\left(t_1+t_2\right)\left(K_0 \dot{+} 2V_s\right)\right)(x,y);\end{aligned} \tag{5.14}$$

$$\int |D_s(t,x,y)|^2\, dx \leq \mathbb{E}_y\left(A(t)^2\right) \sup_{x\in E} \exp\left(-t\left(K_0 \dot{+} 2V_s\right)\right)(x,y). \tag{5.15}$$

Proof. From (5.6) it follows that $D_s(t,x,y)$ is given by

$$\begin{aligned}&D_s(t,x,y) \\ &= \int_0^t \int \exp\left(-uK(s)\right)(x,z)(W(z)-V(z)) \exp\left(-(t-u)K(s)\right)(z,y)dz\,du \\ &= \int_0^t \mathbb{E}_x\left[\exp\left(-\int_0^u V_s(X(\sigma))d\sigma\right)(W(X(u)) - V(X(u)))\right. \\ &\qquad \left. \times \int \exp\left(-\int_0^{t-u} V_s(X(\sigma))d\sigma\right) d\mu_{0,X(u)}^{t,y}\right] du\end{aligned}$$

(Markov property for pinned measure)

$$= \int_0^t \int \exp\left(-\int_0^t V_s(X(\sigma))d\sigma\right) (W(X(u)) - V(X(u)))\, d\mu_{0,x}^{t,y} du. \tag{5.16}$$

The equality in (5.9) is an immediate consequence of (5.16). The equality in (5.10) corresponds to (5.8). For $t_2 = 0$, the equality in (5.11) reduces to the one in (5.9). In order to prove (5.11) in its full generality, we observe that

$$\begin{aligned}
&\int D_s(t_1, x, z) \exp\left(-t_2 K(s)\right)(z, y) dz \\
&= \int_0^{t_1} \int \exp\left(-uK(s)\right)(x, z)(W(z) - V(z)) \exp\left(-(t_1 + t_2 - u)K(s)\right)(z, y) dz\, du.
\end{aligned}$$

Reasoning as in the proof of (5.9) will then prove (5.11). The inequalities in (5.12) and (5.13) follow from the equalities (5.9) and (5.11) respectively by applying the Cauchy-Schwarz inequality. For the proof of the inequality in (5.14) and also for the one in (5.15), it suffices to show that

$$\int dx \int A(t_1)^2 d\mu_{0,x}^{t_1+t_2,y} \leq \left[\exp\left(-t_2 K_0\right) \mathbb{E}_{(\cdot)}\left(A(t_1)^2\right)\right](y). \tag{5.17}$$

In the proof of inequality (5.15), we take $t_1 = t$ and $t_2 = 0$ in (5.17). In order to prove (5.17), we observe the equality

$$\begin{aligned}
&\iint A(t_1)^2 d\mu_{0,x}^{t_1+t_2,y} dx \\
&= 2\int_0^{t_1} du_1 \int_{u_1}^{t_1} du_2 \int dx \int dz_1 \int dz_2\, p_0(u_1, x, z_1) \\
&\qquad \times (W(z_1) - V(z_1))\, p_0(u_2 - u_1, z_1, z_2)\, (W(z_2) - V(z_2))\, p_0(t_1 + t_2 - u_2, z_2, y)
\end{aligned}$$

(make the substitutions $v_1 = t_1 - u_2$, $v_2 = t_1 - u_1$, $z_2 = w_1$, $z_1 = w_2$;
also use the symmetry and Markov property)

$$\begin{aligned}
&= 2\int dw_0 \int_0^{t_1} dv_1 \int_{v_1}^{t_1} dv_2 \int dw_1 \int dw_2 \int dx\, p_0(t_2, y, w_0)\, p_0(v_1, w_0, w_1) \\
&\qquad \times (W(w_1) - V(w_1))\, p_0(v_2 - v_1, w_1, w_2)\, (W(w_2) - V(w_2))\, p_0(t_1 - v_2, w_2, x) \\
&= \int dw_0\, p_0(t_2, y, w_0) \int dx \int \left(\int_0^{t_1} (W(X(v)) - V(X(v)))\, dv\right)^2 d\mu_{0,w_0}^{t,x} \\
&= \int dw_0\, p_0(t_2, y, w_0) \int dx \int \left(\lim_{\epsilon \downarrow 0} \int_0^{t_1-\epsilon} (W(X(v)) - V(X(v)))\, dv\right)^2 d\mu_{0,w_0}^{t,x}
\end{aligned}$$

(Fatou)

$$\begin{aligned}
&\leq \lim_{\epsilon \downarrow 0} \int dw_0\, p_0(t_2, y, w_0) \\
&\mathbb{E}_{w_0}\left[\left(\int_0^{t_1-\epsilon} (W(X(u)) - V(X(u)))\, du\right)^2 \int dx\, p_0(\epsilon, X(t_1 - \epsilon), x)\right]
\end{aligned}$$

$$\leq \int dw_0\, p_0(t_2, y, w_0)\mathbb{E}_{w_0}\left(\int_0^{t_1} (W(X(u)) - V(X(u)))\, du\right)^2$$
$$= \left[\exp\left(-t_2K_0\right)\mathbb{E}_{(\cdot)}\left(A(t_1)^2\right)\right](y).$$

This proves the inequality in (5.14).

As a preliminary result for the Hilbert-Schmidt properties of Theorem 5.6, we obtain the following theorem.

5.4. THEOREM. *Let V and W be Kato-Feller potentials and put*

$$A(t) = \int_0^t (W(X(u)) - V(X(u)))\, du\,.$$

Suppose that the semigroup $\{\exp(-tK_0) : t \geq 0\}$ is L^1-L^∞-smoothing.

(a) If
$$\int \mathbb{E}_x\left[A(t)^2\right] dx < \infty,$$
then the operator $D(t)$ is Hilbert-Schmidt.

(b) If
$$\int dx\left(\mathbb{E}_x\left(A(t/2)^2\right)\right)^{1/2} < \infty,$$
then the operator $D(t)$ is a trace class operator.

(c) Let φ and ψ be Borel measurable functions on E with the property that the integral
$$\int \mathbb{E}_x\left(A(t)^2\right)\left(|\varphi(x)|^4 + |\psi(x)|^4\right) dx$$
is finite. Then the operator $\varphi D(t)\psi$ is Hilbert-Schmidt.

REMARK 1. In part (c) with $\psi \equiv 1$, the assumption that the integral

$$\int \mathbb{E}_x\left(A(t)^2\right)|\varphi(x)|^2\, dx$$

is finite suffices in order to conclude that the operator $\varphi D(t)$ is Hilbert-Schmidt.

REMARK 2. Let T and S be two bounded operators in $L^2(E, m)$. Let φ be a nowhere vanishing Borel function. If we want to estimate the trace norm of TS, we rewrite TS as $T\varphi^{-1}\varphi S$ and we notice that $\|TS\|_{\text{trace}} \leq \left\|T\varphi^{-1}\right\|_{\text{HS}} \|\varphi S\|_{\text{HS}}$.

REMARK 3. Let the process $A(t)$ be as in Theorem 5.4. Suppose that $W - V$ belongs to $L^2(E, m)$. Then

$$\int \mathbb{E}_x\left(A(t)^2\right) dx$$

$$
\begin{aligned}
&= 2\int_0^t dv \int_0^v du \int dx \int dy \int dz p_0(u,x,y) \\
&\qquad p_0(v-u,y,z)\big(W(y)-V(y)\big)\big(W(z)-V(z)\big) \\
&\leq 2\int_0^t dv \int_0^v du \int dx \int dy \int dz p_0(u,x,y) \\
&\qquad p_0(v-u,y,z)\,|W(y)-V(y)|\,|W(z)-V(z)| \\
&\leq 2\int_0^t dv \int_0^v du \int dy \int dz\, p_0(v-u,y,z)\,|W(y)-V(y)|\,|W(z)-V(z)| \\
&= 2\int_0^t du\,(t-u) \int dy \int dz\, p_0(u,y,z)\,|W(y)-V(y)|\,|W(z)-V(z)| \\
&\leq t^2\,\|W-V\|_2^2 .
\end{aligned}
$$

Similarly, we see that

$$
\int \mathbb{E}_x\left(A(t)^2\right)dx \leq 2t\left\|\int_0^t du\exp\left(-uK_0\right)|W-V|\right\|_\infty \|W-V\|_1 ,
$$

in case $W-V$ belongs to $K(E)\cap L^1(E,m)$.

Proof. (a) The Hilbert-Schmidt norm of the operator $D_s(t)$ equals the square root of the double integral $\iint dxdyD_s(t,x,y)^2$. From (5.15) it follows that

$$
\begin{aligned}
&\iint D_s(t,x,y)^2 dx\,dy \\
&\leq \int \mathbb{E}_y\left(A(t)^2\right)dy \sup_{x,y\in E} \exp\left(-t\left(K_0\dot{+}2(1-s)V+2sW\right)\right)(x,y).
\end{aligned}
$$

So that

$$
\begin{aligned}
&\|D(t)\|_{\mathrm{HS}} \leq \int_0^1 \|D_s(t)\|_{\mathrm{HS}}\,ds \\
&\leq \sqrt{\int \mathbb{E}_x\left(A(t)^2\right)dx}\left(\sup_{0\leq s\leq 1}\sup_{x,y\in E}\exp\left(-tK_0\dot{+}2(1-s)V+2sW\right)(x,y)\right)^{1/2}.
\end{aligned}
$$

This proves (a).

(b) For the proof of this assertion, we shall employ the following result on trace class operators. A proof of this fact will be given in Chapter 6: see Theorem 6.1.

Let $K = K_1\circ K_2$ be composition of integral operators K_1 and K_2 with integral kernels k_1 respectively k_2. Suppose that the integral

$$
\int dz\left(\int |k_1(x,z)|^2\,dx\right)^{1/2}\left(\int |k_2(z,x)|^2\,dx\right)^{1/2}
$$

is finite. Then the operator K is a trace class operator and

$$\|K\|_{\text{trace}} \leq \int \left(\int |k_1(x,z)|^2 \, dx \right)^{1/2} \left(\int |k_2(z,x)|^2 \, dx \right)^{1/2} dz. \tag{5.18}$$

A proof of assertion (b) then reads as follows:

$$\|D_s(t)\|_{\text{trace}} \leq 2 \left\| D_s(t/2) \exp\left(-\frac{t}{2} K(s) \right) \right\|_{\text{trace}}$$

$$\leq \int \left(\int D_s(t/2,x,z)^2 dx \right)^{1/2} \left(\int \exp\left(-\frac{t}{2} K(s) \right) (z,x)^2 dx \right)^{1/2} dz$$

(employ (5.15) and Chapman-Kolmogorov)

$$\leq \int dz \left(\mathbb{E}_z \left[A(t/2)^2 \right] \right)^{1/2} \left(\sup_{0 \leq s \leq 1} \sup_{x,y \in E} \exp\left(-\frac{t}{2} \left(K_0 + 2V_s \right) \right) (x,y) \right)^{1/2}$$

$$\times \left(\sup_{0 \leq s \leq 1} \sup_{x \in E} \exp\left(-tK(s) \right) (x,x) \right)^{1/2}.$$

Hence

$$\|D(t)\|_{\text{trace}} \leq \int_0^1 \|D_s(t)\|_{\text{trace}} \, ds \tag{5.19}$$

$$\leq \int \left(\mathbb{E}_z \left[A(t/2)^2 \right] \right)^{1/2} \left(\sup_{0 \leq s \leq 1} \sup_{x \in E} \exp\left(-\frac{t}{2} \left(K_0 + 2V_s \right) \right) (x,x) \right)^{1/2} dz$$

$$\times \left(\sup_{0 \leq s \leq 1} \sup_{x \in E} \exp\left(-tK(s) \right) (x,x) \right)^{1/2}.$$

(c) Again we invoke inequality (5.13). We are going to estimate the integral:

$$\iint dxdy \, |\varphi(x)|^2 \, D_s(t,x,y)^2 \, |\psi(y)|^2$$

(employ (5.13))

$$\leq \left\| \exp\left(-t \left(K_0 \dot{+} 2V_s \right) \right) \right\|_{\infty,1} \iint dxdy \, |\varphi(x)|^2 \int A(t)^2 d\mu_{0,x}^{t,y} \, |\psi(y)|^2$$

$$\leq \frac{1}{2} \left\| \exp\left(-t \left(K_0 \dot{+} 2V_s \right) \right) \right\|_{\infty,1} \iint dxdy \left(|\varphi(x)|^4 + |\psi(y)|^4 \right) \int A(t)^2 d\mu_{0,x}^{t,y}$$

(the function $(x,y) \mapsto \int A(t)^2 d\mu_{0,x}^{t,y}$ is symmetric)

$$= \frac{1}{2} \left\| \exp\left(-t \left(K_0 \dot{+} 2V_s \right) \right) \right\|_{\infty,1} \iint dxdy \left(|\varphi(x)|^4 + |\psi(x)|^4 \right) \int A(t)^2 d\mu_{0,x}^{t,y}$$

$$= \frac{1}{2} \left\| \exp\left(-t \left(K_0 \dot{+} 2V_s \right) \right) \right\|_{\infty,1} \int dx \left(|\varphi(x)|^4 + |\psi(x)|^4 \right) \mathbb{E}_x \left[A(t)^2 \right]. \tag{5.20}$$

An inspection of the proof of (c) shows that Remark 1 is valid as well.

5.5. COROLLARY. *Suppose that the Kato-Feller norm of $W-V$ is finite. If*

$$\int dx\, |W(x)-V(x)| \int_0^t dv \int dy\, p_0(v,x,y) \left(|\varphi(y)|^4 + |\psi(y)|^4\right)$$

is finite, then $\varphi D(t)\psi$ is Hilbert-Schmidt. Moreover, the following estimate is valid:

$$\begin{aligned}
&\|\varphi D(t)\psi\|_{\mathrm{HS}}^2 \\
&\le \sup_{0<s<1} \left\|\exp\left(-t\left(K_0\dot{+}2V_s\right)\right)\right\|_{\infty,1} \sup_{x\in E} \int_0^t dv \left[\exp\left(-vK_0\right)|W-V|\right](x) \\
&\quad\times \int dx\, |W(x)-V(x)| \int_0^t du \left[\exp\left(-uK_0\right)\left(|\varphi|^4+|\psi|^4\right)\right](x).
\end{aligned}$$

If $\psi \equiv 1$, then the latter inequality is valid with $|\varphi|^2$ replacing the function $\frac{1}{2}\left(|\varphi|^4+|\psi|^4\right)$.

Proof. We employ inequality (5.20) in the proof assertion (c) of Theorem 5.4. From this inequality we see that, with $C_s(t) := \left\|\exp\left(-t\left(K_0\dot{+}2V_s\right)\right)\right\|_{\infty,1}$,

$$\begin{aligned}
&\iint dx\, dy\, |\varphi(x)|^2\, D_s(t,x,y)^2\, |\psi(y)|^2 \\
&\le \frac{1}{2} C_s(t) \int dx \left(|\varphi(x)|^4 + |\psi(x)|^4\right) \mathbb{E}_x\left[A(t)^2\right]
\end{aligned}$$

(Markov property)

$$\begin{aligned}
&= C_s(t) \int_0^t du \int_u^t dv \int dx \left[\exp\left(-uK_0\right)(W-V)\exp\left(-(v-u)K_0\right)(W-V)\right](x) \\
&\quad\times \left(|\varphi(x)|^4 + |\psi(x)|^4\right) \\
&\le C_s(t) \int_0^t du \int_0^t dv \int dx \left[\exp\left(-uK_0\right)|W-V|\exp\left(-vK_0\right)|W-V|\right](x) \\
&\quad\times \left(|\varphi(x)|^4 + |\psi(x)|^4\right) \\
&\le C_s(t) \sup_{x\in E} \int_0^t du \left[\exp\left(-uK_0\right)|W-V|\right](x) \\
&\quad\times \int_0^t dv \int dx \left[\exp\left(-vK_0\right)|W-V|\right](x) \left(|\varphi(x)|^4 + |\psi(x)|^4\right).
\end{aligned}$$

From this the result in the Corollary readily follows. An inspection of the proof shows that the statement with $\psi \equiv 1$ is valid as well.

The following theorem contains a general result on Hilbert-Schmidt properties of differences of Feynman-Kac semigroups. As usual we write $D(t) = \exp\left(-t\left(K_0\dot{+}V\right)\right) - \exp\left(-t\left(K_0\dot{+}W\right)\right)$.

5.6. THEOREM. *Fix $t > 0$. Among the assertions*
(i) the function $W - V$ belongs to $L^2(E, m)$;
(i′) the integrals

$$\int_0^{t/2} du\, u \left\|\exp\left(-u\left(K_0 \dot{+} V\right)\right)|W - V|\right\|_2^2 \quad \textit{and}$$
$$\int_0^{t/2} du\, u \left\|\exp\left(-u\left(K_0 \dot{+} W\right)\right)|W - V|\right\|_2^2$$

are finite;
(ii) the operators

$$(V - W)\exp\left(-\frac{t}{2}\left(K_0 \dot{+} V\right)\right) \quad \textit{and} \quad (V - W)\exp\left(-\frac{t}{2}\left(K_0 \dot{+} W\right)\right)$$

are Hilbert-Schmidt;
(iii) the operators

$$D(t/2)\exp\left(-\frac{t}{2}\left(K_0 \dot{+} V\right)\right) \quad \textit{and} \quad D(t/2)\exp\left(-\frac{t}{2}\left(K_0 \dot{+} W\right)\right)$$

are Hilbert-Schmidt;
(iv) the operator $D(t)$ is Hilbert-Schmidt;
the following implications are valid:

$$\begin{array}{ccccc} (i) & \Longrightarrow & (i') & & \\ \Downarrow & & \Downarrow & & \\ (ii) & \Longrightarrow & (iii) & \Longrightarrow & (iv). \end{array}$$

REMARK 1. Suppose that $D(t)$ is a Hilbert-Schmidt operator. Then the essential spectra are preserved:

$$\sigma_{\text{ess}}(K_0 \dot{+} V) = \sigma_{\text{ess}}(K_0 \dot{+} W).$$

We return to this matter in Chapter 8.

REMARK 2. Suppose that $V \geq -a_0$ and $W \geq -b_0$. Then the integrals in (i′) are bounded by

$$\exp\left(\frac{t}{2}\max\left(0, a_0, b_0\right)\right)\int_0^{t/2} du\, u \left\|\exp\left(-uK_0\right)|W - V|\right\|_2^2$$
$$\leq \exp\left(\frac{t}{2}\max\left(0, a_0, b_0\right)\right)\frac{t^2}{4}\left\|W - V\right\|_2^2,$$

if $W - V$ belongs to $L^2(E, m)$ and by

$$\exp\left(\frac{t}{2}\max\left(0,a_0,b_0\right)\right)\left\|\int_0^{t/2} du\, u \exp\left(-uK_0\right)|W-V|\right\|_\infty \|W-V\|_1,$$

in case $W - V$ belongs to $K(E) \cap L^1(E, m)$. Compare this remark with Remark 3 following Theorem 5.4.

Proof. (i) $\Rightarrow$ (ii). The square of the Hilbert-Schmidt norm of the operator

$$(W-V)\exp\left(-\frac{t}{2}\left(K_0\dot{+}V\right)\right)$$

is given by the integral

$$\int dx \exp\left(-t\left(K_0+V\right)\right)(x,x)\left|W(x)-V(x)\right|^2.$$

A similar statement is valid for the operator $(W-V)\exp\left(-\frac{t}{2}\left(K_0\dot{+}W\right)\right)$. The result follows, because the quantities

$$\sup_{x\in E}\exp\left(-t\left(K_0+V\right)\right)(x,x) \quad \text{and} \quad \sup_{x\in E}\exp\left(-t\left(K_0+W\right)\right)(x,x)$$

are finite.

(i) $\Rightarrow$ (i′). Suppose that a, $b > 0$, are such that $K_0\dot{+}V \geq -a$ and $K_0\dot{+}W \geq -b$ in form sense. Then the integrals in (i′) may be estimated as follows:

$$\int_0^{t/2} u\left\|\exp\left(-u\left(K_0\dot{+}V\right)\right)|W-V|\right\|_2^2 du \leq \int_0^{\frac{t}{2}} ue^{au}\left\|W-V\right\|_2^2 du < \infty$$

and

$$\int_0^{\frac{t}{2}} u\left\|\exp\left(-u\left(K_0\dot{+}W\right)\right)|W-V|\right\|_2^2 du \leq \int_0^{t/2} ue^{bu}\left\|W-V\right\|_2^2 du < \infty.$$

This proves (i′).

(i′) $\Rightarrow$ (iii). Since

$$D(t/2)\exp\left(-\frac{t}{2}\left(K_0\dot{+}V\right)\right)$$
$$= \int_0^{t/2}\exp\left(-u\left(K_0\dot{+}W\right)\right)(W-V)\exp\left(-(t-u)\left(K_0\dot{+}V\right)\right)du,$$

we see that the Hilbert-Schmidt norm of the operator

$$D(t/2)\exp\left(-\frac{t}{2}\left(K_0\dot{+}V\right)\right)$$

is given by

$$
\begin{aligned}
&\left\| D(t/2)\exp\left(-\frac{t}{2}\left(K_0\dot{+}V\right)\right)\right\|_{\mathrm{HS}}^2 \\
&= \int_0^{t/2} du_1 \int_0^{t/2} du_2 \int dx \int dy \exp\left(-\left(u_1+u_2\right)\left(K_0+W\right)\right)(x,y) \\
&\quad \times \left(W(y)-V(y)\right)\exp\left(-\left(2t-u_1-u_2\right)\left(K_0+V\right)\right)(y,x)\left(W(x)-V(x)\right).
\end{aligned}
$$

Hence

$$
\begin{aligned}
&\left\| D(t/2)\exp\left(-\frac{t}{2}\left(K_0\dot{+}V\right)\right)\right\|_{\mathrm{HS}}^2 \\
&\leq \sup_{x,y\in E}\ \sup_{t<u<2t} \exp\left(-u\left(K_0+V\right)\right)(x,y) \\
&\quad \times \int_0^{t/2} du_1 \int_{u_1}^{u_1+t/2} du_2 \int dx \int dy \exp\left(-u_2\left(K_0+W\right)\right)(x,y) \\
&\quad \times \left|W(y)-V(y)\right|\left|W(x)-V(x)\right| \\
&= \sup_{x,y\in E}\ \sup_{t<u<2t} \exp\left(-u\left(K_0+V\right)\right)(x,y) \\
&\quad \times \iint dx dy \int_0^t du \min(u,t/2)\exp\left(-u\left(K_0+W\right)\right)(x,y) \\
&\quad \left|W(y)-V(y)\right|\left|W(x)-V(x)\right| \\
&= \sup_{x,y\in E}\ \sup_{t<u<2t} \exp\left(-u\left(K_0+V\right)\right)(x,y) \\
&\quad \times \int_0^t du \min(u,t/2)\left\|\exp\left(-\frac{u}{2}\left(K_0\dot{+}W\right)\right)|W-V|\right\|_2^2 \\
&= \sup_{x,y\in E}\ \sup_{t<u<2t} \exp\left(-u\left(K_0+V\right)\right)(x,y) \\
&\quad \times 2\int_0^{t/2} dv \min(v,t/4)\left\|\exp\left(-v\left(K_0\dot{+}W\right)\right)|W-V|\right\|_2^2 .
\end{aligned}
\tag{5.21}
$$

The Hilbert-Schmidt norm of the operator $D(t/2)\exp\left(-\frac{t}{2}\left(K_0\dot{+}W\right)\right)$ can be estimated in a similar way, and so assertion (iii) readily follows.

(ii) $\Rightarrow$ (iii). From the representation

$$
\begin{aligned}
&D(t/2)\exp\left(-\frac{t}{2}\left(K_0\dot{+}W\right)\right) \\
&= \int_0^{t/2} \exp\left(-u\left(K_0\dot{+}V\right)\right)(W-V) \\
&\quad \exp\left(-\frac{t}{2}\left(K_0\dot{+}W\right)\right)\exp\left(-\left(\frac{t}{2}-u\right)\left(K_0\dot{+}W\right)\right) du,
\end{aligned}
\tag{5.22}
$$

under the assumption that $K_0\dot{+}V \geq -a$ and $K_0\dot{+}W \geq -b$ in form sense, we infer that

$$\begin{aligned}
&\left\| D(t/2)\exp\left(-\frac{t}{2}\left(K_0\dot{+}W\right)\right)\right\|_{\mathrm{HS}} \\
&\leq \int_0^{t/2} \left\| \exp\left(-u\left(K_0\dot{+}V\right)\right)(W-V)\exp\left(-\frac{t}{2}\left(K_0\dot{+}W\right)\right)\right. \\
&\qquad \left.\exp\left(-\left(\frac{t}{2}-u\right)\left(K_0\dot{+}W\right)\right)\right\|_{\mathrm{HS}} du \\
&\leq \int_0^{t/2} \left\|\exp\left(-u\left(K_0\dot{+}V\right)\right)\right\|_{2,2} \left\|(W-V)\exp\left(-\frac{t}{2}\left(K_0\dot{+}W\right)\right)\right\|_{\mathrm{HS}} \\
&\qquad \times \left\|\exp\left(-\left(\frac{t}{2}-u\right)\left(K_0\dot{+}W\right)\right)\right\|_{2,2} du \\
&\leq \int_0^{t/2} \left\|\exp\left(-u\left(K_0\dot{+}V\right)\right)\right\|_{2,2} \left\|\exp\left(-\left(\frac{t}{2}-u\right)\left(K_0\dot{+}W\right)\right)\right\|_{2,2} du \\
&\qquad \times \left\|(W-V)\exp\left(-\frac{t}{2}\left(K_0\dot{+}W\right)\right)\right\|_{\mathrm{HS}} \\
&\leq \int_0^{t/2} du\, e^{au} e^{\left(\frac{1}{2}t-u\right)b} \left\|(W-V)\exp\left(-\frac{t}{2}\left(K_0\dot{+}W\right)\right)\right\|_{\mathrm{HS}} \\
&= \frac{\exp(at/2)-\exp(bt/2)}{a-b}\left\|(W-V)\exp\left(-\frac{t}{2}\left(K_0\dot{+}W\right)\right)\right\|_{\mathrm{HS}} < \infty.
\end{aligned} \tag{5.23}$$

From (5.23) together with a similar estimate for the Hilbert-Schmidt norm of the operator $D(t/2)\exp\left(-\frac{t}{2}\left(K_0\dot{+}V\right)\right)$ assertion (iii) follows.

(iii) $\Rightarrow$ (iv). Because of the identity

$$\begin{aligned}
2D(t) =& D(t/2)\exp\left(-\frac{t}{2}\left(K_0\dot{+}V\right)\right) + \exp\left(-\frac{t}{2}\left(K_0\dot{+}V\right)\right)D(t/2) \\
&+ D(t/2)\exp\left(-\frac{t}{2}\left(K_0\dot{+}W\right)\right) + \exp\left(-\frac{t}{2}\left(K_0\dot{+}W\right)\right)D(t/2),
\end{aligned} \tag{5.24}$$

assertion (iv) follows from (iii). This proves Theorem 5.6.

The following results improve some corresponding results in [82]. The first result improves Theorem 5.5 in [82].

5.7. THEOREM. *Let V and W be Kato-Feller potentials. Suppose that both*

$$\int \exp\left(-2t(K_0+V)\right)(x,x)\left|V(x)-W(x)\right| dx \tag{5.25a}$$

and

$$\int \exp\left(-2t(K_0 + W)\right)(x,x)\,|V(x) - W(x)|\,dx \tag{5.25b}$$

are finite, or that both

$$\int \exp\left(-2t(K_0 + V)\right)(x,x)\,|V(x) - W(x)|^2\,dx \tag{5.26a}$$

and

$$\int \exp\left(-2t(K_0 + W)\right)(x,x)\,|V(x) - W(x)|^2\,dx < \infty \tag{5.26b}$$

are finite. Then the operator $D(t)$ defined by

$$D(t) = \exp\left(-t(K_0 \dot{+} V)\right) - \exp\left(-t(K_0 \dot{+} W)\right)$$

is a Hilbert-Schmidt operator. In the presence of (5.25*a*) *and* (5.25*b*), *the inequality*

$$\begin{aligned}\|D(t)\|_{\mathrm{HS}} \le &\left(2t \int \exp\left(-2t(K_0 + V)\right)(x,x)\,|V(x) - W(x)|\,dx\right)^{1/2}\\ &+\left(2t \int \exp\left(-2t(K_0 + W)\right)(x,x)\,|V(x) - W(x)|\,dx\right)^{1/2},\end{aligned} \tag{5.27}$$

holds true. In the presence of (5.26*a*) *and* (5.26*b*) *the inequality*

$$\begin{aligned}\|D(t)\|_{\mathrm{HS}} \le &\, t\left(\int \exp\left(-2t(K_0 + V)\right)(x,x)\,|V(x) - W(x)|^2\,dx\right)^{1/2}\\ &+t\left(\int \exp\left(-2t(K_0 + W)\right)(x,x)\,|V(x) - W(x)|^2\,dx\right)^{1/2}\end{aligned} \tag{5.28}$$

holds true.

REMARK. From Proposition 5.2 it follows that the operator norm of $D(t)$ may be estimated in the following fashion:

$$\begin{aligned}\|D(t)\|_{2,2} \le &\left\| \int_0^t ds \exp\left(-sK_0\right)|W - V| \exp\left(-(t-s)\left(K_0 \dot{+} 2V\right)\right) 1 \right\|_\infty^{1/2}\\ &\times \left\| \int_0^t ds \exp\left(-sK_0\right)|W - V| \exp\left(-(t-s)\left(K_0 \dot{+} 2W\right)\right) 1 \right\|_\infty^{1/2}\\ &\le \sup_{0<s<t} \left\|\exp\left(-s\left(K_0 \dot{+} 2V\right)\right) 1\right\|_{\infty,1}^{1/2} \sup_{0<s<t} \left\|\exp\left(-s\left(K_0 \dot{+} 2W\right)\right) 1\right\|_{\infty,1}^{1/2}\\ &\times \left\| \int_0^t ds \exp\left(-sK_0\right)|W - V| \right\|_\infty .\end{aligned}$$

Estimates for its Hilbert-Schmidt norm are given by (5.27) and (5.28). Instead of the Kato-Feller norm, the L^2-norm of $W - V$ appears.

Proof. Put $V_0 = \max(V, W)$ and write $D(t) = D_1(t) - D_2(t)$, with

$$D_1(t) = \exp\left(-t(K_0 \dot{+} V)\right) - \exp\left(-t(K_0 \dot{+} V_0)\right) \quad \text{and}$$
$$D_2(t) = \exp\left(-t(K_0 \dot{+} W)\right) - \exp\left(-t(K_0 \dot{+} V_0)\right).$$

We write $\operatorname{trace}(K) = \int k(x,x)dx$, if K is an integral operator with integral kernel $k(x,y)$. The following argument is taken from the proof of Theorem 5.5 in [82]:

$$\begin{aligned}
&\|D_1(t)\|_{\mathrm{HS}}^2 \\
&= \int\int dx\,dy \left[\exp\left(-t\left(K_0 + V\right)\right)(x,y) - \exp\left(-t\left(K_0 + V_0\right)\right)(x,y)\right]^2 \\
&= \operatorname{trace}\left[\exp\left(-2t\left(K_0 \dot{+} V\right)\right) - \exp\left(-t\left(K_0 \dot{+} V\right)\right)\exp\left(-t\left(K_0 \dot{+} V_0\right)\right)\right. \\
&\quad \left. - \exp\left(-t\left(K_0 \dot{+} V_0\right)\right)\exp\left(-t\left(K_0 \dot{+} V\right)\right) + \exp\left(-2t\left(K_0 \dot{+} V_0\right)\right)\right] \\
&\le \operatorname{trace}\left[\exp\left(-2t\left(K_0 \dot{+} V\right)\right) - \exp\left(-2t\left(K_0 \dot{+} V_0\right)\right)\right] \\
&= \int_0^{2t} du \int dx \int dz \exp\left(-u\left(K_0 + V\right)\right)(z,x)\left(V_0(x) - V(x)\right) \\
&\quad \exp\left(-(2t-u)\left(K_0 + V_0\right)\right)(x,z) \\
&\le \int_0^{2t} du \int dx \int dz \exp\left(-u\left(K_0 + V\right)\right)(z,x)\left(V_0(x) - V(x)\right) \\
&\quad \exp\left(-(2t-u)\left(K_0 + V\right)\right)(x,z) \\
&= 2t \int \exp\left(-2t(K_0 + V)\right)(x,x)\left|W(x) - \min(V(x), W(x))\right| dx
\end{aligned} \tag{5.29}$$

and, via a similar argument,

$$\|D_2(t)\|_{\mathrm{HS}}^2 \le 2t \int \exp\left(-2t(K_0 + V)\right)(x,x)\left|V(x) - \min(V(x), W(x))\right| dx. \tag{5.30}$$

From (5.29) and (5.30) the inequality in (5.27) follows. In order to prove (5.28) we proceed as follows (for brevity we keep writing $V_0 = \max(V, W)$):

$$\begin{aligned}
&\|D_1(t)\|_{\mathrm{HS}}^2 \\
&= \int\int dx\,dy \Big(\exp\left(-t(K_0 + V)\right)(x,y) - \exp\left(-t(K_0 + V_0)\right)(x,y)\Big)^2 \\
&\le \int\int dx\,dy \Big(\int_0^t ds \int dz \exp\left(-s(K_0 + V)\right)(x,z)\left(V(z) - V_0(z)\right) \\
&\qquad \times \exp\left(-(t-s)(K_0 + V_0)\right)(z,y)\Big)^2 \\
&= \int\int dz_1 dz_2 \int_0^t \int_0^t du_1 du_2 \exp\left(-(u_1 + u_2)(K_0 + V)\right)(z_1, z_2) \\
&\quad \times \left(V(z_1) - V_0(z_1)\right)\left(V(z_2) - V_0(z_2)\right)\exp\left(-(2t - u_1 - u_2)(K_0 + V_0)\right)(z_1, z_2)
\end{aligned}$$

(Cauchy-Schwarz' inequality)

$$\leq \left(\int \int dz_1 dz_2 \int_0^t \int_0^t du_1 du_2 \exp\left(-(u_1+u_2)(K_0+V)\right)(z_1,z_2) \right.$$

$$\left. |V(z_1) - V_0(z_1)|^2 \exp\left(-(2t-u_1-u_2)(K_0+V_0)\right)(z_1,z_2) \right)^{1/2}$$

$$\times \left(\int \int dz_1 dz_2 \int_0^t \int_0^t du_1 du_2 \exp\left(-(u_1+u_2)(K_0+V)\right)(z_1,z_2) \right.$$

$$\left. |V(z_2) - V_0(z_2)|^2 \exp\left(-(2t-u_1-u_2)(K_0+V_0)\right)(z_1,z_2) \right)^{1/2}$$

$$\leq \left(\int \int dz_1 dz_2 \int_0^t \int_0^t du_1 du_2 \exp\left(-(u_1+u_2)(K_0+V)\right)(z_1,z_2) \right.$$

$$\left. \times |V(z_1) - V_0(z_1)|^2 \exp\left(-(2t-u_1-u_2)(K_0+\min(V,V_0))\right)(z_1,z_2) \right)^{1/2}$$

$$\times \left(\int \int dz_1 dz_2 \int_0^t \int_0^t du_1 du_2 \exp\left(-(u_1+u_2)(K_0+V)\right)(z_1,z_2) \right.$$

$$\left. \times |V(z_2) - V_0(z_2)|^2 \exp\left(-(2t-u_1-u_2)(K_0+\min(V,V_0))\right)(z_1,z_2) \right)^{1/2}$$

$$\leq \int dz \int_0^t \int_0^t du_1 du_2 \exp\left(-2t(K_0+V)\right)(z,z)\, |V(z) - V_0(z)|^2$$

$$\leq t^2 \int dz \exp\left(-2t(K_0+V)\right)(z,z)\, |V(z) - W(z)|^2 . \tag{5.31}$$

Inequality (5.31) together with a similar result for $\|D_2(t)\|_{\mathrm{HS}}$ implies (5.28).

5.8. COROLLARY. *(a) Suppose that the unperturbed semigroup* $\{\exp(-tK_0) : t \geq 0\}$ *is* L^1*-*L^∞*-smoothing. If the function* $W - V$ *belongs to* $L^1(E,m) + L^2(E,m)$*, then the operators* $D(t)$*,* $t \geq 0$*, are Hilbert-Schmidt.*

(b) Fix $t > 0$ *and define* $G(t)$ *by*

$$G(t) = \int \exp\left(-t\left(K_0 + \min(V,W)\right)\right)(x,x)\, |W(x) - V(x)|^2 \, dx.$$

Then the following Hilbert-Schmidt norm estimate holds:

$$\|D(t/2)\|_{\mathrm{HS}}^2 \leq \frac{t^2}{4} G(t).$$

Proof. (a) Write $W - V = F_1 + F_2$, where F_j belongs to $L^j(E,m)$, $j = 1,\ 2$. We also write $D_1(t) = \exp\left(-t\left(K_0 \dot{+} V\right)\right) - \exp\left(-t\left(K_0 \dot{+} V + F_2\right)\right)$ and $D_2(t) =$

$\exp\left(-t\left(K_0\dot{+}W - F_1\right)\right) - \exp\left(-t\left(K_0\dot{+}W\right)\right)$. Then $D(t) = D_1(t) + D_2(t)$. By the previous theorem the operators $D_1(t)$ and $D_2(t)$ are Hilbert-Schmidt and hence $D(t)$ is a Hilbert-Schmidt operator as well.

(b) Note that

$$D(t/2) = D(t/4)\exp\left(-\frac{1}{4}t\left(K_0\dot{+}V\right)\right) + \exp\left(-\frac{1}{4}t\left(K_0\dot{+}W\right)\right)D(t/4).$$

So that it suffices to prove that $\left\|D(t/4)\exp\left(-\frac{1}{4}t\left(K_0\dot{+}V\right)\right)\right\|_{\mathrm{HS}}^2 \leq t^2G(t)/4$ together with a similar estimate for the Hilbert-Schmidt norm of the operator $\exp\left(-\frac{1}{4}t\left(K_0\dot{+}W\right)\right)D(t/4)$. From Duhamel's formula it follows that

$$\begin{aligned}
&\left\|D(t/4)\exp\left(-\frac{t}{4}\left(K_0\dot{+}V\right)\right)\right\|_{\mathrm{HS}}^2 \\
&\leq \int_0^{t/4} du_1 \int_0^{t/4} du_2 \int dz_1 \int dz_2 \exp\left(-(u_1+u_2)\left(K_0+V\right)\right)(z_1,z_2) \\
&\quad \times \left(W(z_2)-V(z_2)\right)\exp\left(-(t-u_1-u_2)\left(K_0+V\right)\right)(z_1,z_2)\left(W(z_1)-V(z_1)\right) \\
&\leq \frac{1}{2}\int_0^{t/4} du_1 \int_0^{t/4} du_2 \int dz_1 \int dz_2 \exp\left(-(u_1+u_2)\left(K_0+V\right)\right)(z_1,z_2) \\
&\quad \times \left|W(z_2)-V(z_2)\right|^2 \exp\left(-(t-u_1-u_2)\left(K_0\dot{+}V\right)\right)(z_1,z_2) \\
&\quad + \frac{1}{2}\int_0^{t/4} du_1 \int_0^{t/4} du_2 \int dz_1 \int dz_2 \exp\left(-(u_1+u_2)\left(K_0+V\right)\right)(z_1,z_2) \\
&\quad \times \exp\left(-(t-u_1-u_2)\left(K_0\dot{+}V\right)\right)(z_1,z_2)\left|W(z_1)-V(z_1)\right|^2 \\
&\leq \frac{1}{2}\int_0^{t/4} du_1 \int_0^{t/4} du_2 \int dz_1 \int dz_2 \exp\left(-(u_1+u_2)\left(K_0+V\right)\right)(z_1,z_2) \\
&\quad \times \exp\left(-(t-u_1-u_2)\left(K_0+V\right)\right)(z_2,z_1)\left|W(z_1)-V(z_1)\right|^2 \\
&\quad + \frac{1}{2}\int_0^{t/4} du_1 \int_0^{t/4} du_2 \int dz_1 \int dz_2 \exp\left(-(u_1+u_2)\left(K_0+V\right)\right)(z_1,z_2) \\
&\quad \times \exp\left(-(t-u_1-u_2)\left(K_0+V\right)\right)(z_1,z_2)\left|W(z_2)-V(z_2)\right|^2 \\
&\leq \frac{t^2}{16}\int dx \exp\left(-t\left(K_0+\min(V,W)\right)\right)(x,x)\left|W(x)-V(x)\right|^2 dx \leq \frac{t^2}{16}G(t).
\end{aligned} \tag{5.32}$$

A combination of inequality (5.32) with a similar estimate for the Hilbert-Schmidt norm of the operator $\exp\left(-\frac{1}{4}t\left(K_0\dot{+}W\right)\right)D(t/4)$ yields the result in assertion (b).

In the following theorem we write $R(a,V)^p$, $p>0$, for the operator

$$R(a,V)^p = \frac{1}{\Gamma(p)}\int_0^\infty t^{p-1}e^{-at}\exp\left(-t(K_0\dot{+}V)\right)dt.$$

Such an operator is called a *power of a resolvent* or *resolvent power*.

5.9. THEOREM. *Let V and W be Kato-Feller potentials with the property that the integral*

$$\int R(a,\min(V,W))^{2p+2}(x,x)\big(W(x)-V(x)\big)^2dx$$

is finite. Then the power resolvent difference $R(a,V)^p - R(a,W)^p$ is a Hilbert-Schmidt operator. Moreover, the following estimate is valid:

$$\|R(a,V)^p - R(a,W)^p\|_{\mathrm{HS}}^2 \le \int R(a,\min(V,W))^{2(p+1)}(x,x)\big(W(x)-V(x)\big)^2dx. \tag{5.33}$$

Proof. As in Proposition 5.2, we write

$$D_s(t) = \int_0^t \exp(-uK(s))\,(W-V)\exp(-(t-u)K(s))\,du = -\frac{\partial}{\partial s}\exp(-tK(s)),$$

where $K(s) = K_0\dot{+}(1-s)V + sW$. Then $D(t) = \int_0^1 D_s(t)ds$: again see Proposition 5.2 equality (5.8). For brevity we write $V_s = (1-s)V + sW$. Then we have

$$\begin{aligned}
R(a,V)^p - R(a,W)^p &= \frac{1}{\Gamma(p)}\int_0^\infty t^{p-1}e^{-at}\left(\exp\left(-t(K_0\dot{+}V)\right) - \exp\left(-t(K_0\dot{+}W)\right)\right)dt\\
&= \frac{1}{\Gamma(p)}\int_0^1 ds\int_0^\infty dt\, t^{p-1}e^{-at}D_s(t)\\
&= \frac{1}{\Gamma(p)}\int_0^1 ds\int_0^\infty dt\, t^{p-1}e^{-at}\int_0^t du\,\exp\left(-u\left(K_0\dot{+}V_s\right)\right)\\
&\qquad\times(W-V)\exp\left(-(t-u)\left(K_0\dot{+}V_s\right)\right).
\end{aligned}\tag{5.34}$$

From (5.34) we get the following estimate:

$$\begin{aligned}
\Gamma(p)\,\|R(a,V)^p - R(a,W)^p\|_{\mathrm{HS}} \le \int_0^1 ds\bigg\|\int_0^\infty dt\, t^{p-1}e^{-at}\int_0^t du\exp\left(-u(K_0\dot{+}V_s)\right)\\
\times(W-V)\exp\left(-(t-u)(K_0\dot{+}V_s)\right)\bigg\|_{\mathrm{HS}}.
\end{aligned}\tag{5.35}$$

So we have to estimate the Hilbert-Schmidt norm

$$\left\|\int_0^\infty dt\, t^{p-1}e^{-at}\int_0^t du\exp\left(-u(K_0\dot{+}V_s)\right)(W-V)\exp\left(-(t-u)(K_0\dot{+}V_s)\right)\right\|_{\mathrm{HS}}. \tag{5.36}$$

The square of the norm in (5.36) is dominated by ($\mathrm{trace}(K) = \int k(x,x)dx$)

$$\mathrm{trace}\bigg\{\int_0^\infty dt_1\int_0^\infty dt_2\,(t_1t_2)^{p-1}e^{-a(t_1+t_2)}\int_0^{t_1}du_1\int_0^{t_2}du_2$$

$$
\begin{aligned}
&\exp\left(-u_1(K_0\dot{+}V_s)\right)(W-V)\exp\left(-(t_1-u_1)(K_0\dot{+}V_s)\right)\\
&\exp\left(-(t_2-u_2)\left(K_0\dot{+}V_s\right)\right)(W-V)\exp\left(-u_2\left(K_0\dot{+}V_s\right)\right)\Big\}\\
&=\operatorname{trace}\Big\{\int_0^\infty dt_1\int_0^\infty dt_2\,(t_1t_2)^{p-1}\,e^{-a(t_1+t_2)}\int_0^{t_1}du_1\int_0^{t_2}du_2\\
&\exp\left(-(u_1+u_2)\left(K_0\dot{+}V_s\right)\right)(W-V)\\
&\quad\exp\left(-(t_1+t_2-u_1-u_2)(K_0\dot{+}V_s)(W-V)\right\}.
\end{aligned}
\tag{5.37}
$$

Next we have

$$
\begin{aligned}
&\operatorname{trace}\{\exp\left(-(u_1+u_2)\left(K_0\dot{+}V_s\right)\right)(W-V)\\
&\quad\exp\left(-(t_1+t_2-u_1-u_2)\left(K_0\dot{+}V_s\right)\right)(W-V)\}\\
&=\int dx\int dy\exp\left(-(u_1+u_2)\left(K_0\dot{+}V_s\right)\right)(x,y)\big(W(y)-V(y)\big)\\
&\qquad\times\exp\left(-(t_1+t_2-u_1-u_2)\left(K_0\dot{+}V_s\right)\right)(y,x)\big(W(x)-V(x)\big)\\
&\le\frac{1}{2}\int dx\int dy\exp\left(-(u_1+u_2)\left(K_0\dot{+}V_s\right)\right)(x,y)\big(W(y)-V(y)\big)^2\\
&\exp\left(-(t_1+t_2-u_1-u_2)\left(K_0\dot{+}V_s\right)\right)(y,x)\\
&\quad+\frac{1}{2}\int dx\int dy\exp\left(-(u_1+u_2)\left(K_0\dot{+}V_s\right)\right)(x,y)\\
&\qquad\exp\left(-(t_1+t_2-u_1-u_2)\left(K_0\dot{+}V_s\right)\right)(y,x)\big(W(x)-V(x)\big)^2\\
&=\int dx\exp\left(-(t_1+t_2)\left(K_0\dot{+}V_s\right)\right)(x,x)\big(W(x)-V(x)\big)^2.
\end{aligned}
\tag{5.38}
$$

So, from (5.37) together with (5.38) we obtain

$$
\begin{aligned}
&\left\|\int_0^\infty dt\,t^{p-1}e^{-at}\int_0^t du\exp\left(-u(K_0\dot{+}V_s)\right)(W-V)\exp\left(-(t-u)(K_0\dot{+}V_s)\right)\right\|_{\mathrm{HS}}^2\\
&\le\int dx\int_0^\infty dt_1\int_0^\infty dt_2\,(t_1t_2)^{p-1}\,e^{-a(t_1+t_2)}t_1t_2\\
&\quad\exp\left(-(t_1+t_2)\left(K_0\dot{+}V_s\right)\right)(x,x)\big(W(x)-V(x)\big)^2\\
&\le\int dx\int dz\int_0^\infty dt_1\int_0^\infty dt_2\,(t_1t_2)^{p}\,e^{-a(t_1+t_2)}\exp\left(-t_1\left(K_0\dot{+}V_s\right)\right)(x,z)\\
&\quad\exp\left(-t_2\left(K_0\dot{+}V_s\right)\right)(z,x)\big(W(x)-V(x)\big)^2\\
&=\int dx\Gamma(p+1)^2R(a,V_s)^{2p+2}(x,x)\big(W(x)-V(x)\big)^2\\
&\le\int dx\Gamma(p+1)^2R(a,\min(V,W))^{2(p+1)}(x,x)\big(W(x)-V(x)\big)^2.
\end{aligned}
\tag{5.39}
$$

From (5.39) inequality (5.33) in the theorem follows.

The following result gives conditions on the Kato-Feller potentials V and W, which ensure that the operator $D(t)$ in Proposition 5.2 is compact.

5.10. THEOREM. *Let V and W be Kato-Feller potentials with the property that, for $a > 0$ sufficiently large, the functions $(aI + K_0)^{-1}|V|$ and $(aI + K_0)^{-1}|W|$ are functions in $C_\infty(E)$. The following operators are compact in $L^2(E, m)$:*

$$\exp\left(-t\left(K_0 \dot{+} V\right)\right) - \exp\left(-t\left(K_0 \dot{+} W\right)\right), \quad t \geq 0;$$
$$\left(aI + K_0 \dot{+} V\right)^{-1} - \left(aI + K_0 \dot{+} W\right)^{-1}, \quad a > 0 \quad \textit{sufficiently large.}$$

REMARK. The condition that the function $(aI + K_0)^{-1}|V|$ belongs to $C_\infty(E)$ should be compared to Weder [251, Lemma III.4] and to [249, Theorem II.9]. More results on relative compactness can be found in Smits [212, Chapter 5], in Weder [250], and in Reed and Simon [194, p. 117, Example 5]. It is also noticed that $(aI + K_0)^{-1}W$ is a function in $C_\infty(E)$, whenever W is a bounded Borel measurable function with the property that for every $\epsilon > 0$ there exists a compact subset K of E for which $|W(x)| \leq \epsilon$ if x does not belong to K (see Definition 1.2).

Proof. Put $V_{k,\ell,m} = V 1_{\{-k \leq V \leq \ell\}} 1_{K_m}$ and similarly $W_{k,\ell,m} = W 1_{\{-k \leq W \leq \ell\}} 1_{K_m}$, where $(K_m : m \in \mathbb{N})$ is a sequence of compact subsets of E with the following properties: $E = \bigcup_{m \in \mathbb{N}} K_m$ and $K_m \subseteq \operatorname{int}(K_{m+1})$. From Theorem 5.7 it follows that the operators

$$\left(aI + K_0 \dot{+} V_{k,\ell,m}\right)^{-1} - \left(aI + K_0 \dot{+} W_{k,\ell,m}\right)^{-1},$$

for $a > 0$ large enough, and

$$\exp\left(-t\left(K_0 \dot{+} V_{k,\ell,m}\right)\right) - \exp\left(-t\left(K_0 \dot{+} W_{k,\ell,m}\right)\right),$$

$t \geq 0$, are compact operators, for all k, ℓ and $m \in \mathbb{N}$. It suffices to prove that the differences

$$\left\|\exp\left(-t\left(K_0 \dot{+} V\right)\right) - \exp\left(-t\left(K_0 \dot{+} V_{k,\ell,m}\right)\right)\right\|_{2,2} \quad \text{and}$$
$$\left\|\exp\left(-t\left(K_0 \dot{+} W\right)\right) - \exp\left(-t\left(K_0 \dot{+} W_{k,\ell,m}\right)\right)\right\|_{2,2}$$

tend to zero, if k, ℓ, and m tend to ∞. Therefore we estimate

$$\left\|\exp\left(-t\left(K_0 \dot{+} V\right)\right) - \exp\left(-t\left(K_0 \dot{+} V_{k,\ell,m}\right)\right)\right\|_{2,2}$$

(symmetry together with Riesz-Thorin interpolation: see Appendix A.40)

$$\leq \left\|\exp\left(-t\left(K_0 \dot{+} V\right)\right) - \exp\left(-t\left(K_0 \dot{+} V_{k,\ell,m}\right)\right)\right\|_{\infty,\infty}$$
$$= \left\|\int_0^t \exp\left(-s(K_0 + V_{k,\ell,m})\right)\left(V - V_{k,\ell,m}\right)\exp\left(-(t-s)(K_0 + V)\right) ds\right\|_{\infty,\infty}$$

$$\leq \sup_{x\in E} \int_0^t \mathbb{E}_x \left(\exp\left(- \int_0^s V_{k,\ell,m}(X(u))du \right) |V(X(s)) - V_{k,\ell,m}(X(s))| \right.$$
$$\left. \times \mathbb{E}_{X(s)} \left(\exp\left(- \int_0^{t-s} V(X(v))dv \right) \right) \right) ds$$

(Markov property together with Cauchy-Schwarz' inequality)

$$\leq \sup_{x\in E} \left(\int_0^t \mathbb{E}_x \left(\exp\left(-2 \int_0^s V_{k,\ell,m}(X(u))du \right) |V(X(s)) - V_{k,\ell,m}(X(s))| \right) ds \right)^{1/2}$$
$$\times \left(\int_0^t \mathbb{E}_x \left(\exp\left(-2 \int_s^t V(X(v))dv \right) |V(X(s)) - V_{k,\ell,m}(X(s))| \right) ds \right)^{1/2}$$

$$\leq \sup_{x\in E} \left(\int_0^t \mathbb{E}_x \left(\exp\left(-4 \int_0^s V_{k,\ell,m}(X(u))du \right) |V(X(s)) - V_{k,\ell,m}(X(s))| \right) ds \right)^{1/4}$$
$$\times \left(\int_0^t \mathbb{E}_x \left(|V(X(s)) - V_{k,\ell,m}(X(s))| \right) ds \right)^{1/4}$$
$$\times \left(\int_0^t \mathbb{E}_x \left(\exp\left(-4 \int_s^t V(X(v))dv \right) |V(X(s)) - V_{k,\ell,m}(X(s))| \right) ds \right)^{1/4}$$
$$\times \left(\int_0^t \mathbb{E}_x \left(|V(X(s)) - V_{k,\ell,m}(X(s))| \right) ds \right)^{1/4}$$

$$\leq \sup_{x\in E} \left(\int_0^t \mathbb{E}_x \left(\exp\left(-4 \int_0^s V_{k,\ell,m}(X(u))du \right) |V(X(s)) - V_{k,\ell,m}(X(s))| \right) ds \right)^{1/4}$$
$$\times \left(\int_0^t \mathbb{E}_x \left(\exp\left(-4 \int_s^t V(X(v))dv \right) |V(X(s)) - V_{k,\ell,m}(X(s))| \right) ds \right)^{1/4}$$
$$\times \left(\int_0^t \mathbb{E}_x \left(|V(X(s)) - V_{k,\ell,m}(X(s))| \right) ds \right)^{1/2}$$

(again Cauchy-Schwarz' inequality)

$$\leq \sup_{x\in E} \left(\mathbb{E}_x \left(\exp\left(+8 \int_0^t \max(-V_{k,\ell,m}(X(u)), 0))du \right) \right) \right)^{1/8}$$
$$\times \sup_{x\in E} \left(\mathbb{E}_x \left(\int_0^t |V(X(s)) - V_{k,\ell,m}(X(s))| \, ds \right)^2 \right)^{1/4}$$
$$\times \sup_{x\in E} \left(\mathbb{E}_x \left(\exp\left(+8 \int_0^t \max(-V(X(u)), 0)du \right) \right) \right)^{1/8}$$
$$\times \sup_{x\in E} \left(\mathbb{E}_x \left(\int_0^t |V(X(s)) - V_{k,\ell,m}(X(s))| \, ds \right) \right)^{1/2}$$

(Markov property)

$$\leq \sup_{x\in E}\left(\mathbb{E}_x\left(\exp\left(+8\int_0^t \max(-V_{k,\ell,m}(X(u)),0))du\right)\right)\right)^{1/8}$$
$$\times \sup_{x\in E}\left(2\mathbb{E}_x\left(\int_0^t |V(X(s_1))-V_{k,\ell,m}(X(s_1))|\right.\right.$$
$$\left.\left.\mathbb{E}_{X(s_1)}\left[\int_{s_1}^t |V(X(s_2-s_1))-V_{k,\ell,m}(X(s_2-s_1))|\,ds_2\right]ds_1\right)\right)^{1/4}$$
$$\times \sup_{x\in E}\left(\mathbb{E}_x\left(\exp\left(+8\int_0^t \max(-V(X(u)),0)du\right)\right)\right)^{1/8}$$
$$\times \sup_{x\in E}\left(\mathbb{E}_x\left(\int_0^t |V(X(s))-V_{k,\ell,m}(X(s))|\,ds\right)\right)^{1/2}$$
$$\leq 2^{1/4}\sup_{x\in E}\left(\mathbb{E}_x\left(\exp\left(8\int_0^t V_-(X(u))du\right)\right)\right)^{1/4}$$
$$\times \sup_{x\in E}\mathbb{E}_x\left(\int_0^t |V(X(s))-V_{k,\ell,m}(X(s))|\,ds\right). \tag{5.40}$$

Since the function $(aI+K_0)^{-1}|V|$ belongs to $C_\infty(E)$ we infer that

$$\lim_{k,\ell,m\to\infty}\sup_{x\in E}\mathbb{E}_x\left(\int_0^t |V(X(s))-V_{k,\ell,m}(X(s))|\,ds\right)=0 \tag{5.41}$$

by Dini's lemma (see Lemma E.1, Appendix E). From (5.41) it follows that the right-hand side of (5.40) tends to zero and hence the claim in the theorem follows.

B. Singular perturbations

In what follows we will be concerned with Hilbert-Schmidt properties of differences of Feynman-Kac semigroups, but now with singular perturbations. That is we study the difference of the Feynman-Kac semigroup $\{\exp(-t(K_0\dot{+}V)) : t \geq 0\}$ and a corresponding Dirichlet semigroup (or, more accurately, pseudo-Dirichlet semigroup) $\{\exp(-t(K_0\dot{+}V)_\Sigma) : t \geq 0\}$ which is obtained by killing the semigroup on a "small" singularity region $\Gamma = E \setminus \Sigma$: see Definition 2.25 and the comments following it. Here "small" means that the set Γ is not necessarily compact, but small for some capacity: the precise conditions will be given in terms of the function h_Σ^{a+V} defined by (cf. Definition 4.19)

$$h_\Sigma^{a+V}(x) = H_\Sigma^{a+V}1(x) = \mathbb{E}_x\left[\exp\left(-\int_0^S (a + V(X(u)))\,du\right) : S < \infty\right].$$

As explained in section B of Chapter 2, the operator H_Σ^{a+V} is defined via the formula

$$H_\Sigma^{a+V}f(x) = \mathbb{E}_x\left[\exp\left(-\int_0^S (a + V(X(u)))\,du\right) f(X(S)) : S < \infty\right],$$

for appropriate functions f. For more details see Definition 2.30, Proposition 4.14, Proposition 4.15, and equation (4.55). That h_Σ^{a+V} is well-defined, at least for a large enough, follows from Proposition 4.20. We kill the Feynman-Kac semigroup $\{\exp(-t(K_0\dot{+}V)) : t \geq 0\}$ on Γ to obtain the corresponding Dirichlet semigroup

$$\{\exp(-t(K_0\dot{+}V)_\Sigma) : t \geq 0\}$$

defined by

$$\left[\exp(-t(K_0\dot{+}V)_\Sigma) f\right](x) = \mathbb{E}_x\left[\exp\left(-\int_0^t V(X(u))du\right) f(X(t)) : S > t\right].$$

Here Σ is the complement of Γ and S is the *penetration time* of Γ: i.e. $S = \inf\left\{s > 0 : \int_0^s 1_\Gamma(X(u))du > 0\right\}$. In most of the results, we also might have used the *hitting time* T instead of S: $T = \inf\{s > 0 : X(s) \in \Gamma\}$. The restriction operator $J = J_\Sigma$ is defined as follows: $Jf = f \restriction_\Sigma$, where $\Sigma = E \setminus \Gamma$. Hence $J^* = Id_{L^2(\Sigma)\to L^2(E)}$, $J^*J = 1_\Sigma$ and $JJ^* = I_{L^2(\Sigma)}$. Denote by $K_\Sigma = (K_0\dot{+}V)_\Sigma$ the Feynman-Kac generator of the Dirichlet semigroup $\{\exp(-t(K_0\dot{+}V)_\Sigma) : t \geq 0\}$. This semigroup is strongly continuous in $L^2(\Sigma, m)$. From formula (2.51) in Corollary 2.32.(b), we infer that $(K_0\dot{+}V)_\Sigma \supseteq J(K_0\dot{+}V)J^*$. Most of the time, especially in proofs, we write J instead of J_Σ. Also see the comments preceding Proposition 2.31. We write $D_\Sigma(t)$ or $D_\Sigma^V(t)$ for the operator

$$D_\Sigma(t) = \exp(-t(K_0\dot{+}V)) - J_\Sigma^*\exp(-t(K_0\dot{+}V)_\Sigma)J_\Sigma. \tag{5.42}$$

We emphasize the fact that the integral kernel of the operator $D_\Sigma(t)$ is given by (cf. Proposition 4.16 and Theorem 2.34.(b))

$$\begin{aligned} &D_\Sigma(t,x,y) \qquad (5.43)\\ &= \mathbb{E}_x\left[\exp\left(-\int_0^S V(X(u))du\right)\exp\left(-(t-S)\left(K_0+V\right)\right)(X(S),y) : S<t\right]. \end{aligned}$$

In Chapter 2, we discussed the relation between H_Σ^{a+V} and the resolvents $\left(aI + K_0\dot{+}V\right)^{-1}$ and $J_\Sigma^*\left(aI + K_0\dot{+}V\right)_\Sigma^{-1} J_\Sigma$. In fact we have Dynkin's formula (see (2.47)):

$$H_\Sigma^{a+V}\left(aI + K_0\dot{+}V\right)^{-1} = \left(aI + K_0\dot{+}V\right)^{-1} - J_\Sigma^*\left(aI + K_0\dot{+}V\right)_\Sigma^{-1} J_\Sigma.$$

For more details and proofs of this and related results, the reader is referred to Proposition 2.31, to Definition 2.30, and to Proposition 4.14.

Equality (5.44) in the next proposition will also be used in the corresponding theorem about trace properties. It is called *Duhamel's formula* for singular perturbations. A proof was given in Proposition 4.15 assertion (b) equality (4.64). In this proof the invertibility of the operator $K_0\dot{+}V + aI$ was used. That is why we make the assumption that $K_0\dot{+}V + aI \geq bI > 0$. Formula (5.45) was stated in Theorem 2.34 equality (2.61).

5.11. PROPOSITION. *Suppose that in form sense $K_0\dot{+}V + aI \geq bI > 0$. Then the following identity is true:*

$$\begin{aligned} D_\Sigma(t) &= H_\Sigma^{a+V}\exp\left(-t\left(K_0\dot{+}V\right)\right) - J^*\exp\left(-t\left(K_0\dot{+}V\right)_\Sigma\right) J H_\Sigma^{a+V} \qquad (5.44)\\ &+\int_0^t ds J^*\exp\left(-s\left(\left(K_0\dot{+}V\right)_\Sigma\right)\right) J H_\Sigma^{a+V}\left(aI + K_0\dot{+}V\right)\exp\left(-(t-s)\left(K_0\dot{+}V\right)\right)\\ &= 1_\Gamma\exp\left(-t\left(K_0\dot{+}V\right)\right) \qquad (5.45)\\ &+\int_0^t ds J^*\left(aI + K_0\dot{+}V\right)_\Sigma\exp\left(-s\left(\left(K_0\dot{+}V\right)_\Sigma\right)\right) J H_\Sigma^{a+V}\exp\left(-(t-s)\left(K_0\dot{+}V\right)\right). \end{aligned}$$

The following theorem contains a general result on Hilbert-Schmidt properties of differences of Feynman-Kac semigroups which are singularly perturbed. In this context, the implication (i′) ⇒ (iv) seems to be the most relevant one. For some results on the operator appearing in (i), the reader is referred to Proposition 6.15 and its Corollary 6.16.

5.12. THEOREM. *Among the assertions*

(i) There exists a real number $\beta \geq \frac{1}{2}$ together with a function $u : E \to \mathbb{R}$ that is identically 1 on Γ with the property that for all $a \in \mathbb{R}$, sufficiently large, the operators

$$\left(aI + K_0\dot{+}V\right)^\beta u\exp\left(-t\left(aI + K_0\dot{+}V\right)\right), \quad t>0,$$

are Hilbert-Schmidt;

(i′) There exists an $a > 0$ such that the integral $\int h_{\Sigma}^{a+V}(x)^2 dx$ is finite;
(ii) There exists a real number a such that the operators

$$H_{\Sigma}^{a+V} \exp\left(-t\left(aI + K_0 \dot{+} V\right)\right), \quad t > 0,$$

are Hilbert-Schmidt;
(iii) The operators

$$D_{\Sigma}(t/2) \exp\left(-\frac{t}{2}\left(K_0 \dot{+} V\right)\right), \quad t > 0,$$

are Hilbert-Schmidt;
(iv) the operators $D_{\Sigma}(t)$, $t > 0$ are Hilbert-Schmidt;
the following implications are valid:

$$\begin{array}{ccccccc} (i) & \Longrightarrow & (ii) & \Longrightarrow & (iii) & \Longleftrightarrow & (iv). \\ & & \Uparrow & & & & \\ & & (i') & & & & \end{array}$$

REMARK. Recall that, the λ-*capacity* (see [108, §3.1. p. 61 ff], Stroock [218], and Definition D.17 in Appendix D) of the set Γ can be defined by

$$\begin{aligned} &\operatorname{cap}(\Gamma) = \inf\left\{\operatorname{cap}(U) : U \supseteq \Gamma, \quad U \text{ open }\right\}, \text{ where} \\ &\operatorname{cap}(U) = \inf\left\{\langle K_0 f, f\rangle + \lambda\langle f, f\rangle : f \geq 1_U \ m\text{-almost everywhere}, \quad U \text{ open }\right\}, \end{aligned}$$

where U is an open subset of E. If the capacity, or more accurately the λ-capacity of the set Γ is finite, then the λ-*equilibrium potential* $v_{\lambda}(x)$, defined by $v_{\lambda}(x) = \mathbb{E}_x\left[\exp(-\lambda T) : T < \infty\right]$, belongs to $\operatorname{dom}(K_0^{1/2})$ and it realizes the infimum in the definition of λ-capacity : i.e.

$$\operatorname{cap}(\Gamma) = \left\langle K_0^{1/2} v_{\lambda}, K_0^{1/2} v_{\lambda}\right\rangle + \lambda\langle v_{\lambda}, v_{\lambda}\rangle = \lambda \int v_{\lambda}(x) dx. \tag{5.46}$$

Also note the fact that the finiteness of $\operatorname{cap}(\Gamma)$ implies that its measure $m(\Gamma)$ is finite. If $T = S$ (e.g. in case $\Gamma^r = (\operatorname{int}(\Gamma))^r$) and if $V \equiv 0$, then $h_{\Sigma}^{a+V} = v_a = v_{a,\Gamma}$. For more details see Proposition 2.24, Definition 2.36, Proposition 2.37, Definition D.17, and the Propositions D.18, D.19. From (i′, it also follows that the difference $D_{\Sigma}^{0}(t) := \exp(-tK_0) - J^* \exp\left(-t\left(K_0\right)_{\Sigma}\right) J$ is a Hilbert-Schmidt operator, whenever the capacity of Γ is finite.

Proof. (i) $\Rightarrow$ (ii). For brevity we write $A(a) = aI + K_0 \dot{+} V$, where $aI + K_0 \dot{+} V \geq bI > 0$ in form sense. The following identities follow because $H_{\Sigma}^{a+V} f = H_{\Sigma}^{a+V}(1_{\Gamma} f)$ depends only on the restriction of the function f to Γ:

$$\begin{aligned} &\left(aI + K_0 \dot{+} V\right)^{1/2} H_{\Sigma}^{a+V} \exp\left(-t\left(aI + K_0 \dot{+} V\right)\right) \\ &= A(a)^{1/2} H_{\Sigma}^{a+V} A(a)^{-1/2} A(a)^{-(\beta-1/2)} A(a)^{\beta} u \exp\left(-tA(a)\right). \end{aligned} \tag{5.47}$$

(By (i) the operator $A(a)^{\beta} u \exp(-tA(a))$ is Hilbert-Schmidt, and hence it is bounded.) It follows that

$$\begin{aligned}&\left\| \left(aI + K_0 \dot{+} V\right)^{1/2} H_{\Sigma}^{a+V} \exp\left(-t\left(aI + K_0 \dot{+} V\right)\right) \right\|_{\mathrm{HS}} \\ &\leq \left\| A(a)^{1/2} H_{\Sigma}^{a+V} A(a)^{-1/2} A(a)^{-(\beta-1/2)} \right\|_{2,2} \left\| \| A(a)^{\beta} u \exp(-tA(a)) \right\|_{\mathrm{HS}} .\end{aligned} \tag{5.48}$$

The operator $A(a)^{1/2} H_{\Sigma}^{a+V} A(a)^{-1/2}$ is a bounded orthogonal projection and the operator $A(a)^{-(\beta-1/2)}$ is bounded as well. Hence the expression in (5.48) is finite and so assertion (ii) follows.

(i′) ⇒ (ii). The integral kernel of the operator

$$H_{\Sigma}^{a+V} \exp\left(-t\left(aI + K_0 \dot{+} V\right)\right)$$

is given by the expression:

$$\mathbb{E}_x\left(\exp\left(-\int_0^S (a + V(X(u)))\,du\right) \exp\left(-t\left(aI + K_0 + V\right)\right)(X(S), y) : S < \infty\right).$$

So that the square of its Hilbert-Schmidt norm is given by the formula

$$\begin{aligned}&\left\| H_{\Sigma}^{a+V} \exp\left(-t\left(aI + K_0 \dot{+} V\right)\right) \right\|_{\mathrm{HS}}^2 = \int dx \int dy \\ &\left(\mathbb{E}_x\left[\exp\left(-\int_0^S (a + V(X(u)))\,du\right) \exp\left(-t\left(aI + K_0 + V\right)\right)(X(S), y) : S < \infty\right]\right)^2\end{aligned}$$

and it is dominated by the quantity:

$$\begin{aligned}&\sup_{z\in\Gamma, y\in E}\left(\exp\left(-t\left(aI + K_0 \dot{+} V\right)\right)(z, y)\right)^2 \\ &\quad\times \int dx \left(\mathbb{E}_x\left(\exp\left(-\int_0^S (a + V(X(u)))\,du\right) : S < \infty\right)\right)^2 .\end{aligned} \tag{5.49}$$

In view of (i′) the expression in (5.49) is finite. This proves the implication (i′) ⇒ (ii).

(ii) ⇒ (iii). Choose a, $b \in \mathbb{R}$, such that $K_0 \dot{+} V + aI \geq bI > 0$. From equality (5.44) in Proposition 5.11, we get

$$\begin{aligned}&D_{\Sigma}(t/2) \exp\left(-\frac{t}{2}\left(K_0 \dot{+} V\right)\right) \\ &= e^{at} H_{\Sigma}^{a+V} \exp\left(-t\left(aI + K_0 \dot{+} V\right)\right)\end{aligned} \tag{5.50}$$

$$- e^{at/2} J^* \exp\left(-\frac{t}{2}\left(K_0 \dot{+} V\right)_\Sigma\right) J H_\Sigma^{a+V} \exp\left(-\frac{t}{2}\left(aI + K_0 \dot{+} V\right)\right)$$
$$+ e^{at} \int_0^{t/2} ds J^* \exp\left(-s\left(aI + \left(K_0 \dot{+} V\right)_\Sigma\right)\right) J$$
$$H_\Sigma^{a+V}\left(aI + K_0 \dot{+} V\right) \exp\left(-\frac{t}{2}\left(aI + K_0 \dot{+} V\right)\right) \exp\left(-\left(\frac{t}{2} - s\right)\left(aI + K_0 \dot{+} V\right)\right).$$

Hence, we infer that

$$\left\| D_\Sigma(t/2) \exp\left(-\frac{t}{2}\left(K_0 \dot{+} V\right)\right)\right\|_{\mathrm{HS}}$$
$$\leq e^{at} \left\| H_\Sigma^{a+V} \exp\left(-t\left(aI + K_0 \dot{+} V\right)\right)\right\|_{\mathrm{HS}}$$
$$+ e^{at/2} \left\| J^* \exp\left(-\frac{t}{2}\left(K_0 \dot{+} V\right)_\Sigma\right) J\right\|_{2,2} \left\| H_\Sigma^{a+V} \exp\left(-\frac{t}{2}\left(aI + K_0 \dot{+} V\right)\right)\right\|_{\mathrm{HS}}$$
$$+ e^{at} \int_0^{t/2} ds \left\| J^* \exp\left(-s\left(aI + \left(K_0 \dot{+} V\right)_\Sigma\right)\right) J\right\|_{2,2}$$
$$\times \left\| H_\Sigma^{a+V} \exp\left(-\frac{t}{4}\left(aI + K_0 \dot{+} V\right)\right)\right\|_{\mathrm{HS}}$$
$$\times \left\| \left(aI + K_0 \dot{+} V\right) \exp\left(-\frac{t}{4}\left(aI + K_0 \dot{+} V\right)\right) \exp\left(-\left(\frac{t}{2} - s\right)\left(aI + K_0 \dot{+} V\right)\right)\right\|_{2,2}$$
$$\leq e^{at} \left\| H_\Sigma^{a+V} \exp\left(-t\left(aI + K_0 \dot{+} V\right)\right)\right\|_{\mathrm{HS}}$$
$$+ e^{at} \left\| H_\Sigma^{a+V} \exp\left(-\frac{t}{2}\left(aI + K_0 \dot{+} V\right)\right)\right\|_{\mathrm{HS}}$$
$$+ e^{at} \int_0^{t/2} ds \left\| J^* \exp\left(-s\left(aI + K_0 \dot{+} V\right)_\Sigma\right) J\right\|_{2,2}$$
$$\times \left\| H_\Sigma^{a+V} \exp\left(-\frac{t}{4}\left(aI + K_0 \dot{+} V\right)\right)\right\|_{\mathrm{HS}}$$
$$\times \left\| \left(aI + K_0 \dot{+} V\right) \exp\left(-\frac{t}{4}\left(aI + K_0 \dot{+} V\right)\right)\right\|_{2,2}$$
$$\times \left\| \exp\left(-\left(\frac{t}{2} - s\right)\left(aI + K_0 \dot{+} V\right)\right)\right\|_{2,2}$$

(employ $\left(aI + K_0 \dot{+} V\right)_\Sigma \geq 0$ and $aI + K_0 \dot{+} V \geq 0$)

$$\leq e^{at} \left\| H_\Sigma^{a+V} \exp\left(-t\left(aI + K_0 \dot{+} V\right)\right)\right\|_{\mathrm{HS}}$$
$$+ e^{at} \left\| H_\Sigma^{a+V} \exp\left(-\frac{t}{2}\left(aI + K_0 \dot{+} V\right)\right)\right\|_{\mathrm{HS}}$$
$$+ e^{at} \frac{t}{2} \left\| H_\Sigma^{a+V} \exp\left(-\frac{t}{4}\left(aI + K_0 \dot{+} V\right)\right)\right\|_{\mathrm{HS}} \sup_{\xi > 0} \xi \exp\left(-\frac{t}{4}\xi\right). \tag{5.51}$$

So, because of (ii), (iii) is a consequence of (5.51).

(iii) $\Rightarrow$ (iv). As in the proof of the implication (iii) implies (iv) of Theorem 5.6, we have the identity

$$\begin{aligned} &2D_\Sigma(t) \\ &= D_\Sigma(t/2)\exp\left(-\frac{t}{2}\left(K_0\dot{+}V\right)\right) + \exp\left(-\frac{t}{2}\left(K_0\dot{+}V\right)\right) D_\Sigma(t/2) \\ &+ D_\Sigma(t/2)J^*\exp\left(-\frac{t}{2}\left(K_0\dot{+}V\right)_\Sigma\right) J + J^*\exp\left(-\frac{t}{2}\left(K_0\dot{+}V\right)_\Sigma\right) JD_\Sigma(t/2). \end{aligned} \tag{5.52}$$

Hence assertion (iv) follows as in Theorem 5.6.

Finally the implication (iv) $\Rightarrow$ (iii) is trivial. This proves the theorem.

A formulation and proof of the following result was given in Proposition 4.16. The corresponding formulae are (4.69) and (4.70).

5.13. PROPOSITION. *Define the pinned measure $\mu_{0,x}^{t,y}$ on the σ-field generated by $X(s)$, $0 \le s < t$, via the equality:*

$$\mu_{0,x}^{t,y}(A) = \mathbb{E}_x\left[1_A p_0(t-s, X(s), y)\right], \quad A \in \mathcal{F}_s, \quad 0 \le s < t. \tag{5.53}$$

Define the martingale $s \mapsto M_V^{t,y}(s)$, $0 \le s < t$, by

$$M_V^{t,y}(s) = \exp\left(-\int_0^s V(X(\sigma))d\sigma\right)\exp\left(-(t-s)\left(K_0+V\right)\right)(X(s),y).$$

Then the function

$$\int_{\{S<t\}} \exp\left(-\int_0^t V(X(\sigma))d\sigma\right) d\mu_{0,x}^{t,y} \tag{5.54}$$

$$= \mathbb{E}_x\left[M_V^{t,y}(S) : S < t\right] \tag{5.55}$$

serves as an integral kernel for the operator $D_\Sigma(t)$.

Instead of (5.54) or (5.55) we write $D_\Sigma(t,x,y)$.

The result of the following theorem is the analogue of the operator equality

$$\begin{aligned} &D_\Sigma(t) + D_\Sigma(t/2)^2 \\ &= \exp\left(-\frac{t}{2}\left(K_0\dot{+}V\right)\right) D_\Sigma(t/2) + D_\Sigma(t/2)\exp\left(-\frac{t}{2}\left(K_0\dot{+}V\right)\right) \end{aligned} \tag{5.56}$$

for the corresponding integral kernels. A proof was given in Proposition 4.18 of Chapter 4 section B.

5.14. PROPOSITION. *The integral kernel $D_\Sigma(t,x,y)$ satisfies the following identities:*

$$\begin{aligned}
&D_\Sigma(t,x,y) + \int D_\Sigma(t/2,x,z) D_\Sigma(t/2,z,y)dz \\
&= \mathbb{E}_x\left[\exp\left(-\int_0^S V(X(\sigma))d\sigma\right)\exp\left(-(t-S)\left(K_0+V\right)\right)(X(S),y) : S < \frac{t}{2}\right] \\
&\quad + \mathbb{E}_y\left[\exp\left(-\int_0^S V(X(\sigma))d\sigma\right)\exp\left(-(t-S)\left(K_0+V\right)\right)(X(S),x) : S < \frac{t}{2}\right]
\end{aligned} \tag{5.57}$$

$$\begin{aligned}
&= \int D_\Sigma(t/2,x,z)\exp\left(-\frac{t}{2}\left(K_0+V\right)\right)(z,y)dz \\
&\quad + \int D_\Sigma(t/2,y,z)\exp\left(-\frac{t}{2}\left(K_0+V\right)\right)(z,x)dz;
\end{aligned} \tag{5.58}$$

$$\begin{aligned}
&D_\Sigma(t,x,y) - \int D_\Sigma(t/2,x,z) D_\Sigma(t/2,z,y)dz \\
&= \int D_\Sigma(t/2,x,z)\exp\left(-\frac{t}{2}\left(K_0+V\right)_\Sigma\right)(z,y)dz \\
&\quad + \int D_\Sigma(t/2,y,z)\exp\left(-\frac{t}{2}\left(K_0+V\right)_\Sigma\right)(z,x)dz.
\end{aligned} \tag{5.59}$$

The following result was established in Proposition 4.20 of Chapter 4.B. See (4.55) for the definition of H_Σ^{a+V} and Definition 4.13 for that of T_Σ^{a+V}.

5.15. PROPOSITION. *Let $1 \le p \le \infty$. Let V be a Kato-Feller potential and choose $a > 0$ and $t > 0$ such that*

$$C(t) := e^{-at}\sup_{x\in E}\mathbb{E}_x\left[\exp\left(-\int_0^t V(X(u))du\right)\right] < 1.$$

If the function $T_\Sigma^{a+V}(t)1$ belongs to $L^p(E,m)$, then the function $H_\Sigma^{a+V}1$ is a member of $L^p(E,m)$ as well. Moreover, the following inequalities are valid:

$$\left\|T_\Sigma^{a+V}(t)1\right\|_p \le \left\|H_\Sigma^{a+V}1\right\|_p \le \frac{1}{1-C(t)}\left\|T_\Sigma^{a+V}(t)1\right\|_p.$$

In particular, for such a choice of a the function $H_\Sigma^{a+V}1$ is bounded and

$$H_\Sigma^{a+V}1(x) \le \frac{1}{1-C(t)}\sup_{x\in E}\mathbb{E}_x\left(\exp\left(\int_0^t V_-(X(u))du\right)\right).$$

That we can choose $a > 0$ and $t > 0$ such that $C(t) < 1$ is a consequence of Khas'minskii's lemma: see Proposition 3.5 and its Corollary 3.6.

We establish a number of Hilbert-Schmidt properties of resolvent and we continue the study of further (weighted) semigroup differences. We begin with a proposition on differences of powers of resolvents. In the following theorem we write, for $q > 0$,

$$R(a)^q := \left(aI + K_0 \dot{+} V\right)^{-q} = \frac{1}{\Gamma(q)} \int_0^\infty s^{q-1} e^{-as} \exp\left(-s\left(K_0 \dot{+} V\right)\right) ds, \text{ and}$$
$$R_\Sigma(a)^q := \left(\left(aI + K_0 \dot{+} V\right)_\Sigma\right)^{-q} = \frac{1}{\Gamma(q)} \int_0^\infty s^{q-1} e^{-as} \exp\left(-s\left(K_0 \dot{+} V\right)_\Sigma\right) ds,$$

for the resolvent of the operators $K_0 \dot{+} V$ and $\left(K_0 \dot{+} V\right)_\Sigma$.

5.16. THEOREM. *Among the assertions*
(i) $H_\Sigma^{a+V} 1$ *belongs to* $L^2(E, m)$ *and* $\left\| 1_\Gamma \left(aI + K_0\right)^{-q} \right\|_{\infty,1} < \infty$;
(ii) $H_\Sigma^{a+V} 1$ *belongs to* $L^2(E, m)$ *and* $\| 1_\Gamma R(a)^q 1_\Gamma \|_{\infty,1} < \infty$;
(iii) The operator $R(a)^q - J^* R_\Sigma(a)^q J$ *is Hilbert-Schmidt;*
(iv) The operator $H_\Sigma^{a+V} R(a)^q$ *is Hilbert-Schmidt;*
the following implications are valid:

$$(i) \implies (ii) \implies (iii) \iff (iv).$$

The corresponding inequalities are:

$$\begin{aligned} \left\| H_\Sigma^{a+V} R(a)^q \right\|_{\mathrm{HS}} &\le \| R(a)^q - J^* R_\Sigma(a)^q J \|_{\mathrm{HS}} \le 2^{\max(q+1/2,1)} \left\| H_\Sigma^{a+V} R(a)^q \right\|_{\mathrm{HS}} \\ &\le 2^{\max(q+1/2,1)} \left\| H_\Sigma^{a+V} 1 \right\|_2 \left\| 1_\Gamma \left(aI + K_0\right)^{-q} \right\|_{\infty,1}^{1/2} \left\| \left(aI + K_0 \dot{+} 2V\right)^{-q} 1 \right\|_\infty^{1/2}. \end{aligned} \tag{5.60}$$

Before we prove this result, we would like to make some comments.

REMARK 1. High enough powers of resolvents smooth out the singularity at $t = 0$ for semigroups of finite dimension. This is a reason why the conditions in (i) are sensible. Since perturbations with Kato-Feller potentials do not change the spectral dimension (see Definition 1.4), the same comment applies to assertion (ii). For some relevant information in this direction the reader may consult Corollary 2.10 assertion (b).

REMARK 2. Since $h_\Sigma^{a+V} = H_\Sigma^{a+V} 1$, it follows from Proposition 5.15 that h_Σ^{a+V} belongs to $L^2(E, m)$ if and only if the function $T_\Sigma^{a+V} 1$ does. Suppose that $\int dx \left(1 + V_-(x)^2\right) \mathbb{P}_x \left[S < 1\right]^2 dx$ is finite. From Lemma 4.21 it follows that the function $T_\Sigma^{a+V} 1$ belongs to $L^2(E, m)$ and hence the same is true for h_Σ^{a+V}.

REMARK 3. The proof of the implication (iii) $\Rightarrow$ (iv) proves the following pointwise inequality in x:

$$\int dy \left[R(a)^q(x, y) - \left[J^* R_\Sigma(a)^q J\right](x, y)\right]^2 \ge \int dy \left[H_\Sigma^{a+V} R(a)^q(x, y)\right]^2. \tag{5.61}$$

Proof. We begin with the implication (iii) $\Longrightarrow$ (iv) and at the same time we prove the inequality

$$\left\| H_{\Sigma}^{a+V} R(a)^q \right\|_{\text{HS}} \leq \left\| R(a)^q - J^* R_{\Sigma}(a)^q J \right\|_{\text{HS}} . \tag{5.62}$$

If $T : L^2(E, m) \to L^2(E, m)$ is an integral operator we write $[T](x, y)$ for its integral kernel. In order to prove (iii) $\Longrightarrow$ (iv), we observe first that the integral kernel of the operator $R(a)^q - J^* R_{\Sigma}(a)^q J$ is given by

$$\begin{aligned} &\frac{1}{\Gamma(q)} \int_0^\infty dt e^{-at} t^{q-1} \\ &\quad \times \mathbb{E}_x \left(\exp\left(-\int_0^S V(X(u)) du \right) \exp\left(-(t - S) \left(K_0 \dot{+} V \right) \right) (X(S), y) : S < t \right). \end{aligned} \tag{5.63}$$

Consequently from Chapman-Kolmogorov's identity we get:

$$\begin{aligned} &\int dy \left\{ R(a)^q (x, y) - \left[J^* R_{\Sigma}(a)^q J \right] (x, y) \right\}^2 \\ &= \int dy \left(\left(aI + K_0 \dot{+} V \right)^{-q} (x, y) - J^* \left(aI + \left(K_0 \dot{+} V \right)_{\Sigma} \right)^{-q} J(x, y) \right)^2 \\ &= \frac{1}{\Gamma(q)^2} \int_0^\infty dt_1 \int_0^\infty dt_2 e^{-a(t_1 + t_2)} (t_1 t_2)^{q-1} \\ &\quad \mathbb{E}_x \otimes \mathbb{E}_x \Bigg[(\omega, \omega') \mapsto \exp\left(-\int_0^{S(\omega)} V(X(u))(\omega) du \right) 1_{[0, t_1)}(S(\omega)) \\ &\qquad \times \exp\left(-\int_0^{S(\omega')} V(X(u))(\omega') du \right) 1_{[0, t_2)}(S(\omega')) \\ &\qquad\quad \times \exp\left(-\left(t_1 + t_2 - S(\omega) - S(\omega') \right) (K_0 + V) \right) \left(X(S)(\omega), X(S)(\omega') \right) \Bigg] \end{aligned}$$

(apply Fubini, substitute $t_1 - S(\omega) = \tau_1$ and $t_2 - S(\omega') = \tau_2$ and apply Fubini again)

$$\begin{aligned} &= \frac{1}{\Gamma(q)^2} \int_0^\infty d\tau_1 \int_0^\infty d\tau_2 \\ &\quad \mathbb{E}_x \otimes \mathbb{E}_x \Bigg[(\omega, \omega') \mapsto \exp\left(-\int_0^{S(\omega)} (a + V(X(u))(\omega)) \, du \right) 1_{[0, \infty)}(S(\omega)) \\ &\qquad \times \exp\left(-\int_0^{S(\omega')} (a + V(X(u))(\omega')) \, du \right) 1_{[0, \infty)}(S(\omega')) \\ &\qquad \times \exp\left(-(\tau_1 + \tau_2)(a + K_0 + V) \right) \left(X(S)(\omega), X(S)(\omega') \right) \\ &\qquad\quad \times (\tau_1 + S(\omega))^{q-1} (\tau_2 + S(\omega'))^{q-1} \Bigg] \end{aligned}$$

(substitute $\tau_1 + \tau_2 = \tau$ and $\tau_1 = \sigma$)

$$
\begin{aligned}
&= \frac{1}{\Gamma(q)^2}\int_0^\infty d\tau \\
&\quad \mathbb{E}_x \otimes \mathbb{E}_x\Bigg[(\omega,\omega') \mapsto \exp\left(-\int_0^{S(\omega)} (a + V(X(u))(\omega))\,du\right) 1_{[0,\infty)}(S(\omega)) \\
&\qquad \times \exp\left(-\int_0^{S(\omega')} (a + V(X(u))(\omega'))\,du\right) 1_{[0,\infty)}(S(\omega')) \\
&\qquad \times \exp\left(-\tau(a + K_0 + V)\right)\left(X(S)(\omega), X(S)(\omega')\right) \\
&\qquad \times \int_0^\tau d\sigma\,(\sigma + S(\omega))^{q-1}\,(\tau - \sigma + S(\omega'))^{q-1}\Bigg] \\
&\geq \frac{1}{\Gamma(q)^2}\int_0^\infty d\tau \\
&\quad \mathbb{E}_x \otimes \mathbb{E}_x\Bigg[(\omega,\omega') \mapsto \exp\left(-\int_0^{S(\omega)} (a + V(X(u))(\omega))\,du\right) 1_{[0,\infty)}(S(\omega)) \\
&\qquad \times \exp\left(-\int_0^{S(\omega')} (a + V(X(u))(\omega'))\,du\right) 1_{[0,\infty)}(S(\omega')) \\
&\qquad \times \exp\left(-\tau(a + K_0 + V)\right)\left(X(S)(\omega), X(S)(\omega')\right) \int_0^\tau d\sigma \sigma^{q-1}\,(\tau - \sigma)^{q-1}\Bigg] \\
&= \frac{1}{\Gamma(2q)}\int_0^\infty \tau^{2q-1} \\
&\quad \mathbb{E}_x \otimes \mathbb{E}_x\Bigg[(\omega,\omega') \mapsto \exp\left(-\int_0^{S(\omega)} (a + V(X(u))(\omega))\,du\right) 1_{[0,\infty)}(S(\omega)) \\
&\qquad \times \exp\left(-\int_0^{S(\omega')} (a + V(X(u))(\omega'))\,du\right) 1_{[0,\infty)}(S(\omega')) \\
&\qquad \times \exp\left(-\tau(a + K_0 + V)\right)\left(X(S)(\omega), X(S)(\omega')\right)\Bigg] d\tau.
\end{aligned}
\tag{5.64}
$$

We also compute the integral $\int \left(H_\Sigma^{a+V} R(a)^q(x,y)\right)^2 dy$. Because of the identity

$$
\begin{aligned}
H_\Sigma^{a+V} R(a)^q(x,y) = \frac{1}{\Gamma(q)}\mathbb{E}_x\Bigg[&\int_0^\infty t^{q-1}\exp\left(-\int_0^S (a + V(X(u)))\,du\right) \\
&\times \exp\left(-t\,(aI + K_0 + V)\right)(X(S), y)dt : S < \infty\Bigg],
\end{aligned}
$$

we infer that

$$\begin{aligned}
&\int \left(H_{\Sigma}^{a+V} R(a)^q(x,y)\right)^2 dy \\
&= \frac{1}{\Gamma(q)^2}\mathbb{E}_x \otimes \mathbb{E}_x \Bigg[(\omega,\omega') \mapsto \int_0^\infty dt_1 \int_0^\infty dt_2 t_1^{q-1} t_2^{q-1} \\
&\times \exp\left(-\int_0^{S(\omega)} (a + V(X(u)(\omega)))\, du\right) \exp\left(-\int_0^{S(\omega')} (a + V(X(u)(\omega')))\, du\right) \\
&\exp\left(-(t_1+t_2)(aI + K_0 + V)\right)(X(S)(\omega), X(S)(\omega'))\, 1_{[0,\infty)}(S(\omega)) 1_{[0,\infty)}(S(\omega'))\Bigg] \\
&= \frac{1}{\Gamma(q)^2}\mathbb{E}_x \otimes \mathbb{E}_x \Bigg[(\omega,\omega') \mapsto \int_0^\infty dt_1 \int_{t_1}^\infty dt_2 t_1^{q-1} (t_2 - t_1)^{q-1} \\
&\times \exp\left(-\int_0^{S(\omega)} (a + V(X(u)(\omega)))\, du\right) \exp\left(-\int_0^{S(\omega')} (a + V(X(u)(\omega')))\, du\right) \\
&\times \exp\left(-t_2 (aI + K_0 + V)\right)(X(S)(\omega), X(S)(\omega')\, 1_{[0,\infty)}(S(\omega)) 1_{[0,\infty)}(S(\omega'))\Bigg] \\
&= \frac{1}{\Gamma(q)^2}\mathbb{E}_x \otimes \mathbb{E}_x \Bigg[(\omega,\omega') \mapsto \int_0^\infty dt_2 \int_0^{t_2} dt_1 t_1^{q-1} (t_2 - t_1)^{q-1} \\
&\times \exp\left(-\int_0^{S(\omega)} (a + V(X(u)(\omega)))\, du\right) \exp\left(-\int_0^{S(\omega')} (a + V(X(u)(\omega')))\, du\right) \\
&\times \exp\left(-t_2 (aI + K_0 + V)\right)(X(S)(\omega), X(S)(\omega')\, 1_{[0,\infty)}(S(\omega)) 1_{[0,\infty)}(S(\omega'))\Bigg] \\
&= \frac{1}{\Gamma(2q)}\mathbb{E}_x \otimes \mathbb{E}_x \Bigg[(\omega,\omega') \mapsto \int_0^\infty dt\, t^{2q-1} \\
&\times \exp\left(-\int_0^{S(\omega)} (a + V(X(u)(\omega)))\, du\right) \exp\left(-\int_0^{S(\omega')} (a + V(X(u)(\omega')))\, du\right) \\
&\times \exp\left(-t (aI + K_0 + V)\right)(X(S)(\omega), X(S)(\omega')\, 1_{[0,\infty)}(S(\omega)) 1_{[0,\infty)}(S(\omega'))\Bigg].
\end{aligned} \tag{5.65}$$

From (5.64) and (5.65) we obtain the inequality:

$$\int dy \left[R(a)^q(x,y) - \left[J^* R_\Sigma(a)^q J\right](x,y)\right]^2 \geq \int dy \left[H_\Sigma^{a+V} R(a)^q(x,y)\right]^2. \tag{5.66}$$

Upon integrating both sides of the inequality (5.66) with respect to x, we get the first inequality in (5.60) and the implication (iii) $\Longrightarrow$ (iv).

(iv) $\Longrightarrow$ (iii). This implication will be proved simultaneously with the second inequality in (5.60). Put

$$M_V^{t,y}(s) = \exp\left(-\int_0^s V(X(u))du\right)\exp\left(-(t-s)\left(K_0+V\right)\right)(X(s),y).$$

From Theorem 5.14 (in fact from Proposition 4.16 together with Proposition 4.18 of Chapter 4.B) we deduce the following pointwise inequality:

$$\begin{aligned}
&R(a)^q(x,y) - \left[J^* R_\Sigma(a)^q J\right](x,y)\\
&= \frac{1}{\Gamma(q)}\int_0^\infty dt\, t^{q-1}e^{-at}D_\Sigma(t,x,y)
\end{aligned}$$

(first equality in Theorem 5.14)

$$\begin{aligned}
&\le \frac{1}{\Gamma(q)}\int_0^\infty dt\, t^{q-1}e^{-at}D_\Sigma(t/2)\exp\left(-\frac{t}{2}\left(K_0\dot{+}V\right)\right)(x,y)\\
&\quad + \frac{1}{\Gamma(q)}\int_0^\infty dt\, t^{q-1}e^{-at}D_\Sigma(t/2)\exp\left(-\frac{t}{2}\left(K_0\dot{+}V\right)\right)(y,x)
\end{aligned}$$

(Theorem 5.14: the equality of (5.57) and (5.58))

$$\begin{aligned}
&= \frac{1}{\Gamma(q)}\int_0^\infty dt\, t^{q-1}e^{-at}\mathbb{E}_x\left[M_V^{t,y}(S) : S < \frac{t}{2}\right]\\
&\quad + \frac{1}{\Gamma(q)}\int_0^\infty dt\, t^{q-1}e^{-at}\mathbb{E}_y\left[M_V^{t,x}(S) : S < \frac{t}{2}\right].
\end{aligned} \tag{5.67}$$

The inequality in (5.67) yields the estimate

$$\begin{aligned}
&\left(\iint dx\, dy\left[R(a)^q(x,y) - \left[J^* R_\Sigma(a)^q J\right](x,y)\right]^2\right)^{1/2}\\
&\le \frac{2}{\Gamma(q)}\left(\int dx\int dy\left(\int_0^\infty dt\, t^{q-1}e^{-at}\mathbb{E}_x\left[\exp\left(-\int_0^S V(X(u))du\right)\right.\right.\right.\\
&\qquad\left.\left.\left.\times\exp\left(-(t-S)\left(K_0+V\right)\right)(X(S),y) : S < \frac{t}{2}\right]\right)^2\right)^{1/2}.
\end{aligned} \tag{5.68}$$

Next we calculate and estimate the following expression:

$$\begin{aligned}
&\int dy\left(\int_0^\infty dt\, t^{q-1}e^{-at}\mathbb{E}_x\left[\exp\left(-\int_0^S V(X(u))du\right)\right.\right.\\
&\qquad\left.\left.\times\exp\left(-(t-S)\left(K_0+V\right)\right)(X(S),y) : S < \frac{t}{2}\right]\right)^2
\end{aligned}$$

$$= \int_0^\infty dt_1 \int_0^\infty dt_2 t_1^{q-1} t_2^{q-1} \mathbb{E}_x \otimes \mathbb{E}_x$$

$$\Bigg[(\omega,\omega') \mapsto \exp\left(-\int_0^{S(\omega)} (a + V(X(u)(\omega)))\,du\right)$$

$$\times \exp\left(-\int_0^{S(\omega')} (a + V(X(u)(\omega')))\,du\right)$$

$$\times \exp\left(-\left(t_1 + t_2 - S(\omega) - S(\omega')\right)(aI + K_0 + V)\right)(X(S)(\omega), X(S)(\omega'))$$

$$\times 1_{[0,t_1/2)}(S(\omega)) 1_{[0,t_2/2)}(S(\omega'))\Bigg]$$

(substitute $\tau_1 = t_1$ and $\tau_2 = t_1 + t_2 - S(\omega) - S(\omega')$)

$$= \mathbb{E}_x \otimes \mathbb{E}_x \Bigg[(\omega,\omega') \mapsto$$

$$\int_{S(\omega)+S(\omega')}^\infty d\tau_2 \int_{2S(\omega)}^{\tau_2+S(\omega)-S(\omega')} d\tau_1 \tau_1^{q-1} \left(\tau_2 + S(\omega) + S(\omega') - \tau_1\right)^{q-1}$$

$$\times \exp\left(-\int_0^{S(\omega)} (a + V(X(u)(\omega)))\,du\right) \exp\left(-\int_0^{S(\omega')} (a + V(X(u)(\omega')))\,du\right)$$

$$\times \exp\left(-\tau_2 (aI + K_0 + V)\right)(X(S)(\omega), X(S)(\omega')) : S(\omega) < \infty,\ S(\omega') < \infty\Bigg]$$

$$\le \mathbb{E}_x \otimes \mathbb{E}_x \Bigg[(\omega,\omega') \mapsto$$

$$\int_{S(\omega)+S(\omega')}^\infty d\tau_2 \int_0^{\tau_2+S(\omega)+S(\omega')} d\tau_1 \tau_1^{q-1} \left(\tau_2 + S(\omega) + S(\omega') - \tau_1\right)^{q-1}$$

$$\times \exp\left(-\int_0^{S(\omega)} (a + V(X(u)(\omega)))\,du\right) \exp\left(-\int_0^{S(\omega')} (a + V(X(u)(\omega')))\,du\right)$$

$$\times \exp\left(-\tau_2 (aI + K_0 + V)\right)(X(S)(\omega), X(S)(\omega')) : S(\omega) < \infty,\ S(\omega') < \infty\Bigg]$$

(for $s > 0$ we have the beta-function $q \mapsto \int_0^s (s-\sigma)^{q-1}\sigma^{q-1} d\sigma = s^{2q-1}\Gamma(q)^2/\Gamma(2q)$)

$$= \frac{\Gamma(q)^2}{\Gamma(2q)} \mathbb{E}_x \otimes \mathbb{E}_x \Bigg[(\omega,\omega') \mapsto \int_{S(\omega)+S(\omega')}^\infty d\tau_2 \left(\tau_2 + S(\omega) + S(\omega')\right)^{2q-1}$$

$$\times \exp\left(-\int_0^{S(\omega)} (a + V(X(u)(\omega)))\,du\right) \exp\left(-\int_0^{S(\omega')} (a + V(X(u)(\omega')))\,du\right)$$

$$\left.\times \exp\left(-\tau_2\left(aI+K_0+V\right)\right)\left(X(S)(\omega),X(S)(\omega')\right) : S(\omega)<\infty,\ S(\omega')<\infty\right]$$

$$\leq \frac{2^{\max(2q-1,0)}\Gamma(q)^2}{\Gamma(2q)}\mathbb{E}_x\otimes\mathbb{E}_x\left[(\omega,\omega')\mapsto\int_0^\infty d\tau_2\tau_2^{2q-1}\right.$$

$$\times\exp\left(-\int_0^{S(\omega)}\left(a+V(X(u)(\omega))\right)du\right)\exp\left(-\int_0^{S(\omega')}\left(a+V(X(u)(\omega'))\right)du\right)$$

$$\left.\times \exp\left(-\tau_2\left(aI+K_0+V\right)\right)\left(X(S)(\omega),X(S)(\omega')\right) : S(\omega)<\infty,\ S(\omega')<\infty\right]$$

$$= 2^{\max(2q-1,0)}\Gamma(q)^2\int\left(\left[H_\Sigma^{a+V}R(a)^q\right](x,y)\right)^2dy. \tag{5.69}$$

The inequalities (5.68) and (5.69) together yield

$$\left(\iint dx\,dy\left[R(a)^q(x,y)-\left[J^*R_\Sigma(a)^qJ\right](x,y)\right]^2\right)^{1/2}$$

$$\leq\frac{2}{\Gamma(q)}\left(\int dx\int dy\left(\int_0^\infty dt\,t^{q-1}e^{-at}\mathbb{E}_x\left[\exp\left(-\int_0^S V(X(u))du\right)\right.\right.\right.$$

$$\left.\left.\left.\times\exp\left(-(t-S)\left(K_0+V\right)\right)\left(X(S),y\right) : S<\frac{t}{2}\right]\right)^2\right)^{1/2}$$

$$\leq 2^{\max(q+1/2,1)}\left(\iint dxdy\left(\left[H_\Sigma^{a+V}R(a)^q\right](x,y)\right)^2\right)^{1/2}. \tag{5.70}$$

This proves the implication (iv) $\Longrightarrow$ (iii) and also the second second inequality in (5.60).

(ii) $\Longrightarrow$ (iv). It suffices to show that the operator $H_\Sigma^{a+V}R(a)^q$ is a Hilbert-Schmidt operator. Consider the identity

$$\int\left(\left[H_\Sigma^{a+V}R(a)^q\right](x,y)\right)^2dy$$

$$=\frac{1}{\Gamma(2q)}\mathbb{E}_x\otimes\mathbb{E}_x\left[(\omega,\omega')\mapsto\int_0^\infty dt\,t^{2q-1}\right.$$

$$\times\exp\left(-\int_0^{S(\omega)}\left(a+V(X(u)(\omega))\right)du\right)\exp\left(-\int_0^{S(\omega')}\left(a+V(X(u)(\omega'))\right)du\right)$$

$$\left.\times\exp\left(-t\left(aI+K_0+V\right)\right)\left(X(S)(\omega),X(S)(\omega')\right)1_{[0,\infty)}(S(\omega))1_{[0,\infty)}(S(\omega'))\right]$$

$$
\begin{aligned}
&= \mathbb{E}_x \otimes \mathbb{E}_x \Bigg[(\omega, \omega') \mapsto R(a)^{2q} \left(X(S)(\omega), X(S)(\omega') \right) \\
&\quad \times \exp\left(- \int_0^{S(\omega)} \left(a + V(X(u)(\omega)) \right) du \right) \exp\left(- \int_0^{S(\omega')} \left(a + V(X(u)(\omega')) \right) du \right) \\
&\quad \times 1_{[0,\infty)}(S(\omega)) 1_{[0,\infty)}(S(\omega')) \Bigg] \\
&\leq \left(\mathbb{E}_x \left[\exp\left(- \int_0^S \left(a + V(X(u)) \right) du \right) : S < \infty \right] \right)^2 \left\| 1_\Gamma R(a)^{2q} 1_\Gamma \right\|_{\infty,1} .
\end{aligned} \tag{5.71}
$$

The assertion in (iv) is a consequence of (5.71).

(i) $\Longrightarrow$ (ii). The (in-)equalities

$$
\begin{aligned}
\left\| 1_\Gamma R(a)^{2q} 1_\Gamma \right\|_{\infty,1} &= \left\| 1_\Gamma R(a)^q R(a)^q 1_\Gamma \right\|_{\infty,1} \leq \left\| 1_\Gamma R(a)^q \right\|_{\infty,2} \left\| R(a)^q 1_\Gamma \right\|_{2,1} \\
&= \left\| 1_\Gamma R(a)^q \right\|_{\infty,2}^2
\end{aligned} \tag{5.72}
$$

in combination with ($x \in \Gamma$ and $f \geq 0$, $f \in L^2(E, m)$)

$$
\begin{aligned}
&\frac{1}{\Gamma(q)} \int_0^\infty dt\, t^{q-1} e^{-at} \mathbb{E}_x \left[\exp\left(- \int_0^t V(X(u)) du \right) f(X(t)) \right] \\
&\leq \left(\frac{1}{\Gamma(q)} \int_0^\infty dt\, t^{q-1} e^{-at} \mathbb{E}_x \left[\exp\left(-2 \int_0^t V(X(u)) du \right) \right] \right)^{1/2} \\
&\quad \times \left(\frac{1}{\Gamma(q)} \int_0^\infty dt\, t^{q-1} e^{-at} \mathbb{E}_x \left[|f(X(t))|^2 \right] \right)^{1/2} \\
&\leq \left\| 1_\Gamma \left(aI + K_0 \dot{+} 2V \right)^{-q} 1 \right\|_\infty^{1/2} \left\| 1_\Gamma \left(aI + K_0 \right)^{-q} \right\|_{\infty,1}^{1/2} \|f\|_2
\end{aligned}
$$

prove the implication (i) $\Longrightarrow$ (ii) and the final inequality in (5.60). This concludes the proof of Theorem 5.16.

We recall the definition of the operators $T_\Sigma^V(s)$, $s \geq 0$, (cf. Definition 2.33):

$$
T_\Sigma^V(s) f(x) = \mathbb{E}_x \left[\exp\left(- \int_0^S V(X(u)) du \right) : S < s \right],
$$

for whatever functions f it makes sense.

5.17. THEOREM. *Suppose that the function $T_\Sigma^V(t/2)1$ belongs to $L^2(E, m)$ and suppose that the semigroup $\{\exp(-tK_0) : t \geq 0\}$ is L^1-L^∞-smoothing. Then the operator $D_\Sigma(t)$ is Hilbert-Schmidt and*

$$
\begin{aligned}
\|D_\Sigma(t)\|_{HS} &\leq 2 \sup_{t<u<2t} \left\| 1_\Gamma \exp\left(-u \left(K_0 \dot{+} V \right) \right) 1_\Gamma \right\|_{\infty,1}^{1/2} \left\| T_\Sigma^V(t/2) 1 \right\|_2 \qquad (5.73) \\
&\leq 2 \sup_{t/2<u<t} \left\| 1_\Gamma \exp(-uK_0) \right\|_{\infty,1}^{1/2} \left\| \exp\left(-u \left(K_0 \dot{+} 2V \right) \right) 1 \right\|_\infty^{1/2} \left\| T_\Sigma^V(t/2) 1 \right\|_2 .
\end{aligned}
$$

Proof. According to equality (5.57) in Proposition 5.14 the integral kernel $D_\Sigma(t,x,y)$ of the operator $D_\Sigma(t)$ obeys the following inequality

$$\begin{aligned}
&D_\Sigma(t,x,y)\\
&\leq \mathbb{E}_x\left[\exp\left(-\int_0^S V(X(\sigma))d\sigma\right)\exp\left(-(t-S)\left(K_0+V\right)\right)(X(S),y) : S<\frac{t}{2}\right]\\
&\quad+\mathbb{E}_y\left[\exp\left(-\int_0^S V(X(\sigma))d\sigma\right)\exp\left(-(t-S)\left(K_0+V\right)\right)(X(S),x) : S<\frac{t}{2}\right].
\end{aligned}$$

Consequently, with

$$M_V^{t,y}(S)=\exp\left(-\int_0^S V(X(\sigma))d\sigma\right)\exp\left(-(t-S)\left(K_0+V\right)\right)(X(S),y),$$

we get

$$\begin{aligned}
&\iint dx\,dy D_\Sigma(t,x,y)^2 \leq 4\int dx\,dy\left(\mathbb{E}_x\left[M_V^{t,y}(S) : S<\frac{t}{2}\right]\right)^2\\
&=4\int dx\mathbb{E}_x\otimes\mathbb{E}_x\Bigg[(\omega,\omega')\mapsto\\
&\quad\exp\left(-\int_0^S V(X(\sigma)(\omega))d\sigma\right)1_{[0,t/2)}(S(\omega))\\
&\quad\times\exp\left(-\int_0^S V(X(\sigma)(\omega'))d\sigma\right)1_{[0,t/2)}(S(\omega'))\\
&\quad\times\exp\left(-\left(2t-S(\omega)-S(\omega')\right)\left(K_0+V\right)\right)\left(X(S)(\omega),X(S)(\omega')\right)\Bigg]\\
&\leq 4\left\|T_\Sigma^V(t/2)1\right\|_2^2\sup_{t<u<2t}\left\|1_\Gamma\exp\left(-u\left(K_0\dot{+}V\right)\right)1_\Gamma\right\|_{\infty,1}.
\end{aligned}\tag{5.74}$$

By virtue of the inequalities

$$\begin{aligned}
&\left\|1_\Gamma\exp\left(-2u\left(K_0\dot{+}V\right)\right)1_\Gamma\right\|_{\infty,1}\leq\left\|1_\Gamma\exp\left(-u\left(K_0\dot{+}V\right)\right)\right\|_{\infty,2}^2\\
&\leq\left\|1_\Gamma\exp\left(-uK_0\right)\right\|_{\infty,1}\left\|1_\Gamma\exp\left(-u\left(K_0\dot{+}2V\right)\right)1\right\|_\infty
\end{aligned}$$

the statement and the inequalities in Theorem 5.17 follow from (5.74).

We would like to include some weighted Hilbert-Schmidt norm estimates. The following lemma will facilitate the proof of Theorem 5.19.

5.18. LEMMA. *Let $D_\Sigma(t,x,y)$ be the integral kernel of the operator $D_\Sigma(t)$. Fix $1 > \epsilon > 0$. The following inequalities are valid:*

$$D_\Sigma(t,x,y) \le \left(\mathbb{E}_x\left[p_0(t-S,X(S),y) : S < t\right]\right)^{1-\epsilon} \left\|\exp\left(-t\left(K_0 \dot{+} \epsilon^{-1}V\right)\right)\right\|_{\infty,1}^{\epsilon}; \tag{5.75}$$

$$\int D_\Sigma(t,x,y)^2 dy \le \left(\mathbb{E}_x\left[p_0(2t-S,X(S),x) : S < t\right]\right)^{1-\epsilon} \left\|\exp\left(-2t\left(K_0 \dot{+} \epsilon^{-1}V\right)\right)\right\|_{\infty,1}^{\epsilon}. \tag{5.76}$$

Proof. Let $\mu_{0,x}^{t,y}$ be the measure on $\mathcal{F}_{t-}$, the σ-field generated by $X(s)$, $0 \le s < t$, determined by the equality $\mu_{0,x}^{t,y}(A) = \mathbb{E}_x\left[p_0(t-s,X(s),y)1_A\right]$, where A is an event in $\mathcal{F}_s$, $s < t$. Then the kernel $D_\Sigma(t,x,y)$ may be written as

$$D_\Sigma(t,x,y) = \int_{\{S<t\}} \exp\left(-\int_0^t V(X(u))du\right) d\mu_{0,x}^{t,y}.$$

For this result the reader is referred to Proposition 4.16 in Chapter 4.B. From Hölder's inequality, it then follows that

$$\begin{aligned} &D_\Sigma(t,x,y) \\ &\le \left(\int_{\{S<t\}} d\mu_{0,x}^{t,y}\right)^{1-\epsilon} \left(\int_{\{S<t\}} \exp\left(-\epsilon^{-1}\int_0^t V(X(u))du\right) d\mu_{0,x}^{t,y}\right)^{\epsilon} \\ &= \left(\mathbb{E}_x\left[p_0(t-S,X(S),y) : S < t\right]\right)^{1-\epsilon} \left\|\exp\left(-t\left(K_0 \dot{+} \epsilon^{-1}V\right)\right)\right\|_{\infty,1}^{\epsilon}. \end{aligned} \tag{5.77}$$

This proves (5.75). In the final equality of (5.77), we employed the fact that $\mu_{0,x}^{t,y}\{S<t\} = \mathbb{E}_x\left[p_0(t-S,X(S),y) : S < t\right]$. This is a consequence of the martingale property of the process $s \mapsto p_0(t-s,X(s),y)$ together with Doob's optional sampling theorem: see Appendix C.4 and C.11. To be more explicit, we have

$$\mu_{0,x}^{t,y}\{S<t\} = \lim_{s\uparrow t} \mu_{0,x}^{t,y}\{S \le s\} = \lim_{s\uparrow t} \mathbb{E}_x\left[p_0(t-s,X(s),y) : S \le s\right]$$

(Doob's theorem on optional sampling)

$$= \lim_{s\uparrow t} \mathbb{E}_x\left[p_0(t-S,X(S),y) : S \le s\right] = \mathbb{E}_x\left[p_0(t-S,X(S),y) : S < t\right]. \tag{5.78}$$

For more information see Definition 2.18 item (e). For the proof of inequality (5.76), we proceed as follows:

$$\begin{aligned} &\int D_\Sigma(t,x,y)^2 dy \le \int D_\Sigma(t,x,y) \exp\left(-t\left(K_0+V\right)\right)(y,x)dy \\ &= \mathbb{E}_x\left[\exp\left(-\int_0^S V(X(u))du\right) \exp\left(-(2t-S)\left(K_0+V\right)\right)(X(S),x) : S < t\right] \\ &= \int_{\{S<t\}} \exp\left(-\int_0^{2t} V(X(u))du\right) d\mu_{0,x}^{2t,x} \end{aligned}$$

(Hölder's inequality)

$$\leq \left(\mu_{0,x}^{2t,x}\{S<t\}\right)^{1-\epsilon}\left(\int \exp\left(-\epsilon^{-1}\int_0^{2t} V(X(u))du\right) d\mu_{0,x}^{2t,x}\right)^{\epsilon}$$
$$\leq \left(\mathbb{E}_x\left[p_0(2t-S,X(S),x) : S<t\right]\right)^{1-\epsilon}\left\|\exp\left(-2t\left(K_0\dot{+}\epsilon^{-1}V\right)\right)\right\|_{\infty,1}^{\epsilon}. \tag{5.79}$$

The inequality in (5.76) coincides with the one in (5.79).

In the following theorem we apply Lemma 5.18 with $\epsilon = 1/2$.

5.19. THEOREM. *Let $\{\exp(-tK_0) : t \geq 0\}$ be L^1-L^∞-smoothing. Suppose that the functions φ and ψ possess one of the properties (a), (b) or (c):*

(a) The integral

$$\int dx\left(|\varphi(x)|^4+|\psi(x)|^4\right)\mathbb{P}_x\{S<t\} \tag{5.80}$$

is finite;

(b) The integral

$$\int dx\left(|\varphi(x)|^4+|\psi(x)|^4\right)\left(\mathbb{E}_x\left(p_0(2t-S,X(S),x) : S<t\right)\right)^{1/2} \tag{5.81}$$

is finite;

(c) The functions $\varphi T_\Sigma(t/2)1$ and $\psi T_\Sigma(t/2)1$ belong to $L^2(E,m)$ and the expressions

$$\sup_{t/2<u<t}\left\|1_\Gamma \exp\left(-u\left(K_0\dot{+}V\right)\right)|\varphi|^2\exp\left(-u\left(K_0\dot{+}V\right)\right)1_\Gamma\right\|_{\infty,1}, \quad \textit{and}$$

$$\sup_{t/2<u<t}\left\|1_\Gamma \exp\left(-u\left(K_0\dot{+}V\right)\right)|\psi|^2\exp\left(-u\left(K_0\dot{+}V\right)\right)1_\Gamma\right\|_{\infty,1}$$

are finite.

Then the operator $\varphi D_\Sigma(t)\psi$ is a Hilbert-Schmidt operator, and

$$\|\varphi D_\Sigma(t)\psi\|_{\mathrm{HS}}^2 \leq \frac{1}{2}\int dx\left(|\varphi(x)|^4+|\psi(x)|^4\right)\mathbb{P}_x\{S<t\}\left\|\exp\left(-t\left(K_0\dot{+}V\right)\right)\right\|_{\infty,1}, \tag{5.82}$$

in the case the functions φ and ψ have property (a). In the presence of property (b), the Hilbert-Schmidt norm of $\varphi D_\Sigma(t)\psi$ may be estimated in the following fashion:

$$\|\varphi D_\Sigma(t)\psi\|_{\mathrm{HS}}^2 \leq \frac{1}{2}\int dx\left(|\varphi(x)|^4+|\psi(x)|^4\right)\left(\mathbb{E}_x\left[p_0(2t-S,X(S),x) : S<t\right]\right)^{1/2}$$
$$\times\left\|\exp\left(-2t\left(K_0\dot{+}2V\right)\right)\right\|_{\infty,1}^{1/2}. \tag{5.83}$$

If the functions φ and ψ possess property (c), then the Hilbert-Schmidt norm of the operator $\varphi D_\Sigma(t)\psi$ satisfies:

$$\|\varphi D_\Sigma(t)\psi\|_{\mathrm{HS}} \tag{5.84}$$
$$\leq \|\varphi T_\Sigma(t/2)1\|_2 \sup_{t/2<u<t}\left\|1_\Gamma \exp\left(-u\left(K_0\dot{+}V\right)\right)|\psi|^2\exp\left(-u\left(K_0\dot{+}V\right)\right)1_\Gamma\right\|_{\infty,1}^{1/2}$$
$$+\|\psi T_\Sigma(t/2)1\|_2 \sup_{t/2<u<t}\left\|1_\Gamma \exp\left(-u\left(K_0\dot{+}V\right)\right)|\varphi|^2\exp\left(-u\left(K_0\dot{+}V\right)\right)1_\Gamma\right\|_{\infty,1}^{1/2}.$$

REMARK 1. Suppose that $\psi \equiv 1$. An inspection of the proof below will show that the conclusions (5.82) and (5.83) still hold if in (5.80) and (5.81) respectively as well as in (5.82) and (5.83) respectively the quantity $|\varphi(x)|^4 + |\psi(x)|^4$ is replaced by $2|\varphi(x)|^2$.

REMARK 2. Notice that the expression in (5.83) is finite whenever

$$\int dx \left(|\varphi(x)|^4 + |\psi(x)|^4\right) \sup_{\frac{1}{2}t<s<t} \sup_{z\in\Gamma} p_0(s,z,x)^{1/2}\mathbb{P}_x(S<t)^{1/2} < \infty. \tag{5.85}$$

This is so because, on $\{S<\infty\}$, $X(S)$ belongs to the closure of Γ, $\mathbb{P}_x$-almost surely.

REMARK 3. Suppose that $f \mapsto \varphi \exp\left(-\frac{t}{4}\left(K_0\dot{+}V\right)\right) f$ is a continuous operator in $L^2(E,m)$. Also suppose that the quantity

$$\sup_{t/4<u<3t/4} \left\|1_\Gamma \exp\left(-uK_0\right)\right\|_{\infty,1}$$

is finite. Then the expression

$$\sup_{t/2<u<t} \left\|1_\Gamma \exp\left(-u\left(K_0\dot{+}V\right)\right)|\varphi|^2 \exp\left(-u\left(K_0\dot{+}V\right)\right)1_\Gamma\right\|_{\infty,1}$$

is finite. This can be seen as follows. We write, with $v = u - t/4$,

$$\begin{aligned}
&1_\Gamma \exp\left(-u\left(K_0\dot{+}V\right)\right)|\varphi|^2 \exp\left(-u\left(K_0\dot{+}V\right)\right)1_\Gamma \\
&= 1_\Gamma \exp\left(-v\left(K_0\dot{+}V\right)\right)\exp\left(-\frac{t}{4}\left(K_0\dot{+}V\right)\right)|\varphi|^2 \exp\left(-\frac{t}{4}\left(K_0\dot{+}V\right)\right) \\
&\quad \times \exp\left(-v\left(K_0\dot{+}V\right)\right)1_\Gamma
\end{aligned}$$

and hence

$$\begin{aligned}
&\sup_{t/2<u<t} \left\|1_\Gamma \exp\left(-u\left(K_0\dot{+}V\right)\right)|\varphi|^2 \exp\left(-u\left(K_0\dot{+}V\right)\right)1_\Gamma\right\|_{\infty,1} \\
&\leq \sup_{t/4<v<3t/4} \left\|1_\Gamma \exp\left(-v\left(K_0\dot{+}V\right)\right)\right\|_{\infty,2} \\
&\qquad \times \left\|\exp\left(-\frac{t}{4}\left(K_0\dot{+}V\right)\right)|\varphi|^2 \exp\left(-\frac{t}{4}\left(K_0\dot{+}V\right)\right)\right\|_{2,2} \\
&\qquad \times \left\|\exp\left(-v\left(K_0\dot{+}V\right)\right)1_\Gamma\right\|_{2,1} \\
&= \sup_{t/4<u<3t/4} \left\|1_\Gamma \exp\left(-u\left(K_0\dot{+}V\right)\right)\right\|_{\infty,2}^2 \left\|\varphi \exp\left(-\frac{t}{4}\left(K_0\dot{+}V\right)\right)\right\|_{2,2}^2.
\end{aligned}$$

This proves our claim, because we assume that

$$\sup_{t/4<u<3t/4} \left\|1_\Gamma \exp\left(-uK_0\right)\right\|_{\infty,1}$$

is finite. Hence, by virtue of

$$\begin{aligned}&\left\|1_\Gamma \exp\left(-v\left(K_0\dot{+}V\right)\right)\right\|_{\infty,2}\\ &\le \left\|1_\Gamma \exp\left(-v\left(K_0\dot{+}2V\right)\right)1\right\|_\infty^{1/2} \left\|1_\Gamma \exp\left(-vK_0\right)\right\|_{\infty,1}^{1/2},\end{aligned}$$

the quantity $\sup_{t/4<u<3t/4}\left\|1_\Gamma \exp\left(-u\left(K_0\dot{+}V\right)\right)\right\|_{\infty,2}$ is finite as well.

Proof. First suppose that the functions φ and ψ have property (a). As a consequence of inequality (5.75) with $\epsilon = \frac{1}{2}$, we get

$$\begin{aligned}&\iint dxdy\left|\varphi(x)\right|^2 D_\Sigma(t,x,y)^2\left|\psi(y)\right|^2 \qquad (5.86)\\ &\le \iint dxdy\left|\varphi(x)\right|^2\left|\psi(y)\right|^2 \mathbb{E}_x\left[p_0(t-S,X(S),y) : S<t\right]\left\|\exp\left(-t\left(K_0\dot{+}2V\right)\right)\right\|_{\infty,1}.\end{aligned}$$

The inequality in (5.82) will follow from (5.86), because the function $(x,y)\mapsto \mathbb{E}_x\left[p_0(t-S,X(S),y) : S<t\right]$ is symmetric and $2\left|\varphi(x)\right|^2\left|\psi(y)\right|^2$ is dominated by $\left|\varphi(x)\right|^4+\left|\psi(y)\right|^4$.

Next suppose that the functions φ and ψ possess property (b). Then as a consequence of (5.76) we may proceed as follows:

$$\begin{aligned}&\iint dx\,dy\left|\varphi(x)\right|^2 D_\Sigma(t,x,y)^2\left|\psi(y)\right|^2\\ &\le \frac{1}{2}\iint dxdy\left(\left|\varphi(x)\right|^4+\left|\psi(x)\right|^4\right) D_\Sigma(t,x,y)^2\\ &\le \frac{1}{2}\int dx\left(\left|\varphi(x)\right|^4+\left|\psi(x)\right|^4\right)\left(\mathbb{E}_x\left[p_0(2t-S,X(S),x) : S<t\right]\right)^{1/2}\\ &\qquad\times\left\|\exp\left(-2t\left(K_0\dot{+}2V\right)\right)\right\|_{\infty,1}^{1/2}. \qquad (5.87)\end{aligned}$$

Finally suppose that the functions φ and ψ possess property (c). Then, as in the situation $\varphi\equiv\psi\equiv 1$, we have

$$\begin{aligned}&\left\|\varphi D_\Sigma(t)\psi\right\|_{\mathrm{HS}}\\ &\le \left\|\varphi D_\Sigma(t/2)\exp\left(-\frac{t}{2}\left(K_0\dot{+}V\right)\right)\psi\right\|_{\mathrm{HS}}+\left\|\varphi\exp\left(-\frac{t}{2}\left(K_0\dot{+}V\right)\right)D_\Sigma(t/2)\psi\right\|_{\mathrm{HS}}\\ &= \left\|\varphi D_\Sigma(t/2)\exp\left(-\frac{t}{2}\left(K_0\dot{+}V\right)\right)\psi\right\|_{\mathrm{HS}}+\left\|\psi D_\Sigma(t/2)\exp\left(-\frac{t}{2}\left(K_0\dot{+}V\right)\right)\varphi\right\|_{\mathrm{HS}}.\end{aligned}$$

So, because the roles of ψ and φ are interchangeable, it suffices to estimate the integral

$$\begin{aligned}
&\iint dxdy\, |\varphi(x)|^2\, |\psi(y)|^2 \\
&\times \left(\mathbb{E}_x\left[\exp\left(-\int_0^S V(X(u))du\right)\exp\left(-(t-S)\left(K_0\dot{+}V\right)\right)(X(S),y) : S<\frac{t}{2}\right]\right)^2 \\
&= \int dx\, |\varphi(x)|^2\, \mathbb{E}_x\otimes\mathbb{E}_x\Bigg[(\omega,\omega')\mapsto \exp\left(-\int_0^{S(\omega)} V(X(u)(\omega))du\right) 1_{[0,t/2)}(S(\omega)) \\
&\times \exp\left(-\int_0^{S(\omega')} V(X(u)(\omega'))du\right) 1_{[0,t/2)}(S(\omega')) \\
&\times \int dy \exp\left(-(t-S(\omega))\left(K_0+V\right)\right)(X(S)(\omega),y)\, |\psi(y)|^2 \\
&\times \exp\left(-(t-S(\omega')\left(K_0+V\right)\right)(y,X(S)(\omega'))\Bigg] \\
&\le \sup_{t/2<u<t}\left\| 1_\Gamma \exp\left(-u\left(K_0\dot{+}V\right)\right) |\psi|^2 \exp\left(-u\left(K_0\dot{+}V\right)\right) 1_\Gamma\right\|_{\infty,1} \|\varphi T_\Sigma(t/2)1\|_2^2 .
\end{aligned} \tag{5.88}$$

Inequality (5.88) together with a similar estimate with the roles of ψ and φ interchanged settles the final claim in Theorem 5.19.

5.20. COROLLARY. *Let φ and ψ be real functions defined on E with the property that the expression*

$$\int dx \left(|\varphi(x)|^4 + |\psi(x)|^4\right)\left(\mathbb{E}_x\left(p_0(2t-S,X(S),x) : S<t\right)\right)^{1/2} \tag{5.89}$$

is finite. Also suppose that the integral $\int_0^\infty \exp(-a_0 t)\, \|\varphi D_\Sigma(t)\psi\|_{2,2}\, dt$ is finite for some $a_0 \in \mathbb{R}$. Then, for Re $z < -a_0$, *the operator*

$$\varphi\left[\left(K_0\dot{+}V - zI\right)^{-1} - J^*\left(\left(K_0\dot{+}V - zI\right)_\Sigma\right)^{-1} J\right]\psi \tag{5.90}$$

is compact.

If, in addition, the operator-valued function in (5.90) is locally bounded on $\mathrm{res}(K_0\dot{+}V) \cap \mathrm{res}\left(\left(K_0\dot{+}V\right)_\Sigma\right)$, *then the operator in (5.90) is compact for* $z \in \mathrm{res}(K_0\dot{+}V) \cap \mathrm{res}\left(\left(K_0\dot{+}V\right)_\Sigma\right)$.

Proof. First we take $a \in \mathbb{C}$ with Re a larger than a_0. From the hypothesis it follows via Theorem 5.19 that the operators $\varphi D_\Sigma(t)\psi$, $t>0$, are Hilbert-Schmidt. In particular they are compact. Let $\{f_\ell : \ell \in \mathbb{N}\}$ be a sequence in $L^2(E,m)$, that weakly converges to 0. Then, for every $t > 0$, $\lim_{\ell\to\infty} \|\varphi D_\Sigma(t)\psi f_\ell\|_2 = 0$. It follows that

$$\limsup_{\ell\to\infty}\left\|\int_0^\infty \exp(-at)\,\varphi D_\Sigma(t)\psi f_\ell dt\right\|_2$$
$$\leq \limsup_{\ell\to\infty}\int_0^\infty \exp(-\operatorname{Re} a\, t)\,\|\varphi D_\Sigma(t)\psi f_\ell\|_2\, dt. \tag{5.91}$$

Since $\text{w-lim}_{\ell\to\infty} f_\ell = 0$, the sequence $\{f_\ell : \ell \in \mathbb{N}\}$ is uniformly bounded. Hence in conjunction with $\int_0^\infty dt \exp(-\operatorname{Re} a\, t)\,\|\varphi D_\Sigma(t)\psi\|_{2,2} < \infty$, we may interchange the limit and integral in (5.91). So that we obtain ($\operatorname{Re}\, a > a_0$):

$$\limsup_{\ell\to\infty}\left\|\int_0^\infty e^{-at}\varphi D_\Sigma(t)\psi f_\ell dt\right\|_2 \leq \int_0^\infty e^{-\operatorname{Re} a\, t}\limsup_{\ell\to\infty}\|\varphi D_\Sigma(t)\psi f_\ell\|_2\, dt = 0. \tag{5.92}$$

This proves the claim in Corollary 5.20 for z in some left half-plane. The result for the other z's is obtained upon employing analytic continuation via Taylor series, to wit

$$\varphi\left(R(a) - J^* R_\Sigma(a)J\right)\psi = \sum_{k=0}^\infty (a_1 - a)^k \varphi\left(R(a_1)^{k+1} - J^* R_\Sigma(a_1)^{k+1}J\right)\psi$$
$$= \sum_{k=0}^\infty (a_1 - a)^k \varphi \sum_{j=0}^k R(a_1)^{k-j}\left(R(a_1) - J^*R_\Sigma(a_1)J\right)\left(J^*R_\Sigma(a_1)J\right)^j\psi, \tag{5.93}$$

for all $a \in \mathbb{C}$ with $\|(a - a_1)R(a_1)\|_{2,2} < 1$. Here, as in Theorem 5.16, $R(a) := \left(aI + K_0\dot{+}V\right)^{-1}$ and $R_\Sigma(a) := \left(aI + K_0\dot{+}V\right)_\Sigma^{-1}$. Also notice the identity

$$\varphi\left(R(a_1)^{k+1} - J^*R_\Sigma(a_1)^{k+1}J\right)\psi = \frac{(-1)^k}{2\pi i}\int_\gamma \frac{\varphi\left(R(z) - J^*R_\Sigma(z)J\right)\psi}{(z - a_1)^{k+1}}dz, \tag{5.94}$$

where γ is for example the boundary of a disc centered at a_1, and contained in the intersection $\operatorname{res}\left(-\left(K_0\dot{+}V\right)\right) \cap \operatorname{res}\left(-(K_0\dot{+}V)_\Sigma\right)$. From (5.94) it also follows that the function For Re $z > a_0$ the operator

$$\varphi\left(\left(K_0\dot{+}V + zI\right)^{-1} - J^*\left(\left(K_0\dot{+}V + zI\right)_\Sigma\right)^{-1}J\right)\psi$$

is compact. The local boundedness of the function

$$z \mapsto \varphi\left(\left(K_0\dot{+}V - zI\right)^{-1} - J^*\left(\left(K_0\dot{+}V - zI\right)_\Sigma\right)^{-1}J\right)\psi$$

on $\operatorname{res}(K_0\dot{+}V) \cap \operatorname{res}\left(\left(K_0\dot{+}V\right)_\Sigma\right)$ together with the representation (5.94) implies the convergence of the series in (5.93) in operator-norm sense. Hence, from the representations (5.93) and (5.94) and the above consideration in (5.92), we infer that the operator $\varphi\left[R(a) - J^* R_\Sigma(a)J\right]\psi$ is compact on a subset Λ of $\mathbb{C}$ containing the left half-plane $\{\mathrm{Re}\ a > a_0\}$, where the operator valued function $z \mapsto \varphi\left[R(z) - J^* R_\Sigma(z)J\right]\psi$ is analytic. Because of the structure of this function, this set Λ may be taken to be the largest open subset of $\mathrm{res}\left(-\left(K_0\dot{+}V\right)\right) \cap \mathrm{res}\left(-\left(K_0\dot{+}V\right)_\Sigma\right)$ on which the mapping $z \mapsto \varphi\left(R(z) - J^* R_\Sigma(z)J\right)\psi$ is locally bounded. Because of our assumptions we may take $-\Lambda := \{-z \in \mathbb{C} : z \in \Lambda\}$ assumptions we may take $-\Lambda := \{-z \in \mathbb{C} : z \in \Lambda\}$ equal to $\mathrm{res}(K_0\dot{+}V) \cap \mathrm{res}\left(\left(K_0\dot{+}V\right)_\Sigma\right)$.

We conclude this chapter with the following compactness result. We suppose that the function $(aI + K_0)^{-1}|V|$ belongs to $C_\infty(E)$ for a sufficiently large. It then follows that $\lim_{x\to\infty}\left[J^*\left(aI + K_0\right)_\Sigma^{-1} J\,|V|\right](x) = 0$. So the other assumption, being that the function $x \mapsto \left[J^*\left(aI + K_0\right)_\Sigma^{-1} J\,|V|\right](x)$ belongs to $C_\infty(E)$, then says that the boundary of Σ is S-regular in the sense that the equality $\mathbb{P}_x(S = 0) = 1$ is true for all $x \in \Gamma$.

5.21. THEOREM. *Suppose that for $a > 0$ sufficiently large the following functions belong to $C_\infty(E)$:*

$$(aI + K_0)^{-1}|V| \quad \textit{and} \quad J^*\left(aI + K_0\right)_\Sigma^{-1} J\,|V|.$$

Then the operators $D_\Sigma(t)$, $t > 0$, are compact.

The proof follows the same lines as that of Theorem 5.10. Because it is so close to that proof, it is omitted.

Chapter 6

Trace Class Properties of Semigroup Differences

A. General trace class criteria

We start with some rather general results on trace norms for integral operators. The proofs of Theorem 6.1, its Corollary 6.2, and Theorem 6.4 are taken from [80].

6.1. THEOREM. *Let k_1 and k_2 be Borel measurable functions on $E \times E$ with the property that the integral*

$$\int \left(\int k_2(x,z)^2 dx\right)^{1/2} \left(\int k_1(z,x)^2 dx\right)^{1/2} dz$$

is finite. Then the function $k(x,y) := \int k_2(x,z)k_1(z,y)dz$ serves as the integral kernel of a trace class (or nuclear) operator K in $L^2(E,m)$. Its trace norm is estimated by

$$\|K\|_{\text{trace}} \leq \int \left(\int k_2(x,z)^2 dx\right)^{1/2} \left(\int k_1(z,x)^2 dx\right)^{1/2} dz. \tag{6.1}$$

Proof. Define the functions h_j, $j = 1, 2$, as follows:

$$h_1(z) = \left(\int k_1(z,x)^2 dx\right)^{1/2} \text{ and } h_2(z) = \left(\int k_2(x,z)^2 dx\right)^{1/2}. \tag{6.2}$$

Let the operator K_1' have integral kernel

$$k_1'(z,y) = (h_2(z)h_1(z))^{1/2}\, h_1(z)^{-1} k_1(z,y)$$

and let the operator K_2' have integral kernel

$$k_2'(x,z)) = k_2(x,z)h_2(z)^{-1}\, (h_2(z)h_1(z))^{1/2}.$$

Since $\iint k_1'(z,y)^2 dydz = \int h_2(z)h_1(z)dz < \infty$, it follows that the operator K_1' is a Hilbert-Schmidt operator. For the same reason, the operator K_2' is also Hilbert-Schmidt and since the integral kernel of $K_2' \circ K_1'$ coincides with the function $k(x,y)$, it follows that $K = K_2' \circ K_1'$ is a trace operator and that its trace norm is dominated be $\|K_2'\|_{\mathrm{HS}} \|K_1'\|_{\mathrm{HS}} = \int h_2(z)h_1(z)dz$. This proves (6.1) and the theorem.

REMARK. Another way of proving Theorem 6.1 reads as follows. The operator K is a trace operator if and only if the supremum

$$\sup \left\{ |\mathrm{trace}(UK)| : U \text{ finite rank and } \|U\|_{2,2} \leq 1 \right\} \tag{6.3}$$

is finite: a proof is given below. Moreover, if this is the case, then $\|K\|_{\mathrm{trace}}$ coincides with this supremum. (If necessary, we may even assume that the finite rank operators U possess continuous integral kernels.) Let U be such a finite rank operator, then

$$\begin{aligned}
|\mathrm{trace}(UK)| &= \left| \int dz \int dx \, [Uk_2(\cdot, z)] \, (x) k_1(z, x) \right| \\
&\leq \int dz \left(\int dx \, |[Uk_2(\cdot, z)] \, (x)|^2 \right)^{1/2} \left(\int dx \, |k_1(z, x)|^2 \right)^{1/2} \\
&\leq \int dz \left(\int dx \, |k_2(x, z)|^2 \right)^{1/2} \left(\int dx \, |k_1(z, x)|^2 \right)^{1/2} \\
&= \int dz h_2(z) h_1(z),
\end{aligned}$$

where the functions h_1 and h_2 are as in the proof above. A proof of (6.3) reads as follows. Let T be a continuous linear operator and let $T = V|T|$ be its polar decomposition. Then $|T| = V^*T$ and so we get:

$$\begin{aligned}
\|T\|_{\mathrm{trace}} &= \mathrm{trace}\,(|T|) \\
&= \sup \{\mathrm{trace}\,(|T|\, P) : P \ \text{ finite rank orthogonal projection}\} \\
&= \sup \{\mathrm{Re} \ \ \mathrm{trace}\,(V^*TP) : P \ \text{ finite rank orthogonal projection}\} \\
&= \sup \{\mathrm{Re} \ \ \mathrm{trace}\,(TPV^*) : P \ \text{ finite rank orthogonal projection}\} \\
&\leq \sup \{\mathrm{Re} \ \ \mathrm{trace}\,(TW) : W \ \text{ finite rank operator of norm } \leq 1\} \\
&\leq \|T\|_{\mathrm{trace}},
\end{aligned}$$

which proves (6.3).

Again consider the functions k_1 and k_2 in Theorem 6.1. Suppose that $\sup_{z \in E} \int |k_2(x, z)|^2 \, dx$ is finite. Then the operator K_2 with kernel k_2 is an operator from $L^1(E, m)$ to $L^2(E, m)$ with norm $\|K_2\|_{2,1} = \sup_{z \in E} \left(\int |k_2(x, z)|^2 \, dx \right)^{1/2}$. Let h_1 be as in the proof of Theorem 6.1 and suppose that h_1 belongs to $L^1(E, m)$. Then we see $\|K_2 \circ K_1\|_{\mathrm{trace}} \leq \|K_2\|_{2,1} \int h_1(z)dz$. Also notice that the function $h_1(z)$ is dominated by $\|K_1\|_{\infty,1}^{1/2} \times \left(\int |k_1(z, x)| \, dx \right)^{1/2}$. Altogether this proves the following corollary to Theorem 6.1.

6.2. COROLLARY. *Let K_2 be an operator from $L^1(E,m)$ to $L^2(E,m)$ with integral kernel k_2 and suppose that K_1 is an operator with integral kernel k_1 with the property that $\int \left(\int dx\,|k_1(z,x)|^2\right)^{1/2} dz$ is finite. Then the composition $k(x,y) = \int k_2(x,z)k_1(z,y)dz$ is the integral kernel of a trace class operator K. Its trace norm satisfies the following inequalities:*

$$\begin{aligned}\|K_2 \circ K_1\|_{\text{trace}} &\leq \|K_2\|_{2,1} \int dz \left(\int |k_1(z,x)|^2\, dx\right)^{1/2} \\ &\leq \|K_2\|_{2,1}\, \|K_1\|_{\infty,1}^{1/2} \int dz \left(\int |k_1(z,x)|\, dx\right)^{1/2}. \qquad (6.4)\end{aligned}$$

In the following (preliminary) result we place ourselves again in the context of general semigroups $\{V(t) : t \geq 0\}$ and $\{W(t) : t \geq 0\}$. Put $D(t) = V(t) - W(t)$.

6.3. PROPOSITION. *Suppose that the operators*

$$V(t)D(2t)V(t), \quad W(t)D(2t)W(t) \quad \text{and} \quad [V(t), W(t)]$$

are trace class. Then the operator $D(4t)$ is a trace operator and

$$\begin{aligned}\|D(4t)\|_{\text{trace}} \leq\; &\|V(t)D(2t)V(t)\|_{\text{trace}} + \|W(t)D(2t)W(t)\|_{\text{trace}} \\ &+ 2\,\|[V(t), W(t)]\|_{\text{trace}}\, \|V(t)W(t)\|_{2,2}\,.\end{aligned}$$

Proof. This assertion is an immediate consequence of equality (5.5).

Next we specialize to differences of self-adjoint positivity preserving semigroups $(V(t) : t \geq 0)$ and $(W(t) : t \geq 0)$. Put $D(t) = V(t) - W(t)$ and note the identity $D(t) = V(t/2)D(t/2) + D(t/2)W(t/2)$. Let $V(t,x,y)$ and $W(t,x,y)$ serve as continuous symmetric integral kernels for the operators $V(t)$ and $W(t)$ respectively. Put $D(t,x,y) = V(t,x,y) - W(t,x,y)$. With this notation, we have the following result. Observe that $\|V(t)\|_{\infty,1} = \sup_{x,y\in E} V(t,x,y)$.

6.4. THEOREM. *Fix $t > 0$.*

(a) Suppose that $\|V(t)\|_{\infty,1}$ and $\|W(t)\|_{\infty,1}$ are finite. Also suppose that the integral $\Phi(y) := \left(\int |D(t/2,x,y)|^2\, dx\right)^{1/2}$ belongs to $L^1(E,m)$. Then the operator $D(t)$ is of trace class and

$$\|D(t)\|_{\text{trace}} \leq \left(\|V(t)\|_{\infty,1}^{1/2} + \|W(t)\|_{\infty,1}^{1/2}\right) \|\Phi\|_1\,. \qquad (6.5)$$

(b) If, in addition, $D(t,x,y) \geq 0$, and the function $(D(t/2)1)^{1/2}$ belongs to $L^1(E,m)$, then again $D(t)$ is a trace operator and

$$\|D(t)\|_{\text{trace}} \leq 2\,\|V(t)\|_{\infty,1}^{1/2}\, \|D(t/2)\|_{\infty,1}^{1/2} \left\|(D(t/2)1)^{1/2}\right\|_1. \qquad (6.6)$$

Proof. (a) As mentioned above $D(t) = V(t/2)D(t/2) + D(t/2)W(t/2)$, and so

$$\|D(t)\|_{\text{trace}} \le \|V(t/2)D(t/2)\|_{\text{trace}} + \|W(t/2)D(t/2)\|_{\text{trace}} .$$

An application of Corollary 6.2 results in

$$\|D(t)\|_{\text{trace}} \le \left(\|V(t/2)\|_{2,1} + \|W(t/2)\|_{2,1}\right) \|\Phi\|_1 . \tag{6.7}$$

Since the operators $V(t/2)$ and $W(t/2)$ are self-adjoint, the following identities are true: $\|V(t/2)\|_{2,1} = \|V(t/2)\|_{\infty,2}$ and $\|W(t/2)\|_{2,1} = \|W(t/2)\|_{\infty,1}$. Next let f be a function in $L^2(E,m)$ and consider

$$\begin{aligned} |[V(t/2)f](x)| &\le \int V(t/2,x,y)\,|f(y)|\,dy \\ &\le \left(\int V(t/2,x,y)^2 dy\right)^{1/2} \|f\|_2 \\ &= (V(t,x,x))^{1/2}\,\|f\|_2 . \end{aligned} \tag{6.8}$$

It follows that $\|V(t/2)\|_{2,1} \le \|V(t)\|_{\infty,1}^{1/2}$ and, of course, a similar statement holds for $\|W(t/2)\|_{2,1}$. This proves (a).
(b) Since $D(t,x,y) \ge 0$, we get $W(t,x,y) \le V(t,x,y)$ and hence

$$\|W(t)\|_{\infty,1} \le \|V(t)\|_{\infty,1} .$$

It also gives $\Phi(y) \le \|D(t/2)\|_{\infty,1}^{1/2} (D(t/2)1(y))^{1/2}$. So (b) follows as well.

We continue with a general inequality for the relevant martingales $M_V^{t,y}$ defined by (see Definition 2.18)

$$M_V^{t,y}(s) = \exp\left(-\int_0^s V(X(u))du\right) \exp\left(-(t-s)\left(K_0+V\right)\right)(X(s),y), \tag{6.9}$$

where $0 \le s < t$. Notice that $M_0^{t,y}(s) = p_0(t-s, X(s), y)$. In terms of this martingale, we define the measure $\mu_{0,x}^{V,t,y}$ on $\mathcal{F}_{t-}$ via the equality

$$\mu_{0,x}^{V,t,y}(A) = \mathbb{E}_x\left[M_V^{t,y}(s)1_A\right], \tag{6.10}$$

where A belongs to $\mathcal{F}_s$ and where $0 \le s < t$. Instead of $\mu_{0,x}^{0,t,y}$, we often write $\mu_{0,x}^{t,y}$. The following theorem states among other things that the measure $\mu_{0,x}^{V,t,y}$ is absolutely continuous with respect to $\mu_{0,x}^{t,y}$. Its Radon-Nikodym derivative is given by the random variable $\exp\left(-\int_0^t V(X(u))du\right)$. For proofs of these results, we refer to Theorem 2.21. If $F \equiv 1$, then (6.11) represents the integral kernel of the operator $\exp\left(-t\left(K_0\dot{+}V\right)\right)$. If $F = 1_{\{S>t\}}$, then (6.11) represents the integral kernel of the operator $\exp\left(-t\left(K_0\dot{+}V\right)_\Sigma\right)$. If $F = 1_{\{S<t\}}$, then (6.11) represents the integral kernel of the difference

$$D_\Sigma(t) = \exp\left(-t\left(K_0\dot{+}V\right)\right) - J_\Sigma^* \exp\left(-t\left(K_0\dot{+}V\right)_\Sigma\right) J_\Sigma.$$

The inequality in (6.12) may be used to compare integral kernels of the perturbed semigroup and the unperturbed one. It follows from (6.11) by using Hölder's inequality. This inequality is a generalization of inequality (5.75) in Lemma 5.18. The assertion in (c) is a consequence of Doob's optional sampling theorem: see Appendix C.11. So we have the following result.

6.5. THEOREM. *Let $F \geq 0$ be a random variable that is measurable with respect to $\mathcal{F}_s$, $0 \leq s < t$. Fix $0 < \epsilon \leq 1$.*

(a) *The following equalities are true:*

$$\int F d\mu_{0,x}^{V,t,y} = \mathbb{E}_x\left[M_V^{t,y}(s)F\right] = \int \exp\left(-\int_0^t V(X(u))du\right) F d\mu_{0,x}^{t,y}. \quad (6.11)$$

(b) *The following inequality is valid:*

$$\mathbb{E}_x\left[M_V^{t,y}(s)F\right] \leq \left(\mathbb{E}_x\left[M_{\epsilon^{-1}V}^{t,y}(s)F\right]\right)^\epsilon \left(\mathbb{E}_x\left[M_0^{t,y}(s)F\right]\right)^{1-\epsilon}. \quad (6.12)$$

(c) *The results in (a) and (b) are also true if s is replaced with a stopping time $S < t$.*

The inequality in the following lemma will always be used without comment.

6.6. LEMMA. *Let $(I, \mathcal{B}, \mu)$ be a non-negative measure space and let*

$$\{A(s) : s \in I\}$$

be a weakly measurable family of linear operators with the property that

$$\int_I \|A(s)\|_{\text{trace}}\, d\mu(s) < \infty.$$

Then the operator $\int_I A(s)d\mu(s)$ exists and is a trace class operator for which

$$\left\|\int_I A(s)d\mu(s)\right\|_{\text{trace}} \leq \int_I \|A(s)\|_{\text{trace}}\, d\mu(s).$$

Proof. Let U be any bounded linear operator and let $(\varphi_j : 1 \leq j \leq n)$ be any (finite) orthonormal sequence. Then the following (in-)equalities are self-explanatory:

$$\operatorname{Re} \sum_{j=1}^n \left\langle U \int_I A(s)d\mu(s)\varphi_j, \varphi_j\right\rangle = \operatorname{Re} \int_I \sum_{j=1}^n \langle UA(s)\varphi_j, \varphi_j\rangle\, d\mu(s)$$

$$\leq \int_I \|UA(s)\|_{\text{trace}}\, d\mu(s) \leq \|U\|_{2,2} \int_I \|A(s)\|_{\text{trace}}\, d\mu(s). \quad (6.13)$$

From (6.13) the assertion and inequality in the lemma readily follow.

B. Regular Perturbations

This section begins with a general result, which is in fact the main result, on trace properties of differences of Feynman-Kac semigroups with regular perturbations.

6.7. THEOREM. *Fix $t > 0$. Among the assertions*

(i) the integrals

$$\int dx \sqrt{\int \left(\exp\left(-\frac{t}{2}(K_0 + V)\right)(x,y)\right)^2 |W(y) - V(y)|^2\, dy}$$

and

$$\int dx \sqrt{\int \left(\exp\left(-\frac{t}{2}(K_0 + W)\right)(x,y)\right)^2 |W(y) - V(y)|^2\, dy}$$

are finite;

(ii) the operators

$$(V - W)\exp\left(-t\left(K_0 \dot{+} V\right)\right) \quad \textit{and} \quad (V - W)\exp\left(-t\left(K_0 \dot{+} W\right)\right)$$

are of trace class;

(iii) the operators

$$D(t/2)\exp\left(-\frac{t}{2}\left(K_0 \dot{+} V\right)\right) \quad \textit{and} \quad D(t/2)\exp\left(-\frac{t}{2}\left(K_0 \dot{+} W\right)\right)$$

are of trace class;

(iv) the operator $D(t)$ is of trace class;

the following implications are valid:

$$(i) \implies (ii) \implies (iii) \implies (iv).$$

REMARK. Suppose that the unperturbed semigroup $\{\exp(-tK_0) : t \geq 0\}$ is L^1-L^∞-smoothing. Then the first integral in (i) is dominated by

$$C(t) \int dx \sqrt{\int p_0(t/2, x, y)\, |W(y) - V(y)|^2\, dy}, \tag{6.14}$$

where

$$C(t) = M_{4V}^{1/2} \exp\left(\frac{t}{8} b_{4V}\right) \sup_{x,y \in E} p_0(t/4, x, y).$$

For these facts see inequality (2.7) in Chapter 2.A. See also assertion (4) in Theorem 2.9 of Chapter 2 and the comment following it. Let $\{Q_j : j \in \mathbb{N}\}$ be a partition of E into Borel sets. Then the integral in (6.14) may be estimated by an expression of the form

$$\sup_{j\in\mathbb{N}} \int dx \sup_{y\in Q_j} p_0(t/2,x,y)^{1/2} \sum_{j=1}^{\infty} \sqrt{\int_{Q_j} dy\, |W(y)-V(y)|^2}.$$

In the case $K_0 = -\frac{1}{2}\Delta$ in $\mathbb{R}^\nu$, the function $p_0(t,x,y)$ is the classical Gaussian density and the sequence $\{Q_j : j\in\mathbb{N}\}$ may be an enumeration of the partition $\{\alpha+Q : \alpha\in\mathbb{A}^\nu\}$, where Q is the cube $Q=(0,1]^\nu$. In this case, the first integral in (i) is dominated by a constant times the expression

$$\sum_{\alpha\in\mathbb{A}^\nu} \sqrt{\int_{\alpha+Q} dy\, |W(y)-V(y)|^2}. \tag{6.15}$$

The space of functions $W-V$ for which the quantity in (6.15) is finite is the *Birman-Solomjak space* $\ell^1\left(L^2(\mathbb{R}^\nu)\right)$.

Proof. (i) $\Rightarrow$ (ii). We shall apply Theorem 6.1 to the operators

$$(V-W)\exp\left(-\frac{1}{2}t\left(K_0\dot{+}V\right)\right)\exp\left(-\frac{1}{2}t\left(K_0\dot{+}V\right)\right)$$

and

$$(V-W)\exp\left(-\frac{1}{2}t\left(K_0\dot{+}W\right)\right)\exp\left(-\frac{1}{2}t\left(K_0\dot{+}W\right)\right).$$

For the first operator we use

$$k_2(x,z) = (V(x)-W(x))\exp\left(-\frac{t}{2}(K_0+V)\right)(x,z)$$

and

$$k_1(z,x) = \exp\left(-\frac{t}{2}\left(K_0+V\right)\right)(z,x).$$

For the second operator we interchange the roles of V and W. From Theorem 6.1 we get

$$\begin{aligned}
&\|K_2\circ K_1\|_{\text{trace}} \\
&\le \int dz \sqrt{\int dx \left|\exp\left(-\frac{t}{2}\left(K_0+V\right)\right)(z,x)\right|^2} \\
&\quad\times \sqrt{\int dx \left(\exp\left(-\frac{t}{2}\left(K_0+V\right)\right)(x,z)\right)^2 |V(x)-W(x)|^2} \\
&\le \sup_{z\in E} \sqrt{\exp\left(-t\left(K_0+V\right)\right)(z,z)} \\
&\quad\times \int dz \sqrt{\int dx \left(\exp\left(-\frac{t}{2}\left(K_0+V\right)\right)(x,z)\right)^2 |V(x)-W(x)|^2}.
\end{aligned}$$

So (i) implies that the operator $(V-W)\exp\left(-t\left(K_0\dot{+}V\right)\right)$ is of trace class. Upon interchanging the role of V and W, from (i) it follows that the operator $(V-W)\exp\left(-t\left(K_0\dot{+}W\right)\right)$ is of trace class as well.

The implications (ii) $\Rightarrow$ (iii) and (iii) $\Rightarrow$ (iv) follow the same lines as the corresponding implications for the Hilbert-Schmidt case. The reader is referred to the proof of Theorem 5.6: the only change that occurs is the fact that the Hilbert-Schmidt norms have to be replaced with trace norms.

The following theorem seems more practical then the corresponding result in [82, Theorem 5.7]. See the remark below.

6.8. THEOREM. *Let V and W be Kato-Feller potentials. Suppose that the quantity $M(t)$, defined by*

$$\begin{aligned} M(t) = \int_0^t ds \int dz\, |V(z) - W(z)| \qquad\qquad (6.16)\\ \times \Big\{ \left(\exp\left(-s(K_0+V)\right)(z,z)\right)^{1/2} \sup_{t\le s\le 2t} \exp\left(-s\left(K_0+W\right)\right)^{1/2}(z,z) \\ + \left(\exp\left(-s(K_0+W)\right)(z,z)\right)^{1/2} \sup_{t\le s\le 2t} \exp\left(-s\left(K_0+V\right)\right)^{1/2}(z,z) \Big\} \end{aligned}$$

is finite. Then $D(t) := \exp\left(-t(K_0+V)\right) - \exp\left(-t(K_0+W)\right)$ *is of trace class and* $\|D(t)\|_{\text{trace}} \le \frac{1}{2}M(t)$.

REMARK. For semigroups of spectral dimension n (see Definition 1.4 in Chapter 1.B), the quantity in (6.16) is finite if $V-W$ belongs to $L^1(E,m)$ and if $n \le 3$. For this see also Proposition 8.13 and the result in Theorem 6.8 is applicable in scattering theory: see Chapter 8.

Proof. Suppose that the unitary operator U satisfies $D(t) = |D(t)|\,U^*$. This is the so-called polar decomposition of the operator $D(t)$. Let $(\varphi_j : j \in \mathbb{N})$ be an orthonormal basis for $L^2(E,m)$. Then the following (in-)equalities are more or less self-explanatory:

$$\begin{aligned} &\|D(t)\|_{\text{trace}} = \text{trace}\left(D(t)U^*\right) = \sum_{j\in\mathbb{N}} \langle D(t)U^*\varphi_j, \varphi_j\rangle \\ &= \sum_{j\in\mathbb{N}} \int dx D(t)U^*\varphi_j(x)\overline{\varphi_j(x)} \\ &= \sum_{j\in\mathbb{N}} \int dx \int_0^t ds \int dz \exp\left(-s(K_0+V)\right)(x,z)\left(V(z)-W(z)\right) \\ &\qquad \times \int \exp\left(-(t-s)(K_0+W)\right)(z,y)\left[U^*\varphi_j\right](y)dy\overline{\varphi_j(x)} \end{aligned}$$

$$= \int dx \int_0^t ds \int dz \exp\left(-s(K_0+V)\right)(x,z)\left(V(z)-W(z)\right)$$
$$\times \sum_{j\in\mathbb{N}} \langle \varphi_j, U\exp\left(-(t-s)(K_0+W)\right)(z,\cdot)\rangle\, \overline{\varphi_j(x)}$$
$$= \int dx \int_0^t ds \int dz \exp\left(-s(K_0+V)\right)(x,z)\left(V(z)-W(z)\right)$$
$$\times \overline{\sum_{j\in\mathbb{N}} \langle U\exp\left(-(t-s)(K_0+W)\right)(z,\cdot), \varphi_j\rangle\, \varphi_j(x)}$$
$$= \int dx \int_0^t ds \int dz \exp\left(-s(K_0+V)\right)(x,z)\left(V(z)-W(z)\right)$$
$$\times \overline{\left[U\exp\left(-(t-s)(K_0+W)\right)(z,\cdot)\right](x)}$$
$$= \int_0^t ds \int dz \left(V(z)-W(z)\right)\langle \exp\left(-s(K_0+V)\right)(\cdot,z),$$
$$U\exp\left(-(t-s)(K_0+W)\right)(z,\cdot)\rangle$$

(Cauchy-Schwarz and Chapman-Kolmogorov)

$$\leq \frac{1}{2}\int_0^{2t} ds \int dz\, |V(z)-W(z)|\left(\exp\left(-s(K_0+V)\right)(z,z)\right)^{1/2}$$
$$\times \left(\exp\left(-(2t-s)(K_0+W)\right)(z,z)\right)^{1/2}. \tag{6.17}$$

From (6.17) the claim in Theorem 6.8 readily follows.

In what follows we write $\operatorname{trace}(T) = \int t(x,x)dx$ whenever T is an integral operator with kernel $t(x,y)$ for which the integral $\int |t(x,x)|\, dx$ is finite.

Remark. Let T be trace class operator with a continuous integral kernel $t(x,y)$. Then its trace is given by $\operatorname{trace}(T) = \int t(x,x)dx$: see e.g. Reed and Simon [191]. In [27] Birman has given a generalization of this result in case T is an operator in $L^2(\mathbb{R}^\nu)$. In particular he proves the following equality: $\operatorname{trace}(T) = \lim_{\lambda\downarrow 0} \frac{1}{V_\nu \lambda^\nu} \int\int_{\{y\in\mathbb{R}^\nu : |y-x|\leq\lambda\}} t(x,y)dx\, dy$. Here V_ν is the volume of the ν-dimensional unit ball.

6.9. Proposition. *Fix $t > 0$ and put*

$$F_1(t) = \int \exp\left(-t\left(K_0+V\right)\right)(x,x)\, |W(x)-V(x)|\, dx,$$
$$F_2(t) = \int \exp\left(-t\left(K_0+W\right)\right)(x,x)\, |W(x)-V(x)|\, dx, \quad \textit{and}$$
$$F_3(t) = \int_0^1 \int \exp\left(-t\left(K_0+(1-s)V+sW\right)\right)(x,x)\, |W(x)-V(x)|\, dx. \tag{6.18}$$

Suppose that $F_j(t)$, $1 \le j \le 3$, is finite. Then the following assertions are valid:

(i) The operator $D(t/2)$ is Hilbert-Schmidt and

$$\|D(t/2)\|_{\mathrm{HS}} \le \sqrt{t}\sqrt{F_1(t)} + \sqrt{t}\sqrt{F_2(t)}. \tag{6.19}$$

(ii) The operator $\exp\left(-\frac{t}{2}\left(K_0\dot{+}V\right)\right)(W-V)\exp\left(-\frac{t}{2}\left(K_0\dot{+}W\right)\right)$ is of trace class and

$$\left\|\exp\left(-\frac{t}{2}\left(K_0\dot{+}V\right)\right)(W-V)\exp\left(-\frac{t}{2}\left(K_0\dot{+}W\right)\right)\right\|_{\mathrm{trace}} \le \sqrt{F_1(t)F_2(t)}. \tag{6.20}$$

(iii) The integral kernel of $\left[\exp\left(-\frac{1}{2}t\left(K_0\dot{+}W\right)\right), \exp\left(-\frac{1}{2}t\left(K_0\dot{+}V\right)\right)\right]$ vanishes on the diagonal, i.e. for $x \in E$,

$$\left[\exp\left(-\frac{1}{2}t\left(K_0\dot{+}W\right)\right), \exp\left(-\frac{1}{2}t\left(K_0\dot{+}V\right)\right)\right](x,x) = 0. \tag{6.21}$$

(iv) The expression $\int |D(t,x,x)|\,dx$ is finite and

$$\int |D(t,x,x)|\,dx \le tF_3(t). \tag{6.22}$$

Here $[A,B] = AB - BA$ denotes the Lie-product of the operators A and B.

Proof. (i) The inequality in (6.19) is a direct consequence of Theorem 5.7.

(ii) For a proof of (6.20), we observe that

$$\begin{aligned}
&\left\|\exp\left(-\frac{1}{2}t\left(K_0\dot{+}V\right)\right)(W-V)\exp\left(-\frac{1}{2}t\left(K_0\dot{+}W\right)\right)\right\|_{\mathrm{trace}} \\
&\le \left\|\exp\left(-\frac{t}{2}\left(K_0\dot{+}V\right)\right)|W-V|^{1/2}\right\|_{\mathrm{HS}} \left\||W-V|^{1/2}\exp\left(-\frac{t}{2}\left(K_0\dot{+}W\right)\right)\right\|_{\mathrm{HS}} \\
&= \sqrt{F_1(t)F_2(t)}.
\end{aligned} \tag{6.23}$$

This inequality proves (ii).

(iii) This assertion follows from

$$\begin{aligned}
&\left[\exp\left(-\frac{t}{2}\left(K_0\dot{+}V\right)\right), \exp\left(-\frac{t}{2}\left(K_0\dot{+}W\right)\right)\right](x,x) \\
&= \int \exp\left(-\frac{t}{2}\left(K_0+V\right)\right)(x,y)\exp\left(-\frac{t}{2}\left(K_0+W\right)\right)(y,x) \\
&\qquad - \int \exp\left(-\frac{t}{2}\left(K_0+W\right)\right)(x,y)\exp\left(-\frac{t}{2}\left(K_0+V\right)\right)(y,x)
\end{aligned}$$

(symmetry)

$$= \int \exp\left(-\frac{t}{2}(K_0+V)\right)(x,y)\exp\left(-\frac{t}{2}(K_0+W)\right)(y,x)$$
$$- \int \exp\left(-\frac{t}{2}(K_0+V)\right)(x,y)\exp\left(-\frac{t}{2}(K_0+W)\right)(y,x) = 0. \tag{6.24}$$

Here we wrote $[T](x,y)$ for the integral kernel of the operator T and $[T,S](x,y)$ for the integral kernel of the operator $[T,S]$. This proves assertion (iii).

(iv) Let $D(t,x,y)$ be the integral kernel of the operator $D(t)$. From equality (5.10) in Proposition 5.3, we see that

$$\begin{aligned} D(t,x,y) &= \int_0^1 D_s(t,x,y)ds \\ &= \int_0^1\int_0^t\int \exp\left(-(t-u)\left(K_0+(1-s)V+sW\right)\right)(x,z)\left(W(z)-V(z)\right) \\ &\quad \times \exp\left(-u\left(K_0+(1-s)V+sW\right)\right)(z,y)dz\,du\,ds. \end{aligned}$$

Hence we get

$$\begin{aligned} &\int |D(t,x,x)|\,dx \\ &\leq \int_0^1\int_0^t\int\int \exp\left(-(t-u)\left(K_0+(1-s)V+sW\right)\right)(x,z)\left|W(z)-V(z)\right| \\ &\times \exp\left(-u\left(K_0+(1-s)V+sW\right)\right)(z,x)dx\,dz\,du\,ds \\ &= \int_0^1\int_0^t\int\int \exp\left(-t\left(K_0+(1-s)V+sW\right)\right)(z,z)\left|W(z)-V(z)\right|dz\,du\,ds \\ &= tF_3(t). \end{aligned} \tag{6.25}$$

Since (6.25) coincides with (6.22), this proves assertion (iv) in Proposition 6.9.

6.10. COROLLARY. *(a) Suppose that the unperturbed semigroup*

$$\{\exp(-tK_0) : t \geq 0\}$$

is L^1-L^∞-smoothing. Also suppose that $W-V$ belongs to $L^1(E,m)$. Then the operator $\exp\left(-\frac{t}{2}(K_0\dot{+}V)\right)(W-V)\exp\left(-\frac{t}{2}(K_0\dot{+}W)\right)$ is a trace operator and

$$\begin{aligned} &\left\|\exp\left(-\frac{t}{2}(K_0\dot{+}V)\right)(W-V)\exp\left(-\frac{t}{2}(K_0\dot{+}W)\right)\right\|_{\text{trace}} \\ &\leq (M_{4V}M_{4W})^{1/4}\exp\left(\frac{t}{4}(b_{4V}+b_{4V})\right)\sup_{x,y\in E} p_0(t,x,y)^{1/2}\left\|W-V\right\|_1. \end{aligned} \tag{6.26}$$

(b) Suppose that $V(t)$ and $W(t)$ are taken to be of the form

$$V(t) = \exp\left(-t\left(K_0 \dot{+} V\right)\right) \quad \textit{and} \quad W(t) = \exp\left(-t\left(K_0 \dot{+} W\right)\right)$$

respectively, where V and W are Kato-Feller potentials. Suppose that the semigroup

$$\{\exp(-tK_0) : t \geq 0\}$$

is L^1-L^∞-smoothing. Also suppose that $V - W$ belongs to $L^1(E, m)$. Then the operators $D(t)$ are trace class if and only if the Lie products $[V(t), W(t)]$ are trace class.

The constants M_V and b_V are chosen such that

$$\mathbb{E}_x\left[\exp\left(-\int_0^t V(X(u))du\right)\right] \leq M_V \exp(tb_V),$$

for all $t > 0$ and for all $x \in E$: see Theorem 2.9 and its Corollary 2.10 in Chapter 2.A. It is perhaps useful to observe that the trace class of the operator mentioned in Corollary 6.10 is important for establishing the completeness of the wave operators in scattering theory: see Proposition 8.13.

Proof. The assertion in (a) immediately follows from inequality (6.20) in assertion (ii) of Proposition 6.9.
(b) This assertion is a consequence of assertion (a) in conjunction with Proposition 6.3. Since the semigroup $\{\exp(-tK_0) : t \geq 0\}$ is L^1-L^∞-smoothing, the same is true for the perturbed semigroups $\{V(t) : t \geq 0\}$ and $\{W(t) : t \geq 0\}$: see Theorem 2.9 and its Corollary 2.10 in Chapter 2.A. From assertion (a) in this corollary, it follows that the operators $V(t)D(2t)V(t)$ and $W(t)D(2t)W(t)$ are trace class operators, because, by assumption, the function $W - V$ belongs to $L^1(E, m)$.

C. Singular Perturbations

The following result is a general result on trace properties of differences of Feynman-Kac semigroups which are singularly perturbed. We write (see Chapter 5, beginning of section B)

$$h_{\Sigma}^{a+V}(x) = H_{\Sigma}^{a+V}1(x) = \mathbb{E}_x\left[\exp\left(-\int_0^S (a+V(X(u)))\,du\right) : S < \infty\right] \quad \text{and}$$

$$D_{\Sigma}(t) = \exp\left(-t\left(K_0 \dot{+} V\right)\right) - J_{\Sigma}^{*} \exp\left(-t\left(K_0 \dot{+} V\right)_{\Sigma}\right) J_{\Sigma}.$$

The following theorem is the trace class counterpart of Theorem 5.12 in Chapter 5.B where Hilbert-Schmidt properties are considered. For some remarks on the operator in (i), see Proposition 6.15 and Corollary 6.16.

6.11. THEOREM. *Among the assertions*

(i) There exists a real number $\beta \geq \frac{1}{2}$ together with a function $u : E \to \mathbb{R}$, that is identically 1 on Γ with the property that, for all sufficiently large real numbers a, the operators

$$\left(aI + K_0 \dot{+} V\right)^{\beta} u \exp\left(-t\left(aI + K_0 \dot{+} V\right)\right), \quad t > 0,$$

are of trace class;

(i′) There exists $a > 0$ such that the integrals

$$\int h_{\Sigma}^{a+V}(x)^2 dx \quad \text{and} \quad \int \sup_{y \in \Gamma} \exp\left(-t(K_0 + V)\right)(y, x) dx$$

are finite;

(i″) There exists $a > 0$ such that the integral $\int h_{\Sigma}^{a+V}(x)^{1/2} dx$ is finite;

(ii) The operators

$$H_{\Sigma}^{a+V} \exp\left(-t\left(aI + K_0 \dot{+} V\right)\right), \quad t > 0,$$

are of trace class;

(iii) The operators

$$D_{\Sigma}(t/2) \exp\left(-\frac{t}{2}\left(K_0 \dot{+} V\right)\right), \quad t > 0,$$

are of trace class;

(iv) The operators $D_{\Sigma}(t)$, $t > 0$ are of trace class;

the following implications are valid:

$$\begin{array}{ccccc} (i) & & & & \\ \Downarrow & & & & \\ (ii) & \Longrightarrow & (iii) & \Longleftrightarrow & (iv). \\ \Uparrow & & \Uparrow & & \\ (i') & & (i'') & & \end{array}$$

REMARK. The main implication is (i′) ⇒ (iv): if

$$\int \sqrt{h_{\Sigma}^{a+V}(x)}dx < \infty,$$

then $D_{\Sigma}(t)$ is a trace class operator. We already know that

$$\int h_{\Sigma}^{a+V}(x)^2 dx < \infty$$

implies that $D_{\Sigma}(t)$ is a Hilbert-Schmidt operator: see Theorem 5.12, implication (i) ⇒ (iv). From equality (5.57) in Proposition 5.14, it follows that

$$\begin{aligned}
&D_{\Sigma}(t,x,y) + \int D_{\Sigma}(t/2,x,z)D_{\Sigma}(t/2,z,y)dz \\
&= \mathbb{E}_x\left[\exp\left(-\int_0^S V(X(\sigma))d\sigma\right)\exp\left(-(t-S)\left(K_0+V\right)\right)(X(S),y) : S < \frac{t}{2}\right] \\
&\quad + \mathbb{E}_y\left[\exp\left(-\int_0^S V(X(\sigma))d\sigma\right)\exp\left(-(t-S)\left(K_0+V\right)\right)(X(S),x) : S < \frac{t}{2}\right],
\end{aligned}$$

and so

$$\begin{aligned}
&D_{\Sigma}(t,x,x) \\
&\leq 2\mathbb{E}_x\left[\exp\left(-\int_0^S V(X(s))ds\right) : S < t/2\right]\sup_{y\in\Gamma}\sup_{t/2<s<t}\exp\left(-s(K_0+V)\right)(y,x) \\
&\leq 2e^{at/2}h_{\Sigma}^{a+V}(x)\sup_{y\in\Gamma}\sup_{t/2<s<t}\exp\left(-s(K_0+V)\right)(y,x).
\end{aligned}$$

Hence we see that the finiteness of the integral $\int h_{\Sigma}^{a+V}(x)dx$ implies that the integral $\int D_{\Sigma}(t,x,x)dx$ is finite as well. If Γ has finite capacity, then the integral $\int h_{\Sigma}^{a+V}(x)dx$ is finite. However there are sets of finite capacity for which $D_{\Sigma}(t)$ is not a trace class operator, i.e. there exist sets Γ for which the integral $\int h_{\Sigma}^{a+V}(x)dx$ is finite, but for which $D_{\Sigma}(t)$ is not a trace class operator. To see this consider the following example. For K_0 we take the generator $K_0 = -\frac{1}{2}\Delta$. Suppose that $D_{\Sigma}(t) = \exp(-tK_0) - J^*\exp(-t(K_0)_{\Sigma})J$ is a trace class operator. Then the operator $1_{\Gamma}\exp(-tK_0)$ is of trace class as well. Consequently the function 1_{Γ} has to be a member of the Birman-Solomjak space $\ell^1(L^2|\mathbb{R}^{\nu}|)$, i.e.

$$\|1_{\Gamma}\|_{\ell^1(L^2)} := \sum_{\alpha\in\mathbb{A}^{\nu}}\sqrt{\int_{\alpha+Q} 1_{\Gamma}(y)dy} < \infty.$$

Here Q is the unit cube in $\mathbb{R}^{\nu}$. For the definition of the Birman-Solomjak space, see the remark following Theorem 6.7. For more on this and other properties on trace operators, the reader should consult Simon [209].

Suppose that Γ is a disjoint union of balls $\Gamma = \bigcup_{j=1}^{\infty} B(x_j, r_j)$, where x_j has integer coefficients and where $r_j \leq \frac{1}{2}$. Then the Birman-Solomjak norm $\|1_\Gamma\|_{\ell^1(L^2)}$ equals a constant times $\sum_{j=1}^{\infty} r_j^{\nu/2}$. A consequence is that for the trace class operator $D_\Sigma(t)$, the sum $\sum_{j=1}^{\infty} r_j^{\nu/2}$ is finite. As usual $\Sigma = \mathbb{R}^\nu \setminus \Gamma$. However there are sets Γ of finite capacity for which this condition is not satisfied. This is so because the capacity of $\Gamma = \bigcup_{j=1}^{\infty} B(x_j, r_j)$ may be estimated by a constant times $\sum_{j=1}^{\infty} r_j^{\nu-2}$. So, if $\nu \geq 5$, and if $r_j = \frac{1}{2} j^{-2/\nu}$, then $\operatorname{cap}(\Gamma)$ is finite and 1_Γ is not a function in $\ell^1\left(L^2\left(\mathbb{R}^\nu\right)\right)$. For details see [81].

Proof. The proof of the implications (i) $\Rightarrow$ (ii) and (ii) $\Rightarrow$ (iii) goes along the same lines as the corresponding proofs in the Hilbert-Schmidt case: see Theorem 5.12. Again the trace norms have to replace the Hilbert-Schmidt norms in the proofs.

(i′) $\Rightarrow$ (ii). We apply Theorem 6.1 with $K_1 = \exp\left(-\frac{t}{2}\left(K_0 \dot{+} V\right)\right)$ and with $K_2 = H_\Sigma^{a+V} \exp\left(-\frac{t}{2}\left(aI + K_0 \dot{+} V\right)\right)$. According to Theorem 6.1 the trace norm of $K_2 \circ K_1$ is dominated by the expression

$$
\begin{aligned}
&\int dz \left(\int dx k_2(x,z)^2\right)^{1/2} \left(\int dx k_1(z,x)^2\right)^{1/2} \\
&= \int dz \left(\int dx \left(\mathbb{E}_x \left[\exp\left(-\int_0^S (a + V(X(u)))\, du\right)\right.\right.\right. \\
&\qquad \left.\left.\left. \times \exp\left(-\frac{t}{2}(aI + K_0 + V)\right)(X(S), z) : S < \infty\right]\right)^2\right)^{1/2} \\
&\qquad \times \left(\int dx \left[\exp\left(-\frac{t}{2}(K_0 + V)\right)(z,x)\right]^2\right)^{1/2} \\
&\leq \left(\int dx \left(h_\Sigma^{a+V}(x)\right)^2\right)^{1/2} \\
&\qquad \times \int dz \sup_{z_1 \in \Gamma} \exp\left(-\frac{t}{2}(aI + K_0 + V)\right)(z_1, z) \\
&\qquad \times \left(\sup_{z_1 \in E} \exp\left(-t\left(K_0 + V\right)\right)(z_1, z_1)\right)^{1/2}.
\end{aligned}
\tag{6.27}
$$

This proves the implication (i′) $\Rightarrow$ (ii).

(i″) $\Rightarrow$ (iii). We apply Theorem 6.1 with $K_1 = \exp\left(-\frac{t}{2}\left(K_0 \dot{+} V\right)\right)$ and with $K_2 = D_\Sigma(t/2)$. By virtue of Theorem 6.1, the trace norm of $K_2 \circ K_1$ is dominated by the expression

$$\int dz \left(\int dx\, k_2(x,z)^2\right)^{1/2} \left(\int dx k_1(z,x)^2\right)^{1/2}$$

$$\leq \int dz \left(\int dx D_\Sigma(t/2,x,z)^2\right)^{1/2} \left(\exp\left(-t\left(K_0 \dot{+} V\right)\right)(z,z)\right)^{1/2}$$

$$\leq \int dz \left(\int dx D_\Sigma(t/2,z,x)\right)^{1/2}$$

$$\left(\sup_{z\in E} \exp\left(-t\left(K_0 \dot{+} V\right)\right)(z,z) \sup_{x,z\in E} D_\Sigma(t/2,x,z)\right)^{1/2}$$

(Feynman-Kac formula)

$$= \int dz \left(\mathbb{E}_z \left[\exp\left(-\int_0^S V(X(u))du\right)\right.\right.$$

$$\left.\left.\mathbb{E}_{X(S)} \left\{\exp\left(-\int_0^{t/2-S} V(X(u))du\right)\right\} : S < t/2\right]\right)^{1/2}$$

$$\times \left(\sup_{z\in E} \exp\left(-t\left(K_0 \dot{+} V\right)\right)(z,z) \sup_{x,z\in E} D_\Sigma(t/2,x,z)\right)^{1/2}$$

$$\leq \int dz \left(\mathbb{E}_z \left[\exp\left(-\int_0^S V(X(u))du\right) : S < t/2\right]\right)^{1/2}$$

$$\times \left(\sup_{y\in\Gamma} \sup_{0<s<t/2} \mathbb{E}_y \left\{\exp\left(-\int_0^s V(X(u))du\right)\right\}\right)^{1/2}$$

$$\times \left(\sup_{z\in E} \exp\left(-t\left(K_0 \dot{+} V\right)\right)(z,z) \sup_{x,z\in E} \exp\left(-\frac{t}{2}\left(K_0 + V\right)\right)(x,z)\right)^{1/2}$$

$$\leq \int dz \left(\mathbb{E}_z \left[\exp\left(-\int_0^S V(X(u))du\right) : S < t/2\right]\right)^{1/2}$$

$$\times \left(\sup_{y\in\Gamma} \sup_{0<s<t/2} \mathbb{E}_y \left\{\exp\left(-\int_0^s V(X(u))du\right)\right\}\right)^{1/2}$$

$$\times \left(\sup_{z\in E} \int dw \exp\left(-\frac{t}{2}\left(K_0 \dot{+} V\right)\right)(z,w) \exp\left(-\frac{t}{2}\left(K_0 \dot{+} V\right)\right)(w,z)\right)^{1/2}$$

$$\times \left(\sup_{z,w\in E} \exp\left(-\frac{t}{2}\left(K_0 + V\right)\right)(z,w)\right)^{1/2}$$

$$\leq \int dz \left(\mathbb{E}_z \left[\exp \left(- \int_0^S V(X(u))du \right) : S < t/2 \right] \right)^{1/2}$$
$$\times \left(\sup_{y \in \Gamma} \sup_{0<s<t/2} \mathbb{E}_y \left\{ \exp \left(- \int_0^s V(X(u))du \right) \right\} \right)^{1/2}$$
$$\times \left(\sup_{z \in E} \int dw \exp \left(-\frac{t}{2} \left(K_0 \dot{+} V \right) \right) (z,w) \right)^{1/2}$$
$$\times \sup_{z,w \in E} \exp \left(-\frac{t}{2} \left(K_0 + V \right) \right) (z,w)$$

(Feynman-Kac formula)

$$= \int dz \left(\mathbb{E}_z \left[\exp \left(- \int_0^S V(X(u))du \right) : S < t/2 \right] \right)^{1/2}$$
$$\times \left(\sup_{y \in \Gamma} \sup_{0<s<t/2} \mathbb{E}_y \left\{ \exp \left(- \int_0^s V(X(u))du \right) \right\} \right)^{1/2}$$
$$\times \left(\sup_{z \in E} \mathbb{E}_z \left\{ \exp \left(- \int_0^{t/2} V(X(u))du \right) \right\} \right)^{1/2} \sup_{z,w \in E} \exp \left(-\frac{t}{2} \left(K_0 + V \right) \right) (z,w)$$

$$\leq \int dz \left(\mathbb{E}_z \left[\exp \left(- \int_0^S V(X(u))du \right) : S < t/2 \right] \right)^{1/2}$$
$$\times \sup_{y \in E} \sup_{0<s\leq t/2} \mathbb{E}_y \left\{ \exp \left(- \int_0^s V(X(u))du \right) \right\} \times \sup_{y \in E} \exp \left(-\frac{t}{2} \left(K_0 + V \right) \right) (y,y).$$

This proves the implication (i″) ⇒ (iii), because

$$\int dz \left(\mathbb{E}_z \left[\exp \left(- \int_0^S V(X(u))du \right) : S < t/2 \right] \right)^{1/2} \leq e^{at/4} \int h_\Sigma^{a+V}(x)^{1/2} dx.$$

(iii) ⇒ (iv). First we shall show that the operator

$$D_\Sigma(t/2) \exp \left(-\frac{t}{2} \left(K_0 \dot{+} V \right)_\Sigma \right)$$

is a trace class operator. By virtue of the identity

$$D_\Sigma(t/2) J^* \exp \left(-\frac{t}{2} \left(K_0 \dot{+} V \right)_\Sigma \right) J = D_\Sigma(t/2) \exp \left(-\frac{t}{2} \left(K_0 \dot{+} V \right) \right) - D_\Sigma(t/2)^2, \tag{6.28}$$

it suffices to prove that trace $\left(D_\Sigma(t/2)^2\right)$ is finite. Since the integral kernel of the operator $D_\Sigma(t/2)\exp\left(-\frac{t}{2}\left(K_0\dot{+}V\right)_\Sigma\right)$ is non-negative, it follows from (6.28) that the trace of the operator $D_\Sigma(t/2)^2$ is dominated by

$$\begin{aligned}&\int dx \int dz\, D_\Sigma(t/2,x,z)\exp\left(-\frac{t}{2}\left(K_0\dot{+}V\right)\right)(z,x)\\&\le \left\|D_\Sigma(t/2)\exp\left(-\frac{t}{2}\left(K_0\dot{+}V\right)\right)\right\|_{\text{trace}}.\end{aligned}$$

So from (iii) it also follows that the operator $D_\Sigma(t/2)\exp\left(-\frac{t}{2}\left(K_0\dot{+}V\right)_\Sigma\right)$ is of trace class as well. The identity (5.52), to wit

$$\begin{aligned}2D_\Sigma(t) &= D_\Sigma(t/2)\exp\left(-\frac{t}{2}\left(K_0\dot{+}V\right)\right) + \exp\left(-\frac{t}{2}\left(K_0\dot{+}V\right)\right)D_\Sigma(t/2)\\&\quad + D_\Sigma(t/2)J^*\exp\left(-\frac{t}{2}\left(K_0\dot{+}V\right)_\Sigma\right)J + J^*\exp\left(-\frac{t}{2}\left(K_0\dot{+}V\right)_\Sigma\right)JD_\Sigma(t/2).\end{aligned}$$

then yields assertion (iv).

To what extent the result in (c) of the following proposition is useful is not at all clear. The right-hand side might be identically infinite.

6.12. PROPOSITION. *The following assertions are valid.*

(a) The following equality is true:

$$\begin{aligned}D_\Sigma(t,x,y) &= \exp\left(-t(K_0+V)\right)(x,y) - \exp\left(-t(K_0+V)_\Sigma\right)(x,y)\\&= \int_0^\infty ds \int_0^t d\tau \int_\Gamma dz \exp\left(-\tau\left(K_0+V+s1_\Gamma\right)\right)(x,z)\\&\qquad \exp\left(-(t-\tau)\left(K_0+V+s1_\Gamma\right)\right)(z,y).\end{aligned}$$

(b) Let $\mu_{0,x}^{t,y}$ be the measure, that pins the motion at x at time 0 and at y at time t. Then

$$\begin{aligned}&\operatorname{trace}\left\{\exp\left(-t\left(K_0\dot{+}V\right)\right) - J_\Sigma^*\exp\left(-t\left(K_0\dot{+}V\right)_\Sigma\right)J_\Sigma\right\}\\&= t\int_0^\infty ds \int_\Gamma dz \exp\left(-t\left(K_0+V+s1_\Gamma\right)\right)(z,z)\\&= t\int_\Gamma dx \int \frac{\exp\left(-\int_0^t V(X(u))du\right)}{\int_0^t 1_\Gamma(X(u))du}\,d\mu_{0,x}^{t,x},\end{aligned}$$

(c) As above, let $D_\Sigma(t) = \exp\left(-t\left(K_0\dot{+}V\right)\right) - J_\Sigma^*\exp\left(-t\left(K_0\dot{+}V\right)_\Sigma\right)J_\Sigma$. If the integral

$$\int_0^{2t} d\tau \left(\frac{\operatorname{trace}\left(D_\Sigma(\tau)\right)}{\tau}\right)^{1/2}\left(\frac{\operatorname{trace}\left(D_\Sigma(2t-\tau)\right)}{2t-\tau}\right)^{1/2}$$

is finite, then the operator $D_\Sigma(t)$ is of trace class and

$$\|D_\Sigma(t)\|_{\text{trace}} \\ \leq \frac{1}{2}\int_0^{2t} d\tau \left(\frac{\text{trace}\,(D_\Sigma(\tau))}{\tau}\right)^{1/2} \left(\frac{\text{trace}\,(D_\Sigma(2t-\tau))}{2t-\tau}\right)^{1/2}.$$

Proof. (a) From general arguments (Feynman-Kac formula and Lebesgue's dominated convergence theorem: see Lemma 7.1 in Chapter 7) it follows that

$$\exp\left(-t(K_0+V)\right)(x,y) - \exp\left(-t\left(K_0+V\right)_\Sigma\right)(x,y)$$
$$= \lim_{\beta\to\infty} \left(\exp\left(-t(K_0+V)\right)(x,y) - \exp\left(-t(K_0+V+\beta 1_\Gamma)\right)(x,y)\right)$$
$$= \lim_{\beta\to\infty} \int_0^\beta ds \int d\mu_{0,x}^{t,y} \exp\left(-\int_0^t \left(V(X(u)) + s1_\Gamma(X(u))\right) du\right) \int_0^t d\tau\, 1_\Gamma\big(X(\tau)\big) d\tau$$

(Markov property)

$$= \int_0^\infty ds \int_0^t d\tau \mathbb{E}_x \left(\exp\left(-\int_0^\tau \left(V(X(u)) + s1_\Gamma(X(u))\right) du\right) 1_\Gamma(X(\tau))\right.$$
$$\left. \int \exp\left(-\int_0^{t-\tau} \left(V(X(u)) + 1_\Gamma(X(u))\right) du\right) d\mu_{0,X(\tau)}^{t-\tau,y}\right)$$
$$= \int_0^\infty ds \int_0^t d\tau \int_\Gamma dz \exp\left(-\tau\left(K_0+V+s1_\Gamma\right)\right)(x,z)$$
$$\exp\left(-(t-\tau)\left(K_0+V+s1_\Gamma\right)\right)(z,y).$$

This proves (a).
(b) From (a) we infer (b), because we have

$$\text{trace}\,(D_\Sigma(t)) = \int dx \left(\exp\left(-t(K_0+V)\right)(x,x) - \exp\left(-t\left(K_0+V\right)_\Sigma\right)(x,x)\right)$$
$$= \int_0^\infty ds \int_0^t d\tau \int_\Gamma dz \int dx \exp\left(-\tau\left(K_0+V+s1_\Gamma\right)\right)(x,z)$$
$$\exp\left(-(t-\tau)\left(K_0+V+s1_\Gamma\right)\right)(z,x)$$

(Chapman-Kolmogorov)

$$= \int_0^\infty ds \int_0^t d\tau \int_\Gamma dz \exp\left(-t\left(K_0+V+s1_\Gamma\right)\right)(z,z)$$
$$= t\int_0^\infty ds \int_\Gamma dz \exp\left(-t\left(K_0+V+s1_\Gamma\right)\right)(z,z)$$

(Feynman-Kac)

$$= t\int_\Gamma dx \int \frac{\exp\left(-\int_0^t V(X(u))du\right)}{\int_0^t 1_\Gamma(X(u))du} d\mu_{0,x}^{t,x}.$$

(c) Let U be a unitary operator. In order to prove (c), it suffices to show that

$$|\operatorname{trace}(UD_\Sigma(t))| \leq \frac{1}{2}\int_0^{2t} d\tau \left(\frac{\operatorname{trace}(D_\Sigma(\tau))}{\tau}\right)^{1/2} \left(\frac{\operatorname{trace}(D_\Sigma(2t-\tau))}{2t-\tau}\right)^{1/2}.$$

From (a) it follows that

$$\begin{aligned}
&|\operatorname{trace}(UD_\Sigma(t))| \\
&= \int_0^\infty ds \int_0^t d\tau \int_\Gamma dz \int dx \,[U\exp(-\tau(K_0+V+s1_\Gamma))(\cdot,z)](x) \\
&\qquad \times \exp(-(t-\tau)(K_0+V+s1_\Gamma))(z,x)
\end{aligned}$$

and hence, by Cauchy-Schwarz' inequality and since U is unitary,

$$\begin{aligned}
&|\operatorname{trace}(UD_\Sigma(t))| \\
&\leq \int_0^t d\tau \int_\Gamma dz \left(\int_0^\infty ds \int dx \,|[U\exp(-\tau(K_0+V+s1_\Gamma))(\cdot,z)](x)|^2\right)^{1/2} \\
&\qquad \times \left(\int_0^\infty ds \int dx\,|\exp(-(t-\tau)(K_0+V+s1_\Gamma))(z,x)|^2\right)^{1/2} \\
&= \int_0^t d\tau \int_\Gamma dz \left(\int_0^\infty ds \int dx\,|\exp(-\tau(K_0+V+s1_\Gamma))(x,z)|^2\right)^{1/2} \\
&\qquad \times \left(\int_0^\infty ds \int dx\,|\exp(-(t-\tau)(K_0+V+s1_\Gamma))(z,x)|^2\right)^{1/2}
\end{aligned}$$

(Chapman-Kolmogorov)

$$\begin{aligned}
&= \int_0^t d\tau \int_\Gamma dz \left(\int_0^\infty ds \exp(-2\tau(K_0+V+s1_\Gamma))(z,z)\right)^{1/2} \\
&\quad \times \left(\int_0^\infty ds \exp(-2(t-\tau)(K_0+V+s1_\Gamma))(z,z)\right)^{1/2}
\end{aligned}$$

(employ (b))

$$\leq \int_0^t d\tau \left(\frac{\operatorname{trace}(D_\Sigma(2\tau))}{2\tau}\right)^{1/2} \left(\frac{\operatorname{trace}(D_\Sigma(2t-2\tau))}{2t-2\tau}\right)^{1/2}.$$

This proves Proposition 6.12.

The following result is the singular analogue of Proposition 6.9.

6.13. PROPOSITION. *(a) Suppose that the integral*

$$\int dx \left(\mathbb{E}_x\left(\exp\left(-\int_0^S V(X(u))du\right) : S < t/4\right)\right)^2$$

is finite. Then the operator $D_\Sigma(t/2)$ is Hilbert-Schmidt and

$$\|D_\Sigma(t/2)\|_{\mathrm{HS}}^2 \leq 4 \sup_{y_1,y_2\in\Gamma} \sup_{t/2<u<t} \exp\left(-u(K_0+V)\right)(y_1,y_2)$$
$$\times \int dx \left(\mathbb{E}_x\left(\exp\left(-\int_0^S V(X(u))du\right) : S < \frac{t}{4}\right)\right)^2. \tag{6.29}$$

(b) Suppose that the integral $\int dx\, \mathbb{E}_x\left(\exp\left(-\int_0^S V(X(u))du\right) : S < t/4\right)$ is finite. Choose real numbers a and b such that, in form sense, $aI+K_0\dot{+}V \geq bI_{L^2(E,m)} > 0$.

Then $\left(aI+K_0\dot{+}V\right)_\Sigma \geq bI_{L^2(\Sigma,m)}$ and the integral $\int h_\Sigma^{a+V}(x)dm(x)$ is finite. In addition, the operator

$$J^*\left(aI+K_0\dot{+}V\right)_\Sigma \exp\left(-\frac{t}{2}\left(K_0\dot{+}V\right)_\Sigma\right) JH_\Sigma^{a+V}\exp\left(-\frac{t}{2}\left(K_0\dot{+}V\right)\right)$$

is a trace operator and, for $0<2\varepsilon<t$,

$$\left\|J^*\left(aI+K_0\dot{+}V\right)_\Sigma \exp\left(-\frac{t}{2}\left(K_0\dot{+}V\right)_\Sigma\right) JH_\Sigma^{a+V}\exp\left(-\frac{t}{2}\left(K_0\dot{+}V\right)\right)\right\|_{\mathrm{trace}}$$
$$\leq \sqrt{\frac{b}{2}}\frac{\exp(2\varepsilon a)}{\varepsilon e\sqrt{\varepsilon e}} \sup_{y\in E\setminus\Sigma, x\in E} \exp\left(-(t-2\varepsilon)(K_0+V)\right)(y,x)$$
$$\times \int h_\Sigma^{a+V}(x)dm(x). \tag{6.30}$$

REMARK 1. An appropriate choice for ε is given by $4a\varepsilon = 3$. For b we may choose $b = \left\|\left(aI+K_0\dot{+}V\right)^{-1}\right\|_{2,2}$. With these choices (6.30) turns into

$$\left\|J^*\left(aI+K_0\dot{+}V\right)_\Sigma \exp\left(-\frac{t}{2}\left(K_0\dot{+}V\right)_\Sigma\right) JH_\Sigma^{a+V}\exp\left(-\frac{t}{2}\left(K_0\dot{+}V\right)\right)\right\|_{\mathrm{trace}}$$
$$\leq a\left\|a\left(aI+K_0\dot{+}V\right)^{-1}\right\|_{2,2}^{1/2}\sqrt{\frac{64}{54}} \sup_{y\in E\setminus\Sigma, x\in E} \exp\left(-\left(t-\frac{3}{2a}\right)(K_0+V)\right)(y,x)$$
$$\times \int h_\Sigma^{a+V}(x)dm(x). \tag{6.31}$$

REMARK 2. Recall

$$h_\Sigma^{a+V}(x) = \mathbb{E}_x\left[\exp\left(-\int_0^S \left(a+V(X(u))\right)du\right) : S < \infty\right].$$

Notice that for a sufficiently large, the function h_Σ^{a+V} is bounded: see Proposition 4.20 in Chapter 4.

REMARK 3. On an appropriate domain we have

$$\begin{aligned}
&J^*\left(aI+K_0\dot{+}V\right)_\Sigma JH_\Sigma^{a+V}\\
&=J^*\left(aI+K_0\dot{+}V\right)_\Sigma JH_\Sigma^{a+V}\left(aI+K_0\dot{+}V\right)^{-1}\left(aI+K_0\dot{+}V\right)\\
&=J^*\left(aI+K_0\dot{+}V\right)_\Sigma J\left\{\left(aI+K_0\dot{+}V\right)^{-1}-J^*\left(aI+K_0\dot{+}V\right)_\Sigma^{-1}J\right\}\left(aI+K_0\dot{+}V\right)\\
&=J^*\left(aI+K_0\dot{+}V\right)_\Sigma J-J^*J\left(aI+K_0\dot{+}V\right)\quad=J^*\left(K_0\dot{+}V\right)_\Sigma J-J^*J\left(K_0\dot{+}V\right)
\end{aligned}$$

and so that the operator $J^*\left(aI+K_0\dot{+}V\right)_\Sigma JH_\Sigma^{a+V}$ does not really depend on a.

Proof. (a) We write

$$D_\Sigma(t/2)=D_\Sigma(t/4)\exp\left(-\frac{t}{4}\left(K_0\dot{+}V\right)\right)+\exp\left(-\frac{t}{4}\left(K_0\dot{+}V\right)_\Sigma\right)D_\Sigma(t/4)$$

and hence we see that

$$\begin{aligned}
&\|D_\Sigma(t/2)\|_{\mathrm{HS}}\\
&\leq\left\|D_\Sigma(t/4)\exp\left(-\frac{t}{4}\left(K_0\dot{+}V\right)\right)\right\|_{\mathrm{HS}}+\left\|\exp\left(-\frac{t}{4}\left(K_0\dot{+}V\right)_\Sigma\right)D_\Sigma(t/4)\right\|_{\mathrm{HS}}\\
&\leq 2\left\|D_\Sigma(t/4)\exp\left(-\frac{t}{4}\left(K_0\dot{+}V\right)\right)\right\|_{\mathrm{HS}}.
\end{aligned}\tag{6.32}$$

Let $(\mathbb{P}'_x,X')$ be an stochastically independent copy of the Markov process $(\mathbb{P}_x,X)$ with the same transition laws. We note the equality

$$\begin{aligned}
&\left\|D_\Sigma(t/4)\exp\left(-\frac{t}{4}\left(K_0\dot{+}V\right)\right)\right\|_{\mathrm{HS}}^2\\
&=\int dx\int dy\\
&\quad\mathbb{E}_x\left[\exp\left(-\int_0^S V(X(u))du\right)\exp\left(-\left(\frac{t}{2}-S\right)(K_0+V)\right)(X(S),y):S<\frac{t}{4}\right]\\
&\times\mathbb{E}'_x\left[\exp\left(-\int_0^{S'} V(X'(u))du\right)\exp\left(-\left(\frac{t}{2}-S'\right)(K_0+V)\right)(X(S'),y):S'<\frac{t}{4}\right]\\
&=\int dx\int dy\\
&\quad\mathbb{E}_x\otimes\mathbb{E}'_x\left[\exp\left(-\int_0^S V(X(u))du\right)\exp\left(-\left(\frac{t}{2}-S\right)(K_0+V)\right)(X(S),y)\right.\\
&\times\exp\left(-\int_0^{S'} V(X'(u))du\right)\exp\left(-\left(\frac{t}{2}-S'\right)(K_0+V)\right)(X(S'),y):\\
&\qquad\qquad\left.S<\frac{t}{4},\ S'<\frac{t}{4}\right]
\end{aligned}$$

$$= \int \mathbb{E}_x \otimes \mathbb{E}'_x \left[\exp\left(-\int_0^S V(X(u))du - \int_0^{S'} V(X'(u))du \right) \right.$$
$$\left. \times \exp\left(-\left(t - S - S'\right)\left(K_0 + V\right)\right)\left(X(S), X(S')\right) : S < \frac{t}{4}, \quad S' < \frac{t}{4} \right] dx$$
$$\leq \int \left(\mathbb{E}_x \left[\exp\left(-\int_0^S V(X(u))du \right) : S < \frac{t}{4} \right] \right)^2 dx$$
$$\times \sup_{y_1, y_2 \in \Gamma} \sup_{t/2 < u < t} \exp\left(-u\left(K_0 + V\right)\right)\left(y_1, y_2\right). \tag{6.33}$$

The inequality in (6.29) results from a combination of (6.32) and (6.33).

(b) First we prove that a lower bound for the operator $aI + K_0 \dot{+} V$ is also a lower bound for $\left(aI + K_0 \dot{+} V\right)_\Sigma$. Let ω_V be the lower bound of the spectrum of $K_0 \dot{+} V$ and let $\omega_{V,\Sigma}$ be the lower bound of the spectrum of $\left(K_0 \dot{+} V\right)_\Sigma$. Then, since

$$\left\| \exp\left(-t\left(K_0 \dot{+} V\right)_\Sigma\right)\right\|_{2,2} \leq \left\|\exp\left(-t\left(K_0 \dot{+} V\right)\right)\right\|_{2,2},$$

it follows that $\omega_{V,\Sigma} \geq \omega_V$. So, if $aI + K_0 \dot{+} V \geq bI_{L^2(E,m)}$, then $a + \omega_V \geq b$ and hence $a + \omega_{V,\Sigma} \geq a + \omega_V \geq b$. Consequently $\left(aI + K_0 \dot{+} V\right)_\Sigma \geq bI_{L^2(\Sigma,m)}$. The fact that the function h_Σ^{a+V} is integrable follows from Proposition 4.20 in Chapter 4. It is only in this proof that we write $A = aI + K_0 \dot{+} V$, $A_\Sigma = J^* \left(aI + K_0 \dot{+} V\right)_\Sigma J$, $H = H_\Sigma^{a+V}$ and $P = P^* = A^{1/2} H A^{-1/2} = P^2$. Rephrasing the problem in terms of this notation, we need to estimate the trace norm of the operator

$$A_\Sigma \exp\left(-\frac{t}{2} A_\Sigma\right) A^{-1/2} P A^{1/2} \exp\left(-\frac{t}{2} A\right).$$

We obtain the following estimates:

$$\left\| A_\Sigma \exp\left(-\frac{t}{2} A_\Sigma\right) A^{-1/2} P A^{1/2} \exp\left(-\frac{t}{2} A\right)\right\|_{\text{trace}}^2$$
$$\leq \left\| A_\Sigma \exp\left(-\frac{t}{2} A_\Sigma\right) A^{-1/2} P\right\|_{\text{HS}}^2 \left\| A^{1/2} \exp\left(-\frac{t}{2} A\right) P\right\|_{\text{HS}}^2$$
$$\leq \left\| A_\Sigma \exp\left(-\frac{t}{2} A_\Sigma\right) A^{-1/2} P A^{-1/2} A_\Sigma \exp\left(-\frac{t}{2} A_\Sigma\right)\right\|_{\text{trace}}$$
$$\times \left\| A^{1/2} \exp\left(-\frac{t}{2} A\right) P A^{1/2} \exp\left(-\frac{t}{2} A\right)\right\|_{\text{trace}}$$
$$\leq \left\| A_\Sigma \exp\left(-\epsilon A_\Sigma\right)\right\|_{L^2(\Sigma,m), L^2(\Sigma,m)}^2$$
$$\times \left\| \exp\left(-\left(\frac{t}{2} - \epsilon\right) A_\Sigma\right) A^{-1/2} P A^{-1/2} \exp\left(-\left(\frac{t}{2} - \epsilon\right) A_\Sigma\right)\right\|_{\text{trace}}$$

$$\times \left\| A^{1/2} \exp\left(-\epsilon A\right) \right\|^2_{L^2(E,m),L^2(E,m)}$$
$$\times \left\| \exp\left(-\left(\frac{t}{2}-\epsilon\right)A\right) P \exp\left(-\left(\frac{t}{2}-\epsilon\right)A\right) \right\|_{\text{trace}}$$

(the operator $A^{-1/2}PA^{-1/2}$ is form positive and possesses a positive integral kernel and the integral kernel of $\exp\left(-uA_\Sigma\right)$ is dominated by the integral kernel of $\exp\left(-uA\right)$)

$$\begin{aligned}
&\le \left\| A_\Sigma \exp\left(-\epsilon A_\Sigma\right) \right\|^2_{L^2(\Sigma,m),L^2(\Sigma,m)} \\
&\quad \times \left\| \exp\left(-\left(\frac{t}{2}-\epsilon\right)A\right) A^{-1/2}PA^{-1/2} \exp\left(-\left(\frac{t}{2}-\epsilon\right)A\right) \right\|_{\text{trace}} \\
&\quad \times \left\| A^{1/2} \exp\left(-\epsilon A\right) \right\|^2_{L^2(E,m),L^2(E,m)} \\
&\quad \times \left\| \exp\left(-\left(\frac{t}{2}-\epsilon\right)A\right) P \exp\left(-\left(\frac{t}{2}-\epsilon\right)A\right) \right\|_{\text{trace}} \\
&\le \left\| A_\Sigma \exp\left(-\epsilon A_\Sigma\right) \right\|^2_{L^2(\Sigma,m),L^2(\Sigma,m)} \left\| A^{-1/2} \right\|^2_{L^2(E,m),L^2(E,m)} \\
&\quad \times \left\| \exp\left(-\left(\frac{t}{2}-\epsilon\right)A\right) P \exp\left(-\left(\frac{t}{2}-\epsilon\right)A\right) \right\|^2_{\text{trace}} \\
&\quad \times \left\| A^{1/2} \exp\left(-\epsilon A\right) \right\|^2_{L^2(E,m),L^2(E,m)} \\
&\le \frac{1}{2be^3\epsilon^3} \left\| \exp\left(-\left(\frac{t}{2}-\epsilon\right)A\right) P \exp\left(-\left(\frac{t}{2}-\epsilon\right)A\right) \right\|^2_{\text{trace}} \\
&\le \frac{1}{2be^3\epsilon^3} \left(\text{trace}\left\{ \exp\left(-\left(\frac{t}{2}-\epsilon\right)A\right) H \exp\left(-\left(\frac{t}{2}-\epsilon\right)A\right) \right\}\right)^2 .
\end{aligned} \tag{6.34}$$

The ultimate inequality in (6.34) follows from spectral theory, to wit

$$\left\| \left(\varepsilon A\right)^{1/2} \exp\left(-\varepsilon A\right) \right\|_{L^2(E,m),L^2(E,m)} \le \sup_{\xi>0} \sqrt{\xi} \exp(-\xi),$$

and similarly for the norm of the operator $A_\Sigma \exp\left(-\varepsilon A_\Sigma\right)$. The inequality in (6.30) is an immediate consequence of (6.34).

In the following result, we discuss to some extent the trace property of the operator $H_\Sigma^{a+V} \exp\left(-t\left(aI + K_0 \dot{+} V\right)\right)$. If throughout the theorem and its proof we replace the trace norms by Hilbert-Schmidt norms, we obtain the corresponding result for Hilbert-Schmidt properties. For the notational conventions we refer to the Propositions 4.14 and 4.15. In Definition 4.13 the operators $T_\Sigma(t)$, $t \ge 0$, are defined s follows:

$$T_\Sigma(t)f(x) = T_\Sigma^V(t)f(x) = \mathbb{E}_x\left[\exp\left(-\int_0^S V(X(u))du\right) f(X(S)) : S \le t\right].$$

6.14. THEOREM. *The following assertions are equivalent:*
(i) For every t_1 and $t_2 > 0$ the supremum

$$\sup_{0<s<t_1} \left\| T_\Sigma(s) \exp\left(-t_2\left(K_0 \dot{+} V\right)\right) \right\|_{\text{trace}} \tag{6.35}$$

is finite;
(ii) For every t_1 and $t_2 > 0$ the supremum

$$\sup_{0<s<t_1} \left\| D_\Sigma(s) \exp\left(-t_2\left(K_0 \dot{+} V\right)\right) \right\|_{\text{trace}} \tag{6.36}$$

is finite;
(iii) For every $t > 0$ and for every $a \in \mathbb{R}$ sufficiently large the operator

$$H_\Sigma^{a+V} \exp\left(-t\left(aI + K_0 \dot{+} V\right)\right)$$

is of trace class.

Proof. (i) $\Rightarrow$ (ii). Equality (4.62) in Proposition 4.15 implies among other things that

$$\begin{aligned}
&D_\Sigma(s) \exp\left(-t_2\left(K_0 \dot{+} V\right)\right) \\
&= -\int_0^s T_\Sigma(u) \exp\left(-t_2'\left(K_0 \dot{+} V\right)\right)\left(K_0 \dot{+} V\right) \exp\left(-\left(s-u+t_2-t_2'\right)\left(K_0 \dot{+} V\right)\right) du \\
&\qquad + T_\Sigma(s) \exp\left(-t_2\left(K_0 \dot{+} V\right)\right)
\end{aligned}$$

for some $0 < t_2' < t_2$. Hence we get

$$\begin{aligned}
&\left\| D_\Sigma(s) \exp\left(-t_2\left(K_0 \dot{+} V\right)\right) \right\|_{\text{trace}} \\
&\leq s \sup_{0<v<s} \left\| T_\Sigma(v) \exp\left(-t_2'\left(K_0 \dot{+} V\right)\right) \right\|_{\text{trace}} \\
&\quad \times \sup_{0<v<s} \left\| \left(K_0 \dot{+} V\right) \exp\left(-\left(v+t_2-t_2'\right)\left(K_0 \dot{+} V\right)\right) \right\|_{2,2} \left\| \exp\left(-(s-v)\left(K_0 \dot{+} V\right)_\Sigma\right) \right\|_{2,2} \\
&\quad + \left\| T_\Sigma(s) \exp\left(-t_2\left(K_0 \dot{+} V\right)\right) \right\|_{\text{trace}} \\
&\left(K_0 \dot{+} V \geq -\omega \text{ in form sense}\right) \\
&\leq s \sup_{0<v<s} \left\| T_\Sigma(v) \exp\left(-t_2'\left(K_0 \dot{+} V\right)\right) \right\|_{\text{trace}} \\
&\quad \times \sup_{0<v<s} \sup_{\xi \geq -\omega} |\xi| \exp\left(-\left(v+t_2-t_2'\right)\xi\right) \sup_{0<v<s} \left\| \exp\left(-v\left(K_0 \dot{+} V\right)_\Sigma\right) \right\|_{2,2} \\
&\quad + \left\| T_\Sigma(s) \exp\left(-t_2\left(K_0 \dot{+} V\right)\right) \right\|_{\text{trace}}.
\end{aligned} \tag{6.37}$$

From (i) it follows that (6.37) is finite uniformly for s in a compact interval.

(ii) $\Rightarrow$ (i). Again we pick $0 < t_2' < t_2$ and use the identity in (4.61) of Proposition 4.15, to wit

$$T_\Sigma(t)f = D_\Sigma(t)f + \int_0^t D_\Sigma(u)(K_0\dot{+}V)f\,du, \quad f \in \operatorname{dom}\left(K_0\dot{+}V\right). \tag{6.38}$$

From (6.38) we see that

$$\begin{aligned}
&\left\|T_\Sigma(s)\exp\left(-t_2\left(K_0\dot{+}V\right)\right)\right\|_{\text{trace}} \\
&\leq \left\|D_\Sigma(s)\exp\left(-t_2\left(K_0\dot{+}V\right)\right)\right\|_{\text{trace}} \\
&\quad + \int_0^s \left\|D_\Sigma(u)\exp\left(-t_2'\left(K_0\dot{+}V\right)\right)\right\|_{\text{trace}} \left\|\left(K_0\dot{+}V\right)\exp\left(-\left(t_2-t_2'\right)\left(K_0\dot{+}V\right)\right)\right\|_{2,2} du \\
&\leq \left\|D_\Sigma(s)\exp\left(-t_2\left(K_0\dot{+}V\right)\right)\right\|_{\text{trace}} \\
&\quad + s\sup_{0<u<s}\left\|D_\Sigma(u)\exp\left(-t_2'\left(K_0\dot{+}V\right)\right)\right\|_{\text{trace}} \sup_{\xi>-\omega}|\xi|\exp\left(-\left(t_2-t_2'\right)\xi\right).
\end{aligned}$$

So (i) follows from (ii).

(iii) $\Rightarrow$ (ii). From equality (4.59) in Proposition 4.14, with a sufficiently large, we see that

$$T_\Sigma^{a+V}(t) = \left(I - J_\Sigma^*\exp\left(-t\left(aI + K_0\dot{+}V\right)_\Sigma\right)J_\Sigma\right)H_\Sigma^{a+V}. \tag{6.39}$$

Here we write T_Σ^{a+V} in order to distinguish it from T_Σ which in fact equals T_Σ^V. From (6.39) we obtain

$$\begin{aligned}
&\sup_{0<s<t_1}\left\|T_\Sigma^{a+V}(s)\exp\left(-t_2\left(aI + K_0\dot{+}V\right)\right)\right\|_{\text{trace}} \\
&= \sup_{0<s<t_1}\left\|\left(I - J_\Sigma^*\exp\left(-s\left(aI + K_0\dot{+}V\right)_\Sigma J_\Sigma\right)\right)H_\Sigma^{a+V}\exp\left(-t_2\left(aI + K_0\dot{+}V\right)\right)\right\|_{\text{trace}} \\
&\leq 2\left\|H_\Sigma^{a+V}\exp\left(-t_2\left(aI + K_0\dot{+}V\right)\right)\right\|_{\text{trace}}.
\end{aligned} \tag{6.40}$$

The quantity in (6.40) is finite and hence (6.35) follows with $T_\Sigma^{a+V}(s)$ replacing $T_\Sigma(s) = T_\Sigma^V(s)$. The equivalence of (i) and (ii) implies that the expression

$$\sup_{0<s<t_1}\left\|D_\Sigma^{a+V}(s)\exp\left(-t_2\left(aI + K_0\dot{+}V\right)\right)\right\|_{\text{trace}} \tag{6.41}$$

is finite. Here $D_\Sigma^{a+V}(s) = \exp(-as)\,D_\Sigma^V(s) = \exp(-as)\,D_\Sigma(s)$. A consequence of (6.41) is that the quantity

$$\sup_{0<s<t_1}\exp\left(-as - at_2\right)\left\|D_\Sigma^V(s)\exp\left(-t_2\left(K_0\dot{+}V\right)\right)\right\|_{\text{trace}} \tag{6.42}$$

is finite. But then the supremum

$$\sup_{0<s<t_1}\left\|D_\Sigma^V(s)\exp\left(-t_2\left(K_0\dot{+}V\right)\right)\right\|_{\text{trace}}$$

is finite. This proves (ii).

(ii) $\Rightarrow$ (iii). Suppose that

$$\sup_{0<s<t_1} \left\| D_\Sigma^V(s) \exp\left(-t_2\left(K_0 \dot{+} V\right)\right) \right\|_{\text{trace}}$$

is finite. Then so is the expression

$$\sup_{0<s<t_1} \left\| D_\Sigma^{a+V}(s) \exp\left(-t_2\left(aI + K_0 \dot{+} V\right)\right) \right\|_{\text{trace}}.$$

From the equivalence of (i) and (ii) we also have that

$$\sup_{0<s<t_1} \left\| T_\Sigma^{a+V}(s) \exp\left(-t_2\left(aI + K_0 \dot{+} V\right)\right) \right\|_{\text{trace}}$$

is finite. Again an appeal to equality (6.39) clearly shows that the expression

$$\left\| \left(I - J^* \exp\left(-t_1\left(aI + K_0 \dot{+} V\right)_\Sigma\right) J\right) \exp\left(-t_2\left(aI + K_0 \dot{+} V\right)\right) \right\|_{\text{trace}}$$

is finite as well. Since for a sufficiently large the operator

$$\left(I - J^* \exp\left(-t_1\left(aI + K_0 \dot{+} V\right)_\Sigma\right) J\right)$$

has a bounded inverse, assertion (iii) is obtained.

We conclude this chapter with some results on operators of the form

$$\left(aI + K_0 \dot{+} V\right)^{1/2} u \exp\left(-2t\left(aI + K_0 \dot{+} V\right)\right). \tag{6.43}$$

This kind of operator occurred in the first assertion of Theorem 5.12 and 6.11. In Proposition 6.15 below we give some sufficient conditions under which this operator is bounded, Hilbert-Schmidt, or of trace class. We assume that the operator K_0 generates a diffusion: see Definition 1.8 and the comments following it. In particular, this means that

$$K_0(f_1 f_2) = f_1 K_0 f_2 + \left(K_0 f_1\right) f_2 - \Gamma_1\left(f_1, f_2\right) \tag{6.44}$$

for f_1 and f_2 in an appropriate large enough algebra $\mathcal{A}$ of functions in $\operatorname{dom}(K_0)$. We assume that u belongs to $\mathcal{A}$. As in Definition 1.7 the carré du champ operator $\Gamma_1 : \mathcal{A} \times \mathcal{A} \to L^2(E, m)$ is given by

$$\left[\Gamma_1\left(f_1, f_2\right)\right](x) = \lim_{s \downarrow 0} \frac{1}{s} \mathbb{E}_x\left[\left(f_1(X(s)) - f_1(X(0))\right)\left(f_2(X(s)) - f_2(X(0))\right)\right] \tag{6.45}$$

for f_1, $f_2 \in \mathcal{A}$.

6.15. PROPOSITION. *Let K_0 and Γ_1 be as in (6.44) and (6.45). Suppose that the operators $f \mapsto u\exp\left(-t\left(K_0\dot{+}V\right)\right)f$, $f \mapsto (K_0 u)\exp\left(-t\left(K_0\dot{+}V\right)\right)f$, and $f \mapsto \left(aI + K_0\dot{+}V\right)^{-1/2}\Gamma_1\left(u, \left(aI + K_0\dot{+}V\right)^{-1/2}\exp\left(-2t\left(K_0\dot{+}V\right)\right)f\right)$ are bounded (or Hilbert-Schmidt, or of trace class respectively), then so is the operator*

$$\left(aI + K_0\dot{+}V\right)^{1/2} u\exp\left(-2t\left(aI + K_0\dot{+}V\right)\right).$$

Proof. Let the operator (Lie-bracket) $[K_0, u]$ be defined by $[K_0, u] = K_0(uf) - uK_0 f$. Then

$$\begin{aligned}&\left(aI + K_0\dot{+}V\right)^{1/2} u - u\left(aI + K_0\dot{+}V\right)^{1/2}\\ &= \left(aI + K_0\dot{+}V\right)^{-1/2}[K_0, u]\left(aI + K_0\dot{+}V\right)^{-1/2},\end{aligned} \tag{6.46}$$

and thus

$$\begin{aligned}&\left\{\left(aI + K_0\dot{+}V\right)^{1/2} u - u\left(aI + K_0\dot{+}V\right)^{1/2}\right\}\exp\left(-2t\left(aI + K_0\dot{+}V\right)\right)\\ &= \left(aI + K_0\dot{+}V\right)^{-1/2}[K_0, u]\left(aI + K_0\dot{+}V\right)^{-1/2}\exp\left(-2t\left(aI + K_0\dot{+}V\right)\right).\end{aligned} \tag{6.47}$$

Since

$$\left\|\left(aI + K_0\dot{+}V\right)^{1/2}\exp\left(-t\left(aI + K_0\dot{+}V\right)\right)\right\|_{2,2} \leq \frac{1}{\sqrt{2et}},$$

the operator in (6.43) is bounded (or Hilbert-Schmidt, or of trace class) whenever the operators

$$u\exp\left(-t\left(K_0\dot{+}V\right)\right) \text{ and}$$
$$\left(aI + K_0\dot{+}V\right)^{-1/2}[K_0, u]\left(aI + K_0\dot{+}V\right)^{-1/2}\exp\left(-2t\left(aI + K_0\dot{+}V\right)\right)$$

possess this property. Since $[K_0, u]f = K_0(uf) - uK_0 f = (K_0 u)f - \Gamma_1(u, f)$, the result in Proposition 6.15 follows.

6.16. COROLLARY. *Suppose that, for every $f \in L^2(E, m)$, the integrals*

$$\iiint dz\,dy\,dx\overline{f(x)}\exp\left(-t\left(K_0 + V\right)\right)(x, z)u(z)^2\exp\left(-t\left(K_0 + V\right)\right)(z, y)f(y),$$

$$\iiint dz\,dy\,dx\overline{f(x)}\exp\left(-t\left(K_0 + V\right)\right)(x, z)\left(\left[K_0 u\right](z)\right)^2$$
$$\exp\left(-t\left(K_0 + V\right)\right)(z, y)f(y),$$

and

$$\int\left[\Gamma_1(u, u)\right](x)\left[\Gamma_1\left(\exp\left(-2t\left(K_0\dot{+}V\right)\right)f, \exp\left(-2t\left(K_0\dot{+}V\right)\right)\overline{f}\right)\right](x)dx$$

are finite. Then the operator in (6.43) is bounded.

Proof. Since, for real functions f_1 and f_2 in $\mathcal{A}$,

$$\left|\Gamma_1(f_1, f_2)\right|^2 \leq \left|\Gamma_1(f_1, f_1)\right| \times \left|\Gamma_1(f_2, f_2)\right|,$$

the corollary follows from Proposition 6.15.

Chapter 7

Convergence of Resolvent Differences

In this chapter we investigate the resolvent convergence of Feller operators additively perturbed by potentials on sets $\Gamma \subset E$ of height β. Let $K_0 \dot{+} V$ be the Feller operator as introduced in Definition 2.8 (see Theorem 2.5.(a) as well). Here K_0 is the free Feller operator (see Definition 1.3) and V is a Kato-Feller potential given in Definition 2.1. The operator $K_0 \dot{+} V$ is perturbed by a potential of the form $\beta 1_\Gamma$. Here β is a positive parameter and Γ is a closed region in E as discussed in the beginning of section C of Chapter 2. This means that we introduce a new self-adjoint operator K_β given by

$$K_\beta = K_0 \dot{+} V + \beta 1_\Gamma \tag{7.1}$$

in $L^2(E,m)$ with $\mathrm{dom}(K_\beta) = \mathrm{dom}\left(K_0 \dot{+} V\right)$. We also introduced the operator $\left(K_0 \dot{+} V\right)_\Sigma$ in Definition 2.25 as the generator of the Dirichlet semigroup on $L^2(\Sigma, m)$:

$$\left[\exp\left(-t\left(K_0 \dot{+} V\right)_\Sigma\right) f\right](x) = \mathbb{E}_x\left[\exp\left(-\int_0^t V(X(u))du\right) f(X(t)) : S > t\right].$$

Here we assume again that $\Gamma^r = (\mathrm{int}(\Gamma))^r$ so that S, the penetration time, equals T, the first hitting time: see Proposition 2.24. Upon writing $K_\Sigma = \left(K_0 \dot{+} V\right)_\Sigma$, there is a natural relation between K_β and K_Σ for such singularity regions Γ given by the strong resolvent convergence. Let $a = -z$ be a large enough real number and denote

$$R_\beta(a) = \left(K_0 \dot{+} V + \beta 1_\Gamma + aI\right)^{-1} \quad \text{and} \tag{7.2}$$

$$R_\Sigma(a) := \left(\left(K_0 \dot{+} V\right)_\Sigma + aI_\Sigma\right)^{-1}. \tag{7.3}$$

Then, for $f \in L^2(E,m)$, we have the existence (in L^2-sense) of the following limits:

$$\lim_{\beta\to\infty} 1_\Gamma R_\beta(a) f = 0 \quad \text{and} \tag{7.4}$$

$$\lim_{\beta\to\infty} 1_\Sigma R_\beta(a) f = J^* R_\Sigma(a) J f \tag{7.5}$$

with $J = J_\Sigma$ and $Jf = f\restriction_\Sigma$.

The main aim of this chapter is to give sufficient conditions for the convergence of $R_\beta(a)$ to $J^* R_\Sigma(a) J$ with respect to different operator norms. The norms we will consider are the usual operator norm, the Hilbert-Schmidt norm and the trace norm. We will also look at the corresponding semigroups and powers of resolvents. First we make a remark about the spectra of the operators K_β and K_Σ.

REMARK. The spectrum of the Feller operator $K_0\dot{+}V$ is contained in an interval of the form $[-\omega_V,\infty)$, $\omega_V \geq 0$, or what amounts to the same thing, $\mathbb{C}\backslash[-\omega_V,\infty) \subseteq \operatorname{res}\left(K_0\dot{+}V\right)$, the resolvent set. If operator norm convergence

$$\text{u-}\lim_{\beta\to\infty} R_\beta(z_0) = J^* R_\Sigma(z_0) J$$

can be established for some $z_0 \in \operatorname{res}(K_\Sigma)$, then this convergence is true for all $z \in \operatorname{res}(K_\Sigma)$ (see e.g. Kato [148, p. 206–207]) because $\operatorname{res}(K_\Sigma) \subseteq \bigcup_{\beta_0>0}\bigcap_{\beta\geq\beta_0}\operatorname{res}(K_\beta)$. We also have $\mathbb{C}\setminus[-\omega_V,\infty) \subseteq \operatorname{res}(K_\Sigma)$.

Before we study the resolvent norm convergences, we consider semigroup differences and we will also give some quantitative estimates in β that guarantee the convergence and that give an indication of how to determine the rate of convergence as well. Prior to the investigation of the different operator norms we would like to mention that we already have convergence of the semigroup kernels. This result was used in the proof of Proposition 6.12.(a).

7.1. LEMMA. *Let K_β and K_Σ be given as above and fix x and y in E. Then the kernels of the semigroups satisfy*

$$\lim_{\beta\to\infty} \exp(-\lambda K_\beta)(x,y) = \exp(-\lambda K_\Sigma)(x,y). \tag{7.6}$$

The convergence in (7.6) is uniform on compact subsets of $\Sigma\times E$.

Proof. Let $\mu_{0,x}^{\lambda,y}$ be the Markov bridge measure determined by the equality

$$\mu_{0,x}^{\lambda,y}(A) = \mathbb{E}_x\left[p_0(\lambda-s,X(s),y)1_A\right], \quad A\in\mathcal{F}_s, \quad s<\lambda.$$

Then it is shown in Chapter 2, Proposition 2.22 and Chapter 4, Proposition 4.16 (see (4.70)), that the difference of the integral kernels in (7.6) is given by the integral

$$\begin{aligned}&\exp(-\lambda K_\beta)(x,y) - \exp(-\lambda K_\Sigma)(x,y)\\ &= \int \exp\left(-\int_0^\lambda V(X(u))du\right)\exp\left(-\beta\int_0^\lambda 1_\Gamma(X(u))du\right)1_{\{S<\lambda\}}d\mu_{0,x}^{\lambda,y}.\end{aligned}$$

Since

$$\begin{aligned}&\lim_{\beta\to\infty}\left[\exp(-\lambda K_\beta)(x,y) - \exp(-\lambda K_\Sigma)(x,y)\right]\\ &= \lim_{\beta\to\infty}\int \exp\left(-\int_0^\lambda V(X(u))du\right)\exp\left(-\beta\int_0^\lambda 1_\Gamma(X(u))du\right)1_{\{S<\lambda\}}d\mu_{0,x}^{\lambda,y}\\ &= \int \exp\left(-\int_0^\lambda V(X(u))du\right)1_{\{\int_0^\lambda 1_\Gamma(X(u))du=0\}}1_{\{S<\lambda\}}d\mu_{0,x}^{\lambda,y}\\ &= \int \exp\left(-\int_0^\lambda V(X(u))du\right)1_{\{S\geq\lambda\}}1_{\{S<\lambda\}}d\mu_{0,x}^{\lambda,y} = 0,\end{aligned}$$

equality (7.6) follows. In order to establish the uniform convergence we write the difference of the integral kernels in the following form:

$$\exp\left(-\lambda K_{\beta}\right)(x,y) - \exp\left(-\lambda K_{\Sigma}\right)(x,y) = \mathbb{E}_x\left[M^{\lambda,y}_{V+\beta 1_{\Gamma}}(S) : S < \lambda\right], \tag{7.7}$$

where $M^{\lambda,y}_{V+\beta 1_{\Gamma}}(s)$, $0 \le s < \lambda$, is the martingale (see formula (2.23) in Definition 2.18, and Proposition 2.20)

$$\begin{aligned} &M^{\lambda,y}_{V+\beta 1_{\Gamma}}(s) \\ &= \exp\left(-\int_0^s \left(V(X(u)) + \beta 1_{\Gamma}(X(u))\right) du\right) \exp\left(-(\lambda - s)\left(K_0 \dot{+} V + \beta 1_{\Gamma}\right)\right)(X(s), y). \end{aligned}$$

The expression in (7.7), as a function of λ, x and y is jointly continuous on $(0,\infty)\times\Sigma\times E$ and decreases to zero. Appendix E, Lemma E.1 yields the desired result.

7.2. PROPOSITION. *Let K_β and K_Σ be as in Lemma 7.1. and put*

$$D_{\Sigma,\beta}(t) = \exp\left(-tK_{\beta}\right) - J^{*}\exp\left(-tK_{\Sigma}\right)J. \tag{7.8}$$

The following assertions are valid:

(a) *The operator norm of $D_{\Sigma,\beta}(t)$ satisfies the following inequality*

$$\begin{aligned} &\left\|D_{\Sigma,\beta}(t)\right\|_{2,2} \qquad (7.9)\\ &\le \sup_{x\in E}\mathbb{E}_x\left[\exp\left(-\int_0^t V(X(u))du\right)\exp\left(-\beta\int_0^t 1_{\Gamma}(X(u))du\right) : S < t\right]; \end{aligned}$$

(b) *The Hilbert-Schmidt norm of $D_{\Sigma,\beta}(t)$ obeys the following inequality*

$$\begin{aligned} &\left\|D_{\Sigma,\beta}(t)\right\|_{\mathrm{HS}} \qquad (7.10)\\ &\le 2\left\|\exp\left(-t\left(K_0 \dot{+} V + \beta 1_{\Gamma}\right)\right)\right\|_{\infty,1}^{1/2} \\ &\quad \left(\int dz\left\{\mathbb{E}_z\left[\exp\left(-\int_0^{t/2}\left(V+\beta 1_{\Gamma}\right)(X(u))du\right) : S < t/2\right]\right\}^2\right)^{1/2}; \end{aligned}$$

(c) *For the Hilbert-Schmidt norm of $D_{\Sigma,\beta}(t)$ there is also an inequality in the other direction which gives a necessary condition on the set Γ provided $D_{\Sigma,\beta}(t)$ is a Hilbert-Schmidt operator. Assume $V \le 0$ so that $V = -V_-$. Then*

$$\left\|D_{\Sigma,\beta}(t)\right\|_{\mathrm{HS}}^2 \ge \exp(-2\beta t)\int_{\Gamma} p_0(2t,x,x)dx;$$

(d) *The trace norm of $D_{\Sigma,\beta}(t)$ satisfies the following inequality:*

$$\begin{aligned}
&\left\|D_{\Sigma,\beta}(t)\right\|_{\text{trace}} \qquad (7.11)\\
&\leq 2\left\|\exp\left(-t\left(K_0\dot{+}V+\beta 1_\Gamma\right)\right)\right\|_{\infty,1}^{1/2}\left\|\exp\left(-\frac{t}{2}\left(K_0\dot{+}V+\beta 1_\Gamma\right)\right)\right\|_{\infty,1}^{1/2}\\
&\int dz\left\{\mathbb{E}_z\left[\exp\left(-\int_0^{t/2}\left(V+\beta 1_\Gamma\right)(X(u))du\right): S<t/2\right]\right\}^{1/2}.
\end{aligned}$$

Before we prove all this we insert some remarks.

REMARKS. In order to simplify the conditions in the previous proposition, we assume that the semigroup $(\exp(-tK_0) : t \geq 0)$ is L^1-L^∞-smoothing with spectral dimension m (see Definition 1.1). Then from Theorem 2.9 we get (M, b, c, and C are positive constants)

$$\begin{aligned}
&\left\|\exp\left(-t\left(K_0\dot{+}V+\beta 1_\Gamma\right)\right)\right\|_{\infty,1} \leq \left\|\exp\left(-t\left(K_0\dot{+}V\right)\right)\right\|_{\infty,1}\\
&\leq M\exp(bt)\sup_{x,y} p_0(t/2,x,y) \leq M\exp(bt)\left\|\exp\left(-\frac{t}{2}K_0\right)\right\|_{\infty,1}\\
&\leq C\exp(ct)t^{-m/2}.
\end{aligned}$$

Again for the sake of simplicity assume for the moment that the function V is bounded from below, i.e. $V(x) \geq -V_0$ with (possibly) $V_0 \geq 0$. Denote by

$$T_{\lambda,\Gamma} := \text{meas}\left\{0 \leq s \leq \lambda,\ X(s) \in \Gamma\right\} = \int_0^\lambda 1_\Gamma(X(s))ds$$

the *spending time* of the trajectory in the singularity region Γ. Then the inequalities in Proposition 7.2 reduce to the following ones:

$$\begin{aligned}
&\left\|D_{\Sigma,\beta}(t)\right\|_{2,2} \qquad (7.12)\\
&\leq \exp\left(tV_0\right)\sup_{x\in E}\mathbb{E}_x\left[\exp\left(-\beta T_{t,\Gamma}\right): T_{t,\Gamma}>0\right];\\
&\left\|D_{\Sigma,\beta}(t)\right\|_{\text{HS}} \qquad (7.13)\\
&\leq 2\sqrt{C}\exp\left(\frac{t}{2}(c+V_0)\right)t^{-m/4}\left(\int dz\left\{\mathbb{E}_z\left[\exp\left(-\beta T_{t/2,\Gamma}\right): T_{t/2,\Gamma}>0\right]\right\}^2\right)^{1/2};\\
&\left\|D_{\Sigma,\beta}(t)\right\|_{\text{trace}} \qquad (7.14)\\
&\leq 2c\exp\left(\frac{t}{4}\left(3c+V_0\right)\right)t^{-m/2}\int dz\left\{\mathbb{E}_z\left[\exp\left(-\beta T_{t/2,\Gamma}\right): T_{t/2,\Gamma}>0\right]\right\}^{1/2}.
\end{aligned}$$

From these inequalities, we easily see that in order to get similar estimates for resolvent differences we are going to need restrictions on the dimension m, if we want to have finiteness of the expressions on the right-hand side. This is caused by the singularity for $t = 0$.

First we notice that the spending time $T_{t,\Gamma}$ is given by $T_{t,\Gamma} = \int_0^t 1_\Gamma(X(s))ds$. Hence the equality $\{S < t\} = \{T_{t,\Gamma} > 0\}$ holds $\mathbb{P}_x$-almost surely. For the Hilbert-Schmidt norm in (7.10), one has a condition of the form

$$\int \left[\mathbb{P}_z \{S < t/2\}\right]^2 dz < \infty,$$

which is in agreement with assertion (i′) of Theorem 5.12 where we assumed the integral $\int h_\Sigma^{a+V}(x)^2 dx$ to be finite. The latter integral is finite if Γ is a set of finite capacity: for this see the remark after Theorem 5.12. For the trace norm in (7.11), the convergence condition is basically that

$$\int \left[\mathbb{P}_z \{S < t/2\}\right]^{1/2} dz < \infty,$$

which is compatible with assertion (ii′) of Theorem 6.11. There we assumed the integral $\int h_\Sigma^{a+V}(x)^{1/2} dx$ to be finite. The right-hand side members in (7.9) and (7.12) tend to zero provided the supremum

$$\sup_{x\in E} \mathbb{E}_x \left[\exp\left(-\beta T_{t,\Gamma}\right) : T_{t,\Gamma} > 0\right]$$

tends to zero as β tends to infinity. In general this kind of condition allows more general sets Γ than Hilbert-Schmidt convergence. For instance if $K_0 = -\frac{1}{2}\Delta$, then Γ can be any unbounded set with some regularity condition on the boundary (see Theorem 8.31 in Chapter 8). In the general context the following result can be proved. Suppose that the following conditions are met by Γ:

$$\inf_{\delta>0} \sup_{y\in\Gamma} \mathbb{P}_y \left\{\int_0^\eta 1_\Gamma(X(u))du \le \delta\right\} = 0, \text{ for all } \eta > 0 \text{ and}$$

$$\inf_{\delta>0} \sup_{x\in\Sigma} \left[\exp\left(-\frac{t}{2}\left(K_0\right)_\Sigma\right) \mathbb{P}_{(\cdot)} \{S < \delta\}\right](x) = 0.$$

Then

$$\lim_{\beta\to\infty} \left\|\exp\left(-tK_\beta\right) - J^* \exp\left(-tK_\Sigma\right) J\right\|_{2,2} = 0.$$

A proof of this assertion reads as follows. We begin with a trivial reduction:

$$\left\|\exp\left(-tK_\beta\right) - J^* \exp\left(-tK_\Sigma\right) J\right\|_{2,2} \le \left\|\exp\left(-tK_\beta\right) - J^* \exp\left(-tK_\Sigma\right) J\right\|_{\infty,\infty}$$

$$= \sup_{x\in E} \int D_{\Sigma,\beta}(t,x,y)dy$$

$$= \sup_x \mathbb{E}_x \left[\exp\left(-\int_0^t \left(V + \beta 1_\Gamma\right)(X(u))du\right) : S < t\right]$$

$$\leq \sup_x \left(\mathbb{E}_x\left[\exp\left(-2\int_0^t V(X(u))du\right) : S < t\right]\right)^{1/2}$$
$$\sup_x \left(\mathbb{E}_x\left[\exp\left(-2\beta\int_0^t 1_\Gamma(X(u))du\right) : S < t\right]\right)^{1/2}.$$

So it suffices to show that the following limit vanishes:

$$\lim_{\beta\to\infty}\sup_x \mathbb{E}_x\left[\exp\left(-\beta\int_0^t 1_\Gamma(X(u))du\right) : S < t\right].$$

Therefore we write, for $\delta > 0$ fixed,

$$\mathbb{E}_x\left[\exp\left(-\beta\int_0^t 1_\Gamma(X(u))du\right) : S < t\right]$$
$$= \mathbb{E}_x\left[\exp\left(-\beta\int_0^t 1_\Gamma(X(u))du\right) : S < t,\ \int_0^t 1_\Gamma(X(u))du \geq \delta\right]$$
$$+ \mathbb{E}_x\left[\exp\left(-\beta\int_0^t 1_\Gamma(X(u))du\right) : S < t,\ \int_0^t 1_\Gamma(X(u))du < \delta\right]$$
$$\leq e^{-\beta\delta} + \mathbb{P}_x\left[\int_0^t 1_\Gamma(X(u))du < \delta,\ S < t\right].$$

The choice of $\delta = M/\beta$, with M fixed but large, makes it clear that in order to obtain our result it suffices to prove that

$$\inf_{\delta>0}\sup_x \mathbb{P}_x\left[\int_0^t 1_\Gamma(X(u))du < \delta,\ S < t\right] = 0.$$

For $\xi \geq 0$, let the symbol T_ξ denote the *stochastic time change*

$$T_\xi = \inf\left\{s > 0 : \int_0^s 1_\Gamma(X(u))du > \xi\right\}.$$

Then $T_0 = S$, the penetration time of Γ, $T_\xi + T_\eta \circ \vartheta_{T_\xi} = T_{\xi+\eta}$, and $\{\eta \leq T_\delta\} = \left\{\int_0^\delta 1_\Gamma(X(u))du \leq \eta\right\}$. The (in-)equalities that follow will show that the equality

$$\inf_{\delta>0}\sup_x \mathbb{P}_x\left[\int_0^t 1_\Gamma(X(u))du < \delta,\ S < t\right] = 0$$

is a consequence of our hypotheses:

$$\mathbb{P}_x\left\{\int_0^t 1_\Gamma(X(u))du \leq \delta,\ S < t\right\} = \mathbb{P}_x\left\{t \leq T_\delta,\ S < t\right\}$$
$$= \mathbb{P}_x\left\{t \leq T_0 + T_\delta \circ \vartheta_{T_0},\ S < t\right\} = \mathbb{P}_x\left\{t \leq S + T_\delta \circ \vartheta_S,\ S < t\right\}$$

(strong time dependent Markov property)

$$
\begin{aligned}
&= \mathbb{E}_x \left\{ \mathbb{P}_{X(S)} \left[t - S \le T_\delta \right],\ S < t \right\} \\
&= \mathbb{E}_x \left\{ \mathbb{P}_{X(S)} \left[t - S \le T_\delta \right],\ S \le t - \eta \right\} \\
&\quad + \mathbb{E}_x \left\{ \mathbb{P}_{X(S)} \left[t - S \le T_\delta \right],\ t - \eta \le S < t \right\} \\
&\le \mathbb{P}_x \left[t - \eta < S < t \right] + \sup_{y \in \Gamma} \mathbb{P}_y \left[\eta \le T_\delta \right] \\
&= \left[\exp\left(- \left(\frac{t}{2} - \eta \right) (K_0)_\Sigma \right) \exp\left(- \frac{t}{2} (K_0)_\Sigma \right) \mathbb{P}_{(\cdot)} \left[S \le \eta \right] \right](x) \\
&\quad + \sup_{y \in \Gamma} \mathbb{P}_y \left[\int_0^\eta 1_\Gamma(X(u)) du \le \delta \right].
\end{aligned}
$$

Altogether we see that the convergence depends to a great extent on the distribution of the spending or occupation time. More precisely for semigroups it depends on the expected value of $\exp\left(-\beta T_{t,\Gamma}\right)$ and for resolvent convergence it depends on its Laplace transform. In Chapter 8 we use the weighted Laplace transform of the occupation time to investigate several spectral theoretical consequences for convergence by large coupling parameters.

Proof of Proposition 7.2. (a) The inequality in (a) directly follows from the symmetry of the integral kernel $D_{\Sigma,\beta}(t,x,y)$ of the operator $D_{\Sigma,\beta}(t)$ which has the representation

$$
D_{\Sigma,\beta}(t,x,y) = \int_{\{S<t\}} \exp\left(- \int_0^t \left(V + \beta 1_\Gamma\right)(X(u)) du \right) d\mu_{0.x}^{t,y} :
$$

see Proposition 4.16 and Lemma 5.18. Its symmetry is proved in Appendix D. For integral operators T with symmetric integral kernels $t(x,y)$, the operator norm can be estimated as follows:

$$
\|T\|_{2,2} \le \sup_{x \in E} \int_E |t(x,y)|\, dy = \|T\|_{\infty,\infty} .
$$

(b) In order to estimate the Hilbert-Schmidt norm of $D_{\Sigma,\beta}(t)$ we employ the following inequality for its integral kernel (see Proposition 4.18 in Chapter 4 formula (4.73)):

$$
\begin{aligned}
&D_{\Sigma,\beta}(t,x,y) \\
&\le \mathbb{E}_x \left[M_{V+\beta 1_\Gamma}^{t,y} : S < t/2 \right] + \mathbb{E}_y \left[M_{V+\beta 1_\Gamma}^{t,x} : S < t/2 \right],
\end{aligned} \tag{7.15}
$$

where, as above, $M_{V+\beta 1_\Gamma}^{t,y}(s)$, $0 \le s < t$, is the martingale

$$
\begin{aligned}
&M_{V+\beta 1_\Gamma}^{t,y}(s) \\
&= \exp\left(- \int_0^s \left(V + \beta 1_\Gamma\right)(X(u)) du \right) \exp\left(-(t-s) \left(K_0 \dot{+} V + \beta 1_\Gamma \right) \right) (X(s), y).
\end{aligned}
$$

(The inequality in (7.15) with $V = 0$ and $\beta = 0$ has been used in the proof of Proposition 3.2.) From inequality (7.15) we see that because the roles of x and y are interchangeable,

$$\|D_{\Sigma,\beta}(t)\|_{\mathrm{HS}}^2 \leq 4 \int\int \left(\mathbb{E}_x\left[M_{V+\beta 1_\Gamma}^{t/2,y}(t/2) : S < t/2\right]\right)^2 dx\, dy. \tag{7.16}$$

Let the Markov process

$$\left\{(\Omega', \mathcal{F}', \mathbb{P}'_x), (X'(t) : t \geq 0), (\vartheta_t : t \geq 0), (E, \mathcal{E})\right\}$$

be a copy of the original Markov process. Then from symmetry, Fubini, the Chapman-Kolmogorov identity, and Doob's optional sampling (Appendix C.11) we obtain

$$\begin{aligned}
&\int dy \left(\mathbb{E}_x\left[M_{V+\beta 1_\Gamma}^{t/2,y}(t/2) : S < t/2\right]\right)^2 \qquad\qquad (7.17)\\
&= \mathbb{E}_x \otimes \mathbb{E}'_x\left[\exp\left(-\int_0^{t/2} (V(X(u)) + \beta 1_\Gamma(X(u)))\, du\right)\right.\\
&\qquad \times \exp\left(-\int_0^{t/2} (V(X'(u))du + \beta 1_\Gamma(X'(u)))\right)\\
&\qquad \left.\times \exp\left(-t\left(K_0 \dot{+} V + \beta 1_\Gamma\right)\right)(X(t/2), X'(t/2)) : S < t/2, \quad S' < t/2\right]\\
&\leq \left\|\exp\left(-t\left(K_0 \dot{+} V + \beta 1_\Gamma\right)\right)\right\|_{\infty,1}\\
&\qquad \times \left(\mathbb{E}_x\left[\exp\left(-\int_0^{t/2} (V + \beta 1_\Gamma)(X(u))du\right) : S < t/2\right]\right)^2.
\end{aligned} \tag{7.18}$$

A combination of the inequalities in (7.16) and (7.18) yields the result in (7.10) of assertion (b).

(c) A proof of inequality (c) reads as follows (see (7.7) and Corollary 4.17):

$$\begin{aligned}
&\left\|\exp(-\lambda K_\beta) - J^* \exp(-tK_\Sigma) J\right\|_{\mathrm{HS}}^2\\
&= \int dx \int dy \left(\mathbb{E}_x\left[\exp\left(-\int_0^S (V + \beta 1_\Gamma)(X(u))du\right)\right.\right.\\
&\qquad\qquad \left.\left.\exp\left(-(t-S)K_\beta\right)(X(S), y) : S < t\right]\right)^2\\
&\geq \int_\Gamma dx \int dy \left(\mathbb{E}_x\left[\exp\left(-\int_0^S (V + \beta 1_\Gamma)(X(u))du\right)\right.\right.\\
&\qquad\qquad \left.\left.\exp\left(-(t-S)K_\beta\right)(X(S), y) : S < t\right]\right)^2
\end{aligned}$$

(on Γ, $S = 0$, $\mathbb{P}_x$-almost surely for m-almost all $x \in \Gamma$)

$$= \int_\Gamma dx \int dy \left(\mathbb{E}_x \left[\exp\left(-tK_\beta\right)(X(0), y)\right]\right)^2$$
$$= \int_\Gamma dx \int dy \left(\exp\left(-tK_\beta\right)(x, y)\right)^2$$
$$= \int_\Gamma dx \exp\left(-2tK_\beta\right)(x, x)$$

(the function V is non-positive)

$$\geq \int_\Gamma dx \exp\left(-2t\left(K_0 + \beta 1_\Gamma\right)\right)(x, x)$$
$$\geq \exp\left(-2\beta t\right) \int_\Gamma dx\; p_0\left(-2t, x, x\right).$$

(d) In order to prove the trace norm estimate, we write:

$$D_{\Sigma,\beta}(t) = D_{\Sigma,\beta}(t/2) J^* \exp\left(-\frac{t}{2}\left(K_0 \dot{+} V + \beta 1_\Gamma\right)_\Sigma\right) J + \exp\left(-\frac{t}{2}\left(K_0 \dot{+} V + \beta 1_\Gamma\right)\right) D_{\Sigma,\beta}(t/2).$$

So from Theorem 6.1 in Chapter 6 we obtain

$$\begin{aligned}
&\left\|D_{\Sigma,\beta}(t)\right\|_{\text{trace}} \\
&\leq \left\|D_{\Sigma,\beta}(t/2) J^* \exp\left(-\frac{t}{2}\left(K_0 \dot{+} V + \beta 1_\Gamma\right)_\Sigma\right) J\right\|_{\text{trace}} \\
&\quad + \left\|\exp\left(-\frac{t}{2}\left(K_0 \dot{+} V + \beta 1_\Gamma\right)\right) D_{\Sigma,\beta}(t/2)\right\|_{\text{trace}} \\
&\leq \int\left(\int \left(D_{\Sigma,\beta}(t/2, x, z)\right)^2 dx \int \left(\exp\left(-\frac{t}{2}\left(K_0 + V + \beta 1_\Gamma\right)_\Sigma\right)(z, x)\right)^2 dx\right)^{1/2} dz \\
&\quad + \int\left(\int \left(\exp\left(-\frac{t}{2}\left(K_0 + V + \beta 1_\Gamma\right)\right)(x, z)\right)^2 dx \int \left(D_{\Sigma,\beta}(t/2, z, x)\right)^2 dx\right)^{1/2} dz \\
&\leq 2\int\left(\int \left(D_{\Sigma,\beta}(t/2, x, z)\right)^2 dx \int \left(\exp\left(-\frac{t}{2}\left(K_0 + V + \beta 1_\Gamma\right)\right)(z, x)\right)^2 dx\right)^{1/2} dz \\
&= 2\int\left(\int \left(D_{\Sigma,\beta}(t/2, x, z)\right)^2 dx \exp\left(-t\left(K_0 + V + \beta 1_\Gamma\right)\right)(z, z)\right)^{1/2} dz \\
&\leq 2\int\left(\int \left(D_{\Sigma,\beta}(t/2, x, z)\right)^2 dx\right)^{1/2} dz \left\|\exp\left(-t\left(K_0 \dot{+} V + \beta 1_\Gamma\right)\right)\right\|_{\infty,1}^{1/2}.
\end{aligned} \tag{7.19}$$

Notice that, on the event $\{S < \infty\}$, the state variable $X(S)$ attains its values $\mathbb{P}_x$-almost surely in the subset Γ^ρ consisting of the S-regular points in Γ. Moreover, the set $\Gamma \setminus \Gamma^\rho$ might not be a polar set, but its m-measure is zero. These results are spelled out in Port and Stone [189] in Theorem 6.3, p. 44, and Proposition 3.7, p. 32, respectively, for Brownian motion and hitting times. They can be extended for symmetric Markov processes: see Dellacherie and Meyer [72], Theorem D.14 and (in particular for S) Remark 2 following Theorem D.16 in Appendix D. In the following argument, the symmetry of the kernel of $D_{\Sigma,\beta}(t)$ plays a crucial role. Next we estimate the integral

$$\begin{aligned}
&\int \left(D_{\Sigma,\beta}(t/2,x,z)\right)^2 dx \\
&= \int dx \left(\mathbb{E}_z\left[M^{t/2,x}_{V+\beta 1_\Gamma}(S) : S < t/2\right]\right)^2 \\
&= \mathbb{E}_z \otimes \mathbb{E}'_z\Bigg[\exp\left(-\int_0^S (V+\beta 1_\Gamma)(X(u))du\right) M^{t-S,X(S)}_{V+\beta 1_\Gamma}(S') : \\
&\qquad\qquad S < t/2,\ S' < t/2\Bigg] \\
&\le \mathbb{E}_z\left\{\exp\left(-\int_0^S (V+\beta 1_\Gamma)(X(u))du\right)\mathbb{E}'_z\left[M^{t-S,X(S)}_{V+\beta 1_\Gamma}(S') : S' < t-S\right] :\right. \\
&\qquad\qquad\qquad\qquad \left. S < t/2\right\}
\end{aligned}$$

(the function $(z,w) \mapsto \mathbb{E}'_z\left[M^{t-S,w}_{V+\beta 1_\Gamma}(S') : S' < t-S\right]$ is symmetric)

$$\begin{aligned}
&= \mathbb{E}_z\left\{\exp\left(-\int_0^S (V+\beta 1_\Gamma)(X(u))du\right)\mathbb{E}'_{X(S)}\left[M^{t-S,z}_{V+\beta 1_\Gamma}(S') : S' < t-S\right] :\right. \\
&\qquad\qquad\qquad\qquad \left. S < t/2\right\} \\
&= \mathbb{E}_z\left\{\exp\left(-\int_0^S (V+\beta 1_\Gamma)(X(u))du\right)\mathbb{E}'_{X(S)}\left[M^{t-S,z}_{V+\beta 1_\Gamma}(S') : S' < t-S\right] :\right. \\
&\qquad\qquad\qquad\qquad \left. S < t/2, \quad X(S) \in \Gamma^\rho\right\}
\end{aligned}$$

(for an S-regular point w of Γ the time S' is $\mathbb{P}_w$-almost surely zero)

$$\begin{aligned}
&= \mathbb{E}_z\left\{\exp\left(-\int_0^S (V+\beta 1_\Gamma)(X(u))du\right)\mathbb{E}'_{X(S)}\left[M^{t-S,z}_{V+\beta 1_\Gamma}(0)\right] :\right. \\
&\qquad\qquad\qquad\qquad \left. S < t/2, \quad X(S) \in \Gamma^r\right\}
\end{aligned}$$

($X'(0) = w$, $\mathbb{P}_w$-almost surely)

$$\leq \mathbb{E}_z\left\{ \exp\left(-\int_0^S (V+\beta 1_\Gamma)(X(u))du \right) \right.$$

$$\left. \times \exp\left(-(t-S)\left(K_0+V+\beta 1_\Gamma\right)\right)(X(S),z) : S < t/2 \right\}$$

(martingale property and Doob's theorem on optional sampling: Theorem C.11)

$$= \mathbb{E}_z\left\{ \exp\left(-\int_0^{t/2} (V+\beta 1_\Gamma)(X(u))du \right) \right.$$

$$\left. \times \exp\left(-\left(\frac{t}{2}\right)\left(K_0+V+\beta 1_\Gamma\right)\right)(X(t/2),z)1_{\{S<t/2\}} \right\}. \tag{7.20}$$

From the inequality in (7.20), we conclude

$$\int \left(D_{\Sigma,\beta}(t/2,x,z)\right)^2 dx$$

$$\leq \left\| \exp\left(-\frac{t}{2}\left(K_0 \dot{+} V + \beta 1_\Gamma\right)\right)\right\|_{\infty,1}$$

$$\times \mathbb{E}_z\left\{ \exp\left(-\int_0^{t/2} (V+\beta 1_\Gamma)(X(u))du \right) : S < t/2 \right\}. \tag{7.21}$$

A combination of (7.19) and (7.21) results in the equality:

$$\left\|D_{\Sigma,\beta}(t)\right\|_{\text{trace}}$$

$$\leq 2\left\|\exp\left(-t\left(K_0 \dot{+} V+\beta 1_\Gamma\right)\right)\right\|_{\infty,1}^{1/2} \left\|\exp\left(-\frac{t}{2}\left(K_0 \dot{+} V+\beta 1_\Gamma\right)\right)\right\|_{\infty,1}^{1/2}$$

$$\times \int \left(\mathbb{E}_z\left[\exp\left(-\int_0^{t/2} (V+\beta 1_\Gamma)(X(u))du \right) : S < t/2 \right] \right)^{1/2} dz.$$

This proves Proposition 7.2.

From Proposition 7.2 we get a first and direct consequence for resolvents using Laplace transforms. In Proposition 7.4 we will relax these conditions in the Hilbert-Schmidt case by avoiding too rough estimates which may happen when using Laplace transforms.

7.3. COROLLARY. *Suppose that there exists an integer $q \geq 1$ such that for some constant $a_0 > 0$ the expression*

$$\int_0^\infty d\lambda \lambda^{2q-6/5} \exp(-a_0\lambda) \left\|\exp\left(-\lambda\left(K_0 \dot{+} V\right)\right)\right\|_{\infty,1}$$

$$\int dz \left[\mathbb{E}_z\left\{ \exp\left(-\int_0^{t/2} V(X(u))du \right) : S < \lambda/2 \right\}\right]^2 \tag{7.22}$$

is finite. Then the following assertions hold true:

(i) *For all* $\beta \geq 0$ *and for all* $z \in \mathbb{C}$, *with* $-z \in \operatorname{res}(K_\beta) \cap \operatorname{res}(K_\Sigma)$, *the resolvent difference* $R_\beta(z) - J^* R_\Sigma(z) J$ *is a compact operator.*

(ii) *For all* $z \in \mathbb{C}$ *with* $-z \in \operatorname{res}(K_\Sigma)$, *the equality*

$$\lim_{\beta\to\infty} \|R_\beta(z) - J^* R_\Sigma(z) J\|_{2,2} = 0$$

is valid. Moreover, the rate of convergence is uniform with respect to $-z$ *in compact subsets of* $\operatorname{res}(K_\Sigma)$.

(iii) *The resolvent power differences* $R_\beta(z_0)^q - J^* R_\Sigma(z_0)^q J$ *are Hilbert-Schmidt operators for* z_0 *in the half-plane* $\{\operatorname{Re} z_0 \geq a_0\}$ *and the identity*

$$\lim_{\beta\to\infty} \|R_\beta(z_0)^q - J^* R_\Sigma(z_0)^q J\|_{\mathrm{HS}} = 0$$

is valid.

(iv) *In order to specialize and simplify the assertion in (iii), assume additionally that the semigroup* $(\exp(-tK_0) : t \geq 0)$ *is* L^1-L^∞*-smoothing with spectral dimension* m. *Moreover, assume that, for some constants* C *and* c, *the inequality*

$$\int \mathbb{P}_z \{S < \lambda\} \, dz \leq C \exp(c\lambda)$$

is satisfied for all $\lambda > 0$. *Then*

$$\lim_{\beta\to\infty} \|R_\beta(z_0)^q - J^* R_\Sigma(z_0)^q J\|_{\mathrm{HS}} = 0$$

is valid for $q > \dfrac{m}{4} + \dfrac{1}{10}$ *(e.g.* q *can be chosen to be equal to one for* $m \leq 3$*) and for* z_0 *with* $\operatorname{Re} z_0$ *large enough. For* $m \leq 3$ *the latter assertion can be extended to all* $z \in \mathbb{C}$ *with* $-z \in \operatorname{res}(K_\Sigma)$:

$$\lim_{\beta\to\infty} \|R_\beta(z) - J^* R_\Sigma(z) J\|_{\mathrm{HS}} = 0, \quad -z \in \operatorname{res}(K_\Sigma).$$

(v) *Again the semigroup* $(\exp(-tK_0) : t \geq 0)$ *is supposed to be* L^1-L^∞*-smoothing with spectral dimension* m. *Moreover, assume that, for some* $\epsilon > 0$,

$$\int (\mathbb{P}_z \{S < \lambda\})^{\frac{1}{2}-\epsilon} \, dz \leq C \exp(c\lambda), \quad \lambda > 0.$$

Then, for $q > m/2$ *and for* z_0 *with* $\operatorname{Re} z_0$ *large enough, the following identity is valid*

$$\lim_{\beta\to\infty} \|R_\beta(z_0)^q - J^* R_\Sigma(z_0)^q J\|_{\mathrm{trace}} = 0.$$

This result can be extended to

$$\lim_{\beta\to\infty} \|R_\beta(z) - J^* R_\Sigma(z) J\|_{\mathrm{trace}} = 0$$

for all $z \in \mathbb{C}$, *with* $z \in \operatorname{res}(-K_\Sigma)$ *provided* $m = 1$.

Proof. We begin with a proof of (iii). The assertion in (iii) is an obvious consequence of Proposition 7.2.(b) because we assume that the left-hand side in (7.10) possesses a finite Laplace transform. The convergence in (iii) for z_0 with $\operatorname{Re} z_0 \geq a_0$ follows by Lebesgue's dominated convergence theorem because the expression

$$\mathbb{E}_z \left[\exp \left(- \int_0^{t/2} (V + \beta 1_\Gamma)(X(u)) du \right) : S < t/2 \right]$$

tends to 0 (as β tends to infinity) for all $z \in E$: see Lemma 7.1.

We show now that from (iii) the assertions (i) and (ii) follow. The equality

$$\lim_{\beta \to \infty} \|R_\beta(z_0)^q - J^* R_\Sigma(z_0)^q J\|_{\mathrm{HS}} = 0$$

implies the convergence of $R_\beta(z_0)^q$ to $J^* R_\Sigma(z_0)^q J$ for the operator norm topology. In order to reduce the power of the resolvents, we may use the representation

$$\begin{aligned} &R_\beta(z_0) - J^* R_\Sigma(z_0) J \\ &= (q-1) \int_0^\infty t^{q-2} \left(R_\beta(z_0 + t)^q - J^* R_\Sigma(z_0 + t)^q J \right) dt, \end{aligned} \tag{7.23}$$

where the integral converges with respect to the operator norm topology. On account of the inequality

$$\|R_\beta(z_0 + t)^q - J^* R_\Sigma(z_0 + t)^q J\|_{2,2} \leq C_1 \left(\operatorname{Re} z_0 + t \right)^{-q}, \tag{7.24}$$

we get the compactness of the resolvent difference $R_\beta(z_0) - J^* R_\Sigma(z_0) J$ and

$$\lim_{\beta \to \infty} \|R_\beta(z_0) - J^* R_\Sigma(z_0) J\|_{2,2} = 0.$$

Via the first resolvent equation this result can be extended to all $z \in \operatorname{res}(K_\Sigma)$. In fact, let d be the distance between $-z_0$ and the spectrum of $K_0 \dot{+} V$. For z in the disc $\{z \in \mathbb{C} : |z - z_0| < d\}$, the Neumann series gives rise to

$$\begin{aligned} &\|R_\beta(z + t) - J^* R_\Sigma(z + t) J\|_{2,2} \\ &\leq \|R_\beta(z_0 + t) - J^* R_\Sigma(z_0 + t) J\|_{2,2} \sum_{k=0}^\infty (k+1) \frac{|z - z_0|^k}{d^k}, \end{aligned} \tag{7.25}$$

which proves (i) and (ii).

In order to prove (iv), we are going to estimate the Hilbert-Schmidt norm:

$$\begin{aligned} &\|R_\beta(z_0)^q - J^* R_\Sigma(z_0)^q J\|_{\mathrm{HS}}^2 \\ &\leq C_2 \int_0^\infty \exp(-a_0 \lambda) \lambda^{2q - 6/5} \|D_{\Sigma,\beta}(\lambda)\|_{\mathrm{HS}}^2 \, d\lambda \end{aligned}$$

$$
\begin{aligned}
&\leq 4C_2 \int_0^\infty \exp\left(-a_0\lambda\right) \lambda^{2q-6/5} \left\|\exp\left(-\lambda\left(K_0\dot{+}V\right)\right)\right\|_{\infty,1} \\
&\quad \times \int \left(\mathbb{E}_z\left[\exp\left(-\int_0^{\lambda/2} \left(V+\beta 1_\Gamma\right)(X(u))du\right) : S < \frac{\lambda}{2}\right]\right)^2 dz\, d\lambda \\
&\leq C_3 \int_0^\infty \exp\left(-a_0\lambda\right) \exp(a_V\lambda)\lambda^{2q-6/5-m/2} \\
&\quad \times \int \mathbb{E}_z\left[\exp\left(-2\int_0^{\lambda/2} V(X(u))du\right)\right] \mathbb{P}_z\left[S < \frac{\lambda}{2}\right] dz\, d\lambda \\
&\leq C_4 \int_0^\infty \exp\left(-\left(a_0 - c_1\right)\lambda\right) \lambda^{2q-6/5-m/2} \int \mathbb{P}_z\left[S < \frac{\lambda}{2}\right] dz\, d\lambda \\
&\leq C_5 \int_0^\infty \exp\left(-\left(a_0 - c_2\right)\lambda\right) \lambda^{2q-6/5-m/2} d\lambda. \qquad (7.26)
\end{aligned}
$$

Here we may choose $C_4 = M_{2V}C_3$ and $c_1 = a_V + \frac{1}{2}b_{2V}$. For the meaning of the constants a_V, b_V and M_V we refer to Chapter 2 formulae (2.7) through (2.10). The expression in (7.26) is finite for $q > \dfrac{m}{4} + \dfrac{1}{10}$. So the convergence of the operator $R_\beta(z_0)^q$ to the operator $R_\Sigma(z_0)^q$ with respect to the Hilbert-Schmidt norm follows. In order to reduce the resolvent powers, we follow the line of proof for the operator norm above. For certain (small) spectral dimensions m, the representation in (7.23) is also true in Hilbert-Schmidt norm sense. This is so because our assumptions are strong enough to ensure that the mapping $t \mapsto R_\beta(z_0+t)^q - J^*R_\Sigma(z_0+t)^qJ$ is continuous with respect to the Hilbert-Schmidt norm. More precisely, we have

$$
\begin{aligned}
&R_\beta(z_0+t)^q - J^*R_\Sigma(z_0+t)^qJ - R_\beta(z_0+t_0)^q + J^*R_\Sigma(z_0+t_0)^qJ \\
&= -q\int_{t_0}^t \left(R_\beta(z_0+s)^{q+1} - J^*R_\Sigma(z_0+s)^{q+1}J\right) ds,
\end{aligned}
$$

for t and t_0 close enough to each other, and so continuity follows. (In fact we have analyticity, using Cauchy's theorem.) It follows that this representation also makes sense in the Hilbert-Schmidt norm sense. It remains to be shown that the integral

$$
\int_0^\infty dt\; t^{q-2} \left\|R_\beta(z_0+t)^q - J^*R_\Sigma(z_0+t)^qJ\right\|_{\mathrm{HS}} \qquad (7.27)
$$

is finite. However from (7.26) we obtain

$$
\left\|R_\beta(z_0+t)^q - J^*R_\Sigma(z_0+t)^qJ\right\|_{\mathrm{HS}} \leq C_6 \left(\frac{1}{a_0 - c_2 + t}\right)^{q - \frac{1}{10} - \frac{m}{4}},
$$

where $\operatorname{Re} z_0 \geq a_0$, $a_0 > c_2$. Therefore the integral in (7.27) is finite provided m satisfies the inequality $-2 + 1/10 + m/4 < -1$ or $m < 3.6$. This proves (iv). The proof of (v) follows the same pattern as that of (iv). From Lemma 6.6 we see, with $a_0 = \operatorname{Re} z_0$,

$$\begin{aligned}
&\|R_\beta(z_0)^q - J^* R_\Sigma(z_0)^q J\|_{\text{trace}} \\
&\le C_7 \int_0^\infty \lambda^{q-1} \exp(-a_0\lambda) \|D_{\Sigma,\beta}(\lambda)\|_{\text{trace}}\, d\lambda \\
&\le C_8 \int_0^\infty \lambda^{q-1} \exp(-a_0\lambda) \exp(c_4\lambda) \\
&\quad \int \left[\mathbb{P}_z\left\{S < \frac{\lambda}{2}\right\}\right]^{\frac{1}{2}(1-\epsilon)} \left[\mathbb{E}_z\left\{\exp\left(-\frac{1}{\epsilon}\int_0^{\lambda/2} V(X(u))du\right)\right\}\right]^{\frac{1}{2}\epsilon} dz\, d\lambda \\
&\le C_9 \int_0^\infty \lambda^{q-1}\lambda^{-\frac{1}{2}m} \exp\left(-\left(a_0 - c_5\right)\lambda\right) d\lambda \\
&= C_9 \left(\frac{1}{a_0 - c_5}\right)^{q-\frac{1}{2}m} \int_0^\infty \sigma^{q-1-\frac{1}{2}m} \exp(-\sigma) d\sigma,
\end{aligned}$$

which is finite for $q > m/2$ and $a_0 > c_5$.

The integral in (7.23) converges in trace norm sense provided the integral

$$\int_0^\infty dt\, t^{q-2} \left(\frac{1}{a_0 + c_5 - t}\right)^{q-\frac{1}{2}m}$$

is finite. This is only the case for $m < 2$. If this happens to be the case, then

$$\lim_{\beta\to\infty} \|R_\beta(z_0) - J^* R_\Sigma(z_0) J\|_{\text{trace}} = 0,$$

for $-\operatorname{Re} z_0 = a_0$ large enough. This result can be extended for all $z \in \operatorname{res}(K_\Sigma)$ using again the Neumann series (7.25).

In the following proposition, we improve the Hilbert-Schmidt result in the previous corollary somewhat. Recall that $-\omega_V$ is the lower bound of the spectrum of $K_0 \dot{+} V$: see formula (2.8).

7.4. PROPOSITION. *Let K_β and K_Σ be as in Lemma 7.1 and define for $q > 0$ and $a > \omega_V$, the operators $R_\beta(a)^q := \left(aI + K_0 \dot{+} V + \beta 1_\Gamma\right)^{-q}$ and $R_\Sigma(a)^q := \left(aI + K_0 \dot{+} V\right)_\Sigma^{-q}$ by*

$$\begin{aligned}
R_\beta(a)^q &= \frac{1}{\Gamma(q)} \int_0^\infty t^{q-1} \exp(-at) \exp\left(-t(K_0 \dot{+} V + \beta 1_\Gamma)\right) dt \quad \textit{and} \\
R_\Sigma(a)^q &= \frac{1}{\Gamma(q)} \int_0^\infty t^{q-1} \exp(-at) \exp\left(-t(K_0 \dot{+} V)_\Sigma\right) dt
\end{aligned}$$

respectively. Assume that the semigroup $(\exp(-tK_0) : t \ge 0)$ is L^1-L^2-smoothing with spectral dimension m.

(a) *Suppose $q > m/4$. The Hilbert-Schmidt norm of the difference $R_\beta(a)^q - J^* R_\Sigma(a)^q J$ is dominated by*

$$2^{q+1}\frac{\Gamma(q-m/4)}{\Gamma(q)} M_{2V}^{1/2} c_0^{1/2} \left\| \left[R_\beta(a_1)^{q-m/4} - J^* R_\Sigma(a_1)^{q-m/4} \right] 1 \right\|_2,$$

where $a_1 = 2a - \frac{1}{2}b_{2V} - \frac{1}{2}a_0$, and where

$$\|\exp\left(-t\left(K_0 \dot{+} 2V\right)\right)\|_{\infty,\infty} \le M_{2V} \exp\left(b_{2V} t\right);$$

$$\|\exp(-tK_0)\|_{\infty,1} \le c_0 t^{-m/2} \exp(t a_0).$$

(b) *Suppose $q > m/2$. The trace (not the trace norm) of the difference $R_\beta(a)^q - J^* R_\Sigma(a)^q J$ is dominated by*

$$2^{q+1+m/2}\frac{\Gamma(q-m/2)}{\Gamma(q)} M_{2V} c_0 \left\| \left[R_\beta(a_1)^{q-m/2} - J^* R_\Sigma(a_1)^{q-m/2} \right] 1 \right\|_1,$$

where a_1, c_0 and M_{2V} are as in (a).

REMARK. For $q = 1$ we have the ordinary resolvents. Notice that the operator $R(a) - J^* R_\Sigma(a) J$ is form positive because we have $R(a) - J^* R_\Sigma(a) J = \int_0^\infty ds R_s(a) 1_\Gamma R_s(a)$ in operator norm sense: see e.g. Proposition 7.5 below. The form positivity of the resolvent difference $R_\beta(a) - J^* R_\Sigma(a) J$ was proved in another way in assertion (c) of Corollary 2.32. As a consequence, it follows that the trace norm of $R_\beta(a) - J^* R_\Sigma(a) J$ coincides with its trace.

Proof. It suffices to prove the results for $\beta = 0$.

(a) Use the inequality

$$\begin{aligned} & D_\Sigma(t,x,y) \\ & \le \mathbb{E}_x \left[\exp\left(- \int_0^{t/2} V(X(u))du \right) \exp\left(-\frac{t}{2}\left(K_0 \dot{+} V\right)\right)(X(t/2),y) : S < \frac{t}{2} \right] \\ & \quad + \mathbb{E}_y \left[\exp\left(- \int_0^{t/2} V(X(u))du \right) \exp\left(-\frac{t}{2}\left(K_0 \dot{+} V\right)\right)(X(t/2),x) : S < \frac{t}{2} \right] \end{aligned}$$

to obtain

$$\begin{aligned} & 0 \le R(a)^q(x,y) - J^* R_\Sigma(a)^q J(x,y) \\ & \le \frac{1}{\Gamma(q)} \int_0^\infty dt\, t^{q-1} \exp(-at) \\ & \quad \left\{ \mathbb{E}_x \left[\exp\left(- \int_0^{t/2} V(X(u))du \right) \exp\left(-\frac{t}{2}\left(K_0 \dot{+} V\right)\right)(X(t/2),y) : S < \frac{t}{2} \right] \right. \\ & \quad \left. + \mathbb{E}_y \left[\exp\left(- \int_0^{t/2} V(X(u))du \right) \exp\left(-\frac{t}{2}\left(K_0 \dot{+} V\right)\right)(X(t/2),x) : S < \frac{t}{2} \right] \right\}. \end{aligned}$$

So that the Hilbert-Schmidt norm of $R(a)^q - J^* R_\Sigma(a)^q J$ is dominated by the square root of

$$\frac{4}{\Gamma(q)^2}\int dx \int dy \int_0^\infty dt_1 \int_0^\infty dt_2\,(t_1t_2)^{q-1}\exp\left(-\left(t_! + t_2\right)a\right)$$
$$\mathbb{E}_x\otimes\mathbb{E}'_x\Bigg[\exp\left(-\int_0^{t_1/2} V(X(u))du - \int_0^{t_2/2} V(X'(u))du\right)$$
$$\exp\left(-\frac{t_1}{2}\left(K_0\dot{+}V\right)\right)(X(t_1/2),y)\exp\left(-\frac{t_2}{2}\left(K_0\dot{+}V\right)\right)(X'(t_2/2),y):$$
$$S<\frac{t_1}{2}, S'<\frac{t_2}{2}\Bigg]$$
$$=\frac{4}{\Gamma(q)^2}\int dx \int_0^\infty dt_1 \int_0^\infty dt_2\,(t_1t_2)^{q-1}\exp\left(-\left(t_! + t_2\right)a\right)$$
$$\mathbb{E}_x\otimes\mathbb{E}'_x\Bigg[\exp\left(-\int_0^{t_1/2} V(X(u))du - \int_0^{t_2/2} V(X'(u))du\right)$$
$$\exp\left(-\frac{t_1+t_2}{2}\left(K_0\dot{+}V\right)\right)(X(t_1/2),X'(t_2/2)):S<\frac{t_1}{2}, S'<\frac{t_2}{2}\Bigg]$$
$$\leq\frac{4}{\Gamma(q)^2}\int dx \int_0^\infty dt_1 \int_0^\infty dt_2\,(t_1t_2)^{q-1}\exp\left(-\left(t_! + t_2\right)a\right)$$
$$\mathbb{E}_x\otimes\mathbb{E}'_x\left[\exp\left(-\int_0^{t_1/2} V(X(u))du - \int_0^{t_2/2} V(X'(u))du\right):S<\frac{t_1}{2},\ S'<\frac{t_2}{2}\right]$$
$$\left\|\exp\left(-\frac{t_1+t_2}{2}\left(K_0\dot{+}V\right)\right)\right\|_{\infty,1}$$
$$\leq\frac{4}{\Gamma(q)^2}\int dx \int_0^\infty dt_1 \int_0^\infty dt_2\,(t_1t_2)^{q-1}\exp\left(-\left(t_! + t_2\right)a\right)$$
$$\mathbb{E}_x\otimes\mathbb{E}'_x\left[\exp\left(-\int_0^{t_1/2} V(X(u))du - \int_0^{t_2/2} V(X'(u))du\right):S<\frac{t_1}{2},\ S'<\frac{t_2}{2}\right]$$
$$\left\|\exp\left(-\frac{t_1}{2}\left(K_0\dot{+}V\right)\right)\right\|_{\infty,2}\left\|\exp\left(-\frac{t_2}{2}\left(K_0\dot{+}V\right)\right)\right\|_{2,1}$$

($K_0\dot{+}V$ is self-adjoint)

$$=\frac{4}{\Gamma(q)^2}\int dx \int_0^\infty dt_1 \int_0^\infty dt_2\,(t_1t_2)^{q-1}\exp\left(-\left(t_! + t_2\right)a\right)$$
$$\mathbb{E}_x\otimes\mathbb{E}'_x\left[\exp\left(-\int_0^{t_1/2} V(X(u))du - \int_0^{t_2/2} V(X'(u))du\right):S<\frac{t_1}{2},\ S'<\frac{t_2}{2}\right]$$
$$\left\|\exp\left(-\frac{t_1}{2}\left(K_0\dot{+}V\right)\right)\right\|_{\infty,2}\left\|\exp\left(-\frac{t_2}{2}\left(K_0\dot{+}V\right)\right)\right\|_{\infty,2}$$

$$= \left(\frac{2}{\Gamma(q)} \int dx \int_0^\infty dt\, t^{q-1} \exp(-at)\, \mathbb{E}_x \left[\exp\left(- \int_0^{t/2} V(X(u))du \right) : S < \frac{t}{2} \right] \right.$$
$$\left. \left\| \exp\left(-\frac{t}{2} (K_0 \dot{+} V) \right) \right\|_{\infty,2} \right)^2 .$$

Consequently, we obtain

$$\begin{aligned} &\|R(a)^q - J^* R_\Sigma(a)^q J\|_{\text{HS}} \\ &\le \frac{2^{q+1}}{\Gamma(q)} \left(\int dx \left[\int_0^\infty dt\; t^{q-1} \exp(-2at) \left\| \exp\left(-t\,(K_0 \dot{+} V)\right) \right\|_{\infty,2} \right. \right. \\ &\qquad \left. \left. \mathbb{E}_x \left\{ \exp\left(- \int_0^t V(X(u))du \right) : S < t \right\} \right]^2 \right)^{1/2} . \end{aligned}$$

Since we have (see Theorem 2.9 and formulae (2.7) through (2.10))

$$\begin{aligned} \left\| \exp\left(-t\,(K_0 \dot{+} V)\right) \right\|_{\infty,2} &\le \left\| \exp\left(-t\,(K_0 \dot{+} 2V)\right) \right\|_{2,2}^{1/2} \left\| \exp\left(-tK_0\right) \right\|_{\infty,1}^{1/2} \\ &\le M_{2V}^{1/2} c_0^{1/2} \exp\left(\frac{1}{2} t b_{2V} \right) \exp\left(\frac{1}{2} a_0 \right) t^{-m/4}, \end{aligned}$$

the result in Proposition 7.4.(a) follows.
(b) As in the proof of (a), we have

$$\begin{aligned} &D_\Sigma(t,x,x) \\ &\le 2\mathbb{E}_x \left[\exp\left(- \int_0^{t/2} V(X(u))du \right) \exp\left(-\frac{t}{2} (K_0 \dot{+} V) \right) (X(t/2),x) : S < t/2 \right] . \end{aligned}$$

If T is an integral operator with kernel $t(x,y)$ for which $\int |t(x,x)|\, dx$ is finite, then we write $\operatorname{trace}(T) = \int t(x,x)dx$. So we have

$$\begin{aligned} &\operatorname{trace}\left(R(a)^q - J^* R_\Sigma(a)^q J\right) \\ &= \frac{1}{\Gamma(q)} \int_0^\infty t^{q-1} \exp(-at) \int D_\Sigma(t,x,x)\, dx\, dt \\ &\le \frac{2}{\Gamma(q)} \int_0^\infty t^{q-1} \exp(-at) \int \\ &\quad \mathbb{E}_x \left[\exp\left(- \int_0^{t/2} V(X(u))du \right) \exp\left(-\frac{t}{2} (K_0 \dot{+} V) \right) (X(t/2),x) : S < t/2 \right] dx\, dt \\ &\le \frac{2}{\Gamma(q)} \int_0^\infty t^{q-1} \left\| \exp\left(-\frac{t}{2} (K_0 \dot{+} V) \right) \right\|_{\infty,1} \exp(-at) \int \left[D_\Sigma\left(\frac{t}{2}\right) 1 \right] (x) dx\, dt \end{aligned}$$

$$\leq \frac{2}{\Gamma(q)} \int_0^\infty t^{q-1} \left\| \exp\left(-\frac{t}{4}\left(K_0 \dot{+} V\right)\right)\right\|^2_{\infty,2} \exp(-at) \int \left[D_\Sigma\left(\frac{t}{2}\right) 1\right](x) dx\, dt$$

(see Theorem 2.9)

$$\leq \frac{2}{\Gamma(q)} \int_0^\infty t^{q-1} \exp(-at) \left\| \exp\left(-\frac{t}{4}\left(K_0 \dot{+} 2V\right)\right)\right\|_{\infty,\infty} \left\| \exp\left(-\frac{t}{4} K_0\right)\right\|_{\infty,1}$$
$$\int \left[D_\Sigma\left(\frac{t}{2}\right) 1\right](x) dx\, dt.$$

Proceeding as in (a) will yield the desired result.

Another possibility to treat the trace norm is contained in the following proposition.

7.5. PROPOSITION. *The resolvent difference $R_\beta(a_0) - J^* R_\Sigma(a_0) J$ is a trace class operator if and only if the integral*

$$\int_\beta^\infty ds \int_\Gamma dx \int_0^\infty d\lambda\, \lambda \exp\left(-a_0\lambda\right) \exp\left(-\lambda K_s\right)(x,x) \quad \textit{is finite,}$$

with $K_s = K_0 \dot{+} V + s 1_\Gamma$. Moreover, we have

$$\int_\Gamma dx \int_0^\infty d\lambda \exp\left(-a_0\lambda\right) \exp\left(-\beta\lambda\right) \exp\left(-\lambda\left(K_0 \dot{+} V\right)\right)(x,x)$$
$$\leq \left\|R_\beta(a_0) - J^* R_\Sigma(a_0) J\right\|_{\text{trace}}$$
$$= \int_\beta^\infty ds \int_\Gamma dx \int_0^\infty d\lambda\, \lambda \exp\left(-a_0\lambda\right) \exp\left(-\lambda K_s\right)(x,x).$$

The real number a_0 is always supposed to be sufficiently large.

Proof. Set $R_s = \left(a_0 I + K_0 \dot{+} V + s 1_\Gamma\right)^{-1}$. Then the derivative of the function R_s exists in operator norm sense because we have

$$\frac{\partial R_s}{\partial s} = -R_s 1_\Gamma R_s.$$

This follows from the identity

$$-\frac{R_{s_2} - R_{s_1}}{s_2 - s_1} = R_{s_1} 1_\Gamma R_{s_1} + (s_1 - s_2) R_{s_1} 1_\Gamma R_{s_2} 1_\Gamma R_{s_1}.$$

Hence we get

$$R_\beta - J^* R_\Sigma J = \int_\beta^\infty R_s 1_\Gamma R_s ds.$$

So we obtain

$$\left\langle \left(R_\beta - J^* R_\Sigma J\right) f, f\right\rangle = \int_\beta^\infty \left\langle 1_\Gamma R_s f, 1_\Gamma R_s f\right\rangle ds \geq 0.$$

As a consequence we see that the operator $R_\beta - J^* R_\Sigma J$ is a form positive operator. So we obtain

$$\begin{aligned}
\|R_\beta - J^* R_\Sigma J\|_{\text{trace}} &= \int_\beta^\infty ds \, \|1_\Gamma R_s\|_{\text{HS}}^2 \\
&= \int_\beta^\infty ds \int_\Gamma dx \int_0^\infty d\lambda \lambda \exp(-a_0\lambda) \exp(-\lambda K_s)(x,x) \\
&\geq \int_\Gamma dx \int_0^\infty d\lambda \lambda \exp(-a_0\lambda) \int_\beta^\infty ds \exp(-\lambda s) \exp\left(-\lambda\left(K_0 \dot{+} V\right)\right)(x,x) \\
&= \int_\Gamma dx \int_0^\infty d\lambda \exp(-a_0\lambda) \exp(-\beta\lambda) \exp\left(-\lambda\left(K_0 \dot{+} V\right)\right)(x,x).
\end{aligned}$$

This proves Proposition 7.5.

In what follows we are mainly interested in the convergence and the rate of convergence of resolvent differences. For one part of the resolvent difference this is relatively easy. The resolvent difference $R_\beta - J^* R_\Sigma J$ is decomposed as follows:

$$R_\beta - J^* R_\Sigma J = 1_\Gamma R_\beta + J^*\left(J R_\beta - R_\Sigma J\right).$$

Moreover, we have the following norm-inequality

$$\|J R_\beta - R_\Sigma J\|_{2,2} \leq \|R_\beta - J^* R_\Sigma J\|_{2,2}.$$

In order to understand $J R_\beta - R_\Sigma J$, we looked at $1_\Gamma R_\beta$ as well as at $R_\beta - J^* R_\Sigma J$. However the considerations for $1_\Gamma R_\beta$ have their own interest and to some extent they are also simpler. Because of the relation

$$R_\beta - J^* R_\Sigma J = \int_\beta^\infty R_s 1_\Gamma R_s ds,$$

lower bounds for $\|R_\beta - J^* R_\Sigma J\|_{2,2}$ provide us with lower bounds for $\|1_\Gamma R_s\|_{2,2}$ and upper bounds for $\|1_\Gamma R_\beta\|_{2,2}$ might determine the rate of convergence for the difference $R_\beta - J^* R_\Sigma J$. Moreover, a detailed study of $\int_0^\infty R_s 1_\Gamma R_s ds$ makes it possible to describe the domain of K_Σ. In the Laplace operator context, this is done in detail by A. Boutet de Monvel, Georgescu and Soffer [39].

7.6. LEMMA. *For $V \equiv 0$ the following inequality is valid:*

$$\|1_\Gamma R_\beta\|_{2,2} \leq \frac{1}{\sqrt{\beta}}.$$

Here $R_\beta = (K_0 + \beta 1_\Gamma + I)^{-1}$.

Proof. The following identities are self-explanatory:

$$\begin{aligned}
I &= R_\beta^{1/2}\left(K_0 + \beta 1_\Gamma + I\right) R_\beta^{1/2} \\
&= R_\beta^{1/2}\left(K_0 + I\right) R_\beta^{1/2} + \beta R_\beta^{1/2} 1_\Gamma R_\beta^{1/2}. \qquad (7.28)
\end{aligned}$$

From (7.28) we see for $f \in L^2(E, m)$

$$\beta \left\| 1_\Gamma R_\beta^{1/2} f \right\|_2^2 = \|f\|_2^2 - \left\| (K_0 + I)^{1/2} R_\beta^{1/2} f \right\|_2^2 \leq \|f\|_2^2 .$$

Hence $\left\| 1_\Gamma R_\beta^{1/2} \right\|_{2,2} \leq \frac{1}{\sqrt{\beta}}$ and so $\|1_\Gamma R_\beta\|_{2,2} \leq \frac{1}{\sqrt{\beta}}$, because $\left\| R_\beta^{1/2} \right\|_{2,2} \leq 1$.

REMARK. Since we have

$$\|R_\beta - J^* R_\Sigma J\|_{2,2} \leq \int_\beta^\infty \|1_\Gamma R_s\|_{2,2}^2 \, ds,$$

the rate of $s^{-1/2}$ for the norm $\|1_\Gamma R_s\|_{2,2}$ is not good enough to find a rate for $\|R_\beta - J^* R_\Sigma J\|_{2,2}$. For subsets Γ of $\mathbb{R}^d$ with smooth enough boundaries and for $K_0 = H_0 = -\frac{1}{2}\Delta$, A. Boutel de Monvel, Georgescu and Soffer [39] discovered that

$$\left\| 1_\Gamma (H_0 + s 1_\Gamma + I)^{-1} \right\|_{2,2} \leq C s^{-3/4}$$

which gives an optimal rate of convergence for

$$\left\| (H_0 + \beta 1_\Gamma + I)^{-1} - J^* \left((H_0)_\Sigma + I \right)^{-1} J \right\|_{2,2} \leq 2C\beta^{-1/2}.$$

This is in agreement with the results obtained by Demuth, Kirsch and McGillivray [78] which to some extent will be described in Chapter 8, section C. In [78] lower bounds are given as well.

As a consequence of Proposition 7.2 and its Corollary 7.3, we may verify the following properties of the operator $1_\Gamma R_\beta$.

7.7. COROLLARY. *Let* $(\exp(-tK_0) : t \geq 0)$ *be an* L^1-L^2*-smoothing free Feller semigroup of spectral dimension* m. *Suppose that* $m(\Gamma) < \infty$. *Then the following assertions are valid:*

(i) *For* $q \in \mathbb{N}$, $\beta \geq 0$, *and* $z \in \operatorname{res}(K_\beta)$ *the operator* $1_\Gamma (K_\beta - zI)^{-q}$ *is compact.*
(ii) *Let* $m \leq 3$. *For* $q \in \mathbb{N}$, $\beta \geq 0$, *and* $z \in \operatorname{res}(K_\beta)$ *the operator* $1_\Gamma (K_\beta - zI)^{-q}$ *is Hilbert-Schmidt. Moreover,*

$$\lim_{\beta \to \infty} \left\| 1_\Gamma (K_\beta - zI)^{-q} \right\|_{\mathrm{HS}} = 0$$

for $z \in \operatorname{res}(K_\Sigma)$.

Proof. The proof is patterned after the proof of Corollary 7.3. In fact it is somewhat easier. We give only a sketch. A Neumann series makes it possible to reduce the problem to a situation where an arbitrary $z \in \operatorname{res}(K_\Sigma)$ is replaced with some z_0 such that $-\operatorname{Re} z_0 = a_0$ large enough. Assertion (ii) follows from

$$\begin{aligned}
&\lim_{\beta\to\infty} \left\| 1_\Gamma \left(K_\beta - z_0 I\right)^{-q} \right\|_{\mathrm{HS}}^2 \\
&\leq \frac{1}{\Gamma(q)} \int_\Gamma dx \int_0^\infty d\lambda \int_0^\infty d\mu \lambda^{q-1}\mu^{q-1} \exp\left(-a_0(\lambda+\mu)\right) \exp\left(-(\lambda+\mu)K_\beta\right)(x,x) \\
&\leq C\, m(\Gamma) \int_0^\infty d\lambda \int_0^\infty d\mu \left(\lambda+\mu\right)^{-m/2} \lambda^{q-1}\mu^{q-1} \exp\left(-\left(a_0 - c\right)\left(\lambda+\mu\right)\right) \\
&\leq C_1\, m(\Gamma) \left(\frac{1}{a_0 - c}\right)^{2q-m/2} \left(\int_0^\infty d\lambda\, \lambda^{q-1-m/4} \exp(-\lambda)\right)^2 . \qquad (7.29)
\end{aligned}$$

The expression in (7.29) is finite for $4q > m$. So, for $m \leq 3$, the operator $1_\Gamma \left(K_\beta - z_0 I\right)^{-1}$ is a Hilbert-Schmidt operator. Then the compactness of $1_\Gamma \left(K_\beta - zI\right)^{-1}$ and the Hilbert-Schmidt property of $1_\Gamma \left(K_\beta - zI\right)^{-1}$ for $z \in \operatorname{res}(K_\beta)$ follow again from the representation

$$1_\Gamma \left(K_\beta + a_0 I\right)^{-1} = (q-1) \int_0^\infty dt\; t^{q-2} 1_\Gamma \left(K_\beta + \left(a_0 + t\right) I\right)^{-q} .$$

Finally, the convergence in (ii) is given by

$$\begin{aligned}
&(\Gamma(q))^2 \left\| 1_\Gamma \left(K_\beta + a_0 I\right)^{-q} \right\|_{\mathrm{HS}}^2 \\
&= \int_0^\infty d\lambda \int_0^\infty d\mu \left(\lambda\mu\right)^{q-1} \exp\left(-a_0\left(\lambda+\mu\right)\right) \\
&\qquad \int_\Gamma \Big[\exp\left(-(\lambda+\mu)K_\beta\right)(x,x) - \exp\left(-(\lambda+\mu)K_\Sigma\right)(x,x) \\
&\qquad\qquad + \exp\left(-(\lambda+\mu)K_\Sigma\right)(x,x) \Big]\, dx.
\end{aligned}$$

From Lemma 7.1 we have

$$\lim_{\beta\to\infty} \int_\Gamma \Big[\exp\left(-(\lambda+\mu)K_\beta\right)(x,x) - \exp\left(-(\lambda+\mu)K_\Sigma\right)(x,x) \Big]\, dx = 0.$$

Assume $S = T$. The set $\Gamma \setminus \Gamma^r$, with $\Gamma^r := \{x \in \Gamma : \mathbb{P}_x \{T = 0\} = 1\}$, is polar and hence has m-measure zero. Therefore $\int_\Gamma \exp\left(-tK_\Sigma\right)(x,x)dx = 0$.

REMARK. For scattering theoretical applications the systems given by K_Σ and K_β establish a two-space scattering situation because the unitary groups $(\exp\left(-itK_\Sigma\right) : t \in \mathbb{R})$ and $(\exp\left(-itK_\beta\right) : t \in \mathbb{R})$ are defined on the spaces $L^2(\Sigma, m)$ and $L^2(E, m)$ respectively. Using for instance trace class conditions one has to verify the inclusion

$$JR_\beta - R_\Sigma J \in \mathcal{B}_1\left(L^2(E,m), L^2(\Sigma,m)\right) .$$

However the results given above for the operators $R_\beta - J^* R_\Sigma J$ and $1_\Gamma R_\beta$ imply the corresponding results for $JR_\beta - R_\Sigma J$.

At the end of this chapter let us mention sandwiched resolvents and semigroups. We will confine ourselves to some rather easy facts. In Chapter 5 we estimated already the Hilbert-Schmidt norm of $\varphi D_\Sigma(t)\psi$: see Theorem 5.19. The purpose here is to study the dependence on β for some norms of the operator $\varphi D_{\Sigma,\beta}(t)\psi$. The estimates in Theorem 5.19 are somewhat too rough for this purpose. Nevertheless the theory developed so far is rich enough to improve these conditions.

7.8. PROPOSITION.
(i) *Suppose that the functions* φ *and* ψ *have the property that the integral*

$$\int dx \left(|\varphi(x)|^4 + |\psi(x)|^4\right) \mathbb{E}_x \left[\exp\left(-2\beta T_{t,\Gamma}\right) : T_{t,\Gamma} > 0\right] \tag{7.30}$$

is finite. Then $\varphi D_{\Sigma,\beta}(t)\psi$ *is a Hilbert-Schmidt operator and*

$$\begin{aligned} \|\varphi D_{\Sigma,\beta}(t)\psi\|_{\mathrm{HS}}^2 \le \frac{1}{2} \left\|\exp\left(-t\left(K_0 \dot{+} 2V\right)\right)\right\|_{\infty,1} \\ \times \int dx \left(|\varphi(x)|^4 + |\psi(x)|^4\right) \mathbb{E}_x \left[\exp\left(-2\beta T_{t,\Gamma}\right) : T_{t,\Gamma} > 0\right]. \end{aligned} \tag{7.31}$$

(ii) *Suppose* $\varphi = \psi$. *Then the operator norm of* $\varphi D_{\Sigma,\beta}(t)\varphi$ *can be estimated as follows:*

$$\begin{aligned} &\|\varphi D_{\Sigma,\beta}(t)\varphi\|_{2,2} \\ &\le \left\|\exp\left(-t\left(K_0 \dot{+} 2V\right)\right)\right\|_{\infty,1} \sup_{x\in E} \left(|\varphi(x)|^4 \mathbb{E}_x \left[\exp\left(-2\beta T_{t,\Gamma}\right) : T_{t,\Gamma} > 0\right]\right). \end{aligned} \tag{7.32}$$

Here $T_{t,\Gamma}$ denotes the spending time as introduced in the remark following Proposition 7.2,

$$T_{t,\Gamma} := \text{meas}\left\{0 \le s \le t,\ X(s) \in \Gamma\right\} = \int_0^t 1_\Gamma(X(s))ds.$$

Proof. (i) The following identities are true (see the proof of Theorem 5.19 and notice $\{S < t\} = \{T_{t,\Gamma} > 0\}$):

$$\begin{aligned} &\|\varphi D_{\Sigma,\beta}(t)\psi\|_{\mathrm{HS}}^2 \\ &= \int dx \int dy \, |\varphi(x)|^2 \, |\psi(y)|^2 \, D_{\Sigma,\beta}(t,x,y)^2 \\ &\le \frac{1}{2} \int dx \int dy \left(|\varphi(x)|^4 + |\psi(y)|^4\right) D_{\Sigma,\beta}(t,x,y)^2 \end{aligned}$$

(the kernel of $D_{\Sigma,\beta}(t)$ is symmetric)

$$\leq \frac{1}{2} \int dx \int dy \left(|\varphi(x)|^4 + |\psi(x)|^4\right)$$

$$\int_{\{S<t\}} \exp\left(-2\int_0^t V(X(u))du\right) d\mu_{0,x}^{t,y} \int_{\{S<t\}} \exp\left(-2\beta\int_0^t 1_\Gamma(X(u))du\right) d\mu_{0,x}^{t,y}$$

$$\leq \frac{1}{2} \left\|\exp\left(-t\left(K_0 \dot{+} 2V\right)\right)\right\|_{\infty,1}$$

$$\int dx \left(|\varphi(x)|^4 + |\psi(x)|^4\right) \mathbb{E}_x\left\{\exp\left(-2\beta T_{t,\Gamma}\right) : T_{t,\Gamma} > 0\right\}.$$

(ii) This assertion follows because the operator $\varphi D_{\Sigma,\beta}(t)\varphi$ has a symmetric kernel.

7.9. COROLLARY. *A sandwiched resolvent difference can be estimated as follows:*

$$\left\|\varphi\left[\left(K_\beta + a_0 I\right)^{-1} - J^*\left(K_\Sigma + a_0 I\right)^{-1} J\right]\varphi\right\|_{2,2}$$

$$\leq \int_0^\infty d\lambda \exp\left(-a_0\lambda\right) \left\|\varphi D_{\Sigma,\beta}(\lambda)\varphi\right\|_{2,2}.$$

Hence the inequalities in (7.31) and (7.32) are also useful to estimate resolvent differences.

Chapter 8

Spectral Properties of Self-adjoint Feller Operators

In the last four chapters we have developed some fundamentals for the spectral theory of Feller operators. In this section we will put these ideas to work in order to investigate some consequences for the different parts of the spectra of the operators $K_0\dot{+}V$ and $(K_0\dot{+}V)_\Sigma$. For the free Feller operator K_0 we make again the general assumptions BASSA as set forth in the conditions A.1.–A.4. in Chapter 1.B. Additionally we assume that $\exp(-tK_0)$ is L^1-L^∞-smoothing (see Definition 1.1). This means that for every $t > 0$ there exists a finite constant c_t such that

$$\sup_{x,y\in E} p_0(t,x,y) \le c_t. \tag{8.1}$$

In order to give quantitative norm estimates, we suppose that the semigroup $\{\exp(-tK_0) : t \ge 0\}$ possesses finite spectral dimension, i.e. we require that there exists a constant m with the property that

$$c_t \le c_0 t^{-m/2} \exp(a_0 t), \quad t > 0, \tag{8.2}$$

where we fix a_0 and $c_0 \ge 0$.

We are mainly interested in perturbation theoretical results. The theory developed so far enables us to study the invariance of the absolutely continuous and/or essential part of the spectra of the pair of operators $K_0\dot{+}V$ and $K_0\dot{+}W$ or the pair $K_0\dot{+}V$ and $(K_0\dot{+}V)_\Sigma$. In addition we are in a position to investigate the bound states. Because we can estimate the norms of the resolvent differences in terms of the Kato-Feller norm or, in the case of singular perturbations, in terms of L^p-conditions for the equilibrium potentials, it is possible to give quantitative estimates for several spectral or scattering data. In particular we will consider the convergence and also the rate of convergence for wave operators, scattering operators and scattering matrices. Moreover, the behavior of spectral measures is studied via the limiting absorption principle. This topic is connected with the perturbation theory for resonances and the determination of resonance free regions. The results given here are not at all complete. We want to give an overview and compendium with the intention of demonstrating how the general theory can be used. We are convinced that the general theory as explained and described in the preceding chapters allows several further consequences for the control of the spectrum or the behavior of eigenfunctions. For instance the possibility of studying semi-classical limits is merely sketched in one example rather than treated rigorously. Large coupling limits are studied in some detail, but it seems possible and worthwhile to find quantitative estimates for thermodynamic limits as well.

In general the results in this chapter are based on some knowledge about the spectrum of the unperturbed generator K_0. The literature concerning the spectra of pseudo-differential (or differential) operators of order less than or equal to two is huge. We will not make an attempt to give a survey here. We mention only two examples. Because the properties of the spectrum $\sigma(K_0)$ depend heavily on the operator K_0, they are not determined by the general conditions BASSA. If one wants to compare $\sigma(K_0)$ with the spectrum of some other, well-known, Feller generator, for example with the negative Laplace operator or its square root, then one is more or less obliged to study "top order perturbations". That is a useful and promising task for the future. As a matter of fact, some results are already known. For instance Ben-Artzi and Devinatz [23] have developed a theory of comparing properties of second order differential operators with, in the end, properties of the Laplace operator. They study operators of the form (see Example 2 in section C of Chapter 1):

$$K_0 = -c(x,y)^2 \rho(x,y) \left[\nabla \frac{1}{\rho(x,y)} \nabla \right]$$

with $x \in \mathbb{R}^\nu$ and $y \in \mathbb{R}$. They compare this operator with

$$\widetilde{K}_0 = -c_0(y)^2 \left[\Delta_x + \rho_0(y) \frac{\partial}{\partial y} \left(\frac{1}{\rho_0(y)} \frac{\partial}{\partial y} \right) \right].$$

Here $c(x,y)$ and $\rho(x,y)$ are short range perturbations with respect to $x \in \mathbb{R}^\nu$ of $c_0(y)$ and $\rho_0(y)$. Moreover, the functions $c(x,y)$ and $\rho(x,y)$ are assumed to be bounded from below and above with finite strictly positive constants and they are supposed to have uniformly bounded derivatives. Using the powerful tool of the limiting absorption principle, the spectrum of K_0 is shown to coincide with $[0, \infty)$ and, up to a possible discrete sequence of eigenvalues of finite multiplicity which can only accumulate at 0 or at ∞, the spectrum of K_0 is purely absolutely continuous. Weder (see [22, Theorem A, page 484, Theorem B, page 485]) proved under more restrictive conditions on the functions $c(x,y)$ and $\rho(x,y)$ that either no positive eigenvalues occur or there are only a countable number of eigenvalues of finite multiplicity which accumulate at the bottom of the spectrum or at ∞. In fact, in [22] the authors Ben-Artzi, Dermanjian, and Guillot prove a limiting absorption principle in Theorem A and B (page 484–485).

In relativistic quantum mechanics and in quantum field theory, the Hamiltonians are in general not partial differential operators, as they are in non-relativistic quantum-mechanics, but they are pseudo-differential operators (see Example 4 in section C of Chapter 1). In [248] Weder studied the spectra of operators with symbol $|p|^\alpha$, $0 < \alpha \leq 2$, and with symbol $\sqrt{|p|^2 + m^2}$: see Example 4 in Chapter 1.C.

For instance he showed the equalities:

$$\sigma\left(\sqrt{|p|^2+m^2}\right)=\sigma_{\text{ess}}\left(\sqrt{|p|^2+m^2}\right)=[m,\infty),$$

where we identify the operator with its symbol. In our case we consider the operator with symbol $\sqrt{|p|^2+m^2}-m$. From the observations above it follows that $\sigma\left(\sqrt{|p|^2+m^2}-m\right)$ coincides with the half-line $[0,\infty)$. Perturbations of these relativistic operators are studied for instance by Herbst [119] and also by Carmona, Masters and Simon [48]. The relativistic stability of matter is investigated by Fefferman and de la Llave [104] and again by Fefferman [103]. In [230] Umeda and Nagase investigate the spectrum of relativistic operators with magnetic vector potentials.

The main objective of this chapter is the investigation of the convergence of spectral data. However, before we start with this, we have to ensure ourselves that the magnitudes involved exist (and are finite). Therefore this chapter contains three sections. The first one gives the qualitative permanence properties of the spectra together with the existence and completeness of wave operators. The second and third section are devoted to explicit quantitative estimates for both regular and singular perturbations.

A. Qualitative spectral results

Due to the Hilbert-Schmidt and trace class considerations of the chapters 5 and 6, some of the spectral consequences for the stability of the essential or the absolutely continuous part of the spectrum are obvious. In order to have a more complete and consistent description we repeat some results of the earlier chapters. We always assume that the unperturbed kernel satisfies BASSA together with condition (8.1) (L^1-L^∞-smoothing) or (8.2) (being of finite spectral dimension). Correspondingly we can estimate the semigroup kernels by (see inequality (3) in Theorem 2.9. together with the comment following it and the inequality in (2.7)):

$$\begin{aligned}
&[\exp(-t(K_0+V))(x,y)]^2\\
&\leq M_{4V}\exp\left(\frac{t}{2}b_{4V}\right)\sup_{x,y\in E}p_0(t/2,x,y).p_0(t,x,y)\\
&\leq c_0M_{4V}\exp\left(\frac{t}{2}(a_0+b_{4V})\right)t^{-m/2}p_0(t,x,y) &(8.3)\\
&\leq c'\exp(ct)\,t^{-m/2}p_0(t,x,y), &(8.4)
\end{aligned}$$

where c and c' are sufficiently large positive constants. In what follows we sometimes write $\mathcal{H}=L^2(E,m)$.

In the first part of Chapter 8.A, we will give sufficient conditions for the existence of bound states, for the invariance of the essential and absolutely continuous spectra, and for the existence and completeness of wave and scattering operators.

Let us begin with the general relations between the recurrence of the semigroup $(\exp(-tK_0) : t \geq 0)$ and the spectral bound of $K_0 \dot{+} V$ including the existence of bound states. For this topic we follow McGillivray and Ouhabaz [178] who investigated this problem for even more general sub-Markovian semigroups. We shall first give some additional properties of free Feller semigroups.

In what follows we write $L^p_+(E, m)$ for the cone of all non-negative functions that belong to $L^p(E, m)$.

8.1. DEFINITION. The semigroup $(\exp(-tK_0) : t \geq 0)$ is called *irreducible* if $[\exp(-tK_0) f](x) > 0$ m-almost everywhere for all $f \in L^2_+(E, m)$, $f \neq 0$. The semigroup $(\exp(-tK_0) : t \geq 0)$ is called *recurrent* if for every $f \in L^1_+(E, m)$, $f \neq 0$, the equality

$$\int_0^\infty [\exp(-tK_0) f](x) dt = \infty$$

is valid for m-almost all x. This is equivalent to saying that for all such functions f the expectation

$$\mathbb{E}_x \left[\int_0^\infty f(X(t)) dt \right] = \infty$$

for m-almost all $x \in E$. The semigroup is called *transient* if for all $f \in L^1(E, m)$, $f \geq 0$, the integral,

$$\int_0^\infty [\exp(-tK_0) f](x) dt < \infty$$

for m-almost all $x \in E$.

Any irreducible semigroup is either recurrent or transient. Since we assume here that $p_0(t, x, y) > 0$ for all x and $y \in E$ (see BASSA, Chapter 1, section B), all the free Feller semigroups are irreducible. For some general information about these topics, the reader may consult Appendix B.9 and B.10. The recurrence and transience of a semigroup is related to its L^2-L^2-type. Let ω_0 and ω_V be the L^2-L^2 types of the semigroup $(\exp(-tK_0) : t \geq 0)$ and of $(\exp(-t(K_0 \dot{+} V)) : t \geq 0)$ respectively (see formula (2.8)): i.e.

$$\left\| \exp\left(-t\left(K_0 \dot{+} V\right)\right) \right\|_{2,2} = \exp(\omega_V t)$$

for all $t \geq 0$. Since K_0 is self-adjoint and form positive the L^2-L^2 type of the unperturbed semigroup ω_0 coincides with the negative spectral bound: i.e.

$$\omega_0 = -\inf \sigma(K_0).$$

In order to understand the relation between ω_V and the recurrence of the semigroup $(\exp(-tK_0) : t \geq 0)$, we have to introduce the notion of L^p-*stability*.

8.2. DEFINITION. The semigroup $(\exp(-tK_0) : t \geq 0)$ is called *L^p-stable* if, for every $f \in L^p(E, m)$,

$$\lim_{t\to\infty} \|\exp(-tK_0) f\|_p = 0.$$

8.3. LEMMA. *Let* $\mathrm{dom}(\mathcal{E}^0)$ *be the form domain of* K_0 *and denote by* $\mathrm{dom}_+(\mathcal{E}^0)$ *the cone* $\mathrm{dom}(\mathcal{E}^0) \bigcap L^2_+(E, m)$. *The semigroup* $(\exp(-tK_0) : t \geq 0)$ *is* L^2*-stable if and only if*

(1) (i) $\langle K_0 u, u\rangle > 0$ *for all* $u \in \mathrm{dom}(\mathcal{E}^0)$, $u \neq 0$,
if and only if
(2) (ii) $\langle K_0 u, u\rangle > 0$ *for all* $u \in \mathrm{dom}_+(\mathcal{E}^0)$, $u \neq 0$.

Proof. If the semigroup $(\exp(-tK_0) : t \geq 0)$ is L^2-stable, then zero is not an eigenvalue of K_0. It follows that $\langle K_0 u, u\rangle > 0$ for all non-zero $u \in \mathrm{dom}(\mathcal{E}^0)$. On the other hand, if $\langle K_0 u, u\rangle > 0$ for all non-zero $u \in \mathrm{dom}(\mathcal{E}^0)$, then zero cannot be an eigenvalue of K_0. And so, in that case the range of K_0 is dense in $L^2(E, m)$. For $f = K_0 g$ we have

$$\|\exp(-tK_0) f\|_2 = \|\exp(-tK_0) K_0 g\|_2 \leq \frac{C}{t} \|g\|_2 .$$

An approximation argument gives the stability. The equivalence of (i) and (ii) remains to be shown. Since $(u, v) \mapsto \langle K_0 u, v\rangle$, $u, v \in \mathrm{dom}(\mathcal{E}^0)$, is a Dirichlet form, we have $\langle K_0 |u|, |u|\rangle \leq \langle K_0 u, u\rangle$ for $u \in \mathrm{dom}(\mathcal{E}^0)$: see assertion (e) of Theorem 1.4.1. in Fukushima, Oshima, and Takeda [110, page 23–24]. Hence, on the one hand, if (ii) holds for all $u \in \mathrm{dom}_+(\mathcal{E}^0)$, we get (i) because $\langle K_0 u, u\rangle \geq \langle K_0 |u|, |u|\rangle > 0$. On the other hand the inclusion $\mathrm{dom}_+(\mathcal{E}^0) \subset \mathrm{dom}(\mathcal{E}^0)$ yields (i) $\Rightarrow$ (ii).

8.4. LEMMA. *Let* $V = V_+$, $V_+ \neq 0$, *be a Kato-Feller potential and suppose that the semigroup* $(\exp(-tK_0) : t \geq 0)$ *is* L^2*-stable. Then the semigroup* $(\exp(-t(K_0 \dot{+} V)) : t \geq 0)$ *is* L^2*-stable.*

Proof. First assume that V is a bounded function, so that $\mathrm{dom}(K_0) = \mathrm{dom}(K_0 \dot{+} V)$. Assume, to arrive at a contradiction, that the semigroup

$$(\exp(-t(K_0 \dot{+} V)) : t \geq 0)$$

is not L^2-stable. By virtue of Lemma 8.3, there exists $v \in \mathrm{dom}_+(\mathcal{E}^0)$, $v \neq 0$, such that $\langle K_0 v, v\rangle + \int V(x) v(x)^2 dm(x) = 0$. Hence $\langle K_0 v, v\rangle = 0$ and so $\exp(-tK_0)v = v$ for all $t \geq 0$. The irreducibility of the semigroup $(\exp(-tK_0) : t \geq 0)$ yields $v(x) > 0$ for m-almost all $x \in E$. Consequently $V \equiv 0$ and so $K_0 = K_0 \dot{+} V$. Thus $\exp(-t(K_0 \dot{+} V)) = \exp(-tK_0)$. This contradicts the non-stability of the semigroup $(\exp(-t(K_0 \dot{+} V)) : t \geq 0)$ because the unperturbed semigroup $(\exp(-tK_0) : t \geq 0)$ is supposed to be stable.

There is an equivalent condition for the recurrence of the semigroup

$$(\exp(-tK_0) : t \geq 0).$$

8.5. PROPOSITION. *Let* $(\exp(-tK_0) : t \geq 0)$ *be the irreducible free Feller semigroup. Then the semigroup* $(\exp(-tK_0) : t \geq 0)$ *is recurrent if and only if the perturbed semigroup* $\left(\exp\left(-t\left(K_0 \dot{+} V\right)\right) : t \geq 0\right)$ *is* L^1*-stable for every non-zero non-negative potential* V*, together with the fact that the unperturbed semigroup* $(\exp(-tK_0) : t \geq 0)$ *is not* L^1*-stable.*

For Schrödinger semigroups the reader is referred to Arendt, Batty and Bénilan [12] and to the same authors for the Laplace operator [10]. For a stochastic approach see Batty [18].

The proof of Proposition 8.5 is given in McGillivray and Ouhabaz [178]. It will not be given here, because we are not going to use it in the sequel.

However, even the equivalent description gives no information about $-\omega_V$ or, what amounts to saying the same thing, about $\inf \sigma\left(K_0 \dot{+} V\right)$. In fact one can have $\omega_V = \omega_0 = 0$ even if the semigroup $\left(\exp\left(-t(K_0 \dot{+} V)\right) : t \geq 0\right)$ is L^p-stable. In order to obtain more information about the spectrum, some additional hypotheses are needed.

8.6. PROPOSITION. *Suppose that the semigroup* $(\exp(-tK_0) : t \geq 0)$ *is recurrent and let* $V = V_+ \neq 0$ *be a non-negative Kato-Feller potential. Assume that* $K_0 \dot{+} V$ *has a compact resolvent. Then* $\omega_V < 0$*, or equivalently,* $\inf \sigma\left(K_0 \dot{+} V\right) > 0$.

Proof. Suppose that $\omega_V = 0$. Since $K_0 \dot{+} V$ has compact resolvent, $0 = \omega_V$ is an eigenvalue of $K_0 \dot{+} V$. By Lemma 8.3 there is a function v, such that $v \geq 0$, $v \neq 0$, and $\exp\left(-t\left(K_0 \dot{+} V\right)\right) v = v$ for all $t \geq 0$. From the inequality

$$\left[\exp\left(-t\left(K_0 \dot{+} V\right)\right) v\right](x) \leq \left[\exp\left(-tK_0\right) v\right](x),$$

we conclude that

$$v(x) \leq \left[\exp\left(-tK_0\right) v\right](x).$$

Together with the inequality $\|\exp(-tK_0)v\|_2 \leq \|v\|_2$ this implies that

$$\left[\exp\left(-tK_0\right) v\right](x) = v(x), \quad m\text{-almost everywhere.}$$

The irreducibility of the semigroup $(\exp(-tK_0) : t \geq 0)$ implies that $v(x) > 0$ for m-almost all $x \in E$. For any $u \in L^2(E, m)$, $u \geq 0$, we see that

$$\begin{aligned} &\left\langle \exp(-tK_0)u - \exp\left(-t\left(K_0 \dot{+} V\right)\right) u, v\right\rangle \\ &= \left\langle u, \exp(-tK_0)v - \exp\left(-t\left(K_0 \dot{+} V\right)\right) v\right\rangle = 0. \end{aligned}$$

Also, note the inequality $\left[\exp(-tK_0)u\right](x) \geq \left[\exp\left(-t\left(K_0 \dot{+} V\right)\right) u\right](x)$, $x \in E$. Consequently $\exp\left(-tK_0\right) u = \exp\left(-t\left(K_0 \dot{+} V\right)\right) u$ for all $u \in L^2(E, m)$. So we conclude $V \equiv 0$, which is a contradiction.

These kinds of considerations enable us to give abstract conditions for the existence of bound states in case K_0 is a general free Feller operator and V is a negative Kato-Feller potential.

8.7. PROPOSITION. *If the semigroup* $(\exp(-tK_0) : t \geq 0)$ *is recurrent and if the Kato-Feller potential is negative, i.e. if* $V = -V_-$, $V_- \geq 0$, $V_- \neq 0$, *then* $\omega_V > 0$, *or equivalently,* $\inf \sigma\left(K_0 \dot{+} V\right) < 0$.

REMARK. Proposition 8.7 implies the existence of at least one bound state, provided we assume additionally that V is relatively compact with respect to K_0. In that case the essential spectra of K_0 and $K_0 \dot{+} V$ coincide and are contained in the positive half line. Conditions for the operator $(aI + K_0)^{-1} - \left(aI + K_0 \dot{+} V\right)^{-1}$ to be compact were given in Chapter 5 for instance.

Proof. Instead of studying a general Kato-Feller potential V, it suffices to consider $V \in L^2(E,m) \cap L^\infty(E,m)$ because otherwise we replace V with $\max(V,-1)1_K$, where K is a subset of E of finite m-measure on which V does not vanish identically. (In fact we may suppose that K is compact.) For a given f_0, $f_0 \geq 0$, $f_0 \neq 0$, we write $v_0 = \exp(-tK_0)f_0$. We suppose that f_0 belongs to $L^2(E,m)$. Then v_0 is an element of $\operatorname{dom}(K_0)$ and, since the semigroup $(\exp(-tK_0) : t \geq 0)$ is irreducible, the function v_0 is almost everywhere positive. In order to arrive at a contradiction, assume that $\omega_V \leq 0$. Then

$$0 \leq \langle K_0 u, u\rangle - \langle V_- u, u\rangle$$

for $u \in \operatorname{dom}(K_0)$. Setting $a = \langle V_- v_0, v_0\rangle$ we obtain

$$\langle V_- v_0, u\rangle^2 \leq a\,\langle V_- u, u\rangle \leq a\,\langle K_0 u, u\rangle$$

for all $u \in \operatorname{dom}(K_0)$. So we get the following estimate:

$$\begin{aligned}
&\left\langle V_- v_0, \int_0^t \exp\left(-sK_0\right) V_- v_0\, ds\right\rangle^2 \\
&\leq a\left\langle K_0 \int_0^t \exp\left(-sK_0\right) V_- v_0\, ds, \int_0^t \exp\left(-sK_0\right) V_- v_0 ds\right\rangle \\
&= a\left\langle V_- v_0 - \exp\left(-tK_0\right) V_- v_0, \int_0^t \exp\left(-sK_0\right) V_- v_0 ds\right\rangle \\
&\leq a\left\langle V_- v_0, \int_0^t \exp\left(-sK_0\right) V_- v_0 ds\right\rangle.
\end{aligned}$$

This means that

$$\left\langle V_- v_0, \int_0^t \exp\left(-sK_0\right) V_- v_0 ds\right\rangle \leq a < \infty, t \geq 0.$$

By hypothesis $V_- v_0 \neq 0$ and therefore

$$\int_0^\infty \exp\left(-sK_0\right) V_- v_0 ds < \infty, \quad m\text{-almost everywhere.}$$

Consequently the semigroup $(\exp(-tK_0) : t \geq 0)$ is transient and so it cannot be recurrent. Thus we obtain $\omega_V > 0$.

REMARK. There is even a converse to Proposition 8.7. If the operators $\exp(-tK_0)$ maps $L^2(E,m)$ to $L^2(E,m)$ (as is presently the case) meaning that $p_0(t,x,y)$ is a Carleman kernel and if $\omega_V > 0$ for all $V = -V_-$, $V_- \geq 0$, $V_- \neq 0$, then the semigroup $(\exp(-tK_0) : t \geq 0)$ is recurrent. The interested reader can find a proof in McGillivray and Ouhabaz [178].

In [177] McGillivray relates the bottom of the spectra of singularly perturbed Hamiltonians to the capacity of the relative complement. We will describe his result presently.

Let K_0 be the free Feller operator with an ultracontractive (also called L^1-L^∞-smoothing) semigroup, i.e. one for which

$$\|\exp(-tK_0)\|_{\infty,1} \leq \left\|\exp\left(-\frac{t}{2}K_0\right)\right\|_{\infty,2}^2 < \infty.$$

The capacity, or more precisely, the 1-capacity of an open subset U of E is defined by

$$\operatorname{cap}(U) = \inf\left\{\left\langle K_0^{1/2}u, K_0^{1/2}u\right\rangle + \langle u,u\rangle : u \geq 1_U,\ u \in \operatorname{dom}\left(K_0^{1/2}\right)\right\}.$$

For an arbitrary subset A, it is defined as the infimum of the capacities of the open sets U containing A:

$$\operatorname{cap}(A) = \inf\left\{\operatorname{cap}(U) : U \supseteq A,\ U \text{open}\right\}.$$

For more on this notion see Definition 2.36, Proposition 2.37 and Definition D.17 in Appendix D. Similarly the 0-capacity of an open set U is given by

$$\operatorname{cap}_{(0)}(U) = \inf\left\{\left\langle K_0^{1/2}u, K_0^{1/2}u\right\rangle : u \geq 1_U,\ u \in \operatorname{dom}\left(K_0^{1/2}\right)\right\}.$$

For an arbitrary set A, we again take the infimum of the 0-capacities of open sets containing A:

$$\operatorname{cap}_{(0)}(A) = \inf\left\{\operatorname{cap}_{(0)}(U) : U \supseteq A,\ U \text{open}\right\}.$$

Let Σ be an open subset of E with the property that $(K_0)_\Sigma$ possesses discrete spectrum only. The smallest eigenvalue λ_Σ is non-negative: $\lambda_\Sigma \geq 0$. Next we cut away a piece of Σ, i.e. we take a compact subset K of Σ. The operator $(K_0)_{\Sigma\setminus K}$ again has a discrete spectrum. Its smallest eigenvalue $\lambda_{\Sigma\setminus K}$ is greater than or equal to λ_Σ. As mentioned earlier, assume that the semigroup $\{\exp(-tK_0) : t \geq 0\}$ is ultracontractive. In [177] McGillivray proved the following relations:

(a) There is a positive constant c_1 such that

$$0 \leq \lambda_{\Sigma\setminus K} - \lambda_\Sigma \leq c_1 \operatorname{cap}(K).$$

(b) If λ_Σ is strictly positive, then there is a constant c_2 such that

$$0 \leq \lambda_{\Sigma\setminus K} - \lambda_\Sigma \leq c_2 \operatorname{cap}_{(0)}(K).$$

(c) Suppose that Σ has finite measure, i.e $m(\Sigma) < \infty$, and suppose that

(1) $\inf_{x,y\in\Sigma} \left(I + K_0\right)^{-1}(x, y) = \inf_{x,y\in\Sigma} \int_0^\infty ds\, e^{-s} p_0(s, x, y) > 0.$

In addition let U be a relatively compact subset with the property that its closure $\overline{U}$ is a (necessarily compact) subset of Σ that does not coincide with Σ. Then there exists a strictly positive constant c_3 with the property that

$$\lambda_{\Sigma\setminus K} - \lambda_\Sigma \geq c_3 \operatorname{cap}_{(0)}(K),$$

for all compact subsets K of U.

(d) If in (c) the hypothesis in (1) is removed, then there exists a strictly positive constant c_4 such that

$$\lambda_{\Sigma\setminus K} - \lambda_\Sigma \geq c_4 \operatorname{cap}(K),$$

The interested reader is invited to read the proof by McGillivray in [177].

Thus we can establish a general framework concerning the existence of bound states and properties of free Feller semigroups. In section C of this chapter we study then one example with a quantitative description of the ground state.

The next objective in the present part is to make sure, that the scattering magnitudes exist. Part (a) of Proposition 8.8 is the same as Theorem 5.7.

8.8. PROPOSITION. *Let V_1 and V_2 be Kato-Feller potentials.*

(1) (a) *Suppose that*

$$\int \exp\left(-2t(K_0 + V_i)\right)(x, x)\, |V_1(x) - V_2(x)|\, dx < \infty,$$

for $i = 1$ and 2, or that

$$\int \exp\left(-2t(K_0 + V_i)\right)(x, x)\, |V_1(x) - V_2(x)|^2\, dx < \infty,$$

for $i = 1$ and 2. Then the operator $D(t)$, defined by

$$D(t) = \exp\left(-t(K_0 \dot{+} V_1)\right) - \exp\left(-t(K_0 \dot{+} V_2)\right),$$

is Hilbert-Schmidt.

According to (8.4) and (8.1) the conditions above are satisfied if the function $V_1 - V_2$ belongs to $L^1(E, m) + L^2(E, m)$ (see Corollary 5.8)

(2) (b) *Suppose that, for some $a \in \mathbb{R}$, the functions*

$$(aI + K_0)^{-1} |V_i| \text{ belong to } C_\infty(E),\ i = 1,\ 2. \tag{8.5}$$

Then the operator $D(t)$ is compact (see Theorem 5.10).

The condition in (8.5) is satisfied if for every $\epsilon > 0$ there exists a compact subset K such that $|V_i(x)| \leq \epsilon$ for $x \notin K$ and for $i = 1, 2$, i.e. if the functions V_i, $i = 1, 2$, belong to

$$L_\epsilon^\infty(E, m) := \left\{ f \in L^\infty(E, m) : \lim_{x \to \triangle} f(x) = 0 \right\}.$$

Here, as usual, we use $\triangle$ to indicate the point at infinity of the locally compact space E.

8.9. COROLLARY. *The essential spectra of K_0 and $K_0 \dot{+} V$ are the same, i.e.*

$$\sigma_{\text{ess}}\left(K_0 \dot{+} V\right) = \sigma_{\text{ess}}(K_0),$$

provided the function V belongs to the space $L^1(E, m) + L^2(E, m) + L_\epsilon^\infty(E, m)$.

A similar result holds for singularly perturbed Feller operators: see Theorem 5.12. In the sequel we write $Jf = f \restriction_\Sigma$.

8.10. PROPOSITION. *Let K_0, $K_0 \dot{+} V$, $\left(K_0 \dot{+} V\right)_\Sigma$ be Feller operators as defined in Chapter 2. Let Γ be a closed singularity region as introduced in Chapter 2.C. Its complement $E \setminus \Gamma$ is denoted by Σ. Suppose $\Gamma^r = (int(\Gamma))^r$. Then the first hitting time T is $\mathbb{P}_x$-almost surely equal to the penetration time S of Γ (see Proposition 2.24). Denote by*

$$h_\Sigma^{a+V}(x) = \mathbb{E}_x\left[\exp\left(-\int_0^S (a + V(X(u)))du\right) : S < \infty\right] \tag{8.6}$$

the $a + V$-equilibrium potential and assume that

$$\int h_\Sigma^{a+V}(x)^2 dx < \infty. \tag{8.7}$$

Then, for any $t > 0$, the operator $D_\Sigma(t)$ given by

$$D_\Sigma(t) = \exp\left(-t\left(K_0 \dot{+} V\right)\right) - J^* \exp\left(-t\left(K_0 \dot{+} V\right)_\Sigma\right) J$$

is a Hilbert-Schmidt operator (with $Jf = f \restriction_\Sigma$).

REMARK. Assume that $V \equiv 0$ in (8.6). Then we can take $a = 1$ and we have to study the 1-equilibrium potential:

$$v_1(x) = \mathbb{E}_x\left[\exp(-S) : S < \infty\right].$$

In case of singular perturbations, the function v_1 plays a role corresponding to the potential V for regular perturbations. This is certainly the case for the description of the conditions under which results like Hilbert-Schmidt and trace class property are true. Because the function v_1 is bounded, the condition in (8.7) implies that

v_1 belongs to $L^1(E,m) \cap L^\infty(E,m)$ which is somehow related to Corollary 8.9. In fact, if V is a Kato-Feller potential, then the function h_Σ^{a+V} is bounded for a sufficiently large (see Proposition 4.20). Moreover, for Markov processes with symmetric transition kernels (see Proposition 2.37 and Proposition D.18), the capacity cap(Γ) equals $\|v_1\|_1 = \|v_1\|_{L^1(E,m)}$, i.e.

$$\operatorname{cap}(\Gamma) = \int_E v_1(x)\,dx.$$

In this context, it would be feasible to introduce the notion of an L^2-capacity of Γ as $\int dx\; v_1(x)^2$ because this kind of L^2-condition implies the Hilbert-Schmidt property of the operators $\exp(-tK_0) - J^* \exp\left(-t\left(K_0\right)_\Sigma\right) J$. But we will not do this. In any case, the Hilbert-Schmidt property of the singularly perturbed semigroup difference $D_\Sigma(t)$ follows for subsets Γ of finite capacity. The obvious implication is the invariance of the essential spectrum:

$$\sigma_{\text{ess}}(K_0) = \sigma_{\text{ess}}\left(\left(K_0\right)_\Sigma\right).$$

Because of the inequality

$$h_\Sigma^{1+V}(x)^2 \le c\,\mathbb{E}_x\left[\exp(-S) : S < \infty\right]$$

with a constant c that depends on the Kato-Feller norm of V_-, the results in Proposition 8.8 and Corollary 8.9 can be combined to yield the following corollary. Here we use Weyl's theorem (see Appendix A, A.28).

8.11. Corollary. *Let V be a Kato-Feller potential belonging to the space $L^1 + L^2 + L^\infty_\epsilon$. If cap(Γ) is finite, then*

$$\sigma_{\text{ess}}(K_0) = \sigma_{\text{ess}}\left((K_0)_\Sigma\right) = \sigma_{\text{ess}}\left((K_0 \dot{+} V)\right) = \sigma_{\text{ess}}\left((K_0 \dot{+} V)_\Sigma\right).$$

Similar considerations may be used to obtain the invariance of the absolutely continuous part of the spectrum. The role played by Weyl's theorem (stating that if the difference of two operators is compact, then they possess the same essential spectra), is now taken over by the existence and completeness of the wave operators. For the sake of completeness they will be defined presently.

8.12. Definition. Let K_0, V_1, V_2, Γ and the restriction operator J be as in Proposition 8.10. Then we define the wave operators $\Omega_\pm\left(K_0 \dot{+} V_2, K_0 \dot{+} V_1\right)$ as

$$\begin{aligned} &\Omega_\pm\left(K_0 \dot{+} V_2, K_0 \dot{+} V_1\right) \\ &= \text{s-}\lim_{t\to\pm\infty} \exp\left(+it\left(K_0 \dot{+} V_2\right)\right) \exp\left(-it\left(K_0 \dot{+} V_1\right)\right) P_{\text{ac}}\left(K_0 \dot{+} V_1\right) \end{aligned} \tag{8.8}$$

and the two-space wave operators $\Omega_\pm\left(\left(K_0 \dot{+} V\right)_\Sigma, J, K_0 \dot{+} V\right)$ as

$$\begin{aligned} &\Omega_\pm\left(\left(K_0 \dot{+} V\right)_\Sigma, J, K_0 \dot{+} V\right) \\ &= \text{s-}\lim_{t\to\pm\infty} \exp\left(+it\left(K_0 \dot{+} V\right)_\Sigma\right) J \exp\left(-it\left(K_0 \dot{+} V\right)\right) P_{\text{ac}}\left(K_0 \dot{+} V\right). \end{aligned} \tag{8.9}$$

The symbol $P_{\mathrm{ac}}(A)$ stands for the orthogonal projection operator onto the absolutely continuous subspace of the self-adjoint operator A. The wave operator in (8.8) is a mapping from $L^2(E,m)$ to $L^2(E,m)$ and the one in (8.9) sends the space $L^2(E,m)$ to $L^2(\Sigma,m)$.

The wave operator $\Omega_+\left(K_0\dot{+}V_2, K_0\dot{+}V_1\right)$ is called *complete* if the operator

$$\Omega_+\left(K_0\dot{+}V_1, K_0\dot{+}V_2\right)$$

exists as well. The wave operator

$$\Omega_+\left(\left(K_0\dot{+}V\right)_\Sigma, J, K_0\dot{+}V\right)$$

is called *complete* if the operator

$$\Omega_+\left(K_0\dot{+}V, J^*, \left(K_0\dot{+}V\right)_\Sigma\right)$$

also exists and if, additionally,

$$\text{s-}\lim_{t\to+\infty}\left(J^*J - I_{L^2(E)}\right)\exp\left(-it\left(K_0\dot{+}V\right)\right)P_{\mathrm{ac}}\left(K_0\dot{+}V\right) = 0 \tag{8.10}$$

and

$$\text{s-}\lim_{t\to+\infty}\left(JJ^* - I_{L^2(\Sigma)}\right)\exp\left(-it\left(K_0\dot{+}V\right)_\Sigma\right)P_{\mathrm{ac}}\left(\left(K_0\dot{+}V\right)_\Sigma\right) = 0. \tag{8.11}$$

The corresponding definitions, with $t\to-\infty$ are used to define the completeness of Ω_-. The scattering system $\left(\left(K_0\dot{+}V\right)_\Sigma, J, K_0\dot{+}V\right)$ is said to be *complete* if Ω_+ and Ω_- are complete.

If the wave operators $\Omega_\pm$ exist, then the *scattering operator* S is defined by

$$\begin{aligned} &S\left(K_0\dot{+}V_2, K_0\dot{+}V_1\right)\\ &= \Omega_+^*\left(K_0\dot{+}V_2, K_0\dot{+}V_1\right)\Omega_-\left(K_0\dot{+}V_2, K_0\dot{+}V_1\right) \end{aligned}\tag{8.12}$$

$$\begin{aligned} &S\left(\left(K_0\dot{+}V\right)_\Sigma, K_0\dot{+}V\right)\\ &= \Omega_+^*\left(\left(K_0\dot{+}V\right)_\Sigma, J, K_0\dot{+}V\right)\Omega_-\left(\left(K_0\dot{+}V\right)_\Sigma, J, K_0\dot{+}V\right). \end{aligned}\tag{8.13}$$

Mathematical scattering theory will not be repeated here. The reader may consult standard textbooks like Amrein, Jauch and Sinha [8], Reed and Simon [193], Baumgärtel and Wollenberg [21] or, for the time-independent scattering theory, Yafaev [253].

Upon employing the results of Chapter 6 (trace class) we can ensure ourselves of the existence and completeness of scattering systems which will be the basis for our quantitative estimates in the second section of the present chapter. For $f\in P_{\mathrm{ac}}(K_0)(\mathcal{H})$, $\mathcal{H} = L^2(E,m)$, the norm $|||\cdot|||$ is defined by

$$|||f||| = \sup_{\lambda\in\sigma_{\mathrm{ac}}(K_0)}\frac{d\left\|E_{K_0}(\lambda)f\right\|^2}{d\lambda},$$

where $E_{K_0}(\cdot)$ is the resolution of the identity corresponding to K_0.

8.13. PROPOSITION. *(a) Let K_0 be the unperturbed Feller operator and let V_1 and V_2 be two Kato-Feller potentials. Suppose that the integral*

$$C(t) := \int \left(\int p_0(t,x,y) \left| V_2(y) - V_1(y) \right|^2 dy \right)^{1/2} dx \tag{8.14}$$

is finite. Then the wave operator $\Omega_+ \left(K_0 \dot{+} V_2, K_0 \dot{+} V_1\right)$ exists and is complete. Consequently the absolutely continuous parts of the spectra are preserved:

$$\sigma_{\mathrm{ac}} \left(K_0 \dot{+} V_1\right) = \sigma_{\mathrm{ac}} \left(K_0 \dot{+} V_2\right).$$

(b) In terms of $|||\cdot|||$, the difference of the wave operators may be estimated as follows:

$$\begin{aligned} &\left\| \left[\Omega_+ \left(K_0 \dot{+} V_2, K_0\right) - \Omega_+ \left(K_0 \dot{+} V_1, K_0\right)\right] f \right\|_2^2 \\ &\leq 16\pi |||f|||^2 \left(C_{V_1}(t) + C_{V_2}(t)\right) C(t), \end{aligned} \tag{8.15}$$

where the constants $C_{V_i}(t)$, $i = 1,\ 2$, are given by

$$\begin{aligned} C_{V_i}(t) = \int_0^{t/2} d\tau \left\| \exp\left(- \left(\frac{t}{2} - \tau \right) \left(K_0 \dot{+} V_1\right) \right) \right\|_{2,2} \left\| \exp\left(-\tau \left(K_0 \dot{+} V_2\right)\right) \right\|_{2,2} \\ \times M_{4V_i}^{1/2} M_{2V_i}^{1/2} \exp\left(\frac{t}{8} b_{2V_i} \right) \exp\left(\frac{t}{16} b_{4V_i} \right) \sup_{w \in E} p_0(t/4, w, w). \end{aligned}$$

Here, the constants M_V and b_V satisfy the inequality

$$\left\| \exp\left(-t \left(K_0 \dot{+} V\right)\right) \right\|_{\infty,\infty} \leq M_V \exp\left(t b_V\right)$$

for all $t > 0$: see inequality (3.48) in Corollary 3.6 (Khas'minskii's lemma). A corresponding result holds for the wave operators $\Omega_- \left(K_0 \dot{+} V_j, K_0\right)$, $j = 1,\ 2$.

Proof. (a) By virtue of Theorem 6.7, the finiteness of the constant $C(t)$ implies the existence and completeness of

$$\Omega_+ \left(\exp\left(- \left(K_0 \dot{+} V_2\right)\right), \exp\left(-K_0 \dot{+} V_1\right)\right).$$

Using the invariance principle, this wave operator is identical to the wave operator $\Omega_+ \left(K_0 \dot{+} V_2, K_0 \dot{+} V_1\right)$.

(b) By virtue of the corollary on page 26 of Reed and Simon [193] together with the invariance principle for the wave operators, we have, for $f \in P_{\mathrm{ac}}(K_0 \dot{+} V_1)\mathcal{H}$ and for $t > 0$, the following Pearson's estimate:

$$\left\| \left(\Omega_+ \left(K_0 \dot{+} V_2, K_0 \dot{+} V_1\right) - I\right) f \right\| \leq 16\pi \left\| D(t) \right\|_{\mathrm{trace}} |||f||| \tag{8.16}$$

where $D(t)$ is the difference

$$\begin{aligned}
D(t) &= \exp\left(-t\left(K_0 \dot{+} V_1\right)\right) - \exp\left(-t\left(K_0 \dot{+} V_2\right)\right) \\
&= \int_0^{t/2} d\tau \exp\left(-\left(\frac{t}{2}-\tau\right)\left(K_0 \dot{+} V_1\right)\right) \exp\left(-\frac{t}{2}\left(K_0 \dot{+} V_1\right)\right) \\
&\qquad (V_2 - V_1) \exp\left(-\tau\left(K_0 \dot{+} V_2\right)\right) \\
&\quad + \int_0^{t/2} d\tau \exp\left(-\left(\frac{t}{2}-\tau\right)\left(K_0 \dot{+} V_1\right)\right) \\
&\qquad (V_2 - V_1) \exp\left(-\frac{t}{2}\left(K_0 \dot{+} V_2\right)\right) \exp\left(-\tau\left(K_0 \dot{+} V_2\right)\right).
\end{aligned}$$

Hence we get

$$\begin{aligned}
&\|D(t)\|_{\mathrm{trace}} \\
&\leq \int_0^{t/2} d\tau \left\|\exp\left(-\left(\frac{t}{2}-\tau\right)\left(K_0 \dot{+} V_1\right)\right)\right\|_{2,2} \left\|\exp\left(-\tau\left(K_0 \dot{+} V_2\right)\right)\right\|_{2,2} \\
&\times \left\|\exp\left(-\frac{t}{2}\left(K_0 \dot{+} V_1\right)\right)(V_2 - V_1)\right\|_{\mathrm{trace}} \\
&+ \int_0^{t/2} d\tau \left\|\exp\left(-\left(\frac{t}{2}-\tau\right)\left(K_0 \dot{+} V_1\right)\right)\right\|_{2,2} \left\|\exp\left(-\tau\left(K_0 \dot{+} V_2\right)\right)\right\|_{2,2} \\
&\times \left\|(V_2 - V_1)\exp\left(-\frac{t}{2}\left(K_0 \dot{+} V_2\right)\right)\right\|_{\mathrm{trace}}.
\end{aligned} \tag{8.17}$$

From Theorem 6.1 it follows that

$$\begin{aligned}
&\left\|\exp\left(-\frac{t}{2}\left(K_0 \dot{+} V_i\right)\right)(V_2 - V_1)\right\|_{\mathrm{trace}} = \left\|(V_2 - V_1)\exp\left(-\frac{t}{2}\left(K_0 \dot{+} V_i\right)\right)\right\|_{\mathrm{trace}} \\
&\leq \int dz \sqrt{\int dx \left[\exp\left(-\frac{t}{4}\left(K_0 + V_i\right)\right)(x,z)\right]^2} \\
&\quad \sqrt{\int dx \left[\exp\left(-\frac{t}{4}\left(K_0 + V_i\right)\right)(z,x)\right]^2 |V_2(x) - V_1(x)|^2} \\
&= \int dz \sqrt{\exp\left(-\frac{t}{2}\left(K_0 + V_i\right)\right)(z,z)} \\
&\quad \sqrt{\int dx \left[\exp\left(-\frac{t}{4}\left(K_0 + V_i\right)\right)(z,x)\right]^2 |V_2(x) - V_1(x)|^2}
\end{aligned}$$

(see the inequalities (3) and (4) in Theorem 2.9)

$$\leq M_{4V_i}^{1/2} M_{2V_i}^{1/2} \exp\left(\frac{t}{8} b_{2V_i}\right) \exp\left(\frac{t}{16} b_{4V_i}\right) \sup_{w \in E} p_0(t/4, w, w)$$
$$\times \int dz \sqrt{\int p_0(t/4, z, x) \left|V_2(x) - V_1(x)\right|^2 dx}.$$

Consequently from (8.17) we get

$$\|D(t)\|_{\text{trace}} \leq \left(C_{V_1}(t) + C_{V_2}(t)\right) \int dz \sqrt{\int dx p_0\left(t/4, z, x\right) \left|V_2(x) - V_1(x)\right|^2},$$

where

$$C_{V_i}(t) = \int_0^{t/2} d\tau \left\|\exp\left(-\left(\frac{t}{2} - \tau\right)\left(K_0 \dot{+} V_1\right)\right)\right\|_{2,2} \left\|\exp\left(-\tau\left(K_0 \dot{+} V_2\right)\right)\right\|_{2,2}$$
$$\times M_{4V_i}^{1/2} M_{2V_i}^{1/2} \exp\left(\frac{t}{8} b_{2V_i}\right) \exp\left(\frac{t}{16} b_{4V_i}\right) \sup_{w \in E} p_0(t/4, w, w).$$

Here, the constants M_V and b_V satisfy the inequality

$$\left\|\exp\left(-t\left(K_0 \dot{+} V\right)\right)\right\|_{\infty,\infty} \leq M_V \exp\left(t b_V\right)$$

for all $t > 0$: see inequality (2.7).

The singular counterpart of the previous proposition reads as follows.

8.14. PROPOSITION. *Let V be a Kato-Feller potential and let Γ be a singularity region with penetration time S. Let h_Σ^{a+V} be the $a+V$-equilibrium potential defined by*

$$h_\Sigma^{a+V}(x) = \mathbb{E}_x \left[\exp\left(-\int_0^S (a + V(X(u)))\, du\right) : S < \infty\right].$$

Let f be a function in $P_{\text{ac}}(K_0 + V)\mathcal{H}$. Then we have the following estimate for the wave operators:

$$\left\|\left[\Omega_\pm\left(\left(K_0 \dot{+} V\right)_\Sigma, J, K_0 \dot{+} V\right) - J\right] f\right\|_2^2 \tag{8.18}$$
$$\leq 32\pi C(t) |||f|||^2 \int dx \sqrt{\mathbb{E}_x \left[\exp\left(-\int_0^S V(X(u)) du\right) : S < t/2\right]}$$
$$\leq 32\pi C(t) |||f|||^2 \exp\left(\frac{at}{4}\right) \int dx \sqrt{h_\Sigma^{a+V}(x)}$$

where the constant $C(t)$ is given by

$$C(t) = \sup_{y \in E} \exp\left(-\frac{t}{2}\left(K_0 + V\right)\right)(y, y)$$
$$\times \sup_{0 < s \leq t/2} \sup_{y \in E} \mathbb{E}_y \left[\exp\left(-\int_0^s V(X(u)) du\right)\right].$$

If the function $V \equiv 0$, or even more generally if $V \geq 0$, the constant $C(t)$ may be taken to be

$$C(t) = \sup_{y \in E} p_0(t/2, y, y).$$

Proof. The proof follows the same pattern as that of Proposition 8.13. However, in order to keep track of the constants, we shall repeat the basic steps here. As in the previous chapters, we write

$$D_\Sigma(t) = \exp\left(-t\left(K_0 \dot{+} V\right)\right) - J^* \exp\left(-t\left(K_0 \dot{+} V\right)_\Sigma\right) J.$$

First we note the identities

$$\begin{aligned} D_\Sigma(t) &= D_\Sigma(t/2)\exp\left(-\frac{t}{2}\left(K_0 \dot{+} V\right)\right) + J^*\exp\left(-\frac{t}{2}\left(K_0 \dot{+} V\right)_\Sigma\right) J D_\Sigma(t/2) \\ &= -D_\Sigma(t/2)^2 + D_\Sigma(t/2)\exp\left(-\frac{t}{2}\left(K_0 \dot{+} V\right)\right) + \exp\left(-\frac{t}{2}\left(K_0 \dot{+} V\right)\right) D_\Sigma(t/2) \\ &= D_\Sigma(t/2) J^*\exp\left(-\frac{t}{2}\left(K_0 \dot{+} V\right)_\Sigma\right) J + \exp\left(-\frac{t}{2}\left(K_0 \dot{+} V\right)\right) D_\Sigma(t/2). \end{aligned}$$

Hence we get the inequality

$$\begin{aligned} &\|D_\Sigma(t)\|_{\text{trace}} \\ &\leq \left\| D_\Sigma(t/2)\exp\left(-\frac{t}{2}\left(K_0 \dot{+} V\right)\right)\right\|_{\text{trace}} + \left\| D_\Sigma(t/2) J^* \exp\left(-\frac{t}{2}\left(K_0 \dot{+} V\right)_\Sigma\right) J\right\|_{\text{trace}}. \end{aligned}$$

From the proof of the implication (i″) ⇒ (iii) of Theorem 6.11, we obtain the inequality

$$\begin{aligned} &\left\| D_\Sigma(t/2)\exp\left(-\frac{t}{2}\left(K_0 \dot{+} V\right)\right)\right\|_{\text{trace}} \\ &\qquad \leq C_1(t)\int dx \sqrt{\mathbb{E}_x\left[\exp\left(-\int_0^S V(X(u))du\right) : S < t/2\right]} \end{aligned}$$

where the constant $C_1(t)$ is given by

$$\begin{aligned} C_1(t) &= \sup_{y \in E} \exp\left(-\frac{t}{2}\left(K_0 + V\right)\right)(y, y) \\ &\qquad \times \sup_{0 < s \leq t/2} \sup_{y \in E} \mathbb{E}_y\left[\exp\left(-\int_0^s V(X(u))du\right)\right]. \end{aligned}$$

In the same way as we proved the implication (i″) ⇒ (iii) in Theorem 6.11, we obtain the inequality

$$\begin{aligned}&\left\| D_\Sigma(t/2)J^* \exp\left(-\frac{t}{2}\left(K_0\dot{+}V\right)_\Sigma\right) J\right\|_{\text{trace}}\\ &\leq C_2(t)\int dx \sqrt{\mathbb{E}_x\left[\exp\left(-\int_0^S V(X(u))du\right) : S < t/2\right]}\end{aligned}$$

where the constant $C_2(t)$ is given by

$$\begin{aligned}C_2(t) &= \sup_{y\in E} \exp\left(-\frac{t}{2}\left(K_0+V\right)_\Sigma\right)(y,y)\\ &\quad\times \sup_{0<s\leq t/2}\sup_{y\in E}\mathbb{E}_y\left[\exp\left(-\int_0^s V(X(u))du\right)\right].\end{aligned}$$

By virtue of the inequality

$$\exp\left(-t\left(K_0+V\right)_\Sigma\right)(x,y) \leq \exp\left(-t\left(K_0+V\right)\right)(x,y),$$

we see that $C_2(t) \leq C_1(t)$. Hence the conclusion in the theorem follows.

The advantages in the latter propositions are that we have explicit estimates for the vector norm of the wave or scattering operators in terms of the regular or singular perturbations. This will be useful for quantitative estimates as we will see in section B of this chapter. On the other hand, in general the integral $\int dx \sqrt{h_\Sigma^{a+V}(x)}$ in (8.18) is not necessarily finite for a singularity region Γ of finite capacity even in case $V \equiv 0$. To see this we take $K_0 = H_0 = -\frac{1}{2}\Delta$ in $L^2(\mathbb{R}^\nu)$ and let $\Gamma = \bigcup_{n=1}^\infty B(x_n, r_n)$ be a union of disjoint balls of radii r_n, $0 < r_n < \frac{1}{2}$, and centers at $x_n = (n, 0, \ldots, 0)$ in $\mathbb{R}^\nu$, $n \in \mathbb{N}$. Let $D_\Sigma(H_0, t)$ denote the operator $D_\Sigma(H_0,t) = \exp(-tH_0) - J^*\exp(-t(H_0)_\Sigma)J$ and suppose that this operator is trace class. Then the operator $1_\Gamma \exp(-tH_0)$ is of trace class as well. Among other things, this implies that $\sum_{n=1}^\infty r_n^{\nu/2} < \infty$. On the other hand, up to a multiplicative constant, the capacity of $\Gamma = \bigcup_{n=1}^\infty B(x_n, r_n)$ is dominated by $\sum_{n=1}^\infty r_n^{\nu-2}$. This means that, for $\nu \geq 5$, there are sets of finite capacity for which the operator $D_\Sigma(H_0,t)$ is not a trace operator. More details of this example can be found in [80]. Some information is given in the remark following Theorem 6.11.

Nevertheless the completeness of the wave operators can be extended to sets of finite capacity (and to regular potentials in $L^1(E, m)$). In what follows we write $\mathcal{B}_1(\mathcal{H}_1, \mathcal{H}_2)$ for the trace class operators from the Hilbert space $\mathcal{H}_1$ to the Hilbert space $\mathcal{H}_2$. The symbol $\mathcal{B}_2(\mathcal{H}_1, \mathcal{H}_2)$ denotes the space of Hilbert-Schmidt operators between $\mathcal{H}_1$ and $\mathcal{H}_2$ and similarly $\mathcal{B}_\infty(\mathcal{H}_1, \mathcal{H}_2)$ stands for the space of compact operators from $\mathcal{H}_1$ to $\mathcal{H}_2$. If, from the context it is clear between what spaces the operators are acting we often write $\mathcal{B}_1$ instead of $\mathcal{B}_1(\mathcal{H}_1, \mathcal{H}_2)$. The same remark applies to $\mathcal{B}_2$ and to $\mathcal{B}_\infty$.

8.15. LEMMA. *Let K_1 and K_2 be self-adjoint semi-bounded operators in not necessarily the same Hilbert spaces $\mathcal{H}_1$ and $\mathcal{H}_2$ respectively. Let $J : \mathcal{H}_1 \to \mathcal{H}_2$ be a bounded identification operator between $\mathcal{H}_1$ and $\mathcal{H}_2$. Then the wave operators $\Omega_\pm(K_2, J, K_1)$ and $\Omega_\pm(K_1, J^*, K_2)$ exist provided that the following conditions are satisfied:*

(i) *The operator*

$$\exp(-K_2)(\exp(-K_2) J - J\exp(-K_1))\exp(-K_1) \\ \text{belongs to } \mathcal{B}_1(\mathcal{H}_1, \mathcal{H}_2); \tag{8.19}$$

(ii) *The operator*

$$\exp(-K_2) J - J\exp(-K_1) \quad \text{belongs to} \quad \mathcal{B}_\infty(\mathcal{H}_1, \mathcal{H}_2). \tag{8.20}$$

The wave operators are complete, if in addition the following conditions are satisfied:

(iii) *The operator*

$$(J^*J - I_{\mathcal{H}_1})\exp(-K_1) \quad \text{belongs to} \quad \mathcal{B}_\infty(\mathcal{H}_1, \mathcal{H}_1); \tag{8.21}$$

(iv) *The operator*

$$(JJ^* - I_{\mathcal{H}_2})\exp(-K_2) \quad \text{belongs to} \quad \mathcal{B}_\infty(\mathcal{H}_2, \mathcal{H}_2). \tag{8.22}$$

Proof. From Pearson's two space trace class criterion (see e.g. Reed and Simon [193, p. 24], it follows that the wave operators

$$\Omega_\pm(\exp(-K_2), \exp(-K_2) J\exp(-K_1), \exp(-K_1))$$

and

$$\Omega_\pm(\exp(-K_1), \exp(-K_1) J^*\exp(-K_2), \exp(-K_2))$$

exist. We consider only the first one. It suffices to prove the existence of wave operators on some dense set in $P_{\text{ac}}(\exp(-K_1))\mathcal{H}_1$. Notice that the range of $\exp(-K_1)$ is dense in $P_{\text{ac}}(\exp(-K_1))\mathcal{H}_1$. Following the arguments in Baumgärtel and Wollenberg [21, p. 347] we get the existence of the strong limit:

$$\text{s-}\lim_{t\to\infty} \exp\left(it\left(e^{-K_2}\right)\right) e^{-K_2} J\exp\left(-it\left(e^{-K_1}\right)\right) P_{\text{ac}}\left(e^{-K_1}\right).$$

The latter limit equals

$$\text{s-}\lim_{t\to\infty} \exp\left(it\left(e^{-K_1}\right)\right) Je^{-K_1}\exp\left(-it\left(e^{-K_1}\right)\right) P_{\text{ac}}\left(e^{-K_1}\right)$$

as long as

$$\text{s-}\lim_{t\to\infty} \left(e^{-K_2}J - Je^{-K_1}\right)\exp\left(-it\left(e^{-K_1}\right)\right) P_{\text{ac}}\left(e^{-K_1}\right) = 0.$$

The latter equality is true because of the Riemann-Lebesgue Lemma and because the operator $e^{-K_2}J - Je^{-K_1}$ is assumed to be compact. So the existence of the strong limit

$$\text{s-}\lim_{t\to\infty} \exp\left(it\left(e^{-K_2}\right)\right) Je^{-K_1} \exp\left(-it\left(e^{-K_1}\right)\right) P_{\mathrm{ac}}\left(e^{-K_1}\right)$$

follows. Hence, again an approximation argument implies the existence of the wave operator $\Omega_+\left(e^{-K_2}, J, e^{-K_1}\right)$. Finally the invariance principle is applicable for the exponential function which implies the existence of $\Omega_-\left(K_2, J, K_1\right)$.

For the completeness we need (8.10) and (8.11). But (8.21) and (8.22) imply (8.10) and (8.11) respectively, again via the Riemann-Lebesgue Lemma and because the range of e^{-K_i} is dense in $\mathcal{H}_i$ for $i = 1, 2$.

In the spirit of Lemma 8.15, we next obtain complete scattering systems if V is a Kato-Feller potential in $L^1(E, m)$ and if the singularity region Γ has finite capacity. These results will be formulated in the following proposition. For convenience we repeat here the definitions of the functions v_λ (see Proposition 2.35, Definition 2.36, Proposition D.18, and the remarks following Proposition 2.34 and Theorem 5.12 respectively) and of h_Σ^{a+V} (see Definition 2.30):

$$v_{\lambda,\Gamma}(x) = \mathbb{E}_x\left(\exp(-\lambda S) : S < \infty\right);$$

$$h_\Sigma^{a+V}(x) = \mathbb{E}_x\left(\exp\left(-\int_0^S (a + V(X(u)))\,du\right) : S < \infty\right).$$

In terms of the harmonic extension operator H_Σ^{a+V} (see Definition 2.30) and the operator $T_\Sigma^V(t)$ (see Definition 2.33), we have the following equalities $v_{\lambda,\Gamma} = h_\Sigma^\lambda = H_\Sigma^\lambda 1$ and $h_\Sigma^{a+V} = a\int_0^\infty e^{-as}T_\Sigma^V(s)1\,ds$: see equality (2.57) in Theorem 2.34. The operators $T_\Sigma^V(t)$ (Definition 2.33) and H_Σ^{a+V} (Definition 2.30) are given by the formulae:

$$\left[T_\Sigma^V(t)f\right](x) = \mathbb{E}_x\left[\exp\left(-\int_0^S V(X(u))du\right) f(X(S)) : S < t\right];$$

$$\left[H_\Sigma^{a+V}f\right](x) = \mathbb{E}_x\left[\exp\left(-\int_0^S (a + V(X(u)))\,du\right) f(X(S)) : S < \infty\right].$$

Suppose $1 \leq p \leq \infty$. Then the function $H_\Sigma^{a+V}1$ belongs to $L^p(E, m)$ (for $a > 0$ sufficiently large) if and only if the function $T_\Sigma^V 1$ belongs to $L^p(E, m)$: see Proposition 4.20.

8.16. PROPOSITION.

(a) *Let K_0 be a free Feller operator and let V be a Kato-Feller potential. Then the scattering system $\left(K_0 \dot{+} V, K_0\right)$ is complete if the function V belongs to $L^1(E, m)$.*

(b) *Let K_0 be a free Feller operator and let Γ be a singularity region of finite capacity, i.e. suppose that $\int dx\, v_{1,\Gamma}(x) < \infty$. Then the corresponding scattering system is complete. Together with (a), this implies that the scattering system $\left((K_0)_\Sigma, J, K_0 \dot{+} V\right)$ is complete as well, provided V belongs to $L^1(E,m)$.*

(c) *Let K_0 be a free Feller operator, Γ a singularity region and V a Kato-Feller potential with the property that the $a+V$-equilibrium potential h_Σ^{a+V} belongs to $L^1(E,m)$. Then the scattering system $\left((K_0 \dot{+} V)_\Sigma, J, K_0 \dot{+} V\right)$ is complete.*

Proof. (a) Here we are in a one space situation. In that case Lemma 8.15 is applicable if the following conditions are satisfied:

$$\exp\left(-\left(K_0 \dot{+} V\right)\right)\left(\exp\left(-\left(K_0 \dot{+} V\right)\right) - \exp\left(-K_0\right)\right)\exp\left(-K_0\right) \quad \text{belongs to} \quad \mathcal{B}_1$$

and the operator

$$\exp\left(-\left(K_0 \dot{+} V\right)\right) - \exp\left(-K_0\right) \quad \text{belongs to} \quad \mathcal{B}_\infty.$$

From Proposition 8.8 (see Theorem 5.7 as well), it follows that the operator $\exp\left(-\left(K_0 \dot{+} V\right)\right) - \exp\left(-K_0\right)$ is Hilbert-Schmidt because the function V belongs to $L^1(E,m)$. From Duhamel's formula we infer the equality

$$\begin{aligned} &\exp\left(-\left(K_0 \dot{+} V\right)\right)\left(\exp\left(-\left(K_0 \dot{+} V\right)\right) - \exp\left(-K_0\right)\right)\exp\left(-K_0\right) \\ &= \int_0^1 ds \exp\left(-(2-s)\left(K_0 \dot{+} V\right)\right) V \exp\left(-(1+s)K_0\right). \end{aligned}$$

The integrand is a trace class operator for any s, $0 \le s \le 1$, using the trace class result in Theorem 6.1 and its consequence in Proposition 6.9 item (ii). To be more precise, from Proposition 6.9.(ii) it follows that the operator $\exp\left(-\left(K_0 \dot{+} V\right)\right) V \exp\left(-K_0\right)$ is a trace class operator. In addition its trace norm can be estimated as follows:

$$\begin{aligned} &\left\|\exp\left(-\left(K_0 \dot{+} V\right)\right) V \exp\left(-K_0\right)\right\|_{\text{trace}} \\ &\le \sqrt{\int dx \exp\left(-2\left(K_0 + V\right)\right)(x,x)\left|V(x)\right| \int dx \exp\left(-2K_0\right)(x,x)\left|V(x)\right|} \end{aligned}$$

(see Theorem 2.9 equality (1) and inequality (3))

$$\le \left\|\exp\left(-\left(K_0 \dot{+} 2V\right)\right)\right\|_{\infty,\infty}^{1/2} \sqrt{\sup_{y\in E} p_0(1,y,y) \sup_{y\in E} p_0(2,y,y)} \int dx \left|V(x)\right|$$

(the function $t \mapsto \sup_{x\in E} p_0(t,x,x)$ decreases: see Corollary 2.10 assertion (a))

$$\begin{aligned} &\le \left\|\exp\left(-\left(K_0 \dot{+} 2V\right)\right)\right\|_{\infty,\infty}^{1/2} \sup_{y\in E} p_0(1,y,y) \int dx \left|V(x)\right| \\ &\le c_1 \left\|V\right\|_1, \end{aligned}$$

where the constant c_1 is given by

$$c_1 = \left\|\exp\left(-\left(K_0\dot{+}2V\right)\right)\right\|_{\infty,\infty}^{1/2} \sup_{x\in E} p_0(1,x,x).$$

Moreover the integral

$$\int_0^1 ds \exp\left(-(2-s)\left(K_0\dot{+}V\right)\right) V \exp\left(-(1+s)K_0\right)$$

is to be interpreted as a strong integral because for $f \in L^2(E,m)$, the function $s \mapsto \exp\left(-(1-s)\left(K_0\dot{+}V\right)\right) T \exp\left(-sK_0\right) f$ is continuous as a mapping from the interval $[0,1]$ to $L^2(E,m)$ whenever T is a continuous linear operator. In the present situation we take $T = \exp\left(-\left(K_0\dot{+}V\right)\right) V \exp\left(-K_0\right)$. The trace norm of this integral can be estimated as follows (a.o. see Lemma 6.6):

$$\begin{aligned}
&\left\|\int_0^1 ds \exp\left(-(2-s)\left(K_0\dot{+}V\right)\right) V \exp\left(-(1+s)K_0\right)\right\|_{\text{trace}} \\
&\leq \int_0^1 ds \left\|\exp\left(-(1-s)\left(K_0\dot{+}V\right)\right)\right\|_{2,2} \\
&\qquad \left\|\exp\left(-\left(K_0\dot{+}V\right)\right) V \exp\left(-K_0\right)\right\|_{\text{trace}} \left\|\exp\left(-sK_0\right)\right\|_{2,2} \\
&\leq c_2 \left\|V\right\|_1,
\end{aligned}$$

where the constant c_2 is given by

$$\begin{aligned}
c_2 = &\left\|\exp\left(-\left(K_0\dot{+}2V\right)\right)\right\|_{\infty,\infty}^{1/2} \sup_{x\in E} p_0(1,x,x) \\
&\times \int_0^1 ds \left\|\exp\left(-(1-s)\left(K_0\dot{+}V\right)\right)\right\|_{2,2} \left\|\exp\left(-sK_0\right)\right\|_{2,2}.
\end{aligned}$$

In form sense, $K_0\dot{+}V \geq -\omega_V I$ ($\omega_V \in \mathbb{R}$), and, for $V = 0$, $\omega_0 \geq 0$ (see formula (2.8) in Chapter 2) so that

$$\begin{aligned}
c_2 &\leq \left\|\exp\left(-\left(K_0\dot{+}2V\right)\right)\right\|_{\infty,\infty}^{1/2} \sup_{x\in E} p_0(1,x,x) \int_0^1 ds \exp\left((1-s)\omega_V\right) \\
&\leq M_{2V}^{1/2} \exp\left(\frac{1}{2} b_{2V}\right) \sup_{x\in E} p_0(1,x,x) \frac{e^{\omega_V}-1}{\omega_V}.
\end{aligned}$$

(b) and (c) The proof of the first part of (b) and that of (c) are about the same. The second assertion in (b) follows via the chain rule for scattering systems. So we only prove the first assertion in (b). We explore Lemma 8.15 for this special situation. We want to apply Lemma 8.15 with $K_1 = K_0$ and with $K_2 = (K_0)_\Sigma$. In the case of singular perturbations, $JJ^* = I_{L^2(\Sigma,m)}$ and hence (8.22) is trivially satisfied. The condition in (8.21) only says that the operator $1_\Gamma \exp(-K_0)$ has to be compact. Since the m-measure of Γ is dominated by its capacity, it follows that its measure is finite. Consequently the operator $1_\Gamma \exp(-K_0)$ is Hilbert-Schmidt and hence (8.21) is satisfied as well. In order to verify condition (8.20), we may use (8.7) in Proposition 8.10 with $a = 1$ and $V \equiv 0$ together with the fact that the operator $1_\Gamma \exp(-K_0)$ is Hilbert-Schmidt.

The condition in (8.19) remains to be proved. Since we have a sandwiched semigroup difference, we may apply Theorem 6.1. Since

$$\begin{aligned}\exp\left(-t\left(K_0\right)_\Sigma\right)(x,y) &\leq \exp\left(-tK_0\right)(x,y)\\ &= \int \exp\left(-\frac{t}{2}K_0\right)(x,z)\exp\left(-\frac{t}{2}K_0\right)(z,y)dz,\end{aligned}$$

we do this with the following kernels:

$$k_1(x,y) = p_0(1/2,x,y);$$
$$k_2(x,y) = \left[\exp\left(-\frac{1}{2}K_0\right)\left\{\exp\left(-\left(K_0\right)_\Sigma\right)J - J\exp\left(-K_0\right)\right\}\exp\left(-\frac{1}{2}K_0\right)\right](x,y).$$

In order to verify (8.19), it suffices to prove that the integral $\int dy\sqrt{\int dx\, k_2(x,y)^2}$ is finite. However, we have the following estimates (see also Theorem 2.34.(b)):

$$\begin{aligned}
\int k_2(x,y)^2dx &= \int dx\left[\int du_1\int du_2 p_0\left(1/2,x,u_1\right)\mu_{0,u_1}^{1,u_2}\left\{S<1\right\}p_0\left(1/2,u_2,y\right)\right]^2\\
&= \int dx\int du_1\int du_2\int du_1'\int du_2' p_0\left(1/2,x,u_1\right)\,p_0\left(1/2,x,u_1'\right)\\
&\quad\times \mu_{0,u_1}^{1,u_2}\left\{S<1\right\}\mu_{0,u_1'}^{1,u_2'}\left\{S<1\right\}p_0\left(1/2,u_2,y\right)\,p_0\left(1/2,u_2',y\right)\\
&\quad\text{(symmetry and Chapman-Kolmogorov)}\\
&= \int du_1\int du_2\int du_1'\int du_2' p_0\left(1,u_1,u_1'\right)\mu_{0,u_1}^{1,u_2}\left\{S<1\right\}\mu_{0,u_1'}^{1,u_2'}\left\{S<1\right\}\\
&\qquad\times p_0\left(1/2,u_2,y\right)\,p_0\left(1/2,u_2',y\right)\\
&\quad\text{(symmetry once more)}\\
&= \int du_1\int du_2\int du_1'\int du_2' p_0\left(1,u_1,u_1'\right)\mu_{0,u_2}^{1,u_1}\left\{S<1\right\}\mu_{0,u_2'}^{1,u_1'}\left\{S<1\right\}\\
&\quad\times p_0\left(1/2,u_2,y\right)\,p_0\left(1/2,u_2',y\right)\\
&\quad\left(\mu_{0,u}^{1,v}\left\{S<1\right\} = \mathbb{E}_u\left[p_0(1-S,X(S),v) : S<1\right]\text{: see Theorem 2.34.(b)}\right)
\end{aligned}$$

$$\begin{aligned}
&= \int du_1 \int du_2 \int du_1' \int du_2' \, p_0\left(1, u_1, u_1'\right) \\
&\quad \times \mathbb{E}_{u_2}\left[p_0(1-S, X(S), u_1) : S < 1\right] \mathbb{E}_{u_2'}\left[p_0(1-S, X(S), u_1') : S < 1\right] \\
&\quad \times p_0\left(1/2, u_2, y\right) p_0\left(1/2, u_2', y\right) \\
&\leq \sup_{u_1, u_1' \in E} p_0\left(1, u_1, u_1'\right) \\
&\quad \times \int du_2 \int du_2' \mathbb{P}_{u_2}\left[S < 1\right] \mathbb{P}_{u_2'}\left[S < 1\right] p_0\left(1/2, u_2, y\right) p_0\left(1/2, u_2', y\right) \\
&= \sup_{u \in E} p_0\left(1, u, u\right) \left(\int \mathbb{P}_v\left[S < 1\right] p_0\left(1/2, v, y\right) dv \right)^2,
\end{aligned}$$

and hence

$$\begin{aligned}
&\int dy \sqrt{\int dx \, k_2(x,y)^2} \leq \sqrt{\sup_{u \in E} p_0(1,u,u)} \int \mathbb{P}_y\left\{S < 1\right\} dy \\
&\leq e \sqrt{\sup_{u \in E} p_0(1,u,u)} \int v_1(y) \, dy \leq c \;\; \operatorname{cap}(\Gamma),
\end{aligned}$$

where $c = e\sqrt{\sup_{x \in E} p_0(1,x,x)}$.

REMARK. Part (a) of Proposition 8.16 corresponds to the situation for Schrödinger operators $-\frac{1}{2}\Delta + V$ where V is relative form bounded with respect to $H_0 = -\frac{1}{2}\Delta$ and with form bound strictly less than 1 and with V a function in $L^1(E, m)$: see Reed and Simon [193, Theorem XI.38, p. 63]. Here the form boundedness is a consequence of the fact that the function V is a Kato-Feller potential: see inequality (2.12) in the proof of Theorem 2.13. Also notice that any bounded function V belongs to the Kato-Feller class. For general free Feller operators K_0, it seems to be impossible to have detailed information on $\exp\left(-itK_0\right) f$, even for functions f in some dense subset of $\mathcal{H}$. This means that even for the existence of wave operators one can only rely on time-independent or trace class methods.

Following the line of the proof of Lemma 8.15, it is obvious that instead of the conditions (8.19) and (8.20) one may insert any continuous function $\varphi(K_i)$ instead of $\exp(-K_i)$. This idea was used by Demuth and Ouhabaz [79], with $\varphi(\lambda) = (a + \lambda)^{-q}$, to prove the existence and completeness of scattering systems with magnetic potentials. Roughly speaking it was shown that $\Omega_{\pm}\left(A(b), -\Delta\right)$ exist and are complete where

$$\begin{aligned}
A(b) = &- \sum_{k,j=1}^{\nu} \left(\frac{\partial}{\partial x_j} - i b_j \right) a_{kj} \left(\frac{\partial}{\partial x_k} - i b_k \right) \\
&+ \sum_{k=1}^{\nu} \left\{ a_k \left(\frac{\partial}{\partial x_k} - i b_k \right) - \left(\frac{\partial}{\partial x_k} - i b_k \right) c_k \right\}.
\end{aligned}$$

The interested reader can find the sufficient conditions for the space dependent variables a_{jk}, a_k, c_k, and for the magnetic field potential $b = (b_1, \dots, b_\nu)$ in [79]. On the other hand, in (b) we proved the completeness of scattering systems for infinitely high potential barriers on sets of finite capacity. The trace class criterion for obstacles has been considered in the literature several times (e.g. by Arsenev [13] or Deift and Simon [71]), but the authors only considered compact singularity regions Γ. However, here we allow unbounded Γ. Again let K_0 be the operator $K_0 = -\frac{1}{2}\Delta$ in $L^2(\mathbb{R}^\nu)$. Then we already mentioned that a singularity region of the form $\Gamma = \bigcup_{n=1}^{\infty} B(x_n, r_n)$ possesses finite capacity whenever the quantity $\sum_{n=1}^{\infty} r_n^{\nu-2}$ is finite. This allows us to take Γ to be a union of balls $B(x_n, r_n)$ for which $\sum_{n=1}^{\infty} r_n = \infty$, at least if $\nu \geq 5$. Such regions may be called star-shaped or better *hedge-hog*-shaped.

It was already known that the existence and completeness of scattering systems were not affected by local changes of the potential in bounded regions. But here we have established a non-local freedom for changing the potential without violating the scattering behavior. This claim is true for any free Feller operator, not only for the Laplace operator, and for any singularity region of finite capacity. Of course all these completeness properties imply the invariance of the absolutely continuous part of the self-adjoint operators involved. However, if for instance we know that $\sigma_{\mathrm{ac}}(K_0) = [0, \infty)$, then it follows already from the existence of the wave operator $\Omega_+ \left((K_0)_\Sigma, J, K_0\right)$ that

$$\sigma_{\mathrm{ac}}(K_0) = \sigma_{\mathrm{ac}}\left((K_0)_\Sigma\right).$$

This is useful in special cases if one is able to study the existence criteria of the time-dependent theory assuming some properties on the function $x \mapsto \exp(-itK_0)\, u(x)$ for functions u in an appropriate dense subset of $\mathcal{H}$. For simple second order differential operators, the method of *stationary phase* is applicable.

B. Quantitative estimates for regular potentials

In the present section of this chapter, we will study the convergence of spectral and scattering data. For regular perturbations the rates are given via the Kato-Feller norms introduced in Definition 2.3 of Chapter 2. For singular perturbations the rate of convergence is based on results in Chapter 7 where we studied the operator norm convergence of $K_0\dot{+}V + \beta 1_\Gamma$ to $\left(K_0\dot{+}V\right)_\Sigma$ in terms of the large coupling parameter β. This will be described in part C of this Chapter.

Here we begin with regular perturbations. The Kato-Feller norm was defined as (see Definition 2.3)

$$\|V\|_{\mathrm{KF}} = \sup_{x\in E} \int_0^1 \int p_0(s,x,y)\,|V(y)|\,dy\,ds = \sup_{x\in E} \mathbb{E}_x\left[\int_0^1 |V(X(u))|\,du\right].$$

Therefore one can investigate the spectral behavior of a sequence of Hamiltonians $\left(K_0\dot{+}V_n : n\in\mathbb{N}\right)$, where the corresponding sequence of potentials $(V_n : n\in\mathbb{N})$ converges with respect to the Kato-Feller norm to a function V. From Proposition 2.4 it follows that the function V is automatically locally integrable and that it is a Kato-Feller potential again. To simplify the notation for the constants in the quantitative estimates, we will establish a stability condition of the Feynman-Kac semigroups with respect to the index n. In this context we also refer to Herbst and Sloan [120].

8.17. STABILITY CONDITIONS. Let $(V_n : n\in\mathbb{N})$ be a sequence of Kato-Feller functions. Then for every $n\in\mathbb{N}$ there exist finite constants M_{V_n} and b_{V_n} (see (2.7) in Chapter 2) such that

$$\sup_{x\in E} \mathbb{E}_x\left[\exp\left(-\int_0^t V_n(X(s))ds\right)\right] \le M_{V_n}\exp\left(b_{V_n}t\right),$$

for all $t\ge 0$. As stability condition we propose the existence of finite positive constants M and b such that, for all $n\in\mathbb{N}$, the following inequalities are satisfied: $M_{V_n}\le M$ and $b_{V_n}\le b$. This means that

$$\sup_{x\in E} \mathbb{E}_x\left\{\exp\left(-\int_0^t V_n(X(s))ds\right)\right\} \le M\exp(bt), \tag{8.23}$$

for all $n\in\mathbb{N}$.

REMARK. The stability condition is satisfied if we can find a constant c, $0<c<1$, together with a common fixed $t>0$ such that the following inequality is true:

$$\sup_{x\in E} \int_0^t \int V_{n,-}(y)p_0(s,x,y)dy\,ds \le c < 1,$$

uniformly in $n\in\mathbb{N}$, or if for a common sufficiently large a the inequality

$$\sup_{x\in E}\left[\left(aI+K_0\right)^{-1}V_{n,-}\right](x) \le c < 1$$

holds uniformly in $n\in\mathbb{N}$. Here $V_{n,-}$ denotes the negative part of V_n.

First we establish some results relating convergence with respect to the Kato-Feller norm of sequences of Kato-Feller functions to stability.

8.18. PROPOSITION. *As above, let $(V_n : n \in \mathbb{N})$ be a sequence of Kato-Feller potentials such that for each $n \in \mathbb{N}$ the inequality*

$$\sup_{x\in E} \mathbb{E}_x \left[\exp\left(-\int_0^t V_n(X(s))ds\right)\right] \leq M_{V_n} \exp\left(b_{V_n} t\right)$$

is satisfied. The following assertions are valid:

(1) (a) *If the sequence $(V_n : n \in \mathbb{N})$ converges with respect to the Kato-Feller norm, then this sequence is stable in the sense that (8.23) is satisfied.*
(2) (b) *If the sequence $(V_n : n \in \mathbb{N})$ is stable with stability constants M and b as in (8.23) and if the sequence $(V_n : n \in \mathbb{N})$ converges in Kato-Feller norm to the function V, then*

$$\sup_{x\in E} \mathbb{E}_x \left\{\exp\left(-\int_0^t V(X(s))ds\right)\right\} \leq M \exp\left(bt\right).$$

REMARK. In Chapter 2, Proposition 2.4 it was proved that every $\|\cdot\|_{\mathrm{KF}}$-Cauchy sequence converges to a Kato-Feller function. So in assertion (a) of Proposition 8.18 we could have said: every $\|\cdot\|_{\mathrm{KF}}$-Cauchy sequence is stable in the sense of (8.23).

Proof. (a) Suppose that the sequence $(V_n : n \in \mathbb{N})$ converges to V with respect to the Kato-Feller norm. For $0 < t \leq 1$ we have

$$\begin{aligned}
\int_0^t \int V_-(y)p_0(u,x,y)dy\,du &= \int_0^t \mathbb{E}_x\left[V_-(X(u))\right]du \\
&= \int_0^t \mathbb{E}_x\left[V_-(X(u)) - V_{n,-}(X(u))\right]du + \int_0^t \mathbb{E}_x\left[V_{n,-}(X(u))\right]du \\
&\leq \|V - V_n\|_{\mathrm{KF}} + \int_0^t \mathbb{E}_x\left[V_{n,-}(X(u))\right]du.
\end{aligned}$$

Fix $1 > \epsilon > 0$ and choose $n = n_\epsilon$ so large that $\|V - V_n\|_{\mathrm{KF}} < \epsilon/2$. Then choose $t_\epsilon > 0$ so small that

$$\sup_{x\in E} \int_0^{t_\epsilon} \mathbb{E}_x\left[V_{n,-}(X(u))\right]du < \epsilon/2.$$

It follows that $\sup_{x\in E} \int_0^{t_\epsilon} \int_E V_-(y)p_0(u,x,y)dy\,du < \epsilon$. From the Markov property and a Dyson expansion we get

$$\begin{aligned}
&\mathbb{E}_x\left[\exp\left(\int_0^{t_\epsilon} V_{n,-}(X(u))du\right)\right] \leq \sum_{k=0}^{\infty}\left[\sup_{x\in E}\mathbb{E}_x\left(\int_0^{t_\epsilon} V_{n,-}(X(u))du\right)\right]^k \\
&= \frac{1}{1 - \sup_{x\in E}\mathbb{E}_x\left[\int_0^{t_\epsilon} V_{n,-}(X(u))du\right]} \to \frac{1}{1 - \sup_{x\in E}\mathbb{E}_x\left[\int_0^{t_\epsilon} V_-(X(u))du\right]},
\end{aligned}$$

as n tends to infinity. So there exists $0 < \alpha_\epsilon < 1$ together with n_ϵ such that for all $n \geq n_\epsilon + 1$ the following inequality is satisfied:

$$\sup_{x\in E} \mathbb{E}_x \left[\exp\left(\int_0^{t_\epsilon} V_{n,-}(X(u))du\right)\right] \leq \frac{1}{1-\alpha_\epsilon}.$$

By Khas'minskii's Lemma (see Corollary 3.6) applied to the functions V_k, $1 \leq k \leq n_\epsilon$, there exist strictly positive (small) numbers t_k and numbers $0 < \alpha_k < 1$ such that

$$\sup_{x\in E} \mathbb{E}_x \left[\exp\left(\int_0^{t_k} V_{k,-}(X(u))du\right)\right] \leq \frac{1}{1-\alpha_k}, \quad 1 \leq k \leq n_\epsilon.$$

Put $t_0 = \min(t_1, \ldots, t_{n_\epsilon}, t_\epsilon)$ and write $\alpha_0 = \max(\alpha_1, \ldots, \alpha_{n_\epsilon}, \alpha_\epsilon)$. Then

$$\sup_{x\in E} \mathbb{E}_x \left[\exp\left(\int_0^{t_0} V_{k,-}(X(u))du\right)\right] \leq \frac{1}{1-\alpha_0}, \quad k \in \mathbb{N}.$$

Applying the Markov property again will yield constants M and b such that

$$\sup_{x\in E} \mathbb{E}_x \left[\exp\left(\int_0^{t} V_{k,-}(X(u))du\right)\right] \leq M\exp(bt), \quad k \in \mathbb{N}, \quad t > 0.$$

This proves assertion (a).

(b) Fix $0 < t_0 < 1$ such that

$$\sup_{k\in\mathbb{N}} \sup_{x\in E} \mathbb{E}_x \left[\int_0^{t_0} V_-(X(u))du + \int_0^{t_0} V_{k,-}(X(u))du\right] < 1/2. \tag{$*$}$$

Then, choose for given $t > 0$, a non-negative integer n such that $nt_0 \leq t < (n+1)t_0$. In fact $n = \lfloor t/t_0 \rfloor$, the integer part of t/t_0. Put $t_j = jt_0$, $1 \leq j \leq n$, $t_{n+1} = t$, $s_j = t_j - t_{j-1}$. If g is a Kato-Feller function, we have, using the triangle inequality and the Markov property,

$$\begin{aligned}
&\left(\mathbb{E}_x\left[\left(\int_0^t g(X(u))du\right)^2\right]\right)^{1/2} \leq \sum_{j=1}^{n+1}\left(\mathbb{E}_x\left[\left(\int_{t_{j-1}}^{t_j} g(X(u))du\right)^2\right]\right)^{1/2} \\
&= \sqrt{2}\sum_{j=1}^{n+1}\left(\mathbb{E}_x\left[\mathbb{E}_{X(t_{j-1})}\left\{\int_0^{s_j} g(X(u))\mathbb{E}_{X(u)}\left(\int_0^{s_j-u} g(X(v))dv\right)du\right\}\right]\right)^{1/2} \\
&\leq (n+1)\sqrt{2}\sup_{x\in E}\mathbb{E}_x\left[\int_0^{t_0} |g(X(v))|\,dv\right].
\end{aligned}$$

So, for fixed $t > 0$, $0 < t_0 < 1$ satisfying $(*)$, and $n = \lfloor t/t_0 \rfloor$, we have

$$\begin{aligned}
&\left[\exp\left(-t\left(K_0\dot{+}V\right)\right)1\right](x) - \left[\exp\left(-t\left(K_0\dot{+}V_k\right)\right)1\right](x) \\
&= \int_0^1 \mathbb{E}_x\left[\exp\left(-\int_0^t ((1-s)V(X(u)) + sV_k(X(u)))\,du\right)\right. \\
&\qquad \left.\int_0^t (V(X(u)) - V_k(X(u)))\,du\right] ds
\end{aligned}$$

(Cauchy-Schwarz' inequality)

$$\begin{aligned}
&\leq \int_0^1 \left(\mathbb{E}_x\left[\exp\left(-2\int_0^t ((1-s)V(X(u)) + sV_k(X(u)))\,du\right)\right]\right)^{1/2} \\
&\qquad \left(\mathbb{E}_x\left[\int_0^t (V(X(u)) - V_k(X(u)))\,du\right]^2\right)^{1/2} ds
\end{aligned}$$

(triangle inequality together with Markov property)

$$\begin{aligned}
&\leq (n+1)\sqrt{2} \\
&\quad \int_0^1 \left(\mathbb{E}_x\left[\exp\left(2\int_0^{(n+1)t_0} ((1-s)V_-(X(u)) + sV_{k,-}(X(u)))\,du\right)\right]\right)^{1/2} \\
&\qquad\qquad \times \sup_{x\in E}\mathbb{E}_x\left[\int_0^{t_0} |V(X(u)) - V_k(X(u))|\,du\right] ds
\end{aligned}$$

(again using Markov property and Dyson expansions)

$$\begin{aligned}
&\leq (n+1)\sqrt{2}\,\|V - V_k\|_{\mathrm{KF}} \\
&\quad \sup_{x\in E}\left(\mathbb{E}_x\left[\exp\left(2\int_0^{t_0} (V_-(X(u)) + V_{k,-}(X(u)))\,du\right)\right]\right)^{(n+1)/2}.
\end{aligned}$$

So from the stability of the sequence $(V_k : k \in \mathbb{N})$ we get

$$\begin{aligned}
&\left[\exp\left(-t\left(K_0\dot{+}V\right)\right)1\right](x) \\
&\leq \left[\exp\left(-\left(K_0\dot{+}V_k\right)\right)1\right](x) + (n+1)\sqrt{2}\,\|V - V_k\|_{\mathrm{KF}} \\
&\quad \sup_{x\in E}\left(\mathbb{E}_x\left[\exp\left(2\int_0^{t_0} (V_-(X(u)) + V_{k,-}(X(u)))\,du\right)\right]\right)^{(n+1)/2} \\
&\leq M\exp(bt) + (n+1)\sqrt{2}\,\|V - V_k\|_{\mathrm{KF}} \\
&\quad \sup_{x\in E}\left(\mathbb{E}_x\left[\exp\left(2\int_0^{t_0} (V_-(X(u)) + V_{k,-}(X(u)))\,du\right)\right]\right)^{(n+1)/2}.
\end{aligned}$$

Next let k tend to infinity to obtain (b).

A fundamental concept in the spectral theory of a self-adjoint operator is its spectral family and the corresponding spectral measure. Let K be a self-adjoint operator. We denote its spectral family (sometimes also called its resolution of the identity) with $E_K(\cdot)$. Let $\triangle = (\lambda_1, \lambda_2)$ be an interval in the spectrum of the operator K where we assume that neither λ_1 nor λ_2 is an eigenvalue of K. Then the spectral measure of $\triangle$ is given by

$$\begin{aligned} &\langle E_K(\triangle)f, g\rangle \\ &= \lim_{\epsilon\to 0}\frac{1}{2\pi i}\int_{\lambda_1}^{\lambda_2}\left\langle\left[(K-(\lambda+i\epsilon)I)^{-1}-(K-(\lambda-i\epsilon)I)^{-1}\right]f, g\right\rangle d\lambda. \end{aligned} \tag{8.24}$$

Our next objective in this section is to find a quantitative estimate for the norm of the difference $E_{K_0\dot{+}V}(\triangle) - E_{K_0\dot{+}W}(\triangle)$ in terms of $\|W-V\|_{\mathrm{KF}}$. Instead of studying the operator norm from one weighted L^2-space, $L^2(\varphi^{-1}dm)$, to another one of the form $L^2(E, \varphi dm)$, we prefer here to treat the sandwiched operator norm in $L^2(E, dm)$, where we sandwich with a multiplication operator of weight φ.

8.19. Definition. Let φ be non-vanishing positive continuous function. The function φ is said to be *admissible* if it has the following properties:

(a) The supremum $\sup\limits_{x\in E}\dfrac{1}{\varphi(x)} \leq 1$ is finite; and
(b) For φ an estimate of the following form is valid:

$$\int p_0(t,x,y)\varphi(y)^\rho dy \leq c_0^2 e^{mt}\varphi(x)^\rho, \tag{8.25}$$

for some $m \in \mathbb{N}$, for all $t \geq 0$ and for all $x \in E$. Most of the time ρ is one of the numbers 2, 4 or 8.

In case of the Laplace operator, a typical choice for the function $\varphi(x)$ is $\varphi(x) = 1 + |x|^\alpha$.

Condition (8.25) will be discussed in more detail in the remarks following Proposition 8.21 below.

The machinery for estimating the sandwiched spectral data quantitatively is based on obtaining quantitative estimates for resolvent and sandwiched resolvent differences at regular resolvent values. The next lemma provides estimates in terms of $V - W$.

8.20. Lemma. *Let V and W be Kato-Feller potentials. Choose constants M_V and b_V such that*

$$\mathbb{E}_x\left[\exp\left(-\int_0^t V(X(u))du\right)\right] \leq M_V \exp(b_V t), \quad x \in E, \quad t > 0. \tag{$*$}$$

Similarly, choose M_W and b_W so that (*) holds with V replaced by W. In addition let φ be an admissible function for which

$$\int p_0(t,x,y)\varphi(y)^\rho dy \leq c_0^2 e^{mt}\varphi(x)^\rho, \tag{8.26}$$

for all $t > 0$, for all $x \in E$ and for $\rho = 4$.

For brevity we write

$$a_0 = \max(b, M, m, 1), \quad M = \max(M_{2V}, M_{4V}, M_{2W}, M_{4W})$$

and

$$b = \max(b_{2V}, b_{4V}, b_{2W}, b_{4W}).$$

Moreover, we suppose that a is a regular resolvent value with $a > 2a_0$.

For such a regular value $a > 2a_0$ the following inequalities are valid:

$$\begin{aligned}
&\left\| \left(aI + K_0 \dot{+} V\right)^{-1} - \left(aI + K_0 \dot{+} W\right)^{-1} \right\|_{2,2} \\
&\leq \frac{M}{a-b} \frac{1}{1-e^{-a}} \|V - W\|_{\mathrm{KF}} \\
&\leq 2\|V - W\|_{\mathrm{KF}}
\end{aligned} \tag{8.27}$$

and

$$\begin{aligned}
&\left\| \varphi \left[\left(aI + K_0 \dot{+} V\right)^{-1} - \left(aI + K_0 \dot{+} W\right)^{-1} \right] \varphi \right\|_{2,2} \\
&\leq c_0 \frac{M^{1/2}}{(a-b)^{1/2}} \frac{1}{(a-m)^{1/2}} \frac{1}{1-e^{-a}} \left\| (V - W)\varphi^2 \right\|_{\mathrm{KF}} \\
&\leq 2c_0 a_0^{-1/2} \left\| (V-W)\varphi^2 \right\|_{\mathrm{KF}}.
\end{aligned} \tag{8.28}$$

In the special case with $W \equiv 0$ the following inequality is obtained:

$$\begin{aligned}
&\left\| \varphi \left[\left(aI + K_0 \dot{+} V\right)^{-1} - (aI + K_0)^{-1} \right] \varphi \right\|_{2,2} \\
&\leq c_0 \frac{M^{1/4}}{(a-b)^{1/4}} \frac{1}{a^{1/4}} \frac{1}{(a-m)^{1/2}} \frac{1}{1-e^{-a}} \left\| R_0(-a) \left| V\varphi^2 \right| \right\|_\infty \\
&\leq 2c_0 a_0^{-1/2} \left\| V\varphi^2 \right\|_{\mathrm{KF}}.
\end{aligned}$$

Proof. From the remark following Proposition 4.2, we have

$$\begin{aligned}
&\left\| \left(aI + K_0 \dot{+} V\right)^{-1} - \left(aI + K_0 \dot{+} W\right)^{-1} \right\|_{2,2} \\
&\leq \frac{M_{2W}^{1/2}}{(a-b_{2V})^{1/2}} \frac{M_{2V}^{1/2}}{(a-b_{2V})^{1/2}} \left\| (aI + K_0)^{-1} |V - W| \right\|_\infty \\
&\leq \frac{M}{a-b} \frac{1}{1-e^{-a}} \|V - W\|_{\mathrm{KF}} \\
&\leq \frac{a_0}{a-a_0} \frac{1}{1-e^{-2a_0}} \|V - W\|_{\mathrm{KF}} \\
&\leq \frac{a_0}{a-a_0} \frac{1}{1-e^{-2}} \|V - W\|_{\mathrm{KF}} \leq 2\|V - W\|_{\mathrm{KF}}.
\end{aligned}$$

According to inequality (4.3) in Proposition 4.2, we also have

$$\begin{aligned}
&\left\|\varphi\left[\left(aI+K_0\dot{+}V\right)^{-1}-\left(aI+K_0\dot{+}W\right)^{-1}\right]\varphi\right\|_{2,2}\\
&\leq\left\|\left(aI+K_0\right)^{-1}|V-W|\left(aI+K_0\dot{+}2V\right)^{-1}\varphi^2\right\|_{\infty}^{1/2}\\
&\times\left\|\left(aI+K_0\right)^{-1}|V-W|\left(aI+K_0\dot{+}2W\right)^{-1}\varphi^2\right\|_{\infty}^{1/2}.
\end{aligned}$$

The first factor squared is smaller than

$$\sup_{x\in E}\int_0^\infty e^{-a\lambda}\,\mathbb{E}_x\left\{|V(X(\lambda))-W(X(\lambda))|\left[R_{2V}(-a)\varphi^2\right](X(\lambda))\right\}d\lambda$$

with $R_V(-a)=\left(aI+K_0\dot{+}V\right)^{-1}$. Next we estimate $|[R_{2V}(-a)\varphi^2](x)|$

$$\begin{aligned}
\left|\left[R_{2V}(-a)\varphi^2\right](x)\right| &\leq\left[\int_0^\infty e^{-a\lambda}\,\mathbb{E}_x\left\{\exp\left(-4\int_0^\lambda V(X(s))ds\right)\right\}d\lambda\right]^{1/2}\\
&\quad\times\left[\int_0^\infty e^{-a\lambda}\,\mathbb{E}_x\left\{\varphi(X(\lambda))^4\right\}d\lambda\right]^{1/2}\\
&\leq\frac{M_{4V}^{1/2}}{(a-b_{4V})^{1/2}}\frac{c_0}{(a-m)^{1/2}}\varphi(x)^2\\
&\leq c_0\frac{M^{1/2}}{(a-b)^{1/2}}\frac{1}{a_0^{1/2}}\varphi(x)^2.
\end{aligned}$$

This implies that

$$\begin{aligned}
&\|\varphi\left[R_V(-a)-R_W(-a)\right]\varphi\|_{2,2}\\
&\leq c_0\frac{M_{4V}^{1/4}}{(a-b_{4V})^{1/4}}\frac{M_{4W}^{1/4}}{(a-b_{4W})^{1/4}}\frac{1}{(a-m)^{1/2}}\left\|R_0(-a)\,|V-W|\,\varphi^2\right\|_\infty\\
&\leq c_0\frac{M^{1/2}}{(a-b)^{1/2}}\frac{1}{(a-m)^{1/2}}\frac{1}{1-e^{-a}}\left\|(V-W)\varphi^2\right\|_{\mathrm{KF}}\\
&\leq 2c_0\frac{a_0^{1/2}}{a-a_0}\left\|(V-W)\varphi^2\right\|_{\mathrm{KF}}\leq 2\frac{c_0}{a_0^{1/2}}\left\|(V-W)\varphi^2\right\|_{\mathrm{KF}}.
\end{aligned}$$

The result in Lemma 8.20 will be extended to resolvent differences in the limiting absorption case, where the resolvent value is given by $a=\lambda\pm i\epsilon$, $\lambda>0$, and $\epsilon\to 0$. As mentioned in the context of equality (8.24), we will investigate the matrix elements of the resolvents by studying the $L^2(E,m)$-operator norms of sandwiched resolvents. The sandwiching operator is given as a multiplication by the function $\varphi(\cdot)$ (see Definition 8.19 and Remark 1 following the next proposition). For the Laplace operator the only relevant estimates of Definition 8.19 are those with $\varphi(x) = \left(1 + |x|^2\right)^{\epsilon+1/2}$, with $\varepsilon > 0$. If we assume that the limiting absorption principle is valid for the unperturbed operator, i.e. for $(K_0 - (\lambda \pm i0)\, I)^{-1}$, it turns out to be possible to make a quantitative study of the limiting absorption case for perturbations by regular potentials.

8.21. PROPOSITION. *Let K_0, V, W and φ be as in Lemma 8.20. Let $\triangle = (\lambda_1, \lambda_2)$ be an open interval in $\mathbb{R}$ and assume that*

$$\sup_{\lambda\in\triangle}\sup_{\epsilon\in(0,1]} \left\|\varphi^{-1}\left(K_0 - (\lambda \pm i\epsilon)\, I\right)^{-1}\varphi^{-1}\right\|_{2,2} =: d_\triangle < \infty. \tag{8.29}$$

Suppose that the pointwise estimate (8.26) in Lemma 8.20 holds, for some fixed c_0 and m, with $\rho = 4$. Let $R_V(-a) = R(a,V)$ denote the resolvent $R_V(-a) = R(a,V) = \left(K_0 \dot{+} V + aI\right)^{-1}$ for a in the resolvent set of $K_0 \dot{+} V$.
(a) Assume that

$$\left\|V\varphi^2\right\|_{\mathrm{KF}} < \frac{1}{6c_0 a_0^{1/2}\left(1 + 3a_0 d_\triangle\right)} \tag{8.30}$$

with $a_0 \geq \max\left(2, \lambda_2, 2m, M_{4V}, 2b_{4V}\right)$. Then

$$\left\|\varphi^{-1} R_V(\lambda + i\epsilon)\varphi^{-1}\right\|_{2,2} \leq \gamma_V^{-1}\left(d_\triangle + \frac{1}{3a_0}\right), \tag{8.31}$$

where $\gamma_V = 1 - 6c_0 a_0^{1/2}\left(1 + 3d_\triangle a_0\right)\left\|V\varphi^2\right\|_{\mathrm{KF}}$.
(b) Assume that (8.30) holds for both V and W. Take $a_0 \geq 2\max(2,\lambda_2,2m,M,2b)$, where M is as in Lemma 8.20. Then

$$\begin{aligned} &\left\|\varphi^{-1}\left[R_V(\lambda \pm i\epsilon) - R_W(\lambda \pm i\epsilon)\right]\varphi^{-1}\right\|_{2,2} \\ &\leq c(V,W,\triangle)\left\|(V - W)\varphi^2\right\|_{\mathrm{KF}}, \end{aligned} \tag{8.32}$$

with the constant

$$c(V,W,\triangle) = 2c_0 a_0^{1/2}\left(1 + \gamma_V^{-1}\left(1 + 3a_0 d_\triangle\right)\right)\left(1 + \gamma_W^{-1}\left(1 + 3a_0 d_\triangle\right)\right). \tag{8.33}$$

REMARK. Condition (8.29) should be compared with corresponding results in Ben-Artzi and Devinatz in [23] and with results in Ben-Artzi, Dermenjian and Guillot in [22] for K_0.

Proof. The proof is based on the Hilbert identity for resolvents. Let R_1 and R_2 be two resolvent families and let a and b be complex numbers for which $R_1(a)$, $R_2(a)$, $R_1(b)$ and $R_2(b)$ are defined. Then the *Hilbert identity* tells us that

$$\begin{aligned} R_1(a) - R_2(a) &= (-aI + K_1)^{-1} - (-aI + K_2)^{-1} \\ &= (1 + (a-b)R_1(a))\,(R_1(b) - R_2(b))\,(1 + (a-b)R_2(a)). \end{aligned} \tag{8.34}$$

(a) From the Hilbert identity for $K_0\dot{+}V$ and $K_0\dot{+}W$ it follows that

$$\begin{aligned}
&\left\|\varphi^{-1}\left[R_V(\lambda+i\epsilon)-R_W(\lambda+i\epsilon)\right]\varphi^{-1}\right\|_{2,2} \\
&\leq \left(1+|\lambda+i\epsilon-a|\left\|\varphi^{-1}R_V(\lambda+i\epsilon)\varphi^{-1}\right\|_{2,2}\right) \\
&\quad\times\left(1+|\lambda+i\epsilon-a|\left\|\varphi^{-1}R_W(\lambda+i\epsilon)\varphi^{-1}\right\|_{2,2}\right) \\
&\quad\times\left\|\varphi\left[R_V(-a)-R_W(-a)\right]\varphi\right\|_{2,2}.
\end{aligned}$$

Upon using (8.29) and taking $W\equiv 0$ we obtain

$$\begin{aligned}
\left\|\varphi^{-1}R_V(\lambda+i\epsilon)\varphi^{-1}\right\|_{2,2} &\leq d_\triangle+\left\|\varphi^{-1}\left[R_V(\lambda+i\epsilon)-R_0(\lambda+i\epsilon)\right]\varphi^{-1}\right\|_{2,2} \\
&\leq d_\triangle+2\left(1+3a_0\left\|\varphi^{-1}R_V(\lambda+i\epsilon)\varphi^{-1}\right\|_{2,2}\right) \\
&\quad\times(1+3d_\triangle a_0)\,c_0 a_0^{-1/2}\left\|V\varphi^2\right\|_{\mathrm{KF}}.
\end{aligned}$$

Consequently

$$\begin{aligned}
\left\|\varphi^{-1}R_V(\lambda+i\epsilon)\varphi^{-1}\right\|_{2,2} &\leq \frac{d_\triangle+2c_0a_0^{-1/2}(1+3d_\triangle a_0)\left\|V\varphi^2\right\|_{\mathrm{KF}}}{1-6c_0a_0^{1/2}(1+3d_\triangle a_0)\left\|V\varphi^2\right\|_{\mathrm{KF}}} \\
&=\frac{1}{\gamma_V}\left(d_\triangle+\frac{2c_0a_0^{-1/2}(1+3d_\triangle a_0)}{6c_0a_0^{1/2}(1+3d_\triangle a_0)}\right) \\
&=\frac{1}{\gamma_V}\left(d_\triangle+\frac{1}{3a_0}\right).
\end{aligned}$$

(b) The estimate in (8.32) follows from the Hilbert identity (8.34), from the estimate in (8.28) and from part (a) of this proposition, i.e. from (8.31), because

$$\begin{aligned}
&\left\|\varphi^{-1}\left[R_V(\lambda+i\epsilon)-R_W(\lambda+i\epsilon)\right]\varphi^{-1}\right\|_{2,2} \\
&\leq\left[1+|\lambda+i\epsilon-a_0|\,\frac{1}{\gamma_V}\left(d_\triangle+\frac{1}{3a_0}\right)\right] \\
&\quad\times\left[1+|\lambda+i\epsilon-a_0|\,\frac{1}{\gamma_W}\left(d_\triangle+\frac{1}{3a_0}\right)\right] \\
&\quad\times 2c_0a_0^{-1/2}\left\|(V-W)\varphi^2\right\|_{KF},
\end{aligned}$$

with $a_0\geq\max\Big(2,2m,\lambda_2,M_{2V},M_{4V},M_{2W},M_{4W},2b_{2V},2b_{4V},2b_{2W},2b_{4W}\Big)$.

The result in (8.32) of the last proposition has several interesting consequences. Some of them will be mentioned here: the convergence of spectral measures, the convergence of local wave operators, the stability of the smoothness property, and last but not least a quantitative estimate for the corresponding scattering matrices.

It is obvious that this list of applications of Proposition 8.21 is not complete. It could be extended with the investigation of pre-wave operators, wave matrices or different kind of definitions of wave operators. For these topics the reader may consult Baumgärtel and Wollenberg [21] or Yafaev [253].

Before we treat in detail some of the applications of Proposition 8.21 we discuss the kind of conditions that were used there.

REMARK 1. The condition in (8.25) or in (8.26) seems very technical, but it is not critical and in many cases rather harmless. Many diffusion processes possess a transition density function $p_0(t,x,y)$ which can be pointwise estimated by the Wiener density function, i.e.: $p_0(t,x,y) \leq \frac{a}{(2\pi t)^{\nu/2}} \exp\left(-b\frac{|x-y|^2}{2t}\right)$, where a and b are some finite positive constants. In this situation (8.25) and (8.26) are satisfied for all polynomially bounded functions.

However, the condition in (8.25) can also be verified for certain translation invariant processes. For weight functions of the form $\varphi(x) := 1 + |x|^\alpha$ for some positive power $\alpha > 0$, it follows that $\varphi(x+y) \leq 2^\alpha\left(\varphi(x)+\varphi(y)\right)$. Having these functions in mind, we assume that $\varphi(x) = g\left(|x|\right)$ (i.e. φ is *radially symmetric*), and that $\varphi(x+y) \leq C\left(\varphi(x)+\varphi(y)\right)$, and that

$$\sup\left\{g(s)\exp\left(-\frac{ms}{2}\right) : s > 0\right\} \tag{8.35}$$

is finite. For such a function φ, and for a translation invariant process (like any Lévy process) with $p_0(t,x,y) = p_0(t,0,x-y)$ a verification of (8.25) reduces to showing that the inequality $\int p_0(t,0,u)\varphi(u)du \leq c\exp(mt)$ holds for some finite constants c and m. Let us consider the process with generator $\sqrt{-\Delta+m^2}-m$ (see Example 4 in section C of Chapter 1). Its density (see equality (1.26)) has the explicit representation:

$$p_0(t,x,y) = te^{mt}\widetilde{p}\left(\sqrt{|x-y|^2+t^2}\right), \quad \text{with}$$

$$\widetilde{p}(r) = \frac{1}{(2\pi)^\nu}\frac{1}{r}\int_{\mathbb{R}^\nu}\exp\left(-r\sqrt{|p|^2+m^2}\right)dp. \tag{8.36}$$

Upon taking a radially symmetric sandwiching function $\varphi(u) = g\left(|u|\right)$ for which the quantity in (8.35) is finite, we obtain:

$$\begin{aligned}&\int_{\mathbb{R}^\nu} du\,(2\pi)^\nu\,\widetilde{p}\left(\sqrt{|u|^2+t^2}\right)\varphi(u)\\ &= \int_{\mathbb{R}^\nu} du \int_{\mathbb{R}^\nu} dp \frac{1}{\sqrt{|u|^2+t^2}}\varphi(u)\exp\left(-\sqrt{\left(|u|^2+t^2\right)\left(|p|^2+m^2\right)}\right)\end{aligned}$$

$$\leq c_\nu^2 \int_0^\infty ds \int_0^\infty d\tau \frac{s^{\nu-1}}{\sqrt{s^2+t^2}} g(s) \tau^{\nu-1} \exp\left(-\sqrt{(s^2+t^2)(\tau^2+m^2)}\right)$$

$$\leq \frac{c_\nu^2}{2} \int_{t^2}^\infty \frac{\sqrt{\lambda - t^2}^{\nu-2}}{\sqrt{\lambda}} g\left(\sqrt{\lambda - t^2}\right) \exp\left(-\frac{m}{2}\sqrt{\lambda}\right)$$

$$\times \int_0^\infty \tau^{\nu-1} \exp\left(-\frac{\tau}{2}\sqrt{\lambda}\right) d\tau \, d\lambda$$

$$= c_\nu^2 2^{\nu-1} \Gamma(\nu) \int_{t^2}^\infty \frac{(\lambda - t^2)^{\frac{1}{2}\nu - 1}}{\lambda^{\frac{1}{2}(\nu+1)}} g\left(\sqrt{\lambda - t^2}\right) \exp\left(-\frac{m}{2}\sqrt{\lambda}\right) d\lambda$$

$$\leq c_\nu^2 2^{\nu-1} \Gamma(\nu) \int_{t^2}^\infty \frac{(\lambda - t^2)^{\frac{1}{2}\nu - 1}}{\lambda^{\frac{1}{2}(\nu+1)}} g\left(\sqrt{\lambda - t^2}\right) \exp\left(-\frac{m}{2}\sqrt{\lambda - t^2}\right) d\lambda$$

$$\leq c_\nu^2 2^{\nu-1} \Gamma(\nu) \sup_{s>0} \left(g(s) \exp\left(-\frac{m}{2}s\right)\right) \int_{t^2}^\infty \frac{(\lambda - t^2)^{\frac{1}{2}\nu - 1}}{\lambda^{\frac{1}{2}(\nu+1)}} d\lambda$$

$$= \frac{1}{t} c_\nu^2 2^{\nu-1} \Gamma(\nu) \sup_{s>0} \left(g(s) \exp\left(-\frac{m}{2}s\right)\right) \int_0^\infty \frac{\sigma^{\frac{1}{2}\nu - 1}}{(1+\sigma)^{\frac{1}{2}(\nu+1)}} d\sigma.$$

In the inequalities above we have used several identities which can be taken from standard reference book on integrals. It holds

$$\int_0^\infty \frac{\sigma^{\frac{1}{2}\nu - 1}}{(1+\sigma)^{\frac{1}{2}(\nu+1)}} d\sigma = \int_0^1 (1-\xi)^{\frac{1}{2}\nu - 1} \xi^{-\frac{1}{2}} d\xi$$

$$= \frac{\Gamma\left(\frac{1}{2}\nu\right)\sqrt{\pi}}{\Gamma\left(\frac{1}{2}\nu + \frac{1}{2}\right)}.$$

The constant c_ν is given by

$$c_\nu = \frac{2\left(\sqrt{\pi}\right)^\nu}{\Gamma\left(\frac{1}{2}\nu\right)},$$

and we used the identity

$$\sqrt{\pi}\Gamma(\nu) = 2^{\nu-1} \Gamma\left(\frac{1}{2}\nu\right) \Gamma\left(\frac{1}{2}\nu + \frac{1}{2}\right).$$

Hence we obtain

$$\int_{\mathbb{R}^\nu} du \int_{\mathbb{R}^\nu} dp \frac{1}{\sqrt{|u|^2 + t^2}} \varphi(u) \exp\left(-\sqrt{\left(|u|^2 + t^2\right)\left(|p|^2 + m^2\right)}\right)$$

$$\leq \frac{1}{t} (4\pi)^\nu \sup_{s>0} \left(g(s) \exp\left(-\frac{m}{2}s\right)\right).$$

The result is applicable with $g(s) = 1 + s^\alpha$, $\alpha > 0$, i.e. for $\varphi(x) = 1 + |x|^\alpha$. Note that for such function φ we have

$$\varphi(x+y) \le 2^\alpha \left(\varphi(x) + \varphi(y)\right).$$

For these functions the above observations mean:

$$\begin{aligned}\int_{\mathbb{R}^\nu} p_0(t,x,y)\varphi(y)dy &\le 2^\alpha \int_{\mathbb{R}^\nu} p_0(t,0,y)\varphi(y)dy + 2^\alpha \int_{\mathbb{R}^\nu} p_0(t,0,y)\varphi(x)dy \\ &\le 2^{\alpha+\nu} e^{mt} \sup_{s>0} \left\{ g(s) \exp\left(-\frac{m}{2}s\right)\right\} + 2^\alpha \varphi(x) \\ &\le C(\alpha) e^{mt} \varphi(x),\end{aligned}$$

where $C(\alpha) = 2^{\alpha+\nu} \sup\left\{(1+s^\alpha)\exp\left(-\frac{1}{2}ms\right) : s > 0\right\}$.

REMARK 2. Instead of condition (8.25) we may even allow the weight function φ to satisfy a weaker estimate of the form

$$\int p_0(t,x,y)\varphi(y)^\rho dy \le c_0^2 \left(t^{-\alpha} + e^{mt}\right) \varphi^\rho(x), \tag{8.37}$$

with $\alpha < 1$. This follows from the proof of Lemma 8.20 where

$$\begin{aligned}&\left|\left[R_{2V}(-a)\varphi^2\right](x)\right| \\ &\le \left[\int_0^\infty d\lambda e^{-a\lambda}\lambda^{-\alpha}\mathbb{E}_x\left\{\exp\left(-4\int_0^\lambda V(X(s))ds\right)\right\}\right]^{1/2} \\ &\quad \times \left[\int_0^\infty d\lambda e^{-a\lambda}\lambda^{\alpha}\mathbb{E}_x\left\{\varphi(X(\lambda))^4\right\}\right]^{1/2} \\ &\le M_{4V}^{1/2}\left[\int_0^\infty d\lambda \exp\left(-\lambda(a-b_{4V})\right)\lambda^{-\alpha}\right]^{1/2} \\ &\quad \times \left[c_0^2 \int_0^\infty d\lambda e^{-a\lambda}\lambda^\alpha\left(\lambda^{-\alpha} + e^{m\lambda}\right)\right]^{1/2} \varphi(x)^2.\end{aligned} \tag{8.38}$$

The verification of the estimate in (8.37) is much simpler for the density (8.36) of the relativistic Hamiltonian.

REMARK 3. For part (b) of Proposition 8.21 it would be sufficient to assume that (8.30) holds for just one of the potentials, for example for V. In this situation an estimate for the operator norm of

$$\varphi^{-1}\left\{R_V(\lambda+i\epsilon) - R_W(\lambda+i\epsilon)\right\}\varphi^{-1}$$

would follow from the inequality

$$\begin{aligned} &\left\|\varphi^{-1}\left\{R_V(\lambda+i\epsilon)-R_W(\lambda+i\epsilon)\right\}\varphi^{-1}\right\|_{2,2} \\ &\leq \left\|\varphi\left[R_V(-a_0)-R_W(-a_0)\right]\varphi\right\|_{2,2} \\ &\quad \left[1+|\lambda+1+a_0|\left\|\varphi^{-1}R_V(\lambda\pm i\epsilon)\varphi^{-1}\right\|_{2,2}\right] \\ &\quad \left[1+|\lambda+1+a_0|\left\|\varphi^{-1}R_V(\lambda\pm i\epsilon)\varphi^{-1}\right\|_{2,2}\right. \\ &\quad \left.+|\lambda+1+a_0|\left\|\varphi^{-1}\left[R_V(\lambda\pm i\epsilon)-R_W(\lambda+i\epsilon)\right]\varphi^{-1}\right\|_{2,2}\right]. \end{aligned} \tag{8.39}$$

This would also imply the convergence in terms of the Kato-Feller norm

$$\left\|(V-W)\varphi^2\right\|_{\mathrm{KF}}.$$

Notice that in all of the considerations above neither the potential V nor the potential W is assumed to be bounded.

REMARK 4. The explicit constant $c(V,W,\triangle)$ in (8.33) looks artificial. However, it is well determined. From the choice of the Kato-Feller potentials we know the values of the constants M_V, M_W, b_V and b_W (at least in principle). Therefore we may choose a_0 explicitly. Of course a_0 should be chosen as small as possible. For instance suppose that $V(x)$ is such that $\left\|V\varphi^2\right\|_{\mathrm{KF}}\leq\frac{1}{2}$: see (8.30). Then $\gamma_V\geq\frac{1}{2}$ and the constant $c(V,W,\triangle)$ is almost completely determined by the value of $d_\triangle$: see (8.29).

Next let us give the following applications of Proposition 8.21: convergence of spectral measures, convergence of local wave operators, and convergence of scattering matrices.

8.22. COROLLARY. *Assume that the hypotheses of Proposition 8.21 and Lemma 8.20 hold. Let $E_{K_0\dot{+}V}(\cdot)$ and $E_{K_0\dot{+}W}(\cdot)$ be the spectral measures of the self-adjoint operators $K_0\dot{+}V$ and $K_0\dot{+}W$ respectively. Let $\triangle=(\lambda_1,\lambda_2)$ be an open interval, satisfying (8.29), with the property that neither λ_1 nor λ_2 is an eigenvalue of the operators $K_0\dot{+}V$ and $K_0\dot{+}W$. Suppose that V and W satisfy (8.30). Then the spectral measures converge in the weighted operator norm from $L^2(E,\varphi^{-1}dm)$ to $L^2(E,\varphi dm)$ provided $\left\|(V-W)\varphi^2\right\|_{\mathrm{KF}}$ tends to zero, and the rate of convergence is given by*

$$\begin{aligned} &\left\|\varphi^{-1}\left[E_{K_0\dot{+}V}(\triangle)-E_{K_0\dot{+}W}(\triangle)\right]\varphi^{-1}\right\|_{2,2} \\ &\leq\frac{|\triangle|}{\pi}c(V,W,\triangle)\left\|(V-W)\varphi^2\right\|_{\mathrm{KF}}, \end{aligned} \tag{8.40}$$

where $c(V,W,\triangle)$ is given in Proposition 8.21.

Proof. This is an obvious consequence of the definition in (8.24) and of the estimate in (8.32).

Next we study the convergence of the wave operators. In (8.15) the norm of the difference of $\Omega_+\left(K_0\dot{+}V,K_0\right)f$ and $\Omega_+\left(K_0\dot{+}W\right)f$, $f\in P_{\mathrm{ac}}(K_0)(\mathcal{H})$, was already estimated. However, the conditions given there do not involve estimates in terms of the Kato-Feller norm. Nevertheless one can use Proposition 8.13 to ensure the existence of wave operators.

8.23. COROLLARY. *Suppose that the wave operators $\Omega_+\left(K_0\dot{+}V,K_0\right)$ and $\Omega_+\left(K_0\dot{+}W,K_0\right)$ exist. Let $\triangle=(\lambda_1,\lambda_2)$ be chosen as in Proposition 8.21. Let $E_0(\cdot)$ denote the spectral measure of K_0. Then the local wave operator*

$$\Omega_+\left(K_0\dot{+}V,K_0\right)E_0(\triangle)\quad \textit{converges to}\quad \Omega_+\left(K_0\dot{+}W,K_0\right)E_0\left(\triangle\right)$$

in the operator norm of the space $L^2(E,\varphi dm)$, provided the Kato-Feller norm $\left\|(V-W)^2\varphi^2\right\|_{\mathrm{KF}}$ and the L^2-L^2-norm $\left\|(V-W)\varphi^{-1}\right\|_{2,2}$ tend to zero. More precisely the following inequality is valid:

$$\begin{aligned}&\left\|\varphi^{-1}\left[\Omega_+\left(K_0\dot{+}V,K_0\right)-\Omega_+\left(K_0\dot{+}W,K_0\right)\right]E_0(\triangle)\right\|_{2,2} \\ &\leq c(V,W,\triangle)\left\{\left\|V\varphi^2\right\|_{\mathrm{KF}}\left\|(W-V)\varphi^{-1}\right\|_{2,2}+\left\|(W-V)^2\varphi^2\right\|_{\mathrm{KF}}^{1/2}\right\},\end{aligned}\tag{8.41}$$

with the constant $c(V,W,\triangle)$ given by

$$c(V,W,\triangle)=2c_0a_0^{1/2}\left|\triangle\right|\left(2+3a_0\gamma_V^{-1}d_\triangle\right).$$

Here the constant $d_\triangle$ is defined in (8.29), $a_0\geq 2\max\left(1,m,\lambda_2,b,M\right)$ (see Lemma 8.20) and the constant γ_V is given in (8.31).

Proof. The pre-wave operator of $\Omega_+\left(K_0\dot{+}V,K_0\right)$ is by definition

$$\Omega_{\epsilon,V}=\epsilon\int_0^\infty dte^{-\epsilon t}\exp\left(it\left(K_0\dot{+}V\right)\right)\exp\left(-itK_0\right).$$

We have

$$\begin{aligned}\left\langle\Omega_{\epsilon,V}E_0(\triangle)f,g\right\rangle&=\epsilon\int_0^\infty dt\int_\triangle e^{-\epsilon t}e^{-it\lambda}\left\langle\exp\left(it\left(K_0\dot{+}V\right)\right)E_0(d\lambda)f,g\right\rangle\\ &=-i\epsilon\int_\triangle\left\langle E_0(d\lambda)f,R_V(\lambda-i\epsilon)g\right\rangle.\end{aligned}\tag{8.42}$$

On the other hand we also have

$$\lim_{\epsilon\to 0}\left\langle\Omega_{\epsilon,V}f,g\right\rangle=\left\langle\Omega_+\left(K_0\dot{+}V,K_0\right)f,g\right\rangle,$$

because $\Omega_+\left(K_0\dot{+}V,K_0\right)$ was assumed to exist as a strong limit of

$$\exp\left(it\left(K_0\dot{+}V\right)\right)\exp\left(-itK_0\right)P_{\mathrm{ac}}(K_0),$$

for t tending to ∞. Therefore, for f and g in $L^2(E,m)$, we get

$$\begin{aligned}&\left|\left\langle\varphi^{-1}\left[\Omega_+\left(K_0\dot{+}V,K_0\right)-\Omega_+\left(K_0\dot{+}W,K_0\right)\right]E_0(\triangle)f,g\right\rangle\right|\\&\le\lim_{\epsilon\downarrow 0}\epsilon\int_\triangle\left|\left\langle E_0(d\lambda)f,\left[R_W(\lambda-i\epsilon)-R_V(\lambda-i\epsilon)\right]\varphi^{-1}g\right\rangle\right|\end{aligned}$$

(since $\epsilon\left\|R_W\left(\lambda-i\epsilon\right)\right\|\le 1$)

$$\le\lim_{\epsilon\downarrow 0}|\triangle|\,\|f\|_2\sup_{\lambda\in\triangle}\left\|(W-V)R_V(\lambda-i\epsilon)\varphi^{-1}\right\|_{2,2}\|g\|_2$$

(using the first resolvent equation)

$$\begin{aligned}&\le|\triangle|\,\|f\|_2\,\|g\|_2\,\|(W-V)R_V(-a_0)\varphi\|_{2,2}\\&\quad\times\lim_{\epsilon\downarrow 0}\sup_{\lambda\in\triangle}\left\{1+(a_0+\lambda_2+1)\left\|\varphi^{-1}R_V(\lambda-i\epsilon)\varphi^{-1}\right\|_{2,2}\right\}.\end{aligned}$$

So that

$$\left\|\varphi^{-1}\left[\Omega_+\left(K_0\dot{+}V,K_0\right)-\Omega_+\left(K_0\dot{+}W,K_0\right)\right]E_0(\triangle)\right\|_{2,2}$$

(see (8.31))

$$\le|\triangle|\left(2+3a_0\gamma_V^{-1}d_\triangle\right)\|(W-V)R_V(-a_0)\varphi\|_{2,2}\,.$$

However,

$$\begin{aligned}&\|(W-V)R_V(-a_0)\varphi\|_{2,2}\\&\le\left\|(W-V)\varphi^{-1}\right\|_{2,2}\|\varphi\left[R_V(-a_0)-R_0(-a_0)\right]\varphi\|_{2,2}\\&\quad+\|\,|W-V|\,R_0(-a_0)\varphi\|_{2,2}\\&\le 2c_0a_0^{-1/2}\left[\left\|V\varphi^2\right\|_{\mathrm{KF}}\left\|\,|W-V|\,\varphi^{-1}\right\|_{2,2}+\left\|\,|W-V|^2\varphi^2\right\|_{\mathrm{KF}}^{1/2}\right].\end{aligned}\tag{8.43}$$

In (8.43) we again used the following inequality:

$$\begin{aligned}&\|\,|W-V|\,R_0(-a_0)\varphi f\|^2_{L^2(E,m)}\\&=\int dx\,|W(x)-V(x)|^2\left(\int_0^\infty d\lambda\,e^{-a_0\lambda}\int dy\,p_0(\lambda,x,y)\varphi(y)\,|f(y)|\right)^2\\&\le\int dx\,|W(x)-V(x)|^2\\&\quad\frac{1}{a_0}\int_0^\infty d\lambda\,e^{-a_0\lambda}\int dy\,p_0(\lambda,x,y)\varphi(y)^2\int du\,p_0(\lambda,x,u)\,|f(u)|^2\end{aligned}$$

$$\leq \frac{c_0^2}{a_0} \int_0^\infty d\lambda \, e^{-(a_0+m)\lambda} \int du \, |f(u)|^2$$
$$\int dx \, p_0(\lambda, x, u) \, |W(x) - V(x)|^2 \, \varphi(x)^2$$
$$\leq \frac{c_0^2}{a_0} \|f\|^2 \sup_{u \in E} \left| \left[R_0 \left(-a_0 - m\right) |W - V|^2 \varphi^2 \right] (u) \right|$$
$$\leq \frac{c_0^2}{a_0} \|f\|^2 \left\| R_0 \left(-a_0 - m\right) |W - V|^2 \varphi^2 \right\|_\infty$$
$$\leq \frac{c_0^2}{a_0} \|f\|^2 \frac{1}{1 - \exp\left(-\left(a_0 + m\right)\right)} \left\| |W - V|^2 \varphi^2 \right\|_{\mathrm{KF}}.$$

REMARK 1. We insert a remark on the *stability of smoothness*. In principle the result of Corollary 8.23 can be used to investigate the stability of the smoothness of operators for such Feller perturbations. A closed linear operator A is called $(K_0\dot{+}V)$*-smooth*, if there exists a constant C_V with the property that

$$\int_{-\infty}^{\infty} \left\| A \exp\left(-it\left(K_0\dot{+}V\right)\right) f \right\|_2^2 dt \leq C_V \|f\|_2^2, \tag{8.44}$$

for all $f \in L^2(E, m)$, for which $\exp\left(it\left(K_0\dot{+}V\right)\right) f$ belongs to $\mathrm{dom}(A)$ for almost all $t \in \mathbb{R}$: see Reed and Simon [194, p. 144]. An appropriate choice of the constant C_V is given by

$$C_V = \frac{1}{2\pi} \sup_{\mu \in \mathbb{C} \setminus \mathbb{R}} \sup_{\substack{f \in \mathrm{dom}(A^*) \\ \|f\|_2 = 1}} \left| \langle A^* f, [R_V(\mu) - R_V(\overline{\mu})] \, A^* f \rangle \right|.$$

This implies that for special classes of operators A we can find a constant C_W which is finite as long as C_V is finite and the norm $\left\| |W - V| \, \varphi^2 \right\|_{\mathrm{KF}}$ is sufficiently small. If that is the case, then A is $(K_0\dot{+}W)$-smooth as well.

REMARK 2. The strong convergence of the wave operators can be obtained easily if the Cook criterion is satisfied for all $K_0\dot{+}V_n$. Again let K_0 be the free Feller operator and let $(V_n : n \in \mathbb{N})$ be a sequence of Kato-Feller potentials satisfying the stability conditions in 8.17. For the sake of simplicity we assume that $\mathrm{dom}(K_0\dot{+}V_n) = D(K_0)$ and that $|V_n| \leq |U|$, for some U for which the Cook criterion is satisfied, i.e. for which for all f in some dense subset of $P_{\mathrm{ac}}(K_0)\mathcal{H}$ the integral

$$\int_{-\infty}^{\infty} \|U \exp\left(-isK_0\right) f\|_2 \, ds$$

is finite. Suppose that for some Kato-Feller function V we have

$$\lim_{n \to \infty} \|V_n - V\|_{\mathrm{KF}} = 0.$$

From the proof of the completeness of the space $KF(E)$ (see Proposition 2.4) it follows that $|V| \leq |U|$, m-almost everywhere.

Then for the wave operators it follows that (see (8.8))

$$\text{s-}\lim_{n\to\infty} \Omega_\pm \left(K_0\dot{+}V_n, K_0\right) = \Omega_\pm \left(K_0\dot{+}V, K_0\right) \tag{8.45}$$

and for the scattering operators (see (8.12))

$$\text{w-}\lim_{n\to\infty} S\left(K_0\dot{+}V_n, K_0\right) = S\left(K_0\dot{+}V, K_0\right). \tag{8.46}$$

If we have completeness, the limit in (8.46) may also be taken in the strong sense.

The convergence in (8.45) follows from the integral representation

$$\begin{aligned}
&\Omega_\pm \left(K_0\dot{+}V, K_0\right) f - f \\
&= i\int_0^\infty ds \exp\left(\pm is\left(K_0\dot{+}V\right)\right) V \exp\left(\mp isK_0\right) f,
\end{aligned}$$

where f varies in a dense subset of $D(K_0) \cap P_{\mathrm{ac}}(K_0)\mathcal{H}$. It suffices to show the convergence in (8.45) for functions f in the space $D\left(K_0^2\right) \bigcap P_{\mathrm{ac}}(K_0)\mathcal{H}$ which is dense in $P_{\mathrm{ac}}(K_0)\mathcal{H}$. Upon choosing $g = (aI + K_0)^2 f$, we get from the intertwining relation ($W = V_n$) the following:

$$\begin{aligned}
&\left\|\Omega_+ \left(K_0\dot{+}V, K_0\right) f - \Omega_+ \left(K_0\dot{+}W, K_0\right) f\right\|_2 \\
&= \left\|\left(aI + K_0\dot{+}V\right)^{-1} \Omega_+ \left(K_0\dot{+}V, K_0\right) (aI + K_0)^{-1} g \right. \\
&\quad \left. - \left(aI + K_0\dot{+}W\right)^{-1} \Omega_+ \left(K_0\dot{+}W, K_0\right) (aI + K_0)^{-1} g\right\|_2 \\
&\leq \int_0^\infty ds \left\|\exp\left(is\left(K_0\dot{+}V\right)\right) \left[(aI + K_0)^{-1} - \left(aI + K_0\dot{+}V\right)^{-1}\right] \exp\left(-isK_0\right) g \right. \\
&\quad \left. - \exp\left(is\left(K_0\dot{+}W\right)\right) \left[(aI + K_0)^{-1} - \left(aI + K_0\dot{+}W\right)^{-1}\right] \exp\left(-isK_0\right) g\right\|_2 \\
&\quad + \left\|\left(aI + K_0\dot{+}V\right)^{-1} (aI + K_0)^{-1} g \right. \\
&\qquad \left. - \left(aI + K_0\dot{+}W\right)^{-1} (aI + K_0)^{-1} g\right\|_2 .
\end{aligned}$$

According to Lemma 8.20 the final term tends to zero as $\|V - W\|_{\mathrm{KF}} \to 0$. The integrand tends to zero because

$$\text{s-}\lim_{n\to\infty} \exp\left(is\left(K_0\dot{+}V_n\right)\right) = \exp\left(is\left(K_0\dot{+}V\right)\right)$$

and because the integrand can be estimated by $\left\|U \exp\left(isK_0\right) (aI + K_0)^{-1} g\right\|_2$.

In a similar way we can also make a study of the scattering operators instead of the wave operators.

REMARK 3. As the final aim of this section we investigate the scattering matrices of the scattering systems given by $\{K_0\dot{+}V, K_0\}$ and $\{K_0\dot{+}W, K_0\}$. We have already introduced the scattering operators in a time-dependent manner. The scattering operators also possess a time-independent representation of the form

$$\begin{aligned}S(K_0\dot{+}V, K_0) &= P_{\mathrm{ac}}(K_0)\\ &+ \operatorname{s-lim}_{\eta\downarrow 0}\lim_{\delta\downarrow 0}\int_{\sigma(K_0)}\left\{(K_0-(\lambda-i\eta)I)^{-1}-(K_0-(\lambda+i\eta)I)^{-1}\right\}\\ &\qquad\left\{V-V\left(K_0\dot{+}V-(\lambda-i\delta)I\right)^{-1}V\right\}dE_{K_0}(\lambda):\end{aligned}$$

see e.g. Amrein, Jauch and Sinha [8], p. 290. Nevertheless having established the convergence of the scattering operator, e.g. as in (8.46), this does not mean that we can say anything about the scattering matrix. On the other hand, from the physical point of view, the scattering matrix is the most interesting object in scattering theory. It determines for example the cross section and it establishes the link with resonance theory. Hence the next objective will be to show that the continuity in terms of the Kato-Feller norms implies the convergence of the corresponding scattering matrices.

Any self-adjoint operator K_0 in a separable Hilbert space $\mathcal{H}$ gives rise to a direct integral decomposition of the form

$$\mathcal{H} = \int_{-\infty}^{\infty} \oplus\mathcal{H}_{K_0}(\lambda)dm(\lambda). \tag{8.47}$$

The reader may for example consult Birman and Solomjak [28]. The spaces $\mathcal{H}_{K_0}(\lambda)$ are called auxiliary Hilbert spaces or fibers of the decomposition (8.47). Any $f \in \mathcal{H}$ admits a decomposition "$f = \int \widetilde{f}(\lambda)dm(\lambda)$", with $\widetilde{f}(\lambda) \in \mathcal{H}_{K_0}(\lambda)$, m-almost everywhere. This decomposition is done in such a way that the scalar product of $f = \int \widetilde{f}(\lambda)dm(\lambda)$ and of $g = \int \widetilde{g}(\lambda)dm(\lambda)$ is given by

$$\langle f, g\rangle = \int_{-\infty}^{\infty}\left\langle \widetilde{f}(\lambda), \widetilde{g}(\lambda)\right\rangle dm(\lambda).$$

The fibers $\mathcal{H}_{K_0}(\lambda)$ are defined only up to unitary equivalence. For any self-adjoint K_0 there exists a so-called diagonal decomposition of $\mathcal{H}$. This means that one can find a special unitary map $\Phi_{K_0} : \mathcal{H} \to \mathcal{H}_{K_0}(\lambda)$ with the property that the spectral map $E_{K_0}(\triangle)$ is represented by the indicator function $1_\triangle(\lambda)$, i.e.

$$\langle E_{K_0}(\triangle)f, g\rangle = \int_\triangle \langle [\Phi_{K_0}f](\lambda), [\Phi_{K_0}g](\lambda)\rangle\, dm(\lambda).$$

This special choice of Φ_{K_0} is called the spectral transformation of K_0. It is diagonal because

$$[\Phi_{K_0}K_0f]_\lambda = \lambda\,[\Phi_{K_0}f]_\lambda\,,$$

where we prefer to write $[\Phi_{K_0} f]_\lambda$ as a notation for the representer of f in $\mathcal{H}_{K_0}(\lambda)$. To any bounded operator $B : \mathcal{H} \to \mathcal{H}$ that commutes with K_0 there corresponds an m-measurable family $B(\lambda) : \mathcal{H}_{K_0}(\lambda) \to \mathcal{H}_{K_0}(\lambda)$ with the property that for every $f \in \mathcal{H}$ the equality

$$[\Phi_{K_0} Bf]_\lambda = B(\lambda) [\Phi_{K_0} f]_\lambda$$

holds. For more details see Yafaev [253, p. 31 ff] or Baumgärtel and Wollenberg [21, p. 189].

If the wave operators $\Omega_\pm \left(K_0 \dot{+} V, K_0\right)$ exist, then they satisfy the intertwining relation

$$\Omega_\pm \left(K_0 \dot{+} V, K_0\right) \exp\left(itK_0\right) = \exp\left(it\left(K_0 \dot{+} V\right)\right) \Omega_\pm \left(K_0 \dot{+} V, K_0\right).$$

Therefore the scattering operator $S = \Omega_+ \Omega_-$ commutes with K_0, i.e.

$$S \exp\left(itK_0\right) = \exp\left(itK_0\right) S.$$

8.24. Definition. Let K_0 be a self-adjoint free Feller operator. Let $\mathcal{H}_{K_0}(\lambda)$, $\lambda \in \sigma_{\mathrm{ac}}(K_0)$, be the fibers of the spectral integral decomposition of $P_{\mathrm{ac}}(K_0)\mathcal{H}$ (see (8.47)). Let Φ_{K_0} be the spectral transformation of K_0 on $\sigma_{\mathrm{ac}}(K_0)$. This is a unitary map from $P_{\mathrm{ac}}(K_0)\mathcal{H}$ onto $L^2\left(\sigma_{\mathrm{ac}}(K_0), \mathcal{H}_{K_0}(\lambda), dm(\lambda)\right)$. Assume that the wave operators $\Omega_\pm \left(K_0 \dot{+} V, K_0\right)$ exist and define S_V by

$$S_V = \Omega_+^* \left(K_0 \dot{+} V, K_0\right) \Omega_- \left(K_0 \dot{+} V, K_0\right).$$

Then the scattering matrix $S_V(\lambda)$ is defined by

$$[\Phi_{K_0} S_V g]_\lambda =: S_V(\lambda) [\Phi_{K_0} g]_\lambda, \tag{8.48}$$

where $[\Phi_{K_0} g]_\lambda$ is the representer of $g \in P_{\mathrm{ac}}(K_0)\mathcal{H}$ in $\mathcal{H}_{K_0}(\lambda)$ with $\lambda \in \sigma_{\mathrm{ac}}(K_0)$, the absolutely continuous part of the spectrum of K_0.

The scattering matrix $S_V(\lambda)$, for some fixed λ, can be represented by

$$S_V(\lambda) = I_{\mathcal{H}_{K_0}(\lambda)} - 2\pi i F_\varphi(\lambda) \left(\varphi^2 V - \varphi V R_V(\lambda - i0) V \varphi\right) F_\varphi^*(\lambda), \tag{8.49}$$

if the expression on the right-hand side exists for that particular λ. The symbol $F_\varphi(\lambda)$ stands for a bounded linear operator that maps the space $L^2(E, \varphi dm)$ onto $\mathcal{H}_{K_0}(\lambda)$. In order to prove (8.65) below, we shall use a concrete representation that can be found in Baumgärtel and Wollenberg [21, p. 65 and pp 387 ff]. The following realization of the space $\mathcal{H}_{K_0}(\lambda)$ will be used:

$$\mathcal{H}_{K_0}(\lambda) = \left\{ f \in L^2(E, m) : \limsup_{s \downarrow 0} \frac{\langle f, E[\lambda, \lambda + s] f\rangle}{s} < \infty \right\}.$$

In fact, from the space $\mathcal{H}_{K_0}(\lambda)$ we divide out the null-space:

$$\left\{ f \in L^2(E, m) : \limsup_{s \downarrow 0} \frac{1}{s} \langle f, E[\lambda, \lambda + s] f \right\} = 0,$$

and so we consider equivalence classes. The operator $F_\varphi(\lambda)$ is defined via the relation $F_\varphi(\lambda) = D_\lambda \varphi^{-1}$, where the operator D_λ is related to the spectral measure $E_{K_0}(\cdot)$ via the formulae (see Baumgärtel and Wollenberg [21, p. 393])

$$D_\lambda f = D(\lambda) f = f + \left\{ h \in L^2(E,m) : \lim_{s\downarrow 0} \frac{\langle h, E_{K_0}[\lambda, \lambda+s]h\rangle}{s} = 0 \right\},$$

$$\langle D(\lambda)f, D(\lambda)f\rangle_{\mathcal{H}_{K_0}(\lambda)} = \lim_{s\downarrow 0} \frac{\langle f, E_{K_0}[\lambda, \lambda+s]f\rangle}{s} = \frac{d\langle f, E_{K_0}(\lambda)f\rangle}{d\lambda}, \text{ and}$$

$$\left\| F_\varphi^*(\lambda) F_\varphi(\lambda) \right\|_{2,2} = \left\| \varphi^{-1} D_\lambda^* D_\lambda \varphi^{-1} \right\|_{2,2} = \left\| \varphi^{-1} \frac{dE_{K_0}(\lambda)}{d\lambda} \varphi^{-1} \right\|_{2,2}, \tag{8.50}$$

for functions f in $\mathcal{H}_{K_0}(\lambda)$. Here φ is a strictly positive function defined on all of E.

Now we are able to formulate conditions on V and W such that the difference of the scattering matrices $S_V(\lambda) - S_W(\lambda)$ tends to zero in operator norm whenever $\left\|(V-W)\varphi^4\right\|_{\mathrm{KF}}$ tends to zero.

8.25. THEOREM. *Let $\{K_0 \dot{+} V, K_0\}$ and $\{K_0 \dot{+} W, K_0\}$ be two scattering systems of Feller operators with scattering matrices $S_V(\lambda)$ and $S_W(\lambda)$ acting in the fiber $\mathcal{H}_{K_0}(\lambda)$ of the direct composition of $P_{\mathrm{ac}}(K_0)\mathcal{H}$. Let φ be an admissible sandwiching function as described in Definition 8.19.*

(1) *Put*

$$v_\varphi = \max \left\{ \sup_{x\in E} |\varphi(x)|^2 |V(x)|, \ \sup_{x\in E} |\varphi(x)|^2 |W(x)| \right\} \tag{8.51}$$

and suppose that $v_\varphi < \infty$.

(2) *Set*

$$d_0(\lambda_0) = \sup_{\epsilon\in(-1,1)} \left\| \varphi^{-1} \left(K_0 - (\lambda_0 + i\epsilon) I\right)^{-1} \varphi^{-1} \right\|_{2,2}$$

and suppose that $d_0(\lambda_0) < \infty$ (see (8.29)). For the sake of simplicity, assume that

$$\max\left(\left\|V\varphi^2\right\|_{\mathrm{KF}}, \left\|W\varphi^2\right\|_{\mathrm{KF}} \right) \le \frac{1}{2} \frac{1}{6c_0 a_0^{1/2} \left(1 + 3a_0 d_0(\lambda_0)\right)}, \tag{8.52}$$

with $a_0 > 2\max(1, \lambda_0, m, M, b)$ (see (8.50)), and where the constant c_0 is given in (8.53) below.

(3) *Suppose that there exist constants c_0 and m with the property that (8.25) is valid, i.e.*

$$\int p_0(t,x,y) \varphi(y)^\rho dy \le c_0^2 e^{mt} \varphi(x)^\rho, \tag{8.53}$$

for $\rho = 2, 4$. Then the operator norms of the scattering matrices in $\mathcal{H}_{K_0}(\lambda_0)$ can be estimated by

$$\left\| S_V(\lambda_0) - S_W(\lambda_0) \right\|^2_{\mathcal{H}_{K_0}(\lambda_0)} \le c \left\| (V-W)\varphi^2 \right\|_{\mathrm{KF}}. \tag{8.54}$$

The constant c can be given explicitly in terms of the constants which are used in the present section B of this chapter.

Altogether this means that we take the assumptions of Lemma 8.20 and of Proposition 8.21 for granted. The only extra hypothesis is the finiteness of v_φ in (8.51).

REMARK. The convergence of wave operators, scattering operators and scattering matrices was considered by Kato [149] using totally different methods. He applied his general results to Schrödinger operators $-\frac{1}{2}\Delta + V$. He considered continuity with respect to the L^∞-norm, i.e. with respect to $\|V - W\|_\infty$. The continuity in V for scattering systems was also considered by Brüning and Gesztesy [42].

Proof. From (8.52) together with the definitions we get the inequalities $\gamma_V^{-1} \le 2$ and $\gamma_W^{-1} \le 2$. Hence, upon invoking Proposition 8.21 we infer that

$$\left\|\varphi^2 V \varphi^{-1} R_V(\lambda_0 \pm i0)\,\varphi^{-1} V \varphi^2\right\|_{2,2} \le v_\varphi^2 \left(d_0(\lambda_0) + \frac{1}{3a_0}\right).$$

Here $R_V(a)$ is given by $R_V(a) = \left(K_0 \dot{+} V - aI\right)^{-1}$. The same argument applies to $R_W(\lambda_0 \pm i0)$. Therefore the representation in (8.49) is available and may be used for $S_V(\lambda_0)$ as well as for $S_W(\lambda_0)$. Consequently we have to consider the inequality

$$\begin{aligned}
&\frac{1}{2\pi}\left\|\left(S_V(\lambda_0) - S_W(\lambda_0)\right) f\right\|_2 && (8.55)\\
&\le \left\|F_\varphi(\lambda_0)\varphi^2(V - W)F_\varphi^*(\lambda_0) f\right\|_2 && (\mathrm{I})\\
&\quad + \left\|F_\varphi(\lambda_0)\varphi V\left(R_V(\lambda_0 - i0) - R_W(\lambda_0 - i0)\right) V\varphi F_\varphi^*(\lambda_0) f\right\|_2 && (\mathrm{II})\\
&\quad + \left\|F_\varphi(\lambda_0)\varphi(V - W)R_W(\lambda_0 - i0)V\varphi F_\varphi^*(\lambda_0) f\right\|_2 && (\mathrm{III})\\
&\quad + \left\|F_\varphi(\lambda_0)\varphi W R_W(\lambda_0 - i0)(V - W)\varphi F_\varphi^*(\lambda_0) f\right\|_2 && (\mathrm{IV}).
\end{aligned}$$

Next we estimate every term in (8.55) separately. We write $\varphi^2(V - W) = \varphi^2 U$. We also employ the (in-)equalities (see (8.50)):

$$\begin{aligned}
&\left\|F_\varphi(\lambda_0)\right\|^2_{L^2(E,\varphi^{-1}dm)\to\mathcal{H}_{K_0}(\lambda_0)} = \left\|F_\varphi^*(\lambda_0)F_\varphi(\lambda_0)\right\|_{L^2(E,\varphi^{-1}dm)\to L^2(E,\varphi dm)}\\
&= \left\|\varphi^{-1}D_{\lambda_0}^* D_{\lambda_0}\varphi^{-1}\right\|_{2,2} = \left\|\varphi^{-1}\frac{dE_{K_0}(\lambda)}{d\lambda}\right|_{\lambda=\lambda_0}\left.\varphi^{-1}\right\|_{2,2}\\
&\le \frac{1}{2\pi}\left\|\varphi^{-1}\left[R_0(\lambda_0 - i0) - R_0(\lambda_0 + i0)\right]\varphi^{-1}\right\|_{2,2}\\
&\le \frac{d_0(\lambda_0)}{\pi}. && (8.56)
\end{aligned}$$

For the first term in (8.55), we obtain the following estimate:

$$\begin{aligned}
&\left\langle F_\varphi(\lambda_0)\varphi^2 U F_\varphi^*(\lambda_0) f, F(\lambda_0)\varphi^2 U F_\varphi^*(\lambda_0) f\right\rangle^{1/2}\\
&= \left\langle \varphi^2 U \varphi^{-1} D_{\lambda_0}^* D_{\lambda_0}\varphi^{-1}\varphi^2 U\varphi^{-1} D_{\lambda_0}^* f, \varphi^{-1} D_{\lambda_0}^* f\right\rangle^{1/2}\\
&= \Big\langle \varphi^2 U\varphi^{-1}\left(aI + K_0\right)^{-1}\varphi\varphi^{-1}\left(aI + K_0\right)\frac{dE_{K_0}(\lambda_0)}{d\lambda_0}\left(aI + K_0\right)\varphi^{-1}\\
&\qquad \varphi\left(aI + K_0\right)^{-1}\varphi^{-1}\varphi^2 U\varphi^{-1} D_{\lambda_0}^* f, \varphi^{-1} D_{\lambda_0}^* f\Big\rangle^{1/2}
\end{aligned}$$

$$\begin{aligned}
&\leq \left\|\varphi^2 U \varphi^{-1} (aI+K_0)^{-1} \varphi\right\|_{2,2} \left\|\varphi^{-1} D^*_{\lambda_0}\right\|_{2,2} \|f\|_2 \\
&\quad \times \left\| \varphi^{-1} (aI+K_0) \left.\frac{dE_{K_0}(\lambda)}{d\lambda}\right|_{\lambda=\lambda_0} (aI+K_0)\varphi^{-1} \right\|_{2,2}^{1/2} \\
&\leq (\lambda_0 + a) \frac{d_0(\lambda_0)}{\pi} \|f\|_2 \left\| \varphi |V-W| (aI+K_0)^{-1} \varphi \right\|_{2,2}. \qquad (8.57)
\end{aligned}$$

In order to estimate term (II) in (8.55), we write $R = R_V(\lambda_0 - i0) - R_W(\lambda_0 - i0)$. Then we get

$$\begin{aligned}
&\left\langle F_\varphi(\lambda_0)\varphi V R V \varphi F^*_\varphi(\lambda_0) f, F_\varphi(\lambda_0)\varphi V R V \varphi F^*_\varphi(\lambda_0) f\right\rangle^{1/2} \\
&= \left\langle \varphi V \varphi^{-1} \left.\frac{dE_{K_0}(\lambda)}{d\lambda}\right|_{\lambda=\lambda_0} \varphi^{-1}\varphi V R V \varphi\varphi^{-1} D^*_{\lambda_0} f, R V \varphi \varphi^{-1} D^*_{\lambda_0} f \right\rangle^{1/2} \\
&= \left\langle \varphi^2 V \varphi^{-1} \left.\frac{dE_{K_0}(\lambda)}{d\lambda}\right|_{\lambda=\lambda_0} \varphi^{-1} V \varphi^2 \varphi^{-1} R \varphi^{-1} V \varphi^2 \varphi^{-1} D^*_{\lambda_0} f, \right. \\
&\qquad\qquad \left. \varphi^{-1} R \varphi^{-1} V \varphi^2 \varphi^{-1} D^*_{\lambda_0} f \right\rangle^{1/2} \\
&\leq \frac{v_\varphi^2 d_0(\lambda_0)}{\pi} \|f\|_2 \left\|\varphi^{-1} R \varphi^{-1}\right\|_{2,2} \\
&\leq \frac{v_\varphi^2 d_0(\lambda_0)}{\pi} c(V,W,\lambda_0) \|f\|_2 \left\|(V-W)\varphi^2\right\|_{\mathrm{KF}}; \qquad (8.58)
\end{aligned}$$

see (8.32) of Proposition 8.21.

The estimates for the terms (III) and (IV) are similar. Here we consider only (III). In what follows we use a.o. the equality in (8.50):

$$\begin{aligned}
&\left\langle F_\varphi(\lambda_0)\varphi U R_W(\lambda_0 - i0) V \varphi F^*_\varphi(\lambda_0) f, F_\varphi(\lambda_0)\varphi U R_W(\lambda_0 - i0) V \varphi F^*_\varphi(\lambda_0) f\right\rangle^{1/2} \\
&= \left\langle \varphi^2 U \varphi^{-1} (aI+K_0)^{-1} \varphi\varphi^{-1} (aI+K_0) \left.\frac{dE_{K_0}(\lambda)}{d\lambda}\right|_{\lambda=\lambda_0} (aI+K_0)\varphi^{-1} \right. \\
&\quad \varphi (aI+K_0)^{-1} \varphi^{-1} U \varphi^2 \varphi^{-1} R_W(\lambda_0 - i0) \varphi^{-1}\varphi^2 V F^*_\varphi(\lambda_0) f, \\
&\quad \left. \varphi^{-1} R_W(\lambda_0 - i0)\varphi^{-1}\varphi^2 V F^*_\varphi(\lambda_0) f \right\rangle^{1/2} \\
&\leq \frac{d_0(\lambda_0)}{\pi} v_\varphi \left(d_0(\lambda_0) + \frac{1}{3a_0}\right)(a+\lambda_0) \left\|\varphi^2 |V-W| \varphi^{-1} (aI+K_0)^{-1}\varphi\right\|_{2,2} \|f\|_2. \qquad (8.59)
\end{aligned}$$

A combination of the estimates in (8.55) through (8.59) yields the following inequality:

$$\|S_V(\lambda_0) - S_W(\lambda_0)\| \leq c\left\{\left\|\varphi\,|V-W|\,(aI+K_0)^{-1}\,\varphi\right\|_{2,2} + \left\|(V-W)\varphi^2\right\|_{\mathrm{KF}}\right\}. \tag{8.60}$$

From the Stein interpolation theorem (Theorem A.40 in Appendix A) we infer that

$$\begin{aligned}
&\left\|\varphi\,|W-V|\,(aI+K_0)^{-1}\varphi\right\|_{2,2} \\
&\leq \sup_{\xi\in\mathbb{R}}\left\|\varphi^{2-2i\xi}\,|W-V|\,(aI+K_0)^{-1}\,\varphi^{2i\xi}\right\|_{1,1}^{1/2} \\
&\qquad\times \sup_{\xi\in\mathbb{R}}\left\|\varphi^{-2i\xi}\,|W-V|\,(aI+K_0)^{-1}\,\varphi^{2+2i\xi}\right\|_{\infty,\infty}^{1/2} \\
&\leq \left\|\varphi^2\,|W-V|\,(aI+K_0)^{-1}\right\|_{1,1}^{1/2} \times \left\||W-V|\,(aI+K_0)^{-1}\,\varphi^2\right\|_{\infty,\infty}^{1/2}
\end{aligned}$$

(see Proposition 2.2 in Chapter 2)

$$\begin{aligned}
&\leq \left\||W-V|\,(aI+K_0)^{-1}\,\varphi^2\right\|_{\infty,\infty} \\
&\leq c'\left\|(V-W)\varphi^2\right\|_{\mathrm{KF}}.
\end{aligned} \tag{8.61}$$

Here we applied the Stein interpolation theorem to the function $T : z \mapsto |W-V|\,\varphi^{2-2z}\,(aI+K_0)^{-1}\,\varphi^{2z}$. On the line $\mathrm{Re}\, z = 1$ we estimate $\|T(z)\|_{1,1}$ and on the line $\mathrm{Re}\, z = 0$ we estimate $\|T(z)\|_{\infty,\infty}$. A combination of (8.60) and (8.61) yields

$$\|S_V(\lambda_0) - S_W(\lambda_0)\| \leq c(c'+1)\left\|(W-V)\varphi^2\right\|_{\mathrm{KF}}.$$

This proves assertion (3) in Theorem 8.25.

Another approach to studying scattering matrices is contained in the next proposition.

8.26. PROPOSITION. *Suppose that the wave operators $\Omega_\pm\left(K_0\dot{+}V, K_0\right)$ and $\Omega_\pm\left(K_0\dot{+}W, K_0\right)$ exist and that the operators $S_V(\lambda)$ and $S_W(\lambda)$ are given according to Definition 8.24. Fix a point λ_0 in the absolutely continuous part of the spectrum of K_0. In addition, assume that φ is a strictly positive function with the following properties:*

(i) *For some real number $a > 0$, there exists a constant $k(a)$ such that, for all $x \in E$,*

$$\left[\left(aI + K_0\dot{+}2V\right)^{-1}\varphi^2\right](x) \leq k(a)\varphi(x)^2 \text{ and}$$
$$\left[\left(aI + K_0\dot{+}2W\right)^{-1}\varphi^2\right](x) \leq k(a)\varphi(x)^2.$$

(ii) *The operator B defined by*

$$B = B(\lambda_0) = \varphi^{-1} \frac{dE_{K_0}(\lambda)}{d\lambda} \varphi^{-1} \Big|_{\lambda=\lambda_0} \tag{8.62}$$

is bounded on $P_{ac}(K_0)\mathcal{H}$;

(iii) *The following operator norms are bounded:*

$$\left\| \varphi^{-1} R_V(\lambda_0 - i0) \varphi^{-1} \right\|_{2,2} \quad \text{and} \quad \left\| \varphi^{-1} R_W(\lambda_0 - i0) \varphi^{-1} \right\|_{2,2}. \tag{8.63}$$

(iv) *The Kato-Feller norm of the function $(V - W)\varphi^2$ is finite and there exists an open interval I_0 containing λ_0 such that the following norms are finite:*

$$\|\varphi E_{K_0}(I_0) V \varphi\|_{2,2} \text{ and } \|\varphi E_{K_0}(I_0) W \varphi\|_{2,2}. \tag{8.64}$$

Then the norm of the difference of the scattering matrices can be estimated by

$$\begin{aligned} &\|S_V(\lambda_0) - S_W(\lambda_0)\|_{2,2} \\ &\le c_1 \|\varphi(V - W) E_{K_0}(I_0)\varphi\|_{2,2} + c_2 \left\| \varphi^{-1} \left(R_V(\lambda_0 - i0) - R_W(\lambda_0 - i0) \right) \varphi^{-1} \right\|_{2,2} \\ &\le c_1 \|\varphi(V - W) E_{K_0}(I_0)\varphi\|_{2,2} + c_2' \left\| (W - V)\varphi^2 \right\|_{KF}. \end{aligned} \tag{8.65}$$

REMARK. The interval I_0 in assertion (iv) should not contain exceptional sets like eigenvalues or thresholds of $K_0 \dot{+} V$ nor of $K_0 \dot{+} W$: see Reed and Simon [194, p. 99 ff]. This can be expressed by

$$\begin{aligned} &\operatorname{ran}\left(E_{K_0 \dot{+} V}(I_0)\right) \subseteq P_{ac}\left(K_0 \dot{+} V\right)\mathcal{H}, \text{ and} \\ &\operatorname{ran}\left(E_{K_0 \dot{+} W}(I_0)\right) \subseteq P_{ac}\left(K_0 \dot{+} W\right)\mathcal{H}. \end{aligned}$$

Before we proceed with the proof of this proposition, we mention the following corollary. It follows from Proposition 8.26 with $\mathbb{R}$ replacing I_0: compare with Theorem 8.25.

8.27. COROLLARY. *Suppose that φ has the properties (i), (ii) and (iii) of Proposition 8.26 and suppose that (iv) is replaced with*

(iv′) *The supremum norm of the function $(V - W)\varphi^2$ is finite.*

Then the norm of the difference of the scattering matrices can be estimated as follows:

$$\|S_V(\lambda_0) - S_W(\lambda_0)\| \le c \left\| (V - W)\varphi^2 \right\|_\infty. \tag{8.66}$$

REMARK. The proof follows more or less the same pattern as that of Theorem 8.25. Nevertheless it is somewhat more sophisticated. It is not clear whether one can find weaker conditions than the ones in Theorem 8.25 under which the conclusion of Proposition 8.26 is valid.

Proof. For brevity we write $F(\lambda)$ instead of $F_\varphi(\lambda)$. We employ the identity (see (8.55))

$$\begin{aligned} S_V(\lambda) - S_W(\lambda) = -2\pi i F(\lambda) \\ \times \{\varphi(V-W)\varphi - \varphi V R_V(\lambda - i0)V\varphi + \varphi V R_W(\lambda - i0)\varphi V \\ -\varphi V R_W(\lambda - i0)\varphi V + \varphi W R_W \lambda - i0)\varphi V \\ -\varphi W R_W(\lambda - i0)\varphi V + \varphi W R_W(\lambda - i0)\varphi W\} F^*(\lambda). \end{aligned} \tag{8.67}$$

Let $\{E(\lambda) = E_{K_0}(\lambda) : \lambda \in \mathbb{R}\}$ be the spectral decomposition corresponding to K_0. We use the same notation as in the proof of Theorem 8.25. In addition let φ be a strictly positive function that has the properties (i), (ii), (iii) and (iv) of Proposition 8.26. So $F(\lambda_0)f$ is defined by $F(\lambda_0)f = D(\lambda_0)\varphi^{-1}f$. Hence we have by assumption

$$\left\| \varphi^{-1} \frac{dE(\lambda)}{d\lambda}\bigg|_{\lambda=\lambda_0} \varphi^{-1} \right\|_{2,2} = \|B\|_{2,2} < \infty. \tag{8.68}$$

Following (8.50) we note the following equalities:

$$\begin{aligned} \left\|\varphi^{-1}D(\lambda_0)^*\right\|^2_{\mathcal{H}_{K_0}(\lambda_0), L^2(E,m)} = \left\|D(\lambda_0)\varphi^{-1}\right\|^2_{L^2(E,m)^2, \mathcal{H}_{K_0}(\lambda_0)} \\ = \left\|\varphi^{-1}D(\lambda_0)^*D(\lambda_0)\varphi^{-1}\right\|_{L^2(E,m), L^2(E,m)} = \|B\|^2_{L^2(E,m), L^2(E,m)} . \end{aligned} \tag{8.69}$$

For brevity we write

$$D_{V,W}(\lambda_0) = R_V(\lambda_0 - i0) - R_W(\lambda_0 - i0),$$

and we fix $f \in \mathcal{H}_{K_0}(\lambda_0)$. Then, by (8.67) together with the triangle inequality we obtain

$$\begin{aligned} &\frac{1}{2\pi}\left(\langle (S_V(\lambda_0) - S_W(\lambda_0))\, f, (S_V(\lambda_0) - S_W(\lambda_0))\, f\rangle\right)^{1/2} \\ &\le \langle F(\lambda_0)\varphi\,(V-W)\,E(I_0)\varphi F^*(\lambda_0)f, F(\lambda_0)\varphi\,(V-W)\,E(I_0)\varphi F^*(\lambda_0)f\rangle^{1/2}_{\lambda_0} \\ &\quad + \langle F(\lambda_0)E(I_0)V\,(R_V(\lambda_0 - i0) - R_W(\lambda_0 - i0))\,VE(I_0)\varphi F^*(\lambda_0)f, \\ &\quad\ F(\lambda_0)\varphi E(I_0)V\,(R_V(\lambda_0 - i0) - R_W(\lambda_0 - i0))\,VE(I_0)\varphi F^*(\lambda_0)f\rangle^{1/2}_{\lambda_0} \\ &\quad + \langle F(\lambda_0)\varphi E(I_0)\,(V-W)\,\varphi R_W(\lambda_0 - i0)VE(I_0)\varphi F^*(\lambda_0)f, \\ &\quad\quad F(\lambda_0)\varphi E(I_0)\,(V-W)\,\varphi R_W(\lambda_0 - i0)VE(I_0)\varphi F^*(\lambda_0)f\rangle^{1/2}_{\lambda_0} \\ &\quad + \langle F(\lambda_0)\varphi E(I_0)WR_W(\lambda_0 - i0)\,(V-W)\,E(I_0)\varphi F^*(\lambda_0)f, \\ &\quad\ F(\lambda_0)\varphi E(I_0)WR_W(\lambda_0 - i0)\,(V-W)\,E(I_0)\varphi F^*(\lambda_0)f\rangle^{1/2}_{\lambda_0} \end{aligned}$$

(definition of $F(\lambda_0)$ and equality (8.62))

$$
\begin{aligned}
&= \left\langle \varphi\,(V-W)\,E(I_0)\varphi\varphi^{-1}D(\lambda_0)^* f, B\varphi\,(V-W)\,E(I_0)\varphi\varphi^{-1}D(\lambda_0)^* f\right\rangle^{1/2}\\
&\quad + \left\langle \varphi V E(I_0)\varphi\varphi^{-1}D_{V,W}(\lambda_0-i0)\varphi^{-1}\varphi V E(I_0)\varphi\varphi^{-1}D(\lambda_0)^* f,\right.\\
&\quad\quad \left. B\varphi V E(I_0)\varphi\varphi^{-1}D_{V,W}(\lambda_0-i0)\varphi^{-1}\varphi V E(I_0)\varphi\varphi^{-1}D(\lambda_0)^* f\right\rangle^{1/2}\\
&\quad + \left\langle \varphi E(I_0)\,(V-W)\,\varphi\varphi^{-1}R_W(\lambda_0-i0)\varphi^{-1}\varphi V E(I_0)\varphi\varphi^{-1}D(\lambda_0)^* f,\right.\\
&\quad\quad \left. B\varphi E(I_0)\,(V-W)\,\varphi\varphi^{-1}R_W(\lambda_0-i0)\varphi^{-1}\varphi V E(I_0)\varphi\varphi^{-1}D(\lambda_0)^* f\right\rangle^{1/2}\\
&\quad + \left\langle \varphi E(I_0) W\varphi\varphi^{-1}R_W(\lambda_0-i0)\varphi^{-1}\varphi\,(V-W)\,E(I_0)\varphi\varphi^{-1}D(\lambda_0)^* f,\right.\\
&\quad\quad \left. B\varphi E(I_0) W\varphi\varphi^{-1}R_W(\lambda_0-i0)\varphi^{-1}\varphi\,(V-W)\,E(I_0)\varphi\varphi^{-1}D(\lambda_0)^* f\right\rangle^{1/2}\\
&\le \|B\|_{2,2}^{1/2}\,\|\varphi\,(V-W)\,E(I_0)\varphi\|_{2,2}\,\left\|\varphi^{-1}D(\lambda_0)^* f\right\|_{2,2}\\
&\quad + \|B\|_{2,2}^{1/2}\,\|\varphi V E(I_0)\varphi\|_{2,2}\\
&\quad \times \left\|\varphi^{-1}\left(R_V(\lambda_0-i0)-R_W(\lambda_0-i0)\right)\varphi^{-1}\right\|_{2,2}\left\|\varphi^{-1}D(\lambda_0)^* f\right\|_{2,2}\\
&\quad + \|B\|_{2,2}^{1/2}\,\|\varphi E(I_0)\,(V-W)\,\varphi\|_{2,2}\\
&\quad \times \left\|\varphi^{-1}R_W(\lambda_0-i0)\varphi^{-1}\right\|_{2,2}\|\varphi V E(I_0)\varphi\|_{2,2}\left\|\varphi^{-1}D(\lambda_0)^* f\right\|_{2,2}\\
&\quad + \|B\|_{2,2}^{1/2}\,\|\varphi E(I_0) W\varphi\|_{2,2}\left\|\varphi^{-1}R_W(\lambda_0-i0)\varphi^{-1}\right\|_{2,2}\\
&\quad \times \|\varphi E(I_0)\,(V-W)\,\varphi\|_{2,2}\left\|\varphi^{-1}D(\lambda_0)^* f\right\|_{2,2}
\end{aligned}
$$

(by (8.69))

$$
\begin{aligned}
&\le \|B\|_{2,2}^{1/2}\,\|\varphi\,(V-W)\,E(I_0)\varphi\|_{2,2}\,\langle f,f\rangle_{\lambda_0}^{1/2}\\
&\quad + \|B\|_{2,2}^{1/2}\,\|\varphi V E(I_0)\varphi\|_{2,2}\\
&\quad \times \left\|\varphi^{-1}\left(R_V(\lambda_0-i0)-R_W(\lambda_0-i0)\right)\varphi^{-1}\right\|_{2,2}\langle f,f\rangle_{\lambda_0}^{1/2}\\
&\quad + \|B\|_{2,2}^{1/2}\,\|\varphi E(I_0)\,(V-W)\,\varphi\|_{2,2}\\
&\quad \times \left\|\varphi^{-1}R_W(\lambda_0-i0)\varphi^{-1}\right\|_{2,2}\|\varphi V E(I_0)\varphi\|_{2,2}\,\langle f,f\rangle_{\lambda_0}^{1/2}\\
&\quad + \|B\|_{2,2}^{1/2}\,\|\varphi E(I_0) W\varphi\|_{2,2}\left\|\varphi^{-1}W(\lambda_0-i0)\varphi^{-1}\right\|_{2,2}\\
&\quad \times \|\varphi E(I_0)\,(V-W)\,\varphi\|_{2,2}\,\langle f,f\rangle_{\lambda_0}^{1/2}. \qquad (8.70)
\end{aligned}
$$

From (8.70) the first inequality in (8.65) follows. In order to obtain the second, we use the identity (see (8.34))

$$
\begin{aligned}
&\varphi^{-1}\left(R_V(\lambda_0-i0)-R_W(\lambda_0-i0)\right)\varphi^{-1}\\
&= \left\{\varphi^{-2}+(\lambda_0+a)\varphi^{-1}R_W(\lambda_0-i0)\varphi^{-1}\right\}\\
&\quad \varphi\left\{\left(aI+K_0\dot{+}V\right)^{-1}-\left(aI+K_0\dot{+}W\right)^{-1}\right\}\varphi\\
&\quad \left\{\varphi^{-2}+(\lambda_0+a)\varphi^{-1}R_V(\lambda_0-i0)\varphi^{-1}\right\}. \qquad (8.71)
\end{aligned}
$$

From inequality (4.3) in Proposition 4.2, with $\beta = 0$, $\alpha = 1$, $\gamma_0 = 1/2$, and with φ replacing ψ we infer that

$$\begin{aligned}
&\left\|\varphi\left\{\left(aI+K_0\dot{+}V\right)^{-1}-\left(aI+K_0\dot{+}W\right)^{-1}\right\}\varphi\right\|_{2,2} \\
&\leq \left\|\left(aI+K_0\right)^{-1}|V-W|\left(aI+K_0\dot{+}2V\right)^{-1}\varphi^2\right\|_\infty^{1/2} \\
&\quad\times\left\|\left(aI+K_0\right)^{-1}|V-W|\left(aI+K_0\dot{+}2W\right)^{-1}\varphi^2\right\|_\infty^{1/2}.
\end{aligned} \tag{8.72}$$

So from (8.70), from (8.71) together with (8.72) it follows that

$$\begin{aligned}
&\frac{1}{2\pi}\left\|S_V(\lambda_0)-S_W(\lambda_0)\right\|_{2,2} \\
&\leq \|B\|_{2,2}^{1/2}\left\|\varphi\left(V-W\right)E(I_0)\varphi\right\|_\infty \\
&\quad+\|B\|_{2,2}^{1/2}\left\|\varphi VE(I_0)\varphi\right\|_{2,2} \\
&\qquad\times\left(\varphi^{-2}+|\lambda_0+a|\left\|\varphi^{-1}R_V(\lambda_0-i0)\varphi^{-1}\right\|_{2,2}\right) \\
&\qquad\times\left(\varphi^{-2}+|\lambda_0+a|\left\|\varphi^{-1}R_W(\lambda_0-i0)\varphi^{-1}\right\|_{2,2}\right) \\
&\qquad\times\left\|\left(aI+K_0\right)^{-1}|V-W|\left(aI+K_0\dot{+}2V\right)^{-1}\varphi^2\right\|_\infty^{1/2} \\
&\qquad\times\left\|\left(aI+K_0\right)^{-1}|V-W|\left(aI+K_0\dot{+}2W\right)^{-1}\varphi^2\right\|_\infty^{1/2} \\
&\quad+\|B\|_{2,2}^{1/2}\left\|\varphi E(I_0)\left(V-W\right)\varphi\right\|_{2,2}\left\|\varphi^{-1}R_W(\lambda_0-i0)\varphi^{-1}\right\|_{2,2}\left\|\varphi VE(I_0)\varphi\right\|_{2,2} \\
&\quad+\|B\|_{2,2}^{1/2}\left\|\varphi E(I_0)W\varphi\right\|_{2,2}\left\|\varphi^{-1}R_W(\lambda_0-i0)\varphi^{-1}\right\|_{2,2}\left\|\varphi\left(V-W\right)E(I_0)\varphi\right\|_{2,2}.
\end{aligned}$$

Consequently we obtain the following inequalities:

$$\begin{aligned}
&\frac{1}{2\pi}\left\|S_V(\lambda_0)-S_W(\lambda_0)\right\|_{2,2} \\
&\leq \|B\|_{2,2}^{1/2}\left\|\varphi\left(V-W\right)E(I_0)\varphi\right\|_\infty \\
&\quad+\|B\|_{2,2}^{1/2}\left\|\varphi VE(I_0)\varphi\right\|_{2,2}\left(\varphi^{-1}+|\lambda_0+a|\left\|\varphi^{-1}R_V(\lambda_0-i0)\varphi^{-1}\right\|_{2,2}\right) \\
&\qquad\times\left(\varphi^{-1}+|\lambda_0+a|\left\|\varphi^{-1}R_W(\lambda_0-i0)\varphi^{-1}\right\|_{2,2}\right) \\
&\qquad\times\left\|\left(aI+K_0\right)^{-1}|V-W|\left(aI+K_0\dot{+}2V\right)^{-1}\varphi^2\right\|_\infty^{1/2} \\
&\qquad\times\left\|\left(aI+K_0\right)^{-1}|V-W|\left(aI+K_0\dot{+}2W\right)^{-1}\varphi^2\right\|_\infty^{1/2} \\
&\quad+\|B\|_{2,2}^{1/2}\left\|\varphi\left(V-W\right)E(I_0)\varphi\right\|_{2,2} \\
&\qquad\times\left\|\varphi^{-1}R_W(\lambda_0-i0)\varphi^{-1}\right\|_{2,2}\left\|\varphi VE(I_0)\varphi\right\|_{2,2} \\
&\quad+\|B\|_{2,2}^{1/2}\left\|\varphi E(I_0)W\varphi\right\|_{2,2}\left\|\varphi^{-1}R_W(\lambda_0-i0)\varphi^{-1}\right\|_{2,2}\left\|\varphi\left(V-W\right)E(I_0)\varphi\right\|_{2,2}
\end{aligned}$$

$$
\begin{aligned}
&\leq \|B\|_{2,2}^{1/2} \|\varphi (V - W) E(I_0)\varphi\|_\infty \\
&\quad + k(a) \|B\|_{2,2}^{1/2} \|\varphi V E(I_0)\varphi\|_{2,2} \\
&\qquad \times \left(\varphi^{-2} + |\lambda_0 + a| \left\|\varphi^{-1} R_V(\lambda_0 - i0)\varphi^{-1}\right\|_{2,2}\right) \\
&\qquad \times \left(\varphi^{-2} + |\lambda_0 + a| \left\|\varphi^{-1} R_W(\lambda_0 - i0)\varphi^{-1}\right\|_{2,2}\right) \\
&\qquad \times \left\|(aI + K_0)^{-1} |V - W| \varphi^2\right\|_\infty \\
&\quad + \|B\|_{2,2}^{1/2} \|\varphi (V - W) E(I_0)\varphi\|_{2,2} \\
&\qquad \times \left\|\varphi^{-1} R_W(\lambda_0 - i0)\varphi^{-1}\right\|_{2,2} \|\varphi V E(I_0)\varphi\|_{2,2} \\
&\quad + \|B\|_{2,2}^{1/2} \|\varphi E(I_0) W \varphi\|_{2,2} \left\|\varphi^{-1} R_W(\lambda_0 - i0)\varphi^{-1}\right\|_{2,2} \\
&\qquad \times \|\varphi (V - W) E(I_0)\varphi\|_{2,2} .
\end{aligned}
$$

Hence the result in Proposition 8.26 follows.

C. Quantitative estimates for singular potentials in terms of the weighted Laplace transform of the occupation time (for large coupling parameters)

Many quantum mechanical systems are described by self-adjoint Feller operators K, where the height of the potential plays a very crucial role for the spectral properties of K. In several models, the potential is relatively high over a certain region Γ of $\mathbb{R}^d$, compared to the height of the potential on the complement $\Sigma = \mathbb{R}^d \setminus \Gamma$. The choice of Γ is given in a natural way in these models. This will become obvious as well from the examples that follow. Let $W = W_+ - W_-$ be the decomposition of a Kato-Feller potential into its positive and negative part. We consider models in which W_+ is large on a set $\Gamma \subset \mathbb{R}^d$, i.e. we decompose the positive part of the potential W:

$$W_+ = 1_\Gamma W_+ + 1_\Sigma W_+, \quad \text{and set} \quad U = 1_\Gamma W_+, \quad \operatorname{supp}(U) = \Gamma,$$

where we fix a possibly large value U_0 and determine Γ in such a way that $U(x) \geq U_0$, $x \in \Gamma$. Introducing $V = 1_\Sigma W_+ - W_-$, the physical systems are described by self-adjoint Feller operators of the form

$$\hbar^2 K_0 + V + MU.$$

Very often these real models are compared with the artificial one where either $\hbar$ tends to zero or M tends to infinity. In our context, the artificial systems are described by the Dirichlet-Feller operators $\left(\hbar^2 K_0 + V\right)_\Sigma$ (see Definition 2.25). The main topic in this section is a quantitative estimate for resolvent differences in terms of $\hbar$ (for small $\hbar$) or of M (for large M). With $Jf = f\restriction_\Sigma$ the operator norm of the resolvent difference can be estimated as follows:

$$\left\| J\left(K_0 \dot{+} \frac{1}{\hbar^2}V + \frac{M}{\hbar^2}U + \frac{a}{\hbar^2}I\right)^{-1} - \left(\left(K_0 \dot{+} \frac{1}{\hbar^2}V\right)_\Sigma + \frac{a}{\hbar^2}I\right)^{-1} J \right\|_{2,2} \tag{8.73}$$

$$\leq \int_0^\infty \exp\left(-\frac{a\lambda}{\hbar^2}\right)$$
$$\left\| J \exp\left(-\lambda\left(K_0 \dot{+} \frac{1}{\hbar^2}V + \frac{M}{\hbar^2}U\right)\right) - \exp\left(-\lambda\left(K_0 \dot{+} \frac{1}{\hbar^2}V\right)_\Sigma\right) J \right\|_{2,2} d\lambda$$

$$\leq \int_0^\infty \exp\left(-\frac{a\lambda}{\hbar^2}\right)$$
$$\sup_{x \in E} \mathbb{E}_x \left\{ \exp\left(-\frac{1}{\hbar^2}\int_0^\lambda V(X(s))ds\right) \exp\left(-\frac{M}{\hbar^2}\int_0^\lambda U(X(s))ds\right), T_{\lambda,\Gamma} > 0 \right\} d\lambda.$$

Here $T_{\lambda,\Gamma} = \int_0^\lambda 1_\Gamma(X(u))du$. Since $S < \lambda$ if and only if $T_{\lambda,\Gamma} := \int_0^\lambda 1_\Gamma(X(u))du > 0$, we see that events $\{S < \lambda\}$ and $\{T_{\lambda,\Gamma} > 0\}$ coincide. Here S is the penetration time of Γ: $S = \inf\{\lambda > 0 : T_{\lambda,\Gamma} > 0\}$. For the sake of simplicity we suppose that $V \geq -V_0$, $V_0 > 0$. Because of the inequality $U(x) \geq U_0 1_\Gamma(x)$, the final term is smaller than

$$\int_0^\infty \exp\left(-\lambda\left(a - V_0\right)\right) \sup_{x\in E} \mathbb{E}_x \left\{\exp\left(-\frac{MU_0}{\hbar^2} T_{\lambda,\Gamma}\right), T_{\lambda,\Gamma} > 0\right\} d\lambda, \tag{8.74}$$

assuming $0 < \hbar \leq 1$. In expressions like (8.73) and (8.74), studying the semi-classical limit $\hbar \to 0$ is equivalent to studying the large coupling limit $M \to \infty$. The main ingredient for a quantitative resolvent estimate is the behavior of the quantity

$$\sup_{x\in E} \mathbb{E}_x \left\{\exp\left(-\beta T_{\lambda,\Gamma}\right), T_{\lambda,\Gamma} > 0\right\},$$

i.e. of the Laplace transform of the occupation time of the trajectories in Γ for large values of the parameter β.

Before we start the study of these quantitative resolvent estimates, we will illustrate that in different physical models one can find regions with high potentials in a natural way.

Examples

Solid state physics: In solid state physics, the potential of a crystal is typically of the form

$$V(x) = \sum_{\substack{m_i \in \mathbb{A} \\ i=1,2,3}} W(x_1 + m_1 d, x_2 + m_2 d, x_3 + m_3 d),$$

with $x = (x_1, x_2, x_3) \in \mathbb{R}^3$ and with a lattice constant $d \in \mathbb{R}^3$. If the atomic potential W has a positive maximum with the property that $W(y) \geq W_0 > 0$ for y in a ball $B \subset \mathbb{R}^3$ centered let us say at zero, then V is larger than W_0 for a series of balls centered at the lattice points. In this case the set Γ is unbounded in $\mathbb{R}^3$ consisting of an infinite union of a sequence of balls.

In Figure 1 a crystal potential is shown. The possible sets Γ for that potential are given in Figure 2.

N-body systems: In N-body systems with two-body interactions, the total potential V is given as a sum of two-body potentials V_ℓ, i.e.

$$V(x) = \sum_\ell V_\ell\left(u_\ell\right),$$

where $x = (x_1, \dots, x_N)$ belongs to $\mathbb{R}^{3N}$, $x_j \in \mathbb{R}^3$, where ℓ labels the possible two-body interactions, and where each u_ℓ is a linear combination of x_j. Consequently every u_ℓ is a vector in $\mathbb{R}^3$. For an easier explanation, assume $N = 3$ and so we

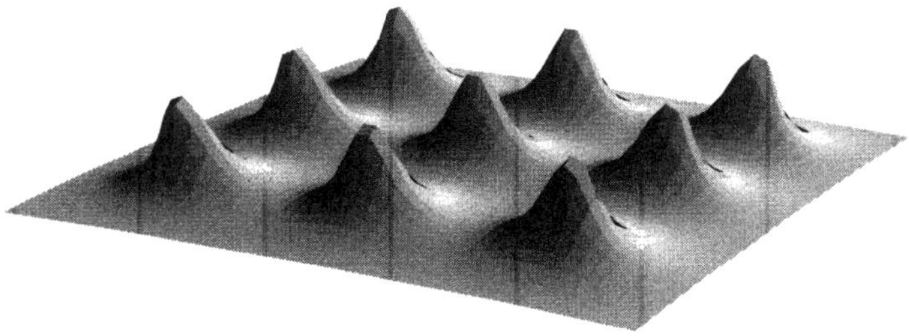

Fig. 1

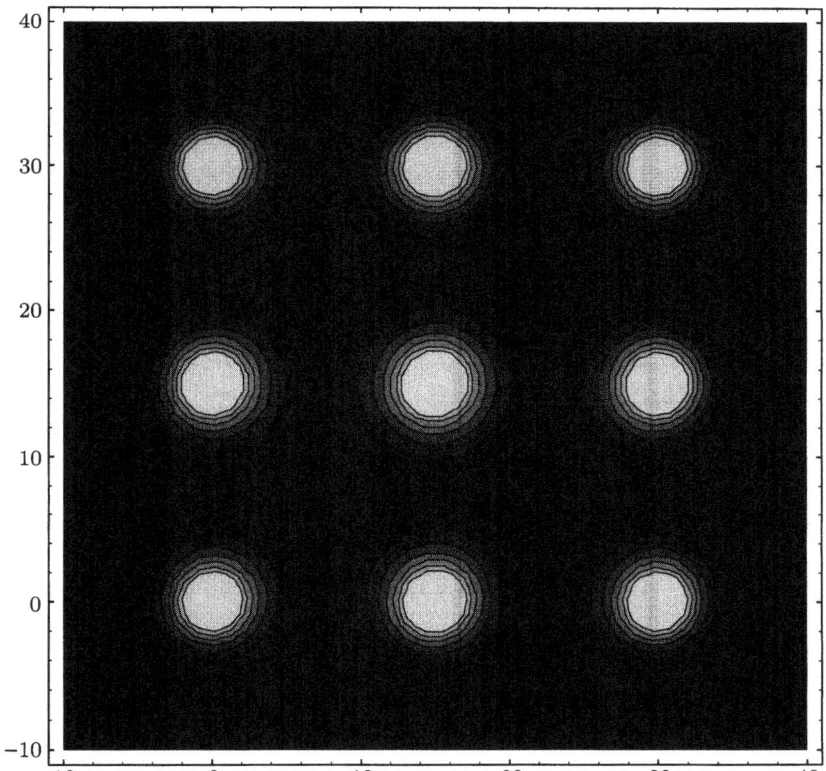

Fig. 2

consider a three-body potential V. In relative coordinates u_1 and u_2, the potential V is given via the formula

$$V(u_1, u_2) = V_1(u_1) + V_2(u_2) + V_3(u_1 + u_2).$$

If $V_1(u_1)$ is for instance larger than V_0 for all u_1 in a ball $B \subset \mathbb{R}^3$, then V is larger than V_0 on $B \times \mathbb{R}^3$. In that case, Γ consists of unbounded intersecting sheets in $\mathbb{R}^6$ possessing edges and corners.

A simple example of a three-body potential is given in Figure 3. Here V_3 is assumed to be zero.

Atomic systems: In atomic systems, the potentials have different signs corresponding to the attractive or repulsive forces. For instance, the Helium atom is described by the potential

$$W(u_1, u_2) = -\frac{a_1}{|u_1|} - \frac{a_2}{|u_2|} + \frac{a_3}{|u_1 - u_2|},$$

where a_1, a_2 and a_3 are positive constants. One sees easily that

$$\begin{aligned} W(u) &= +\infty \quad \text{for} \quad u \in \left\{(u_1, u_2) \in \mathbb{R}^6 \setminus \{(0,0)\} : u_1 = u_2\right\}, \\ W(u) &= -\infty \quad \text{for} \quad u \in \left\{(0, u_2) \in \mathbb{R}^6 : u_2 \in \mathbb{R}^3 \setminus \{0\}\right\}, \text{ and} \\ W(u) &= -\infty \quad \text{for} \quad u \in \left\{(u_1, 0) \in \mathbb{R}^6 : u_1 \in \mathbb{R}^3 \setminus \{0\}\right\}. \end{aligned}$$

The (natural) value of $W(0)$ depends on the values of the constants a_1, a_2 and a_2. This example shows that there are regions in $\mathbb{R}^6$ in a neighborhood of the subset $\{0\} \times \mathbb{R}^3 \cup \mathbb{R}^3 \times \{0\}$ where W is large. For Γ we may choose appropriate cones in $\mathbb{R}^6$.

In these examples, physical models give rise to a certain region Γ where the potentials are high. In fact they have height β with β large. On the other hand, from the mathematical point of view, the Dirichlet-Feller or Dirichlet-Schrödinger operators are well studied and their spectral properties are better understood. The systems in which the potential is infinite over Γ are somewhat artificial. However, they can be used as comparison systems provided we have a quantitative estimate between the system with finite β and the one with infinite β. In that case we will be able to extend the information about the artificial system governed by the Hamiltonian $(K_0 \dot{+} V)_\Sigma$ to the realistic one determined by the Hamiltonian $K_0 \dot{+} V + \beta 1_\Gamma$.

In Chapter 7, we investigated the convergence of the resolvent differences

$$J\left(K_0 \dot{+} V + \beta 1_\Gamma + aI\right)^{-1} - \left(\left(K_0 \dot{+} V + aI\right)_\Sigma\right)^{-1} J$$

with respect to different operator norms. The key for both the convergence studied in Chapter 7 and the quantitative estimate which is the topic of this section is the *Laplace transform* of *the occupation time*, i.e.

$$\sup_{x \in E} \mathbb{E}_x \left\{\exp\left(-\beta T_{\lambda,\Gamma}\right)\right\} =: \rho(\beta, \lambda).$$

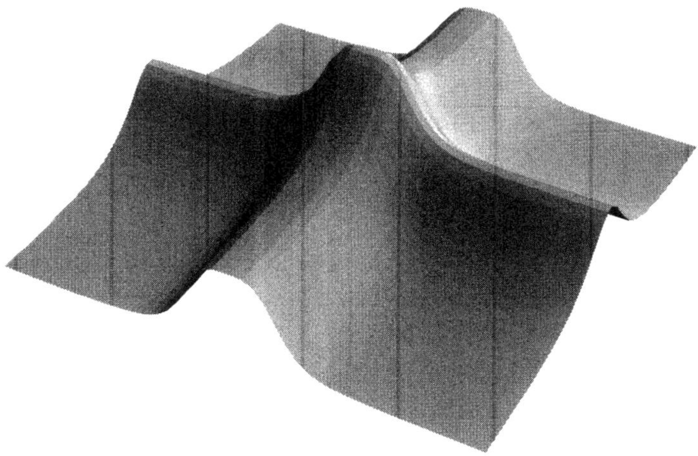

Fig. 3

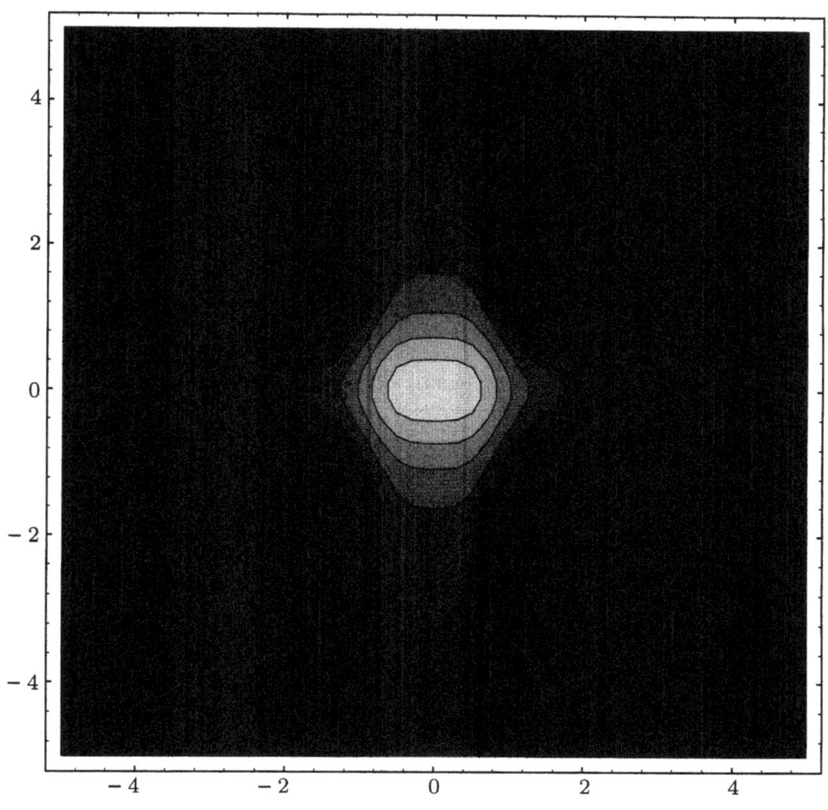

Fig. 4

The function $\rho(\beta,\lambda)$ depends heavily on the process under consideration. There is no hope to find for general strong Markov processes upper and lower bounds on $\rho(\beta,\lambda)$. Hence we will formulate the quantitative results in the present chapter in terms of $\rho(\beta,\lambda)$ or in terms of

$$\sup_{x\in E}\varphi(x)^2\left[\mathbb{E}_x\left\{\exp\left(-\beta T_{\lambda,\Gamma}\right)\right\}\right]^{1/2}=:\rho_\varphi(\beta,\lambda),$$

where φ is an admissible function: see Definition 8.19. For the resolvent differences we have to integrate with respect to $d\lambda$ and the quantitative estimates depend on the respective expressions

$$\int_0^\infty d\lambda\exp(-\lambda)\rho(\beta,\lambda)=:\rho(\beta)\text{ and}$$
$$\int_0^\infty d\lambda\exp(-\lambda)\rho_\varphi(\beta,\lambda)=:\rho_\varphi(\beta).$$

This means that the quantitative estimate is reduced to the study of the behavior of $\rho(\beta)$. This is a difficult task, but it is not impossible. In any case the behavior of $\rho(\beta)$ is sufficiently well known if the underlying Markov process is Brownian motion, also called the Wiener process.

In the present book, our intention was to allow general strong Markov processes with the Feller property. In our general setup we did not confine ourselves to the Wiener process in $\mathbb{R}^\nu$. Nevertheless here it seems worthwhile to explain the results from Demuth, Kirsch and McGillivray [78], in order to convince the reader that for special processes, probably for instance canonical diffusion processes, the study of the Laplace transform of the occupation time, $\rho(\beta,\lambda)$ or $\rho(\beta)$, can be carried out in a successful way.

C.1. Estimates for the Laplace transform of the occupation time for Wiener processes

For the moment, let

$$\left\{(\Omega,\mathcal{F},\mathbb{P}_x),(X(t),t\geq 0),(\vartheta_t:t\geq 0),\left(\mathbb{R}^d,\mathcal{B}^d\right)\right\}$$

be the Wiener process. Its free generator is $H_0=-\frac{1}{2}\Delta$. We will compare the systems given by $H_\beta=H_0+\beta 1_\Gamma$ and by the Dirichlet Laplacian $H_\Sigma=\frac{1}{2}(-\Delta)_\Sigma$, where $\Sigma=\mathbb{R}^d\setminus\Gamma$. In order to study the Laplace transform of the occupation time, we need some geometric conditions on the boundary $\partial\Gamma$. The main assumption we make is that both Γ and Σ should satisfy the *uniform cone condition.*

8.28. DEFINITION. A set $\Omega\subset\mathbb{R}^d$ satisfies the *uniform cone condition* if there exist an angle α and a height h such that for every $y\in\partial\Omega$ there is a cone C, with vertex at y, of angle α and of height h contained in Ω, i.e. $C\subset\Omega$.

For arbitrary uniform Lipschitz domains (see Definition 8.30 below), the estimates of the occupation time Laplace transform can be reduced to the case where Γ is itself a cone C of infinite height and angle α. In that case, the following estimate is valid (see [78], Theorem 2.3 and Proposition 2.5):

$$c_1 \left(\frac{1}{\beta\lambda}\right)^{A(C)} \leq \mathbb{E}_0 \left\{\exp\left(-\beta T_{\lambda,C}\right)\right\} \leq c_2 \left(\frac{\log(\beta\lambda)}{\beta\lambda}\right)^{A(C)}, \tag{8.75}$$

where $\beta\lambda > 2$ and where c_1 and c_2 are positive finite constants which can be given explicitly. The number $A(C)$ is called the *cone constant* which will be defined below.

8.29. DEFINITION. Let C be a cone with vertex at the origin of $\mathbb{R}^d$. Let S^{d-1} be the unit sphere in $\mathbb{R}^d$ centered at the origin. Let F be the closed subset of S^{d-1} given by $F := S^{d-1} \cap C$. Denote by $F^{\text{compl}} := S^{d-1} \setminus F$ the complement of F in S^{d-1}. Let $\lambda_1\left(F^{\text{compl}}\right)$ be the lowest eigenvalue of the Laplace-Beltrami operator $-\frac{1}{2}\Delta$ on F^{compl} with Dirichlet boundary condition on ∂F. Then the cone constant $A(C)$ is given by

$$A(C) := \frac{\lambda_1\left(F^{\text{compl}}\right)}{\sqrt{\left(\frac{d}{2}-1\right)^2 + 2\lambda_1\left(F^{\text{compl}}\right)} + \frac{d}{2} - 1},$$

for $d \geq 2$.

Set $C = C_\alpha$ in order to emphasize that the cone has the cone angle α. The cone C_π is the half space. The cone constant of the half space is $A\left(C_\pi\right) = \dfrac{1}{2}$, i.e. the lowest eigenvalue in that case is equal to $\frac{1}{2}(d-1)$. The function $\alpha \mapsto A\left(C_\alpha\right)$ is monotone in α, i.e. if $\alpha \leq \alpha'$ or if $C_\alpha \subseteq C_{\alpha'}$, then $A\left(C_\alpha\right) \leq A\left(C_{\alpha'}\right)$. In the latter, we only have equality if $\alpha = \alpha'$. Hence $A\left(C_\alpha\right) < \frac{1}{2}$ if $\alpha < \pi$.

The inequalities in (8.75) are crucial for obtaining the upper and lower bounds of the semigroup or resolvent differences. For this matter we have to classify the sets Γ. To start we assume that they are *uniformly Lipschitz.*

8.30. DEFINITION. Let Γ be a closed subset in $\mathbb{R}^d$. Define for $\ell > 0$ the set $\partial_\ell\Gamma$ via the equality

$$\partial_\ell\Gamma = \left\{x \in \mathbb{R}^d : \text{dist}\left(x, \partial\Gamma\right) \leq \ell\right\}.$$

The set Γ is said to be uniformly Lipschitz if there is a collection of open balls $\{B_k : k \in \mathbb{N}\}$ of radius r covering $\partial_\ell\Gamma$ such that the following conditions are satisfied:

- the intersection $\partial\Gamma \cap B_k$ is locally a graph of a Lipschitz function with Lipschitz constant L independent of k;

– there exists $\delta > 0$ such that

$$\operatorname{dist}\left(\partial_\ell \Gamma, \mathbb{R}^d \setminus \bigcup_{k=1}^{\infty} B_k\right) \geq \delta > 0;$$

– each $x \in \partial_\ell \Gamma$ belongs to at most m balls B_k, where m does not depend on x.

In any uniform Lipschitz set Γ, there exists an interior cone C_{int} such that for each $y \in \partial\Gamma$ there is a rigid transformation S_y (i.e. a combination of translations and rotations) of $\mathbb{R}^d$ such that $S_y C \cap B(y,r) \subset \Gamma$ and $S_y \{0\} = y$. In other words, C_{int} plays the role of the cone in the definition of uniform cone condition.

The main result concerning the quantitative estimates (see [78], Theorem 3.3 where the precise definition of the function $u_\gamma(\lambda)$ can also be found) is the following statement.

8.31. THEOREM. *Let Γ be a uniform Lipschitz set in $\mathbb{R}^d$ and let A_{int} be the cone constant of the interior cone C_{int}. Then for any $\gamma \in (0,1)$ and for any $\beta > 1$ the following inequality is valid:*

$$\|J \exp(-\lambda H_\beta) - \exp(-\lambda H_\Sigma) J\|_{2,2} \leq u_\gamma(\lambda)\beta^{-\gamma A_{\text{int}}},$$

where, for $0 < \gamma < 1$, $u_\gamma(\lambda)$ is a function in λ for which

$$\int_0^\infty \exp(-a\lambda)\, u_\gamma(\lambda) d\lambda =: c_\gamma(a) < \infty,$$

for any $a > 0$. The latter implies the inequality

$$\left\| J (H_\beta + aI)^{-1} - (H_\Sigma + aI_\Sigma)^{-1} J \right\|_{2,2} \leq c_\gamma(a)\beta^{-\gamma A_{\text{int}}},$$

for $\beta > 1$. If $\partial\Gamma$ is a smooth boundary, the constant γ an be chosen such that γA_{int} is arbitrarily close to $\frac{1}{2}$.

Finally in [78] lower bounds for the resolvent differences are given as well. This is an advantage of the stochastic spectral analysis because it enables us to find local estimates from below. For lower estimates, the cone constant is taken from the best possible cone covering Γ.

8.32. DEFINITION. Let y belong to $\partial\Gamma$. A *covering cone* C_{cov} (with vertex in y) is a cone covering Γ, i.e. $C_{\text{cov}}^{\text{compl}} \setminus \{y\} \subset \Sigma$. Let Γ_0 be a part of the boundary $\partial\Gamma$ with a strictly positive measure. A *uniform* covering cone (with respect to Γ_0) is a covering cone which has the same size for every $y \in \Gamma_0$. This means that for each $y \in \Gamma_0$ there is a rigid transformation S_y of $\mathbb{R}^d$ such that $S_y C_{\text{cov}}^{\text{compl}} \setminus \{y\} \subset \Sigma$.

In general the covering cone is larger than a half space and hence its cone constant is larger than $1/2$.

The following theorem is taken from [78], Theorem 4.2, where the precise definition of the function $k(\lambda)$ is given as well.

8.33. THEOREM. *Let Γ_0 be a subset of $\partial\Gamma$ with strictly positive surface measure. Let C_{cov} be the uniform covering cone with respect to Γ_0. Suppose $d \geq 2$, $\lambda \geq 1$, and $\rho > 1$. Then*

$$\|J \exp\left(-\lambda H_\beta\right) - \exp\left(-\lambda H_\Sigma\right) J\|_{2,2} \geq k(\lambda)\beta^{-\rho A_{\text{cov}}},$$

for β sufficiently large. The function $k(\lambda)$ possesses a finite Laplace transform. It follows that, for each $a > 0$, there exists a positive constant c such that for sufficiently large β the inequality

$$\left\| J\left(H_\beta + aI\right)^{-1} - \left(H_\Sigma + aI_\Sigma\right)^{-1} J \right\|_{2,2} \geq c\beta^{-\rho A_{\text{cov}}}, \quad \rho > 1,$$

is valid.

Summarizing these results for the Wiener process shows that one has upper and lower bounds for the operator norm of the resolvent differences in terms of the large coupling parameter β. This lower bound is mainly due to the inequalities in (8.75), the explicit estimates for the Laplace transform of the occupation time.

REMARK. In the following section we generalize these kind of estimates to Feller operators. These estimates can not be so explicit like in the Wiener case. The final dependence of β of the Laplace transform of the occupation time is determined by the underlying process.

The results are only applied to the convergence of wave operators and scattering matrices. Nevertheless, there are many further possibilities to use these estimates for quantitative considerations. For physical systems it is important to find similar estimates for the scattering phase, for the cross section and, some more difficult, for the spectral shift function. These are open questions left to the interested reader for further research.

In this context it seems to be worthwhile to study also the "visibility" of obstacle scattering systems. With visibility we mean quantitative estimates between the scattered and the free system in terms of the geometry of the obstacle. If the difference of the Feller semigroups is trace class we know not only the completeness of the scattering system. In this case it is also possible to analyse the distance of the wave operator with the trivial situation of a free dynamics. This difference is given in terms of the trace norm of the semigroup difference (see also section 8.A.). The most general proof is due to Pearson, hence we call it Pearson's estimate (see e.g. [193] page 24). The idea is to use then Pearson's estimate in order to find a quantitative description of the visibility explained above.

C.2. Quantitative large coupling estimates for Feller operators in terms of the weighted Laplace transform of the occupation time

In the present subsection of this chapter we investigate again general Markov processes. In this case the rate of convergence is given in terms of $\rho(\beta)$ or of $\rho_\varphi(\beta)$, respectively (see Definition 8.34 below).

The major part of this final subsection will describe the scattering phenomena in terms of $\rho(\beta)$ or $\rho_\varphi(\beta)$. We will follow the treatment in Section 8.B where we studied the quantitative behavior for regular perturbations. Once we have the existence of the wave operators we can give quantitative estimates for their operator norms. The convergence of the scattering matrices is investigated and possible resonance free regions are described.

Let us start with a short example on the behavior of eigenfunctions which depicts the way that these kinds of estimates work. Let Ω be a compact subset of $\mathbb{R}^d$ and let K_0 be a free Feller operator which generates a strong Markov process having a continuous symmetric density $p_0(t,x,y)$ for which

$$\sup_{x,y\in E} p_0(t,x,y) \le c_t < \infty.$$

We only consider positive Kato-Feller potentials of the form

$$V_+ = V_+ 1_{\Omega\setminus\Gamma} + V_+ 1_\Gamma$$

with $\Gamma \subset \Omega$, $V_+ 1_\Gamma \ge 1_\Gamma$. Let us consider the operator $\left(\hbar^2 K_0 + V_+\right)_\Omega$. This is the operator $\hbar^2 K_0 + V_+$ in $L^2(\Omega)$ with Dirichlet boundary conditions on $\partial\Omega$. The lowest eigenvalue of $\left(\hbar^2 K_0 + V_+\right)_\Omega$ is denoted by $\hbar^2\lambda_0$. Let $\psi_\hbar$ be the corresponding normalized eigenfunction, i.e.

$$\left(\hbar^2 K_0 + V_+\right)_\Omega \psi_\hbar = \hbar^2 \lambda_0 \psi_\hbar,$$

with $\|\psi_\hbar\|_{L^2(\Omega)} = 1$. Then $\psi_\hbar$ is a bounded function. Let c_1 be an upper bound for $\psi_\hbar$. We obtain the asymptotics of $\psi_\hbar$ via

$$\begin{aligned}\psi_\hbar(x) &= e^{\lambda_0}\left[\exp\left(-\left(K_0 \dot{+} \frac{1}{\hbar^2}V_+\right)_\Omega\right)\psi_\hbar\right](x)\\ &\le c_1 e^{\lambda_0}\mathbb{E}_x\left\{\exp\left(-\frac{1}{\hbar^2}\int_0^1 V_+(X(s))ds\right)\right\}\\ &\le c_1 e^{\lambda_0}\mathbb{E}_x\left\{\exp\left(-\frac{1}{\hbar^2}T_{1,\Gamma}\right)\right\}.\end{aligned}$$

This inequality implies that the asymptotics of $\psi_\hbar(0)$ is governed by the Laplace transform of the occupation time of Γ, i.e. of $\rho\left(\frac{1}{\hbar^2},1\right)$.

For a translation invariant process, we have a uniform estimate on certain subsets of Γ. For this we take a subset $\widetilde{\Gamma}$ of Γ such that $\operatorname{dist}\left(\widetilde{\Gamma}, \Omega \setminus \Gamma\right) = d > 0$. Let $B_d(x)$ be a ball of radius d centered at x. For $x \in \widetilde{\Gamma}$ we have

$$\begin{aligned}
|\psi_\hbar(x)| &\leq c_1 e^{\lambda_0} \mathbb{E}_x \left\{ \exp\left(-\frac{1}{\hbar^2} \int_0^1 1_{B_d(x)}(X(s))\, ds \right) \right\} \\
&= c_1 e^{\lambda_0} \mathbb{E}_0 \left\{ \exp\left(-\frac{1}{\hbar^2} \int_0^1 1_{B_d(0)}(X(s))\, ds \right) \right\} \\
&= c_1 e^{\lambda_0} \mathbb{E}_0 \left\{ \exp\left(-\frac{1}{\hbar^2} T_{1, B_d(0)} \right) \right\}.
\end{aligned}$$

The next and main part of this section follows the considerations in section 7.B. There we investigated the quantitative estimate of the spectral magnitudes in terms of regular perturbations. However, here we neglect the regular perturbation and we compare systems given by $K_\beta = K_0 + \beta 1_\Gamma$ and $K_\Sigma = (K_0)_\Sigma$ in terms of the large coupling constant β. In particular, we study spectral measures, resolvent differences, limiting absorptions limits, scattering matrices, existence of resonance free regions.

We begin with the sandwiched resolvent difference. For this we now introduce in a rigorous way the Laplace transform of the occupation time, also called spending time.

8.34. Definition. (*Laplace transform of the occupation time*) Let Γ be a closed singularity region with $\Gamma^r = (\operatorname{int}(\Gamma))^r$ (see Proposition 2.24), so that the first hitting time and the penetration time of Γ coincide. Let

$$T_{\lambda,\Gamma} = \int_0^\lambda 1_\Gamma(X(u))du = \operatorname{meas}\{\lambda \geq s \geq 0 : X(s) \in \Gamma\},$$

the *occupation* or *spending* time of the trajectory in Γ. Let $\varphi(\cdot)$ be an admissible function. According to Definition 8.19 this means that φ is a non-vanishing function mapping E to $\mathbb{R}_+$, with $\sup_{x \in E} \dfrac{1}{\varphi(x)} \leq 1$ and

$$\int_E p_0(t,x,y)\varphi(y)^\rho dy \leq c_0^2 \exp(2mt)\, \varphi(x)^\rho \tag{8.76}$$

(see (8.25)), for some $m > 0$, $c_0 \geq 1$, and for some positive integer ρ. (Most of the time $\rho = 2$ or 4.) Suppose that for every $\lambda > 0$ and for every $\epsilon > 0$ there exists a compact subset K_ϵ with the property that for $x \notin K_\epsilon$ the inequality

$$\varphi(x)^2 \left[\mathbb{P}_x\{T_{\lambda,\Gamma} > 0\}\right]^{1/2} < \epsilon \tag{8.77}$$

is valid. Then we define the following Laplace transforms of the occupation time:

$$\sup_{x\in E}\left[\mathbb{E}_x\left\{\exp\left(-\beta T_{\lambda,\Gamma}\right) : T_{\lambda,\Gamma} > 0\right\}\right] =: \rho(\beta,\lambda)$$

and

$$\int_0^\infty \exp\left(-m\lambda\right)\rho(\beta,\lambda)d\lambda =: \rho(\beta),$$

where m is taken from (8.76). Moreover, we define the weighted Laplace transform of the occupation time:

$$\sup_{x\in E}\varphi(x)^2\left[\mathbb{E}_x\left\{\exp\left(-\beta T_{\lambda,\Gamma}\right) : T_{\lambda,\Gamma} > 0\right\}\right]^{1/2} =: \rho_\varphi(\beta,\lambda)$$

and

$$\int_0^\infty \exp(-m\lambda)\rho_\varphi(\beta,\lambda)d\lambda =: \rho_\varphi(\beta).$$

Notice that ρ_1 does not coincide with ρ.

8.35. LEMMA. *The quantities $\rho(\beta)$ and $\rho_\varphi(\beta)$ tend to zero as β tends to infinity.*

Proof. Let $\epsilon > 0$ be given. According to (8.77) the quantity

$$\sup_{\beta>0}\varphi(x)^2\left[\mathbb{E}_x\left\{\exp\left(-\beta T_{\lambda,\Gamma}\right) : T_{\lambda,\Gamma} > 0\right\}\right]^{1/2}$$

is less than ϵ for $x \notin K_\epsilon$. We know that $\lim_{\beta\to\infty}\mathbb{E}_x\left\{\exp\left(-\beta T_{\lambda,\Gamma}\right) : T_{\lambda,\Gamma} > 0\right\} = 0$ for each $x \in E$ separately. From Dini's lemma (Appendix E, Lemma E.1) we conclude that

$$\lim_{\beta\to\infty}\sup_{x\in K_\epsilon}\varphi(x)^2\left[\mathbb{E}_x\left\{\exp\left(-\beta T_{\lambda,\Gamma}\right) : T_{\lambda,\Gamma} > 0\right\}\right]^{1/2} = 0.$$

Lebesgue's dominated convergence theorem yields the second result in Lemma 8.35. The same arguments imply that $\lim_{\beta\to\infty}\rho(\beta) = 0$ with an obvious modification of the assumptions in (8.77) and so on.

In principle, any estimate in terms of $\rho_\varphi(\beta)$ is a quantitative estimate in terms of the large coupling parameter β as long as one is able to find quantitative estimates for these Laplace transforms. For the Wiener process this was explained in subsection C.1 of this chapter. However, for other processes this work is still to be carried out. Put $K_\beta = K_0 + \beta 1_\Gamma$ and $K_\Sigma = (K_0)_\Sigma$. We estimate the norm of the resolvent difference:

$$\begin{aligned}
&\left\|J\left(K_\beta + aI\right)^{-1} - \left(K_\Sigma + aI_\Sigma\right)^{-1}J\right\|_{2,2} \\
&\le \int_0^\infty d\lambda\exp\left(-a\lambda\right)\sup_{x\in\Sigma}\mathbb{E}_x\left\{\exp\left(-\beta T_{\lambda,\Gamma}\right) : T_{\lambda,\Gamma} > 0\right\} \\
&\le \int_0^\infty d\lambda\exp\left(-a\lambda\right)\rho(\beta,\lambda) \le \rho(\beta),
\end{aligned} \tag{8.78}$$

for $a > m$. Of course, (8.78) was the original starting point in the study of the different Laplace transforms of the occupation time.

In order to find estimates for the differences of the spectral measures, the first step is to consider sandwiched resolvent differences.

8.36. LEMMA. *Let K_β, K_Σ, φ be as above. Then, for $a > 2m$,*

$$\left\|\varphi^{-1}\left[(K_\beta + aI)^{-1} - J^*\left(K_\Sigma + aI_\Sigma\right)^{-1} J\right]\varphi^{-1}\right\|_{2,2} \leq c_0\rho_\varphi(\beta).$$

Proof. The mapping

$$\varphi^{-1}\left[(K_\beta + aI)^{-1} - J^*\left(K_\Sigma + aI_\Sigma\right)^{-1} J\right]\varphi^{-1}$$

is an integral operator with a symmetric kernel. Therefore its operator norm is smaller than

$$\begin{aligned}
&\sup_x \int_0^\infty d\lambda \exp\left(-a\lambda\right)\varphi(x)\mathbb{E}_x\left\{\exp\left(-\beta T_{\lambda,\Gamma}\right)\varphi\left(X(\lambda)\right) : T_{\lambda,\Gamma} > 0\right\}\\
&\leq \frac{1}{2}\int_0^\infty d\lambda \exp\left(-a\lambda\right)\sup_x \varphi(x)^2\mathbb{E}_x\left\{\exp\left(-\beta T_{\lambda,\Gamma}\right) : T_{\lambda,\Gamma} > 0\right\}\\
&\quad + \frac{1}{2}\int_0^\infty d\lambda \exp\left(-a\lambda\right)\sup_x \left[\mathbb{E}_x\left\{\exp\left(-2\beta T_{\lambda,\Gamma}\right) : T_{\lambda,\Gamma} > 0\right\}\right]^{1/2}\\
&\qquad\qquad \times \left[\mathbb{E}_x\left\{\varphi\left(X(\lambda)\right)^4\right\}\right]^{1/2}\\
&\leq \frac{1}{2}\int_0^\infty d\lambda \exp\left(-a\lambda\right)\rho_\varphi(\beta,\lambda)\\
&\quad + \frac{1}{2}\int_0^\infty d\lambda \exp\left(-a\lambda\right)c_0\exp\left(m\lambda\right)\sup_x \varphi(x)^2\left[\mathbb{E}_x\left\{\exp\left(-2\beta T_{\lambda,\Gamma}\right)\right\} : T_{\lambda,\Gamma} > 0\right]^{1/2}\\
&\leq \frac{1}{2}\int_0^\infty d\lambda \exp\left(-m\lambda\right)\rho_\varphi(\beta,\lambda)\\
&\quad + \frac{c_0}{2}\int_0^\infty d\lambda \exp\left(-(a-m)\lambda\right)\rho_\varphi(\beta,\lambda)\\
&\leq c_0\int_0^\infty d\lambda \exp\left(-m\lambda\right)\rho_\varphi(\beta,\lambda) = c_0\rho_\varphi(\beta),
\end{aligned}$$

using $a > 2m$, $\rho_\varphi(2\beta,\lambda) < \rho_\varphi(\beta,\lambda)$.

This last lemma provides the basis to treat the *limiting absorption* case. See for example Proposition 8.21.

8.37. THEOREM. *Let K_0, $K_\beta = K_0 + \beta 1_\Gamma$, K_Σ and φ be given as above. Let $\triangle = (\lambda_1, \lambda_2)$ be an open interval in $\mathbb{R}$ and assume that*

$$\sup_{\lambda\in\triangle}\sup_{\epsilon\in(0,1]}\left\|\varphi^{-1}J^*\left(K_\Sigma - (\lambda \pm i\epsilon)\,I_\Sigma\right)^{-1}J\varphi^{-1}\right\|_{2,2} =: d_{\triangle,\Sigma} < \infty.$$

Fix β_0 *such that for all* $\beta \geq \beta_0$ *the inequality*

$$\rho_\varphi(\beta) \leq \frac{1}{2a_0c_0} \frac{1}{1 + 3a_0 d_{\triangle,\Sigma}} \tag{8.79}$$

is valid with $a_0 > \max\{\lambda_2, 2m\}$. *Here* c_0 *is defined as in (8.76). Such a* β_0 *exists because* $\rho_\varphi(\beta)$ *tends to zero for large* β. *Then*

$$\begin{aligned} &\left\| \varphi^{-1} \left[(K_\beta - (\lambda \pm i\epsilon) I)^{-1} - J^* (K_\Sigma - (\lambda \pm i\epsilon) I_\Sigma)^{-1} J \right] \varphi^{-1} \right\|_{2,2} \\ &\leq 2c_0 (1 + 3a_0 d_{\triangle,\Sigma})^2 \rho_\varphi(\beta) \end{aligned} \tag{8.80}$$

and as a consequence

$$\begin{aligned} &\left\| \varphi^{-1} (K_\beta - (\lambda \pm i\epsilon) I)^{-1} \varphi^{-1} \right\|_{2,2} \\ &\leq d_{\triangle,\Sigma} + 2c_0 (1 + 3a_0 d_{\triangle,\Sigma})^2 \rho_\varphi(\beta). \end{aligned} \tag{8.81}$$

In what follows (8.79) will be referred to as the *boundedness condition.*

Proof. This result follows immediately from Proposition 8.21.

8.38. COROLLARY. *Let* $E_\beta(\cdot)$ *and* $E_\Sigma(\cdot)$ *be the spectral measures of* K_β *and* K_Σ *respectively. Moreover, suppose that for* $\beta \geq \beta_0$ *the boundedness condition in (8.79) is satisfied. Then the operator norm of the sandwiched difference of these spectral measures satisfies the following inequality:*

$$\begin{aligned} &\left\| \varphi^{-1} \left[E_\beta(\triangle) - J^* E_\Sigma(\triangle) J \right] \varphi^{-1} \right\|_{2,2} = \left\| \left[E_\beta(\triangle) - J^* E_\Sigma(\triangle) J \right] \right\|_{L^2(\varphi^{-1}dm), L^2(\varphi dm)} \\ &\leq \frac{c_0}{\pi} |\triangle| (1 + 3a_0 d_{\triangle,\Sigma})^2 \rho_\varphi(\beta) \end{aligned}$$

with $a_0 \geq \max(\lambda_2, 2m)$.

Proof. This assertion follows from the definition in (8.24) together with the inequality in (8.80).

In Propositions 8.14 and 8.16 we studied the existence (and completeness) of the wave operators. Now we assume that they exist and we shall give a quantitative estimate for their operator norms.

8.39. COROLLARY. *Suppose that the wave operators*

$$\Omega_+ (K_\beta, K_0) \quad \text{and} \quad \Omega_+ (K_\Sigma, J, K_0)$$

exist. Let the interval $\triangle = (\lambda_1, \lambda_2)$ *be chosen as in Theorem 8.23. Denote with* $E_0(\cdot)$ *the spectral measure of* K_0. *Let* φ *be an admissible function. Then the weighted* L^2*-norm for the local wave operators can be estimated as follows:*

$$\begin{aligned} &\left\| \varphi^{-1} \left[\Omega_+ (K_\beta, K_0) - J^* \Omega_+ (K_\Sigma, J, K_0) \right] E_0(\triangle) \right\|_{2,2} \\ &\leq c_0 |\triangle| (1 + \lambda_2 + 3m)^2 \rho_\varphi(\beta). \end{aligned} \tag{8.82}$$

For the definition of c_0 and m see (8.76). The constant in (8.82) is not the optimal one. Here we content ourselves with the existence of some well-determined constant.

Proof. From the existence of the wave operators the following inequalities are readily deduced (compare with (8.42)):

$$
\begin{aligned}
&\left|\left\langle\varphi^{-1}\left[\Omega_{+}\left(K_\beta, K_0\right)-J^* \Omega_{+}\left(K_\Sigma, J, K_0\right) J\right] E_0(\triangle) f, g\right\rangle\right| \\
&\leq \lim_{\epsilon \downarrow 0} \epsilon\left|\int_0^{\infty} d t \int_{\triangle} \exp (-\epsilon t) \exp (-i t \lambda)\right. \\
&\qquad\left.\left[\left\langle\exp \left(i t K_\beta\right) E_0(d \lambda) f, \varphi^{-1} g\right\rangle-\left\langle J^* \exp \left(i t K_\Sigma\right) J E_0(d \lambda) f, \varphi^{-1} g\right\rangle\right]\right| \\
&\leq \lim_{\epsilon \downarrow 0} \epsilon \int_{\triangle}\left|\left\langle E_0(d \lambda) f,\left[(K_\beta-(\lambda-i \epsilon) I)^{-1}-J^*\left(K_\Sigma-(\lambda-i \epsilon) I_\Sigma\right)^{-1} J\right] \varphi^{-1} g\right\rangle\right| \\
&\leq \lim_{\epsilon \downarrow 0} \epsilon|\triangle|\|f\|_2\|g\|_2 \\
&\qquad \sup_{\lambda \in \triangle}\left\|\left[(K_\beta-(\lambda-i \epsilon) I)^{-1}-J^*\left(K_\Sigma-(\lambda-i \epsilon) I_\Sigma\right)^{-1} J\right] \varphi^{-1}\right\|_{2,2} .
\end{aligned}
\tag{8.83}
$$

Upon setting

$$
D_\beta(a)=\left(K_\beta+a I\right)^{-1}-J^*\left(K_\Sigma+a I_\Sigma\right)^{-1} J
$$

for some $a \geq 2m$, the limit in (8.83) can be estimated by

$$
\begin{aligned}
&|\triangle|\|f\|_2\|g\|_2 \lim_{\epsilon \downarrow 0}\left\{\epsilon\left\|D_\beta(a) \varphi^{-1}\right\|_{2,2}\right. \\
&\quad+\left(\lambda_2+a+1\right)\left\|D_\beta(a)\right\|_{2,2} \epsilon\left\|(K_\beta-(\lambda-i \epsilon) I)^{-1}\right\|_{2,2} \\
&\quad+\left(\lambda_2+a+1\right)\left\|D_\beta(a)\right\|_{2,2} \epsilon\left\|J^*\left(K_\Sigma-(\lambda-i \epsilon) I_\Sigma\right)^{-1} J\right\|_{2,2} \\
&\quad+\left(\lambda_2+a+1\right)^2 \epsilon\left\|(K_\beta-(\lambda-i \epsilon) I)^{-1}\right\|_{2,2} \\
&\qquad\left.\times\left\|D_\beta(a) J^*\left(K_\Sigma-(\lambda-i \epsilon) I_\Sigma\right)^{-1} J \varphi^{-1}\right\|_{2,2}\right\} \\
&\leq|\triangle|\|f\|_2\|g\|_2\left\{2\left(\lambda_2+a+1\right) \rho(\beta)+\left(\lambda_2+a+1\right)^2\left\|D_\beta(a) \varphi\right\|_{2,2} d_{\triangle, \Sigma}\right\} \\
&\leq|\triangle|\|f\|_2\|g\|_2\left(2+d_{\triangle, \Sigma}\right) c_0\left(\lambda_2+a+1\right)^2 \rho_{\varphi}(\beta),
\end{aligned}
$$

for $a > 2m$. Of course, the final estimate is somewhat rough. For instance we used $\rho(\beta) \leq \rho_\varphi(\beta)$.

In the next application we investigate the scattering matrices for two space scattering systems $\{K_0+\beta 1_\Gamma, K_0\}$ and $\{K_\Sigma, J, K_0\}$. Assuming that the invariance principle for our special choice of Γ holds it suffices to study the scattering matrices for the systems $\{R_\beta, R_0\}$ and $\{R_\Sigma, J, R_0\}$, with the resolvents

$$\begin{aligned} R_\beta &:= \left(K_0 + \beta 1_\Gamma + aI\right)^{-1}, \\ R_\Sigma &:= \left(K_\Sigma + aI\right)^{-1}, \text{ and} \\ R_0 &:= \left(K_0 + aI\right)^{-1}, \quad a > 0. \end{aligned}$$

According to Definition 8.12, we let

$$\begin{aligned} S_\beta &:= \Omega_+^* \left(R_\beta, R_0\right) \Omega_- \left(R_\beta, R_0\right); \text{ and} \\ S_\Sigma &:= \Omega_+^* \left(R_\Sigma, J, R_0\right) \Omega_- \left(R_\Sigma, J, R_0\right). \end{aligned}$$

The *scattering matrices* were introduced as follows (see equation (8.48)):

$$\left[\Phi_{R_0} S_\beta g\right]_\lambda =: S_\beta \left(\lambda, R_\beta, R_0\right) \left[\Phi_{R_0} g\right]_\lambda$$

and

$$\left[\Phi_{R_0} S_\Sigma g\right]_\lambda =: S_\Sigma(\lambda, R_\Sigma, R_0) \left[\Phi_{R_0} g\right]_\lambda ,$$

where $\left[\Phi_{R_0} g\right]_\lambda$ is the representer of $g \in P_{\mathrm{ac}}(R_0)\mathcal{H}$ in $\mathcal{H}_{R_0}(\lambda)$, $\lambda \in (0, 1/a)$.

In a general two-space scattering system $\{R_1, J, R_2\}$, where R_1 and R_2 are self-adjoint operators in not necessarily the same Hilbert spaces $\mathcal{H}_1$ and $\mathcal{H}_2$ and where J is a bounded identification operator from $\mathcal{H}_1$ to $\mathcal{H}_2$, the *scattering matrix* $S(\lambda, R_2, R_1)$ is given by (see equality (8.49))

$$\begin{aligned} &S\left(\lambda, R_2, R_1\right) \\ &:= I - 2\pi i F_\varphi(\lambda) \left[T^* - T^* \varphi^{-1} J^* \left(R_2 - (\lambda + i0)\, I_{\mathcal{H}_2}\right)^{-1} J \varphi T\right] F_\varphi^*(\lambda), \end{aligned} \tag{8.84}$$

$$\text{with} \quad T = \varphi^{-1} J^* \left(R_2 J - J R_1\right) \varphi, \tag{8.85}$$

as long as the right-hand side of (8.84) exists.

In our situation for the obstacle scattering system, we have

$$\begin{aligned} &S_\Sigma(\lambda, R_\Sigma, R_0) \\ &= I - 2\pi i F_\varphi(\lambda) \Big\{ \varphi \left(J^* R_\Sigma J - R_0 J^*\right) J \varphi \\ &- \varphi \left(J^* R_\Sigma - R_0 J^*\right) J \varphi \varphi^{-1} J^* \left(R_\Sigma - (\lambda + i0)\, I_\Sigma\right)^{-1} J \varphi^{-1} \varphi J^* \left(R_\Sigma J - J R_0\right) \varphi \Big\} F_\varphi^*(\lambda) \end{aligned} \tag{8.86}$$

for an appropriate $\lambda \in \triangle \subset \sigma_{\mathrm{ac}}(R_0)$. Here $F_\varphi(\lambda)$ maps $L^2(E)$ into $\mathcal{H}_{R_0}(\lambda)$ and corresponds to the spectral measure of the self-adjoint operator $R_0 = \left(K_0 + aI\right)^{-1}$, $a > 0$, via

$$\left\| F_\varphi^*(\lambda) F_\varphi(\lambda) \right\|_{2,2} = \left\| \varphi^{-1} \frac{dE_{R_0}(\lambda)}{d\lambda} \varphi^{-1} \right\|_{2,2} :$$

compare with (8.50). Here φ is an admissible function in $L^2(E, m)$.

A corresponding expression will follow for $S_\beta(\lambda)$, to wit

$$\begin{aligned}
&S_\beta(\lambda, R_\beta, R_0)\\
&= I - 2\pi i F_\varphi(\lambda)\Big\{\varphi\,(R_\beta - R_0)\,\varphi\\
&\quad - \varphi\,(R_\beta - R_0)\,\varphi\,\varphi^{-1}\,(R_\beta - (\lambda + i0)\,I)^{-1}\,\varphi^{-1}\,\varphi\,(R_\beta - R_0)\,\varphi\Big\}F_\varphi^*(\lambda).
\end{aligned}$$

If we consider the difference of these scattering matrices, there is a formal difficulty because the operator

$$\varphi R_0 J^* J \varphi - \varphi R_0 \varphi = \varphi R_0 1_\Gamma \varphi$$

is in general not zero. Using the following lemma, will circumvent this difficulty.

8.40. LEMMA. *Suppose that the wave operators $\Omega_\pm\,(R_\beta, R_0)$ exist. Let Γ be a set of finite measure. Then the following identities are true:*

$$\Omega_\pm\,(R_\beta, R_0) = \Omega_\pm\,(R_\beta, J^*J, R_0)\,. \tag{8.87}$$

Proof. Let f belong to $L^2(E, m)$. It suffices to prove the equality

$$\text{s-}\lim_{t\to\infty}\,(I - J^*J)\exp\left(-itR_0\right)P_{\mathrm{ac}}(R_0)f = 0. \tag{8.88}$$

This may be done in the same way as we proved Proposition 8.16.(b). We have $I - J^*J = 1_\Gamma$. Since $\text{s-}\lim_{s\downarrow 0}\exp\left(-sK_0\right)P_{\mathrm{ac}}(R_0)f = P_{\mathrm{ac}}(R_0)f$ the result in (8.88) will follow as soon as we have shown that

$$\text{s-}\lim_{t\to\infty} 1_\Gamma \exp\left(-sK_0\right)\exp\left(-itR_0\right)P_{\mathrm{ac}}(R_0)f = 0 \tag{8.89}$$

holds. By the Riemann-Lebesgue lemma, we have

$$\text{w-}\lim_{t\to\infty}\exp\left(-itR_0\right)P_{\mathrm{ac}}(R_0)f = 0. \tag{8.90}$$

Because of the inequality

$$\iint \left|1_\Gamma(x)p_0(s,x,y)\right|^2 dxdy \le m(\Gamma)\sup_{x,y\in E} p_0(2s,x,y), \tag{8.91}$$

the operator $1_\Gamma \exp\left(-sK_0\right)$ is a Hilbert-Schmidt operator. Hence (8.88) follows.

Therefore the scattering matrix $S_\beta\,(\lambda, R_\beta, R_0)$ can be written as

$$\begin{aligned}
&S_\beta(\lambda, R_\beta, R_0) \qquad\qquad (8.92)\\
&= I - 2\pi i F_\varphi(\lambda)\Big\{\varphi\,(1_\Sigma R_\beta - R_0)\,1_\Sigma\varphi\\
&\quad - \varphi\,(1_\Sigma R_\beta - R_0)\,1_\Sigma\varphi\,\varphi^{-1}\,(R_\beta - (\lambda + i0)\,I)^{-1}\,\varphi^{-1}\,\varphi 1_\Sigma\,(R_\beta 1_\Sigma - R_0)\,\varphi\Big\}F_\varphi^*(\lambda).
\end{aligned}$$

Employing the identity $1_\Sigma = J^*J$, we obtain

$$\begin{aligned} &S_\beta(\lambda, R_\beta, R_0) - S_\Sigma(\lambda, R_\Sigma, R_0) \\ &= -2\pi i F_\varphi(\lambda) \\ &\Big[\varphi 1_\Sigma (R_\beta - J^* R_\Sigma J) 1_\Sigma \varphi \\ &\quad - \varphi 1_\Sigma (R_\beta - J^* R_\Sigma J) 1_\Sigma \varphi \varphi^{-1} (R_\beta - (\lambda + i0) I)^{-1} \varphi^{-1} \\ &\qquad \times \varphi 1_\Sigma (R_\beta 1_\Sigma - R_0) \varphi \\ &\quad - \varphi (J^* R_\Sigma J - R_0 1_\Sigma) \varphi \\ &\qquad \times \varphi^{-1} \Big[(R_\beta - (\lambda + i0) I)^{-1} - J^* (R_\Sigma - (\lambda + i0) 1_\Sigma)^{-1} J\Big] \varphi^{-1} \\ &\qquad \times \varphi 1_\Sigma (R_\beta 1_\Sigma - R_0) \varphi \\ &\quad - \varphi (J^* R_\Sigma J - R_0 1_\Sigma) \varphi \varphi^{-1} J^* (R_\Sigma - (\lambda + i0) 1_\Sigma)^{-1} J \varphi^{-1} \\ &\qquad \times \varphi 1_\Sigma (R_\beta - J^* R_\Sigma J) 1_\Sigma \varphi\Big] F_\varphi^*(\lambda). \end{aligned} \tag{8.93}$$

Very often $F_\varphi(\lambda)$ is a Hilbert-Schmidt operator. Hence the scattering matrix difference in (8.93) is a trace class operator. Its trace norm is governed by

$$\begin{aligned} &\|\varphi 1_\Sigma (R_\beta - J^* R_\Sigma J) 1_\Sigma \varphi\|_{2,2}, \text{ and} \\ &\left\|\varphi^{-1}\Big[(R_\beta - (\lambda + i0) I)^{-1} - J^* (R_\Sigma - (\lambda + i0) 1_\Sigma)^{-1} J\Big] \varphi^{-1}\right\|_{2,2}. \end{aligned}$$

8.41. THEOREM. *Let K_0 be a free Feller operator. Put $K_\beta = K_0 + \beta 1_\Gamma$ and let K_Σ be the corresponding Dirichlet Feller operator. Suppose that Γ is chosen in such a way that the wave operators $\Omega_\pm(K_\beta, K_0)$ and $\Omega_\pm(K_\Sigma, J, K_0)$ exist and that the invariance principle holds in the strong sense, i.e.*

$$\begin{aligned} \Omega_\pm(K_\beta, K_0) &= \Omega_\mp(R_\beta, R_0), \quad \textit{and} \\ \Omega_\pm(K_\Sigma, J, K_0) &= \Omega_\mp(R_\Sigma, J, R_0). \end{aligned}$$

Here $R_0 = (K_0 + aI)^{-1}$, $R_\beta = (K_\beta + aI)^{-1}$, and $R_\Sigma = (K_\Sigma + aI_\Sigma)^{-1}$, $a > 0$.

Assume that $F_\varphi(\lambda)$ (see (8.50)) is an operator from $L^2(E, m)$ to $\mathcal{H}_{R_0}(\lambda)$, with $\lambda \in \sigma_{ac}(R_0)$. Here $\mathcal{H}_{R_0}(\lambda)$ is a fiber of the spectral resolution with respect to the self-adjoint operator $R_0 = (K_0 + aI)^{-1}$. The spectral measure $E_{R_0}(\cdot)$ and $F_\varphi(\lambda)$ satisfy the equality

$$\|F_\varphi^*(\lambda) F_\varphi(\lambda)\|_{2,2} = \left\|\varphi^{-1} \frac{dE_{R_0}(\lambda)}{d\lambda} \varphi^{-1}\right\|_{2,2}.$$

The function φ is admissible in the sense of Definition 8.19. Assume that $F_\varphi(\lambda)$ is a Hilbert-Schmidt operator. According to Definition 8.34, Lemma 8.36, and Theorem 8.37 the difference of the scattering matrices given in (8.93) is of trace class.

Hence its trace norm can be estimated as follows in terms of the weighted Laplace transform of the occupation time:

$$\begin{aligned}&\left\|S_\beta\left(\lambda, R_\beta, R_0\right)-S_\Sigma\left(\lambda, R_\Sigma, R_0\right)\right\|_{\text{trace}} \\ &\leq c(\lambda)\left\|F_\varphi(\lambda)\right\|_{\mathrm{HS}}^2 \rho_\varphi(\beta),\end{aligned} \tag{8.94}$$

with $\lambda \in \triangle$, $\triangle \subset [0, 1/a]$. The interval $\triangle$ is chosen in such a way that the representations in (8.84) and (8.86) are valid. The constant $c(\lambda)$ can be made explicit.

Proof. The following estimates follow from (8.93):

$$\begin{aligned}&\left\|S_\beta\left(\lambda, R_\beta, R_0\right)-S_\beta\left(\lambda, R_\beta, R_0\right)\right\|_{\text{trace}} \\ &\leq 2\pi\left\|F_\varphi(\lambda)\right\|_{\mathrm{HS}}^2 \\ &\Big[\left\|\varphi 1_\Sigma\left(R_\beta-J^* R_\Sigma J\right) 1_\Sigma \varphi\right\|_{2,2} \\ &+\left\|\varphi 1_\Sigma\left(R_\beta-J^* R_\Sigma J\right) 1_\Sigma \varphi\right\|_{2,2}\left\|\varphi^{-1}\left(R_\beta-(\lambda+i0) I\right)^{-1} \varphi^{-1}\right\|_{2,2} \\ &\qquad\left\|\varphi 1_\Sigma\left(R_\beta 1_\Sigma-R_0\right) \varphi\right\|_{2,2} \\ &+\left\|\varphi\left(J^* R_\Sigma J-R_0 1_\Sigma\right) \varphi\right\|_{2,2} \\ &\left\|\varphi^{-1}\left[\left(R_\beta-(\lambda+i0) I\right)^{-1}-J^*\left(R_\Sigma-(\lambda+i0) 1_\Sigma\right)^{-1} J\right] \varphi^{-1}\right\|_{2,2} \\ &\left\|\varphi 1_\Sigma\left(R_\beta 1_\Sigma-R_0\right) \varphi\right\|_{2,2} \\ &+\left\|\varphi\left(J^* R_\Sigma J-R_0 1_\Sigma\right) \varphi\right\|_{2,2}\left\|\varphi^{-1} J^*\left(R_\Sigma-(\lambda+i0) 1_\Sigma\right)^{-1} J \varphi^{-1}\right\|_{2,2} \\ &\left.\left\|\varphi 1_\Sigma\left(R_\beta-J^* R_\Sigma J\right) 1_\Sigma \varphi\right\|_{2,2}\right].\end{aligned} \tag{8.95}$$

From Lemma 8.36 we get the inequality

$$\left\|\varphi 1_\Sigma\left(R_\beta-J^* R_\Sigma J\right) 1_\Sigma \varphi\right\|_{2,2} \leq\left\|\varphi\left(R_\beta-J^* R_\Sigma J\right) \varphi\right\|_{2,2} \leq c_0 \rho_\varphi(\beta).$$

Using the identity

$$\left((H+a I)^{-1}-z I\right)^{-1}=-(a+\zeta) I-(a+\zeta)^2(H-\zeta I)^{-1},$$

with $-a$, $\zeta \in \operatorname{res}(H)$, $z=\dfrac{1}{a+\zeta}$, we obtain the bound

$$\begin{aligned}&\left\|\varphi^{-1}\left[\left(R_\beta-(\lambda+i0) I\right)^{-1}-J^*\left(R_\Sigma-(\lambda+i0) 1_\Sigma\right)^{-1} J\right] \varphi^{-1}\right\|_{2,2} \\ &=\frac{1}{\lambda^2}\left\|\varphi^{-1}\left[\left(K_\beta-(\mu-i0) I\right)^{-1}-J^*\left(K_\Sigma-(\mu-i0) 1_\Sigma\right)^{-1} J\right] \varphi^{-1}\right\|_{2,2} \\ &\leq c_1 \rho_\varphi(\beta).\end{aligned}$$

Here we used Theorem 8.37 and we wrote $\mu = \lambda^{-1} - a$, $\lambda \in \triangle$ and $0 \notin \triangle$. This implies that

$$\left\| \varphi^{-1} \left(R_\beta - (\lambda + i0) I \right)^{-1} \varphi^{-1} \right\|_{2,2} \leq c_2,$$

uniformly in β, because we have assumed the finiteness of the expression

$$\left\| \varphi^{-1} \left(R_\Sigma - (\lambda + i0) I_\Sigma \right)^{-1} \varphi^{-1} \right\|_{2,2}.$$

Finally we have

$$\|\varphi R_\beta 1_\Gamma \varphi\|_{2,2} = \|\varphi 1_\Gamma R_\beta \varphi\|_{2,2} \leq \rho_\varphi(\beta) \leq \rho_\varphi(0).$$

Hence we get

$$\begin{aligned} &\|\varphi 1_\Sigma \left(R_\beta 1_\Sigma - R_0 \right) \varphi\|_{2,2} \leq \|\varphi \left(R_\beta 1_\Sigma - R_0 \right) \varphi\|_{2,2} \\ &\leq \rho_\varphi(0) + \|\varphi \left(R_\beta - R_0 \right) \varphi\|_{2,2} \\ &\leq 2\rho_\varphi(0). \end{aligned}$$

Altogether we obtain

$$\begin{aligned} &\|S_\beta \left(\lambda, R_\beta, R_0 \right) - S_\Sigma \left(\lambda, R_\Sigma, R_0 \right)\|_{\text{trace}} \\ &\leq 2\pi \|F_\varphi(\lambda)\|_{\text{HS}}^2 \\ &\quad \left(c_0 \rho_\varphi(\beta) + 4 c_0 c_2 \rho_\varphi(0) \rho_\varphi(\beta) + 4 c_1 \rho_\varphi^2(0) \rho_\varphi(\beta) \right) \\ &\leq c(\lambda) \|F_\varphi(\lambda)\|_{\text{HS}}^2 \rho_\varphi(\beta), \end{aligned}$$

which proves Theorem 8.41.

In order to obtain the scattering matrices for the original systems $\{K_\beta, K_0\}$ and $\{K_\Sigma, J, K_0\}$, one may employ the relation

$$S\left(\mu, K_\beta, K_0\right) = S\left(\lambda, R_\beta, R_0\right)^{-1},$$

where $\lambda^{-1} = a + \mu$, $0 < \mu < 1$, $0 < \lambda \leq a^{-1}$. For this formula see Kato [149, p. 149].

Upon using (8.92) and (8.86) and choosing $a > 0$ large enough, it will follow that the inverses of $S_\beta \left(\lambda, R_\beta, R_0 \right)$ and $S_\Sigma \left(\lambda, R_\Sigma, R_0 \right)$ exist because we have

$$\|\varphi \left(J^* R_\Sigma - R_0 J^* \right) J \varphi\|_{2,2} \leq \frac{c}{a}, \quad a > 0.$$

Hence we see that

$$\begin{aligned} &S_\beta \left(\mu, K_\beta, K_0 \right) - S_\Sigma \left(\mu, K_\Sigma, K_0 \right) \qquad (8.96) \\ &= S_\beta \left(\lambda, R_\beta, R_0 \right)^{-1} \left(S_\Sigma \left(\lambda, R_\Sigma, R_0 \right) - S_\beta \left(\lambda, R_\beta, R_0 \right) \right) S_\Sigma \left(\lambda, R_\beta, R_0 \right)^{-1}. \end{aligned}$$

The operator in (8.96) is trace class whenever $S_\Sigma \left(\lambda, R_\Sigma, R_0 \right) - S_\beta \left(\lambda, R_\beta, R_0 \right)$ is trace class. Choosing a large enough yields the following corollary.

8.42. COROLLARY. *Let the assumptions be as in Theorem 8.41. The trace norms of the differences of the scattering matrices of the systems* $\{K_\beta, K_0\}$ *and* $\{K_\Sigma, K_0\}$ *can be estimated as follows:*

$$\|S_\beta(\mu, K_\beta, K_0) - S_\Sigma(\mu, K_\Sigma, K_0)\|_{\text{trace}} \leq c\rho_\varphi(\beta).$$

Another application of Theorem 8.37 is determining *resonance free regions* for the physical system represented by $K_\beta = K_0 + \beta 1_\Gamma$. The notion of "resonances" is not at all uniquely established in the literature. Here we call a value z_0 in the second sheet of the lower half plane a *resonance* or a *resonance value* if the meromorphically continued sandwiched resolvent

$$\varphi^{-1}(K_\beta - zI)^{-1}\varphi^{-1} =: r_\beta(z)$$

has a pole at z_0. The starting point however is that the artificial system has an analytic continuation. The latter means that we assume

$$\varphi^{-1}J^*(K_\Sigma - zI_\Sigma)^{-1}J\varphi^{-1} =: r_\Sigma(z)$$

to be holomorphically continuable into a region G_Σ of the second lower half plane. This assumption is motivated by results for $K_0 = -\frac{1}{2}\Delta$ in $L^2(\mathbb{R}^d)$ and $\varphi(x) = \exp(|x|)$, for which Bardos, Lebeau and Rauch [17] showed that the poles of $r_\Sigma(z)$ lie in the region

$$\Im z \leq -c|z|^{1/3},$$

if Γ is a ball or a convex set. This property implies the existence of a resonance free region G_Σ. Because we have a quantitative estimate on the difference between $r_\Sigma(z)$ and $r_\beta(z)$ in terms of the weighted Laplace transform (here the weight function is $\exp(-|x|)$) of the occupation time we can determine a region G_β which is resonance free for $r_\beta(z)$.

Although it is not absolutely necessary, in the following proposition we neglect the regular part of the potential V. We also fix a weight φ and we suppose that we are in a general state space situation of a second countable locally compact Hausdorff space E.

8.43. PROPOSITION. *Let* $K_\beta = K_0 + \beta 1_\Gamma$ *and* K_Σ *be given as above. Suppose that the operator valued function* $r_\Sigma(z)$ *defined by*

$$r_\Sigma(z) = \varphi^{-1}J^*(K_\Sigma - zI_\Sigma)^{-1}J\varphi^{-1}$$

possesses an analytic continuation from a bounded set G *in the upper half plane to a bounded set, denoted by* G_Σ *in the lower half plane. Assume that any poles of* $r_\Sigma(z)$ *lie outside of* G_Σ, *i.e. assume that*

$$\|r_\Sigma(z)\|_{2,2} =: d_\Sigma(z) < \infty$$

and that

$$\sup_{z\in G\cup G_\Sigma} \|r_\Sigma(z)\|_{2,2} =: d_\Sigma < \infty.$$

Let Γ and φ be chosen in such a way that $\rho_\beta(\beta)$ tends monotonically to zero as β tends to infinity (see Definition 8.34 and Lemma 8.35) where $\varphi_\varphi(\beta)$ was given by

$$\rho_\varphi(\beta) = \sup_{x\in E} \varphi(x)^2 \left[\mathbb{E}_x \left\{\exp\left(-\beta T_{\lambda,\Gamma}\right) : T_{\lambda,\Gamma} > 0\right\}\right]^{1/2}.$$

Put $r_\beta(z) := \varphi^{-1}\left(K_\beta - zI\right)^{-1}\varphi^{-1}$. Then $r_\beta(z)$ is holomorphically continuable into G_Σ if β is sufficiently large.

A quantitative way of expressing this is the following. Choose a constant β_0 such that for all $\beta \geq \beta_0$ the following estimate, (compare it with the inequality in (8.79)), is valid

$$\rho_\varphi(\beta) \leq \frac{1}{16a^2c_0}\frac{1}{d_\Sigma(z)}. \tag{8.97}$$

Here a is a positive constant that is chosen large enough, i.e.

$$a \geq \max\left\{2m, \frac{1}{d_\Sigma}, g\right\}, \tag{8.98}$$

with $g = \sup_{z\in G\cup G_\Sigma}|z|$. The constants c_0 and m are taken from (8.76) with $\rho = 4$ (see the proof of Lemma 8.35). Then the sandwiched resolvent difference can be estimated by the weighted Laplace transform of the occupation time:

$$\|r_\beta(z) - r_\Sigma(z)\|_{2,2} \leq 32a^2 d_\Sigma^2 c_0 \rho_\varphi(\beta), \quad \beta \geq \beta_0, \tag{8.99}$$

for all $z \in G \cup G_\Sigma$ for which condition (8.97) is true. For every z in the region

$$G_\beta := \left\{z \in G\cup G_\Sigma : \quad d_\Sigma(z)\rho_\varphi(\beta) \leq \left(16a^2c_0\right)^{-1}\right\},$$

the operator $r_\beta(z)$ is bounded and the function $z \mapsto r_\beta(z)$ is holomorphic in G_β. Thus, G_β is a resonance free region for the system described by $K_\beta = K_0 + \beta 1_\Gamma$.

REMARKS. As an example assume $r_\beta(z)$ to have a simple pole at z_0 with residue R. The assumption in (8.97) says mainly that

$$\frac{\|R\|_{2,2}}{|z - z_0|}\rho_\varphi(\beta) < c.$$

The region where $r_\beta(z)$ cannot have a pole is

$$G_\beta \subseteq \left\{z \in G_\Sigma : |z - z_0| > c\,\|R\|_{2,2}\,\rho_\varphi(\beta)\right\}$$

or, what amounts to saying the same thing,

$$\operatorname{dist}\{G_\beta, \mathbb{C}\setminus G_\Sigma\} \geq c\rho_\varphi(\beta).$$

If $K_0 = -\frac{1}{2}\Delta$, if Γ is a bounded uniformly Lipschitz set (for the definition see Definition 8.30) and if $\varphi(x) = \exp(|x|)$, it can be shown (using Theorem 8.31) that

$$\rho_\varphi(\beta) \leq c\beta^{-\gamma A_{\text{int}}},$$

with $\gamma < 1$ and A_{int} the interior cone constant (see Definition 8.29 and Theorem 8.31) of Γ. The constant c depends on γ, m and c_0. In this situation, we have

$$\|r_\beta(z)\|_{2,2} \leq d_\Sigma + 32a^2 d_\Sigma^2 c_0 c\beta^{-\gamma A_{\text{int}}}$$

and

$$G_\beta = \left\{z \in G \cup G_\Sigma : d_\Sigma(z) \leq \widetilde{c}\beta^{\gamma A_{\text{int}}}\right\},$$

where $\widetilde{c}$ is some other constant which can be computed explicitly.

Proof of Proposition 8.43. For brevity we write

$$D_{\beta,\varphi}(a) = \varphi\left[(K_\beta + aI)^{-1} - J^*\left(K_\Sigma + aI_\Sigma\right)^{-1} J\right]\varphi.$$

The Hilbert identity (8.34) yields the following equality

$$\begin{aligned}
&r_\beta(z) - r_\Sigma(z)\\
&= \left[\varphi^{-2} + (z-a)r_\Sigma(z)\right] D_{\beta,\varphi}(a)\\
&\quad \left[\varphi^{-2} + (z-a)\left(r_\beta(z) - r_\Sigma(z)\right) + (z-a)r_\Sigma(z)\right].
\end{aligned}$$

Hence,

$$\begin{aligned}
&\left[r_\beta(z) - r_\Sigma(z)\right]\\
&\quad\times\left[1 - (z-a)\left(\varphi^{-2} + (z-a)r_\Sigma(z)\right) D_{\beta,\varphi}(a)\right]\\
&= \left[\varphi^{-2} + (z-a)r_\Sigma(z)\right] D_{\beta,\varphi}(a)\left[\varphi^{-2} + (z-a)r_\Sigma(z)\right].
\end{aligned} \tag{8.100}$$

Since $r_\Sigma(z)$ has a holomorphic extension into G_Σ the right-hand side of (8.100) can be holomorphically continued to G_Σ. This implies that the left-hand side admits an analytic continuation. The continuation

$$I - (z-a)\left(\varphi^{-2} + (z-a)r_\Sigma(z)\right) D_{\beta,\varphi}(a), \tag{8.101}$$

$z \in G_\Sigma$, is invertible if

$$\left\|(z-a)\left(\varphi^{-2} + (z-a)r_\Sigma(z)\right) D_{\beta,\varphi}(a)\right\|_{2,2} < 1. \tag{8.102}$$

From condition (8.97), it follows that the inequality in (8.102) is true, because we have (see Corollary 8.38 and use the fact that $|z-a| \le g+a \le 2a$; see (8.98))

$$|z-a| \left\|\varphi^{-2} + (z-a) r_\Sigma(z)\right\|_{2,2} \|D_{\beta,\varphi}(a)\|_{2,2}$$
$$\le 2a\,(1+2ad_\Sigma(z))\,c_0\rho_\varphi(\beta)$$

(use $1 \le a\, d_\Sigma(z)$)

$$\le 8a^2 c_0 d_\Sigma(z)\rho_\varphi(\beta)$$

(apply the boundedness assumption (8.97))

$$\le \frac{1}{2}.$$

Therefore the holomorphic continued operator family in (8.101) has a bounded inverse with operator norm smaller than two. As a consequence of (8.100), we have

$$\begin{aligned}\|r_\beta(z) - r_\Sigma(z)\|_{2,2} &\le 2\left\|\varphi^{-2} + (z-a) r_\Sigma(z)\right\|_{2,2}^2 \|D_{\beta,\varphi}(a)\|_{2,2} \\ &\le 2\,(1+2ad_\Sigma)^2\, c_0\rho_\varphi(\beta) \\ &\le 32 c_0 a^2 d_\Sigma^2 \rho_\varphi(\beta),\end{aligned}$$

as advertized in (8.99). The boundedness of $r_\beta(z)$ in G_β is easily established. If β tends to infinity, then G_β tends to G_Σ.

REMARK. Since we intend to give an accessible presentation, our constants are not chosen optimally. Our only aim is to give a procedure that leads to quantitative estimates. If one has precise estimates for $\rho_\varphi(\beta)$ as in the case of the Laplace operator, we can easily optimize the constants above.

Appendix A

Spectral Theory

The objective of this appendix is to give a brief overview of the spectral theory of linear self-adjoint operators in separable Hilbert spaces which outlines notions used in the present book. The ultimate aim is to introduce and explain the different and relevant parts of the spectrum. The reader who is not very familiar with spectral theory can take this appendix as a guideline to consult the literature on this well-established subject. The theory explained here is taken from standard text books. It mainly follows the introduction by Amrein [7]. The interested reader may also consult Amrein, Jauch, Sinha [8], Kato [148], Reed and Simon [191, 192, 193, 194], or Weidmann [252].

Self-adjoint operators will be explained in the items A.1 through A.17. The spectra of self-adjoint operators will be described in A.18 through A.39.

A.1. Linear operators

Let $\mathcal{H}$ be a separable complex Hilbert space. A linear operator from $\mathcal{H}$ to $\mathcal{H}$ is an ordered pair $\{\mathrm{dom}(K), K\}$ where $\mathrm{dom}(K)$ is a linear subspace in $\mathcal{H}$ and where K is a linear map from $\mathrm{dom}(K)$ into $\mathcal{H}$. The subspace $\mathrm{dom}(K)$ is called the *domain* of the operator K. In other words, to each $f \in \mathrm{dom}(K)$ there is associated a vector $Kf \in \mathcal{H}$ in such a way that, for f and $g \in \mathrm{dom}(K)$ and for scalars $\alpha \in \mathbb{C}$,

$$K\,(f + \alpha g) = Kf + \alpha K g.$$

Very often, in fact most of the time, the linear operator $\{\mathrm{dom}(K), K\}$ is just denoted by K. We often write K is an operator in $\mathcal{H}$.

A.2. Range of a linear operator

The range of a linear operator K, $\mathrm{ran}(K)$, is the image of its domain $\mathrm{dom}(K)$ under K, i.e.

$$\mathrm{ran}(K) = \{g \in \mathcal{H}:\; g = Kf \text{ for some } f \in \mathrm{dom}(K)\}\,.$$

A.3. Null space of a linear operator

The null space or kernel of a linear operator K, $\mathrm{nul}(K)$, is defined as

$$\mathrm{nul}(K) = \{f \in \mathrm{dom}(K):\; Kf = 0\}\,.$$

Of course, here 0 denotes the zero-vector in $\mathcal{H}$.

A.4. Equality of two linear Operators

Two operators K_1 and K_2 with domain and range in $\mathcal{H}$ are equal if

(1) $\mathrm{dom}(K_1) = \mathrm{dom}(K_2)$, and

(2) $K_1 f = K_2 f$ for all $f \in \mathrm{dom}(K_1)$.
In this case we write $K_1 = K_2$.

A.5. Restriction and extension of a linear operator

A linear operator K_2 is an extension of an operator K_1 if

(1) $\mathrm{dom}(K_1) \subseteq \mathrm{dom}(K_2)$, and

(2) $K_1 f = K_2 f$ for all $f \in \mathrm{dom}(K_1)$. On the other hand, in the same situation, K_1 is called a restriction of K_2. This is expressed by

$$K_2 \supseteq K_1 \text{ or } K_1 \subseteq K_2 \text{ respectively.}$$

A.6. Sum of two Operators

The sum of two operators K_1 and K_2, $K_1 + K_2$, is defined by:

(1) $\mathrm{dom}\,(K_1 + K_2) := \mathrm{dom}(K_1) \cap \mathrm{dom}(K_2)$, and

(2) $(K_1 + K_2)\, f := K_1 f + K_2 f$ for f belonging to $\mathrm{dom}\,(K_1 + K_2)$.

A.7. Inverse operator

A linear operator K is called invertible if its null space consists of the zero vector, i.e. if $\mathrm{nul}(K) = \{0\}$. In that case K^{-1}, the inverse of K, is defined by

(1) $\mathrm{dom}(K^{-1}) = \mathrm{ran}(K)$, and

(2) $K^{-1}(Kf) = f$, for $f \in \mathrm{dom}(K)$.

A.8. Closed linear operators

Let $\{f_n\}$, $f_n \in \mathrm{dom}(K)$, and $\{Kf_n\}$ be Cauchy sequences for the strong or $\|\cdot\|$-topology, i.e. suppose that the sequence $\{(f_n, Kf_n)\}$ is a Cauchy sequence in the Hilbert space $\mathcal{H} \times \mathcal{H}$ equipped with the norm

$$\|(f,g)\|^2 := \|f\|^2 + \|g\|^2 \,,$$

for $(f,g) \in \mathcal{H} \times \mathcal{H}$. Set $f = \text{s-}\lim_{n\to\infty} f_n$. The linear operator K is said to be closed if f belongs to $\mathrm{dom}(K)$ and if $Kf = \text{s-}\lim_{n\to\infty} Kf_n$.

A.9. Graph of an operator

The graph $G(K)$ of an operator K is the subset of $\mathcal{H} \times \mathcal{H}$ consisting of all elements of the form (f, Kf), where f belongs to $\mathrm{dom}(K)$. A linear operator K is closed if and only if its graph $G(K)$ is a closed linear subspace of $\mathcal{H} \times \mathcal{H}$.

A.10. Closable operator

A linear operator is said to be closable if there exists a closed linear operator, denoted here by $\overline{K}$, such that its graph $G\left(\overline{K}\right)$ is the closure of $G(K)$, so that $G\left(\overline{K}\right) = \overline{G(K)}$. The operator $\overline{K}$ is the smallest closed linear extension of K. In fact the operator K is closable if and only if for every pair of strong Cauchy sequences $\{f_n\}$ and $\{Kf_n\}$ for which $\text{s-}\lim_{n\to\infty} f_n = 0$ we also have $\text{s-}\lim_{n\to\infty} Kf_n = 0$.

A.11. Bounded operators and $\mathcal{B}(\mathcal{H})$

A linear operator K is bounded if there is a finite number M such that

$$\|Kf\| \leq M \|f\|$$

for all f in the domain of K. Its norm is given by the formula:

$$\|K\| := \sup \left\{ \frac{\|Kf\|}{\|f\|} : f \in \operatorname{dom}(K),\ f \neq 0 \right\}.$$

Let $\mathcal{B}(\mathcal{H})$ be the set of all bounded operators K on $\mathcal{H}$ for which $\operatorname{dom}(K) = \mathcal{H}$. Any bounded linear operator K has a unique extension $\overline{K}$ with $\operatorname{dom}(\overline{K}) = \overline{\operatorname{dom}(K)}$, the closure of $\operatorname{dom}(K)$. In particular, if K is defined on a dense subspace of $\mathcal{H}$, then $\overline{K}$ belongs to $\mathcal{B}(\mathcal{H})$. A closed linear operator is bounded if it is defined on the whole Hilbert space $\mathcal{H}$ (*closed graph theorem*). In the present book we do not distinguish between a bounded operator K and its bounded extension $\overline{K} \in \mathcal{B}(\mathcal{H})$. If we consider bounded linear operators we always mean operators in $\mathcal{B}(\mathcal{H})$.

A.12. Adjoint operator

A linear operator K is said to be densely defined if its domain is a dense subspace of $\mathcal{H}$. For a densely defined operator K its adjoint K^* is defined as follows. A vector g belongs to the domain of K^* if there exists a vector $h \in \mathcal{H}$ such that

$$\langle Kf, g \rangle = \langle f, h \rangle, \quad \text{for all } f \in \operatorname{dom}(K).$$

Then K^* is a linear mapping from $\operatorname{dom}(K^*)$ to $\mathcal{H}$ given by $K^*g = h$, $g \in \operatorname{dom}(K^*)$. This means

$$\langle Kf, g \rangle = \langle f, K^*g \rangle, \quad \text{for all } f \in \operatorname{dom}(K) \text{ and } g \in \operatorname{dom}(K^*).$$

The operator K^* has a closed graph. If the operator K is closable, then the domain of K^* is dense and $\overline{K} = K^{**}$.

A.13. Unitary operators

A bounded linear operator $K \in \mathcal{B}(\mathcal{H})$ is called unitary if $K^* = K^{-1}$; i.e. if $KK^* = K^*K = I$.

A.14. Symmetric operators

A linear operator K is called symmetric if $\operatorname{dom}(K)$ is dense in $\mathcal{H}$ and if $K \subseteq K^*$, i.e. if

$$\langle Kf, g \rangle = \langle f, Kg \rangle$$

for all vectors f and g belonging to $\operatorname{dom}(K) \subseteq \operatorname{dom}(K^*)$.

A.15. Self-adjoint operators

A linear operator K is called self-adjoint if $K^* = K$, i.e. if K is symmetric and if $\operatorname{dom}(K) = \operatorname{dom}(K^*)$. If $\operatorname{dom}(K) = \mathcal{H}$ and if K is symmetric, then from the closed graph theorem it follows that K is bounded and hence self-adjoint. So that a symmetric everywhere defined operator is bounded and selfadjoint.

A.16. Essentially self-adjoint operators

A symmetric operator K which possesses only one self-adjoint extension is said to be essentially self-adjoint. If $K \subseteq K^*$ and if $K^{**} = K^*$, then K is essentially self-adjoint. If the operator K is symmetric, then it is self-adjoint if and only if $\operatorname{ran}(K + iI) = \operatorname{ran}(K - iI) = \mathcal{H}$: see for instance Amrein [7, p. 28]. A linear operator is *form positive* if $\langle f, Kf \rangle \geq 0$ for all f in the domain of K. For brevity we write $K \geq 0$ (or $0 \leq K$), if the operator K is form positive. A symmetric, form positive linear operator K is essentially self-adjoint if $\operatorname{ran}(K + I)$ is dense in $\mathcal{H}$.

A.17. Dissipative operator

Let K be a linear operator with domain and range in $C_\infty(E)$ (see Definition 1.2), the space of continuous functions vanishing at infinity. The operator K is said to be *dissipative* if for all $\lambda > 0$ and for all $f \in \operatorname{dom}(K)$ the inequality

$$\|\lambda f + Kf\|_\infty \geq \lambda \|f\|_\infty$$

holds. Further properties of dissipative operators are explained in Appendix C.

A.18. Relatively bounded operators

Let K_1 and K_2 be two linear operators with $\operatorname{dom}(K_1) \subseteq \operatorname{dom}(K_2)$. The operator K_2 is called K_1-bounded, if there exist two finite positive numbers α and β for which the inequality

$$\|K_2 f\| \leq \alpha \|K_1 f\| + \beta \|f\|$$

is true for all f belonging to the domain of K_1. If the operators K_1 and K_2 are closed, the existence of such constants is guaranteed by the closed graph theorem. The infimum of all possible α is called the (relative) K_1-bound of the operator K_2. A.20.

Suppose that the operator K_1 is self-adjoint. Then K_2 is K_1-bounded if the operator $K_2 (K_1 - zI)^{-1}$ belongs to $\mathcal{B}(\mathcal{H})$ for some $z \in \operatorname{res}(K_1)$: see A.20.

A.19. Kato-Rellich theorem

Let K_1 be a self-adjoint and let K_2 be a symmetric operator with a K_1-bound strictly smaller than one. Then the operator $K_1 + K_2$ is self-adjoint and its domain coincides with that of K_1. It is noted that the conditions in this result can be weakened. See for instance Reed and Simon [192, p. 162 ff.].

A.20. Resolvent set: $\operatorname{res}(K)$

Let K be a closed linear operator. The complex number z is called a regular point if the operator $K - zI$ is invertible with an inverse $(K - zI)^{-1}$ defined everywhere on $\mathcal{H}$ that belongs to $\mathcal{B}(\mathcal{H})$. From the closed graph theorem it follows that a point z is regular for the closed linear operator K if and only if the null space of $K - zI$ is $\{0\}$ and if its range $\operatorname{ran}(K - zI)$ coincides with $\mathcal{H}$. The set of all regular points constitutes the resolvent set of K denoted by $\operatorname{res}(K)$.

A.21. Resolvent

Correctly defined, the resolvent of an operator K is the operator-valued function $z \mapsto (K - zI)^{-1}$ mapping the set $\operatorname{res}(K)$, which is an open subset of $\mathbb{C}$, into $\mathcal{B}(\mathcal{H})$. Sometimes, for a fixed value of $z \in \operatorname{res}(K)$, the operator $(K - zI)^{-1}$ is called resolvent of K as well. For a shorter notation we sometimes write $(K - zI)^{-1}$ for the resolvent.

A.22. Spectrum: $\sigma(K)$

Let K be a closed linear operator. By definition the complement of its resolvent set $\operatorname{res}(K)$ is the spectrum of K: $\sigma(K) = \mathbb{C} \setminus \operatorname{res}(K)$. If K is a self-adjoint operator, then its spectrum is a closed subset of the real axis. In general, it is a closed subset of $\mathbb{C}$. If K is a bounded linear operator in $\mathcal{B}(\mathcal{H})$, then its spectrum is nonempty and compact.

A.23. Eigenvalue

Let K be a linear operator. If there exists a non-zero vector $f \in \operatorname{dom}(K)$ and a complex number z for which $Kf = zf$, then z is called an eigenvalue of K and f an associated eigenvector. If K is self-adjoint, then the eigenvalues are real. Since $\mathcal{H}$ is separable, a self-adjoint operator possesses at most countably many eigenvalues.

A.24. Pure point and continuous spectra

Let K be a self-adjoint operator. Denote by $\mathcal{H}_{\mathrm{p}}(K)$ the closure of the subspace of $\mathcal{H}$ spanned by all eigenvectors of K. Denote its complement with $\mathcal{H}_{\mathrm{c}}(K)$, i.e.

$$\mathcal{H} = \mathcal{H}_{\mathrm{p}}(K) \oplus \mathcal{H}_{\mathrm{c}}(K).$$

The subspace $\mathcal{H}_{\mathrm{c}}(K)$ is called the subspace of continuity. This name will become more transparent after introducing the spectral family corresponding to K (see A.35). Set

$$K_{\mathrm{p}} = K \restriction \operatorname{dom}(K) \cap \mathcal{H}_{\mathrm{p}}(K),$$
$$K_{\mathrm{c}} = K \restriction \operatorname{dom}(K) \cap \mathcal{H}_{\mathrm{c}}(K),$$

the restrictions of K to $\mathcal{H}_{\mathrm{p}}(K)$ and to $\mathcal{H}_{\mathrm{c}}(K)$ respectively. The restrictions K_{p} and K_{c} are self-adjoint operators. The operator K_{p} possesses pure point spectrum, and the operator K_{c} has no eigenvalues, it has a purely continuous spectrum (see also A.38). Then the pure point spectrum of K is defined by $\overline{\sigma_{\mathrm{p}}(K)} = \sigma(K_{\mathrm{p}})$, and its continuous spectrum by $\sigma_{\mathrm{c}}(K) = \sigma(K_{\mathrm{c}})$.

A.25. Discrete and essential spectrum

The discrete spectrum of a self-adjoint operator K, $\sigma_{\mathrm{d}}(K)$, is the set of all isolated eigenvalues with finite multiplicity. The essential spectrum is defined as its complement:

$$\sigma_{\mathrm{ess}}(K) = \sigma(K) \setminus \sigma_{\mathrm{d}}(K).$$

If a self-adjoint operator has compact resolvent, then its spectrum is purely discrete (see A.26).

A.26. Compact operators

A bounded operator K is called compact if for every bounded sequence $\{f_n\}$ in $\mathcal{H}$ the sequence $\{Kf_n\}$ contains a Cauchy sequence. Since every bounded sequence contains a subsequence that converges with respect to the weak topology, a linear operator K is compact if and only if the sequence $\{Kf_n\}$ converges in the strong sense, whenever the sequence $\{f_n\}$ converges in the weak sense. Notice that a sequence $\{f_n\}$ converges weakly if and only if, for every vector $f \in \mathcal{H}$, the limit $\lim_{n\to\infty}\langle f_n, f\rangle$ exists in $\mathbb{C}$. This is a consequence of the Banach-Steinhaus theorem and the reflexifity of $\mathcal{H}$. If K is compact and self-adjoint, it possesses a pure point spectrum. In such a case, all non-zero eigenvalues have finite multiplicity and are isolated. The complex number 0 is either an accumulation point of the set of eigenvalues or an eigenvalue with infinite multiplicity, i.e. $\sigma_{\mathrm{ess}}(K) = \{0\}$: see Weidmann [252, p. 197].

A.27. Compact perturbations

Assume that K_1 and K_2 are two self-adjoint linear operators with the property that the operator $K_2 - K_1$ is compact. Then their essential spectra coincide: $\sigma_{\mathrm{ess}}(K_2) = \sigma_{\mathrm{ess}}(K_1)$.

A.28. Relative compactness, Weyl's Theorem

Let K_1 be a closed linear operator with $\mathrm{res}\,(K_1) \neq \emptyset$. Another linear operator K_2 is called relatively K_1-compact if the following conditions are met:

(1) $\mathrm{dom}(K_1) \subseteq \mathrm{dom}(K_2)$, and

(2) the operator $K_2\,(K_1 - zI)^{-1}$ is compact for some $z \in \mathrm{res}\,(K_1)$.

Let K_1 be a self-adjoint linear operator in $\mathcal{H}$ and let B be a symmetric relatively compact perturbation of K_1. Then $K_2 := K_1 + B$ is self-adjoint, and

$$\sigma_{\mathrm{ess}}\,(K_1) = \sigma_{\mathrm{ess}}\,(K_2):$$

see Reed and Simon [194, p. 113 ff].

A.29. Spectral representation of compact operators

Let $\{\mu_k\}_1^\infty$ be the sequence of non-zero eigenvalues of a compact operator K. Denote with E_k the associated eigenprojections, so that every E_k is linear a projection on the null space of $\mu_k I - K$. Then K admits the spectral representation

$$K = \sum_{k=1}^{\infty} \mu_k E_k.$$

This sum converges with respect to the operator norm. Every operator E_k may be written in the form

$$E_k = \frac{1}{2\pi i} \int_{|z-\mu_k|=\rho_k} (zI - K)^{-1} \, dz,$$

where $0 < \rho_k$ is so small that μ_k is the only point that the open disc $\{z \in \mathbb{C} : |z - \mu_k| < \rho_k\}$ has in common with the spectrum of K. Let $\{\lambda_j\}$ be the sequence of eigenvalues, counting multiplicities, with associated normalized eigenvectors $\{\varphi_j\}$, then

$$K = \sum_{j=1}^{\infty} \lambda_j \, (\cdot, \varphi_j) \, \varphi_j.$$

Some authors use the phrase "repeated eigenvalues" instead of "multiplicities of eigenvalues": see e.g. Kato [148].

A.30. Hilbert-Schmidt operators

Let $\{\varphi_j\}_{j=1}^{\infty}$ be complete orthonormal family in $\mathcal{H}$. The compact operator K is called a Hilbert-Schmidt operator if

$$\|K\|_{\mathrm{HS}}^2 := \sum_{j=1}^{\infty} \|K\varphi_j\|^2$$

is finite. The latter expression does not depend on the particular choice of the complete orthonormal sequence $\{\varphi_j\}_{j=1}^{\infty}$. If K is a self-adjoint Hilbert-Schmidt operator, then its Hilbert-Schmidt norm $\|K\|_{\mathrm{HS}}$ is determined by

$$\|K\|_{\mathrm{HS}}^2 = \sum_{j=1}^{\infty} |\lambda_j|^2 ,$$

where $\{\lambda_j\}_{j=1}^{\infty}$ is an enumeration of the eigenvalues of K, where multiplicities are counted.

A.31. Trace class operators

Let K be a compact operator and let $\{\alpha_j\}_{j=1}^{\infty}$ be its singular values. This means that each α_j is non-negative and that any α_j^2 is a non-zero eigenvalue of K^*K. Then K is said to be a trace class operator if

$$\|K\|_{\mathrm{trace}} := \sum_{j=1}^{\infty} \alpha_j$$

is finite. Let $|K| := \sqrt{K^*K}$ be the unique positive square root of K^*K. Then K is a trace class operator if and only if $|K|$ is so. Moreover we have

$$\|K\|_{\mathrm{trace}} = \||K|\|_{\mathrm{trace}} = \||K^*|\|_{\mathrm{trace}} .$$

If K is a self-adjoint trace class operator, then its trace norm is given by

$$\|K\|_{\text{trace}} = \sum_{j=1}^{\infty} |\lambda_j|\,,$$

where λ_j, $j = 1, 2, \ldots$ are the eigenvalues of K counting the multiplicities. Let $\{\varphi_j\}_{j=1}^{\infty}$ be any complete orthonormal sequence in $\mathcal{H}$ and let K be any trace class operator in $\mathcal{H}$. Its trace is given by

$$\operatorname{trace}(K) = \sum_{j=1}^{\infty} \langle K\varphi_j, \varphi_j \rangle\,.$$

This definition does not depend on the particular choice of the complete orthonormal sequence $\{\varphi_j\}_{j=1}^{\infty}$. Hence, if K is a self-adjoint trace class operator, then its trace is given by

$$\operatorname{trace}(K) = \sum_{j=1}^{\infty} \lambda_j.$$

A.32. Spectral family

A spectral family is a family of projection operators $\{E(\lambda) : \lambda \in \mathbb{R}\}$ with the following properties:

(1) The family $\{E(\lambda) : \lambda \in \mathbb{R}\}$ is non-decreasing, i.e., in form sense we have

$$E(\lambda) \leq E(\mu), \quad \text{if } \lambda \leq \mu.$$

(2) $\text{s-}\lim_{\varepsilon\downarrow 0} E(\lambda + \varepsilon) = E(\lambda)$, i.e. the mapping $\lambda \mapsto E(\lambda)f$ is continuous from the right for every $\lambda \in \mathbb{R}$ and for every $f \in \mathcal{H}$.

(3) $\text{s-}\lim_{\lambda\downarrow -\infty} E(\lambda) = 0$.

(4) $\text{s-}\lim_{\lambda\uparrow \infty} E(\lambda) = I$.

A.33. Spectral theorem

To each self-adjoint operator K there is associated a unique spectral family $\{E_K(\lambda) : \lambda \in \mathbb{R}\}$ in such a way that

$$\operatorname{dom}(K) = \left\{ f \in \mathcal{H} : \int_{-\infty}^{\infty} \lambda^2 \, d\langle E_K(\lambda)f, f\rangle < \infty \right\}$$

and

$$\langle Kf, g\rangle = \int_{-\infty}^{\infty} \lambda \, d\langle E_K(\lambda)f, g\rangle\,,$$

where f belongs to $\operatorname{dom}(K)$ and g belongs to $\mathcal{H}$. This formula is abbreviated as

$$K = \int_{-\infty}^{\infty} \lambda \, dE_K(\lambda).$$

In fact one can prove that the integral

$$Kf = \int_{-\infty}^{\infty} \lambda \, dE_K(\lambda)f$$

converges in the strong sense in $\mathcal{H}$.

A.34. Functional calculus

Let $\{E_K(\lambda) : \lambda \in \mathbb{R}\}$ be the spectral family corresponding to the self-adjoint operator K. Let $\Phi(\cdot)$ be a continuous function from $\mathbb{R}$ to $\mathbb{C}$. Then one may define a new operator $\Phi(K)$ by

$$\operatorname{dom}(\Phi(K)) = \left\{ f \in \mathcal{H} : \int_{-\infty}^{\infty} |\Phi(\lambda)|^2 \, d\langle E_K(\lambda)f, f\rangle < \infty \right\} \quad \text{and}$$
$$\langle \Phi(K)f, g\rangle = \int_{-\infty}^{\infty} \Phi(\lambda)\, d\langle E_K(\lambda)f, g\rangle, \quad f, \quad g \in \operatorname{dom}(\Phi(K)),$$

which briefly is written as

$$\Phi(K) = \int_{-\infty}^{\infty} \Phi(\lambda)\, dE_K(\lambda).$$

If, in addition, $\Phi(\cdot)$ is a bounded function, then the operator $\Phi(K)$ is bounded and has norm

$$\|\Phi(K)\| = \sup\{|\Phi(\lambda)| : \lambda \in \operatorname{supp}\{E_K(\lambda)\}\},$$

where $\operatorname{supp}\{E_K(\lambda)\}$ denotes the support of the spectral measure associated to the spectral family: see A.36.

A.35. Subspace of continuity

In A.24 we already introduced $\mathcal{H}_c(K)$ as the complement of $\mathcal{H}_p(K)$, the closed subspace spanned by the eigenvectors of K. Let $\{E_K(\lambda) : \lambda \in \mathbb{R}\}$ be the spectral family corresponding to the self-adjoint operator K. Then the vector f belongs to $\mathcal{H}_c(K)$ if and only if the scalar product

$$\langle f, E_K(\lambda)f\rangle = \|E_K(\lambda)f\|^2$$

is continuous as a function of λ. This result explains the choice of the name for $\mathcal{H}_c(K)$ being the continuous subspace of $\mathcal{H}$ with respect to K. Correspondingly the space $\mathcal{H}_p(K)$ is called the subspace of discontinuity.

A.36. Spectral measure

Any spectral family $\{E_K(\lambda) : \lambda \in \mathbb{R}\}$ determines a spectral measure $B \mapsto E_K(B)$, B Borel subset of $\mathbb{R}$, as follows. For a half open interval $(a, b]$, $a < b$, we define $E_K\big((a, b]\big)$ as

$$E_K\big((a, b]\big) = E(b) - E(a).$$

Since, for every $f \in \mathcal{H}$, the mapping $B \mapsto E_K(B)f$ should be a $\mathcal{H}$-valued measure, we are obliged to use the following definitions:

$$E_K\big([a, b]\big) := E(b) - E(a-0);$$
$$E_K\big([a, b)\big) := E(b-0) - E(a-0);$$
$$E_K\big((a, b)\big) := E(b-0) - E(a).$$

This definition can be extended to all Borel subsets of $\mathbb{R}$. In fact if $U = \bigcup_{j=1}^{\infty} (a_j, b_j)$, where the open intervals (a_j, b_j), $j \in \mathbb{N}$, are mutually disjoint, is any open subset of $\mathbb{R}$, then

$$E_K(U) = \text{s-}\lim_{n\to\infty} \sum_{j=1}^{n} E_K\big((a_j, b_j)\big),$$

and if B is a Borel subset of $\mathbb{R}$ and if f belongs to $\mathcal{H}$, then

$$\langle E_K(B)f, f\rangle := \inf\left\{\langle E_K(U)f, f\rangle : U \supseteq B,\ U \text{ open}\right\}.$$

It also follows that

$$\langle E_K(B)f, f\rangle := \sup\left\{\langle E_K(C)f, f\rangle : C \subseteq B,\ C \text{ compact}\right\}$$

for $f \in \mathcal{H}$. In this way one obtains a projection valued mapping $\triangle \mapsto E_K(\triangle)$, $\triangle$ Borel subset of $\mathbb{R}$, with the following properties:

$$\begin{aligned} E_K(\triangle)^2 &= E_K(\triangle) = E_K(\triangle)^*\,; \\ E_K(\triangle \cup \triangle')^2 &= E_K(\triangle) + E_K(\triangle')\,, \text{ if } \triangle \text{ and } \triangle' \text{ are disjoint;} \\ E_K\left(\bigcup_{j=1}^{\infty} \triangle_j\right) &= \sum_{j=1}^{\infty} E_K(\triangle_j)\,, \text{ strong convergence,} \end{aligned}$$

if the intersection $\triangle_j \cap \triangle_k = \emptyset$, for $k \neq j$. To this operator-valued strongly σ-additive measure together with any vector in $\mathcal{H}$ there corresponds a numerical measure m_f:

$$m_f(\triangle) := \langle E_K(\triangle)f, f\rangle = \|E_K(\triangle)f\|^2\,.$$

This is a non-negative, countably additive (which is the same as σ-additive) measure defined on the Borel field of $\mathbb{R}$. The support of the spectral family $\{E_K(\lambda) : \lambda \in \mathbb{R}\}$ is defined as the complement of the largest open subset U for which $E_K(U) = 0$.

A.37. Absolutely and singularly continuous subspaces

The Lebesgue decomposition theorem states that any measure on $\mathbb{R}$ has a unique decomposition into an absolutely continuous and a singular part with respect to Lebesgue measure $|\cdot|$. A vector $f \in \mathcal{H}$ is called absolutely continuous with respect to the self-adjoint operator K if the measure $\triangle \mapsto m_f(\triangle) = \|E_K(\triangle)f\|^2$ (see A.36) is absolutely continuous with respect to the Lebesgue measure on $\mathbb{R}$, i.e. if $|\triangle| = 0$ implies $m_f(\triangle) = 0$. The set of all absolutely continuous vectors is a closed subspace denoted by $\mathcal{H}_{\mathrm{ac}}(K)$. For $f \in \mathcal{H}_{\mathrm{ac}}(K)$ and for $g \in \mathcal{H}$, the function $\lambda \mapsto \langle E_K(\lambda)f, g\rangle$ is absolutely continuous.

A vector f is called singular with respect to the self-adjoint operator K if the measure $\triangle \mapsto m_f(\triangle)$ is singular with respect to the Lebesgue measure. This means that there exists a Borel set $\triangle_0$ of Lebesgue measure zero such that $m_f(\triangle) = m_f(\triangle \cap \triangle_0)$, for any Borel subset $\triangle$ of $\mathbb{R}$, or what amounts to saying the same thing, $E_K(\triangle_0) f = f$. The set of all singular vectors is denoted by $\mathcal{H}_s(K)$. Both subsets $\mathcal{H}_{ac}(K)$ and $\mathcal{H}_s(K)$ are closed subspaces of $\mathcal{H}$. Moreover these subspaces are complementary and orthogonal:

$$\mathcal{H} = \mathcal{H}_{ac}(K) \oplus \mathcal{H}_s(K).$$

In A.24 we already saw the decomposition

$$\mathcal{H} = \mathcal{H}_c(K) \oplus \mathcal{H}_p(K).$$

Upon comparing these decompositions one may prove the following inclusions:

$$\mathcal{H}_{ac}(K) \subseteq \mathcal{H}_c(K) \text{ and } \mathcal{H}_p(K) \subseteq \mathcal{H}_s(K).$$

In turn this gives rise to define the singular continuous subspace:

$$\mathcal{H}_{sc}(K) := \mathcal{H}_c(K) \ominus \mathcal{H}_{ac}(K).$$

A.38. Absolutely and singularly continuous spectra

The subspaces $\mathcal{H}_{ac}(K)$, $\mathcal{H}_{sc}(K)$, $\mathcal{H}_s(K)$ reduce the self-adjoint operator K. Set

$$\begin{aligned} K_{ac} &:= K \restriction \operatorname{dom}(K) \cap \mathcal{H}_{ac}(K), \\ K_{sc} &:= K \restriction \operatorname{dom}(K) \cap \mathcal{H}_{sc}(K), \text{ and} \\ K_s &:= K \restriction \operatorname{dom}(K) \cap \mathcal{H}_s(K). \end{aligned}$$

Then the absolutely continuous spectrum of K is defined by $\sigma_{ac}(K) = \sigma(K_{ac})$, its singularly continuous spectrum by $\sigma_{sc}(K) = \sigma(K_{sc})$, and its singular spectrum by $\sigma_s(K) = \sigma(K_s)$.

A.39. Invariance of the absolutely continuous spectra

The Birman-Kato theorem reads as follows. Let K_1 and K_2 be two self-adjoint operators in the Hilbert space $\mathcal{H}$ that are bounded below in the sense that $K_1 \geq -\omega_1 I$ and $K_2 \geq -\omega_2 I$ for some finite constants ω_1 and ω_2. Let the operator

$$(K_2 - zI)^{-p} - (K_1 - zI)^{-p}$$

be a trace class operator for some non-real z and for some $p \in \mathbb{N}$. Then

$$\sigma_{ac}(K_1) = \sigma_{ac}(K_2).$$

We conclude this appendix with a formulation of the Stein and the Riesz-Thorin interpolation theorem, a proof of which can be found in Reed and Simon [192, Theorem IX.21, page 40]. It was used in the proof of Theorem 2.5 and in Lemma 4.21.

A.40. Stein interpolation

Let $(E_j, \mathcal{A}_j, m_j)$, $j = 0, 1$, be σ-finite measure spaces, and fix $1 \leq p_0, p_1, q_0, q_1 \leq \infty$. Set

$$\frac{1}{p_t} = \frac{1-t}{p_0} + \frac{t}{p_1}, \quad \frac{1}{q_t} = \frac{1-t}{q_0} + \frac{t}{q_1}.$$

Let

$$T(z) : L^{p_0}(E_0, \mathcal{A}_0, m_0) + L^{p_1}(E_0, \mathcal{A}_0, m_0) \to L^{q_0}(E_1, \mathcal{A}_1, m_1) + L^{q_1}(E_1, \mathcal{A}_1, m_1)$$

be a uniformly bounded family of continuous linear operators on the strip $\{0 \leq \operatorname{Re} z \leq 1\}$, which is holomorphic on its interior. In addition, suppose

$$M_0 := \sup_{\xi \in \mathbb{R}} \|T(i\xi)\|_{q_0, p_0} < \infty, \quad \text{and} \quad M_1 := \sup_{\xi \in \mathbb{R}} \|T(1+i\xi)\|_{q_1, p_1} < \infty.$$

Then, for each $0 < t < 1$, $T(t) : L^{p_t}(E_0, \mathcal{A}_0, m_0) \to L^{q_t}(E_0, \mathcal{A}_0, m_0)$ and

$$\|T(t)\|_{q_t, p_t} \leq M_0^{1-t} M_1^t.$$

The space $L^{p_0}(E_0, \mathcal{A}_0, m_0) + L^{p_1}(E_0, \mathcal{A}_0, m_0)$ is supplied with the norm

$$f \mapsto$$
$$\inf \left\{ \|f_0\|_{p_0} + \|f_1\|_{p_1} : f = f_0 + f_1, f_0 \in L^{p_0}(E_0, \mathcal{A}_0, m_0),\ f_1 \in L^{p_1}(E_0, \mathcal{A}_0, m_0) \right\}.$$

If $T(z) = T$ for all z, $0 \leq \operatorname{Re} z \leq 1$, then we get the Riesz-Thorin interpolation theorem.

Appendix B

Semigroup Theory

In this appendix we define several notions having their origin in operator semigroup theory. We confine ourselves to properties which are used in the present book. The literature on the subject is huge. The material in this appendix is taken from the following textbooks: Bliedtner and Hansen [34], Clement et. al. [54], Davies [65], Fukushima et al [110], Goldstein [116], Jacob [143], Ma and Röckner [170], Nagel et. al. [182], Reed and Simon [192], and Robinson [198].

B.1. Strongly continuous, contractive semigroups

A family of bounded linear operators $\{P(t) : t \geq 0\}$ is called a *strongly continuous semigroup* in a Banach space $(X, \|\cdot\|)$, if it possesses the following properties:

(i) $P(s + t) = P(s)P(t)$ for all s, $t \geq 0$;
(ii) $P(0) = I$;
(iii) For every $f \in X$ the mapping $t \mapsto P(t)f$ is continuous from $[0, \infty)$ to $(X, \|\cdot\|)$.
In the presence of (i) and (ii) property (iii) is equivalent to
(iii′) For every $x \in X$ and for every $x^* \in X^*$,the dual space of X, the following identity is valid:

$$\lim_{t\downarrow 0} \langle P(t)x, x^* \rangle = \langle x, x^* \rangle .$$

For this see e.g. Yosida [255, p. 233].

For every strongly continuous semigroup $\{P(t) : t \geq 0\}$ there exist constants $M \geq 1$ and $b \in \mathbb{R}$ such that

$$\|P(t)\| \leq Me^{bt}, \quad t \geq 0.$$

The infimum of all possible real numbers b is called the *type* or *growth rate* of the semigroup under consideration. The semigroup $\{P(t) : t \geq 0\}$ is called a *contraction* semigroup, if the above inequality is valid for $M = 1$ and $b = 0$, i.e. if

$$\|P(t)\| \leq 1, \text{ for } t \geq 0.$$

If X is a Hilbert space and if every operator $P(t)$ is self-adjoint, then M may be chosen to be equal to one.

B.2. Generators

To every semigroup $\{P(t) : t \geq 0\}$ one may associate an operator $\{\text{dom}(K), K\}$, with domain $\text{dom}(K)$, which is called the (infinitesimal) *generator* of the semigroup. Its domain is defined by

$$\text{dom}(K) = \left\{ f \in X : \lim_{t\downarrow 0} \frac{1}{t} (f - P(t)f) \text{ exists in } X \right\}.$$

Then

$$Kf := \lim_{t \downarrow 0} \frac{1}{t}(f - P(t)f),$$

for $f \in \mathrm{dom}(K)$. The generator K is densely defined and has a closed graph. It possesses the following properties:

(i) $f \in \mathrm{dom}(K)$ implies $P(t)f \in \mathrm{dom}(K)$ for every $t \geq 0$;

(ii) If f belongs to $\mathrm{dom}(K)$, then the mapping $t \mapsto P(t)f$ is differentiable on $[0, \infty)$ in the strong sense. For $f \in \mathrm{dom}(K)$ the following identities hold:

$$\frac{d}{dt}P(t)f = -KP(t)f = -P(t)Kf.$$

Therefore we say that K generates the semigroup $\{P(t) : t \geq 0\}$ and we write $P(t) = \exp(-tK)$.

(iii) For any $g \in X$, and for $t > 0$, the vector $\int_0^t P(s)g\,ds$ belongs to $\mathrm{dom}(K)$, and

$$K \int_0^t P(s)g\,ds = g - P(t)g.$$

B.3. Resolvents

Let K be the generator of the strongly continuous semigroup $\{P(t) : t \geq 0\}$ on a Banach space X. Suppose that, for some $\lambda \in \mathbb{C}$, the X-valued integral $\int_0^\infty e^{-\lambda s}P(s)f\,ds$ exists for all $f \in X$. Then $-\lambda$ belongs to the resolvent set of K (see Appendix A.20) and

$$(\lambda I + K)^{-1} f = \int_0^\infty e^{-\lambda s} \exp(-sK)\, f\,ds.$$

The Hille-Yosida theorem gives sufficient and necessary conditions, in order that a closed linear operator generates a strongly continuous semigroup.

A closed linear operator K on a Banach space X generates a contraction semigroup if and only if

(i) The interval $(-\infty, 0)$ is a subset of $\mathrm{res}(K)$;

(ii) The inequality $a\left\|(aI + K)^{-1}\right\| \leq 1$ holds for all $a > 0$.

If (i) and (ii) are satisfied, then the entire open half-plane $\{\lambda \in \mathbb{C} : \mathrm{Re}\,\lambda < 0\}$ is contained in the resolvent set of K and equality (B.3) holds on that open half-plane.

B.4. Myadera perturbations

Let $\{\exp(-tK_0) : t \geq 0\}$ be a strongly continuous semigroup in a Banach space $(X, \|\cdot\|)$. Let B be a linear operator in X with a dense domain $\mathrm{dom}(B)$. Suppose that the following conditions are satisfied:

(i) $\operatorname{dom}(B) \subseteq \operatorname{dom}(K_0)$, $\exp(-tK_0)\operatorname{dom}(B) \subseteq \operatorname{dom}(B)$, $t > 0$, and for all $f \in \operatorname{dom}(B)$, the function $t \mapsto B\exp(-tK_0)f$ is continuous from $[0,\infty)$ to $(X, \|\cdot\|)$.

(ii) There exist constants $\alpha > 0$ and γ, $0 \leq \gamma < 1$, such that for all $f \in \operatorname{dom} B$ the inequality

$$\int_0^\alpha \|B\exp(-sK_0)f\|\, ds \leq \gamma \|f\|$$

is valid.

Then the following assertions can be proved:

(a) The operator B is $(K_0 \restriction \operatorname{dom}(B))$-bounded with $(K_0 \restriction \operatorname{dom}(B))$-bound less than or equal to γ. This means that for every $\varepsilon > 0$ there exists a finite constant $M(\varepsilon)$ such that for every $f \in \operatorname{dom}(B)$ the inequality

$$\|Bf\| \leq M(\varepsilon)\|f\| + \|K_0 f\|$$

is valid.

(b) The operator B extends to a unique K_0-bounded linear operator $\widehat{B}$, with $\operatorname{dom}\left(\widehat{B}\right) = \operatorname{dom}(K_0)$, satisfying the condition

$$\limsup_{\lambda\to\infty} \left\|\widehat{B}\left(\lambda I + K_0\right)^{-1}\right\| \leq \gamma.$$

(c) For all $f \in \operatorname{dom}(K_0)$ the inequality

$$\int_0^\alpha \left\|\widehat{B}\exp(-sK_0)f\right\| ds \leq \gamma\|f\|$$

holds.

(d) If the operator B is closable, then its closure extends $\widehat{B}$.

(e) The operator $K_0 + \widehat{B}$ coincides with the closure of $K_0 + B$ and it is the generator of a strongly continuous semigroup $\{\exp(-t(K_0 + B)) : t \geq 0\}$. There exist constants $M \geq 1$ and b depending on K_0, α, and γ, such that for all $t \geq 0$,

$$\|\exp(-t(K_0 + B))\| \leq M\exp(bt).$$

If the operator K_0 equals $H_0 := -\frac{1}{2}\Delta$ in $L^1(\mathbb{R}^\nu)$, and if B denotes a multiplication operator by a function W, then

$$\int_0^\alpha \|W\exp(-sH_0)f\|_1\, ds \leq \gamma\|f\|$$

if and only if

$$\sup_{x\in\mathbb{R}^\nu} \int_0^\alpha ds \int dy\, p_{0,\nu}(s,x,y)\,|W(y)| \leq \gamma.$$

Such a function W could be called a *Myadera potential*. This result should also be compared with section A of Chapter 2, where Kato functions are introduced. For details the reader is referred to Voigt [247, Theorem 1.1. p. 169–170].

B.5. Trotter product formula

In many textbooks some version of the Trotter product formula is given. A rather general result is given by Kato [151].

Let K_0 and B be non-negative self-adjoint operators in a separable Hilbert space $\mathcal{H}$. Put $\mathcal{D} := \operatorname{dom}\left(K_0^{1/2}\right) \cap \operatorname{dom}\left(B^{1/2}\right)$. Let K be the self-adjoint linear operator associated to the closed quadratic form:

$$f \mapsto \left\|K_0^{1/2} f\right\|^2 + \left\|B^{1/2} f\right\|^2, \quad f \in \mathcal{D},$$

and suppose that K is densely defined in $\overline{\mathcal{D}}$, the closure of $\mathcal{D}$. Then the strong limit

$$\text{s-}\lim_{n\to\infty}\left(\exp\left(-\frac{t}{n}K_0\right)\exp\left(-\frac{t}{n}B\right)\right)^n$$

exists. Moreover this limit is given by

$$\text{s-}\lim_{n\to\infty}\left(\exp\left(-\frac{t}{n}K_0\right)\exp\left(-\frac{t}{n}B\right)\right)^n = \exp\left(-tK\right)P,$$

where P is the orthogonal projection from $\mathcal{H}$ onto $\overline{\mathcal{D}}$. (Kato proved this for even more general functions than the exponential function.)

B.6. Duhamel's formula

Let K_0 be the infinitesimal generator of the strongly continuous semigroup $\{\exp(-tK_0) : t \geq 0\}$. Let B be a bounded linear operator that is everywhere defined on X. Then the operator $K := K_0 + B$ is a closed linear operator with $\operatorname{dom}(K) = \operatorname{dom}(K_0)$, that generates a strongly continuous semigroup $\{\exp(-tK) : t \geq 0\}$. In such a case the perturbed semigroup and the unperturbed semigroup are related as follows:

$$\begin{aligned}
\exp(-tK)f &= \exp(-tK_0) - \int_0^t \frac{d}{ds}\left[\exp(-(t-s)K)\exp(-sK_0)f\right]ds \\
&= \exp(-tK_0) - \int_0^t \left[\exp(-(t-s)K)(K-K_0)\exp(-sK_0)f\right]ds \\
&= \exp(-tK_0) - \int_0^t \left[\exp(-(t-s)K)B\exp(-sK_0)f\right]ds.
\end{aligned}$$

If the operator B is not bounded, then one has to ensure that $\operatorname{ran}(\exp(-sK_0))$ is contained in the domain of K. In the Hilbert space context, we may interpret Duhamel's formula in form sense as follows. Suppose $K = K^*$ and $K_0^* = K_0$ are not necessarily bounded self-adjoint operators in a Hilbert space $\mathcal{H}$. Then, for f belonging to $\operatorname{dom}(K)$ and g to $\operatorname{dom}(K_0)$ we have

$$\begin{aligned}
\langle \exp(-tK)f, g\rangle \langle \exp(-tK_0)f, g\rangle = \int_0^t \Big(&\langle \exp(-(t-s)K)f, \exp(-sK_0)K_0 g\rangle \\
&- \langle \exp(-(t-s)K)Kf, \exp(-sK_0)g\rangle\Big)ds.
\end{aligned}$$

B.7. Sub-Markovian and positivity preserving semigroups

Let $(E, \mathcal{E}, m)$ be a σ-finite measure space. Let $\{P(t) : t \geq 0\}$ be a strongly continuous semigroup defined on $L^2(E, m)$. The semigroup is called *sub-Markovian* (or sometimes just Markovian), if for all $f \in L^2(E, m)$, for which $0 \leq f \leq 1$, we have $0 \leq P(t)f \leq 1$. The generator of a sub-Markovian semigroup is called a *Dirichlet* operator: see e.g. Ma and Röckner [170, p. 31]. The operator $P(t)$ is called *positivity preserving* if $P(t)f \geq 0$ m-almost everywhere, whenever the function $f \in L^2(E, m)$ is so.

In the remainder of this appendix we mainly consider symmetric strongly continuous semigroups.

B.8. Symmetric and conservative (sub-)Markov semigroups

The strongly continuous semigroup $\{P(t) : t \geq 0\}$ on $L^2(E, m)$ is called *symmetric* if the linear bounded operators $P(t)$, $t \geq 0$, are symmetric. This property is equivalent to the self-adjointness of its generator K. It is a symmetric sub-Markov semigroup if $0 \leq P(t)f \leq 1$, m-almost every, for all $f \in L^2(E, m)$ for which $0 \leq f \leq 1$. Let the (form) positive operator K in $L^2(E, m)$ be the generator of the symmetric sub-Markov semigroup $\{\exp(-tK) : t \geq 0\}$. This semigroup can be regarded as a positivity preserving contractive semigroup in the spaces $L^p(E, m)$, $1 \leq p \leq \infty$. For $1 \leq p < \infty$, it is strongly continuous: see e.g. Davies [65, p. 22].

The semigroup $\{\exp(-tK) : t \geq 0\}$ is called conservative if $\exp(-tK)1 = 1$ for all $t \geq 0$. Let the semigroup $\{\exp(-tK) : t \geq 0\}$ be a symmetric sub-Markov semigroup in $L^2(E, m)$, $1 \leq p \leq \infty$. Denote its resolvent by

$$(aI + K)^{-1} f = \int_0^\infty e^{-as} \exp(-sK) f \, ds.$$

Then the semigroup $\{\exp(-tK) : t \geq 0\}$ is conservative if and only if any one of the following equivalent conditions is satisfied:

(i) $P(t)1 = 1$ for some $t > 0$;
(ii) $a(aI + K)^{-1} 1 = 1$ for some $a > 0$;
(iii) $a(aI + K)^{-1} 1 = 1$ for all $a > 0$.

Additional results on conservative semigroups may be found for instance in Jacob [143, p. 100 ff].

B.9. Irreducible semigroups

Let $\{\exp(-tK) : t \geq 0\}$ be a symmetric positivity preserving semigroup on the space $L^2(E, m)$. A Borel subset Ω of E is called *invariant* if for every $f \in L^2(E, m)$ of the form $f = f 1_\Omega$ the equality $\exp(-tK)f = (\exp(-tK)f) 1_\Omega$ is m-almost every valid for all $t \geq 0$ simultaneously. The semigroup $\{\exp(-tK) : t \geq 0\}$ is called irreducible if the only invariant sets are E and $\emptyset$ (up to m-negligible sets). If every operator $\exp(-tK)$ is an integral operator with a strictly positive integral kernel, then the semigroup is irreducible: see Davies [65, p. 24].

B.10. Transient and recurrent semigroups

Let $\{\exp(-tK) : t \geq 0\}$ be a symmetric sub-Markov semigroup in the space $L^2(E, m)$. Set

$$S(t)g := \int_0^t \exp(-sK)\, g\, ds, \quad \text{for } g \in L^2(E, m).$$

The semigroup $\{\exp(-tK) : t \geq 0\}$ is called *transient* if, for every $f \in L^1(E, m)$, $f \geq 0$, the limit

$$Gf := L^1(E, m)\text{-}\lim_{t \to \infty} S(t)f$$

exists. Roughly speaking, the operator G is the resolvent at the point zero. The integral kernel of the operator G, if it exists, is called Green's function.

The semigroup $\{\exp(-tK) : t \geq 0\}$ is called *recurrent* if, for any $f \in L^1(E, m)$, $f \geq 0$, either $Gf(x) = 0$ for m-almost all $x \in E$ or $Gf(x) = \infty$ for m-almost all $x \in E$.

If the semigroup $\{\exp(-tK) : t \geq 0\}$ is irreducible (see B.9), then it is either transient or recurrent with $Gf(x) = \infty$ m-almost everywhere, whenever $f \geq 0$ belongs to $L^1(E, m)$ and f does not vanish m-almost everywhere.

B.11. Feller semigroups

A family $\{P(t) : t \geq 0\}$ of operators defined on $L^\infty(E, \mathcal{E})$, the space of bounded Borel functions, is a *Feller semigroup* on $C_\infty(E)$ (see Definition 1.2) if it possesses the following properties:

(i) It leaves $C_\infty(E)$ invariant: $P(t)C_\infty(E) \subseteq C_\infty(E)$ for $t \geq 0$;
(ii) It is a semigroup: $P(s+t) = P(s) \circ P(t)$ for all s, $t \geq 0$, and $P(0) = I$;
(iii) It consists of contraction operators: $\|P(t)f\|_\infty \leq \|f\|_\infty$ for all $t \geq 0$ and for all $f \in C_\infty(E)$;
(iv) It is positivity preserving: $f \geq 0$, $f \in C_\infty(E)$, implies $P(t)f \geq 0$;
(v) It is continuous in $t = 0$: $\lim_{t \downarrow 0} [P(t)f](x) = f(x)$, for all $f \in C_\infty(E)$ and for all $x \in E$.
In the presence of (iii) and (ii), property (v) is equivalent to:
(v′) $\lim_{t \downarrow 0} \|P(t)f - f\|_\infty = 0$ for all $f \in C_\infty(E)$. So that a Feller semigroup is in fact *strongly continuous* in the sense that, for every $f \in C_\infty(E)$,

$$\lim_{s \to t} \|P(s)f - P(t)f\|_\infty = 0.$$

It is noted that the properties (iii) and (iv) can be combined into one:
(iii′) For all $f \in C_\infty(E)$, for which $0 \leq f \leq 1$, we have $0 \leq P(t)f \leq 1$.

B.12. Kernels of semigroups, Theorem of Dunford-Pettis

Let $\{\exp(-tK) : t \geq 0\}$ be a symmetric sub-Markov semigroup on $L^p(E, m)$, $1 \leq p < \infty$. Suppose that every operator $\exp(-tK)$ is a bounded operator from

$L^p(E,m)$ to $L^\infty(E,m)$. Then there exists a family of kernels $\{K_t(x,y) : t > 0\}$ such that

$$\exp(-tK)f(x) = \int_E K_t(x,y)f(y)dm(y) \ m\text{-almost everywhere and}$$

$$\|\exp(-tK)\|_{\infty,p} = \sup_{x\in E}\left(\int_E |K_t(x,y)|^q\, dm(y)\right)^{1/q}, \text{ for } p > 1,$$

$$\|\exp(-tK)\|_{\infty,1} = \sup_{x,y\in E} K_t(x,y),$$

with $1 = \frac{1}{p} + \frac{1}{q}$. In the main text, of the book we have used the notation the $P(t) = \exp(-tK)$ and $K_t(x,y) = \exp(-tK)(x,y)$.

Remark. In the main text of the book, the symbol $\|f\|_\infty$ denotes the essential supremum of the function $f \in L^\infty(E,\mathcal{E})$. Also, if we write $\sup_{x\in E}|f(x)|$ we in fact mean to take the essential supremum of the function f. Of course if f is a continuous function, then the notions of supremum and essential supremum coincide. In particular this is the case for most of the integral kernels considered here, e.g. $\exp(-t(K_0 + V))(x,y)$: see Theorem 2.5.(a).

B.13. Ultracontractivity

The symmetric sub-Markovian semigroup $\{\exp(-tK) : t \geq 0\}$ is called *ultracontractive* if every operator $\exp(-tK)$, $t > 0$, is a bounded operator from $L^2(E,m)$ to $L^\infty(E,m)$, i.e. if $\|\exp(-tK)\|_{\infty,2} < \infty$, for $t > 0$. Consequently (see B.12) every operator $\exp(-tK)$ is an integral operator: see Davies [65, p. 59 ff]. Since, by symmetry,

$$\begin{aligned}\|\exp(-2tK)\|_{\infty,1} &= \|\exp(-tK)\exp(-tK)\|_{\infty,1}\\ &\leq \|\exp(-tK)\|_{\infty,2}\|\exp(-tK)\|_{2,1} \leq \|\exp(-tK)\|_{\infty,2}^2,\end{aligned}$$

we see that a symmetric, ultracontractive semigroup is L^1-L^∞-smoothing.

B.14. Kernel estimates

Let K be a strictly elliptic operator on $L^2(\mathbb{R}^\nu)$. Then the operator $\exp(-tK)$, $t > 0$, possesses an integral kernel $\exp(-tK)(x,y)$ for which, for $m\otimes m$-almost all $(x,y) \in \mathbb{R}^\nu \times \mathbb{R}^\nu$ and for all $t > 0$, $0 \leq \exp(-tK)(x,y) \leq ct^{-\nu/2}$. Upper and lower bounds are studied e.g. by Davies [65, p. 82 ff].

B.15. Dyson expansions

Let K_0, and $K = K_0 + B$ as in B.6. From Duhamel's formula together with induction we get the following Dyson expansion:

$$\begin{aligned}
&\exp\left(-t\left(K_0+B\right)\right)-\exp\left(-tK_0\right)\\
&=\sum_{k=1}^{n}(-1)^k\int_{\substack{0<s_1,\dots,s_k\\ s_1+\dots+s_k<t}}\int ds_1\dots ds_k\exp\left(-s_1K_0\right)B\exp\left(-s_2K_0\right)\cdots\\
&\quad\exp\left(-s_kK_0\right)B\exp\left(-\left(t-\left(s_1+\dots+s_k\right)\right)K_0\right)\\
&\quad+(-1)^{n+1}\int_{\substack{0<s_1,\dots,s_{n+1}\\ s_1+\dots+s_{n+1}<t}}\int ds_1\dots ds_{n+1}\exp\left(-s_1K_0\right)B\exp\left(-s_2K_0\right)\cdots\\
&\quad\exp\left(-s_{n+1}K_0\right)B\exp\left(-\left(t-\left(s_1+\dots+s_{n+1}\right)\right)K_0\right)\\
&=\sum_{k=1}^{\infty}(-1)^k\int_{0<s_1,\dots,s_k,\ s_1+\cdots s_k<t}\int ds_1\dots ds_k\exp\left(-s_1K_0\right)B\exp\left(-s_2K_0\right)\cdots\\
&\quad\exp\left(-s_kK_0\right)B\exp\left(-\left(t-\left(s_1+\dots+s_k\right)\right)K_0\right).
\end{aligned}$$

At the level of integral kernels, this formula reads as follows:

$$\begin{aligned}
&\exp\left(-t\left(K_0+B\right)\right)(x,y)-\exp\left(-tK_0\right)(x,y)\\
&=\sum_{k=1}^{n}(-1)^k\int_{\substack{0<s_1,\dots,s_k\\ s_1+\dots+s_k<t}}\int\int\dots\int ds_1\dots ds_k dx_1\dots dx_k\\
&\quad\exp\left(-s_1K_0\right)(x,x_1)\left[B\exp\left(-s_2K_0\right)(\cdot,x_2)\right](x_1)\cdots\\
&\quad\exp\left(-s_kK_0\right)(x_{k-1},x_k)\left[B\exp\left(-\left(t-\left(s_1+\dots+s_k\right)\right)K_0\right)(\cdot,y)\right](x_k).
\end{aligned}$$

If B is multiplication by a bounded Borel function V, then we get the Feynman-Kac formula. From the Markov property (see Appendix C.14) we get

$$\begin{aligned}
&\left[\exp\left(-s_1K_0\right)V\exp\left(-s_2K_0\right)\dots\exp\left(-s_kK_0\right)V\exp\left(-\left(t-s_1-\dots-s_k\right)K_0\right)f\right](x)\\
&=\mathbb{E}_x\left[\prod_{j=1}^{k}V\left(X\left(\sum_{i=1}^{j}s_i\right)\right)\times f(X(t))\right],
\end{aligned}$$

and hence

$$\begin{aligned}
&\left[\exp\left(-t\left(K_0\dot{+}V\right)\right)f\right](x)=\sum_{k=0}^{\infty}\frac{(-1)^k}{k!}\mathbb{E}_x\left[\left(\int_0^t V(X(s))ds\right)^k\times f(X(t))\right]\\
&=\mathbb{E}_x\left[\exp\left(-\int_0^t V(X(s))ds\right)f(X(t))\right].
\end{aligned}$$

B.16. Spectral theory of semigroups

In the book edited by Nagel and written by Arendt et. al. [182] one may find three articles by G. Greiner dealing with spectral theory of positive semigroups on Banach spaces, spectral theory of positivity preserving semigroups on $C_\infty(E)$, and spectral theory of positive semigroups on Banach lattices.

Appendix C

Markov Processes, Martingales and Stopping Times

Let $(\Omega, \mathcal{F}, \mathbb{P})$ be a probability space and let $\{\mathcal{F}_t : t \geq 0\}$ be an increasing family of σ-fields in $\mathcal{F}$. If necessary we suppose that the filtration $\{\mathcal{F}_t : t \geq 0\}$ is right continuous, i.e. $\mathcal{F}_t = \bigcap_{s>t} \mathcal{F}_s$, or complete in the sense that, for every $t > 0$, the σ-field $\mathcal{F}_t$ contains all $A \in \mathcal{F}$, with $\mathbb{P}(A) = 0$.

Let $\{H(t) : t \geq 0\}$ be a collection of $\mathbb{R}$-valued functions defined on Ω and measurable with respect to $\mathcal{F}$ and the Borel field of $\mathbb{R}$. Of course each function $H(t)$ on its own is called a *random variable*. Such a family is called a (real-valued) stochastic process. For more details on the items mentioned in C.1 and C.2, the reader is referred to Sharpe [205, Definition 5.2 page 20]. For more details on (semi-)martingale theory, the reader is referred to Dellacherie and Meyer [72, Chapter V (case of discrete time) and Chapter VI (case of continuous time)], to Durrett [93, Chapter §2.2., 2.3., 2.4.], to Gihman and Skorohod [111, part III, Chapter 1, §1], to Revuz and Yor [195, Chapter II], to Ethier and Kurtz [99, Chapter II], or to Karatzas and Shreve [147, Chapter 1].

C.1. Adapted processes

(a) The process $\{H(t) : t \geq 0\}$ is said to be *adapted* or *non-anticipating* if, for every $t \geq 0$, the variable $H(t)$ is measurable with respect to $\mathcal{F}_t$.

(b1) The symbol Λ denotes the σ-field ($=$ σ-algebra) of subsets of $[0,\infty) \times \Omega$, which is generated by the adapted processes which are right-continuous and which possess left limits. These are the so-called *cadlag processes.*

(b2) The process $\{H(t) : t \geq 0\}$ is said to be *optional* if the function $(t, \omega) \to H(t, \omega)$ is measurable with respect to Λ.

(c1) The symbol Π denotes the σ-field of subsets of $[0,\infty) \times \Omega$, which is generated by the adapted processes which are left-continuous adapted processes.

(c2) The process $\{H(t) : t \geq 0\}$ is said to be *predictable* if the function $(t, \omega) \to H(t, \omega)$ is measurable with respect to Π.

(d) A mapping $T : \Omega \to [0, \infty]$ is called an $\mathcal{F}_t$*-stopping* time if for all $t \geq 0$ the event $\{T \leq t\}$ belongs to the σ-field $\mathcal{F}_t$. The corresponding σ-field $\mathcal{F}_T$ is defined by

$$\mathcal{F}_T = \bigcap_{t \geq 0} \{A \in \mathcal{F} : A \cap \{T \leq t\} \in \mathcal{F}_t\}.$$

C.2. Properties of adapted processes

(a) The σ-field Π (see (c1) in C.1) is generated by the set

$$\{(a, b] \times A : 0 \leq a < b, A \in \mathcal{F}_a\}.$$

(b) A mapping $T : \Omega \to [0, \infty]$ is an $\mathcal{F}_t$-stopping time if and only if the process $t \mapsto 1_{[0,t]} \circ T$ is adapted.

(c) The field Π is a subfield of Λ.

Proof. (a) Let A belong to $\mathcal{F}_a$. The variable $\omega \mapsto 1_{(a,b]}(s)1_A(\omega)$ is measurable with respect to $\mathcal{F}_s$ and the function $s \mapsto 1_{(a,b]}(s)1_A(\omega)$ is left continuous. This proves that $(a, b] \times A$ belongs to Π.

Conversely, let F be adapted and left continuous. Set

$$F_n(s, \omega) = \sum_{k=0}^{\infty} F\left(k2^{-n}, \omega\right) 1_{(k2^{-n},(k+1)2^{-n}]}(s).$$

Then, by left continuity, $\lim_{n\to\infty} F_n(s, \omega) = F(s, \omega)$, $\mathbb{P}$-almost surely. Moreover the processes $\{F_n(t) : t \geq 0\}$ are adapted and are measurable with respect to the σ-field generated by $\{(a, b] \times A : 0 \leq a < b, A \in \mathcal{F}_a\}$.

(b) This assertion is easy to prove.

(c) Since $1_{(a,b]}(s) \times 1_A(\omega) = \lim\limits_{n\to\infty} 1_{[a_n,b_n)}(s)1_A(\omega)$, where $a_n \downarrow a$ and where $b_n \downarrow b$, it follows that $\Pi \subseteq \Lambda$. Here we employ assertion (a).

REMARK. Compare C.2. with Exercise 7 page 43 of Chung [49] where a similar result is stated.

C.3. Martingales

Let $\{X(t) : t \geq 0\}$ be a real-valued adapted process. We define the following "martingale"-like processes.

(a) The family $\{X(t) : t \geq 0\}$ is a *martingale* if $\mathbb{E}(|X(t)|) < \infty$, $t \geq 0$, and if, for every $t > s \geq 0$, $\mathbb{E}(X(t) \mid \mathcal{F}_s) = X(s)$, $\mathbb{P}$-almost surely.

(b) The family $\{X(t) : t \geq 0\}$ is a *submartingale* if $\mathbb{E}(|X(t)|) < \infty$, $t \geq 0$, and if, for every $t > s \geq 0$, $\mathbb{E}(X(t) \mid \mathcal{F}_s) \geq X(s)$, $\mathbb{P}$-almost surely.

(c) The family $\{X(t) : t \geq 0\}$ is a *supermartingale* if $\mathbb{E}(|X(t)|) < \infty$, $t \geq 0$, and if, for every $t > s \geq 0$, $\mathbb{E}(X(t) \mid \mathcal{F}_s) \leq X(s)$, $\mathbb{P}$-almost surely.

(d) The family is $\mathbb{P}$-almost surely of *bounded variation* (on $[0, t]$) if

$$\sup\left\{\sum_{j=1}^{n} |X(t_j) - X(t_{j-1})| : 0 \leq t_0 < t_1 < \ldots < t_n < \infty\right\} < \infty,$$
$$\left(\sup\left\{\sum_{j=1}^{n} |X(t_j) - X(t_{j-1})| : 0 \leq t_0 < t_1 < \ldots < t_n \leq t\right\} < \infty,\right)$$

$\mathbb{P}$-almost surely. The supremum is taken of all possible subdivisions of the interval $[0, \infty)$ ($[0, t]$).

(e) The family is a *local martingale* if there exists an increasing sequence of stopping times $(T_n : n \in \mathbb{N})$ for which $\lim_{n\to\infty} T_n = \infty$, $\mathbb{P}$-almost surely, and for which the processes $\{X(T_n \wedge t)(t) : t \geq 0\}$, $n = 1, 2, \ldots$ are martingales with respect to the filtration $\{\mathcal{F}_{T_n \wedge t} : t \geq 0\}$.

(f) Let T be a stopping time. The process $\{X(t) : t \geq 0\}$ is a *local martingale on* $[0, T)$ if there exists an increasing sequence of stopping times $(T_n : n \in \mathbb{N})$ for which $\lim_{n\to\infty} T_n = T$, $\mathbb{P}$-almost surely, and for which the processes $\{X(T_n \wedge t)(t) : t \geq 0\}$, $n = 1, 2, \ldots$ are martingales with respect to the filtration $\{\mathcal{F}_{T_n \wedge t} : t \geq 0\}$.

(g) The definition of "local submartingale", "local supermartingale" and "being locally $\mathbb{P}$-almost surely of bounded variation" are now self-explanatory.

(h) The process $\{X(t) : t \geq 0\}$ is called a *semi-martingale* if it can be written in the form $X(t) = M(t) + A(t)$, where $\{M(t) : t \geq 0\}$ is a martingale and where $\{A(t) : t \geq 0\}$ is an adapted process which is of bounded variation, $\mathbb{P}$-almost surely, on $[0, t]$ for every $t > 0$, and for which $\mathbb{E}\,|A(t)| < \infty$, $t \geq 0$.

(i) The process $\{X(t) : t \geq 0\}$ is of class (DL) if for every $t > 0$ the family

$$\{X(\tau) : 0 \leq \tau \leq t, \tau \text{ is a } (\mathcal{F}_t)\text{-stopping time}\}$$

is uniformly L^1-integrable.

A family $\{f_\alpha : \alpha \in \mathcal{A}\}$ of functions in $L^1(\Omega, \mathcal{F}, \mathbb{P})$ is *uniformly L^1-integrable* or *equi-integrable* if for every $\varepsilon > 0$ there exists a function $f \in L^1(\Omega, \mathcal{F}, \mathbb{P})$, $f \geq 0$, such that

$$\sup_{\alpha \in \mathcal{A}} \int_{\{|f_\alpha| \geq f\}} |f_\alpha| \, d\mathbb{P} \leq \varepsilon.$$

Since $\mathbb{P}$ is a finite positive measure, the latter is equivalent to

$$\lim_{m\to\infty} \sup_{\alpha \in \mathcal{A}} \int_{\{|f_\alpha| \geq m\}} |f_\alpha| \, d\mathbb{P} = 0.$$

REMARK 1. Let $\{X(t) : t \geq 0\}$ be a semi-martingale. The decomposition $X(t) = M(t) + A(t)$, where $\{M(t) : t \geq 0\}$ is a martingale and where for every $t > 0$ the process $\{A(t) : t \geq 0\}$ is $\mathbb{P}$-almost surely of bounded variation and where $\{A(t) : t \geq 0\}$ is predictable and right continuous $\mathbb{P}$-almost surely, is unique provided $A(0) = 0$, $\mathbb{P}$-almost surely. This follows from the fact that a right-continuous martingale which is predictable and of bounded variation is necessarily constant which, in turn, follows from the uniqueness part of the Doob-Meyer decomposition: see e.g. Karatzas and Shreve [147, Theorem 4.10, p. 24] or Ethier and Kurtz [99, Theorem 5.1, page 74].

REMARK 2. An $\mathcal{F}_t$-martingale $\{M(t) : t \geq 0\}$ is of class (DL). An increasing adapted process $\{A(t) : t \geq 0\}$ in $L^1(\Omega, \mathcal{F}, \mathbb{P})$ is also of class (DL). Hence, the sum $\{M(t) + A(t) : t \geq 0\}$ is of class (DL). If $\{X(t) : t \geq 0\}$ is a submartingale and if μ is a real number, then the process $\{\max(X(t), \mu) : t \geq 0\}$ is a submartingale of class (DL). Processes of class (DL) are important in the Doob-Meyer decomposition theorem.

C.4. Examples of martingales

We continue with some examples of martingales, submartingales and the like.

EXAMPLE 1. Let $T : \Omega \to [0, \infty]$ be a stopping time. The process $t \mapsto 1_{\{T<t\}}$, $t \geq 0$, is left continuous and so it is predictable. It also follows that the process $t \mapsto 1_{\{T \geq t\}}$, $t \geq 0$, is predictable.

EXAMPLE 2. Let I be an open interval in $\mathbb{R}$ and let $\varphi : I \to (-\infty, \infty)$ be an increasing convex function. If $\{X(t) : t \geq 0\}$ is a submartingale with values in I, then the process $\{\varphi(X(t)) : t \geq 0\}$ is also a submartingale. Let $t > s \geq 0$. Then by the Jensen inequality and the monotonicity of φ it follows that

$$\mathbb{E}\left(\varphi(X(t)) \mid \mathcal{F}_s\right) \geq \varphi\left(\mathbb{E}\left(X(t) \mid \mathcal{F}_s\right)\right) \geq \varphi\left(X(s)\right).$$

EXAMPLE 3. Let $(B(t), \mathbb{P}_0)$ be one-dimensional Brownian motion starting at 0. Then $\{B(t) : t \geq 0\}$ is a martingale. Since the definition of martingale also makes sense for vector valued processes, we also see that an $\mathbb{R}^\nu$-valued Brownian motion is a martingale.

EXAMPLE 4. Let $(B(t), \mathbb{P}_0)$ be $\mathbb{R}^\nu$-valued Brownian motion starting at 0. The process $\left\{|B(t)|^2 - \nu t : t \geq 0\right\}$ is a martingale.

EXAMPLE 5. Let $\{X(t), \mathbb{P}_x\}$ be a (strong) Markov process (see C.14) such that

$$\mathbb{E}_x\left(f(X(t))\right) = \int p_0(t, x, y) f(y) dm(y), \quad f \geq 0,$$

where the density $p_0(t, x, y)$ satisfies the Chapman-Kolmogorov identity:

$$p_0(s + t, x, y) = \int p_0(s, x, z) p_0(t, z, y) dm(z).$$

For the definition of a Markov process, we refer the reader to Definition C.14. The process $\{p(t - s, X(s), y) : 0 \leq s < t\}$ is a martingale on $[0, t)$. For example, for $X(t)$ we may take $B(t)$, ν-dimensional Brownian motion. Then

$$p_0(t, x, y) = \frac{1}{\left(\sqrt{2\pi t}\right)^\nu} \exp\left(-\frac{|x - y|^2}{2t}\right).$$

A proof of this martingale property reads as follows ($0 \leq s_1 < s_2$):

$$\begin{aligned}
& \mathbb{E}_x\left[p_0\left(t - s_2, X(s_2), y\right) \mid \mathcal{F}_{s_1}\right] \\
\text{(Markov property)} \quad & = \mathbb{E}_{X(s_1)}\left\{p_0\left(t - s_2, X(s_2 - s_1), y\right)\right\} \\
& = \int p_0\left(s_2 - s_1, X(s_1), z\right) p_0\left(t - s_2, z, y\right) dz \\
\text{(Chapman-Kolmogorov)} \quad & = p_0\left(t - s_1, X(s_1), y\right).
\end{aligned}$$

EXAMPLE 6. Let $\{X(t) : t \geq 0\}$ be a right-continuous martingale and let T be a stopping time. The process $\{X(T \wedge t) : t \geq 0\}$ is a martingale with respect to $\{\mathcal{F}_t : t \geq 0\}$ and also with respect to the filtration $\{\mathcal{F}_{T\wedge t} : t \geq 0\}$.

EXAMPLE 7. Let Y be an integrable random variable. The process $s \mapsto \mathbb{E}[Y \mid \mathcal{F}_s]$ is a martingale. If $\{M(s) : s \geq 0\}$ is a uniformly integrable martingale, then $M(s) = \mathbb{E}[Y \mid \mathcal{F}_s]$ $\mathbb{P}$-almost surely, where $Y = \lim_{s\to\infty} M(s)$. Such martingales are called *closed*.

We want to discuss an inequality on the second moment of a martingale.

C.5. Second moments of martingales

Let $\{M(t) : t \geq 0\}$ be a continuous martingale with $M(0) = 0$. Then

$$\mathbb{E}\left(M(t)^2\right) \leq \mathbb{E}\left(M^*(t)^2\right) \leq 4\mathbb{E}\left(M(t)^2\right).$$

Here $M^*(t) = \sup_{0\leq s\leq t} |M(s)|$.

Proof. Define for $\xi > 0$ the stopping time T_ξ by

$$T_\xi = \inf\{t > 0 : M^*(t) \geq \xi\}.$$

Then $\{M^*(t) > \xi\} \subseteq \{T_\xi < t\}$ and $\{T_\xi < t\} \subseteq \{M^*(t) \geq \xi\}$. Hence, since $|M(t)|$ is a submartingale, upon using Doob's optional sampling (see C.11), we obtain

$$\mathbb{E}\left(M^*(t)^2\right) = \int_0^\infty \mathbb{P}\left(M^*(t)^2 > \lambda\right) d\lambda$$

(make the substitution $\lambda = \xi^2$)

$$= 2\int_0^\infty \xi\, \mathbb{P}\left(M^*(t) > \xi\right) d\xi \leq 2\int_0^\infty \xi\, \mathbb{P}\left(T_\xi < t\right) d\xi$$

$$= 2\int_0^\infty \mathbb{E}\left(|M(T_\xi)| : T_\xi < t\right) d\xi$$

(Doob's optional sampling)

$$\leq 2\int_0^\infty \mathbb{E}\left(|M(t)| : T_\xi < t\right) d\xi$$

$$\leq 2\int_0^\infty \mathbb{E}\left(|M(t)| : M^*(t) \geq \xi\right) d\xi$$

$$= 2\mathbb{E}\left(|M(t)|\, M^*(t)\right)$$

(Cauchy-Schwarz' inequality)

$$\leq 2\left(\mathbb{E}\left(M(t)^2\right)\right)^{1/2}\left(\mathbb{E}\left(M^*(t)^2\right)\right)^{1/2}.$$

Consequently $\mathbb{E}\left(M^*(t)^2\right) \leq 4\mathbb{E}\left(M(t)^2\right)$.

REMARK. The method of proof works very well if $\mathbb{E}\left(M^*(t)^2\right)$ is finite. If this is not the case we may use a localization technique (with stopping times), i.e. we may consider the martingale $M(T_\xi \wedge t)$, $t \geq 0$, where we let ξ tend to infinity (compare with Example 6 in C.4.).

Next we will see how to transfer properties of discrete time semi-martingales to continuous time semi-martingales. Most of the results in C.6, C.7 and C.8 of this Appendix are taken from Bhattacharaya and Waymire [26, §13 of Chapter I, and §5 of Chapter IV, theoretical complements]. We begin with an upcrossing inequality for a discrete time submartingale.

C.6. Upcrossings

Consider a sequence $\{Z_n : n \in \mathbb{N}\}$ of real-valued random variables and sigma-fields $\mathcal{F}_1 \subset \mathcal{F}_2 \subset \cdots$, such that, for every $n \in \mathbb{N}$, the variable Z_n is $\mathcal{F}_n$-measurable: see the introduction to this appendix and see C.12. An *upcrossing* of an interval (a, b) by $\{Z_n\}$ is a passage to a value equal to or exceeding b from an value equal to or below a at an earlier time. Define the random variables X_n, $n \in \mathbb{N}$, by $X_n = \max(Z_n - a, 0)$. If the process $\{Z_n\}$ is a submartingale, then so is the process $\{X_n\}$. The upcrossings of $(0, b-a)$ by $\{X_n\}$ are the upcrossings of the interval (a, b) by $\{Z_n\}$. We define the successive upcrossing times η_{2k}, $k \in \mathbb{N}$, of $\{X_n\}$ as follows:

$$\begin{aligned}
\eta_1 &= \inf\{n \geq 1 : X_n = 0\}; \\
\eta_2 &= \inf\{n \geq \eta_1 : X_n \geq b - a\}; \\
\eta_{2k+1} &= \inf\{n \geq \eta_{2k} : X_n = 0\}; \\
\eta_{2k+2} &= \inf\{n \geq \eta_{2k+1} : X_n \geq b - a\}.
\end{aligned}$$

Then every η_k is an $\{\mathcal{F}_n\}$-stopping time. Fix $N \in \mathbb{N}$ and put $\tau_k = \min(\eta_k, N)$. Then every τ_k is also a stopping time and $\tau_k = N$ for $k > \lfloor N/2 \rfloor$ (the largest integer smaller than or equal to $N/2$). It follows that $X_{\tau_{2k}} = X_N$ for $k > \lfloor N/2 \rfloor$ and we also have $\eta_k \geq k$ and so $k \leq \tau_k \leq N$. Let $U_N(a, b)$ be the number of upcrossings of (a, b) by the process $\{Z_n\}$ at time N. That means

$$U_N(a, b) = \sup\{k \geq 1 : \eta_{2k} \leq N\}$$

with the convention that the supremum over the empty set is 0. Notice that $U_N(a, b)$ is also the number of upcrossings of the interval $(0, b-a)$ by $\{X_n\}$ in time N.

C.7. Upcrossing inequality

Let $\{Z_n\}$ be an $\{\mathcal{F}_n\}$-submartingale. For each pair (a, b), $a < b$, the expected number of upcrossings of (a, b) by $Z_1, \ldots, Z_N$ satisfies the inequality

$$\begin{aligned}
\mathbb{E}(U_N(a, b)) &\leq \frac{\mathbb{E}(\max(Z_N - a, 0) - \max(Z_1 - a, 0))}{b - a} \\
&\leq \min\left(\frac{\mathbb{E}(|Z_N|) + |a|}{b - a}, \frac{\max(Z_N - Z_1, 0)}{b - a}\right).
\end{aligned}$$

C.8. Submartingale convergence theorem

Let $\{Z_n\}$ be a submartingale with the property that $\sup_{n\in\mathbb{N}} \mathbb{E}(|Z_n|) < \infty$. Then the sequence $\{Z_n\}$ converges almost surely to an integrable random variable Z_∞. Moreover we have $\mathbb{E}(|Z_\infty|) \leq \liminf_{n\to\infty} \mathbb{E}(|Z_n|)$.

REMARK. In general we do not have $\mathbb{E}(|Z_\infty|) = \lim_{n\to\infty} \mathbb{E}(|Z_n|)$. This is true only in case the submartingale Z_n, $n \in \mathbb{N}$, is uniformly integrable: see Scheffé's theorem (Theorem E.3 in Appendix E). In fact there exist martingales $\{M_n : n \in \mathbb{N}\}$ such that $M_n \geq 0$, such that $\mathbb{E}(M_n) = 1$, $n \in \mathbb{N}$, and such that $M_\infty = \lim_{n\to\infty} M_n = 0$, $\mathbb{P}$-almost surely. To be specific, let $\{B(s) : s \geq 0\}$ be ν-dimensional Brownian motion starting at $x \in \mathbb{R}^\nu$ and let $p_{0,\nu}(t,x,y)$ be the corresponding transition density. Fix $t > 0$ and $y \neq x$ and put

$$M_n = \frac{p_{0,\nu}(t/n, B(t-t/n), y)}{p_{0,\nu}(t,x,y)}. \tag{C.1}$$

The process $\{M_n : n \in \mathbb{N}\}$ defined in (C.1) is a $\mathbb{P}_x$-martingale with respect to the sigma-fields $\mathcal{F}_n$ generated by $B(s)$, $0 \leq s \leq t - t/n$.

As a corollary we obtain the following.

COROLLARY. *A non-negative martingale $\{Z_n\}$ converges almost surely to a finite limit Z_∞. Also $\mathbb{E}(Z_\infty) \leq \mathbb{E}(Z_1)$.*

REMARK. Convergence properties for supermartingales $\{Z_n\}$ are obtained from the submartingale results applied to $\{-Z_n\}$. Since a semi-martingale is a difference of two submartingales, we also have convergence results for semi-martingales.

C.9. Stochastically continuous processes

A continuous time process $\{X(t) : t \in \mathbb{R}\}$ is said to be *stochastically continuous* at t_0 if for every $\epsilon > 0$

$$\lim_{t\to t_0} \mathbb{P}(|X(t) - X(t_0)| > \epsilon) = 0.$$

REMARK 1. Brownian motion possesses almost surely continuous sample paths and is stochastically continuous for every $t \geq 0$. On the other hand a Poisson process is stochastically continuous, but its sample paths are step functions with unit jumps: see Bhattacharya and Waymire [26, Chapter IV, Example 1, pp 261-262. In fact, for $t > s$, $X(t) \geq X(s)$ $\mathbb{P}$-almost surely and $\mathbb{P}(X(t) - X(s) = n) = e^{-\lambda(t-s)} \dfrac{(\lambda(t-s))^n}{n!}$. Hence, for $t > s$ and $\epsilon > 0$,

$$\mathbb{P}(X(t) - X(s) \geq \epsilon) \leq e^{-\lambda(t-s)} \sum_{n=1}^{\infty} \frac{(\lambda(t-s))^n}{n!} = 1 - e^{-\lambda(t-s)}.$$

REMARK 2. In fact all our Markov processes are stochastically continuous. This follows from the fact that they arise from Feller semigroups on a second countable locally compact space: see §2.2. of Chapter 2 in Chung [49], and in particular Proposition 2 of §2.2. (page 50).

C.10. Stochastic equivalence and regularity of sample paths

Let $\{X(t) : t \geq 0\}$ be a submartingale or a supermartingale that is stochastically continuous at each $t \geq 0$. Then there exists a process $\left\{\widetilde{X}(t) : t \geq 0\right\}$ with the following properties:

(i) (*stochastic equivalence*) $\left\{\widetilde{X}(t)\right\}$ is equivalent to $\{X(t)\}$ in the sense that

$$\mathbb{P}\left(\widetilde{X}(t) = X(t)\right) = 1 \quad \text{for every} \quad t \geq 0;$$

(ii) (*sample path regularity*) with probability 1 the sample paths of the process $\left\{\widetilde{X}(t) : t \geq 0\right\}$ are bounded on compact intervals $[a, b]$, $a < b < \infty$, are right-continuous, and possess left-hand limits at each $t > 0$ (in other words $\left\{\widetilde{X}(t) : t \geq 0\right\}$ is *cadlag*).

Proof. Fix $T > 0$ in $\mathbb{Q}$ and let $\mathbb{Q}_T := \mathbb{Q} \cap [0, T]$ denote the set of rational numbers in $[0, T]$. Write $\mathbb{Q}_T = \cup_{n=1}^{\infty} R_n$, where R_n is a finite subset of $[0, T]$ and where $T \in R_1 \subset R_2 \subset R_2 \subset \cdots$. By Doob's maximality inequality we have

$$\mathbb{P}\left(\max_{t \in R_n} |X(t)| > \lambda\right) \leq \frac{\mathbb{E}\,|X(T)|}{\lambda}, \quad n = 1, 2, \ldots$$

and hence

$$\mathbb{P}\left(\sup_{t \in \mathbb{Q}_T} |X(t)| > \lambda\right) \leq \lim_{n \to \infty} \mathbb{P}\left(\max_{t \in R_n} |X(t)| > \lambda\right) \leq \frac{\mathbb{E}\,|X(T)|}{\lambda}, \quad n = 1, 2, \ldots$$

In particular, the paths of $\{X(t) : t \in \mathbb{Q}_T\}$ are bounded with probability 1. Let (c, d) be any interval in $\mathbb{R}$ and let $U^{\{T\}}(c, d)$ denote the number of upcrossings of (c, d) by the process $\{X(t) : t \in \mathbb{Q}_T\}$. Then $U^{\{T\}}(c, d)$ is the limit of the number $U^{\{n\}}(c, d)$ of upcrossings of (c, d) by $\{X(t) : t \in R_n\}$ as n tends to ∞. By the upcrossing inequality we have

$$\mathbb{E}\left(U^{\{n\}}(c, d)\right) \leq \frac{\mathbb{E}\left(|X(T)|\right) + |c|}{d - c}. \tag{C.2}$$

Since $U^{\{n\}}(c, d)$ is increasing, it follows from (C.2) that

$$\mathbb{E}\left(U^{\{T\}}(c, d)\right) \leq \frac{\mathbb{E}\left(|X(T)|\right) + |c|}{d - c}$$

and hence that $U^{\{T\}}(c, d)$ is almost surely finite. Taking unions over all intervals (c, d), with c, $d \in \mathbb{Q}$, and $c < d$, it follows with probability 1 that the process $\{X(t) : t \in \mathbb{Q}_T\}$ has only finitely many upcrossings on any interval. In particular, left- and right-hand limits must exist at each $t < T$, $\mathbb{P}$-almost surely. To construct a right-continuous version of $\{X(t)\}$, we define $\left\{\widetilde{X}(t) : t \geq 0\right\}$ as follows: $\widetilde{X}(t) = \lim_{s \downarrow t, s \in \mathbb{Q}} X(s)$ for $t < T$. That this process $\left\{\widetilde{X}(t)\right\}$ is stochastically equivalent to $\{X(t)\}$ follows from the stochastic continuity of the process $\{X(t)\}$. Further details are left to the reader.

Next we prove Doob's optional sampling for continuous time submartingales (that are right-continuous). A similar result holds for martingales (where the inequality sign in (C.3) is replaced with an equality) and for super-martingales (where the inequality is reversed). For discrete submartingales the result will be taken for granted. For more details and other proofs, see e.g. Theorem 2.13 of Chapter II in Ethier and Kurtz [99, page 61] and Bhattacharaya and Waymire [26, Proposition 13.9. page 52, and Theorem 4.1, page 544] for the discrete case.

C.11. Doob's optional sampling theorem

Let $\{X(t) : t \geq 0\}$ be a right-continuous submartingale of class (DL) (see C.3, (i)) and let T be a stopping time. Suppose $t \geq s$. Then

$$\mathbb{E}\left(X\left(\min(t, T)\right) \mid \mathcal{F}_s\right) \geq X\left(\min(s, T)\right), \quad \mathbb{P}\text{-almost surely.} \tag{C.3}$$

Proof. Put $s_n = 2^{-n}\lceil 2^n s\rceil$, $t_n = 2^{-n}\lceil 2^n t\rceil$, and $T_n = 2^{-n}\lceil 2^n T\rceil$. If A belongs to $\mathcal{F}_s$, then A also belongs to $\mathcal{F}_{s_n}$ for all $n \in \mathbb{N}$. From Doob's optional sampling for discrete time submartingales, we infer that

$$\mathbb{E}\left(X\left(\min(t_n, T_n)\right) 1_A\right) \geq \mathbb{E}\left(X\left(\min(s_n, T_n)\right) 1_A\right)$$

upon using the (DL)-property. Let n tend to ∞ and use the right-continuity of the process $\{X(t) : t \geq 0\}$ to obtain

$$\mathbb{E}\left(X\left(\min(t, T)\right) 1_A\right) \geq \mathbb{E}\left(X\left(\min(s, T)\right) 1_A\right).$$

C.12. Skorohod space

Throughout this book E denotes a second countable locally compact Hausdorff space and $E^{\triangle}$ is its one-point compactification or, if E is compact, $\triangle$ is an isolated point and $E^{\triangle} = E \bigcup\{\triangle\}$. Define the path space Ω as follows. The sample path space Ω is a subset of $\left(E^{\triangle}\right)^{[0,\infty]}$ with the following properties:

(i) If ω belongs to Ω, if there is a $t \geq 0$ such that $\omega(t) = \triangle$, and if $s \geq t$, then $\omega(s) = \triangle$;

(ii) Put $\zeta(\omega) = \inf\{s > 0 : \omega(s) = \triangle\}$ for $\omega \in \Omega$. If ω belongs to Ω, then ω possesses left limits in $E^{\triangle}$ on the interval $[0, \zeta]$ and it is right-continuous on $[0, \infty]$;

(iii) If ω belongs to Ω and $t \geq 0$ is such that $\omega(t)$ belongs to E, then the closure of the set $\{\omega(s) : 0 \leq s \leq t\}$ is a compact subset of E. Equivalently, if $t > 0$ is such that $\omega(t-) = \triangle$ and if $s \geq t$, then $\omega(s) = \triangle$.

The path space Ω is also written as $\Omega = D\left([0,\infty], E^{\triangle}\right)$. It is called the *Skorohod space* of cadlag sample paths from $[0,\infty]$ to $E^{\triangle}$.

The random variable ζ defined in (ii) is called the *life time* of ω. A path $\omega \in \Omega$ is said to be *cadlag* on its life time. This means that on the half open interval $[0,\zeta(\omega))$, ω attains its values in E and is cadlag (continu à droit et limité à gauche). We also define the *state variables* $X(t) : \Omega \to E^{\triangle}$ by $X(t)(\omega) = X(t,\omega) = \omega(t)$, $t \geq 0$, $\omega \in \Omega$. The *translation* or *shift operators* are defined in the following way: $[\vartheta_t(\omega)](s) = \omega(s+t)$, s, $t > 0$ and $\omega \in \Omega$. The largest subset of $\left(E^{\triangle}\right)^{[0,\infty]}$ with the properties (i), (ii) and (iii) is denoted by as $D\left([0,\infty], E^{\triangle}\right)$. Let $\mathcal{F}$ be a σ-field on Ω. A function $Y : \Omega \to \mathbb{C}$ is called a *random variable* if it is measurable with respect to $\mathcal{F}$. Of course, $\mathbb{C}$ is supplied with its Borel field. The so-called *state space* E is also equipped with its Borel field $\mathcal{E}$ and $E^{\triangle}$ is also equipped with its Borel field $\mathcal{E}^{\triangle}$. The sample path $\omega_{\triangle}$ is given by $\omega_{\triangle}(s) = \triangle$, $s \geq 0$. Unless specified otherwise, we write $\Omega = D([0,\infty], E^{\triangle})$. In addition let $\mathcal{F}$ be a σ-field on Ω and let $\{\mathcal{F}_t : t \geq 0\}$ be a filtration on Ω. Suppose $\mathcal{F}_t \subseteq \mathcal{F}$, $t \geq 0$ and suppose that every state variable $X(t)$, $t \geq 0$, is measurable with respect to $\mathcal{F}_t$. (This is the case where e.g. $\mathcal{F}_t$ is the σ-field generated by $\{X(s) : s \leq t\}$.) Equipped with a suitable metric the space Ω is a separable and complete. In other words it is a Polish space. For more details on the Skorohod space, the reader is referred to Ethier and Kurtz [99, §5 through §10 of Chapter III].

C.13. Dissipative operators, martingales, Feller semigroups

Let K be a linear operator with domain and range in $C_\infty(E)$ (the space $C_\infty(E)$ is defined in Definition 1.2).

(i) The operator K is said to be *dissipative* if, for all $\lambda > 0$ and for all $f \in \operatorname{dom}(K)$,
$$\|\lambda f + Kf\|_\infty \geq \lambda \|f\|_\infty .$$
(ii) The operator K is said to satisfy the *maximum principle* if for every $f \in \operatorname{dom}(K)$ with $\sup\{\operatorname{Re} f(x) : x \in E\}$ strictly positive, there exists $x_0 \in E$ with the property that
$$\operatorname{Re} f(x_0) = \sup\{\operatorname{Re} f(x) : x \in E\} \quad \text{and} \quad \operatorname{Re} Kf(x_0) \geq 0.$$
(iii) The martingale problem is said to be *uniquely solvable* or *well-posed* for the operator K, if for every $x \in E$ there exists a unique probability $\mathbb{P} = \mathbb{P}_x$ which satisfies the following:
   (a) For every $f \in \operatorname{dom}(K)$ the process $f(X(t)) - f(X(0)) + \int_0^t Kf(X(s))ds$, $t \geq 0$, is a $\mathbb{P}$-martingale;
   (b) $\mathbb{P}(X(0) = x) = 1$.

(iv) The operator K is said to solve the *martingale problem maximally* if for K the martingale problem is uniquely solvable and if its closure $\overline{K}$ is maximal for this property. This means that, if K_1 is any linear operator with domain and range in $C_\infty(E)$ which extends $\overline{K}$, and for which the martingale problem is uniquely solvable, then $\overline{K}$ coincides with K_1.

(v) The operator K is said to be the (infinitesimal) *generator* of a Feller semigroup (see below) $\{P(t) : t \geq 0\}$ if $K = \text{s-}\lim_{t\downarrow 0} \frac{I - P(t)}{t}$. This means that a function f belongs to $\text{dom}(K)$ whenever $Kf := \lim_{t\downarrow 0} \frac{f - P(t)f}{t}$ exists in $(C_\infty(E), \|\cdot\|_\infty)$.

An operator which satisfies the maximum principle is dissipative (see e.g. [231, p. 14] and can be regarded as kind of a generalized second order derivative operator. A prototype of such an operator is the Laplace operator, an operator for which the martingale problem is uniquely solvable is closable.

As in Appendix B.11 we define a *Feller semigroup* as follows.

DEFINITION. A family $\{P(t) : t \geq 0\}$ of operators defined on $L^\infty(E, \mathcal{E})$ is a *Feller semigroup* on $C_\infty(E)$ if it possesses the following properties:

(i) It leaves $C_\infty(E)$ invariant: $P(t)C_\infty(E) \subseteq C_\infty(E)$ for $t \geq 0$;
(ii) It is a semigroup: $P(s + t) = P(s) \circ P(t)$ for all s, $t \geq 0$, and $P(0) = I$;
(iii) It consists of positivity preserving contraction operators: $0 \leq f \leq 1$ implies $0 \leq P(t)f \leq 1$, $t > 0$;
(iv) It is continuous for $t = 0$: $\lim_{t\downarrow 0} [P(t)f](x) = f(x)$, for all $f \in C_\infty(E)$ and for all $x \in E$.

In the presence of (iii) and (ii), property (iv) is equivalent to:

(iv′) $\lim_{t\downarrow 0} \|P(t)f - f\|_\infty = 0$ for all $f \in C_\infty(E)$. So that a Feller semigroup is in fact *strongly continuous* in the sense that, for every $f \in C_\infty(E)$,

$$\lim_{s\to t} \|P(s)f - P(t)f\|_\infty = 0.$$

It is perhaps useful to observe that $C_\infty(E)$ equipped with the supremum-norm $\|\cdot\|_\infty$ is a Banach space (in fact it is a Banach algebra). We need one more definition. Let $\{P(t) : t \geq 0\}$ be a Feller semigroup. For each open subset U of E, define the *transition probability* $P(t, x, U)$, $t \geq 0$, $x \in E$, by

$$P(t, x, U) = \sup\left\{[P(t)u](x) : 0 \leq u \leq 1_U,\ u \in C_\infty(E)\right\}.$$

This transition function can be extended to all Borel subsets by writing

$$P(t, x, K) = \inf\left\{P(t, x, U) : U \text{ open } U \supseteq K\right\},$$

for K a compact subset of E. If B is a Borel subset of E, then we write

$$\begin{aligned} P(t, x, B) &= \inf\left\{P(t, x, U) : U \supseteq B, U \text{ open }\right\} \\ &= \sup\left\{P(t, x, K) : K \subseteq B, K \text{ compact }\right\}. \end{aligned}$$

It then follows that the mapping $B \mapsto P(t, x, B)$ is a Borel measure on $\mathcal{E}$, the Borel field of E. The Feller semigroup is said to be *conservative* if, for all $t \geq 0$ and for all $x \in E$, $P(t, x, E) = 1$.

C.14. Strong Markov process and Feller semigroups

Next suppose that, for every $x \in E$, a probability measure $\mathbb{P}_x$ on $\mathcal{F}$ is given. Suppose that for every bounded random variable $Y : \Omega \to \mathbb{R}$ the equality $\mathbb{E}_x\left(Y \circ \vartheta_t \mid \mathcal{F}_t\right) = \mathbb{E}_{X(t)}(Y)$ holds P_x-almost surely for all $x \in E$ and for all $t \geq 0$: for the definition of the shift operator ϑ_t see C.12. Then the process

$$\left\{(\Omega, \mathcal{F}, \mathbb{P}_x), (X(t) : t \geq 0), (\vartheta_t : t \geq 0), (E, \mathcal{E})\right\}$$

is called a *Markov process*. If the fixed time t may be replaced with a stopping time T, the process $\left\{(\Omega, \mathcal{F}, \mathbb{P}_x), (X(t) : t \geq 0), (\vartheta_t : t \geq 0), (E, \mathcal{E})\right\}$ is called a *strong Markov process*. By definition $\mathbb{P}_\triangle(A) = 1_A(\omega_\triangle) = \delta_{\omega_\triangle}(A)$. Here A belongs to $\mathcal{F}$. If the process $\left\{(\Omega, \mathcal{F}, \mathbb{P}_x), (X(t) : t \geq 0), (\vartheta_t : t \geq 0), (E, \mathcal{E})\right\}$ is a Markov process, then we write (see C.13)

$$P(t, x, B) = \mathbb{P}_x(X(t) \in B), \quad t \geq 0, \quad B \in \mathcal{E}, \quad x \in E,$$

for the corresponding *transition function*. The operator family $\{P(t) : t \geq 0\}$ is defined by $[P(t)f](x) = \mathbb{E}_x\left(f(X(t))\right)$, $f \in C_\infty(E)$. A recent book on Markov processes is Ethier and Kurtz [99]. An elementary theory for diffusions is given in Durrett [93]. On this topic, the books of Stroock and Varadhan [218] and Ikeda and Watanabe [137] are of interest as well.

We shall mainly be interested in the case that the function $P(t)f$ is a member of $C_\infty(E)$ whenever f is so. In the following theorem $\mathcal{F}$ is the σ-field generated by $\{X(s) : s \geq 0\}$ and $\mathcal{F}_t$ is the σ-field generated by the *past* or *full history*, i.e. $\mathcal{F}_t = \sigma\left\{X(s) : 0 \leq s \leq t\right\}$. If T is a stopping time we write $\mathcal{F}_{T+} = \bigcap_{t \geq 0} \left\{A \in \mathcal{F} : A \cap \{T \leq t\} \in \mathcal{F}_{t+}\right\}$.

THEOREM. *Let $(\Omega, \mathcal{F}, \mathbb{P}_x)$, $x \in E$, be probability spaces with the following properties:*

(a) *For every $\alpha > 0$, for every $t \geq 0$ and for every open subset U of E, the set $\{x \in E : \mathbb{P}_x(X(t) \in U) > \alpha\}$ is open;*
(b) *For every $\alpha > 0$, for every $t \geq 0$ and for every compact subset K of E, the set $\{x \in E : \mathbb{P}_x(X(t) \in K) \geq \alpha\}$ is compact;*
(c) *For every open subset U of E and for every $x \in U$, the equality*

$$\lim_{t \downarrow 0} \mathbb{P}_x\left(X(t) \in U\right) = 1 \quad \textit{is valid.}$$

The following assertions are equivalent:

(i) *For all $t \geq 0$ and for all bounded random variables $Y : \Omega \to \mathbb{C}$ the equality*

$$\mathbb{E}_x\left(Y \circ \vartheta_t \mid \mathcal{F}_t\right) = \mathbb{E}_{X(t)}(Y)$$

holds $\mathbb{P}_x$-almost surely for all $x \in E$;

(ii) *For all finite tuples* $0 \leq t_1 < t_2 < \ldots < t_n < \infty$ *together with Borel subsets* $B_1, \ldots, B_n$ *of* E *the equality*

$$\begin{aligned} &\mathbb{P}_x\left(X(t_1) \in B_1, \ldots, X(t_n) \in B_n\right) \\ &= \int_{B_1} \cdots \int_{B_{n-1}} \int_{B_n} P(t_n - t_{n-1}, x_{n-1}, dx_n) P(t_{n-1} - t_{n-2}, x_{n-2}, dx_{n-1}) \\ &\qquad \ldots P(t_2 - t_1, x_1, dx_2) P(t_1, x, dx_1) \end{aligned} \tag{C.4}$$

is valid for all $x \in E$ *(here* $\mathbb{P}_x(X(t) \in B) = P(t, x, B)$*);*

(iii) *For every* $(\mathcal{F}_{t+})$*-stopping time* T *and for every bounded random variable* $Y : \Omega \to \mathbb{C}$ *the equality*

$$\mathbb{E}_x\left(Y \circ \vartheta_T \mid \mathcal{F}_{T+}\right) = \mathbb{E}_{X(T)}\left(Y\right),$$

holds $\mathbb{P}_x$*-almost surely on* $\{T < \infty\}$ *for all* $x \in E$*;*

(iv) *Let* $\mathcal{B}$ *be the Borel field of* $[0, \infty)$*. For every bounded function* $F : [0, \infty) \times \Omega \to \mathbb{C}$*, which is measurable with respect to* $\mathcal{B} \otimes \mathcal{F}$*, and for every* $(\mathcal{F}_{t+})$*-stopping time* T *the equality*

$$\mathbb{E}_x\left(\{\omega \mapsto F\left(T(\omega), \vartheta_T(\omega)\right)\} \mid \mathcal{F}_T\right) = \left\{\omega' \mapsto \mathbb{E}_{X(T(\omega'))}\left\{\omega \mapsto F\left(T(\omega'), \omega\right)\right\}\right\} \tag{C.5}$$

holds $\mathbb{P}_x$*-almost surely* $\{T < \infty\}$ *for all* $x \in E$*.*

We will not prove this theorem nor its extension mentioned in subsection C.15. below. It is proved in Blumenthal and Getoor [35] and in Sharpe [205].

REMARK 1. Suppose the probability spaces $(\Omega, \mathcal{F}, \mathbb{P}_x)$, $x \in E$, possess the properties (a) and (b). Define the operator $P(t) : L^\infty(E, \mathcal{E}) \to L^\infty(E, \mathcal{E})$, $t \geq 0$, by $[P(t)f](x) = \mathbb{E}_x[f(X(t))]$, $x \in E$. Then the operators $P(t)$ leave the space $C_\infty(E)$ invariant.

REMARK 2. Suppose that (a), (b), and (c) in the previous theorem are satisfied. From Theorem 9.4, page 46 of Blumenthal and Getoor [35], it follows that there exists a strong Markov process (in fact a Hunt process)

$$\left\{\left(\Omega, \mathcal{F}, \mathbb{P}_x\right), \left(X(t) : t \geq 0\right), \left(\vartheta_t : t \geq 0\right), \left(E, \mathcal{E}\right)\right\}$$

with the properties mentioned in (i) through (iv): see C.15. In fact a more general theorem is proved in Sharpe [205, Theorem 7.4, page 31].

REMARK 3. The equivalent properties mentioned in assertion (i) and (ii) are called the *Markov property* with respect to the filtration $(\mathcal{F}_t)_{t \geq 0}$. The property mentioned in assertion (iii) is called the *strong Markov property*. In subsection C.15. below the Markov property is ameliorated to a larger filtration $\left(\widetilde{\mathcal{F}}_t\right)_{t \geq 0}$. We shall not prove this result. Equality (C.5) is called the *strong time-dependent Markov property*.

Since our strong Markov process is related to a Feller semigroup $\{P(t) : t \geq 0\}$ in $C_\infty(E)$ via the identity $[P(t)f](x) = \mathbb{E}_x[f(X(t))]$, $f \in C_\infty(E)$, it is also called a Feller process (see e.g. Chung [50, page 50]). Since $\mathbb{P}_x[X(0) = x] = 1$, $x \in E$, it is called a *normal* strong Markov process. Since, in addition, one may prove the *quasi-left continuity* of the sample paths, meaning that $\lim_{n\to\infty} X(T_n) = X(T)$, $\mathbb{P}_x$-almost surely, on the event $\{T < \infty\}$, whenever $(T_n)_{n\in\mathbb{N}}$ is an sequence of stopping times which increases to the stopping time T. Because of this additional property, such processes are called *Hunt processes*: see e.g. Blumenthal and Getoor [35, page 45], Chung [49, page 75], or Sharpe [205, §47 in Chapter V].

C.15. Markov property and larger filtrations

Let $\{(\Omega, \mathcal{F}, \mathbb{P}_x), (X(t) : t \geq 0), (\vartheta_t : t \geq 0), (E, \mathcal{E})\}$ be a strong Markov process with state space E. For a probability measure μ on $(E, \mathcal{E})$, the measure $\mathbb{P}_\mu$ is defined by $\mathbb{P}_\mu(A) = \int \mathbb{P}_x(A)d\mu(x)$, $A \in \mathcal{F}$. Let $\mathcal{F}^\mu$ be the completion of $\mathcal{F}$ with respect to $\mathbb{P}_\mu$:

$$\mathcal{F}^\mu = \big\{A \subseteq \Omega : \text{there exist subsets } A_1, A_2 \in \mathcal{F}, A_1 \subseteq A \subseteq A_2, \\ \text{such that } \mathbb{P}_\mu(A_2 \setminus A_1) = 0\big\}.$$

Define the σ-fields $\mathcal{F}_t^\mu$, $t \geq 0$, by

$$\mathcal{F}_t^\mu = \{A \in \mathcal{F}^\mu : \text{ there exists } A_0 \in \mathcal{F}_t \text{ such that } \mathbb{P}_\mu(A \setminus A_0) = \mathbb{P}_\mu(A_0 \setminus A) = 0\}.$$

For $t \geq 0$ we define $\widetilde{\mathcal{F}}_t$ by

$$\widetilde{\mathcal{F}}_t := \bigcap \{\mathcal{F}_t^\mu : \mu \text{ probability measure on } (E, \mathcal{E})\},$$

and the σ-field $\widetilde{\mathcal{F}}$ by

$$\widetilde{\mathcal{F}} := \bigcap \{\mathcal{F}^\mu : \mu \text{ probability measure on } (E, \mathcal{E})\}.$$

Put $\mathcal{F}_{t+} = \bigcap_{s>t} \mathcal{F}_s$. It can be proved that $\mathcal{F}_{t+} \subseteq \mathcal{F}_t^\mu$ and that $\bigcap_{s>t} \mathcal{F}_s^\mu = \mathcal{F}_t^\mu$. We also have $\bigcap_{s>t} \widetilde{\mathcal{F}}_s = \widetilde{\mathcal{F}}_t$. So the filtrations $(\mathcal{F}_t^\mu : t \geq 0)$ and $\left(\widetilde{\mathcal{F}}_t : t \geq 0\right)$ are right-continuous. Moreover, all results concerning the strong Markov property remain valid for $\left(\widetilde{\mathcal{F}}_t\right)$-stopping times, which is very important. In fact hitting times of Borel subsets need not be $(\mathcal{F}_t)$-stopping times, but they are $\left(\widetilde{\mathcal{F}}_t\right)$-stopping times. These results are proved in Chapter 2 of Blumenthal and Getoor [35]. A convenient formulation of the strong Markov property reads as follows. For every bounded $\widetilde{\mathcal{F}}$-measurable random variable $Y : \Omega \to \mathbb{C}$ the identity

$$\mathbb{E}_\mu\left[Y \circ \vartheta_T \mid \widetilde{\mathcal{F}}_T\right] = \mathbb{E}_{X(T)}[Y], \quad \text{holds } \mathbb{P}_\mu\text{-almost surely on } \{T < \infty\},$$

where T is an $\left(\widetilde{\mathcal{F}}_t\right)$-stopping time and where

$$\widetilde{\mathcal{F}}_T = \bigcap_{t \geq 0} \left\{ A \in \widetilde{\mathcal{F}} : \; A \cap \{T \leq t\} \in \widetilde{\mathcal{F}}_t \right\}.$$

REMARK. Next fix a probability measure μ on $\mathcal{E}$, fix s and $t \geq 0$, and fix $A \in \widetilde{\mathcal{F}}_t$ as well. Put

$$\mu_s(B) = \int P(s, x, B) d\mu(x) = \int \mathbb{P}_x \{X(s) \in B\} \, d\mu(x), \;\; B \in \mathcal{E}.$$

Then, by the Markov property, $\mathbb{P}_\mu \left[\vartheta_s^{-1}(A)\right] = \mathbb{P}_{\mu_s}(A)$, and there exists $A_s \in \mathcal{F}_t$ such that $\mathbb{P}_{\mu_s}(A \triangle A_s) = 0$. Hence

$$\mathbb{P}_\mu \left[\vartheta_s^{-1}(A) \triangle \vartheta_s^{-1}(A_s)\right] = \mathbb{P}_\mu \left[\vartheta_s^{-1}(A \triangle A_s)\right] = \mathbb{P}_{\mu_s}[A \triangle A_s] = 0.$$

Hence $\vartheta_s^{-1}(A)$ belongs to $\widetilde{\mathcal{F}}_{s+t}$. Here we used the fact that $\vartheta_s^{-1}(A_s)$ belongs to $\mathcal{F}_{s+t}$. These observations show that

$$\vartheta_s^{-1}\left(\widetilde{\mathcal{F}}_t\right) \subseteq \widetilde{\mathcal{F}}_{s+t}, \quad s, \; t \geq 0.$$

Here, by definition, $A \triangle A'$ is the symmetric difference of A and A', i.e. $A \triangle A' = (A \cap (\Omega \setminus A')) \cup (A' \cap (\Omega \setminus A))$.

C.16. Markov processes and Feller semigroups

Let $\{P(t) : t \geq 0\}$ be a Feller semigroup with corresponding transition probabilities $P(t, x, B)$: see C.13. Then there exists a collection of probabilities $(\mathbb{P}_x)_{x \in E}$ on the σ-field $\mathcal{F}$ generated by the state variables $\{X(t) : t \geq 0\}$ defined on the Skorohod space $\Omega := D\left([0, \infty], E^\triangle\right)$ (see C.12.) for which

$$[P(t)f](x) = \mathbb{E}_x [f(X(t)] = \int f(y) P(t, x, dy), \;\; f \in C_\infty(E), \; x \in E,$$

and for which

$$\begin{aligned} &\mathbb{E}_x \left(f\left(X(t_1), \ldots, X(t_n)\right)\right) = \int f\left(X(t_1), \ldots, X(t_n)\right) d\mathbb{P}_x \\ &= \int \ldots \int f(x_1, x_2, \ldots, x_n) P(t_n - t_{n-1}, x_{n-1}, dx_n) P(t_{n-1} - t_{n-2}, x_{n-2}, dx_{n-1}) \\ &\qquad \ldots P(t_2 - t_1, x_1, dx_2) P(t_1, x, dx_1), \end{aligned} \tag{C.6}$$

where f is any bounded complex or non-negative Borel measurable function defined on $E^\triangle \times \ldots \times E^\triangle$ that vanishes outside of $E \times \ldots \times E$. Let the measure spaces $(\Omega, \mathcal{F}, \mathbb{P}_x)_{x \in E}$ be as in (C.6). Then the process

$$\{(\Omega, \mathcal{F}, \mathbb{P}_x), (X(t) : t \geq 0), (\vartheta_t : t \geq 0), (E, \mathcal{E})\}$$

is a strong Markov process: see C.14.

REMARK. The proof of this result is quite technical. The first part follows from a well-known theorem of Kolmogorov on projective systems of measures: see C.17. In the second part we must show that the indicated path space has full measure, so that no information is lost. Proofs are omitted. They can be found in for example Blumenthal and Getoor [35, Theorem 9.4, p. 46]. A closely related result is proved in Ethier and Kurtz [99, Theorem 2.7, page 169].

We mention an important extension theorem of Kolmogorov, which, among other things allows for the construction of Markov processes. The result is copied from Proposition 1.14 in Chapter 1.E. For a proof see e.g. Bhattacharaya and Waymire [26, pages 92–93].

C.17. Kolmogorov's extension theorem

Let I be some index set and let $(\Omega_j, \mathcal{A}_j)$, $j \in I$, be a collection of measurable spaces. In practice I will be the time axis $[0,\infty)$ or $[0,\infty]$ and each Ω_j, $j \in I$, will either be E or $E^\triangle$, with the respective Borel fields $\mathcal{E}$ or $\mathcal{E}^\triangle$. Put $\Omega_J = \prod_{j\in J}\Omega_j$ (Cartesian product) and supply Ω_J with the product field $\mathcal{A}_J$. A collection $\{(\Omega_J, \mathcal{A}_J, \mathbb{P}_J) : J \subseteq I,\ J \text{ finite}\}$ is called a *projective* or *consistent* family of probability measures on $\Omega = \Omega_I = \prod_{j\in I}\Omega_j$, if for any pair of of finite subsets $J_1 \subseteq J_2$ in I, and for any $A \in \mathcal{A}_{J_1}$, the equality

$$\mathbb{P}_{J_1}(A) = \mathbb{P}_{J_2}\left\{(\omega_j)_{j\in J_2} : (\omega_j)_{j\in J_1} \in A\right\}$$

holds. If $p_{J_1}^{J_2} : \Omega_{J_2} \to \Omega_{J_1}$ is the mapping $p_{J_1}^{J_2}\left((\omega_j)_{j\in J_2}\right) = (\omega_j)_{j\in J_1}$, $J_1 \subseteq J_2$, then the latter can be rewritten as

$$\mathbb{P}_{J_1}(A) = \mathbb{P}_{J_2}\left\{\left(p_{J_1}^{J_2}\right)^{-1}(A)\right\}.$$

THEOREM. *Let I be some index set. Let $\{(\Omega_J, \mathcal{A}_J, \mathbb{P}_J) : J \subseteq I,\ J \text{ finite}\}$ be a projective system of probability measures. Suppose that each Ω_i is a metrizable and σ-compact Hausdorff space endowed with its Borel field $\mathcal{A}_i$. Then there exists a unique probability measure $\mathbb{P}_I$ on $(\Omega_I, \mathcal{A}_I)$, such that*

$$\mathbb{P}_I\left(p_J^{-1}(A)\right) = \mathbb{P}_J(A)$$

for every $J \subseteq I$, J finite, and for every $A \in \mathcal{A}_J$.

Here $\Omega_J = \prod_{j\in J}\Omega_j$ and, in our situation $\Omega_i = E^\triangle$ for all $i \in I = [0,\infty]$.

C.18. Markov transition function

For $x \in E$, $t \geq 0$, and for B a Borel subset of $E^\triangle$ we write

$$N(t,x,B) = P(t,x,B\cap E) + \delta_\triangle(B)\left(1 - P(t,x,E)\right), \quad \text{where}$$
$$P(t,x,B) = \mathbb{P}_x\left[X(t) \in B\right],\ t \geq 0,\ x \in E,\ B \in \mathcal{E},$$

denotes the transition function of our Markov process. The advantage of the transition function $N(t,x,B)$ is that it has total mass 1: $N\left(t,x,E^{\triangle}\right)=1$. So, instead of applying Kolmogorov's extension theorem to the sub-Markov transition function $P(t,x,B)$ it is used for $N(t,x,B)$. For more details see e.g. Karatzas and Shreve [147, Theorem 2.2, p. 50]. We also write $\delta_{\triangle}(B)=1_B(\triangle)=1$ if $\triangle$ belongs to B and zero otherwise.

A relevant theorem which we mention without proof reads as follows.

C.19. Feller semigroups and the martingale problem

The following theorem says that densely defined linear operators in $C_\infty(E)$ which maximally solve the martingale problem are generators of Feller semigroups, and conversely, that the martingale problem is maximally solvable for generators of Feller semigroups. For more information on the martingale problem, the reader may consult Ethier and Kurtz [99, §3 through §7 of Chapter III].

THEOREM. *Let K_0 be a linear operator with domain* $\operatorname{dom}(K_0)$ *and range* $\operatorname{ran}(K_0)$ *in $C_\infty(E)$. Let Ω be the path space $\Omega=D\left([0,\infty],E^{\triangle}\right)$. The following assertions are equivalent:*

(i) *The operator K_0 is closable and its closure generates a Feller semigroup;*

(ii) *The operator K_0 solves the martingale problem maximally and its domain* $\operatorname{dom}(K_0)$ *is dense in $C_\infty(E)$;*

(iii) *The operator K_0 satisfies the maximum principle, its domain* $\operatorname{dom}(K_0)$ *is dense in $C_\infty(E)$ and there exists $\lambda_0>0$ such that the range $R(\lambda_0 I+K_0)$ is dense in $C_\infty(E)$.*

REMARKS. The hard part in (iii) is usually the range property: there exists $\lambda_0>0$ such that the range $R(\lambda_0 I+K_0)$ is dense in $C_\infty(E)$. The theorem outlines, in conjunction with the results on Feller semigroups and Markov processes, the relationships which exist between the unique solvability of the martingale problem, the strong Markov property, and densely defined operators satisfying the maximum principle together with the range property. However if K_0 is in fact a second order differential operator, then we want to read off the range property from the coefficients. There do exist results in this direction. The interested reader is referred to the literature: Stroock and Varadhan [218] and also Ikeda and Watanabe [137]. A proof of the previous result can be found in van Casteren [239]. A proof of the equivalence of (i) and (iii) can be found in [231, Theorem 2.2, p. 14]. The proof of (ii) $\Rightarrow$ (i) relies on this result.

The following result was proved very recently. The a priori hypothesis in the following theorem could be phrased as the operator $\lambda_0 I+K_0$ possesses the *Korovkin* property: compare with §2 in Chapter V of Schaefer [202].

THEOREM. *Let K_0 be a densely defined linear operator with domain and range in $C_\infty(E)$. Suppose that, for some $\lambda_0 > 0$, the subspace $S(\lambda_0)$, defined by*

$$S(\lambda_0)= \bigcap_{x_0\in E}\bigcap_{\varepsilon>0}\Big\{g\in C_\infty(E):\sup\{h_1(x_0):(\lambda_0 I+K_0)h_1\le \operatorname{Re}g+\varepsilon, h_1\in\operatorname{dom}(K_0)\}$$
$$\ge \inf\{h_2(x_0):(\lambda_0 I+K_0)h_2\ge \operatorname{Re}g-\varepsilon, h_2\in\operatorname{dom}(K_0)\}\Big\}.$$

The following assertions are equivalent:

(a) *The martingale problem is well-posed for the operator K_0.*
(b) *The operator K_0 extends to the generator of a Feller semigroup.*

A proof of this result appeared in Odimboleko and van Casteren [185].

C.20. First entry and hitting times

We shall define some important stopping times. Let A be an arbitrary subset of $E^\triangle$. Define the first *entry* time D_A and the (first) *hitting* time τ_A of A respectively by

$$D_A := \inf\{t\ge 0 : X(t)\in A\} \quad \text{and} \quad \tau_A := \inf\{t > 0 : X(t)\in A\}.$$

The hitting time is the most important one. Proofs of the results in C.21 through C.23 can be found in §10 and 11 of Chapter I in Blumenthal and Getoor [35]. We begin with some properties of first entry and hitting times.

C.21. Proposition. *Fix $t\ge 0$ and let A, B, and A_n, $n\in\mathbb{N}$, be subsets of $E^\triangle$. The following assertions hold true:*

(i) $t + D_A\circ\vartheta_t = \inf\{s\ge t : X(s)\in A\}$;
(ii) $t + \tau_A\circ\vartheta_t = \inf\{s> t : X(s)\in A\}$;
(iii) $t + D_A\circ\vartheta_t = D_A$ *on* $\{D_A\ge t\}$;
(iv) $t + \tau_A\circ\vartheta_t = \tau_A$ *on* $\{\tau_A> t\}$;
(v) $D_A\le\tau_A$ *and* $D_A=\tau_A$ *on* $\{X(0)\notin A\}$;
(vi) $A\subseteq B$ *implies* $D_A\ge D_B$ *and* $\tau_A\ge\tau_B$;
(vii) $D_{\cup A_n} = \inf D_{A_n}$ *and* $\tau_{\cup A_n} = \inf\tau_{A_n}$;
(viii) $D_{A\cap B}\ge \max(D_A, D_B)$ *and* $\tau_{A\cap B}\ge\max(\tau_A,\tau_B)$.

The proof of this proposition is easy. The following proposition is somewhat more difficult to prove.

C.22 Proposition.

(i) *For any compact subset A of $E^\triangle$ the identities*

$$\lim_{s\downarrow 0}(s + D_A\circ\vartheta_s) = \inf_{s>0}(s + D_A\circ\vartheta_s) = \inf_{s>0}(s+\tau_A\circ\vartheta_s) = \tau_A \quad \textit{are true};$$

(ii) *Let O be an open subset of $E^\triangle$. Then $\tau_O = D_O$ is an $(\mathcal{F}_{t+})$-stopping time;*
(iii) *Let B be a closed subset of $E^\triangle$. Then τ_B and D_B are $\left(\widetilde{\mathcal{F}}_t\right)$-stopping times;*
(iv) *Let B be a closed subset of $E^\triangle$ and suppose that the paths are $\mathbb{P}_x$-almost surely continuous for all $x\in E$. Then D_B and τ_B are $(\mathcal{F}_{t+})$-stopping times.*

Next we formulate some approximation theorems for stopping times. These will be used in Appendix D.

C.23. Theorem. *Let B be a Borel subset of $E^{\triangle}$ and let μ be a probability measure on $\mathcal{E}$. The following assertions are valid:*

(i) *There exists an increasing sequence $(K_n : n \in \mathbb{N})$ of compact subsets of B such that D_{K_n} decreases to D_B, $\mathbb{P}_\mu$-almost surely;*

(ii) *There exists an decreasing sequence $(U_n : n \in \mathbb{N})$ of open subsets, each containing B, such that τ_{U_n} increases to D_B, $\mathbb{P}_\mu$-almost surely;*

(iii) *There exists an increasing sequence $(K_n : n \in \mathbb{N})$ of compact subsets of B such that τ_{K_n} decreases to τ_B, $\mathbb{P}_\mu$-almost surely.*

C.24. Hitting times and Choquet capacity theory

The following result is probably the most important in this appendix. Its proof relies on Choquet capacity theory.

THEOREM. *Let B be any Borel subset of $E^{\triangle}$. Then D_B and τ_B are $\left(\widetilde{\mathcal{F}}_t\right)$-stopping times.*

DEFINITION. Let $\mathcal{K}$ be the collection of all compact subsets of E. A function $\varphi : \mathcal{K} \to \mathbb{R}$ is called a *Choquet capacity* if it possesses the following properties:

(i) If A and B belong to $\mathcal{K}$ and if $A \subseteq B$, then $\varphi(A) \subseteq \varphi(B)$;

(ii) For every $A \in \mathcal{K}$ and for every $\varepsilon > 0$, there exists an open subset U of E, $U \supseteq A$, such that for every $B \in \mathcal{K}$ for which $A \subseteq B \subseteq U$, the inequality $\varphi(B) - \varphi(A) < \varepsilon$ holds;

(iii) For all compact A and B in $\mathcal{K}$, the inequality

$$\varphi(A \cup B) + \varphi(A \cap B) \leq \varphi(A) + \varphi(B)$$

is valid.

Given such a capacity φ, we define an inner capacity $\varphi_*(A)$ of a subset A of E by

$$\varphi_*(A) := \sup \{\varphi(K) : K \subseteq A,\ K \in \mathcal{K}\},$$

and we define an outer capacity $\varphi^*(A)$ of an arbitrary subset A of E by

$$\varphi^*(A) := \inf \{\varphi_*(U) : U \supseteq A,\ U \text{ open}\}.$$

A subset A of E is said to be *capacitable* if $\varphi_*(A) = \varphi^*(A)$, and by definition this common value is called the *capacity* of A and it is denoted by $\varphi(A)$.

C. 25. Theorem. *Every Borel measurable (more generally every analytic) subset of E is capacitable.*

This theorem is used to prove that hitting times of Borel subsets are $\left(\widetilde{\mathcal{F}}_t\right)$-stopping times. The following proposition shows how entry times and Choquet capacities are related.

C.26. Proposition. *Fix $t > 0$ and define, for a probability measure μ on $\mathcal{E}$ the function $\varphi : \mathcal{K} \to [0,1]$ by*

$$\varphi(K) = \mathbb{P}_\mu \{D_K \le t\}, \quad K \in \mathcal{K}.$$

Then φ is a Choquet capacity.

By assertion (iii) of the proposition in C.22 the event $\{D_K \le t\}$ belongs to $\widetilde{\mathcal{F}}_t$ and so φ is well-defined. For more details on the results in the reader is referred to §10 in Chapter I of Blumenthal and Getoor [35].

REMARK. The reader is advised to compare the following definitions: Dirichlet capacity, Newton capacity (see e.g. Definition 2.36, and Proposition 2.37, Definition D.17, and Proposition D.19.(b) in Appendix D), and the Choquet capacity, defined in Proposition C.26. If in Proposition C.26. the measure μ is replaced with the measure m, and if $\varphi(B)$ is finite, then the Choquet capacity is comparable to the Newton capacity: see Lemma 4.21. The two are also comparable in the case when the penetration time S of Γ coincides with the hitting time $\tau_\Gamma = T$. This can be seen as follows. Since, for a closed subset B, $\{D_B \le t\} = \{\tau_B \le t\} \cup \{X(0) \in B\}$, and since $X(0) = x$, $\mathbb{P}_x$-almost surely, we infer $\mathbb{P}_x \{D_B \le t\} = \mathbb{P}_x \{\tau_B \le t\}$ for all $x \in B^r \cup (E \setminus B)$. Here $B^r = \{x \in E : \mathbb{P}_x \{\tau_B = 0\} = 1\}$ is the collection of regular points of the set B. From Proposition D.7 it follows that $m(B \setminus B^r) = 0$. Hence, we infer that

$$\int \mathbb{P}_x \{\tau_B \le t\} \, dx = \int_{B^r \cup (B \setminus B^r)} \mathbb{P}_x \{D_B \le t\} \, dx + \int_{B \setminus B^r} \mathbb{P}_x \{\tau_B \le t\} \, dx$$
$$= \int \mathbb{P}_x \{D_B \le t\} \, dx.$$

Consequently, the set B has finite capacity if and only if $\int \mathbb{P}_x \{D_B \le t\} \, dx$ is finite: see Proposition D.19.(b).

Appendix D

Dirichlet Kernels, Harmonic Measures, Capacities

In this appendix we treat three topics. First we prove the symmetry and the joint continuity of the Dirichlet kernels. The results are given in Theorem D.4 and D.5. For the proof we need the theorems and lemmas up to Lemma D.11. The second part deals with harmonic measures and its relation to the λ-equilibrium potential (Definition D.12 up to Theorem D.16). Finally we study the λ-Dirichlet and λ-Newton capacity (see Definition D.13). They turn out to coincide: see Propostition D.18 and D.19, provided the underlying Markov process is conservative. The main results in this appendix are Theorem D.4, Theorem D.5, and the equivalence of the assertions (i), (iii) and (iv) in Theorem D.14. Theorem D.4 says that the Dirichlet kernel is symmetric, and Theorem D.5 provides us with the joint continuity of the Dirichlet kernel. Both of these results were used in the proof of Theorem 2.5 (see Proposition 3.3 and its proof). The results from Theorem D.14 were a.o. employed in the proofs of equality (2.58) in Theorem 2.34, of Corollary 4.17, and of Proposition 7.2. In these proofs we used assertion (iv) of Theorem D.14 for penetration times instead of hitting times: see Remark 2 following the proof of Theorem D.16. Theorem D.16 says that sets of the form $B \setminus B^r$ are polar. Assertion (7) in Proposition D.18 compares several notions of λ-capacity. Finally, Theorem D.21 deals with Doob's Theorem on regular points.

Throughout this appendix we employ the following notation. Let U be an open subset of the second countable locally compact space E and let $T = \tau_{E \setminus U}$ be the exit time from U: $T = \inf\{s > 0 : X(s) \in E \setminus U\}$. In this section we will employ the martingale property of the process $s \mapsto p_0(t-s, X(s), y)$, $0 \le s < t$: see Example 5 in C.4. of Appendix C. Only in Theorem D.1, in Definition D.17, in Propositions D.18 through D.20, and Theorem D.21, we will use the capitals T and T_0 for hitting times. In the other part we use the symbol τ_B for the hitting time of B. In the main text of the book we have used T for the exit time of an open subset Σ of E (see e.g. the equalities in (2.39) of Chapter 2.C) and S for the penetration time of its complement (see Chapter 1, E.5 and also (2.39) in Chapter 2).

D.1. Continuity and symmetry of Dirichlet kernels

We begin with a relevant convergence property for hitting times.

D.1. THEOREM. *Let K be a compact subset of U and suppose $0 < t_0 < t_1 < \infty$. Let T be the exit time from U. Then the following equalities are valid:*

(a)
$$\inf_{\varepsilon>0} \sup_{x\in K} \mathbb{P}_x \{T < \varepsilon\} = 0;$$

(b)
$$\inf_{\varepsilon>0} \sup_{t\in[t_0,t_1]} \sup_{x\in K} \sup_{y\in K} \mathbb{E}_x \left[p_0(t-T, X(T), y) : T < \varepsilon\right] = 0.$$

Proof. Choose a compact subset K_0 of U and an open subset U_0 such that its closure is compact and such that the following inclusions are valid:

$$K \subseteq \operatorname{int}(K_0) \subseteq K_0 \subseteq U_0 \subseteq \overline{U}_0 \subseteq U.$$

Let u and v be functions in $C_\infty(E)$ such that

$$1_K \leq u \leq 1_{\operatorname{int}(K_0)} \quad \text{and such that} \quad 1_{K_0} \leq v \leq 1_{U_0}.$$

Put $T_0 = \inf\{s > 0 : X(s) \in E \setminus U_0\}$. Since $U_0 \subseteq U$ we see $T_0 \leq T$, $\mathbb{P}_x$-almost surely. So that, for $x \in K$, we get

$$\begin{aligned}
\mathbb{P}_x \{T < \varepsilon\} &\leq \mathbb{P}_x \{T_0 < \varepsilon\} \\
&\leq \mathbb{P}_x \{T_0 < \varepsilon,\ X(\varepsilon) \in \operatorname{int}(K_0)\} + \mathbb{P}_x \{T_0 < \varepsilon,\ X(\varepsilon) \notin \operatorname{int}(K_0)\} \\
&\leq \mathbb{P}_x \{T_0 < \varepsilon,\ X(\varepsilon) \in K_0\} + \mathbb{P}_x \{X(\varepsilon) \notin \operatorname{int}(K_0)\}
\end{aligned}$$

(strong Markov property)

$$= \mathbb{E}_x \left[\mathbb{P}_{X(T_0)} \{X(\varepsilon - T_0) \in K_0\}, T_0 < \varepsilon\right] + 1 - \mathbb{P}_x \{X(\varepsilon) \in \operatorname{int}(K_0)\}$$

(paths are right continuous, so that $X(T_0)$ belongs to $E^\triangle \setminus U_0$)

$$\begin{aligned}
&\leq \sup_{0\leq s\leq\varepsilon} \sup_{y\notin U_0} \mathbb{P}_y \{X(s) \in K_0\} \mathbb{P}_x \{T_0 < \varepsilon\} + 1 - \mathbb{P}_x \{X(\varepsilon) \in \operatorname{int}(K_0)\} \\
&\leq \sup_{0\leq s\leq\varepsilon} \sup_{y\notin U_0} \mathbb{E}_y \{v(X(s))\} + u(x) - \mathbb{E}_x \{u(X(\varepsilon))\} \\
&\leq \sup_{0\leq s\leq\varepsilon} \sup_{y\notin U_0} \{[\exp(-sK_0)\, v](y) - v(y)\} + u(x) - [\exp(-\varepsilon K_0)\, u](x).
\end{aligned}$$

Hence we obtain:

$$\sup_{x\in K} \mathbb{P}_x \{T < \varepsilon\} \leq \sup_{0\leq s\leq\varepsilon} \left\|\exp(-sK_0)\, v - v\right\|_\infty + \left\|u - \exp(-\varepsilon K_0)\, u\right\|_\infty .$$

By the strong continuity of the semigroup $\{\exp(-tK_0) : t \geq 0\}$ in $C_\infty(E)$, the claim (a) in Theorem D.1 follows.

For a proof of item (b) we proceed in a similar manner. To be specific we write ($x \in K$, $y \in K$)

$$\mathbb{E}_x \{p_0(t - T, X(T), y) : T < \varepsilon\}$$

(the process $s \mapsto p_0(t - s, X(s), y)$ is a $\mathbb{P}_x$-martingale on $[0, t)$)

$$\begin{aligned}
&\leq \mathbb{E}_x \{p_0(t - T_0, X(T_0), y) : T_0 < \varepsilon\} \\
&\leq \mathbb{E}_x \{p_0(t - T_0, X(T_0), y) : T_0 < \varepsilon,\ X(\varepsilon) \in \operatorname{int}(K_0)\} \\
&\quad + \mathbb{E}_x \{p_0(t - T_0, X(T_0), y) : T_0 < \varepsilon,\ X(\varepsilon) \notin \operatorname{int}(K_0)\} \\
&\leq \mathbb{E}_x \{p_0(t - T_0, X(T_0), y) : T_0 < \varepsilon,\ X(\varepsilon) \in K_0\} \\
&\quad + \mathbb{E}_x \{p_0(t - T_0, X(T_0), y) : T_0 < \varepsilon,\ X(\varepsilon) \notin \operatorname{int}(K_0)\}
\end{aligned}$$

(strong Markov property and

martingale property of the process $s \mapsto p_0(t - s, X(s), y)$)

$$\begin{aligned}
&\leq \mathbb{E}_x \left[p_0(t - T_0, X(T_0), y) \mathbb{P}_{X(T_0)} \{X(\varepsilon - T_0) \in K_0\}, T_0 < \varepsilon\right] \\
&\quad + \mathbb{E}_x \left[p_0(t - \varepsilon, X(\varepsilon), y)\right] - \mathbb{E}_x \{p_0 (t - \varepsilon, X(\varepsilon), y) : X(\varepsilon) \in \operatorname{int}(K_0)\}
\end{aligned}$$

(paths are right continuous, so that $X(T_0)$ belongs to $E^{\triangle} \setminus U_0$)

$$\begin{aligned}
&\leq \sup_{0 \leq s \leq \varepsilon} \sup_{z \notin U_0} \mathbb{P}_z \{X(s) \in K_0\} \mathbb{E}_x \{p_0(t - T_0, X(T_0), y) : T_0 < \varepsilon\} \\
&\quad + p_0(t, x, y) u(x) - \mathbb{E}_x \{p_0(t - \varepsilon, X(\varepsilon), y) u(X(\varepsilon))\} \\
&\leq \sup_{0 \leq s \leq \varepsilon} \sup_{z \notin U_0} \left[\mathbb{E}_z \{v(X(s))\} - v(z)\right] p_0(t, x, y) \\
&\quad + p_0(t, x, y) u(x) - \mathbb{E}_x \{p_0(t - \varepsilon, X(\varepsilon), y) u(X(\varepsilon))\} \\
&\leq \sup_{0 \leq s \leq \varepsilon} \sup_{z \notin U_0} \{[\exp(-sK_0) v](z) - v(z)\} p_0(t, x, y) \\
&\quad + p_0(t, x, y) u(x) - [\exp(-\varepsilon K_0) p_0(t - \varepsilon, \cdot, y) u(\cdot)](x).
\end{aligned}$$

Since, $0 < \varepsilon_0 < t_0$, the family

$$\{p_0(t - \varepsilon, \cdot, y) u(\cdot) : t \in [t_0, t_1],\ y \in K,\ 0 \leq \varepsilon \leq \varepsilon_0\}$$

is a compact subset of $C_\infty(E)$ (in the uniform norm topology), the conclusion in (b) follows. This is so because of the following argument. Suppose that the sequence $(\varepsilon_n : n \in \mathbb{N})$ decreases to 0. Then, since the function $p_0(t, x, y)$ is continuous on $(0, \infty) \times E \times E$, the expression

$$\sup_{(t,x,y) \in [t_0, t_1] \times K \times K} p_0(t, x, y)$$

is finite. Moreover since the unperturbed kernel is jointly continuous, there exists a sequence $((t_n, y_n) : n \in \mathbb{N})$ in $[t_0, t_1] \times K$ for which

$$\begin{aligned}
&\sup_{(t,x,y) \in [t_0, t_1] \times K \times K} \left|p_0(t, x, y) u(x) - [\exp(-\varepsilon_n K_0) p_0(t - \varepsilon_n, \cdot, y) u(\cdot)](x)\right| \\
&= \sup_{x \in K} \left|p_0(t_n, x, y_n) u(x) - [\exp(-\varepsilon_n K_0) p_0(t_n - \varepsilon_n, \cdot, y_n) u(\cdot)](x)\right|.
\end{aligned}$$

Then, upon choosing a subsequence, we may and do assume that the sequence $((t_n, y_n) : n \in \mathbb{N})$ converges to $(t, y) \in [t_0, t_1] \times K$. It then follows that the sequence $(p_0(t_n - \varepsilon_n, \cdot, y_n)u(\cdot) : n \in \mathbb{N})$ converges uniformly to the function $p_0(t, \cdot, y)u(\cdot)$. Since the sequence $(p_0(t_n, \cdot, y_n)u(\cdot) : n \in \mathbb{N})$ has the same limit, the conclusion in (b) follows.

D.2. PROPOSITION. *Assume that the function* $(t,x) \mapsto \int p_0(t,x,y)dy = \mathbb{P}_x[X(t) \in E]$ *is upper semi-continuous. Then, for every bounded Borel measurable function f defined on E, the function* $(t, x) \mapsto [\exp(-tK_0) f](x)$ *is continuous.*

REMARK. The proof of the joint continuity of the function $(t, x) \mapsto \int_E p_0(t, x, y)dy$ can be reduced to showing the separate continuity of these functions. This claim can be read of from the equality

$$\int p_0(t, x, y)dy = \int p_0(s, x, z) \int p_0(t - s, z, y)\, dy\, dz,$$

together with Scheffé's theorem (Appendix E, Theorem E.3) on uniform integrability (also see Bauer [19, p. 105, Corollary 2.12.5]).

Proof. We begin with $f \equiv 1$. Then

$$\exp(-tK_0)\, 1 = \sup_{n \in \mathbb{N}} \exp(-tK_0)\, u_n,$$

where the sequence $(u_n : n \in \mathbb{N})$ is a sequence in $C_\infty(E)$ that pointwise increases to the function that is identically 1. Moreover each function u_n is supposed to be non-negative. Since the semigroup $\{\exp(-tK_0) : t \geq 0\}$ leaves the space $C_\infty(E)$ invariant, the function $\exp(-tK_0)\, 1$ is lower semi-continuous, i.e. for every $\alpha \in \mathbb{R}$, the set $\{(t, x) \in (0, \infty) \times E : \exp(-tK_0)\, 1(x) > \alpha\}$ is open. Since, by assumption, the function $\exp(-tK_0)\, 1$ is upper semi-continuous, it is continuous. Next, let K be a compact subset of E. Then

$$\int_K p_0(t, x, y)dy = \inf \{\exp(-tK_0)\, v(x) : v \in C_\infty(E),\ v \geq 1_K\},$$

and so the function $\exp(-tK_0)\, 1_K(x) = \int_K p_0(t, x, y)dy$ is upper semi-continuous. It is also lower semi-continuous because it is the limit of a sequence of lower semi-continuous functions that converges uniformly on compact subsets. In fact, let $(U_n : n \in \mathbb{N})$ be a decreasing sequence of open subsets with compact closure, each containing K, for which

$$m(U_n \setminus K) \leq 2^{-n} \left(1 + \sup_{(t,x,y) \in [t_1,t_2] \times K \times U_1} p_0(t, x, y)\right)^{-1}.$$

Here $0 < t_1 < t_2 < \infty$ are fixed, but arbitrary. Then

$$0 \leq \int_{U_n} p_0(t, x, y)dy - \int_K p_0(t, x, y)dy \leq 2^{-n}.$$

Each of the functions $(t,x) \mapsto \int_{U_n} p_0(t,x,y)dy$ is lower semi-continuous and consequently its limit inherits the same property. Next let B be an arbitrary Borel subset of E. Then

$$\int_B p_0(t,x,y)dy = \sup_{K\subseteq B, K \text{ compact}} \int_K p_0(t,x,y)dy,$$

and so the function $(t,x) \mapsto \int_B p_0(t,x,y)dy$ is lower semi-continuous. By the same token the function $(t,x) \mapsto \int_{E\setminus B} p_0(t,x,y)dy$ is lower semi-continuous. From the equality

$$\int_B p_0(t,x,y)dy = \int_E p_0(t,x,y)dy - \int_{E\setminus B} p_0(t,x,y)dy$$

it follows that the function $(t,x) \mapsto \int_B p_0(t,x,y)dy$ is lower as well as upper semi-continuous. Consequently it is continuous. If f is a bounded Borel measurable function, then it can uniformly be approximated by finite linear combinations of the form $f_n = \sum_{j=1}^n \alpha_j 1_{B_j}$, $\alpha_j \in \mathbb{C}$. The continuity of $\exp(-tK_0) f$ will then be a consequence of that of the functions $\exp(-tK_0) f_n$. This completes the proof of Proposition D.2.

Let B be a Borel subset of E. The formulas in (D.1) and (D.2) below should be compared with Theorem 2.21.(b), with Proposition 2.22 (equality (2.36), and with the comments following Proposition 3.1. Notice the identity

$$p_0(t,x,y) - \mathbb{E}_x\left[p_0(t-\tau_B, X(\tau_B), y) : \tau_B < t\right] = \mathbb{E}_x\left[M_0^{t,y}(\tau_B) : \tau_B < \infty\right].$$

and set

$$q_B(t,x,y) = p_0(t,x,y) - \mathbb{E}_x\left[p_0(t-\tau_B, X(\tau_B), y) : \tau_B < t\right] \quad \text{and} \tag{D.1}$$

$$Q_B(t)f(x) = \int q_B(t,x,y)f(y)dy, \tag{D.2}$$

for f a bounded Borel measurable function. The space of bounded Borel measurable functions is denoted by $L^\infty(E,\mathcal{E})$. Here τ_B is the hitting time of B: $\tau_B = \inf\{s > 0 : X(s) \in B\}$. We write $\tau_B(\omega) = \zeta(\omega)$, if the path $X(t,\omega)$ does not hit B. Next we want to prove the following theorems.

D.3. THEOREM. *Let B be a Borel subset of E and let $q_B(t,x,y)$ and Q_B be as in (D.2). For $B \in \mathcal{E}$ the following assertions are true:*

(a) $\mathbb{E}_x\left[f(X(t)) : \tau_B > t\right] = \int q_B(t,x,y)f(y)dy = \left[Q_B(t)f\right](x)$, $t > 0$, $f \in L^\infty(E,\mathcal{E})$, *and* $x \in E$;

(b) $\int q_B(t-\varepsilon,x,z)p_0(\varepsilon,z,y)dz \downarrow q_B(t,x,y)$, *if ε decreases to 0, and if $t > 0$, x, $y \in E$;*

(c) $\int p_0(\varepsilon,x,z)q_B(t-\varepsilon,z,y)dz \downarrow q_B(t,x,y)$, *if ε decreases to 0, and if $t > 0$, x, $y \in E$;*

(d) $q_B(t,x,y) \geq 0$ *for* $t > 0$, x, $y \in E$;
(e) $Q_B(s+t) = Q_B(s) \circ Q_B(t)$, *for* s, $t \geq 0$;
(f) $q_B(s+t,x,y) = \int q_B(s,x,z)q_B(t,z,y)dz$, *for* s, $t > 0$, *and for* x, $y \in E$;
(g) *If* C *is another Borel subset of* E, *for which* $B \subseteq C$, *then*

$$q_B(t,x,y) = q_C(t,x,y) + \mathbb{E}_x\left[q_B\left(t-\tau_C, X(\tau_C), y\right) : \tau_C < t\right], \tag{D.3}$$

for $t > 0$ *and for* x, $y \in E$;
(h) *Fix* $x \in E$ *and let* $(B_n : n \in \mathbb{N})$ *be an increasing sequence of Borel subsets of* E *with the property that* τ_{B_n} *decreases to* τ_B, $\mathbb{P}_x$*-almost surely. Then* $q_{B_n}(t,x,y)$ *decreases to* $q_B(t,x,y)$, $y \in E$ *as* n *goes to* ∞.
(i) *For fixed* x *and* y *in* E *the function* $t \mapsto q_B(t,x,y)$ *is continuous.*
(j) *Let* ε_1 *and* ε_2 *be strictly positive real numbers for which* $\varepsilon_1 + \varepsilon_2 < t$. *For* x *and* y *in* E *the identity*

$$\begin{aligned}
&\int\int p_0\left(\varepsilon_1, x, z_1\right) q_B\left(t-\varepsilon_1-\varepsilon_2, z_1, z_2\right) p_0\left(\varepsilon_2, z_2, y\right) dz_1\, dz_2 - q_B(t,x,y) \\
&= \mathbb{E}_x\left[p_0\left(t-\tau_B, X(\tau_B), y\right) : \tau_B < \varepsilon_1\right] \\
&\quad + \mathbb{E}_x\left[p_0\left(t-\tau_B, X(\tau_B), y\right) : t-\varepsilon_2 < \tau_B < t\right] \\
&\quad - \mathbb{E}_x\left[p_0\left(t-\tau_{\varepsilon_1}, X(\tau_{\varepsilon_1}), y\right) : \tau_B < \varepsilon_1, \tau_{\varepsilon_1} < t-\varepsilon_2\right]
\end{aligned}$$

holds true. Moreover,

$$\begin{aligned}
&\lim_{\varepsilon_1\downarrow 0}\lim_{\varepsilon_2\downarrow 0}\int\int p_0\left(\varepsilon_1, x, z_1\right) q_B\left(t-\varepsilon_1-\varepsilon_2, z_1, z_2\right) p_0\left(\varepsilon_2, z_2, y\right) dz_1\, dz_2 \\
&= \lim_{\varepsilon_2\downarrow 0}\lim_{\varepsilon_1\downarrow 0}\int\int p_0\left(\varepsilon_1, x, z_1\right) q_B\left(t-\varepsilon_1-\varepsilon_2, z_1, z_2\right) p_0\left(\varepsilon_2, z_2, y\right) dz_1\, dz_2 \\
&= q_B(t,x,y).
\end{aligned}$$

(k) *Fix* $t > 0$. *The function* $x \mapsto \mathbb{P}_x\left[\tau_B < t\right]$ *is lower semi-continuous, i.e. for every* $\alpha \in \mathbb{R}$, *the set* $\{x \in E : \mathbb{P}_x\left[\tau_B < t\right] > \alpha\}$ *is open.*

In (j) we wrote $\tau_\varepsilon = \varepsilon + \tau_B \circ \vartheta_\varepsilon$. We note that τ_B is a terminal stopping time so that on the event $\{\tau_B > \varepsilon\}$ the identity $\tau_\varepsilon = \tau_B$ holds: see Definition 2.18. The limits in (j) are infima.

REMARK 1. Only the proof of (g) depends on the symmetry of the function $(t,x,y) \mapsto p_0(t,x,y)$ in the sense that equality (D.10) below is established using the fact that sets of the form $B \setminus B^r$ are polar. The "polarity" of such sets depends on the symmetry of the function $p_0(t,x,y) = p_0(t,y,x)$.

REMARK 2. Notice the equality $\lim_{\varepsilon\downarrow 0}\left(\varepsilon + \tau_B \circ \vartheta_\varepsilon\right) = \tau_B$, $\mathbb{P}_x$-almost surely.

Proof. (a) The equality $Q_B(t)f(x) = \int q_B(t,x,y)f(y)dy$ is in fact a definition: see (D.2). Since τ_B is a terminal stopping time (see Definition 2.18), we have by virtue of Lemma 2.19 $\mathbb{P}_x[\tau_B = t] = 0$, for $t > 0$ and for $x \in E$. From the definition of $q_B(t,x,y)$, (see (D.1)), we get via the Markov property

$$\begin{aligned}
&\int q_B(t,x,y)f(y)dy \\
&= \int p_0(t,x,y)f(y)dy - \mathbb{E}_x\left[\int p_0(t-\tau_B, X(\tau_B), y)f(y)dy : \tau_B < t\right] \\
&= \int p_0(t,x,y)f(y)dy - \mathbb{E}_x\left[\mathbb{E}_{X(\tau_B)}\left\{f\left(X\left(t-\tau_B\right)\right)\right\} : \tau_B < t\right] \\
&= \mathbb{E}_x\left[f(X(t))\right] - \mathbb{E}_x\left[f\left(X\left(t\right)\right) : \tau_B < t\right] \\
&= \mathbb{E}_x\left[f(X(t)) : \tau_B \geq t\right] = \mathbb{E}_x\left[f(X(t)) : \tau_B > t\right].
\end{aligned}$$

This proves (a).

(b) and (d) From (a) we infer among other things that $q_B(t,x,z) \geq 0$ for m-almost all $z \in E$. So for $0 < \varepsilon < t$ and for x, $y \in E$ the following (in-)equalities hold true:

$$\begin{aligned}
0 &\leq \int q_B(t-\varepsilon, x, z)\, p_0(\varepsilon, z, y)\, dz \\
&= \int p_0(t-\varepsilon, x, z)\, p_0(\varepsilon, z, y)\, dz \\
&\quad - \mathbb{E}_x\left[\int p_0(t-\varepsilon-\tau_B, X(\tau_B), z)\, p_0(\varepsilon, z, y)\, dz : \tau_B < t-\varepsilon\right] \\
&= p_0(t,x,y) - \mathbb{E}_x\left[p_0(t-\tau_B, X(\tau_B), y) : \tau_B < t-\varepsilon\right]. \qquad \text{(D.4)}
\end{aligned}$$

Hence (b) follows and so does (d).

(c) For $0 < \varepsilon < t$, and for x, $y \in E$, the following sequence of equalities is valid:

$$\begin{aligned}
&\int p_0(\varepsilon, x, z) q_B(t-\varepsilon, z, y) dz \\
&= \int p_0(\varepsilon, x, z)\left\{p_0(t-\varepsilon, z, y) - \mathbb{E}_z\left[p_0(t-\varepsilon-\tau_B, X(\tau_B), y) : \tau_B < t-\varepsilon\right]\right\} dz \\
&= p_0(t,x,y) - \mathbb{E}_x\left[\mathbb{E}_{X(\varepsilon)}\left\{p_0(t-\varepsilon-\tau_B, X(\tau_B), y) : \tau_B < t-\varepsilon\right\}\right]
\end{aligned}$$

(Markov property)

$$\begin{aligned}
&= p_0(t,x,y) - \mathbb{E}_x\left[p_0\left(t-\varepsilon-\tau_B\circ\vartheta_\varepsilon, X(\varepsilon+\tau_B\circ\vartheta_\varepsilon), y\right) : \varepsilon+\tau_B\circ\vartheta_\varepsilon < t\right] \\
&= p_0(t,x,y) - \mathbb{E}_x\left[p_0\left(t-\varepsilon-\tau_B\circ\vartheta_\varepsilon, X(\varepsilon+\tau_B\circ\vartheta_\varepsilon), y\right) : \varepsilon+\tau_B\circ\vartheta_\varepsilon < t,\ \tau_B > \varepsilon\right] \\
&\qquad - \mathbb{E}_x\left[p_0\left(t-\varepsilon-\tau_B\circ\vartheta_\varepsilon, X(\varepsilon+\tau_B\circ\vartheta_\varepsilon), y\right) : \varepsilon+\tau_B\circ\vartheta_\varepsilon < t,\ \tau_B \leq \varepsilon\right] \\
&= p_0(t,x,y) - \mathbb{E}_x\left[p_0\left(t-\tau_B, X(\tau_B), y\right) : \varepsilon < \tau_B < t\right] \\
&\qquad - \mathbb{E}_x\left[p_0\left(t-\varepsilon-\tau_B\circ\vartheta_\varepsilon, X(\varepsilon+\tau_B\circ\vartheta_\varepsilon), y\right) : \varepsilon+\tau_B\circ\vartheta_\varepsilon < t,\ \tau_B \leq \varepsilon\right] \\
&= q_B(t,x,y) + \mathbb{E}_x\left[p_0(t-\tau_B, X(\tau_B), y) : \tau_B \leq \varepsilon\right] \\
&\qquad - \mathbb{E}_x\left[p_0\left(t-\varepsilon-\tau_B\circ\vartheta_\varepsilon, X(\varepsilon+\tau_B\circ\vartheta_\varepsilon), y\right) : \varepsilon+\tau_B\circ\vartheta_\varepsilon < t,\ \tau_B \leq \varepsilon\right]
\end{aligned}$$

(Markov property)

$$= q_B(t,x,y) + \mathbb{E}_x\left[p_0(t-\tau_B, X(\tau_B), y) : \tau_B \le \varepsilon\right]$$
$$- \mathbb{E}_x\left[\mathbb{E}_{X(\varepsilon)}\left\{p_0\left(t-\varepsilon-\tau_B, X(\tau_B), y\right) : \tau_B < t-\varepsilon\right\}, \tau_B \le \varepsilon\right]. \tag{D.5}$$

From assertion (b), we get $q_B(t,x,y) \ge 0$ for all $t > 0$ and for all x and y in E. Hence

$$0 \le \mathbb{E}_z\left[p_0(t-\varepsilon-\tau_B, X(\tau_B), y) : \tau_B < t-\varepsilon\right] \le p_0(t-\varepsilon, z, y),$$

and so

$$\mathbb{E}_{X(\varepsilon)}\left[p_0(t-\varepsilon-\tau_B, X(\tau_B), y) : \tau_B < t-\varepsilon\right] \le p_0(t-\varepsilon, X(\varepsilon), y). \tag{D.6}$$

Since the function $p_0(s,z,y)$ is continuous on $(0,\infty)\times E\times E$, and since $\tau_\varepsilon := \varepsilon + \tau_B\circ\vartheta_\varepsilon \downarrow \tau_B$, if $\varepsilon \downarrow 0$, we infer that

$$\begin{aligned}
&\liminf_{\varepsilon\downarrow 0} \mathbb{E}_x\left[p_0\left(t-\tau_\varepsilon, X(\tau_\varepsilon), y\right) : \tau_\varepsilon < t,\ \tau_B \le \varepsilon\right] \\
&\ge \mathbb{E}_x\left[\liminf_{\varepsilon\downarrow 0} p_0\left(t-\tau_\varepsilon, X(\tau_\varepsilon), y\right) 1_{\{\tau_\varepsilon < t\}} 1_{\{\tau_B \le \varepsilon\}}\right] \\
&= \mathbb{E}_x\left[p_0(t, X(0), y) : \tau_B = 0\right] = p_0(t,x,y)\mathbb{P}_x\left[\tau_B = 0\right].
\end{aligned} \tag{D.7}$$

From the Markov property and from (D.6) together with the continuity of the function $(s,z) \mapsto p_0(s,z,y)$, we derive the following:

$$\begin{aligned}
&\limsup_{\varepsilon\downarrow 0} \mathbb{E}_x\left[p_0\left(t-\tau_\varepsilon, X(\tau_\varepsilon), y\right) : \tau_\varepsilon < t,\ \tau_B \le \varepsilon\right] \\
&\le \limsup_{\varepsilon\downarrow 0} \mathbb{E}_x\left[\mathbb{E}_{X(\varepsilon)}\left\{p_0\left(t-\varepsilon-\tau_B, X(\tau_B), y\right) : \tau_B < t-\varepsilon\right\},\ \tau_B \le \varepsilon\right] \\
&\le \limsup_{\varepsilon\downarrow 0} \mathbb{E}_x\left[p_0(t-\varepsilon, X(\varepsilon), y) : \tau_B \le \varepsilon\right]
\end{aligned}$$

(the process $s \mapsto p_0(t-s, X(s), y)$ is a martingale)

$$\begin{aligned}
&\le \limsup_{\varepsilon\downarrow 0} \mathbb{E}_x\left[p_0(t-\tau_B, X(\tau_B), y) : \tau_B \le \varepsilon\right] \\
&= \mathbb{E}_x\left[p_0(t, X(0), y) : \tau_B = 0\right] = p_0(t,x,y)\mathbb{P}_x\left[\tau_B = 0\right].
\end{aligned} \tag{D.8}$$

From (D.7) and (D.8) the assertion in (c) follows except for the monotonicity. This can be achieved as follows. First we note the identity

$$\begin{aligned}
&\mathbb{E}_x\left[p_0(t-\tau_B, X(\tau_B), y) : \tau_B \le \varepsilon\right] \\
&\qquad - \mathbb{E}_x\left[p_0\left(t-\varepsilon-\tau_B\circ\vartheta_\varepsilon, X\left(\varepsilon+\tau_B\circ\vartheta_\varepsilon\right), y\right) : \varepsilon+\tau_B\circ\vartheta_\varepsilon < t,\ \tau_B \le \varepsilon\right] \\
&\quad = \mathbb{E}_x\left[p_0\left(t-\tau_B, X(\tau_B), y\right) : \tau_B < t\right] - \mathbb{E}_x\left[p_0\left(t-\tau_\varepsilon, X\left(\tau_\varepsilon\right), y\right) : \tau_\varepsilon < t\right],
\end{aligned}$$

where, as above, $\tau_\varepsilon = \varepsilon + \tau_B\circ\vartheta_\varepsilon = \tau_B$, on the event $\{\tau_B > \varepsilon\}$. Fix $0 < \varepsilon_2 < \varepsilon_1$. Then the equalities

$$\begin{aligned}
&\mathbb{E}_x\left[p_0(t-\tau_{\varepsilon_2}, X(\tau_{\varepsilon_2}), y) : \tau_{\varepsilon_2} < t\right] - \mathbb{E}_x\left[p_0(t-\tau_{\varepsilon_1}, X(\tau_{\varepsilon_1}), y) : \tau_{\varepsilon_1} < t\right] \\
&= \lim_{\eta\uparrow t}\left\{\mathbb{E}_x\left[p_0(t-\tau_{\varepsilon_2}, X(\tau_{\varepsilon_2}), y) : \tau_{\varepsilon_2} \le \eta\right] - \mathbb{E}_x\left[p_0(t-\tau_{\varepsilon_1}, X(\tau_{\varepsilon_1}), y) : \tau_{\varepsilon_1} \le \eta\right]\right\} \\
&= \lim_{\eta\uparrow t} \mathbb{E}_x\left[p_0(t-\eta, X(\eta), y) :\ \tau_{\varepsilon_2} \le \eta < \tau_{\varepsilon_1}\right] \ge 0
\end{aligned}$$

imply the decreasing character of the limit in (c).

(e) This semigroup property is a consequence of the Markov property and the fact that τ_B is a terminal stopping time. Suppose f belongs to $L^\infty(E,\mathcal{E})$. Then

$$\begin{aligned}
&Q_B(s)Q_B(t)f(x) = \mathbb{E}_x\left[Q_B(t)f(X(s) : \tau_B > s\right] \\
&= \mathbb{E}_x\left[\mathbb{E}_{X(s)}\left\{f(X(t)) : \tau_B > t\right\} : \tau_B > s\right] \\
&= \mathbb{E}_x\left[f(X(t+s)) : s + \tau_B \circ \vartheta_s > s+t,\ \tau_B > s\right] = \mathbb{E}_x\left[f(X(t+s)) : \tau_B > s+t\right].
\end{aligned}$$

(f) From (a) and (e) we infer, for $f \in L^\infty(E,\mathcal{E})$, that

$$\begin{aligned}
&\int q_B(s+t,x,y)f(y)dy = \left[Q_B(s+t)f\right](x) = \left[Q_B(s)\left(Q_B(t)f\right)\right](x) \\
&= \int q_B(s,x,z)\left[Q_B(t)f\right](z)dz = \int q_B(s,x,z)\int q_B(t,z,y)f(y)dy\,dz \\
&= \int\left[\int q_B(s,x,z)q_B(t,z,y)dz\right]f(y)dy.
\end{aligned}$$

So that, for s, $t > 0$, and $x \in E$ fixed, the equality

$$q_B(s+t,x,y) = \int q_B(s,x,z)q_B(t,z,y)dz \tag{D.9}$$

holds for m-almost all $y \in E$. Whence, for $0 < \varepsilon < t$, and x, $y \in E$, we get

$$\begin{aligned}
&\int q_B(s+t-\varepsilon,x,u)p_0(\varepsilon,u,y)du \\
&= \int q_B(s,x,z)\left[\int q_B(t-\varepsilon,z,u)p_0(\varepsilon,u,y)du\right]dz.
\end{aligned}$$

In the latter we let ε decrease to 0 to obtain (D.9) for all s, $t > 0$ and for all x, $y \in E$. Here we use (b) and (d).

(g) Here we use the identity

$$\tau_B = \tau_C + \tau_B \circ \vartheta_{\tau_C}, \tag{D.10}$$

on the event $\{\tau_C < \infty\}$, $\mathbb{P}_x$-almost surely. A proof of this equality will be provided while proving Theorem D.16: see (D.45). From (D.10) we get

$$\begin{aligned}
&q_B(t,x,y) \\
&= p_0(t,x,y) - \mathbb{E}_x\left[p_0(t-\tau_B, X(\tau_B), y) : \tau_B < t\right] \\
&= p_0(t,x,y) - \mathbb{E}_x\left[p_0(t-\tau_C-\tau_B\circ\vartheta_{\tau_C}, X(\tau_B)\circ\vartheta_{\tau_C}, y) : \tau_C + \tau_B\circ\vartheta_{\tau_C} < t\right]
\end{aligned}$$

(strong Markov property)

$$\begin{aligned}
&= p_0(t,x,y) - \mathbb{E}_x\left[\mathbb{E}_{X(\tau_C)}\left\{p_0(t-\tau_C-\tau_B, X(\tau_B), y) : \tau_B < t-\tau_C\right\} : \tau_C < t\right] \\
&= q_C(t,x,y) + \mathbb{E}_x\left[q_B(t-\tau_C, X(\tau_C), y) : \tau_C < t\right].
\end{aligned}$$

Whence equality (D.3) follows.

(h) Let x and $(B_n : n \in \mathbb{N})$ be as in (h). Since τ_{B_n} decreases to τ_B, since paths are right-continuous, and since the function $(s,z) \mapsto p_0(s,z,y)$ is continuous on $(0,\infty)\times E$, we conclude that

$$\begin{aligned}&\limsup_{n\to\infty} q_{B_n}(t,x,y)\\ &= p_0(t,x,y) - \liminf_{n\to\infty} \mathbb{E}_x\left[p_0(t-\tau_{B_n}, X(\tau_{B_n}), y) : \tau_{B_n} < t\right]\end{aligned}$$

(Fatou's lemma)

$$\begin{aligned}&\le p_0(t,x,y) - \mathbb{E}_x\left[\liminf_{n\to\infty} p_0(t-\tau_{B_n}, X(\tau_{B_n}), y) 1_{\{\tau_{B_n}<t\}}\right]\\ &= p_0(t,x,y) - \mathbb{E}_x\left[p_0(t-\tau_B, X(\tau_B), y) : \tau_B < t\right]\\ &\le q_{B_{n+1}}(t,x,y) \le q_{B_n}(t,x,y).\end{aligned}$$

So that (h) follows.

(i) Let $(t_n : n \in \mathbb{N})$ be sequence of strictly positive numbers that converges to $t>0$. Fix $0 < s < \inf\{t_n : n \in \mathbb{N}\}$. We may and do assume that $\inf t_n + s > \sup t_n$. Fix $\eta < t - s$ and fix η_0 such that $\sup_n t_n - s < \eta_0 < \inf_n t_n$. Ultimately we let η increase to $t-s$. The following identities then follow from (f) and the martingale property of the process $u \mapsto p_0(t_n - u, X(u), y)$:

$$\begin{aligned}q_B(t_n,x,y) &= \int q_B(t_n - s, x, z) q_B(s,z,y)dz\\ &= \int p_0(t_n - s, x, z) q_B(s,z,y)dz\\ &\quad - \int \mathbb{E}_x\left[p_0(t_n - s - \tau_B, X(\tau_B), z) : \tau_B < t_n - s\right] q_B(s,z,y)dz\\ &= \int \mathbb{E}_x\left[p_0\left(t_n - s - \eta, X(\eta), z\right) : \tau_B > \eta\right] q_B(s,z,y)dz \qquad \text{(D.11)}\\ &\quad - \int \mathbb{E}_x\left[p_0(t_n - s - \tau_B, X(\tau_B), z) : \eta < \tau_B < t_n - s\right] q_B(s,z,y)dz.\end{aligned}$$

Since

$$\begin{aligned}&\int \mathbb{E}_x\left[p_0(t_n - s - \tau_B, X(\tau_B), z) : \eta < \tau_B < t_n - s\right] q_B(s,z,y)dz\\ &\le \int \mathbb{E}_x\left[p_0(t_n - s - \tau_B, X(\tau_B), z) : \eta < \tau_B < t_n - s\right] p_0(s,z,y)dz\\ &= \mathbb{E}_x\left[p_0(t_n - \tau_B, X(\tau_B), z) : \eta < \tau_B < t_n - s\right]\\ &= \mathbb{E}_x\left[p_0(t_n - \eta_0, X(\eta_0), z) : \eta < \tau_B < t_n - s\right]. \qquad \text{(D.12)}\end{aligned}$$

From (D.11) and (D.12), we infer that

$$\begin{aligned}&\int \mathbb{E}_x\left[p_0(t_n - s - \eta, X(\eta), z) : \tau_B > \eta\right] q_B(s,z,y)dz\\ &\quad - \mathbb{E}_x\left[p_0(t_n - \eta_0, X(\eta_0), y) : \eta < \tau_B < t_n - s\right] \le q_B(t_n,x,y)\\ &\le \int \mathbb{E}_x\left[p_0(t_n - s - \eta, X(\eta), z) : \tau_B > \eta\right] q_B(s,z,y)dz. \qquad \text{(D.13)}\end{aligned}$$

From Scheffé's theorem on uniform integrability (see Appendix E, Theorem E.3 or Bauer [19, p. 105, Corollary 2.12.5]), it follows that the sequences

$$(p_0(t_n - \eta_0, X(\eta_0), y) : n \in \mathbb{N}) \quad \text{and} \quad (p_0(t_n - \eta, X(\eta), y) : n \in \mathbb{N})$$

are uniformly $\mathbb{P}_x$-integrable. By virtue of the inequality

$$\begin{aligned} &\int p_0(t_n - s - \eta, X(\eta), z) q_B(s, z, y) dz \\ &\leq \int p_0(t_n - s - \eta, X(\eta), z) p_0(s, z, y) dz = p_0(t_n - \eta, X(\eta), y), \end{aligned}$$

the sequence $\left(p_0(t_n - s - \eta, X(\eta), z) 1_{\{\tau_B > \eta\}} : n \in \mathbb{N}\right)$ is uniformly integrable with respect to the measure determined by

$$(A, C) \mapsto \mathbb{P}_x(A) \int q_B(s, z, y) 1_C(z) dz.$$

Here A belongs to $\mathcal{F}_\eta$ and C is a Borel subset of E. These observations together with (D.13) yield the inequalities

$$\begin{aligned} &\int \mathbb{E}_x \left[p_0(t - s - \eta, X(\eta), z) : \tau_B > \eta\right] q_B(s, z, y) dz \\ &\quad - \mathbb{E}_x \left[p_0(t - \eta_0, X(\eta_0), y) : \eta < \tau_B < t - s\right] \\ &\leq \liminf_{n \to \infty} q_B(t_n, x, y) \leq \limsup_{n \to \infty} q_B(t_n, x, y) \\ &\leq \int \mathbb{E}_x \left[p_0(t - s - \eta, X(\eta), z) : \tau_B > \eta\right] q_B(s, z, y) dz. \end{aligned} \tag{D.14}$$

Next we notice that the expression

$$\begin{aligned} &\mathbb{E}_x \left[p_0(t - s - \eta, X(\eta), z) : \tau_B > \eta\right] \\ &= p_0(t - s, x, z) - \mathbb{E}_x \left[p_0(t - s - \tau_B, X(\tau_B), z) : \tau_B \leq \eta\right] \end{aligned}$$

decreases to $q_B(t - s, x, z)$ as η increases to $t - s$ and that the expression

$$\mathbb{E}_x \left[p_0(t - \eta_0, X(\eta_0), y) : \eta < \tau_B < t - s\right]$$

decreases to 0 as η increases to $t - s$. So (D.14) together with assertion (f) results in (i).

(j) Let ε_1 and ε_2 be strictly positive real numbers for which $\varepsilon_1 + \varepsilon_2 < t$. For x and y in E the following identities follow from the Markov property:

$$
\begin{aligned}
&\int\int p_0(\varepsilon_1, x, z_1)\, q_B(t - \varepsilon_1 - \varepsilon_2, z_1, z_2)\, p_0(\varepsilon_1, z_2, y)\, dz_1\, dz_2 \\
&\quad - q_B(t, x, y) \\
&= \int\int p_0(\varepsilon_1, x, z_1)\, p_0(t - \varepsilon_1 - \varepsilon_2, z_1, z_2)\, p_0(\varepsilon_1, z_2, y)\, dz_1\, dz_2 \\
&\quad - p_0(t, x, y) + \mathbb{E}_x\left[p_0(t - \tau_B, X(\tau_B), y) : \tau_B < t\right] \\
&\quad - \int\int p_0(\varepsilon_1, x, z_1)\, \mathbb{E}_{z_1}\left[p_0(t - \varepsilon_1 - \varepsilon_2 - \tau_B, X(\tau_B), z_2) : \tau_B < t - \varepsilon_1 - \varepsilon_2\right] \\
&\qquad p_0(\varepsilon_2, z_2, y)\, dz_1\, dz_2
\end{aligned}
$$

(Chapman-Kolmogorov and Fubini's theorem)

$$
\begin{aligned}
&= \mathbb{E}_x\left[p_0(t - \tau_B, X(\tau_B), y) : \tau_B < t\right] \\
&\quad - \int\int p_0(\varepsilon_1, x, z_1)\, \mathbb{E}_{z_1}\left[p_0(t - \varepsilon_1 - \tau_B, X(\tau_B), y) : \tau_B < t - \varepsilon_1 - \varepsilon_2\right] dz_1 \\
&= \mathbb{E}_x\left[p_0(t - \tau_B, X(\tau_B), y) : \tau_B < t\right] \\
&\quad - \mathbb{E}_x\left[\mathbb{E}_{X(\varepsilon_1)}\left\{p_0(t - \varepsilon_1 - \tau_B, X(\tau_B), y) : \tau_B < t - \varepsilon_1 - \varepsilon_2\right\}\right]
\end{aligned}
$$

(Markov property)

$$
\begin{aligned}
&= \mathbb{E}_x\left[p_0(t - \tau_B, X(\tau_B), y) : \tau_B < t\right] \\
&\quad - \mathbb{E}_x\left[p_0(t - \tau_{\varepsilon_1}, X(\tau_{\varepsilon_1}), y) : \tau_{\varepsilon_1} < t - \varepsilon_2\right] \\
&= \mathbb{E}_x\left[p_0(t - \tau_B, X(\tau_B), y) : \tau_B < t\right] \\
&\quad - \mathbb{E}_x\left[p_0(t - \tau_{\varepsilon_1}, X(\tau_{\varepsilon_1}), y) : \tau_B < \varepsilon_1,\ \tau_{\varepsilon_1} < t - \varepsilon_2\right] \\
&\quad - \mathbb{E}_x\left[p_0(t - \tau_{\varepsilon_1}, X(\tau_{\varepsilon_1}), y) : \varepsilon_1 < \tau_B,\ \tau_{\varepsilon_1} < t - \varepsilon_2\right] \\
&= \mathbb{E}_x\left[p_0(t - \tau_B, X(\tau_B), y) : \tau_B < \varepsilon_1\right] \\
&\quad + \mathbb{E}_x\left[p_0(t - \tau_B, X(\tau_B), y) : t - \varepsilon_2 < \tau_B < t\right] \\
&\quad - \mathbb{E}_x\left[p_0(t - \tau_{\varepsilon_1}, X(\tau_{\varepsilon_1}), y) : \tau_B < \varepsilon_1, \tau_{\varepsilon_1} < t - \varepsilon_2\right].
\end{aligned}
$$

This proves the first part of (j). For the second part we use the pinned measure $\mu_{0,x}^{t,y}$ introduced in Chapter 2, section B: see the comments following Propositions 2.7 and 2.17 and see Definition 2.18.(e). In terms of this pinned measure, we have

$$
\begin{aligned}
&\int\int p_0(\varepsilon_1, x, z_1)\, q_B(t - \varepsilon_1 - \varepsilon_2, z_1, z_2)\, p_0(\varepsilon_2, z_2, y)\, dz_1\, dz_2 \\
&\quad - q_B(t, x, y) \\
&= \mu_{0,x}^{t,y}\left[\tau_B < \varepsilon_1,\ \tau_{\varepsilon_1} + \varepsilon_2 \geq t\right] + \mu_{0,x}^{t,y}\left[t - \varepsilon_2 < \tau_B < t\right].
\end{aligned}
$$

Since τ_{ε_1} decreases to τ_B as ε_1 decreases to 0, the claim about the limits in (j) follows.

(k) Since

$$\tau_B = \inf \{\varepsilon + \tau_B \circ \vartheta_\varepsilon : \varepsilon > 0\} = \lim_{\varepsilon \downarrow 0} (\varepsilon + \tau_B \circ \vartheta_\varepsilon),$$

we see, using the Markov property, that

$$\mathbb{P}_x [\tau_B < t] = \mathbb{P}_x \left[\bigcup_{\varepsilon > 0} \{\varepsilon + \tau_B \circ \vartheta_\varepsilon < t\} \right]$$

$$= \sup_{t > \varepsilon > 0} \mathbb{P}_x [\tau_B \circ \vartheta_\varepsilon < t - \varepsilon]$$

(Markov property)

$$= \sup_{t > \varepsilon > 0} \mathbb{P}_x \left[\mathbb{P}_{X(\varepsilon)} \{\tau_B < t - \varepsilon\} \right]$$

$$= \sup_{t > \varepsilon > 0} \int p_0(\varepsilon, x, y) \mathbb{P}_y \{\tau_B < t - \varepsilon\} \, dy.$$

So that the function $x \mapsto \mathbb{P}_x [\tau_B < t]$ is a supremum of continuous functions. Hence it is lower semi-continuous. This completes the proof of Theorem D.3.

D.4. THEOREM. *Let B be a Borel subset of E. Suppose that the function $p_0(t, x, y)$ is symmetric and continuous on $(0, \infty) \times E \times E$. Then the function $q_B(t, x, y)$ is symmetric as well.*

D.5. THEOREM. *If, in Theorem D.3, $B = E \setminus U$ where U is an open subset of E, then the function $q_{E \setminus U}(t, x, y)$ is continuous on $(0, \infty) \times U \times U$.*

In order to prove Theorem D.4 and D.5, a number of lemmas and propositions are required (Lemma D.6 through Lemma D.11). From now on we will use the symmetry of the integral kernel $p_0(t, x, y)$ to its full extent.

D.6. LEMMA. *Let f be a function in $L^1(E, m)$ and let g belong to $L^\infty(E, m) \cap L^1(E, m)$. Then*

$$\lim_{t \downarrow 0} \int [\exp(-tK_0) f](x) g(x) dx = \int f(x) g(x) dx.$$

Proof. Fix $\varepsilon > 0$ and let f and g be as in the lemma. Choose $f_1 \in C_\infty(E)$ such that

$$\|f - f_1\|_1 \|g\|_\infty < \frac{1}{3}\varepsilon.$$

By the symmetry of $p_0(t,x,y)$, it follows that

$$\begin{aligned}
&\int \left[\exp\left(-tK_0\right) f\right](x) g(x) dx - \int f(x) g(x) dx \\
&= \int \left[\exp\left(-tK_0\right)(f - f_1)\right](x) g(x) dx + \int \left(f_1(x) - f(x)\right) g(x) dx \\
&\quad + \int \left(\left[\exp\left(-tK_0\right) f_1\right](x) - f_1(x)\right) g(x) dx \\
&= \int \left[(f(y) - f_1(y)) \exp\left(-tK_0\right) g\right](y) dy + \int \left(f_1(x) - f(x)\right) g(x) dx \\
&\quad + \int \left(\left[\exp\left(-tK_0\right) f_1\right](x) - f_1(x)\right) g(x) dx.
\end{aligned}$$

Hence

$$\begin{aligned}
&\left|\int \left[\exp\left(-tK_0\right) f\right](x) g(x) dx - \int f(x) g(x) dx\right| \\
&\leq \|f - f_1\|_1 \left\|\exp\left(-tK_0\right) g\right\|_\infty + \|f_1 - f\|_1 \|g\|_\infty \\
&\quad + \left\|\exp\left(-tK_0\right) f_1 - f_1\right\|_\infty \|g\|_1 \\
&\leq 2\|f_1 - f\|_1 \|g\|_\infty + \left\|\exp\left(-tK_0\right) f_1 - f_1(x)\right\|_\infty \|g\|_1 \\
&\leq \frac{2}{3}\varepsilon + \left\|\exp\left(-tK_0\right) f_1 - f_1\right\|_\infty \|g\|_1 .
\end{aligned}$$

Since $\lim_{t\downarrow 0} \left\|\exp\left(-tK_0\right) f_1 - f_1\right\|_\infty = 0$, the conclusion in Lemma D.6 follows: see Assumption A.3 in BASSA (Chapter 1).B.

D.7. PROPOSITION. *Let B be a Borel subset of E. Then $m\left(B \setminus B^r\right) = 0$.*

Here B^r is the collection of regular points of the set B with respect to τ_B:

$$B^r = \left\{x \in E : \mathbb{P}_x\left[\tau_B = 0\right] = 1\right\}.$$

Proof. Let K be a compact subset of the set $B \setminus B^r$. It suffices to show that $m(K) = 0$. From Lemma D.6 it follows that

$$\begin{aligned}
m(K) &= \int 1_K(x) 1_K(x) dx = \lim_{t\downarrow 0} \int \left[\exp\left(-tK_0\right) 1_K\right](x) 1_K(x) dx \\
&= \lim_{t\downarrow 0} \int_K \mathbb{P}_x\left[X(t) \in K\right] dx \leq \liminf_{t\downarrow 0} \int_K \mathbb{P}_x\left[X(t) \in B\right] dx \\
&\leq \liminf_{t\downarrow 0} \int_K \mathbb{P}_x\left[\tau_B \leq t\right] dx = \int_K \mathbb{P}_x\left[\tau_B = 0\right] dx = 0.
\end{aligned}$$

The ultimate equality follows because $K \cap B^r = \emptyset$.

In the following lemma the symmetry of the function $p_0(t,x,y)$ is used in an essential way.

D.8. LEMMA. *Let $f_1, \ldots, f_m$ be bounded Borel measurable functions defined on E. Let $0 < t_1 < \cdots < t_m < \infty$ be such that*

$$t_1 = t_2 - t_1 = t_3 - t_2 = \cdots = t_m - t_{m-1}.$$

For $x \in E$ the equality

$$\begin{aligned}&\mathbb{E}_x\left[f_1(X(t_1))\ldots f_m(X(t_m))\right]\\&= \int f_m(z)\mathbb{E}_z\left[f_{m-1}(X(t_1))\ldots f_1(X(t_{m-1}))p_0(t_1, X(t_{m-1}), x)\right]dz\end{aligned}$$

is valid.

Proof. Let $f_1, \ldots, f_m$ and $t_1, \ldots, t_m$ be as in the lemma. By the definition of finite dimensional distribution we have, for $x = x_0 \in E$,

$$\begin{aligned}&\mathbb{E}_x\left[\prod_{j=1}^{m} f_j(X(t_j))\right]\\&= \int \cdots \int \prod_{j=1}^{m} f_j(x_j) \prod_{j=1}^{m} p_0(t_1, x_{m-j}, x_{m-j+1})dx_m \ldots x_1\end{aligned}$$

(rename integration variables and apply Fubini's theorem)

$$\begin{aligned}&= \int f_m(z) \int \cdots \int \prod_{j=1}^{m-1} f_{m-j}(x_j)p_0(t_1, x_1, z)\\&\prod_{j=1}^{m-2} p_0\left(t_1, x_{j+1}, x_j\right) p_0(t_1, x, x_{m-1})dx_{m-1} \ldots dx_1 dz\end{aligned}$$

(again an appeal to Fubini's theorem plus symmetry)

$$\begin{aligned}&= \int f_m(z) \int \cdots \int \prod_{j=1}^{m-1} f_{m-j}(x_j)p_0(t_1, x_{m-1}, x)\\&\prod_{j=1}^{m-2} p_0(t_1, x_{m-j-1}, x_{m-j})p_0(t_1, z, x_1)dx_{m-1} \ldots dx_1 dz\\&= \int f_m(z)\mathbb{E}_z\left[f_{m-1}(X(t_1))\ldots f_1(X(t_{m-1}))p_0(t_1, X(t_{m-1}), x)\right]dz.\end{aligned}$$

This proves Lemma D.8.

D.9. LEMMA. *Let U be an open subset of E and f be a bounded Borel function defined on E. Fix $t > 0$ and fix $x \in E$.*

(1) (a) *The following equalities are valid:*

$$\mathbb{E}_x\left[f(X(t)) : \tau_U > t\right] = \int f(z)q_U(t, x, z)dz = \int f(z)q_U(t, z, x)dz. \tag{D.15}$$

(2) (b) *Fix $x \in E$ and fix $t > 0$. The equality*

$$q_U(t,x,z) = q_U(t,z,x) \tag{D.16}$$

holds for m-almost all $z \in E$.

Here m-almost all is meant in the sense that

$$m\{z \in E : q_U(t,x,z) \neq q_U(t,z,x)\} = 0.$$

Proof. The first equality in (D.15) follows from Theorem D.3 assertion (a). For the second equality in (D.15), we proceed as follows. Define the finite set $J_n(t)$ by

$$J_n(t) = \left\{k2^{-n}t : 1 \le k \le 2^n\right\}$$

and notice the inclusion $J_{n+1}(t) \supset J_n(t)$. Also notice that the union $\bigcup_{n=1}^{\infty} J_n(t)$ is dense in the interval $[0,t]$. Since, by Lemma 2.19, $\mathbb{P}_x(\tau_U = t) = 0$ for $x \in E$ and for $t > 0$, by right continuity of paths, we get

$$\begin{aligned}
&\mathbb{E}_x\left[f(X(t)) : \tau_U > t\right] = \mathbb{E}_x\left[f(X(t)) : \tau_U \ge t\right] \\
&= \lim_{n\to\infty} \mathbb{E}_x\left[f(X(t)) : s \in J_n(t) \Rightarrow X(s) \notin U\right] \\
&= \lim_{n\to\infty} \mathbb{E}_x\left[\prod_{s\in J_n(t)\setminus\{t\}} 1_{U^c}(X(s)) 1_{U^c}(X(t)) f(X(t))\right].
\end{aligned} \tag{D.17}$$

Next we apply Lemma D.8 to obtain that

$$\begin{aligned}
&\mathbb{E}_x\left[\prod_{s\in J_n(t)\setminus\{t\}} 1_{U^c}(X(s)) 1_{U^c}(X(t)) f(X(t))\right] \\
&= \int 1_{U^c}(z) f(z) \mathbb{E}_z\left[\left(\prod_{s\in J_n(t)\setminus\{t\}} 1_{U^c}(X(s))\right) p_0\left(2^{-n}t, X\left(t-2^{-n}t\right), x\right)\right] dz \\
&= \int_{U^c} f(z) \mathbb{E}_z\left[p_0\left(2^{-n}t, X\left(t-2^{-n}t\right), x\right) : \bigcap_{s\in J_n(t)\setminus\{t\}} \{X(s) \notin U\}\right] dz \\
&= \int_{U^c} (z) f(z) \mathbb{E}_z\left[p_0\left(2^{-n}t, X\left(t-2^{-n}t\right), x\right)\right] dz \\
&\quad - \int f(z) \mathbb{E}_z\left[p_0\left(2^{-n}t, X\left(t-2^{-n}t\right), x\right) : \bigcup_{s\in J_n(t)\setminus\{t\}} \{X(s) \in U\}\right] dz \\
&= \int_{U^c} f(z) p_0(t,z,x) dz - \int_{U^c} f(z) \mathbb{E}_z\left[p_0\left(2^{-n}t, X\left(t-2^{-n}t\right), x\right) : A_n(t)\right] dz,
\end{aligned} \tag{D.18}$$

where

$$A_n(t) := \bigcup_{s\in J_n(t)\setminus\{t\}} \{X(s) \in U\} = \bigcup_{s\in J_n(t)\setminus\{t\}} \{X(s) \in U\} \cap \{\tau_U \leq t - 2^{-n}t\}. \tag{D.19}$$

The event $A_n(t)$ belongs to the σ-field $\mathcal{F}_{\tau_U}$. So, by virtue of the martingale property of the process $s \mapsto p_0(t-s, X(s), x)$ with respect to the measure $\mathbb{P}_z$, we obtain

$$\begin{aligned}&\mathbb{E}_z\left[p_0\left(2^{-n}t, X\left(t-2^{-n}t\right), x\right) : A_n(t)\right]\\ &= \mathbb{E}_z\left[p_0\left(2^{-n}t, X\left(t-2^{-n}t\right), x\right) : A_n(t), \tau_U \leq t-2^{-n}t\right]\\ &= \mathbb{E}_z\left[p_0\left(t-\tau_U, X(\tau_U), x\right) : A_n(t)\right].\end{aligned} \tag{D.20}$$

Since $A_n(t)$ increases to the event $\{\tau_U < t\}$, by (D.17) through (D.20) it follows that

$$\begin{aligned}&\mathbb{E}_x\left[f(X(t)) : \tau_U > t\right]\\ &= \int_{U^c} f(z)p_0(t,z,x)dz - \lim_{n\to\infty}\int_{U^c} f(z)\mathbb{E}_z\left[p_0(t-\tau_U, X(\tau_U), x) : A_n(t)\right]dz\\ &= \int_{U^c} f(z)p_0(t,z,x)dz - \int_{U^c} f(z)\mathbb{E}_z\left[p_0(t-\tau_U, X(\tau_U), x) : \tau_U < t\right]dz.\end{aligned} \tag{D.21}$$

Since $\mathbb{P}_z\{\tau_U = 0] = 1$, for $z \in U \subseteq U^r$, we see that

$$\mathbb{E}_z\left[p_0\left(t-\tau_U, X(\tau_U), x\right) : \tau_U < t\right] = p_0(t,z,x), \quad z \in U.$$

So from (D.21) we infer that

$$\begin{aligned}\mathbb{E}_x\left[f(X(t)) : \tau_U > t\right] &= \int f(z)\left(p_0(t,z,x) - \mathbb{E}_z\left[p_0\left(t-\tau_U, X(\tau_U), x\right) : \tau_U < t\right]\right)dz\\ &= \int f(z)q_U(t,z,x)dz.\end{aligned}$$

This is the second equality in (D.15). The assertion in (b) is an immediate consequence of (a).

D.10. LEMMA. *Equip $E \times E$ with the product measure $m \otimes m$. Let U be an open subset of E. For $m \otimes m$-almost all pairs $(x,z) \in E \times E$, the equality*

$$q_U(t,z,x) = q_U(t,x,z)$$

is true for all $t > 0$.

Proof. Let A and C be sets in the Borel field $\mathcal{E}$. From assertion (b) of Lemma D.9 and Fubini's theorem, it follows that

$$\begin{aligned}&\int_A \mathbb{P}_z\left[\tau_U > t, X(t) \in C\right]dz = \int_A\left(\int_C q_U(t,z,x)dx\right)dz\\ &= \int_C\left(\int_A q_U(t,z,x)dz\right)dx = \int_C\left(\int_A q_U(t,x,z)dz\right)dx.\end{aligned} \tag{D.22}$$

Hence

$$\int_C \left(\int_A q_U(t,z,x)dz \right) dx = \int_C \left(\int_A q_U(t,x,z)dz \right) dx. \qquad \text{(D.23)}$$

For every fixed $t > 0$ the $m \otimes m$-almost everywhere equality in Lemma D.10 follows from (D.23). Assertion (i) in Theorem D.3 makes the exceptional set independent of $t > 0$.

REMARK. For U open in E, the identity in (D.22) and assertion (a) of Theorem D.3 yields the following identity:

$$\int_A \mathbb{P}_z \left[\tau_U > t, X(t) \in C \right] dz = \int_C \mathbb{P}_x \left[\tau_U > t, X(t) \in A \right] dx, \quad A,\ C \in \mathcal{E}. \quad \text{(D.24)}$$

D.11. LEMMA. *Let B be closed subset of E. Then, for all $t > 0$ and for $m \otimes m$-almost all pairs $(x,y) \in E \times E$, the equality*

$$q_B(t,x,y) = q_B(t,y,x) \qquad \text{(D.25)}$$

is valid.

Proof. Let B be a closed subset of E and choose a sequence of open subsets $(U_n : n \in \mathbb{N})$ of E for which $U_n \supseteq \overline{U_{n+1}}$ and for which $B = \bigcap_{n=1}^{\infty} U_n = \bigcap_{n=1}^{\infty} \overline{U_n}$. Put $\tau = \sup_n \tau_{U_n} = \lim_{n\to\infty} \tau_{U_n}$. Fix $x \in E$ and set $D_C = \inf\{s \geq 0 : X(s) \in C\}$, $C \in \mathcal{E}$. Since $D_{U_n} = \tau_{U_n}$, and since $U_n \supseteq B$, we have $\tau \leq D_B$. On the other hand, let $\omega \in \Omega$ be such that $\tau(\omega) < \infty$ and such that $X(\tau(\omega))(\omega) = \lim_{n\to\infty} X\left(\tau_{U_n}(\omega)\right)(\omega)$. Then $X(\tau)(\omega) = X\left(\tau(\omega)\right)(\omega)$ belongs to $\bigcap_{n=1}^{\infty} \overline{U_n} = B$. So that, by the quasi-left continuity of our process $\{(\Omega, \mathcal{F}, \mathbb{P}_x), (X(t) : t \geq 0), (\vartheta_t : t \geq 0), (E, \mathcal{E})\}$, $D_B \leq \tau$, $\mathbb{P}_x$-almost surely on the event $\{\tau < \infty\}$. Since always $\tau \leq D_B$, we infer that $\tau = D_B$ $\mathbb{P}_x$-almost surely. From (D.24) it follows that

$$\int_A \mathbb{P}_z \left[\tau_{U_n} > t, X(t) \in C \right] dz = \int_C \mathbb{P}_x \left[\tau_{U_n} > t, X(t) \in A \right] dx, \quad A,\ C \in \mathcal{E}.$$

So that, by taking limits,

$$\int_A \mathbb{P}_z \left[D_B > t, X(t) \in C \right] dz = \int_C \mathbb{P}_x \left[D_B > t, X(t) \in A \right] dx, \quad A,\ C \in \mathcal{E}. \quad \text{(D.26)}$$

The left-hand side of (D.26) may be rewritten as follows:

$$\begin{aligned} \int_A \mathbb{P}_z \left[D_B > t, X(t) \in C \right] dz &= \int_{A \cap B^c} \mathbb{P}_z \left[D_B > t, X(t) \in C \right] dz \qquad \text{(D.27)} \\ &+ \int_{A \cap (B \setminus B^r)} \mathbb{P}_z \left[D_B > t, X(t) \in C \right] dz + \int_{A \cap B \cap B^r} \mathbb{P}_z \left[D_B > t, X(t) \in C \right] dz. \end{aligned}$$

It is not so difficult to prove that $\tau_B = D_B$ $\mathbb{P}_x$-almost surely if and only if x belongs to $B^c \cup (B^r \cap B) = B^c \cup B^r$. By Proposition D.7 the middle term of the right-hand side of (D.27) vanishes. So we obtain

$$\begin{aligned}\int_A \mathbb{P}_z\left[D_B > t, X(t) \in C\right] dz &= \int_{A\cap B^c} \mathbb{P}_z\left[\tau_B > t, X(t) \in C\right] dz \\ &+ \int_{A\cap(B\backslash B^r)} \mathbb{P}_z\left[\tau_B > t, X(t) \in C\right] dz + \int_{A\cap B\cap B^r} \mathbb{P}_z\left[\tau_B > t, X(t) \in C\right] dz \\ &= \int_A \mathbb{P}_z\left[\tau_B > t, X(t) \in C\right] dz.\end{aligned}$$

Similarly the right-hand side of equality (D.26) is treated to obtain

$$\int_A \mathbb{P}_z\left[\tau_B > t, X(t) \in C\right] dz = \int_C \mathbb{P}_x\left[\tau_B > t, X(t) \in A\right] dx.$$

So that, by assertion (a) in Theorem D.3 and Fubini's theorem,

$$\int_A \left(\int_C q_B(t,z,x)dx\right) dz = \int_C \left(\int_A q_B(t,x,z)dz\right) dx = \int_A \left(\int_C q_U(t,x,z)dx\right) dz.$$

Hence $q_B(t,x,z) = q_B(t,z,x)$ for $m \otimes m$-almost all pairs $(x,z) \in E \times E$. The continuity of the function $t \mapsto q_B(t,x,z)$ for $t > 0$ together with the separability of $(0,\infty)$, implies that the exceptional set may be chosen independent of $t > 0$.

Proof of Theorem D.4. First let B be a closed subset of E. Fix $(t,x,y) \in (0,\infty) \times E \times E$ and let ε_j, $j = 1, 2$, be strictly positive real numbers for which $\varepsilon_1 + \varepsilon_2 < t$. Lemma D.11 implies the following identity:

$$\begin{aligned}&\int\int p_0\left(\varepsilon_1, x, z_1\right) q_B\left(t-\varepsilon_1-\varepsilon_2, z_1, z_2\right) p_0\left(\varepsilon_2, z_2, y\right) dz_1\, dz_2 \\ &= \int\int p_0\left(\varepsilon_2, y, z_2\right) q_B\left(t-\varepsilon_1-\varepsilon_2, z_2, z_1\right) p_0\left(\varepsilon_1, z_1, x\right) dz_1\, dz_2. \qquad \text{(D.28)}\end{aligned}$$

An application of the second assertion in (j) of Theorem D.3 yields the equality $q_B(t,x,y) = q_B(t,y,x)$. So Theorem D.3 is now proved for closed sets B. Next let B be an arbitrary Borel subset of E and let x and y be points in E. Then we may choose two increasing sequences of compact subsets $(B_{n,1} : n \in \mathbb{N})$ and $(B_{n,2} : n \in \mathbb{N})$ of B such that both $\tau_{B_{n,1}}$ and $\tau_{B_{n,2}}$ decrease $\mathbb{P}_x$-almost surely and $\mathbb{P}_y$-almost surely to τ_B respectively. Put $B_n = B_{n,1} \cup B_{n,2}$ and let τ_n be the first hitting time of B_n. Then

$$\tau_n = \tau_{B_n} = \min\left(\tau_{B_{n,1}}, \tau_{B_{n,2}}\right)$$

decreases to τ_B, $\mathbb{P}_x$- as well as $\mathbb{P}_y$-almost surely. An application of assertion (h) of Theorem D.3 together with the symmetry $q_{B_n}(t,x,y) = q_{B_n}(t,y,x)$ yields the identities

$$q_B(t,x,y) = \inf_{n\in\mathbb{N}} q_{B_n}(t,x,y) = \inf_{n\in\mathbb{N}} q_{B_n}(t,y,x) = q_B(t,y,x). \tag{D.29}$$

This is the content of Theorem D.4.

Proof of Theorem D.5. Let $B = E\setminus U$ be a closed subset of E and hence U is open. From assertion (j) in Theorem D.3, we see that

$$\begin{aligned} 0 &\le \int\int p_0(\varepsilon_1,x,z_1)\, q_B(t-\varepsilon_1-\varepsilon_2,z_1,z_2)\, p_0(\varepsilon_2,z_2,y) - q_B(t,x,y) \\ &\le \mathbb{E}_x\left[p_0(t-\tau_B,X(\tau_B),y) : \tau_B < \varepsilon_1\right] \\ &\quad + \mathbb{E}_x\left[p_0(t-\tau_B,X(\tau_B),y) : t-\varepsilon_2 < \tau_B < t\right]. \end{aligned} \tag{D.30}$$

Next we consider the final term in (D.30). From the Markov property we see

$$\begin{aligned} &\mathbb{E}_x\left[p_0(t-\tau_B,X(\tau_B),y) : t-\varepsilon_2 < \tau_B < t\right] \\ &= \mathbb{E}_x\left[p_0(\varepsilon_2-\tau_B\circ\vartheta_{t-\varepsilon_2}, X(\tau_B)\circ\vartheta_{t-\varepsilon_2},y) : \tau_B\circ\vartheta_{t-\varepsilon_2} < \varepsilon_2,\ t-\varepsilon_2<\tau_B\right] \\ &= \mathbb{E}_x\left[\mathbb{E}_{X(t-\varepsilon_2)}\left\{p_0(\varepsilon_2-\tau_B,X(\tau_B),y) : \tau_B<\varepsilon_2\right\} : t-\varepsilon_2<\tau_B\right] \\ &= \int q_B(t-\varepsilon_2,x,z)\mathbb{E}_z\left[p_0(\varepsilon_2-\tau_B,X(\tau_B),y) : \tau_B<\varepsilon_2\right] dz \end{aligned}$$

(symmetry of $\mathbb{E}_z\left[p_0(\varepsilon_2-\tau_B,X(\tau_B,y),y) : \tau_B<\varepsilon_2\right]$: see Theorem D.4)

$$\begin{aligned} &= \int q_B(t-\varepsilon_2,x,z)\mathbb{E}_y\left[p_0(\varepsilon_2-\tau_B,X(\tau_B),z) : \tau_B<\varepsilon_2\right] dz \\ &\le \int p_0(t-\varepsilon_2,z,x)\mathbb{E}_y\left[p_0(\varepsilon_2-\tau_B,X(\tau_B),z) : \tau_B<\varepsilon_2\right] dz \\ &= \mathbb{E}_y\left[p_0(t-\tau_B,X(\tau_B),x) : \tau_B<\varepsilon_2\right]. \end{aligned} \tag{D.31}$$

From (D.30) and (D.31) we infer the inequality

$$\begin{aligned} 0 &\le \int\int p_0(\varepsilon_1,x,z_1)\, q_B(t-\varepsilon_1-\varepsilon_2,z_1,z_2)\, p_0(\varepsilon_2,z_2,y) \\ &\qquad - q_B(t,x,y) \\ &\le \mathbb{E}_x\left[p_0(t-\tau_B,X(\tau_B),y) : \tau_B<\varepsilon_1\right] \\ &\quad + \mathbb{E}_y\left[p_0(t-\tau_B,X(\tau_B),x) : \tau_B<\varepsilon_2\right]. \end{aligned} \tag{D.32}$$

The function

$$\int\int p_0(\varepsilon_1,x,z_1)\, q_B(t-\varepsilon_1-\varepsilon_2,z_1,z_2)\, p_0(\varepsilon_2,z_2,y),$$

is jointly continuous in t, x and y. From equality (b) in Theorem D.1 together with (D.32), it follows that continuity on $(0,\infty)\times U\times U$ is inherited by the function $q_B(t,x,y)$. This proves Theorem D.5.

D.2. Harmonic measures and equilibrium potentials

Let B be a Borel subset of B. In what follows we want to prove that $B \backslash B^r$ is a polar subset of E: in other words the process is not sensitive for such a set. This result will be shown to hold in Theorem D.16. For this we need hitting distributions.

D.12. DEFINITION. Let x be in E and let B be a Borel subset of E. The hitting distribution for the process

$$\{(\Omega, \mathcal{F}, \mathbb{P}_x), (X(t) : t \geq 0), (\vartheta_t : t \geq 0), (E, \mathcal{E})\}$$

is defined by

$$h_B(x, A) = \mathbb{P}_x [X(\tau_B) \in A, \tau_B < \infty]. \tag{D.33}$$

If U is an open subset of E we also write $H_U(x, A) = h_{E\backslash U}(x, A)$. The mapping $A \mapsto H_U(x, A)$ is a measure on $B := E \setminus U$. It is called the *harmonic measure* corresponding to our Markov process.

D.13. THEOREM. *Let x and B be as in Definition D.12. The following assertions are true:*

(i) *On the event $\{\tau_B < \infty\}$ the state variable $X(\tau_B)$ attains its values $\mathbb{P}_x$-almost surely in $B \cup B^r$;*
(ii) *The measure $h_B(x, \cdot)$ is concentrated on $B \cup B^r$.*

Proof. Assertion (ii) is a consequence of (i). So we only need to show (i). If $\tau_B(\omega)$ is finite and if $X(\tau_B)(\omega)$ does not belong to B, then a simple calculation shows that

$$\tau_B(\omega) + \tau_B \circ \vartheta_{\tau_B}(\omega) = \tau_B(\omega).$$

Here we use the fact that, for $0 < t \leq \tau_B(\omega)$, $X(\tau_B)(\omega)$ does not belong to B. Consequently, by the strong Markov property, we see that

$$\begin{aligned}
&\mathbb{P}_x [X(\tau_B) \notin B, \tau_B < \infty] = \mathbb{P}_x [\tau_B \circ \vartheta_{\tau_B} = 0, X(\tau_B) \notin B, \tau_B < \infty] \\
&= \mathbb{E}_x \left[\mathbb{P}_{X(\tau_B)} \{\tau_B = 0\} : X(\tau_B) \notin B, \tau_B < \infty\right] \\
&= \mathbb{E}_x [X(\tau_B) \in B^r, X(\tau_B) \notin B, \tau_B < \infty].
\end{aligned}$$

Hence we get

$$\begin{aligned}
\mathbb{P}_x [\tau_B < \infty] &= \mathbb{P}_x [\tau_B < \infty, X(\tau_B) \in B] + \mathbb{P}_x [\tau_B < \infty, X(\tau_B) \notin B] \\
&= \mathbb{P}_x [\tau_B < \infty, X(\tau_B) \in B] + \mathbb{P}_x [\tau_B < \infty, X(\tau_B) \in B^r \setminus B] \\
&= \mathbb{P}_x [\tau_B < \infty, X(\tau_B) \in B \cup B^r].
\end{aligned}$$

This proves (i) and also (ii).

D.14. THEOREM. *The following assertions are equivalent:*

(i) *For every Borel set B in E the set $B \setminus B^r$ is a polar set;*
(ii) *For every compact subset K of E and for every $x \in E$, the measure $h_K(x, \cdot)$ is concentrated on K^r;*
(iii) *For every Borel subset B of E and for every $x \in E$, the measure $h_B(x, \cdot)$ is concentrated on B^r;*
(iv) *For every pair of Borel subsets A and B of E for which $B \subseteq A$ and for every $x \in E$, the equality*

$$\tau_B = \tau_A + \tau_B \circ \vartheta_{\tau_A} \tag{D.34}$$

holds $\mathbb{P}_x$-almost surely on the event $\{\tau_A < \infty\}$;
(v) *For every compact subset K of E with $K^r = \emptyset$ and for every Borel subset A of E with $K \subseteq A^r$, the identity*

$$\tau_K = \tau_A + \tau_K \circ \vartheta_{\tau_A} \tag{D.35}$$

holds $\mathbb{P}_x$-almost surely on the event $\{\tau_A < \infty\}$ for all $x \in E$;
(vi) *For every compact subset K of E with $K^r = \emptyset$, the identity*

$$\tau_K \circ \vartheta_{\tau_K} = 0$$

holds $\mathbb{P}_x$-almost surely on the event $\{\tau_K < \infty\}$ for all $x \in E$.

Proof. (i) $\Rightarrow$ (ii). Let K be a compact subset of E. Then

$$\begin{aligned} h_K(x, K \setminus K^r) &= \mathbb{P}_x\left[X(\tau_K) \in K \setminus K^r : \tau_K < \infty\right] \\ &\le \mathbb{P}_x\left[\bigcup_{0<s<\infty} \{X(s) \in K \setminus K^r\}\right] \le \mathbb{P}_x\left[\tau_{K \setminus K^r} < \infty\right], \end{aligned}$$

for all $x \in E$. The equality

$$\mathbb{P}_x\left[\tau_{K \setminus K^r} < \infty\right] = 0$$

follows, since by (i) the set $K \setminus K^r$ is polar. Since, by Theorem D.13 the measure $h_K(x, \cdot)$ is concentrated on $K \cup K^r$, assertion (ii) follows.

(ii) $\Rightarrow$ (i). Fix $x \in E$ and let B be a Borel subset of E. Then there exists an increasing sequence of compact subsets $(K_n : n \in \mathbb{N})$ in $B \setminus B^r$ with the property that τ_{K_n} decreases to $\tau_{B \setminus B^r}$ $\mathbb{P}_x$-almost surely on the event $\left\{\tau_{B \setminus B^r} < \infty\right\}$. So that

$$\mathbb{P}_x\left[\tau_{B \setminus B^r} < \infty\right] = \sup_{n \in \mathbb{N}} \mathbb{P}_x\left[\tau_{K_n} < \infty\right]. \tag{D.36}$$

Since, for all $n \in \mathbb{N}$, $K_n \subseteq B \setminus B^r \subseteq B$, it follows that $K_n^r \subseteq B^r$ and $K_n^r \subseteq K_n \subseteq B \setminus B^r$. Hence $K_n^r = \emptyset$. Since, by (ii) the measures $h_{K_n}(x, \cdot)$ are concentrated on K_n^r, it follows that $h_{K_n}(x, \cdot) \equiv 0$ for all $n \in \mathbb{N}$. Hence we see that

$$0 = h_{K_n}(x, K_n) = \mathbb{P}_x\left[X(\tau_{K_n}) \in K_n, \tau_{K_n} < \infty\right] = \mathbb{P}_x\left[\tau_{K_n} < \infty\right].$$

Consequently, by (D.36), $\mathbb{P}_x\left[\tau_{B\setminus B^r}<\infty\right]=0$. So that $B\setminus B^r$ is a polar set.
(i) $\Rightarrow$ (iii). See the proof of (i) $\Rightarrow$ (ii).
(iii) $\Rightarrow$ (iv). Let B and A be Borel subsets of E for which $B\subseteq A$. Then

$$\tau_A\leq\tau_B\leq\tau_A+\tau_B\circ\vartheta_{\tau_A},$$

and, on the event $\{\tau_A<\tau_B\}$, the equalities

$$\tau_A+\tau_B\circ\vartheta_{\tau_A}=\inf\{s>\tau_A:X(s)\in B\}=\tau_B$$

are readily verified. Next fix $x\in E$. Then

$$\begin{aligned}&\mathbb{P}_x\left[\tau_A+\tau_B\circ\vartheta_{\tau_A}>\tau_B,\tau_A<\infty\right]\\&=\mathbb{P}_x\left[\tau_A+\tau_B\circ\vartheta_{\tau_A}>\tau_B,\tau_A<\tau_B\right]+\mathbb{P}_x\left[\tau_A+\tau_B\circ\vartheta_{\tau_A}>\tau_B,\tau_A=\tau_B,\tau_A<\infty\right]\\&=\mathbb{P}_x\left[\tau_B\circ\vartheta_{\tau_B}>0,\tau_A=\tau_B,\tau_A<\infty\right]\end{aligned}$$

(strong Markov property)

$$\begin{aligned}&=\mathbb{P}_x\left[\mathbb{P}_{X(\tau_B)}\{\tau_B>0\}:\tau_A=\tau_B,\tau_A<\infty\right]\\&=\mathbb{P}_x\left[X(\tau_B)\notin B^r,\tau_A=\tau_B,\tau_A<\infty\right]\\&\leq\mathbb{P}_x\left[X(\tau_B)\notin B^r,\tau_B<\infty\right]=h_B\left(x,E\setminus B^r\right)\end{aligned}$$

(by (iii))

$$=h_B\left(x,B^r\setminus B^r\right)=h_B\left(x,\emptyset\right)=0.$$

This proves (iv).
(iv) $\Rightarrow$ (vi). Set $B=A=K$, with K compact, in (iv).
(vi) $\Rightarrow$ (i). Let B be a Borel subset of E and let K be a compact subset of $B\setminus B^r$. Then $K^r\subset B^r$ and also $K^r\subseteq K$. So that

$$K^r\subseteq B^r\cap K\subseteq B^r\cap(B\setminus B^r)=\emptyset.$$

By the strong Markov property we have

$$\begin{aligned}\mathbb{P}_x\left[\tau_K\circ\vartheta_{\tau_K}>0,\tau_K<\infty\right]&=\mathbb{E}_x\left[\mathbb{P}_{X(\tau_K)}\{\tau_K>0\},\tau_K<\infty\right]\\&=\mathbb{P}_x\left[X(\tau_K)\in K\setminus K^r,\tau_K<\infty\right]\end{aligned}$$

(the set K^r is empty and $X(\tau_K)$ belongs to K $\mathbb{P}_x$-almost surely)

$$=\mathbb{P}_x\left[\tau_K<\infty\right].$$

So, by (vi), $\mathbb{P}_x\left[\tau_K<\infty\right]=0$. Hence any compact subset of $B\setminus B^r$ is polar. Thus $B\setminus B^r$ is a polar set because

$$\mathbb{P}_x\left[\tau_{B\setminus B^r}<\infty\right]=\sup\left\{\mathbb{P}_x\left[\tau_K<\infty\right]:K\subseteq B\setminus B^r,\ K\text{ compact}\right\}.$$

Hitherto we proved the equivalence of (i), (ii), (iii), (iv), and (vi). In Theorem D.16 we will establish assertion (v) for the symmetric case.

(i) $\Rightarrow$ (v). Assertions (i) and (iv) are equivalent and, trivially, (iv) implies (v).

(v) $\Rightarrow$ (i). Fix $\lambda > 0$. We begin by noticing that

$$\begin{aligned} B \setminus B^r &= \{x \in B : \mathbb{E}_x [\exp(-\lambda \tau_B)] < 1\} \\ &= \bigcup_{n=1}^{\infty} \left\{x \in B : \mathbb{E}_x [\exp(-\lambda \tau_B)] \leq 1 - \frac{1}{n}\right\} = \bigcup_{n=1}^{\infty} B_n, \end{aligned}$$

where

$$B_n = \left\{x \in B : \mathbb{E}_x [\exp(-\lambda \tau_B)] \leq 1 - \frac{1}{n}\right\}.$$

Let K be a subset of B_n. Then $\tau_K \geq \tau_{B_n} \geq \tau_B$. So that for $x \in K$ the inequalities

$$\mathbb{E}_x [\exp(-\lambda \tau_K)] \leq \mathbb{E}_x [\exp(-\lambda \tau_B)] \leq 1 - \frac{1}{n}$$

are valid. But then, using (v), K is a polar set, as we will show below. Consequently every set B_n is polar. But then, the set B is polar as well because the sequence τ_{B_n} decreases to τ_B $\mathbb{P}_x$-almost surely for all $x \in E$.

So let K be a compact subset of $B \setminus B^r$ with the property that

$$M := \sup \{\mathbb{E}_x [\exp(-\lambda \tau_K) : \tau_K < \infty] : x \in K\} < 1.$$

As in the proof of (vi) implies (i) the set K^r is empty. Let $\epsilon > 0$ and set

$$A := \{y \in E : \mathbb{E}_y [\exp(-\lambda \tau_K) : \tau_K < \infty] \leq M + \epsilon\}.$$

Then

$$A = \left\{y \in E : \lambda \int_0^{\infty} e^{-\lambda s} \mathbb{P}_y [\tau_K < s] \, ds \leq M + \epsilon\right\}.$$

So from Theorem D.3, assertion (k) it follows that A is a closed subset of E. Next we prove that K is a subset of A^r. For $t > 0$ and $x \in E$ we write

$$\begin{aligned} &\mathbb{E}_x \left[e^{-\lambda t} \mathbb{E}_{X(t)} \{\exp(-\lambda \tau_K) : \tau_K < \infty\} : \tau_K > t\right] \\ &= \mathbb{E}_x \left[e^{-\lambda t} \exp(-\lambda \tau_K \circ \vartheta_t) : \tau_K \circ \vartheta_t < \infty, \tau_K > t\right] \\ &= \mathbb{E}_x [\exp(-\lambda \tau_K) : t < \tau_K < \infty] \leq \mathbb{E}_x [\exp(-\lambda \tau_K)]. \end{aligned}$$

Hence

$$\mathbb{E}_x \left[\mathbb{E}_{X(t)} \{\exp(-\lambda \tau_K) : \tau_K < \infty\} : \tau_K > t\right] \leq e^{\lambda t} \mathbb{E}_x [\exp(-\lambda \tau_K)].$$

Since

$$(M+\epsilon)\, 1_{\{\tau_A>t\}} \leq \mathbb{E}_{X(t)}\left\{\exp\left(-\lambda\tau_K\right)\right\} 1_{\{\tau_A>t\}} \leq \mathbb{E}_{X(t)}\left\{\exp\left(-\lambda\tau_K\right)\right\} 1_{\{\tau_K>t\}},$$

we infer that

$$(M+\epsilon)\, \mathbb{P}_x\left[\tau_A > t\right] \leq \mathbb{E}_x\left[\mathbb{E}_{X(t)}\left\{\exp\left(-\lambda\tau_K\right)\right\} : \tau_A > t\right] \leq e^{\lambda t}\mathbb{E}_x\left[\exp\left(-\lambda\tau_K\right)\right].$$

So that, upon letting t tend to 0 from above,

$$(M+\epsilon)\, \mathbb{P}_x\left[\tau_A > 0\right] \leq \mathbb{E}_x\left[\exp\left(-\lambda\tau_K\right)\right].$$

So that, for $x \notin A^r$, the inequality

$$(M+\epsilon) \leq \mathbb{E}_x\left[\exp\left(-\lambda\tau_K\right)\right]$$

follows. Hence, if x does not belong to A^r, it does not belong to K either. Whence $K \subseteq A^r$. By (v) we obtain

$$\mathbb{E}_x\left[\exp\left(-\lambda\tau_K\right)\right] = \mathbb{E}_x\left[\exp\left(-\lambda\left(\tau_A + \tau_K \circ \vartheta_{\tau_A}\right)\right) : \tau_A < \infty\right]$$

(strong Markov property)

$$= \mathbb{E}_x\left[\exp\left(-\lambda\tau_A\right)\mathbb{E}_{X(\tau_A)}\left\{\exp\left(-\lambda\tau_K\right)\right\} : \tau_A < \infty\right]$$

(since $X\left(\tau_A\right)$ belongs to A)

$$\leq \mathbb{E}_x\left[\exp\left(-\lambda\tau_A\right) : \tau_A < \infty\right] \sup_{y\in A}\mathbb{E}_y\left\{\exp\left(-\lambda\tau_K\right)\right\} \leq M+\epsilon.$$

Hence

$$\sup_{x\in E}\mathbb{E}_x\left[\exp\left(-\lambda\tau_K\right)\right] \leq M+\epsilon.$$

Since $\epsilon > 0$ was arbitrary, we see that

$$\sup_{x\in E}\mathbb{E}_x\left[\exp\left(-\lambda\tau_K\right)\right] \leq \sup_{x\in K}\mathbb{E}_x\left[\exp\left(-\lambda\tau_K\right)\right].$$

Next let $(U_n : n \in \mathbb{N})$ be a decreasing sequence of open subsets of E, each containing K, such that $\overline{U_{n+1}} \subseteq U_n$ and such that $K = \bigcap_{n\in\mathbb{N}} U_n$. Let x be in E. Again invoking assertion (v) yields the equality

$$\tau_K = \tau_{U_n} + \tau_K \circ \vartheta_{U_n}$$

$\mathbb{P}_x$-almost surely on the event $\{\tau_{U_n} < \infty\}$. Another application of the strong Markov property results in

$$\mathbb{E}_x\left[\exp\left(-\lambda\tau_K\right)\right] = \mathbb{E}_x\left[\exp\left(-\lambda\left(\tau_{U_n} + \tau_K \circ \vartheta_{\tau_{U_n}}\right)\right) : \tau_{U_n} < \infty\right]$$

(strong Markov property)

$$= \mathbb{E}_x \left[\exp\left(-\lambda\tau_{U_n}\right) \mathbb{E}_{X(\tau_{U_n})} \left\{ \exp\left(-\lambda\tau_K\right) \right\} : \tau_{U_n} < \infty \right]$$

(since $X(\tau_{U_n})$ belongs to $\overline{U_n}$)

$$\leq \mathbb{E}_x \left[\exp\left(-\lambda\tau_{U_n}\right) : \tau_{U_n} < \infty \right] \sup_{y \in E} \mathbb{E}_y \left\{ \exp\left(-\lambda\tau_K\right) \right\}$$

$$\leq M \mathbb{E}_x \left[\exp\left(-\lambda\tau_{U_n}\right) \right].$$

Put $\tau = \sup_n \tau_{U_n}$. Since $\tau_{U_n} = D_{U_n} \leq D_K$, where

$$D_C = \inf \left\{ s \geq 0 : X(s) \in C \right\},$$

we see that $\tau \leq D_K$. Since, by quasi-left continuity, $X(\tau) = \lim_{n\to\infty} X(\tau_{U_n})$ $\mathbb{P}_x$-almost surely on the event $\{\tau < \infty\}$, we also have that $X(\tau)$ belongs to K $\mathbb{P}_x$-almost surely. Consequently $D_K \leq \tau$, $\mathbb{P}_x$-almost surely on the event $\{\tau < \infty\}$. Hence we get

$$\mathbb{E}_x \left[\exp\left(-\lambda\tau_K\right) \right] \leq M \mathbb{E}_x \left[\exp\left(-\lambda D_K\right) \right].$$

For $x \in K^c \cup K^r$ the equality $\tau_K = D_K$ holds $\mathbb{P}_x$-almost surely. So for such $x \in E$ the inequality

$$(1 - M) \mathbb{E}_x \left[\exp\left(-\lambda\tau_K\right) \right] \leq 0$$

is valid. Whence, for $x \in K^c \cup K^r$, the equality $\mathbb{E}_x \left[\exp\left(-\lambda\tau_K\right) \right] = 0$ follows. So that $\mathbb{P}_x \left[\tau_K < \infty \right] = 0$ for $x \in K^c \cup K^r$. Since $K^r = \emptyset$ and since $\operatorname{int}(K) \subseteq K^r$, we see that K^c is dense in E. Once again using the result in (k) of Theorem D.3 yields the equality $\mathbb{P}_x \left[\tau_K < \infty \right] = 0$ for all $x \in E$. Hence K is a polar set. This proves the implication (v) $\Rightarrow$ (i). This proves Theorem D.14.

Next we introduce the *λ-potential* $g^\lambda(x, y)$ and the λ-equilibrium measure $\mu_B^\lambda(\cdot)$ defined for $\lambda > 0$ as follows:

$$g^\lambda(x, y) = \int_0^\infty e^{-\lambda t} p_0(t, x, y) dt; \tag{D.37}$$

$$\begin{aligned} \mu_B^\lambda(A) &= \lambda \int \mathbb{E}_y \left[\exp\left(-\lambda\tau_B\right) : X\left(\tau_B\right) \in A, \ \tau_B < \infty \right] dy \\ &= \lambda^2 \int_0^\infty e^{-\lambda s} \int \mathbb{P}_y \left[X(\tau_B) \in A, \ \tau_B < s \right] dy\, ds, \ \ A \in \mathcal{E}. \end{aligned} \tag{D.38}$$

From assertion (i) in Theorem D.13, it follows that the measure $\mu_B^\lambda(\cdot)$ is concentrated on $B \cup B^r$. From the right continuity of paths, it follows that the support of $\mu_B^\lambda(\cdot)$ is contained in the closure of B. The measure $\mu_B^\lambda(\cdot)$ is a Radon measure in the sense that $\mu_B^\lambda(K) < \infty$ for every compact subset K of E. This will be a consequence of the following proposition.

D.15. PROPOSITION. *Let ζ be the life time of the Markov process*

$$\left\{ \left(\Omega, \mathcal{F}, \mathbb{P}_x\right), \left(X(t) : t \geq 0\right), \left(\vartheta_t : t \geq 0\right), \left(E, \mathcal{E}\right) \right\}.$$

Let B be a Borel subset of E. The following assertions are true:

(a) *On the event $\{\tau_B < \infty\}$ the strict inequality $\tau_B < \zeta$ holds $\mathbb{P}_x$-almost surely;*

(b) *On the event $\{\tau_B < \infty\}$ the equality $\zeta = \tau_B + \zeta \circ \vartheta_{\tau_B}$ holds $\mathbb{P}_x$-almost surely;*

(c) *For x, $y \in E$, the equality*

$$\int g^\lambda(x,y)d\mu_B^\lambda(y) = \mathbb{E}_x\left[\exp\left(-\lambda\tau_B\right) - \exp\left(-\lambda\zeta\right) : \tau_B < \infty\right]$$

holds;

(d) *If μ is a non-negative Borel measure on $\mathcal{E}$ for which the integrals*

$$\int g^\lambda(x,y)d\mu(x) = \int\int_0^\infty e^{-\lambda t}p_0(t,x,y)dt\,d\mu(x)$$

are finite for m-almost all $y \in E$, then $\mu(K)$ is finite for all compact subsets K of E;

(e) *The λ-equilibrium measures μ_B^λ, $\lambda > 0$, are Radon measures;*

(f) *Let A be Borel subset of E. The following identity holds*

$$\begin{aligned}&\mathbb{E}_x\left[\exp\left(-\lambda\tau_A\right)g^\lambda\left(X\left(\tau_A\right),y\right) : \tau_A < \infty\right]\\ &= \mathbb{E}_y\left[\exp\left(-\lambda\tau_A\right)g^\lambda\left(X\left(\tau_A\right),x\right) : \tau_A < \infty\right],\end{aligned} \tag{D.39}$$

for all $\lambda > 0$ and for all x and y in E.

The assertions in (c), (e) and (f) rely on the symmetry of the function $p_0(t,x,y)$.

REMARK. The expression in assertion (d) of Proposition D.15 is closely related to the Newton capacity: compare with Definition 2.36.(b).

Proof. (a) On the event $\{\tau_B \geq \zeta\}$ we would have $\triangle = X(\tau_B) \in B \cup B^r \subseteq E$ $\mathbb{P}_x$-almost surely. Since $\triangle$ does not belong to E, it follows that $\tau_B < \zeta$, $\mathbb{P}_x$-almost surely.

(b) If $\omega \in \Omega$ is such that $\zeta(\omega) > \tau_B(\omega)$, then the identity $\zeta(\omega) = \tau_B(\omega) + \zeta \circ \vartheta_{\tau_B}(\omega)$ easily follows. So the assertion in (b) follows from (a).

(c) Let λ be strictly positive. Then, by definition and standard results on image measures, we see that

$$\begin{aligned}&\int g^\lambda(x,y)d\mu_B^\lambda(y) = \lambda\int\mathbb{E}_y\left[\exp\left(-\lambda\tau_B\right)g^\lambda\left(x,X\left(\tau_B\right)\right) : \tau_B < \infty\right]dy\\ &= \lambda\int\mathbb{E}_y\left[\exp\left(-\lambda\tau_B\right)\int_0^\infty e^{-\lambda t}p_0\left(t,x,X\left(\tau_B\right)\right)dt : \tau_B < \infty\right]dy\\ &= \lambda\int\mathbb{E}_y\left[\int_{\tau_B}^\infty e^{-\lambda t}p_0\left(t-\tau_B,x,X\left(\tau_B\right)\right)dt : \tau_B < \infty\right]dy\\ &= \lambda\int_0^\infty e^{-\lambda t}\int\mathbb{E}_y\left[p_0\left(t-\tau_B,x,X\left(\tau_B\right)\right) : \tau_B \leq t\right]dy\,dt\end{aligned}$$

(symmetry of the function $p_0(t,x,y)$)

$$= \lambda \int_0^\infty e^{-\lambda t} \int \mathbb{E}_y \left[p_0 \left(t - \tau_B, X\left(\tau_B\right), x\right) : \tau_B < t\right] dy\, dt$$

(Theorem D.4)

$$= \lambda \int_0^\infty e^{-\lambda t} \int \mathbb{E}_x \left[p_0 \left(t - \tau_B, X\left(\tau_B\right), y\right) : \tau_B < t\right] dy\, dt$$

(Fubini's theorem)

$$= \lambda \int_0^\infty e^{-\lambda t} \mathbb{E}_x \left[\int p_0 \left(t - \tau_B, X\left(\tau_B\right), y\right) dy : \tau_B < t\right] dt$$

(assertion (a) in Theorem D.3 with $f = 1_E$)

$$= \lambda \int_0^\infty e^{-\lambda t} \mathbb{P}_x \left[X(t) \in E, \tau_B < t\right] dt$$

$$= \lambda \int_0^\infty e^{-\lambda t} \mathbb{P}_x \left[\tau_B \leq t\right] dt - \lambda \int_0^\infty e^{-\lambda t} \mathbb{P}_x \left[\tau_B < t, X(t) = \triangle\right] dt$$

$$= \mathbb{E}_x \left[\exp\left(-\lambda \tau_B\right) : \tau_B < \infty\right] - \lambda \mathbb{E}_x \left[\int_{\tau_B}^\infty e^{-\lambda t} 1_\triangle(X(t)) dt : \tau_B < \infty\right]$$

(by (a) on the event $\{\tau_B < \infty\}$, $\zeta > \tau_B$, $\mathbb{P}_x$-almost surely)

$$= \mathbb{E}_x \left[\exp\left(-\lambda \tau_B\right) : \tau_B < \infty\right] - \lambda \mathbb{E}_x \left[\int_{\zeta}^\infty e^{-\lambda t} 1_\triangle(X(t)) dt : \tau_B < \infty\right]$$

$$= \mathbb{E}_x \left[\exp\left(-\lambda \tau_B\right) : \tau_B < \infty\right] - \mathbb{E}_x \left[\exp\left(-\lambda \zeta\right) : \tau_B < \infty\right].$$

This proves (c).

(d) Assume that K_0 is a compact subset of E for which $\mu\left(K_0\right) = \infty$, where μ is a Borel measure with the property that for m-almost every $y \in E$ the integral

$$R(\lambda)^* \mu(y) := \int g^\lambda(x, y) d\mu(x) \tag{D.40}$$

is finite. (We wrote $\mu \mapsto R(\lambda)^* \mu$, because it can be regarded as the adjoint of the operator $f \mapsto R(\lambda) f(x) = \int g^\lambda(x, y) f(y) dy$.) Define the subset E^f of E by

$$E^f := \left\{y \in E : R(\lambda)^* \mu(y) < \infty\right\}. \tag{D.41}$$

Then $m\left(E \setminus E^f\right) = 0$. Let $\Lambda \subseteq E^f$ be a finite subset. Then

$$\infty > \sum_{y \in \Lambda} R(\lambda)^* \mu(y) \geq \inf_{x \in K_0} \left\{\sum_{y \in \Lambda} \int_0^\infty e^{-\lambda t} p_0(t, x, y) dt\right\} \mu\left(K_0\right). \tag{D.42}$$

Since, by hypothesis, $\mu(K_0) = \infty$, we see that

$$\inf_{x \in K_0} \left\{\sum_{y \in \Lambda} \int_0^\infty e^{-\lambda t} p_0(t, x, y) dt\right\} = 0. \tag{D.43}$$

Since the function

$$x \mapsto \sum_{y\in\Lambda}\int_0^\infty e^{-\lambda t}p_0(t,x,y)dt$$

is lower semi-continuous, it follows from (D.43) that the set

$$\bigcap_{y\in\Lambda}\left\{x\in K_0 : \int_0^\infty e^{-\lambda t}p_0(t,x,y)dt = 0\right\} \tag{D.44}$$

is a nonempty compact subset of K_0. Since Λ is an arbitrary finite subset of E^f, we infer by a compactness argument that

$$\bigcap_{y\in E^f}\left\{x\in K_0 : \int_0^\infty e^{-\lambda t}p_0(t,x,y)dt = 0\right\} \neq \emptyset.$$

Let x_0 be a point in this intersection. Then

$$\begin{aligned}0 &= \int_{E^f}\int_0^\infty e^{-\lambda t}p_0(t,x_0,y)dt\,dy = \int_0^\infty e^{-\lambda t}\left(\int_{E^f}p_0(t,x_0,y)dy\right)dt\\ &= \int_0^\infty e^{-\lambda t}\left(\int_E p_0(t,x_0,y)dy\right)dt.\end{aligned}$$

Hence, for almost all $t>0$, the integral $\int_E p_0(t,x_0,y)dy$ vanishes. Since $1 = \lim_{t\downarrow 0}\int_E p_0(t,x_0,y)dy$, it would follow that $1=0$. This is clearly a contradiction. This proves (d).

(e) Apply (d) to the measure $\mu_B^\lambda(\cdot)$.

(f) Let A be a Borel subset of E and let x and y belong to E. Then by Fubini's theorem and by Theorem D.4, we have

$$\begin{aligned}&\mathbb{E}_x\left[\exp\left(-\lambda\tau_A\right)g^\lambda\left(X\left(\tau_A\right),y\right) : \tau_A<\infty\right]\\ &= \mathbb{E}_x\left[\int_0^\infty \exp\left(-\lambda\left(t+\tau_A\right)\right)p_0\left(t,X\left(\tau_A\right),y\right)dt : \tau_A<\infty\right]\\ &= \mathbb{E}_x\left[\int_{\tau_A}^\infty e^{-\lambda t}p_0\left(t-\tau_A,X\left(\tau_A\right),y\right)dt : \tau_A<\infty\right]\\ &= \int_0^\infty e^{-\lambda t}\mathbb{E}_x\left[p_0\left(t-\tau_A,X\left(\tau_A\right),y\right)dt : \tau_A<t\right]dt\\ &= \int_0^\infty e^{-\lambda t}\left(p_0(t,x,y)-q_A(t,x,y)\right)dt = \int_0^\infty e^{-\lambda t}\left(p_0(t,y,x)-q_A(t,y,x)\right)dt\\ &= \mathbb{E}_y\left[\exp\left(-\lambda\tau_A\right)g^\lambda\left(X\left(\tau_A\right),x\right) : \tau_A<\infty\right].\end{aligned}$$

D.16. THEOREM. *Let $p_0(t,x,y)$ be as in Theorem D.3. In particular suppose that the function $(t,x,y) \mapsto p_0(t,x,y) = p_0(t,y,x)$ is continuous on $(0,\infty)\times E\times E$. For every Borel subset B of E the set $B\setminus B^r$ is polar and, if B and A are Borel subsets for which $B\subseteq A$, then*

$$\tau_B = \tau_A + \tau_B \circ \vartheta_{\tau_A} \tag{D.45}$$

$\mathbb{P}_x$-almost surely on the event $\{\tau_A < \infty\}$.

Proof. From assertion (v) in Theorem D.14 it follows that it suffices to prove that equation (D.45) holds for all closed subsets B of E contained in A^r. Put $\tau := \tau_A + \tau_B \circ \vartheta_A$. Then $\tau_B \le \tau$. For fixed $\lambda > 0$ we have

$$\begin{aligned}
&\mathbb{E}_x\left[\exp(-\lambda\tau_B) : \tau_B < \infty\right] - \mathbb{E}_x\left[\exp(-\lambda\tau) : \tau < \infty\right] \\
&= \mathbb{E}_x\left[\exp(-\lambda\tau_B) - \exp(-\lambda\zeta) : \tau_B < \infty\right] - \mathbb{E}_x\left[\exp(-\lambda\tau) - \exp(-\lambda\zeta) : \tau < \infty\right] \\
&\quad + \mathbb{E}_x\left[\exp(-\lambda\zeta) : \tau_B < \infty\right] - \mathbb{E}_x\left[\exp(-\lambda\zeta) : \tau < \infty\right].
\end{aligned} \tag{D.46}$$

Next we apply the strong Markov property to obtain

$$\begin{aligned}
&\mathbb{E}_x\left[\exp(-\lambda\tau_B) : \tau_B < \infty\right] - \mathbb{E}_x\left[\exp(-\lambda\tau) : \tau < \infty\right] \\
&\quad - \mathbb{E}_x\left[\exp(-\lambda\zeta) : \tau_B < \infty\right] + \mathbb{E}_x\left[\exp(-\lambda\zeta) : \tau < \infty\right] \\
&= \mathbb{E}_x\left[\exp(-\lambda\tau_B) - \exp(-\lambda\zeta) : \tau_B < \infty\right] \\
&\quad - \mathbb{E}_x\left[\exp(-\lambda\tau_A)\,\mathbb{E}_{X(\tau_A)}\left\{\exp(-\lambda\tau_B) - \exp(-\lambda\zeta) : \tau_B < \infty\right\} : \tau_A < \infty\right]
\end{aligned}$$

(symmetry: Proposition D.15. (c))

$$= \int g^\lambda(x,y)d\mu_B^\lambda(y) - \mathbb{E}_x\left[\exp(-\lambda\tau_A)\int g^\lambda\left(X(\tau_A),y\right)d\mu_B^\lambda(y) : \tau_A < \infty\right]$$

$$= \int g^\lambda(x,y)d\mu_B^\lambda(y) - \int \mathbb{E}_x\left[\exp(-\lambda\tau_A)\,g^\lambda\left(X(\tau_A),y\right) : \tau_A < \infty\right]d\mu_B^\lambda(y)$$

(symmetry: Proposition D.15.(f) equality (D.39))

$$= \int g^\lambda(x,y)d\mu_B^\lambda(y) - \int \mathbb{E}_y\left[\exp(-\lambda\tau_A)\,g^\lambda\left(X(\tau_A),x\right) : \tau_A < \infty\right]d\mu_B^\lambda(y)$$

(the measure $\mu_B^\lambda(\cdot)$ is concentrated on $B\cup B^r \subset A^r$)

$$= \int g^\lambda(x,y)d\mu_B^\lambda(y) - \int_{A^r} \mathbb{E}_y\left[\exp(-\lambda\tau_A)\,g^\lambda\left(X(\tau_A),x\right) : \tau_A < \infty\right]d\mu_B^\lambda(y)$$

(for $y\in A^r$, $\tau_A = 0$, $\mathbb{P}_y$-almost surely)

$$= \int g^\lambda(x,y)d\mu_B^\lambda(y) - \int_{A^r} \mathbb{E}_y\left[g^\lambda\left(X(0),x\right)\right]d\mu_B^\lambda(y)$$

$$= \int g^\lambda(x,y)d\mu_B^\lambda(y) - \int \mathbb{E}_y\left[g^\lambda(y,x)\right]d\mu_B^\lambda(y) = 0. \tag{D.47}$$

The final equality in (D.47) follows from another appeal to the symmetry: $g^\lambda(x,y) = g^\lambda(y,x)$. As a consequence the expression in (D.47) is zero. The equality in (D.45) follows.

REMARK 1. The symmetry of the density $p_0(t,x,y)$ was used in Theorem D.3 assertion (g). However, for $B \subseteq C$, the inequality $q_B(t,x,y) \geq q_C(t,x,y)$ remains true in the non-symmetric case as well. This follows from the representation

$$q_C(t,x,y) = \mu_{0,x}^{t,y}\left[\tau_C \geq t\right] \leq \mu_{0,x}^{t,y}\left[\tau_B \geq t\right] = q_B(t,x,y).$$

In the proof of Theorem D.4 and Theorem D.5, the symmetry was also used. The same is true for the results in Theorem D.6, Proposition D.7, Lemma D.8, Lemma D.9, Lemma D.10, Lemma D.11, assertions (c), (e) and (f) of Proposition D.15, and Theorem D.16. In fact, the use of symmetry is very substantial in the proof of Theorem D.16. This theorem in conjunction with Proposition D.15 and Theorems D.4 and D.5 is among the most relevant results. A nice problem would be to prove similar result in the non-symmetric case (i.e. for non-selfadjoint generators).

REMARK 2. Set $S = \inf\left\{s > 0 : \int_0^s 1_{\{E\setminus U\}}(X(u))du > 0\right\}$: the (first) penetration time of U^c. It is noted that hitting times and penetration times coincide for open subsets. Theorem D.1 remains valid with S replacing T. This is so because $S \geq T$ so that $\{S < \varepsilon\} \subseteq \{T < \varepsilon\}$, and hence

$$\begin{aligned}
&\mathbb{E}_x\left[p_0\left(t-S, X(S), y\right) : S < \varepsilon\right] = \mu_{0,x}^{t,y}\left\{S < \varepsilon\right\} \leq \mu_{0,x}^{t,y}\left\{T < \varepsilon\right\} \\
&= \mathbb{E}_x\left[p_0\left(t-T, X(T), y\right) : T < \varepsilon\right]
\end{aligned}$$

and $\mathbb{P}_x\left\{S < \varepsilon\right\} \leq \mathbb{P}_x\left\{T < \varepsilon\right\}$. Set $\sigma_B = \inf\left\{s > 0 : \int_0^s 1_B(X(u))du > 0\right\}$ and let

$$\begin{aligned}
p_B(t,x,y) &= \mu_{0,x}^{t,y}\left[\sigma_B \geq t\right] = p_0(t,x,y) - \mu_{0,x}^{t,y}\left[\sigma_B < t\right] \text{ and} \\
\left[P_B(t)f\right](x) &= \mathbb{E}_x\left[f(X(t)) : \sigma_B > t\right] = \int p_B(t,x,y)f(y)dy.
\end{aligned}$$

Compare this with Proposition 2.22. Then Theorem D.3 remains true for σ_B replacing τ_B, $p_B(t,x,y)$ instead of $q_B(t,x,y)$, and $P_B(t)$ substituted for $Q_B(t)$. In fact, for none of the assertions in Theorem D.3, including (g), the symmetry of $p_0(t,x,y)$ will be required. Since, for $B \subseteq C$, the equality $\sigma_B = \sigma_C + \sigma_B \circ \vartheta_{\sigma_C}$ holds $\mathbb{P}_x$-almost surely on the event $\{\sigma_B < \infty\}$, the equality in (g) follows from the Markov property. Next, suppose as usual, that the function $p_0(t,x,y)$ is symmetric and continuous on $(0,\infty) \times E \times E$. Theorem D.4 remains valid and so does Theorem D.5 with σ_B instead of τ_B. The reason for this is the following:

$$\begin{aligned}
&\int\int p_B(t,x,y)f(y)\overline{g(x)}dx\,dy \\
&= \lim_{\beta\to\infty}\left\langle \exp\left(-t\left(K_0 \dot{+} \beta 1_B\right)\right)f, g\right\rangle = \lim_{\beta\to\infty}\left\langle f, \exp\left(-t\left(K_0 \dot{+} \beta 1_B\right)\right)g\right\rangle \\
&= \int\int p_B(t,y,x)f(y)\overline{g(x)}dx\,dy. \qquad \text{(D.48)}
\end{aligned}$$

So that $p_B(t,x,y) = p_B(t,y,x)$ for $m \otimes m$-almost all $(x,y) \in E \times E$. An appeal to Theorem D.3 assertion (j) with $p_B(t,z_1,z_2)$ in place of $q_B(t,z_1,z_2)$ will yield the everywhere symmetry of $p_B(t,x,y)$. The proof of Theorem D.5 can be repeated with σ_B replacing τ_B and $p_B(t,x,y)$ instead of $q_B(t,x,y)$. It is not entirely clear how some of the other results have to be modified. But, Lemma D.7. is true for the set $B^\rho := \{x \in E : \mathbb{P}_x\{\sigma_B = 0\} = 1\}$ instead of B^r. We will prove this. Since $\sigma_B \circ \vartheta_{\sigma_B} = 0$, $\mathbb{P}_x$-almost surely, it follows from the Markov property that

$$\mathbb{E}_x\left[\mathbb{P}_{X(\sigma_B)}\{\sigma_B = 0\} : \sigma_B < \infty\right] = \mathbb{P}_x\{\sigma_B \circ \vartheta_{\sigma_B} = 0, \sigma_B < \infty] = \mathbb{P}_x[\sigma_B < \infty].$$

Hence, on the event $\{\sigma_B < \infty\}$, $X(\sigma_B)$ belongs to B^ρ, $\mathbb{P}_x$-almost surely. Let K be a compact subset of $B \setminus B^\rho$. Then Lemma D.6. implies that

$$\begin{aligned}
m(K) &= \int 1_K(x)1_K(x)dx = \lim_{t\downarrow 0}\int_K \frac{1}{t}\mathbb{E}_x\left[\int_0^t 1_K(X(s))ds\right]dx \\
&\le \liminf_{t\downarrow 0}\int_K \frac{1}{t}\mathbb{E}_x\left[\int_0^t 1_B(X(s))ds : \sigma_B \le t\right]dx \\
&\le \liminf_{t\downarrow 0}\int_K \mathbb{P}_x[\sigma_B \le t]\,dx = \int_K \mathbb{P}_x[\sigma_B = 0]\,dx = 0.
\end{aligned}$$

In the very last equality we employ the fact that the set $K \cap B^\rho$ is empty. Hence $m(B \setminus B^\rho) = 0$. The result in Theorem D.13 remains valid with B^ρ replacing $B \cup B^r$ provided the definition of $h_B(x,A)$ (see formula (D.33)) is appropriately modified, i.e. replacing τ_B with σ_B. In fact $h_B(x,A)$ should be replaced with

$$(x,A) \mapsto \mathbb{P}_x[X(\sigma_B) \in A, \sigma_B < \infty].$$

Except for the first, all the assertions in Theorem D.14 remain valid provided that B^ρ replaces B^r. The first assertion is probably not true for B^ρ instead of B^r. Define the λ-equilibrium measures $\nu_B^\lambda(\cdot)$ as follows:

$$\begin{aligned}
\nu_B^\lambda(A) &= \lambda\int \mathbb{E}_y\left[\exp(-\lambda\sigma_B) : X(\sigma_B) \in A,\ \sigma_B < \infty\right]dy \\
&= \lambda^2\int_0^\infty e^{-\lambda s}\int \mathbb{P}_y[X(\sigma_B) \in A,\ \sigma_B < s]\,dy\,ds,\ A \in \mathcal{E}.
\end{aligned} \tag{D.49}$$

Proposition D.15 remains true with σ_B instead of τ_B and with $\nu_B^\lambda(\cdot)$ replacing $\mu_B^\lambda(\cdot)$. Equality (D.45) in Theorem D.16 remains valid with σ_A and σ_B instead of τ_A respectively τ_B. Again the set $B \setminus B^\rho$ might not be polar. If $\sigma_B = \tau_B$ $\mathbb{P}_x$-almost surely for all $x \in E$, then $B^r = B^\rho$, and hence $B \setminus B^\rho$ is a polar set in such a case.

D.3. Capacities

In the sequel we define, as usual,

$$\begin{aligned}\mathcal{E}^{\lambda}(f,g) &= \left\langle (\lambda I + K_0)^{1/2} f, (\lambda I + K_0)^{1/2} g\right\rangle = \left\langle K_0^{1/2} f, K_0^{1/2} g\right\rangle + \lambda \langle f, g\rangle \\ &= \lim_{t\downarrow 0} \frac{1}{t} \left\langle \left(I - \exp\left(-t\left(\lambda I + K_0\right)\right)\right) f, g\right\rangle, \end{aligned} \tag{D.50}$$

for all functions f, $g \in \operatorname{dom}\left(\mathcal{E}^{\lambda}\right)$. We finish this appendix with some results on Dirichlet and Newton capacities. For more information on Dirichlet capacities, the reader is referred to Fukushima et. al. [110, Chapter 2]. For results on classical Newton potential theory and capacities, the reader may consult Port and Stone [189]. Another source of information is Doob [92, Part I Chapter 13]. We fix a closed subset Γ of E with hitting time T and we set $g^{\lambda}(x,y) = \int_0^{\infty} e^{-\lambda s} p_0(s,x,y)ds$: see (D.37).

D.17. DEFINITION. Let Γ be a closed subset of E.

(a) The λ-Dirichlet capacity of Γ is defined by

$$\lambda\text{-Dirichlet-cap}(\Gamma) = \inf_{U\supseteq\Gamma, U \text{open}} \inf\left\{\mathcal{E}^{\lambda}(f,f) : f \in \operatorname{dom}\left(\mathcal{E}^{\lambda}\right),\ f \geq 1 \text{ on } U\right\}. \tag{D.51}$$

The set Γ is said to have *finite* λ-Dirichlet capacity if the expression in (D.51) is finite.

(b) The set Γ is said to be of *finite relative λ-Newton* capacity if there exists a positive Borel measure μ with the following properties:

(i) for every $x \in \Gamma^r$, the inequality $\int g^{\lambda}(x,y)d\mu(y) \geq \lambda \int g^{\lambda}(x,y)dy$ is valid;

(ii) the integral $\int\int g^{\lambda}(x,y)d\mu(y)\, dx$ is finite.

Its relative λ-Newton capacity is defined by the quantity

$$\inf\left\{\lambda \int\int g^{\lambda}(x,y)d\mu(y)\, dx : \int g^{\lambda}(x,y)d\mu(y) \geq \lambda \int g^{\lambda}(x,y)dy \text{ for all } x \in \Gamma^r\right\}. \tag{D.52}$$

(c) The set Γ is said to have *finite λ-Newton* capacity if there exists a positive Borel measure μ, with the following properties:

(i) For every $x \in \Gamma^r$, either the inequality $\int g^{\lambda}(x,y)d\mu(y) \geq 1$, or for every $x \in \Gamma^r$ the inequality $\int g^{\lambda}(x,y)d\mu(y) \geq \lambda \int g^{\lambda}(x,y)dy$ is valid;

(ii) Either the integral $\int\int g^{\lambda}(x,y)d\mu(y)\, dx$, or the total mass $\mu(E)$ is finite.

In this case, its λ-Newton capacity is given by the expression

$$\inf\left\{\lambda \int\int g^{\lambda}(x,y)d\mu(y)\,dx : \quad \int g^{\lambda}(x,y)d\mu(y) \geq 1 \text{ for all } x \in \Gamma^{r}\right\}$$
$$= \inf\left\{\mu(E) : \quad \int g^{\lambda}(x,y)d\mu(y) \geq \lambda \int g^{\lambda}(x,y)dy \text{ for all } x \in \Gamma^{r}\right\}.$$

(d) In Proposition D.19.(b), we shall prove that the expression in (D.51) is finite for some $\lambda > 0$ if and only if (D.51) is finite for all $\lambda > 0$. We say that Γ has finite *Dirichlet capacity* if (D.51) is finite for some $\lambda > 0$ (or equivalently for all $\lambda > 0$). To the knowledge of the authors it is not known whether a similar statement is true for the expression in (D.52). So we don't know whether or not the notion of finite relative *Newton capacity* is λ-independent. In assertion (7) of Proposition D.18, we shall prove that the λ-Newton capacity of Γ is closely related to its λ-Dirichlet capacity.

D.18. PROPOSITION. *Let $e_{\lambda,\Gamma}$ be a function in the domain of $\mathcal{E}^{\lambda}$ with $e_{\lambda,\Gamma} \geq 1$ on Γ^{r} and for which*

$$\mathcal{E}^{\lambda}\left(e_{\lambda,\Gamma}, e_{\lambda,\Gamma}\right) = \inf_{U \supseteq \Gamma, U \, open} \inf\left\{\mathcal{E}^{\lambda}(f,f) : f \in \operatorname{dom}\left(\mathcal{E}^{\lambda}\right),\ f \geq 1 \text{ on } U\right\}. \quad \text{(D.53)}$$

Set

$$v_{\lambda,\Gamma}(x) = \mathbb{E}_{x}\left[\exp\left(-\lambda T\right) : T < \infty\right], \quad \text{(D.54)}$$

let $C(\Gamma)$ be the constant

$$C(\Gamma) = \inf\left\{\mu(E) : \int g^{\lambda}(x,y)d\mu(y) \geq \lambda \int g^{\lambda}(x,y)dy \text{ for all } x \in \Gamma^{r}\right\}, \quad \text{(D.55)}$$

and let the measure μ_{Γ}^{λ} be given by

$$\mu_{\Gamma}^{\lambda}(B) = \lambda \int \mathbb{E}_{x}\left[\exp\left(-\lambda T\right) : X(T) \in B, T < \infty\right] dx. \quad \text{(D.56)}$$

The following assertions hold true:

(1) *For every $v \in \operatorname{dom}\left(\mathcal{E}^{\lambda}\right)$, $v \geq 0$ on Γ, the inequality $\mathcal{E}^{\lambda}\left(e_{\lambda,\Gamma}, v\right) \geq 0$ is valid;*

(2) *For every $s \geq 0$ the inequality $\exp\left(-s\left(\lambda I + K_0\right)\right) e_{\lambda,\Gamma} \leq e_{\lambda,\Gamma}$ is valid quasi-almost everywhere. The continuous functions $\exp\left(-s\left(\lambda I + K_0\right)\right) e_{\lambda,\Gamma}$ increase if s decreases to 0. For the function $v_{\lambda,\Gamma}$ a similar statement is true;*

(3) *The function $e_{\lambda,\Gamma}$ coincides quasi-almost everywhere with $v_{\lambda,\Gamma}$. The function $e_{\lambda,\Gamma}$ may be redefined in such a way that, for all $x \in E$,*

$$e_{\lambda,\Gamma}(x) = v_{\lambda,\Gamma}(x) = \mathbb{E}_{x}\left[\exp\left(-\lambda T\right) : T < \infty\right];$$

(4) *The measure μ_Γ^λ is concentrated on Γ^r, its total mass equals the λ-Newton capacity of Γ, and*

$$v_{\lambda,\Gamma}(x) = e_{\lambda,\Gamma}(x) = \int g^\lambda(x,y)d\mu_\Gamma^\lambda(y) + \mathbb{E}_x\left[\exp(-\lambda\zeta) : \zeta < \infty\right].$$

If ν_Γ^λ is a measure, concentrated on Γ^r, for which $\int g^\lambda(x,y)d\nu_\Gamma^\lambda(y) = \int g^\lambda(x,y)d\mu_\Gamma^\lambda(y)$ for all $x \in \Gamma^r$, then $\nu_\Gamma^\lambda = \mu_\Gamma^\lambda$;

(5) *Let the measure $\nu_\triangle$ be determined by the equality*

$$\int K_0 f(x)dx = \int f(x)d\nu_\triangle(x), \quad f\,, K_0 f \in L^1(E,m).$$

Then $\nu_\triangle$ is a positive Radon measure and

$$\int_0^\infty g^\lambda(x,y)d\nu_\triangle(y) = \mathbb{E}_x\left[\exp(-\lambda\zeta) : \zeta < \infty\right];$$

(6) *Let μ be a Borel measure on E for which $\int g^\lambda(x,y)d\mu(y) = 1$ for all $x \in \Gamma^r$. Then $\mu = \mu_\Gamma^\lambda + \nu_\triangle$ and*

$$\int \mathbb{E}_x\left[\exp(-\lambda T) - \exp(-\lambda\zeta) : T < \infty\right] d\mu(x) = \mu_\Gamma^\lambda(E);$$

(7) *The λ-Dirichlet capacity and the λ-Newton capacity are related as follows:*

λ*-Newton capacity of* Γ

$= $ *relative* λ*-Newton capacity of* $\Gamma + \lambda \int \mathbb{E}_x\left[\exp(-\lambda\zeta) : \zeta < \infty\right] dx$

$= \mu_\Gamma^\lambda(E) - \int v_{\lambda,\Gamma}(x)d\nu_\triangle(x)$

$= \lambda \int \mathbb{E}_x\left[\exp(-\lambda T) : T < \infty\right] dx - \int v_{\lambda,\Gamma}(x)d\nu_\triangle(x)$

$= \lambda \int v_{\lambda,\Gamma}(x)dx = \mathcal{E}^\lambda\left(v_{\lambda,\Gamma}, v_{\lambda,\Gamma}\right) - \int v_{\lambda,\Gamma}(x)d\nu_\triangle(x)$

$= \inf_{U\supseteq\Gamma, U\,\mathit{open}} \inf\left\{\mathcal{E}^\lambda(f,f) : f \in \operatorname{dom}\left(\mathcal{E}^\lambda\right),\ f \geq 1 \text{ on } U\right\} - \int v_{\lambda,\Gamma}(x)d\nu_\triangle(x)$

$= $ λ*-Dirichlet capacity of* $\Gamma - \int v_{\lambda,\Gamma}(x)d\nu_\triangle(x).$

REMARK. In technical terms, assertion (2) says that the functions $e_{\lambda,\Gamma}$ and $v_{\lambda,\Gamma}$ are λ-*excessive*.

Proof. (1) First suppose that $v > 0$ in a neighborhood of Γ. For $\varepsilon > 0$, the function $e_{\lambda,\Gamma} + \varepsilon v$ belongs to the domain of $\mathcal{E}^\lambda$ and is larger than or equal to 1 on a neighborhood of Γ. From the definition of $e_{\lambda,\Gamma}$, it then follows that

$$\mathcal{E}^\lambda \left(e_{\lambda,\Gamma} + \varepsilon v, e_{\lambda,\Gamma} + \varepsilon v\right) \geq \mathcal{E}^\lambda \left(e_{\lambda,\Gamma}, e_{\lambda,\Gamma}\right).$$

But then we get $\mathcal{E}^\lambda \left(e_{\lambda,\Gamma}, v\right) \geq 0$. If the function v is not strictly larger then 0 on Γ, then we consider $e_{\lambda,\Gamma} + \varepsilon v + \delta w$, where $\delta > 0$ and where $w \in \operatorname{dom}\left(\mathcal{E}^\lambda\right)$ is strictly positive on E.
(2) Suppose $v \geq 0$ belongs to $\operatorname{dom}\left(\mathcal{E}^\lambda\right)$. Then

$$\begin{aligned}
&\langle e_{\lambda,\Gamma} - \exp\left(-s\left(\lambda I + K_0\right)\right) e_{\lambda,\Gamma}, v\rangle \\
&= \left\langle \left(\lambda I + K_0\right) \int_0^s \exp\left(-\sigma\left(\lambda I + K_0\right)\right) e_{\lambda,\Gamma} d\sigma, v\right\rangle \\
&= \mathcal{E}^\lambda \left(\int_0^s \exp\left(-\sigma\left(\lambda I + K_0\right)\right) e_{\lambda,\Gamma} d\sigma, v\right) \\
&= \int_0^s \mathcal{E}^\lambda \left(\exp\left(-\sigma\left(\lambda I + K_0\right)\right) e_{\lambda,\Gamma}, v\right) d\sigma \\
&= \int_0^s \mathcal{E}^\lambda \left(e_{\lambda,\Gamma}, \exp\left(-\sigma\left(\lambda I + K_0\right)\right) v\right) d\sigma \geq 0.
\end{aligned}$$

So the claim in (2) follows for the function $e_{\lambda,\Gamma}$. To get a similar statement for $v_{\lambda,\Gamma}$, we proceed as follows. From the Markov property we get

$$\begin{aligned}
&\left[\exp\left(-s\left(\lambda I + K_0\right)\right) v_{\lambda,\Gamma}\right](x) = \exp\left(-\lambda s\right) \mathbb{E}_x \left[\mathbb{E}_{X(s)} \left\{\exp\left(-\lambda T\right) : T < \infty\right\}\right] \\
&= \exp\left(-\lambda s\right) \mathbb{E}_x \left[\exp\left(-\lambda T \circ \vartheta_s\right) : T \circ \vartheta_s < \infty\right]
\end{aligned}$$

($s + T \circ \vartheta_s \geq T$: Appendix C.20)

$$\leq \mathbb{E}_x \left[\exp\left(-\lambda T\right) : T < \infty\right] = v_{\lambda,\Gamma}(x).$$

(3) Put $Y(t) = \exp\left(-\lambda t\right) e_{\lambda,\Gamma}(X(t))$. We shall prove $Y(t)$ is a supermartingale: see Appendix C.3. Therefore we consider, for $t > s$,

$$\begin{aligned}
&\mathbb{E}_x \left[\exp\left(-\lambda t\right) e_{\lambda,\Gamma}(X(t)) \mid \mathcal{F}_s\right] \\
\text{(Markov property)} \quad &= \exp\left(-\lambda t\right) \mathbb{E}_{X(s)} \left[e_{\lambda,\Gamma}\left(X(t-s)\right)\right] \\
&= \exp\left(-\lambda s\right) \left[\exp\left(-(t-s)\left(\lambda I + K_0\right)\right) e_{\lambda,\Gamma}\right] \left(X(s)\right) \\
\text{(Assertion (2))} \quad &\leq \exp\left(-\lambda s\right) e_{\lambda,\Gamma}\left(X(s)\right).
\end{aligned} \tag{D.57}$$

Indeed, from (D.57) it follows that the process $Y(t)$ is a supermartingale. Next we invoke Doob's optional sampling (Appendix C.11) to justify the following argument:

$$\begin{aligned} v_{\lambda,\Gamma}(x) &= \mathbb{E}_x\left[\exp\left(-\lambda T\right) : T < \infty\right] \\ &\leq \mathbb{E}_x\left[\exp\left(-\lambda T\right) e_{\lambda,\Gamma}(X(T)) : T < \infty\right] \\ &\leq \mathbb{E}_x\left[\exp\left(-\lambda 0\right) e_{\lambda,\Gamma}(X(0)) : T < \infty\right] \leq e_{\lambda,\Gamma}(x). \end{aligned}$$

Hence $v_{\lambda,\Gamma} \leq e_{\lambda,\Gamma}$ quasi-almost everywhere. Next we see

$$\left\langle \left(I - \exp\left(-s\left(\lambda I + K_0\right)\right)\right) v_{\lambda,\Gamma}, v_{\lambda,\Gamma}\right\rangle$$

(Assertion (2) for $v_{\lambda,\Gamma}$ together with $v_{\lambda,\Gamma} \leq e_{\lambda,\Gamma}$ quasi-almost everywhere)

$$\begin{aligned} &\leq \left\langle \left(I - \exp\left(-s\left(\lambda I + K_0\right)\right)\right) v_{\lambda,\Gamma}, e_{\lambda,\Gamma}\right\rangle \\ &= \left\langle v_{\lambda,\Gamma}, \left(I - \exp\left(-s\left(\lambda I + K_0\right)\right)\right) e_{\lambda,\Gamma}\right\rangle \end{aligned}$$

(Assertion (2) for $e_{\lambda,\Gamma}$ together with $v_{\lambda,\Gamma} \leq e_{\lambda,\Gamma}$ quasi-almost everywhere)

$$\leq \left\langle e_{\lambda,\Gamma}, \left(I - \exp\left(-s\left(\lambda I + K_0\right)\right)\right) e_{\lambda,\Gamma}\right\rangle.$$

Set $\mathcal{E}^{\lambda,s}(f,g) = \dfrac{1}{s}\left\langle \left(I - \exp\left(-s\left(\lambda I + K_0\right)\right)\right) f, g\right\rangle$. According to the parallelogram law, we see that

$$\begin{aligned} &\mathcal{E}^{\lambda,s}\left(e_{\lambda,\Gamma} - v_{\lambda,\Gamma}, e_{\lambda,\Gamma} - v_{\lambda,\Gamma}\right) \\ &= 2\mathcal{E}^{\lambda,s}\left(e_{\lambda,\Gamma}, e_{\lambda,\Gamma}\right) + 2\mathcal{E}^{\lambda,s}\left(v_{\lambda,\Gamma}, v_{\lambda,\Gamma}\right) - 4\mathcal{E}^{\lambda,s}\left(\frac{1}{2}\left(e_{\lambda,\Gamma} + v_{\lambda,\Gamma}\right), \frac{1}{2}\left(e_{\lambda,\Gamma} + v_{\lambda,\Gamma}\right)\right) \\ &\leq 2\mathcal{E}^{\lambda,s}\left(e_{\lambda,\Gamma}, e_{\lambda,\Gamma}\right) + 2\mathcal{E}^{\lambda,s}\left(e_{\lambda,\Gamma}, e_{\lambda,\Gamma}\right) - 4\mathcal{E}^{\lambda,s}\left(e_{\lambda,\Gamma}, e_{\lambda,\Gamma}\right) = 0. \end{aligned} \tag{D.58}$$

From the triangle inequality it follows that

$$\left|\sqrt{\mathcal{E}^{\lambda,s}\left(v_{\lambda,\Gamma}, v_{\lambda,\Gamma}\right)} - \sqrt{\mathcal{E}^{\lambda,s}\left(e_{\lambda,\Gamma}, e_{\lambda,\Gamma}\right)}\right|^2 \leq \mathcal{E}^{\lambda,s}\left(e_{\lambda,\Gamma} - v_{\lambda,\Gamma}, e_{\lambda,\Gamma} - v_{\lambda,\Gamma}\right),$$

and hence $\lim_{s\downarrow 0} \mathcal{E}^{\lambda,s}\left(\left(v_{\lambda,\Gamma}, v_{\lambda,\Gamma}\right)\right.$ exist and equals $\mathcal{E}^{\lambda}\left(e_{\lambda,\Gamma}, e_{\lambda,\Gamma}\right)$. It follows that $v_{\lambda,\Gamma}$ belongs to $\operatorname{dom}\left(\mathcal{E}^{\lambda}\right)$. Consequently $e_{\lambda,\Gamma}(x) = v_{\lambda,\Gamma}(x)$ quasi-almost everywhere. Finally, redefine $e_{\lambda,\Gamma}$ by

$$e_{\lambda,\Gamma}(x) = \sup_{s>0}\left[\exp\left(-s\left(\lambda I + K_0\right)\right) e_{\lambda,\Gamma}\right](x).$$

(4) The fact that μ_Γ^λ is concentrated on Γ^r follows from Theorem D.16. From Theorem D.4 we obtain the following symmetry:

$$q_\Gamma(s,x,z) = q_\Gamma(s,z,x) = p_0(s,z,x) - \mathbb{E}_z\left[p_0\left(s-T, X(T), x\right) : T < s\right].$$

The following identity follows from this symmetry (see Proposition D.15.(c)):

$$\int g^\lambda(x,y)d\mu_\Gamma^\lambda(y) = \mathbb{E}_x\left[\exp\left(-\lambda T\right) - \exp\left(-\lambda\zeta\right) : T < \infty\right]. \tag{D.59}$$

Except for the assertions concerning the equality in (D.55) and the measure ν_Γ^λ, (D.59) in combination with (3) proves (4). So we still have to prove the assertions about ν_Γ^λ. First we notice, that for $y \in E$ fixed the process $t \mapsto Y^\lambda(t,y) := \exp\left(-\lambda t\right) g^\lambda(X(t),y)$ is a supermartingale. Therefore the Markov property together with the identity of Chapman-Kolmogorov yields, for $t > s$,

$$\begin{aligned}
&\mathbb{E}_x\left[e^{-\lambda t} g^\lambda(X(t),y) \mid \mathcal{F}_s\right] = e^{-\lambda t}\mathbb{E}_{X(s)}\left[g^\lambda(X(t-s),y)\right] \\
&= e^{-\lambda t}\int_0^\infty e^{-\lambda\sigma}\mathbb{E}_{X(s)}\left[p_0(\sigma, X(t-s),y)\right] d\sigma \\
&= e^{-\lambda t}\int_0^\infty e^{-\lambda\sigma}\int p_0(t-s,X(s),z)p_0(\sigma,z,y)dz\, d\sigma \\
&= e^{-\lambda s}\int_0^\infty e^{-\lambda(\sigma+t-s)}\int p_0(\sigma+t-s,X(s),z)d\sigma \\
&= e^{-\lambda s}\int_{t-s}^\infty e^{-\lambda\sigma}\int p_0(\sigma,X(s),z)d\sigma \le e^{-\lambda s} g^\lambda(X(s),y).
\end{aligned}$$

Hence, for all x and $y \in E$, the process $Y^\lambda(t,y)$ is a $\mathbb{P}_x$-supermartingale. Let μ be any positive Borel measure on E for which

$$\int g^\lambda(x,y)d\mu(y) \ge \lambda\int g^\lambda(x,y)dy \text{ for all } x \in \Gamma^r.$$

By Doob's optional sampling (Appendix C.11) we infer that

$$\begin{aligned}
&\int g^\lambda(x,y)d\mu(y) = \int \mathbb{E}_x\left[Y^\lambda(0,y)\right] d\mu(y) \\
&\ge \int \mathbb{E}_x\left[Y^\lambda(T,y) : T < \infty\right] d\mu(y) \ge \lambda\int \mathbb{E}_x\left[Y^\lambda(T,y) : T < \infty\right] dy
\end{aligned}$$

(employ the equality $\lambda\int g^\lambda(x,y)dy = \int g^\lambda(x,y)d\mu_\Gamma^\lambda$

$$= \int \mathbb{E}_x\left[Y^\lambda(T,y) : T < \infty\right] d\mu_\Gamma^\lambda(y)$$

(use the symmetry $\mathbb{E}_x\left[Y^\lambda(T,y) : T < \infty\right] = \mathbb{E}_y\left[Y^\lambda(T,x) : T < \infty\right]$)

$$= \int \mathbb{E}_y\left[Y^\lambda(T,x) : T < \infty\right] d\mu_\Gamma^\lambda(y)$$

(the measure μ_Γ^λ is concentrated on Γ^r)

$$= \int \mathbb{E}_y\left[Y^\lambda(0,x)\right] d\mu_\Gamma^\lambda(y) = \int g^\lambda(x,y) d\mu_\Gamma^\lambda(y). \tag{D.60}$$

Hence we obtain the inequality

$$\int g^\lambda(x,y) d\mu(y) \geq \int g^\lambda(x,y) d\mu_\Gamma^\lambda(y). \tag{D.61}$$

Let $\nu_\triangle$ be the measure considered in (5). (Its existence does not depend on assertion (4)). From what is proved above (see (D.60)), we obtain (upon invoking (D.59) together with equality

$$\int_0^\infty g^\lambda(x,y) d\nu_\triangle(y) = \mathbb{E}_x\left[\exp\left(-\lambda\zeta\right) : \zeta < \infty\right], \tag{D.62}$$

which we need twice), the (in-)equalities ($\nu := \mu_\Gamma^\lambda + \nu_\triangle$):

$$\begin{aligned}
\mu(E) &\geq \int \mathbb{E}_y\left[\exp\left(-\lambda T\right) : T < \infty\right] d\mu(y) \\
&= \int\int g^\lambda(x,y) d\mu(y) d\nu(x) \geq \int\int g^\lambda(x,y) d\mu_\Gamma^\lambda(y) d\nu(x) \\
&= \int \mathbb{E}_y\left[\exp\left(-\lambda T\right) : T < \infty\right] d\mu_\Gamma^\lambda(y) = \mu_\Gamma^\lambda(E).
\end{aligned} \tag{D.63}$$

Consequently $C(\Gamma) \geq \mu_\Gamma^\lambda(E)$, and hence, since $C(A) \leq \mu_\Gamma^\lambda(\Gamma)$, the equality $C(A) = \mu_\Gamma^\lambda(\Gamma)$ follows. Next, let ν_Γ^λ be a measure with the properties indicated in (4). From the arguments in (D.61), we obtain the inequality

$$\int g^\lambda(x,y) d\nu_\Gamma^\lambda(y) \geq \int g^\lambda(x,y) d\mu_\Gamma^\lambda(y), \quad x \in E.$$

Since the roles of ν_Γ^λ and μ_Γ^λ can be exchanged in (D.60), we obtain the equality

$$\int g^\lambda(x,y) d\nu_\Gamma^\lambda(y) = \int g^\lambda(x,y) d\mu_\Gamma^\lambda(y), \quad x \in E. \tag{D.64}$$

As a consequence we see that for all $f \geq 0$, Borel measurable,

$$\int \left(\lambda I + K_0\right)^{-1} f d\mu_\Gamma^\lambda = \int \left(\lambda I + K_0\right)^{-1} f d\nu_\Gamma^\lambda, \tag{D.65}$$

and hence $\mu_\Gamma^\lambda = \nu_\Gamma^\lambda$. This completes the proof of (4).

(5) First we prove that $\nu_\triangle$ is positive. Therefore we pick $f \geq 0$, such that $K_0 f$ belongs to $L^1(E,m)$. Then ($\langle\cdot,\cdot\rangle$ denotes the duality between $L^1(E,m)$ and $L^\infty(E,m)$)

$$\int f d\nu_\triangle = \langle K_0 f, 1\rangle = \lim_{\eta\downarrow 0}\frac{1}{\eta}\left\langle\left(I-\exp\left(-\eta K_0\right)\right)f, 1\right\rangle$$
$$= \lim_{\eta\downarrow 0}\frac{1}{\eta}\left\langle f, \left(I-\exp\left(-\eta K_0\right)\right)1\right\rangle \geq 0. \tag{D.66}$$

The positivity of $\nu_\triangle$ follows from (D.66). The existence of $\nu_\triangle$ can be proved as follows. First define $\nu_\triangle(K)$, $K \subseteq E$ compact, via the equality

$$\nu_\triangle(K)$$
$$=\inf\left\{\langle K_0 f, 1\rangle : f\geq 1 \text{ on } K, f\in \operatorname{dom}(K_0), K_0 f \in L^1(E,m)\cap L^2(E,m)\right\}.$$

From the arguments in (D.66), we will see that $\nu_\triangle(K) < \infty$. A more or less standard extension procedure leads to a genuine Borel measure $\nu_\triangle$: see e.g. Stroock [220, Chapter 7]. We continue by showing that, for a compact subset K of E, $\nu_\triangle(K)$ is finite. This will imply that $\nu_\triangle$ is a Radon measure. There exists a function $u \geq 0$, $u \in C_\infty(E)\cap L^1(E,m)$, such that $u \geq 2$ on K. Choose $\alpha > 0$ so large that $\alpha\left(\alpha I + K_0\right)^{-1} u \geq 1$ on K. Then

$$\nu_\triangle(K) \leq \alpha\int\left(\alpha I + K_0\right)^{-1} u\, d\nu_\triangle = \alpha\int K_0\left(\alpha I + K_0\right)^{-1} u(x)dx$$
$$= \alpha\int\left[\left(u - \left(\alpha I + K_0\right)^{-1} u\right)\right](x)dx \leq \alpha\int u(x)dx < \infty.$$

Hence it follows that $\nu_\triangle$ is a Radon measure. In order to prove the remaining assertion in (5), we write

$$\int g^\lambda(x,y)\left[\left(I-\exp\left(-\eta K_0\right)\right)1\right](y)dy$$
$$= \int_0^\infty e^{-\lambda s}\int p_0(s,x,y)\left[\left(I-\exp\left(-\eta K_0\right)\right)1\right](y)dy\, ds$$
$$= \int_0^\infty e^{-\lambda s}\left[\exp\left(-sK_0\right)\left(I-\exp\left(-\eta K_0\right)\right)1\right](x)ds$$
$$= \left[\left(\lambda I + K_0\right)^{-1} K_0\int_0^\eta \exp\left(-\sigma K_0\right)1 d\sigma\right](x)$$
$$= \int_0^\eta\left[\exp\left(-\sigma K_0\right)\left(I - \lambda\left(\lambda I + K_0\right)^{-1}\right)1\right](x)d\sigma$$
$$= \eta\int_0^1\left[\exp\left(-\sigma\eta K_0\right)\left(I - \lambda\left(\lambda I + K_0\right)^{-1}\right)1\right](x)d\sigma. \tag{D.67}$$

Since $\exp(-\eta K_0) 1 \leq 1$, the integrand in (D.67) increases as η decreases to 0. It follows that

$$\begin{aligned}
&\int g^\lambda(x,y)d\nu_\triangle(y) = \lim_{\eta\downarrow 0}\frac{1}{\eta}\Big(\int g^\lambda(s,x,y)\left[(I-\exp(-\eta K_0))\,1\right](y)dy \\
&= \left[\left(I-\lambda(\lambda I+K_0)^{-1}\right)1\right](x) = 1-\lambda\int_0^\infty e^{-\lambda s}\int p_0(s,x,y)dy\,ds \\
&= 1-\lambda\int_0^\infty e^{-\lambda s}\mathbb{P}_x\left[X(s)\in E\right]ds = 1-\lambda\int_0^\infty e^{-\lambda s}\mathbb{P}_x\left[\zeta>s\right]ds \\
&= 1-\mathbb{E}_x\left[1-\exp(-\lambda\zeta)\right] = \mathbb{E}_x\left[\exp(-\lambda\zeta)\right]. \qquad \text{(D.68)}
\end{aligned}$$

From (D.68) assertion (5) follows.
(6) Let μ be as in (6). Then

$$\begin{aligned}
&\mathbb{E}_x\left[\exp(-\lambda T) : T<\infty\right] = \int \mathbb{E}_x\left[\exp(-\lambda T)\,g^\lambda(X(T),y) : T<\infty\right]d\mu(y) \\
&= \int \mathbb{E}_x\left[\int_0^\infty \exp(-\lambda(s+T))\,p_0(s,X(T),y)ds : T<\infty\right]d\mu(y) \\
&= \int\int_0^\infty e^{-\lambda s}\mathbb{E}_x\left[p_0(s-T,X(T),y)ds : T<s\right]ds\,d\mu(y)
\end{aligned}$$

(symmetry: Theorem D.4)

$$= \int\int_0^\infty e^{-\lambda s}\mathbb{E}_y\left[p_0(s-T,X(T),x)ds : T<s\right]ds\,d\mu(y). \qquad \text{(D.69)}$$

From (D.69) we conclude that

$$\begin{aligned}
\mu_\Gamma^\lambda(E) &= \lambda\int\int\int_0^\infty e^{-\lambda s}\mathbb{E}_y\left[p_0(s-T,X(T),x) : T<s\right]ds\,d\mu(y)\,dx \\
&= \lambda\int\int_0^\infty e^{-\lambda s}\mathbb{E}_y\left[\mathbb{P}_{X(T)}\left\{X(s-T)\in E\right\} : s>T\right]ds\,d\mu(y)
\end{aligned}$$

(strong Markov property)

$$\begin{aligned}
&= \lambda\int\int_0^\infty e^{-\lambda s}\mathbb{P}_y\left[X(s)\in E,\ s>T\right]ds\,d\mu(y) \\
&= \lambda\int\int_0^\infty e^{-\lambda s}\mathbb{P}_y\left[\zeta>s>T\right]ds\,d\mu(y) \\
&= \int \mathbb{E}_y\left[\exp(-\lambda T)-\exp(-\lambda\zeta) : T<\infty\right]d\mu(y). \qquad \text{(D.70)}
\end{aligned}$$

Assertion (6) is the same as (D.70).

(7) We begin by showing that the relative λ-Newton capacity of Γ is given by the integral

$$\lambda \int \mathbb{E}_y \left[\exp(-\lambda T) - \exp(-\lambda\zeta) : T < \infty\right] dy. \tag{D.71}$$

From Proposition D.15.(c), we get

$$\int g^\lambda(x,y) d\mu_\Gamma^\lambda(y) = \mathbb{E}_x \left[\exp(-\lambda T) - \exp(-\lambda\zeta) : T < \infty\right]. \tag{D.72}$$

We also have

$$\lambda \int g^\lambda(x,y) dy = \mathbb{E}_x \left[1 - \exp(-\lambda\zeta)\right]. \tag{D.73}$$

Hence, using (D.72) and (D.73), we obtain

$$\int g^\lambda(x,y) d\mu_\Gamma^\lambda(y) = \lambda \int g^\lambda(x,y) dy$$

for $x \in \Gamma^r$. Hence, from Proposition D.15.(c) it follows that the relative λ-Newton capacity of Γ is less than or equal to

$$\lambda \int\int g^\lambda(x,y) d\mu_\Gamma^\lambda(y) dx = \lambda \int \mathbb{E}_x \left[\exp(-\lambda T) - \exp(-\lambda\zeta) : T < \infty\right] dx. \tag{D.74}$$

On the other hand, we also have, for μ as in the definition of relative λ-Newton capacity,

$$\lambda \int \mathbb{E}_x \left[\exp(-\lambda T) - \exp(-\lambda\zeta) : T < \infty\right] dx$$

(assertion (c) of Proposition D.15)

$$\begin{aligned} &= \lambda \int\int g^\lambda(x,y) d\mu_\Gamma^\lambda(y)\, dx \leq \int \lambda \int g^\lambda(x,y) dx\, d\mu_\Gamma^\lambda(y) \\ &\leq \int\int g^\lambda(x,y) d\mu(x)\, d\mu_\Gamma^\lambda(y) \end{aligned}$$

(again assertion (c) of Proposition D.15)

$$= \int \mathbb{E}_x \left[\exp(-\lambda T) - \exp(-\lambda\zeta) : T < \infty\right] d\mu(x). \tag{D.75}$$

From (D.69) and (D.74) we infer that the relative λ-Newton capacity of Γ is the integral in (D.71): cf. Definition D.17.(b).

Let μ be the measure $\mu = \mu_\Gamma^\lambda + \nu_\triangle$. From assertion (4) and (5) it follows that

$$\int g^\lambda(x,y)d\mu(y) = \mathbb{E}_x\left[\exp(-\lambda T) : T < \infty\right]. \tag{D.76}$$

Since the expression in (D.76) is 1 on Γ^r, and since

$$\begin{aligned}
&\int\int g^\lambda(x,y)d\mu(y)\,dx = \int\int g^\lambda(x,y)d\mu_\Gamma^\lambda(y)\,dx + \int\int g^\lambda(x,y)d\nu_\triangle(y)\,dx \\
&= \int \mathbb{E}_x\left[\exp(-\lambda T) - \exp(-\lambda\zeta) : T < \infty\right]dx + \int \mathbb{E}_x\left[\exp(-\lambda\zeta)\right]dx \\
&= \int \mathbb{E}_x\left[\exp(-\lambda T) : T < \infty\right]dx,
\end{aligned} \tag{D.77}$$

we see that the λ-Newton capacity is dominated by the integral

$$\lambda\int \mathbb{E}_x\left[\exp(-\lambda T) : T < \infty\right]dx. \tag{D.78}$$

Next we consider

$$\begin{aligned}
&\lambda\int \mathbb{E}_x\left[\exp(-\lambda T) : T < \infty\right]dx \\
&\leq \lambda\int \mathbb{E}_x\left[\exp(-\lambda T)\int g^\lambda\left(X(T),y\right)d\mu(y) : T < \infty\right]dx
\end{aligned}$$

(Fubini and symmetry: Theorem D.4)

$$\begin{aligned}
&= \lambda\int \mathbb{E}_y\left[\exp(-\lambda T)\int g^\lambda\left(X(T),x\right)dx : T < \infty\right]d\mu(y) \\
&= \lambda\int e^{-\lambda s}\int \mathbb{E}_y\left[\int p_0(s-T,X(T),x)dx : T < s\right]d\mu(y)\,ds \\
&= \lambda\int e^{-\lambda s}\int \mathbb{E}_y\left[\mathbb{P}_{X(T)}\left\{X(s-T)\in E\right\} : T < s\right]d\mu(y)\,ds
\end{aligned}$$

(strong Markov property)

$$\begin{aligned}
&= \lambda\int e^{-\lambda s}\int \mathbb{P}_y\left[X(s)\in E,\ T < s\right]d\mu(y)\,ds \\
&= \int \mathbb{E}_y\left[\exp(-\lambda T) - \exp(-\lambda\zeta) : T < \infty\right]d\mu(y) \\
&\leq \int \mathbb{E}_y\left[1 - \exp(-\lambda\zeta)\right]d\mu(y) = \lambda\int\int g^\lambda(x,y)d\mu(y)\,dx.
\end{aligned} \tag{D.79}$$

From (D.78) together with (D.79) we see that the λ-Newton capacity of Γ is the integral in (D.78).

In order to complete the proof of (7), the equality

$$\lambda \int v_{\lambda,\Gamma}(x)dx + \int v_{\lambda,\Gamma}(x)d\nu_{\triangle}(x) = \mathcal{E}^{\lambda}\left(v_{\lambda,\Gamma}, v_{\lambda,\Gamma}\right) \tag{D.80}$$

remains to be shown. Let $1_\Gamma \leq v \leq 1$ be any function in the domain of $K_0^{1/2}$. From the proof of Lemma 3.1.1. in Fukushima [108, p. 62] (and also from assertion (1)), it follows that

$$\left\langle K_0^{1/2} v_{\lambda,\Gamma}, K_0^{1/2} v_{\lambda,\Gamma}\right\rangle + \lambda\left\langle v_{\lambda,\Gamma}, v_{\lambda,\Gamma}\right\rangle = \left\langle K_0^{1/2} v_{\lambda,\Gamma}, K_0^{1/2} v\right\rangle + \lambda\left\langle v_{\lambda,\Gamma}, v\right\rangle.$$

So that, upon letting $v = v_{\lambda,\Gamma'}$ with Γ' open and relatively compact tend to 1, we get

$$\begin{aligned}
&\left\langle K_0^{1/2} v_{\lambda,\Gamma}, K_0^{1/2} v_{\lambda,\Gamma}\right\rangle + \lambda\left\langle v_{\lambda,\Gamma}, v_{\lambda,\Gamma}\right\rangle - \int v_{\lambda,\Gamma}(x)d\nu_{\triangle}(x)\\
&= \sup_{\Gamma'\supseteq\Gamma}\left\{\left\langle K_0^{1/2} v_{\lambda,\Gamma}, K_0^{1/2} v_{\lambda,\Gamma'}\right\rangle + \lambda\left\langle v_{\lambda,\Gamma}, v_{\lambda,\Gamma'}\right\rangle\right\} - \int v_{\lambda,\Gamma}(x)d\nu_{\triangle}(x)\\
&= \sup_{\substack{\alpha>0,\beta>0\\ \Gamma'\supseteq\Gamma}}\left\{\left\langle K_0^{1/2}\alpha\left(\alpha I+\lambda I+K_0\right)^{-1} v_{\lambda,\Gamma}, K_0^{1/2}\beta\left(\beta I+\lambda I+K_0\right)^{-1} v_{\lambda,\Gamma'}\right\rangle\right.\\
&\qquad \left. + \lambda\left\langle \alpha\left(\alpha I+\lambda I+K_0\right)^{-1} v_{\lambda,\Gamma}, \beta\left(\beta I+\lambda I+K_0\right)^{-1} v_{\lambda,\Gamma'}\right\rangle\right\} - \int v_{\lambda,\Gamma}(x)d\nu_{\triangle}(x)\\
&= \sup_{\substack{\alpha>0,\beta>0\\ \Gamma'\supseteq\Gamma}}\left\langle \alpha\left(\alpha I+\lambda I+K_0\right)^{-1} v_{\lambda,\Gamma}, \left(\lambda I+K_0\right)\beta\left(\beta I+\lambda I+K_0\right)^{-1} v_{\lambda,\Gamma'}\right\rangle\\
&- \int v_{\lambda,\Gamma}(x)d\nu_{\triangle}(x)\\
&= \lim_{\alpha\to\infty}\lim_{\beta\to\infty}\left\{\left\langle \alpha\left(\alpha I+\lambda I+K_0\right)^{-1} v_{\lambda,\Gamma}, \beta\left(I-\beta\left(\beta I+\lambda I+K_0\right)^{-1}\right)1\right\rangle\right.\\
&\left. - \left\langle \alpha\left(\alpha I+\lambda I+K_0\right)^{-1} v_{\lambda,\Gamma}, (\beta+\lambda)\left(I-(\beta+\lambda)\left(\beta I+\lambda I+K_0\right)^{-1}\right)1\right\rangle\right\}\\
&= \lim_{\alpha\to\infty}\lim_{\beta\to\infty}\left\langle \alpha\left(\alpha I+\beta I+K_0\right)^{-1} v_{\lambda,\Gamma}, 2\lambda\beta\left(\lambda I+\beta I+K_0\right)^{-1}1-\lambda 1\right\rangle\\
&= \lambda\left\langle v_{\lambda,\Gamma}, 1\right\rangle = \lambda^2\int_0^\infty\int e^{-\lambda s}\int\int p_0(s,x,y)dx\, d\mu_{\lambda,\Gamma}dy\, ds = \lambda\mu_{\lambda,\Gamma}(E).
\end{aligned}$$

Here we used the following two facts:

$$\begin{aligned}
&\left(\lambda I+K_0\right)\left(\alpha I+\lambda I+K_0\right)^{-1} v_{\lambda,\Gamma}(x) = v_{\lambda,\Gamma}(x) - \alpha\left(\alpha I+\lambda I+K_0\right)^{-1} v_{\lambda,\Gamma}(x)\\
&= \alpha\int_0^\infty e^{-\alpha s}\mathbb{E}_x\left[\exp\left(-\lambda T\right) - \exp\left(-\lambda\left(s+T\circ\vartheta_s\right)\right) : T<\infty\right] ds \geq 0,
\end{aligned}$$

and, for $x \in E$ fixed, the function

$$\begin{aligned}\alpha \mapsto & \alpha\left[\left(\alpha I+\lambda I+K_0\right)^{-1} v_{\lambda,\Gamma}\right](x) \\ & =\int_0^{\infty} e^{-s}\mathbb{E}_x\left[\exp\left(-\lambda\left(s\alpha^{-1}+T\circ\vartheta_{s\alpha^{-1}}\right)\right): T<\infty\right] ds\end{aligned} \tag{D.81}$$

increases to $v_{\lambda,\Gamma}(x)$ if α increases to ∞. These properties follow from the definition of $v_{\lambda,\Gamma}$, from §20 in Appendix C, Proposition C.22.(i), and from the Markov property.

The equalities in Assertion (7) of Proposition D.18 imply the results in the following corollary. In case our Markov process is conservative or a diffusion, the measure $\nu_\triangle$ is identically zero: compare with Proposition 1.5.

D.19. PROPOSITION. *Suppose that $\nu_\triangle(B)=0$, $B\in\mathcal{E}$, i.e. $\int K_0 f(x)dx=0$, for $f\in \operatorname{dom}(K_0)\cap L^1(E,m)$.*

(1) *(a) Let Γ be a closed subset of E. Then its λ-Newton capacity coincides with its λ-Dirichlet capacity. So that if one of them is finite (infinite), then the other is and they are equal.*
(2) *(b) The set Γ possesses finite λ-Dirichlet capacity for some $\lambda>0$, if and only if it possesses finite λ-Dirichlet capacity for all $\lambda>0$, if and only if $\int \mathbb{P}_x\left[T\le 1\right]dx<\infty$.*
(3) *(c) Suppose $\int \mathbb{E}_x\left[\exp\left(-\lambda\zeta\right)\right]dx<\infty$. Then Γ has finite relative Newton capacity if and only if it possesses finite Dirichlet capacity for some (or equivalently for all) $\lambda>0$.*

Proof. The assertions in (a) and (c) are a consequence of Assertion (7) in Proposition D.18.

The assertion in (b) can be proved as follows. Since T is a terminal stopping time, we have $T=s+T\circ\vartheta_s$ on the event $\{T\ge s\}$, $\mathbb{P}_x$-almost surely (see Definition 2.18 and Lemma 2.19). From the Markov property we get

$$\begin{aligned}\mathbb{E}_x\left[\exp\left(-\lambda T\right): T<\infty\right] & \le \sum_{k=1}^{\infty}\mathbb{E}_x\left[\exp\left(-\lambda T\right): k-1\le T<k\right] \\ \le \sum_{k=1}^{\infty} & e^{-\lambda(k-1)}\mathbb{P}_x\left[T\circ\vartheta_{k-1}<1, k-1<T\right] \\ = \sum_{k=1}^{\infty} & e^{-\lambda(k-1)}\mathbb{E}_x\left[\mathbb{P}_{X(k-1)}\left\{T<1\right\}: k-1<T\right] \\ \le \sum_{k=1}^{\infty} & e^{-\lambda(k-1)}\mathbb{E}_x\left[\mathbb{P}_{X(k-1)}\left\{T<1\right\}\right].\end{aligned} \tag{D.82}$$

Hence we infer that

$$\int \mathbb{E}_x \left[\exp\left(-\lambda T\right) : T < \infty\right] dx \leq \sum_{k=1}^{\infty} e^{-\lambda(k-1)} \int \mathbb{E}_x \left[\mathbb{P}_{X(k-1)} \left\{T < 1\right\}\right] dx$$

$$\leq \sum_{k=1}^{\infty} e^{-\lambda(k-1)} \int \mathbb{P}_y \left\{T < 1\right\} dy. \tag{D.83}$$

From (D.83), we see that the finiteness of the integral $\int \mathbb{P}_x \left[T \leq 1\right] dx$ implies that the set Γ has finite capacity. The inequality

$$\mathbb{E}_x \left[\exp\left(-\lambda T\right) : T < \infty\right] \geq e^{-\lambda} \mathbb{P}_x \left[T \leq 1\right]$$

implies the converse statement.

The following proposition states that every compact subset of E has finite Dirichlet capacity.

D.20. PROPOSITION. *If Γ is a compact subset of E with λ-equilibrium potential $v_{\lambda,\Gamma}$, then $\int v_{\lambda,\Gamma}(x)dx < \infty$. In fact the inequality*

$$\lambda \int v_{\lambda,\Gamma}(x)\, dx \leq \frac{1}{\sup_{\mu\in P(\Gamma^r)} \inf_{x\in\Gamma^r} \int g^{\lambda}(x,y)d\mu(y)} \tag{D.84}$$

holds true. If the Markov process has an infinite life time, then there is equality in (D.84).

Here $v_{\lambda,\Gamma}(x) = \mathbb{E}_x \left[\exp(-\lambda T) : T < \infty\right]$, $P(\Gamma^r)$ denotes the collection of probability measures on E that do not charge $E\backslash\Gamma^r$, and g^{λ} is the function (λ-potential) $g^{\lambda}(x,y) = \int_0^{\infty} e^{-\lambda s} p_0(s,x,y)ds$.

REMARK. Since $g^{\lambda}(x,y) = \sup_{\eta>0} \int_{\eta}^{\infty} e^{-\lambda s} p_0(s,x,y)ds$ and since the function $(s,x,y) \mapsto p_0(s,x,y)$ is continuous, we see that the function $g^{\lambda}(x,y)$ is lower semi-continuous. Notice that the set Γ^r of regular points of a set Γ is a subset of Γ provided that the set Γ is closed. This follows from the right continuity of paths. Since $\lambda \int g^{\lambda}(x,y)dx \leq 1$, the function $g^{\lambda}(x,y)$ is $m\otimes m$-almost everywhere finite. If $m(U) > 0$ for every nonempty open subset of E, then by lower semi-continuity we see that $g^{\lambda}(x,y)$ is finite for all x and y.

Proof. We first show that the quantity $\sup_{\mu\in P(\Gamma^r)} \inf_{x\in\Gamma^r} \int g^{\lambda}(x,y)d\mu(y)$ is strictly positive. Suppose the contrary. Then, by the lower semi-continuity of the function g^{λ}, the set $\left\{x \in \overline{\Gamma^r} : \int g^{\lambda}(x,y)d\mu(y) = 0\right\}$ is a closed nonempty subset of the compact set $\overline{\Gamma^r}$. This is true for every $\mu \in P(\Gamma^r)$. So, by compactness of the set Γ, there exists a point $x_0 \in \overline{\Gamma^r}$ such that $\int g^{\lambda}(x_0,y)d\mu(y) = 0$ for every $\mu \in P(\Gamma^r)$. Consequently, upon inserting $\mu = \delta_y$, $y \in \Gamma^r$, we get $g^{\lambda}(x_0,y) = 0$ for all $y \in \Gamma^r$. Using the lower semi-continuity of g^{λ}, we see that $g^{\lambda}(x_0,x_0) = 0$.

Hence $p_0(s, x_0, x_0) = 0$ for all $s > 0$. Since $\int \left(p_0(s, x_0, y)\right)^2 dy = p_0(2s, x_0, x_0) = 0$, we infer that $p_0(s, x_0, y) = 0$ for all $s > 0$ and for m-almost all $y \in E$. It follows that $1 = \lim_{s\downarrow 0} \int p_0(s, x_0, y)dy = 0$. Clearly this is a contradiction.

By symmetry it follows that, for $y \in \Gamma^r$,

$$\begin{aligned}
&\mathbb{E}_x\left[e^{-\lambda T} g^\lambda(X(T), y) : T < \infty\right] = \int_0^\infty e^{-\lambda s}\mathbb{E}_x\left[p_0(s-T, X(T), y) : T < s\right] ds \\
&= \int_0^\infty e^{-\lambda s}\mathbb{E}_y\left[p_0(s-T, X(T), x) : T < s\right] ds \\
&= \int_0^\infty e^{-\lambda s}\mathbb{E}_y\left[p_0(s-T, X(T), x) : T = 0\right] ds = g^\lambda(y, x) = g^\lambda(x, y).
\end{aligned}$$

Since $\Gamma \setminus \Gamma^r$ is a polar set, we have $\mathbb{P}_x\left[X(T) \in \Gamma \setminus \Gamma^r : T < \infty\right] = 0$ and hence, for $\mu \in P(\Gamma^r)$,

$$\int g^\lambda(x, y)d\mu(y) = \mathbb{E}_x\left[e^{-\lambda T}\int g^\lambda(X(T), y)d\mu(y) : X(T) \in \Gamma^r,\ T < \infty\right].$$

Notice that $X(T)$ belongs to $\Gamma \cup \Gamma^r$, $\mathbb{P}_x$-almost surely on the event $\{T < \infty\}$ (see Theorem D.13 in Appendix D). Hence we see that

$$\begin{aligned}
&\inf_{z\in\Gamma^r}\int g^\lambda(z, y)d\mu(y) \times \lambda\int v_{\lambda,\Gamma}(x)dx \\
&\leq \lambda\int \mathbb{E}_x\left[e^{-\lambda T}\int g^\lambda(X(T), y)d\mu(y) : X(T) \in \Gamma^r,\ T < \infty\right] dx \\
&= \lambda\int\int g^\lambda(x, y)dx\, d\mu(y) = \int \mathbb{E}_y\left[1 - \exp\left(-\lambda\zeta\right)\right] d\mu(y) \\
&\leq 1 - \inf_{y\in\Gamma^r}\mathbb{E}_y\left[\exp\left(-\lambda\zeta\right)\right]. \qquad \text{(D.85)}
\end{aligned}$$

Consequently

$$1 - \sup_{x\in\Gamma^r}\mathbb{E}_x\left[\exp\left(-\lambda\zeta\right)\right] = \inf_{x\in\Gamma^r}\mathbb{E}_x\left[\exp\left(-\lambda T\right) - \exp\left(-\lambda\zeta\right) : T < \infty\right]$$

(assertion (c) in Proposition D.15)

$$\begin{aligned}
&= \inf_{x\in\Gamma^r}\frac{\int g^\lambda(x, y)d\mu_\Gamma^\lambda(y)}{\mu_\Gamma^\lambda\left(\Gamma^r\right)} \times \lambda\int v_{\lambda,\Gamma}(x)dx \\
&\leq \sup_{\mu\in P(\Gamma^r)}\inf_{x\in\Gamma^r}\int g^\lambda(x, y)d\mu(y) \times \lambda\int v_{\lambda,\Gamma}(x)dx \\
&\leq \sup_{\mu\in P(\Gamma^r)}\int \mathbb{E}_y\left[1 - \exp\left(-\lambda\zeta\right)\right] d\mu(y) = 1 - \inf_{y\in\Gamma^r}\mathbb{E}_y\left[\exp\left(-\lambda\zeta\right)\right]. \qquad \text{(D.86)}
\end{aligned}$$

The inequality in (D.84) follows from (D.86). If $\zeta \equiv \infty$, then the extremities in (D.86) are equal, and hence we have equality in (D.78). This proves Proposition D.20.

We have included these results on capacity theory because they have some importance in connection with spectral theory, in particular with Hilbert-Schmidt and trace properties.

Let $f : \Gamma \to \mathbb{C}$ be a bounded Borel measurable function which is continuous at $x_0 \in \Gamma^r$. As a final result in this Appendix D, we prove Doob's Theorem on regular points: see Doob [92, Part 3, Chapter II, §2, pp 720-721].

D.21. THEOREM. *(Doob) (a) Let $f : \Gamma \to \mathbb{C}$ be a bounded Borel measurable function which is continuous at $x_0 \in \Gamma^r$. Then*

$$\lim_{x \to x_0} \mathbb{E}_x \left[f(X(T)) : T < \infty \right] = f(x_0). \tag{D.87}$$

(b) If $\lim_{x \to x_0, x \in \Sigma} H_\Sigma(x, U) = 1$ for all open sets U containing x_0, then x_0 belongs to Γ^r.

Following the notation of (D.33) we have

$$H_\Sigma(x, B) = h_\Gamma(x, B) = \mathbb{P}_x \left\{ X(T) \in B, T < \infty \right\},$$

where, as usual, $\Sigma = E \setminus \Gamma$ and where T is the hitting time of Γ. The mapping $B \mapsto H_\Sigma(x, B)$ is called the harmonic measure of Σ.

REMARK. A result like the one in (a) can also be proved for functions of the form $x \mapsto \mathbb{E}_x \left[\exp(-\lambda T) f(X(T)) : T < \infty \right]$, and $x \mapsto \mathbb{E}_x \left[f(X(t)) : T > t \right]$. The authors don't know whether it is also true if we replace T with S, the penetration time of Γ.

Proof. (a) Fix $\delta > 0$, $U \subseteq E$ open, and suppose $x_0 \in U$. Then we have, for $X(0) = x \in U$ $\mathbb{P}_x$-almost surely and $\tau_{E \setminus U} = \inf \left\{ s > 0 : X(s) \in E \setminus U \right\}$,

$$\begin{aligned} &\mathbb{P}_x \left[X(T) \in U, T < \delta \right] \\ &\geq \mathbb{P}_x \left[T < \delta, \bigcap_{0 \leq s < \delta} (X(s) \in U) \right] \geq \mathbb{P}_x \left[T < \delta, \bigcap_{0 < s < \delta} (X(s) \in U) \right] \\ &\geq \mathbb{P}_x \left[T < \delta, \tau_{E \setminus U} \geq \delta \right] \geq \mathbb{P}_x \left[T < \delta \right] - \mathbb{P}_x \left[\tau_{E \setminus U} < \delta \right]. \end{aligned} \tag{D.88}$$

From the equality $\mathbb{P}_x \left[T < \delta \right] = \sup_{0 < s < \delta} \mathbb{E}_x \left[\mathbb{P}_{X(s)} \left\{ T < \delta - s \right\} \right]$, we infer that the function $\mathbb{P}_x \left[T < \delta \right]$ is lower semicontinuous. From (D.88) we then obtain

$$\begin{aligned} &\liminf_{x \to x_0} \mathbb{P}_x \left[X(T) \in U, T < \delta \right] \geq \liminf_{x \to x_0} \mathbb{P}_x \left[T < \delta \right] - \limsup_{x \to x_0} \mathbb{P}_x \left[\tau_{E \setminus U} < \delta \right] \\ &\geq 1 - \limsup_{x \to x_0} \mathbb{P}_x \left[\tau_{E \setminus U} < \delta \right] = 1 - 0 = 1, \end{aligned} \tag{D.89}$$

where in the last line we used assertion (a) of Proposition D.1. From (D.89) we infer that

$$\begin{aligned}
&\limsup_{x\to x_0} \mathbb{P}_x\left[X(T)\in E\setminus U, T<\infty\right] \\
&= \limsup_{x\to x_0} \left(\mathbb{P}_x\left[T<\infty\right] - \mathbb{P}_x\left[X(T)\in U, T<\infty\right]\right) \\
&\le 1 - \liminf_{x\to x_0} \mathbb{P}_x\left[X(T)\in U, T<\infty\right] = 1-1 = 0. \qquad \text{(D.90)}
\end{aligned}$$

Next let the Borel measurable function f be continuous at x_0. Without loss of generality we assume that $0 \le f \le 1$. Then we get, using (D.89) and (D.90),

$$\begin{aligned}
&\liminf_{x\to x_0} \mathbb{E}_x\left[f(X(T)) : T<\infty\right] \\
&= \liminf_{x\to x_0}\left(\int_0^{f(x_0)} \mathbb{P}_x\left[f(X(T))>\xi : T<\infty\right] d\xi \right. \\
&\quad \left. + \int_{f(x_0)}^1 \mathbb{P}_x\left[f(X(T))\ge\xi : T<\infty\right] d\xi\right) \\
&\ge \int_0^{f(x_0)} \liminf_{x\to x_0} \mathbb{P}_x\left[f(X(T))>\xi : T<\infty\right] d\xi \\
&\quad + \int_{f(x_0)}^1 \liminf_{x\to x_0} \mathbb{P}_x\left[f(X(T))\ge\xi : T<\infty\right] d\xi \ge \int_0^{f(x_0)} 1 d\xi = f(x_0). \qquad \text{(D.91)}
\end{aligned}$$

We may apply the same reasoning to the function $1-f$, to get

$$\liminf_{x\to x_0} \mathbb{E}_x\left[1-f(X(T)) : T<\infty\right] \ge 1-f(x_0).$$

Hence Theorem D.21 follows for $0 \le f \le 1$. A general complex valued function which is continuous at x_0 may be written as a linear combination of 4 real valued functions, each of which is between zero and 1 and continuous at x_0.

(b) Let $x_0 \in E$ satisfy the hypothesis of (b). We shall prove that

$$\mathbb{P}_{x_0}\left[X(T)=x_0,\ T<\infty\right]=1.$$

Since the set $\Gamma\setminus\Gamma^r$ is polar (see (i) of Theorem D.14), we then infer that

$$\mathbb{P}_{x_0}\left[X(T)=x_0,\ X(T)\in\Gamma^r, T<\infty\right]=1.$$

Hence x_0 belongs to Γ^r. In order to prove $\mathbb{P}_{x_0}\left[X(T)=x_0,\ T<\infty\right]=1$, we fix $\varepsilon>0$, we pick an arbitrary open subset U containing x_0, and we choose an open neighborhood V of x_0 such that

$$\mathbb{P}_x\left[X(T)\in U,\ T<\infty\right]\ge 1-\varepsilon. \qquad \text{(D.92)}$$

By the right continuity of paths, there exists $t > 0$ such that

$$\mathbb{P}_{x_0}\left[\bigcap_{0\le s\le t}(X(s)\in U\cap V)\right]\ge 1-\varepsilon. \tag{D.93}$$

Then we write the following string of (in-)equalities:

$$\begin{aligned}
&\mathbb{P}_{x_0}\left[X(T)\in U,\ T<\infty\right]\\
&=\mathbb{P}_{x_0}\left[X(T)\in U,\ T\le t\right]+\mathbb{P}_{x_0}\left[X(T)\in U,\ \infty>T\ge t\right]\\
&=\mathbb{P}_{x_0}\left[X(T)\in U,\ T\le t\right]+\mathbb{P}_{x_0}\left[X(T)\circ\vartheta_t\in U,\ \infty>T\circ\vartheta_t>0,T>t\right]
\end{aligned}$$

(Markov property)

$$\begin{aligned}
&=\mathbb{P}_{x_0}\left[X(T)\in U,\ T\le t\right]+\mathbb{E}_{x_0}\left[\mathbb{P}_{X(t)}\left\{X(T)\in U,\ \infty>T>0\right\}:T>t\right]\\
&\ge\mathbb{P}_{x_0}\left[X(T)\in U\cap V,\ T\le t\right]\\
&\quad+\mathbb{E}_{x_0}\left[\mathbb{P}_{X(t)}\left\{X(T)\in U,\ \infty>T>0\right\}:X(t)\in U\cap V,\ T>t\right]
\end{aligned}$$

($\mathbb{P}_y\left[X(0)=y\right]=1$, $y\in E$, and $0<t<T$ implies $X(t)\in\Gamma^c=E^\triangle\setminus\Gamma$)

$$\begin{aligned}
&\ge\mathbb{P}_{x_0}\left[\bigcap_{0<s\le t}(X(s)\in U\cap V),\ T\le t\right]\\
&\quad+\mathbb{E}_{x_0}\left[\mathbb{P}_{X(t)}\left\{X(T)\in U,\ \infty>T>0\right\}:X(t)\in U\cap V\cap\Gamma^c,\ T>t\right]
\end{aligned}$$

(apply inequality (D.92) to $X(t)\in V\cap\Gamma^c$)

$$\begin{aligned}
&\ge\mathbb{P}_{x_0}\left[\bigcap_{0<s\le t}(X(s)\in U\cap V),\ T\le t\right]+(1-\varepsilon)\mathbb{P}_{x_0}\left[X(t)\in U\cap V,\ T>t\right]\\
&\ge(1-\varepsilon)\mathbb{P}_{x_0}\left[\bigcap_{0<s\le t}(X(s)\in U\cap V),\ T\le t\right]\\
&\quad+(1-\varepsilon)\mathbb{P}_{x_0}\left[\bigcap_{0<s\le t}(X(s)\in U\cap V),\ T>t\right]\\
&=(1-\varepsilon)\mathbb{P}_{x_0}\left[\bigcap_{0<s\le t}(X(s)\in U\cap V)\right]\ge(1-\varepsilon)^2
\end{aligned} \tag{D.94}$$

Here we used (D.93). Since $\varepsilon>0$ was arbitrary, from (D.94) we infer that

$$\mathbb{P}_{x_0}\left[X(T)\in U,\ T<\infty\right]=1.$$

Finally we let U shrink to x_0 to obtain $\mathbb{P}_{x_0}\left[X(T)=x_0,\ T<\infty\right]=1$. Consequently x_0 belongs to Γ^r. This concludes the proof of Theorem D.21.

Appendix E

Dini's Lemma, Scheffé's Theorem, Monotone Class Theorem

In this appendix we formulate and discuss three theorems: Dini's Lemma, Scheffé's Theorem, and the Monotone Class Theorem. Another proof of Dini's lemma can be found in Stroock [220, Lemma 7.1.23, p. 146].

E.1. LEMMA. *(Dini)Let $(f_n : n \in \mathbb{N})$ be a sequence of continuous functions on the locally compact Hausdorff space E. Suppose that $f_n(x) \geq f_{n+1}(x) \geq 0$ for all $n \in \mathbb{N}$ and for all $x \in E$. If $\lim_{n\to\infty} f_n(x) = 0$ for all $x \in E$, then, for all compact subsets K of E, $\lim_{n\to\infty} \sup_{x\in K} f_n(x) = 0$. If the function f_1 belongs to $C_\infty(E)$, then $\lim_{n\to\infty} \sup_{x\in E} f_n(x) = 0$.*

Proof. We only prove the second assertion.
Fix $\eta > 0$ and consider $\bigcap_{n\in\mathbb{N}} \{x \in E : f_n(x) \geq \eta\}$. Since, by assumption the function f_1 belongs to $C_\infty(E)$, and since $\lim_{n\to\infty} f_n(x) = 0$, $x \in E$, it follows that the intersection $\bigcap_{n\in\mathbb{N}} \{x \in E : f_n(x) \geq \eta\}$ is empty. As a consequence $E = \bigcup_{n\in\mathbb{N}} \{f_n < \eta\}$. Let $\varepsilon > 0$ and set $K = \{f_1 \geq \varepsilon\}$. The subset K is compact. By the preceding argument, there exist $n_\varepsilon \in \mathbb{N}$ for which $K \subseteq \{f_{n_\varepsilon} < \varepsilon\}$. For $n \geq n_\varepsilon$, we have $0 \leq f_n(x) \leq \varepsilon$ for all $x \in E$.

In Definition E.2 and in Theorem E.3 of this appendix, $(E, \mathcal{E}, m)$ may be any measure space with $m(B) \geq 0$ for $B \in \mathcal{E}$.

E.2. DEFINITION. A collection of functions $\{f_j : j \in J\}$ in $L^1(E, \mathcal{E}, m)$ is *uniformly L^1-integrable* if for every $\varepsilon > 0$ there exists $g \in L^1(E, \mathcal{E}, \mu)$, $g \geq 0$ for which

$$\sup_{j\in J} \int_{\{|f_j|\geq g\}} |f_j| \, dm \leq \varepsilon.$$

REMARK 1. If the collection $\{f_j : j \in J\}$ is uniformly L^1-integrable, and if $\{g_j : j \in J\}$ is a collection for which $|g_j| \leq |f_j|$ m-almost everywhere for all $j \in J$, then the collection $\{g_j : j \in J\}$ is uniformly L^1-integrable as well.

REMARK 2. Cauchy sequences in $L^1(E, \mathcal{E}, m)$ are uniformly L^1-integrable.

REMARK 3. Let $f \geq 0$ be a function in $L^1(\mathbb{R}^\nu, \mathcal{B}, m)$, where m is the Lebesgue measure. Suppose $\int f(x)dm(x) = 1$ and $\lim_{n\to\infty} n^\nu f(nx) = 0$ for all $x \neq 0$. Set $f_n(x) = n^\nu f(nx)$, $n \in \mathbb{N}$. Then the sequence $\{f_n : n \in \mathbb{N}\}$ is not uniformly L^1-integrable. This will follow from Theorem E.3 below.

A version of Scheffé's theorem reads as follows. Our proof uses the arguments in the proof Theorem 3.3.5 (Lieb's version of Fatou's lemma) in Stroock [220, p. 54]. Another proof can be found in Bauer [19, Theorem 2.12.4, p. 103].

E.3. THEOREM. *(Scheffé) Let* $(f_n : n \in \mathbb{N})$ *be a sequence in* $L^1(E, \mathcal{E}, m)$. *If* $\lim_{n\to\infty} f_n(x) = f(x)$, *m-almost everywhere, then the sequence* $(f_n : n \in \mathbb{N})$ *is uniformly* L^1*-integrable if and and only* $\lim_{n\to\infty} \int |f_n(x)|\, dm(x) = \int |f(x)|\, dm(x)$.

Proof. Consider the m-almost everywhere pointwise inequality

$$0 \leq |f_n - f| + |f| - |f_n| \leq 2\,|f|\,. \tag{E.1}$$

First suppose that the sequence $\{f_n : n \in \mathbb{N}\}$ is uniformly L^1-integrable. Then, by Fatou's lemma,

$$\int |f(x)|\, dm(x) = \int \liminf |f_n(x)|\, dm(x) \leq \liminf \int |f_n(x)|\, dm(x)$$

(choose $g \in L^1(E, m)$ such that $\int_{\{|f_n| \geq g\}} |f_n(x)|\, dm(x) \leq 1$)

$$\begin{aligned} &\leq \liminf \int_{\{|f_n| \geq g\}} |f_n(x)|\, dm(x) + \int_{\{|f_n| \leq g\}} |f_n(x)|\, dm(x) \\ &\leq 1 + \int g(x) dm(x). \end{aligned} \tag{E.2}$$

From (E.2) we see that the function f belongs to $L^1(E, m)$. From Lebesgue's dominated convergence theorem in conjunction with (E.2), we infer that

$$\lim_{n\to\infty} \int (|f_n - f| + |f| - |f_n|)\, dm = 0. \tag{E.3}$$

Since the sequence $\{f_n{:}n{\in}\mathbb{N}\}$ is uniformly L^1-integrable, and since $\lim_{n\to\infty} f_n(x) = f(x)$ m-almost all x, we see that $\lim_{n\to\infty} \int |f_n - f|\, dm = 0$. So from (E.3) we get

$$\lim_{n\to\infty} \int |f_n|\, dm = \int |f| < \infty. \tag{E.4}$$

Conversely, suppose (E.4) holds. Then f belongs to $L^1(E, m)$. Again we may invoke Lebesgue's dominated convergence theorem to conclude (E.3) from (E.1). Again using (E.4) implies $\lim_{n\to\infty} \int |f_n - f|\, dm = 0$. An appeal to Remark 2 yields the desired result.

Our presentation of the monotone class theorems is taken from Blumenthal and Getoor [35, pages 5–7]. For other versions of this theorem see e.g. Sharpe [205, pages 364–366]. In Theorems E.5 and E.6 and Propositions E.7 and E.8, we give closely related versions of this theorem.

E.4. DEFINITION. Let Ω be a set and let $\mathcal{S}$ be a collection of subsets of Ω. Then $\mathcal{S}$ is a *Dynkin* system if it has the following properties:

(a) $\Omega \in \mathcal{S}$;
(b) if A and B belong to $\mathcal{S}$ and if $A \supseteq B$, then $A \setminus B$ belongs to $\mathcal{S}$;
(c) if $(A_n : n \in \mathbb{N})$ is an increasing sequence of elements of $\mathcal{S}$, then the union $\bigcup_{n=1}^{\infty} A_n$ belongs to $\mathcal{S}$.

The following result on Dynkin systems is well-known.

E.5. THEOREM. *Let $\mathcal{M}$ be a collection of subsets of of Ω which is stable (or closed) under finite intersections. The Dynkin system generated by $\mathcal{M}$ coincides with the σ-field generated by $\mathcal{M}$.*

E.6. THEOREM. *Let Ω be a set and let $\mathcal{M}$ be a collection of subsets of of Ω which is stable (or closed) under finite intersections. Let $\mathcal{H}$ be a vector space of real valued functions on Ω satisfying*

(i) *the constant function 1 belongs to $\mathcal{H}$ and 1_A belongs to $\mathcal{H}$ for all $A \in \mathcal{M}$ and*
(ii) *if $(f_n : n \in \mathbb{N})$ is an increasing sequence of non-negative functions in $\mathcal{H}$ such that $f = \sup_{n\in\mathbb{N}} f_n$ is finite (bounded), then f belongs to $\mathcal{H}$.*

Then $\mathcal{H}$ contains all real valued functions (bounded) functions on Ω, that are $\sigma(\mathcal{M})$ measurable.

Proof. Put $\mathcal{D} = \{A \subseteq \Omega : 1_A \in \mathcal{H}\}$. Then by (i) Ω belongs to $\mathcal{D}$ and $\mathcal{D} \supseteq \mathcal{M}$. If A and B are in $\mathcal{D}$ and if $B \supseteq A$, then $B \setminus A$ belongs to $\mathcal{D}$. If $(A_n : n \in \mathbb{N})$ is an increasing sequence in $\mathcal{D}$, then $1_{\cup A_n} = \sup_n 1_{A_n}$ belongs to $\mathcal{D}$ by (ii). Hence $\mathcal{D}$ is a Dynkin system that contains $\mathcal{M}$. Since $\mathcal{M}$ is closed under finite intersection, it follows, by Theorem 2.8, that $\mathcal{D} \supseteq \sigma(\mathcal{M})$. If $f \geq 0$ is measurable with respect to $\sigma(\mathcal{M})$, then

$$f = \sup_n 2^{-n} \sum\nolimits_{j=1}^{n2^n} 1_{\{f \geq j2^{-n}\}}. \tag{E.5}$$

Since $1_{\{f\geq j2^{-n}\}}$, j, $n \in \mathbb{N}$, belong to $\sigma(\mathcal{M})$, we see that f belongs to $\mathcal{H}$. Here we employed the fact that $\sigma(\mathcal{M}) \subseteq \mathcal{D}$. If f is $\sigma(\mathcal{M})$-measurable, then we write f as a difference of two non-negative $\sigma(\mathcal{M})$-measurable functions.

The previous theorems (Theorem E.5 and Theorem E.6) are used in the following form. Let Ω be a set and let $(E_i, \mathcal{E}_i)_{i\in I}$ be a family of measurable spaces indexed by an arbitrary set I. For each $i \in I$, let $\mathcal{S}_i$ denote a collection of subsets of E_i which is closed under finite intersection and which generates the σ-field $\mathcal{E}_i$. Let $f_i : \Omega \to E_i$ be a map from Ω to E_i. In our presentation of the Markov property, the spaces E_i are all the same space and the maps f_i, $i \in I$, are the state variables $X(t)$, $t \geq 0$. In this context the following two propositions follow.

E.7. PROPOSITION. *Let $\mathcal{M}$ be the collection of all sets of the form $\bigcap_{i\in J} f_i^{-1}(A_i)$, $A_i \in \mathcal{S}_i$, $i \in J$, $J \subseteq I$, J finite. Then $\mathcal{M}$ is a collection of subsets of Ω which is stable under finite intersection and $\sigma(\mathcal{M}) = \sigma(f_i : i \in I)$.*

E.8. PROPOSITION. *Let $\mathcal{H}$ be a vector space of real-valued functions on Ω such that*

(i) *the constant function 1 belongs to $\mathcal{H}$,*
(ii) *if $(h_n : n \in \mathbb{N})$ is an increasing sequence of non-negative functions in $\mathcal{H}$ such that $h = \sup_n h_n$ is finite (bounded), then h belongs to $\mathcal{H}$, and*
(iii) *$\mathcal{H}$ contains all products of the form $\prod_{i\in J} 1_{A_i}\circ f_i$, $J \subseteq I$, J finite, and $A_i \in \mathcal{S}_i$, $i \in J$.*

Under these assumptions $\mathcal{H}$ contains all real-valued functions (bounded) functions in $\sigma(f_i : i \in I)$.

E.9. DEFINITION. Theorems E.5 and E.6 together with Propositions E.7 and E.8 are called the *monotone class theorems*.

Other theorems and results on integration theory not explained in the book can be found in any textbook on the subject. In particular this is true for Fatou's lemma and Fubini's theorem on the interchange of the order of integration. Proofs of these results can be found in Bauer [19] and Stroock [220]. The same references contain proofs of the Radon-Nikodym theorem. This theorem may be phrased as follows.

E.10. THEOREM. *(Radon-Nikodym) If a measure μ on some σ-finite measure space $(E, \mathcal{E}, m)$ is absolutely continuous with respect to m, then there exists a function $f \in L^1(E, \mathcal{E}, m)$ such that $\mu(A) = \int_A f(x)dm(x)$ for all subsets $A \in \mathcal{E}$.*

The measure μ is said to be absolutely continuous with respect to m if $m(A) = 0$ implies $\mu(A) = 0$. The measure m is said to be σ-finite if there exists an increasing sequence $(E_n : n \in \mathbb{N})$ in $\mathcal{E}$ such that $E = \bigcup_{n\in\mathbb{N}} E_n$ and for which $m(E_n) < \infty$, $n \in \mathbb{N}$. A very important application is the existence of *conditional expectations*. This can be seen as follows.

E.11. COROLLARY. *Let $(\Omega, \mathcal{F}, \mathbb{P})$ be a probability space, let $\mathcal{F}_0$ be a sub-field of $\mathcal{F}$, and let $Y : \Omega \to [0, \infty]$ be a $\mathcal{F}$-measurable function (random variable) in $L^1(\Omega, \mathcal{F}, \mathbb{P})$. Then there exists a function $G \in L^1(E, \mathcal{F}_0, m)$ such that $E[Y1_A] = \mu(A) = \mathbb{E}[G1_A]$ for all $A \in \mathcal{F}_0$.*

By convention the random variable G is written as $G = \mathbb{E}\left[Y \mid \mathcal{F}_0\right]$.

Proof. Put $m(A) = \mathbb{E}[Y1_A]$, $A \in \mathcal{F}$, and let μ be the restriction of m to $\mathcal{F}_0$. If for some $A \in \mathcal{F}_0$, $m(A) = 0$, then $\mu(A) = 0$. The Radon-Nikodym theorem yields the existence of a function $G \in L^1(E, \mathcal{F}_0, m)$ such that $E[Y1_A] = \mu(A) = \mathbb{E}[G1_A]$ for all $A \in \mathcal{F}_0$.

References

[1] R. Aebi, *Schrödinger diffusion processes*, Birkhäuser, Basel, 1996.

[2] S. Agmon, *Spectral properties of Schrödinger operators and scattering theory*, Annali della Scuola Normale Superiore di Pisa, no. II, **2** (1975), 151–218.

[3] M. Aizenman and B. Simon, *Brownian motion and Harnack inequality for Schrödinger operators*, Comm. Pure and Applied Math. **XXXV** (1982), 209–273.

[4] S. Albeverio, Ph. Blanchard, and Zhiming Ma, *Feynman-Kac semigroups in terms of signed smooth measures*, Random Partial Differential Equations (Oberwolfach 1989), International Series Numer. Math., vol. **102**, Birkhäuser, Basel, 1991, pp. 1–31.

[5] S. Albeverio, F. Gesztesy, R. Hoegh-Krohn and H. Holden, *Solvable models in quantum mechanics*, Springer Verlag, Berlin, 1988.

[6] S. Albeverio and Zhiming Ma, *Perturbation of Dirichlet forms, lower semiboundedness, closability and form cores*, J. Funct. Anal. **99** (1991), 332–356.

[7] W.O. Amrein, *Non-relativistic quantum dynamics*, Math. Phys., vol. **2**, Reidel, Dordrecht, Boston, London, 1981.

[8] W.O. Amrein, J.M. Jauch and K.B. Sinha, *Scattering theory in quantum mechanics*, Benjamin, New York, 1977.

[9] H. Araki, *On an inequality of Lieb and Thirring*, Letters in Math. Phys. **19** (1990), 167–170.

[10] W. Arendt and C.J.K. Batty, *Tauberian theorems and stability of one-parameter semigroups*, Trans. Amer. Math. Soc. **306** (1988), 837–852.

[11] W. Arendt and C.J.K. Batty, *Absorption semigroups and Dirichlet boundary conditions*, Math. Annalen **295** (1993), 427–448.

[12] W. Arendt, C.J.K. Batty and Ph. Bénilan, *Asymptotic stability of Schrödinger semigroups*, Math. Zeitschrift (1992), 511–518.

[13] A.A. Arsenev, *Singular potentials and resonances*, Publishing House of Moscow University, Moscow, 1974. (Russian)

[14] R. Azencott, P. Baldi, A. Bellaiche, C. Bellaiche, P. Bougerol, M. Chaleyat-Maurel, L. Elie, and J. Granara, *Géodésiques et diffusions en temps petit, Séminaire de probabilités*, Astérisque, vol. **84–85**, Soc. Math. de France, Université de Paris VII,, 1981.

[15] D. Bakry, *L'hypercontractivité et son utilisation en théorie de semigroupes*, Lectures on probability theory (P. Bernard, ed.), Lecture Notes in Math., vol. **1581**, Springer Verlag, Berlin, 1994, pp. 1–114, Ecole d'été de Saint-Flour XXII-1992.

[16] D. Bakry and D. Michel, *Inégalités de Sobolew et minorations du semi-groupe de chaleur*, Ann. de la Facultés des Sciences de Toulouse no. 2 **XI** (1990), 23–66.

[17] C. Bardos, G. Lebeau and J. Rauch, *Scattering frequencies and Gevrey 3 singularities*, Inventiones Math. **90** (1987), 77–114.

[18] C.J.K. Batty, *Asymptotic stability of Schrödinger semigroups: path integral methods*, Math. Annalen **292** (1992), 457–492.

[19] H. Bauer, *Probability theory and elements of measure theory*, Holt, Rinehart and Winston Inc., New York, 1972.

[20] H. Baumgärtel and M. Demuth, *Decoupling by a projection*, Rep. Math. Physics **15** (1979), 173–186.

[21] H. Baumgärtel and M. Wollenberg, *Mathematical scattering theory*, Birkhäuser, Basel, 1983.

[22] M. Ben-Artzi, Y. Dermenjian and J.-C. Guillot, *Acoustic waves in perturbed stratisfied fluids: a spectral theory*, Comm. Part. Diff. Equat. **14** (1989), 479–517.

[23] M. Ben-Artzi and A. Devinatz, *The limiting absorption principle for partial differential operators*, Memoirs of the Amer. Math. Soc. no 364, vol. **66**, 1987.

[24] Yu.M. Berezanskii, *Expansions in eigenfunctions of self-adjoint operators*, Translations of Mathematical Monographs, vol. **63**, Amer. Math. Soc., Providence, 1968, Original title: Razioženic po sobstvennyns funkcijam samosoprjažennyh operatorov, published by Naukova Dumka, Kiev, 1965.

[25] C. Berg and G. Forst, *Potential theory on locally compact abelian groups*, Ergebnisse der Mathematik und ihre Grenzgebiete, vol. **87**, Springer Verlag, 1975.

[26] R.N. Bhattacharya and E.C. Waymire, *Stochastic processes with applications*, Wiley Series in Probability and Mathematical Statistics, John Wiley & Sons, New York, 1990, A Wiley-Interscience Publication.

[27] M.Sh. Birman, *A proof of the Fredholm trace formula as an application of a simple embedding for kernels of integral operators of trace class in $L^2(\mathbb{R}^m)$*, Preprint Linköping University (1989).

[28] M.S. Birman and M.A. Solomyak, *Spectral theory of self-adjoint operators in Hilbert spaces*, Reidel, Dordrecht, 1987.

[29] M.Sh. Birman and M.Z. Solomyak, *Double Stieltjes operator integrals I*, Problemy Mat. Fyz. **1** (1966), 33–67.

[30] M.Sh. Birman and M.Z. Solomyak, *Double Stieltjes operator integrals II*, Problemy Mat. Fyz. **2** (1967), 26–60.

[31] M.Sh. Birman and M.Z. Solomyak, *Double Stieltjes operator integrals III*, Problemy Mat. Fyz. **6** (1973), 27–53.

[32] M.Sh. Birman and M.Z. Solomyak, *Operator integration, perturbations, and commutators*, Zap. Nauchn. Sem. Leningrad, Otdel. Inst. Steklov (LOMI) **170** (1989), 34–66.

[33] J.-M. Bismut, *Large deviations and the Malliavin calculus*, Birkhäuser, Basel, 1984.

[34] J. Bliedtner and W. Hansen, *Potential theory*, Universitext, Springer Verlag, Berlin, 1980.

[35] R.M. Blumenthal and R.K. Getoor, *Markov processes and potential theory*, Pure and Appl. Math., vol. **29**, Academic Press, New York, 1986, a series of monographs and textbooks.

[36] J.-M. Bony, P. Courrège and P. Priouret, *Semi-group de Feller sur une variété à bord compacte et problèmes aux limites intégro-différentiels du second ordre donnant lieu au principle du maximum*, Ann. Inst. Fourier (Grenoble) **18** (1986), 369–521.

[37] N. Bouleau and F. Hirsch, *Dirichlet forms and analysis on Wiener space*, De Gruyter Studies in Mathematics, vol. **14**, de Gruyter, New York, 1991.

[38] N. Bouleau and D. Lamberton, *Residual risks and hedging strategies in Markovian markets*, Stoch. Processes and their Applications **33** (1989), 131–150.

[39] A. Boutet de Monvel, V. Georgescu and A. Soffer, *N-body Hamiltonians with hard-core interactions*, Rev. Math. Phys. **6** (1994), 515–596.

[40] M. Bramanti, *Potential theory for stationary Schrödinger operators: a survey of results obtained with non-probabilistic methods*, Le Mathematiche - Fasc. I **XLVII** (1992), 25–61.

[41] J.F. Brasche, P. Exner, Yu.A. Kuperin and P.Šeba, *Schrödinger operators with singular interactions*, J. Math. Anal. Appl. **184** (1994), 112–139.

[42] E. Brüning and F. Gesztesy, *Continuity of wave and scattering operators*, J. Math. Phys. **24** (1983), 1516–1528.

[43] E.A. Carlen, *Conservative diffusions*, Comm. Math. Phys. **94** (1984), 293–315.

[44] E.A. Carlen, S. Kusuoka and D.W. Stroock, *Upper bounds for symmetric Markov transition functions*, Ann. Inst. Henri Poincaré (Probabilités et Statistiques), Suppl. au no. 2 **23** (1987), 245–287.

[45] E.A. Carlen and E.H. Lieb, *Optimal hypercontractivity for Fermi fields and related non-commutative integration inequalities*, Comm. Math. Phys. **155** (1993), 27–46.

[46] R. Carmona, *Probabilistic construction of Nelson processes*, Probabilistic methods in mathematical physics, Proceedings of the Taniguchi International Symposium on Stochastic Methods in Mathematical Physics, Katata/Kyoto, 1985, Academic Press, Boston Mass., 1987, pp. 55–81.

[47] R. Carmona and J. Lacroix, *Spectral theory of random Schrödinger operators*, Probability Theory and its Applications, Birkhäuser, Basel, 1990.

[48] R. Carmona, W.C. Masters and B. Simon, *Relativistic Schrödinger operators: Asymptotic behaviour of eigenfunctions*, J. Funct. Anal. **91** (1990), 117–142.

[49] K.L. Chung, *Lectures from Brownian motion to Markov processes*, Die Grundlehren der mathematischen Wissenschaften in Einzeldarstellungen, vol. **249**, Springer Verlag, Berlin, 1982.

[50] K.L. Chung, *Probability methods in potential theory*, Potential theory - Surveys and problems, Lecture Notes in Math., vol. **1344**, Springer Verlag, Berlin, 1988, pp. 42–54.

[51] K.L. Chung and K.H. Rao, *Feynman-Kac functional and the Schrödinger equation*, Seminar on Stochastic Processes 1981 (E. Cinlar, K.L. Chung and R.K. Getoor, eds.), Birkhäuser, Basel, 1981, pp. 1–29.

[52] K.L. Chung and K.H. Rao, *General gauge theorem for multiplicative functionals*, Trans. Amer. Math. Soc. **306** (1988), 819–836.

[53] K.L. Chung and Z. Zhao, *From Brownian motion to Schrödinger's equation*, Die Grundlehren der mathematischen Wissenschaften in Einzeldarstellungen, vol. **312**, Springer Verlag, Berlin, 1995.

[54] Ph. Clément and H.J.A.M. Heijmans, *One-parameter semigroups*, CWI-Monographs, North-Holland, Amsterdam, 1987.

[55] T. Coulhon, *Dimensions of continuous and discrete semigroups on the L^p-spaces*, Semigroup Theory and Evolution Equations, Lecture Notes in Pure and Appl. Math., vol. **135**, Marcel Dekker, New York, 1991, pp. 93–100.

[56] T. Coulhon, *Inégalités de Gagliardo-Nirenberg pour les semi-groupes d'opérateurs et applications*, Potential Anal. **1** (1992), 343–353.

[57] T. Coulhon, *Noyau de la chaleur et discrétisation d'une variété riemannienne*, Israel J. Math. **80** (1992), 289–300.

[58] T. Coulhon, *Dimension à l'infini d'un semi-groupe analytique*, Bull. Sci. Math. (2), no. 4 **114** (1990), 485–480.

[59] T. Coulhon and L. Saloff-Coste, *Théorèmes de Sobolev pour les semi-groupes d'opéra teurs et applications aux groupes de Lie unimodulaires*, C.R. Acad. Sci. Paris, t. 309, Série I (1989), 289–294.

[60] T. Coulhon and L. Saloff-Coste, *Semi-groupes d'opérateurs et espaces fonctionnels sur les groupes de Lie*, J. Approx. Theory **65** (1991), 176–199.

[61] T. Coulhon, L. Saloff-Coste and N. Varopoulos, *Analysis and geometry on groups*, Cambridge University Press, Cambridge, 1992.

[62] M. Cranston, E.Fabes and Z. Zhao, *Conditional gauge and potential theory for the Schrödinger operator*, Trans. Amer. Math. Soc. **307** (1988), 171–194.

[63] H.L. Cycon, R.G. Froese, W. Kirsch and B. Simon, *Schrödinger operators with applications to quantum mechanics and global geometry*, Springer Verlag, Berlin, 1987.

[64] A. Das, *Field theory: a path integral approach*, Lecture Notes in Physics, vol. **52**, World Scientific, Singapore, 1993.

[65] E.B. Davies, *Pointwise bounds on the space and time derivatives of heat kernels*, J. Operator Theory **21** (1989), 367–378.

[66] E.B. Davies, *Gaussian upper bounds for the heat kernels of some second-order operators on Riemannian manifolds*, J. Funct. Anal. **80** (1988), 16–32.

[67] E.B. Davies, *Heat kernels and spectral theory*, Cambridge University Press, Cambridge, 1989.

[68] E.B. Davies, *Heat kernel bounds, conservation of probability and the Feller property*, J. d'Anal. Math. **58** (1992), 99–119.

[69] E.B. Davies, *The state of the art for heat kernel bounds on negatively curved manifolds*, Bull. London Math. Soc. **25** (1993), 289–292.

[70] E.B. Davies, *Spectral theory and differential operators*, Cambridge Studies in Advanced Mathematics, vol. **42**, Cambridge University Press, Cambridge, 1995.

[71] P. Deift and B. Simon, *On the decoupling of finite singularities from the question of asymptotic completeness in the two-body quantum systems*, J. Funct. Anal. **23** (1976), 218–238.

[72] C. Dellacherie and P.-A. Meyer, *Probabilités et Potentiel, Chapitre I–IV/Chapitre V–VIII*, Hermann, Paris, 1975/1980.

[73] M. Demuth, *On spectral properties for regular and singular generators of Schrödinger semigroups*, Symposium Partial Differential Equations, Teubner-Texte Math., vol. **112**, Teubner, Leipzig, 1989, pp. 52–62, Proceedings Conference Holzhau, 1988.

[74] M. Demuth, *Spectral properties of Schrödinger operators of large coupling and semiclassical asymptotics*, Functional Analysis and Related Topics (S. Koshi, ed.), Proceedings Conference Sapporo, 1990, World Scientific, Singapore, 1991, pp. 77–80.

[75] M. Demuth, *On large coupling operator norm convergences of resolvent differences*, J. of Math. Phys. **32** (1991), 1522–1530.

[76] M. Demuth, *Perturbations of spectral measures for Feller operators*, Operator Calculus and Spectral Theory, Operator Theory: Advances and Applications, vol. **57**, Birkhäuser, Basel, 1992, pp. 27–37, Proceedings Conference Lambrecht, 1991.

[77] M. Demuth, F. Gesztesy, J. van Casteren and Z. Zhao, *Finite capacities in spectral theory, in Operator Theory: Advances and Applications* **78** (1995), Birkhäuser, 89–97.

[78] M. Demuth, W. Kirsch and I. McGillivray, *Schrödinger semigroups - geometric estimates in terms of the occupation time*, Comm. Part. Diff. Equat. **20** (1995), 37–57.

[79] M. Demuth and E. Ouhabaz, *Scattering for Schrödinger operators with magnetic fields*, Math. Nachrichten (1997) (to appear).

[80] M. Demuth, P. Stollmann, G. Stolz and J.A. van Casteren, *Trace norm estimates for products of integral operators and diffusion semigroups*, Integral Equat. and Op. Theory **23** (1995), 145–153.

[81] M. Demuth, P. Stollmann and J.A. van Casteren, *Non trace class operators with finite trace*, Seminar Notes in Functional Analysis, Dep. of Mathematics, Louisiana State University, 1994, pp. 119–128.

[82] M. Demuth and J.A. van Casteren, *On spectral theory for selfadjoint Feller generators*, Rev. Math. Phys. **1** (1989), 325–414.

[83] M. Demuth and J.A. van Casteren, *On potential continuity in stochastic spectral analysis*, Rigorous Results in Quantum Dynamics (J. Dittrich and P. Exner, eds.), World Scientific, Singapore, 1991, pp. 11–15, Proceedings Conference Liblice, CSSR 1990.

[84] M. Demuth and J.A. van Casteren, *Perturbations of generalized Schrödinger operators in stochastic spectral analysis*, Schrödinger operators: The quantum mechanical many body problem (E. Balslev, ed.), Lecture Notes in Physics, vol. **403**, Springer Verlag, Berlin, 1992, pp. 1–15, Proceedings Conference Aarhus 1991.

[85] M. Demuth and J.A. van Casteren, *On generalized Schrödinger semigroups and Feynman-Kac formulae*, Schrödinger Operators Standard and Non-standard (P. Exner and P. Šeba, eds.), World Scientific, Singapore, 1989, pp. 320–331, Proceedings Conference Dubna, 1988.

[86] M. Demuth and J.A. van Casteren, *A Hilbert-Schmidt property of resolvent differences of singularly perturbed generalized Schrödinger semigroups*, Evolution Equations, Control theory, and Biomathematics, Pure and Applied Mathematics, vol. **155**, Marcel Dekker, New York, 1994, pp. 117–144, Proceedings Conference Han-sur-Lesse 1991.

[87] M. Demuth and J.A. van Casteren, *Framework and results of stochastic spectral analysis*, Operator Theory: Advances and Applications, vol. **70**, Birkhäuser, Basel, 1994, pp. 123–132.

[88] M. Demuth and J.A. van Casteren, *Some results in stochastic spectral analysis*, Partial Differential Equations - Models in Physics and Biology, vol. **82**, Akademie Verlag, Berlin, 1994, pp. 56–81.

[89] M. Demuth and J.A. van Casteren, *Completeness of scattering systems with obstacles of finite capacity*, Preprint Max-Planck-Institut (1994), Proceedings of the IWOTA conference, 1995, Regensburg (to appear).

[90] J. Dereziński, *The Mourre estimate for dispersive N-body Schrödinger operators*, Trans. of the Amer. Math. Soc. **317** (1990), 773–798.

[91] J. Diestel and J.J. Uhl, *Vector Measures*, Math. Surveys, vol. **15**, Am. Math. Soc., Providence RI, 1977.

[92] J.L. Doob, *Classical potential theory and its probabilistic counterpart*, Die Grundlehren der Mathematischen Wissenschaften in Einzeldarstellungen, vol. **262**, Springer Verlag, 1983.

[93] R. Durrett, *Brownian motion and martingales in analysis*, Advanced Books and Software, Wadsworth, Belmont, 1984.

[94] E.B. Dynkin, *Markov processes I, II*, Die Grundlehren der Mathematischen Wissenschaften in Einzeldarstellungen, vol. **121–122**, Springer Verlag, Berlin, 1965.

[95] K.D. Elworthy, *Geometric aspects of diffusions on manifolds*, École d'été de Probabilités de Saint-Flour XV–XVII, 1985–1987 (P.L. Hennequin, ed.), Lecture Notes in Math., vol. **1362**, Springer Verlag, Berlin, 1988.

[96] K.D. Elworthy, *Stochastic differential equations on manifolds*, London Math. Soc. Lecture Notes in Math., vol. **70**, Cambridge University Press, Cambridge, 1982.

[97] K.D. Elworthy and A. Truman, *Classical Mechanics, the diffusion (heat) equation, and the Schrödinger equation on a Riemannian manifold*, J. Math. Phys. **22** (1981), 2144–2166.

[98] K.D. Elworthy, A. Truman and K. Watling, *The semi-classical expansion for a charged particle on a curved space background*, J. Math. Phys. **26** (1985), 198–202.

[99] S.N. Ethier and T.G. Kurtz, *Markov processes characterization and convergence*, Wiley Series in Probability and Statistics, John Wiley & Sons, New York, 1985.

[100] S. Fang, *Analyse quasi-sûre et l'estimation du noyau de la chaleur pour temps petit*, Trans. of the Amer. Math. Soc. **339** (1993), 221–241.

[101] Yu.B. Farforovskaya, *On the difference* $f(B) - f(A)$ *for unbounded self-adjoint operators in perturbation theory*, Zap. Nauchn. sem. Leningrad, Otdel. Inst. Steklov (LOMI) **107** (1982), 169–177, 232.

[102] Yu.B. Farforovskaya, *Functions of operators and their commutators in perturbation theory*, Functional analysis and operator theory (J. Zemánek, ed.), Banach Center Publications; Polish Academy of Sciences, Institute of Mathematics Warsaw, vol. **30**, 1994, pp. 147–159, Papers from the thirty-ninth Semester in Warsaw, March 2 - May 30, 1992.

[103] C. Fefferman, *The N-body problem in quantum mechanics*, Comm. Pure Appl. Math. **39** (1986), 67–109.

[104] C. Fefferman and R. de la Llave, *Relativistic stability of matter I*, Rev. Math. Iberoamerica **2** (1986), 119–223.

[105] D. Feyel and A. de La Pradelle, *Étude de l'équation* $\frac{1}{2}\Delta u - u\mu = 0$, Ann. Inst. Fourier, Grenoble **38** (1988,), 199–218.

[106] R. Froese, I. Herbst, M. Hoffmann-Ostenhof and T. Hoffman-Ostenhof, *On the absence of positive eigenvalues for one-body Schrödinger operators*, J. d'Analyse Math. **41** (1982), 272–284.

[107] M. Fukushima, *On the L^p-estimate of resolvents of Markov processes*, Publ. Res. Inst. Math. Sci. **13** (1977), 277–284.

[108] M. Fukushima, *Dirichlet forms and Markov processes*, North-Holland Math. Library, vol. **23**, North-Holland Publishing Company, Amsterdam, 1980.

[109] M. Fukushima, N. Jacob and H. Kaneko, *On $(r, 2)$-capacities for a class of elliptic differential operators*, Mathematische Annalen, no. 2 **293** (1992), 343–348.

[110] M. Fukushima, Y. Oshima, and M Takeda, *Dirichlet forms and symmetric Markov processes*, Studies in Mathematics, vol. **19**, de Gruyter, Berlin, New York, 1994.

[111] I.I. Gihman and A.V. Skorohod, *The theory of stochastic process I, II, and III*, Die Grundlehren der mathematischen Wissenschaften, vol. **210, 218, 232**, Springer Verlag, Berlin, 1974, 1975, 1979.

[112] J. Ginibré, *Some applications of functional integration in statistical mechanics and quantum field theory*, Statistical mechanics and quantum field theory, Les Houches 1970 (C. DeWitt-Morette and R. Storm, eds.), Gordon and Breach, New York, London, Paris, 1971, pp. 327–427.

[113] I.M. Glazman, *Direct methods of qualitative spectral analysis of singular differential operators*, Daniel Davey and Co., New York, 1966, Russian title: Prjamye metody kačestvennogo spektral'nogo analiza singularnyh differencial'nyh operatorov, published by Fizmagiz, Moscow, 1963.

[114] J. Glover, K.H. Rao, H. Silic and R. Song, *Γ-potentials*, Classical and modern potential theory and applications (K. GownSankaran et. al., eds.), Kluwer Academic Publisher, 1994, pp. 217–232.

[115] J. Gohberg and M. Krein, *Theory and applications of Volterra operators in Hilbert space*, Translations of Mathematical Monographs, vol. **24**, Amer. Math. Soc., Providence R.I., 1970.

[116] J. Goldstein, *Semigroups of linear operators an applications*, Oxford Mathematical Monographs, Clarendon Press, New York, Oxford, 1985.

[117] B. Helffer and J. Sjöstrand, *Equation de Schrödinger avec champ magnétique et équation de Harper*, Schrödinger operators, Lecture Notes in Physics, vol. **345**, Springer Verlag, Berlin, 1989.

[118] R. Hempel and J. Voigt, *The spectrum of a Schrödinger operator in $L_p(\mathbb{R}^\nu)$ is p-independent*, Comm. Math. Phys. **104** (1986), 243–250.

[119] I.W. Herbst, *Spectral theory of the operator $\left(p^2+m^2\right)^{1/2} - \frac{Ze^2}{r}$*, Comm. Math. Phys. **53** (1977), 285–294.

[120] I.W. Herbst and A.D. Sloan, *Perturbations of translation invariant positivity preserving semigroups in $L^2(\mathbb{R}^N)$*, Trans. Amer. Math. Soc. **236** (1978), 325–360.

[121] I.W. Herbst and Z. Zhao, Sobolev spaces, Kac regularity and the Feynman-Kac formula, Seminar on Stochastic Processes (E. Cinlar, K.L. Chung, R.K. Getoor and J. Glover, eds.), Birkhäuser, Boston, Basel, 1988, pp. 171–191, Proceedings Conference Princeton, 1987.

[122] P.D. Hislop and I.M. Sigal, *Introduction to spectral theory; with applications to Schrödinger operators*, Applied Math. Sciences, vol. **113**, Springer Verlag, New York, 1996.

[123] W. Hoh, *The martingale problem for a class of pseudo differential operators*, Math. Ann. **300** (1994), 121–147.

[124] W. Hoh, *Feller semigroups generated by pseudo differential operators*, Dirichlet forms and Stochastic Processes (Z. Ma, M. Röckner and J.A. Yau, eds.), de Gruyter, Berlin, New York, 1995, pp. 199–206, Proceedings Conference Bejing 1993.

[125] W. Hoh and N. Jacob, *Some Dirichlet forms generated by pseudo-differential operators*, Bull. Sci. Math. (2) **116** (1992), 383–398.

[126] D. Hundertmark, *Zur Theorie der Schrödinger Operatoren mit Magnetfeld*, Ph.D.-thesis, Institut für Theoretische Physik I, Universität Erlangen-Nürnberg, 1991.

[127] T. Ichinose, *Essential selfadjointness of the Weyl quantized relativistic Hamiltonian*, Ann. Inst. Henri Poincaré **51** (1889), 265–298.

[128] T. Ichinose, *The nonrelativistic limit problem for a relativistic spinless particle in an electromagnetic field*, J. Funct. Anal. **73** (1987), 233–257.

[129] T. Ichinose, *Path integral for a Weyl quantized relativistic Hamiltonian and the nonrelativistic limit problem*, Differential Equations and Mathematical Physics, Lecture Notes in Math., vol. **1285**, Springer Verlag, Berlin, 1988, pp. 205–210.

[130] T. Ichinose, *Kato's inequality and essential self-adjointness for Weyl quantized relativistic Hamiltonian*, Proc. Japan Acad. **64A** (1988), 367–369.

[131] T. Ichinose and H. Tamura, *Imaginary-time path integral for a relativistic spinless particle in a electromagnetic field*, Comm. Math. Phys. **105** (1986), 239–257.

[132] T. Ichinose and H. Tamura, *Path integral for Weyl quantized relativistic Hamiltonian*, Proc. Japan Acad. **62A** (1986), 91–93.

[133] T. Ichinose and W. Ichinose, *On the self-adjointness of the relativistic Hamiltonian with a negative scalar potential*, Rev. Math. Phys. **7** (1995), 709–721.

[134] T. Ichinose and H. Tamura, *Path integral for the relativistic quantun mechanics*, Progress Theor. Phys. Supplement **92** (1987), 144–175.

[135] T. Ichinose and T. Tsuchida, *On essential self-adjointness of the Weyl quantized relativistic Hamiltonian*, Forum Math. **5** (1993), 539–559.

[136] N. Ikeda and H. Matsumoto, *Short time asymptotics of the traces of heat kernels of Schrödinger operators with magnetic fields*, Bull. Sci. Math. (2) **116** (1992), 53–66.

[137] N. Ikeda and S. Watanabe, *Stochastic differential equations and diffusion processes, second edition*, North-Holland, Amsterdam, 1989.

[138] A.A. Ilyin, *Lieb-Thirring inequalities on the N-sphere and in the plane, and some applications*, Proc. London Math. Soc., vol. **67**, 1993, pp. 159–182.

[139] Y. Ishikawa, *A remark on the existence of a diffusion process with non-local boundary conditions*, J. Math. Soc. Japan, no. 1 **42** (1990), 171–184.

[140] N. Jacob, *A class of elliptic differential operators generating symmetric Dirichlet forms*, Potential Anal. **1** (1992), 221–232.

[141] N. Jacob, *A class of Feller semigroups generated by pseudo-differential operators*, Math. Zeitschrift **215** (1994), 151–166.

[142] N. Jacob, *Further pseudo-differential operators generating Feller semigroups and Dirichlet forms*, Rev. Mat. Iberoamerica **9** (1993), 373–407.

[143] N. Jacob, *Pseudo-differential operators and Markov processes: A mathematical survey* (1996), Akademie Verlag, Berlin.

[144] B. Jefferies, *Perturbations of Schrödinger semigroups generated by stochastic integrals*, J. Funct. Anal. (1988), 98–111.

[145] A. Jensen, *High energy resolvent estimates for Schrödinger operators*, Ideas and Methods in Quantum and Statistical Physics, Cambridge Univ. Press, Cambridge, 1992, pp. 254–260, Proceedings Conference Oslo, 1988.

[146] K. Jörgens and J. Weidmann, *Spectral properties of Hamiltonian operators*, Springer Verlag, Berlin, 1973.

[147] I. Karatzas and S.E. Shreve, *Brownian motion and stochastic calculus*, Graduate texts in Mathematics, Springer Verlag, Berlin, 1988.

[148] T. Kato, *Perturbation theory for linear operators*, Springer Verlag, Berlin, 1976.

[149] T. Kato, *Monotonicity theorems in scattering theory*, Hadronic Journal **1** (1978), 134–154.

[150] T. Kato, *On the Trotter-Lie product formula*, Proc. Japan Acad. **50** (1974), 694–698.

[151] T. Kato, *Trotter's product formula for an arbitrary pair of self-adjoint contraction semigroups*, Topics in Functional Analysis, Advances in Mathematics Supplementary Studies, vol. **3**, 1978, pp. 185–195.

[152] R.Z. Khas'minskii, *On positive solutions of the equation* $\mathcal{A}u + Vu = 0$, Theory of probability and its applications IV, no. 3, 1959, pp. 309–318.

[153] J. Kisyński, *On jumps of paths of Markov processes*, Probability measures on Groups VIII, Lecture Notes in Math., vol. **1210**, Springer Verlag, Berlin, 1985, pp. 130–145.

[154] J. Kisyński, *Local uniqueness of Feller processes with integrodifferential generators*, Stochastic Systems and Optimization (Warsaw 1988), Lecture Notes in Control and Information Sciences, vol. **136**, Springer Verlag, Berlin, 1989, pp. 98–111.

[155] J. Kisyński, *Metrization of* $D_E[0.1]$ *by Hausdorff distance between graphs*, Annales Polonici Math. **LI** (1990), 195–203.

[156] J. Kisyński, *On Markov processes generated by boundary systems of Ventcel*, Classical Analysis (T. Mazur, ed.), World Scientific, 1992., pp. 53–99, Proceedings Kazimierz Dolny, 1991 Singapore.

[157] A.N. Kochubeĭ, *Singular parabolic equations and Markov processes*, Math. USSR Izvestiya **24** (1985), 73–97.

[158] A.N. Kochubeĭ, *Parabolic pseudodifferential equations, hypersingular integrals and Markov processes*, Math. USSR Izvestiya **33** (1989), 233–259.

[159] H. König, *Eigenvalue distribution of compact operators*, Operator Theory: Advances and Applications, vol. **16**, Birkhäuser, Basel, 1986.

[160] S. Kusuoka and D. Stroock, *Applications of the Malliavin calculus, Part II*, J. Fac. Sci. Univ. Tokyo Sect. IA, Math. **32** (1985), 1–76.

[161] H.H. Kuo, *Gaussian measures in Banach spaces*, Lecture Notes in Math., vol. **463**, Springer Verlag, Berlin, 1975.

[162] R. Lavine, *Absolute continuity of positive spectrum for Schrödinger operators with long-range potentials*, J. Funct. Anal., **12** (1973), 30–54.

[163] R. Léandre, *Minoration en temps petit de la densité d'une diffusion déférée*, J. Funct. Anal. **74** (1987), 399–414.

[164] L.F. Lemmens, *Functional integration of a Schrödinger semigroup for excitations on a lattice*, Nieuw Archief voor Wiskunde **13** (1995), 409–420.

[165] L.F. Lemmens, F. Brosens, and J.T. Devreese, *Many-body diffusions and path integrals for identical particles*, Physical Review E **53** (1996), 4467–4476.

[166] M. Liao, *The Dirichlet problem of a discontinuous Markov process*, Acta Math. Sinica **5** (1989), 9–15.

[167] E. Lieb and W. Thirring, *Inequalities for the moments of the eigenvalues of the Schrödinger Hamiltonian and their relation to Sobolev inequalities*, Studies in Mathematical Physics (E.H. Lieb, B. Simon and A.S. Wightman, eds.), Essays in honor of Valentine Bargmann, Princeton University Press, Princeton, N.J., 1976, pp. 269–304.

[168] T.M. Liggett, *Interacting particle systems*, Die Grundlehren der Mathematischen Wissenschaften in Einzeldarstellungen, vol. **276**, Springer Verlag, Berlin, New York, 1985.

[169] H.P. Lotz, *Semigroups on L^∞ and H^∞*, One-parameter semigroups of positive operators (R. Nagel, ed.), Lecture Notes in Math., vol. **1184**, Springer Verlag, Berlin, 1986, pp. 54–59.

[170] Z. Ma and M. Röckner, *Introduction to the theory of (non-symmetric) Dirichlet forms*, Universitext, Springer Verlag, Berlin, 1992.

[171] H. Matsumoto, *The short time asymptotics of the traces of the heat kernels for magnetic Schrödinger operators*, J. Math. Soc. Japan **42** (1990), 677–689.

[172] H. Matsumoto, *Classical and non-classical eigenvalue asymptotics for magnetic Schrödinger operators*, J. Funct. Anal. **95** (1991), 460–482.

[173] H. Matsumoto, *The asymptotic distributions of the eigenvalues for the Schrödinger operators with magnetic fields*, Asymptotic problems in probability theory: Wiener functionals and asymptotics (Sanda/Kyoto, 1990), Pitman Research Notes Math. Ser., vol. **284**, Longman Sci. Tech., Harlow, 1993, pp. 169–193.

[174] H. Matsumoto, *On the integrated density of states for the Schrödinger operators with certain random electromagnetic potentials*, J. Math. Soc. Japan **45** (1993), 197–214.

[175] H. Matsumoto, *Semiclassical asymptotics of eigenvalues for Schrödinger operators with magnetic fields*, J. Funct. Anal. **129** (1995), 168–190.

[176] I. McGillivray, *A Feller property for some degenerate elliptic operators*, Dirichlet Forms and Stochastic Processes (Z. Ma, M. Röckner and J. Yau, eds.), Proceedings Conference Bejing, 1993, de Gruyter, Berlin, 1995, pp. 271–282.

[177] I. McGillivray, *Capacitary estimates for Dirichlet eigenvalues*, J. Funct. Anal. **139** (1996), 244–259.

[178] I. McGillivray and E. Ouhabaz, *Some spectral properties of recurrent semigroups*, Arch. Math. **66** (1996), 233–242.

[179] M. Métivier and J. Pellaumail, *Stochastic integration*, Probability Theory and Statistics Monographs and Textbooks, Academic Press, New York, 1980.

[180] E. Mourre, *Absence of singular continuous spectrum for certain operators*, Comm. Math. Phys. **78** (1981), 391–408.

[181] M. Nagasawa, *Schrödinger equations and diffusion theory*, Monographs in Mathematics, vol. **86**, Birkhäuser, Basel, 1993.

[182] R. Nagel (editor), W. Arendt, A. Grabosch, G. Greiner, U. Groh, H.P. Lotz, U. Moustakas, R. Nagel, F. Neubrander, U. Schlotterbeck, *One-parameter semigroups of positive operators*, Lecture Notes in Math., vol. **1184**, Springer Verlag, Berlin, 1986.

[183] S. Nakamura, *Scattering theory for the shape resonance model I. Non-resonant energies*, Ann. Inst. Henri Poincaré **50** (1989), 115–131.

[184] J. Nash, *Continuity of solutions of parabolic and elliptic equations*, Amer. J. Math. **80** (1958), 931–954.

[185] O. Odimboleko and J.A. van Casteren, *On the uniqueness of the martingale problem*, Intern. J. Math. **7** (1996), 775–810.

[186] M.M.H. Pang, *Resolvent estimates for Schrödinger operators in $L^p(\mathbb{R}^N)$ and the theory of exponentially bounded C-semigroups*, Semigroup Forum **41** (1990), 97–114.

[187] V.V. Peller, *Hankel operators in the perturbation theory of unbounded self-adjoint operators*, Analysis and Partial Differential Equations, Lecture Notes in Math., vol. **122**, Marcel Dekker, New York, 1990, pp. 529–544.

[188] P. Perry, I.M. Sigal and B. Simon, *Spectral analysis of N-body Schrödinger operators*, Annals of Math. **114** (1981), 519–567.

[189] S.C. Port and C.J. Stone, *Brownian motion and classical potential theory*, Academic Press, New York, 1978.

[190] N.I. Portenko, Generalized diffusion processes, Proceedings of the third Japan-USSR Symposium on Probability Theory (G. Maruyama and J.V. Prokhorov, eds.), Lecture Notes in Math., vol. **550**, Springer Verlag, Berlin, 1976, pp. 500–523.

[191] M. Reed and B. Simon, *Methods of Modern Mathematical Physics I: Functional Analysis*, Academic Press, New York, 1975.

[192] M. Reed and B. Simon, *Methods of Modern Mathematical Physics II: Fourier Analysis, Self-Adjointness*, Academic Press, New York, 1975.

[193] M. Reed and B. Simon, *Methods of Modern Mathematical Physics III: Scattering Theory*, Academic Press, New York, 1979.

[194] M. Reed and B. Simon, *Methods of Modern Mathematical Physics IV: Analysis of operators*, Academic Press, New York, 1978.

[195] D. Revuz and M. Yor, *Continuous martingales and Brownian motion*, Die Grundlehren der mathematischen Wissenschaften in Einzeldarstellungen, vol. **293**, Springer Verlag, Berlin, 1991.

[196] D. Robert and H. Tamura, *Semi-classical estimates for resolvents and asymptotics for total scattering cross-sections*, Ann. Inst. Henri Poincaré **46** (1987), 415–442.

[197] D. Robert and H. Tamura, *Asymptotic behavior of scattering amplitudes in semi-classical and low energy limits*, Ann. Inst. Fourier (Grenoble) **39** (1989), 155–192.

[198] D.W. Robinson, *Elliptic operators and Lie groups*, Oxford Mathematical Monographs, Clarendon press, Oxford, New York, 1991.

[199] G. Roepstorff, *Path integral approach to quantum physics: an introduction*, Springer Verlag, Berlin, 1994.

[200] J.-P. Roth, *Opérateurs carré du champ et formule de Lévy-Kinchine sur les espaces localement compacts*, C.R. Acad. Sci. Paris Ser. **A 278** (1974), 1103–1106.

[201] K. Sato and T. Ueno, *Multi-dimensional diffusion and the Markov process on the boundary*, J. Math. Kyoto University **4** (1965), 529–605.

[202] H.H. Schaefer, *Banach lattices and positive operators*, Die Grundlehren der mathematischen Wissenschaften in Einzeldarstellungen, vol. **215**, Springer Verlag, Berlin, 1974.

[203] M. Schechter, *Spectra of partial differential operators*, North-Holland, Amsterdam, 1986.

[204] R. Seeley, *Contraction semigroups for diffusions with drift*, Trans. Amer. Math. Soc. **283** (1984), 717–728.

[205] M. Sharpe, *General theory of Markov processes*, Pure and applied mathematics, vol. **133**, Academic Press, New York, 1988.

[206] B. Simon, Schrödinger semigroups, Bulletin (New Series) Amer. Math. Soc. **7** (1982), 447–526.

[207] B. Simon, *Erratum: Schrödinger semigroups*, Bull. (New Series) Amer. Math. Soc., no. 2 **11** (1984), 426.

[208] B. Simon, *Functional integration and quantum physics*, Academic Press, New York, 1979.

[209] B. Simon, *Trace Ideals and their Applications*, London Math. Soc. Lecture Notes Series, vol. **35**, Cambridge, 1979.

[210] B. Simon, *Semiclassical analysis of low lying eigenvalues, I, Non-degenerate minima: Asymptotic expansions*, Ann. Inst. Henri Poincaré **38** (1983), 295–307.

[211] B. Simon, *Semiclassical analysis of low lying eigenvalues, II. Tunneling*, Annals of Math. **120** (1984), 89–118.

[212] L. Smits, *Perturbations of Markov semigroups by additive functionals*, Ph.D.-thesis, University of Antwerp, Antwerp, 1991.

[213] L. Smits and J.A. van Casteren, *Semigroups defined by additive processes*, Trends in Semigroups and Evolution Equations, Proceedings Conference Delft, 1989, Marcel Dekker, New York, 1991, pp. 463–482.

[214] R.M. Song, *Probabilistic approach to the Dirichlet problem of perturbed stable processes*, Probab. Theory Related Fields **95** (1993), 371–389.

[215] R. Song, *Feynman-Kac semigroups with discontinuous additive functionals*, J. Theoret. Probab. **8** (1995), 727–762.

[216] P. Stollmann, *Scattering at obstacles of finite capacity*, J. Funct. Anal. **121** (1994), 416–425.

[217] P. Stollmann and J. Voigt, Perturbation of Dirichlet forms by measures, Potential Anal. **5** (1996), 109–138.

[218] D. Stroock, *The Kac approach to potential theory: Part I*, J. Math. and Mech. **16** (1967), 819–852.

[219] D.W. Stroock, *Probability theory: an analytic view*, Cambridge University Press, Cambridge, 1993.

[220] D.W. Stroock, *A concise introduction to the theory of integration*, second edition, Birkhäuser, Boston, 1994.

[221] D.W. Stroock and S.R.S. Varadhan, *Multidimensional diffusion processes*, Springer Verlag, Berlin, 1979.

[222] K.-T. Sturm, *Störung von Hunt-Prozessen durch signierte additive Funktionale*, Ph.-D. thesis, Universität Erlangen-Nürnberg, 1989.

[223] K.-T. Sturm, *Gauge theorems for resolvents with application to Markov processes*, Probab. Theory Related Fields **89** (1991), 387–406.

[224] K.-T. Sturm, *Heat kernel bounds on manifolds*, Math. Ann. **292** (1992), 149–162.

[225] K.-T. Sturm, *Measures charging no polar sets and additive functionals of Brownian motion*, Forum Math. **4** (1992), 257–297.

[226] K.-Th. Sturm, *Schrödinger semigroups on manifolds*, J. Funct. Anal. **118** (1993), 309–350.

[227] K. Taira, *Semigroups and boundary value problems*, Duke Math. J. **49** (1982), 287–320.

[228] K. Taira, *Diffusion processes and partial differential equations*, Academic Press, New York, 1988.

[229] K. Taira, *On the existence of Feller semigroups with boundary conditions*, Memoirs of the Am. Math. Soc., vol. **267**, Providence, 1991.

[230] T. Umeda and H. Nagase, *Spectra of relativistic Schrödinger operators with magnetic vector potentials*, Osaka J. Math. **30** (1993), 839–853.

[231] J.A. van Casteren, *Generators of strongly continuous semigroups*, Research Notes in Math., vol. **115**, Pitman, London, 1985.

[232] J.A. van Casteren, *Integral kernels and Schrödinger type equations*, Functional integration with emphasis on the Feynman integral, Supplemento ai Rendiconti del Circolo Matematico di Palermo, serie II, no. 17, 1987, pp. 392–421, Proceedings Sherbrooke, Canada, 1986.

[233] J.A. van Casteren, *Integral kernels and the Feynman-Kac formalism*, Aspects of Positivity in Functional Analysis (R. Nagel, U.Schlotterbeck and M.P.H. Wolff, eds.), Proceedings Conference Tübingen 1985, North-Holland, Amsterdam, 1986, pp. 179–185.

[234] J.A. van Casteren, *Pointwise inequalities for Schrödinger semigroups*, Trends in Semigroup Theory and Applications (Ph. Clément, S. Invernizzi, E. Mitidieri, I.I. Vrabie, eds.), Lecture Notes in Pure and Applied Mathematics, vol. **116**, Marcel Dekker, New York, 1989, pp. 67–94, Proceedings Conference Trieste 1988.

[235] J.A. van Casteren, *A pointwise inequality for generalized Schrödinger semigroups*, Symposium "Partial Differential Operators" Holzhau 1988 (B.-W. Schulze and H. Triebel, eds.), Teubner Texte zur Mathematik, Band, vol. **112**, Teubner Verlagungsgesellschaft, Leipzig, 1989, pp. 298–312, Proceedings Conference Holzhau 1988.

[236] J.A. van Casteren, *On generalized Schrödinger semigroups*, Markov Processes and Control Theory (H. Langer and V. Nollau, eds.), Mathematical Research, vol. **54**, Akademie Verlag, Berlin, 1989, pp. 16–39, Proceedings Conference ISAM 88, Gaussig, 1988.

[237] J.A. van Casteren, *Sur des différences de semigroupes de Feynman-Kac: une propriété de trace*, Série I, C.R. Acad. Sci. Paris, t. **314** (1992), 997–1002.

[238] J.A. van Casteren, *A trace class property of singularly perturbed generalized Schrödinger semigroups*, Operator Calculus and Spectral Theory, Operator Theory: Advances Applications, vol. **57**, Birkhäuser, Basel, 1992, pp. 329–347, Proceedings International Symposium Lambrecht, Pfalz, Germany, December 8–14, 1991.

[239] J.A. van Casteren, *On martingales and Feller semigroups*, Results in Mathematics **21** (1992), 274–288.

[240] M. van den Berg, *Gaussian bounds for the Dirichlet heat kernel*, J. Funct. Anal. **88** (1990), 267–278.

[241] M. van den Berg, *On the trace of the difference of Schrödinger heat semigroups*, Proc. Royal Soc. Edinburgh A **119** (1991), 169–175.

[242] M. van den Berg and E.B. Davies, *Heat flow out of regions in* $\mathbb{R}^m$, Math. Zeitschrift, no. 4 **202** (1989), 463–482.

[243] N. Varopoulos, *Hardy-Littlewood theory for semigroups*, J. Funct. Anal. **63** (1985), 240–260.

[244] N. Varopoulos, *Analysis on Lie groups*, J. Funct. Anal. **76** (1988), 346–410.

[245] N. Varopoulos, *Small time Gaussian estimates of heat diffusion kernels I, the semigroup technique*, Bull. Sci. Math. Série II **113** (1989), 263–277.

[246] N. Varopoulos, *Small time Gaussian estimates of heat diffusion kernels II, the theory of large deviations*, J. Funct. Anal. **93** (1990), 1–33.

[247] J. Voigt, *Absorption semigroups, their generators, and Schrödinger semigroups*, J. Funct. Anal. **67** (1986), 167–205.

[248] R. Weder, *Spectral analysis of pseudo-differential operators*, J. Funct. Anal. **20** (1975), 319–337.

[249] R. Weder, *The unified approach to spectral analysis*, Comm. Math. Phys. **60** (1978), 291–299.

[250] R. Weder, *The unified approach to spectral analysis II*, Proc. Amer. Math. Soc. **75** (1979), 81–84.

[251] R. Weder, *Second order operators in the uniform norm*, Comm. Partial Differential Equations **3** (1978), 381–406.

[252] J. Weidmann, *Linear operators in Hilbert space*, Graduate texts in Mathematics, vol. **68**, Springer-Verlag, Berlin, 1980.

[253] R.D. Yafaev, *Mathematical scattering theory, general theory*, Translations of Mathematical Monographs, vol. **105**, Am. Math. Soc., Providence R.I., 1992.

[254] M. Yor, *Some aspects of Brownian motion part I: Some special functionals*, Lectures in Mathematics ETH Zürich, Birkhäuser, Basel, 1992.

[255] K. Yosida, *Functional analysis*, Die Grundlehren der Mathematischen Wissenschaften in Einzeldarstellungen, vol. **123**, Springer Verlag, Berlin, 1980, sixth edition.

[256] V.A. Zagrebnov, *The Trotter-Lie product formula for Gibbs semigroups*, J. Math. Phys. **29** (1988), 888–891.

[257] Zh. Zhou, *The contractivity of the free Hamiltonian semigroup in the* L^p*-space of entire functions*, J. Funct. Anal. **96** (1991), 407–425.

Index of Symbols

Constants

(σ)-Fields

Martingales

Measures

Norms, inner-products

Operators

Operator spaces

Points, sets and spaces

Processes

Stopping times

(i) Throughout the Chapters 1 through 8 of the book the symbol T denotes the hitting time of the set Γ, and S denotes its penetration time: see (2.39) on page 87 and page 52. In the Appendices C and D we write τ_B for the (first) hitting time of the Borel set B, and σ_B for its penetration time. The symbol D_B denotes the entry time of B. For more explanation on this see section C.20 (p. 370) in Appendix C and the comments on the pages 373, 403, 404 of Appendix D.

(ii) If the symbol T has a subscript, then T is some specific stopping time (this depends very much on the context, e.g. see p. 9, p. 357).

Subject Index

General remark:
The most relevant pages of reference to a subject are written in bold face.

S

MIX
Papier aus verantwortungsvollen Quellen
Paper from responsible sources
FSC® C105338

If you have any concerns about our products,
you can contact us on
ProductSafety@springernature.com

In case Publisher is established outside the EU,
the EU authorized representative is:
Springer Nature Customer Service Center GmbH
Europaplatz 3, 69115 Heidelberg, Germany

Printed by Libri Plureos GmbH
in Hamburg, Germany